Mathematics for Physics

Mathematics for
Physics

Michael M. Woolfson
Department of Physics
University of York

Malcolm S. Woolfson
School of Electrical and Electronic Engineering
University of Nottingham

UNIVERSITY PRESS

OXFORD

UNIVERSITY PRESS

Great Clarendon Street, Oxford OX2 6DP

Oxford University Press is a department of the University of Oxford.
It furthers the University's objective of excellence in research, scholarship,
and education by publishing worldwide in

Oxford New York

Auckland Cape Town Dar es Salaam Hong Kong Karachi
Kuala Lumpur Madrid Melbourne Mexico City Nairobi
New Delhi Shanghai Taipei Toronto

With offices in

Argentina Austria Brazil Chile Czech Republic France Greece
Guatemala Hungary Italy Japan Poland Portugal Singapore
South Korea Switzerland Thailand Turkey Ukraine Vietnam

Oxford is a registered trade mark of Oxford University Press
in the UK and in certain other countries

Published in the United States
by Oxford University Press Inc., New York

© Michael M Woolfson & Malcolm S Woolfson

British Library Cataloguing in Publication Data

Data available

Library of Congress Cataloging in Publication Data

Data available

Typeset by Newgen Imaging Systems (P) Ltd., Chennai, India
Printed in Great Britain
on acid-free paper by
Antony Rowe Ltd, Chippenham, Wiltshire

ISBN 978–0–19–928929–5

10 9 8 7 6 5 4 3 2

Contents

4 Differentiation

5 Integration

6 Complex numbers

7 Ordinary differential equations

8 Matrices I and determinants

9 Vector algebra

13 Coordinate systems and multiple integration

14 Distributions

15 Hyperbolic functions

16 Vector analysis

17   Fourier analysis

18   Introduction to digital signal processing

19   Numerical methods for ordinary differential equations

20 Applications of partial differential equations

21 Quantum mechanics I: Schrödinger wave equation and observations

30 Linear equations

31 Numerical solution of equations

32 Signals and noise

33 Digital filters

34 Introduction to estimation theory

35 Linear programming and optimization

36 Laplace transforms

37 Networks

38 Simulation with particles

39 Chaos and physical calculations

Preface

Mathematics is a subject of immense appeal. It covers the range from the most abstract of individual topics to the most applied – from logic and Fermat's last theorem to statistics and the solution of differential equations. But more than being a subject in its own right, complete and without the need to call on knowledge from outside its own boundaries, it is the essential language of science. As Richard Feynman, Nobel Laureate in Physics in 1965, wrote, 'To those that do not know mathematics it is difficult to get across a real feeling as to the beauty, the deepest beauty of nature. If you want to learn about nature, to appreciate nature, it is necessary to understand the language she speaks in.' Many professional physicists may not appreciate mathematics to the full, in the way that professional mathematicians do, but they should all understand its importance. Without a language you cannot communicate and, in fact, there are some that argue that without a language you cannot even think clearly.

For those coming into physics at an undergraduate level it is helpful to motivate the learning of mathematics by demonstrating the applications alongside the development of the mathematical theory. This is the main distinguishing feature of this text. Not only do we apply the mathematics in the context of physics, in close proximity to developing it, but we also have some chapters, primarily on quantum mechanics, which show how several mathematical techniques come together in the development of particular areas of physics.

This is a textbook designed for the twenty-first century, where electronic systems, including computers, exist and systems of interest include radar, nuclear reactors, and signals from spacecraft, including the global positioning system (GPS). All of these are dealt with in the text, at an appropriate level.

It is sometimes mistakenly felt that mathematics is for theoretical physicists only. Not only is this untrue, but there are some areas of mathematics that are mainly of interest to experimentalists. Many experimental physicists need to use signal-processing algorithms to process data and extract meaningful information from noisy data. In this book, there are three chapters that describe the detection, filtering, and estimation of signals in noise. Although there are many specialist textbooks describing such processing techniques, we feel that a gentle introduction to these mathematical methods may motivate the experimentalist to read further.

There are many specialist degree courses in computational physics that deal with quite advanced applications of computational techniques to physical problems. However, even for those following courses on experimental or theoretical physics it is important to appreciate the usefulness of computational techniques. We provide 23 computer programs, applicable to different areas of problem-solving, which are described in outline form, mostly in the appendices, and are available on the web coded in FORTRAN, C, and MATLAB. These can be applied to problems as varied as finding eigenfunctions for a simple quantum-mechanical system, studying the behaviour of a cluster of stars, determining the equation of state of a liquid, estimating a constant from a stream of noisy data, or designing a nuclear reactor. In using these programs the reader may need to call on some form of graphical output. These are readily available – for example, GNUPLOT is a flexible and easy-to-use package that can be downloaded from the Internet.

However, even in the twenty-first century, the basic mathematics that has served scientists for hundreds of years is still the essential core of the scientist's mathematical toolkit and this is dealt with fully. Some of the more important and fundamental equations are highlighted, and particular attention should be paid to the understanding of the material so indicated.

In the body of each chapter there are exercises, related to the topic that they immediately follow and designed to assist the understanding of the material. For the most part they are simple in nature and the given solutions are often just the answer – although a derivation may be given where appropriate. At the end of each chapter there are problems of a more searching nature that should be considered as part of the teaching material. Detailed solutions are given for these problems and, even if the reader is unable to solve the problem completely, then reading the solution should not only indicate how the problem should be tackled but also help in understanding the material with which the chapter deals.

M. M. Woolfson
M. S. Woolfson
York and Nottingham

April 2006

Useful formulae and relationships

A common first step in applying mathematics to physical problems is that one or more mathematical relationships are found, derived from the physics of the experimental situation. These relationships are then mathematically manipulated to give a final result that can be regarded as the conclusion from, or the outcome of, the experiment. The analysis that leads from the initial equations to the final result may involve transformations of expressions from one form to another, approximations of various kinds or the representation of positions in some convenient coordinate system. In this chapter you will meet a large selection of relationships involving angles, some important approximations and functions that frequently occur in physical contexts – the natural logarithm and the exponential e. You will also be introduced to two coordinate systems for defining positions in a plane.

1.1 Relationships for triangles

The best-known relationship relating to triangles must be Pythagoras' theorem for a right-angled triangle, which most students can quote in words as 'The square on the hypotenuse equals the sum of the squares on the other two sides'. For the right-angled triangle shown in Figure 1.1 this gives the equation

$$h^2 = p^2 + b^2. \tag{1.1}$$

A right-angled triangle is convenient for defining various trigonometric functions such as the following, involving the angle θ in the figure:

$$\sin \theta = \frac{p}{h}; \quad \operatorname{cosec} \theta = \frac{1}{\sin \theta} = \frac{h}{p} \tag{1.2a}$$

$$\cos \theta = \frac{b}{h}; \quad \sec \theta = \frac{1}{\cos \theta} = \frac{h}{b} \tag{1.2b}$$

$$\tan \theta = \frac{\sin \theta}{\cos \theta} = \frac{p}{b}; \quad \cot \theta = \frac{1}{\tan \theta} = \frac{b}{p}. \tag{1.2c}$$

From Pythagoras' theorem and the definitions (1.2) we can deduce that

$$\sin^2 \theta + \cos^2 \theta = \frac{p^2}{h^2} + \frac{b^2}{h^2} = 1. \tag{1.3a}$$

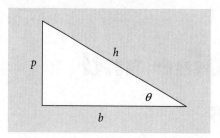

Figure 1.1 A right-angled triangle.

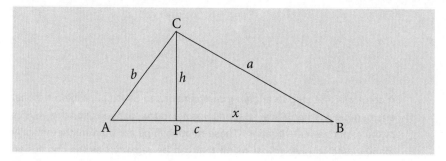

Figure 1.2 A general triangle with a perpendicular from apex C to side AB.

Dividing both sides of (1.3a) by $\cos^2 \theta$ gives

$$\frac{\sin^2 \theta}{\cos^2 \theta} + 1 = \frac{1}{\cos^2 \theta}$$

or

$$\tan^2 \theta + 1 = \sec^2 \theta. \tag{1.3b}$$

However, there are also important and quite simple relationships that link the sides and angles of a general (**scalene**) triangle, to deduce which we have only to apply the definitions of the sine and cosine of an angle.

1.1.1 The sine relationship

We consider the triangle ABC shown in Figure 1.2 with a perpendicular CP from the apex C to the side AB.

The sides are of length a, b, and c as shown in the figure, and the perpendicular is of length h. The angle at point A is A with similar notation for the other angles.

From the figure it is clear that

$$a \sin B = b \sin A = h$$

or

$$\frac{a}{\sin A} = \frac{b}{\sin B} = \frac{c}{\sin C}. \tag{1.4}$$

The last term in (1.4) was deduced from symmetry and can easily be confirmed by dropping a perpendicular from one of the other apices. Time and effort can often be saved by the use of symmetry in mathematical derivations, as is illustrated by this example.

>
> **Exercise 1.1**
> Find the other side and angles of a triangle with $a=5$, $b=6$, $B=50°$.

1.1.2 The cosine relationship

In Figure 1.2 the distance PB is marked as x, but this was not used in the derivation of (1.3). We shall use it now. From Pythagoras' theorem

$$a^2 = h^2 + x^2 \tag{1.5}$$

and we also have $h = b \sin A$ and $x = c - b \cos A$.

Substituting for h and x in (1.5),

$$a^2 = b^2 \sin^2 A + c^2 + b^2 \cos^2 A - 2bc \cos A.$$

Since $\sin^2 A + \cos^2 A = 1$ we find

$$a^2 = b^2 + c^2 - 2bc \cos A. \tag{1.6}$$

which is the required cosine relationship.

Three angles give the shape of the triangle, but not its size, so that only the relative lengths of the sides can be found. If one side and two angles are given then the third angle follows and the given side fixes the size of the triangle, so both other sides can be found. If three sides are given then all the angles can be found from (1.6). If two sides and the included angle are given then clearly the triangle is completely defined. However, if the angle is not included then, in general, the triangle is not uniquely defined. This is illustrated in Figure 1.3. The sides a and b and the angle B are given but there are two possible triangles – ABC and A'BC.

>
> **Exercise 1.2**
> Find the angles of a triangle with $a=5$, $b=6$, $c=8$.

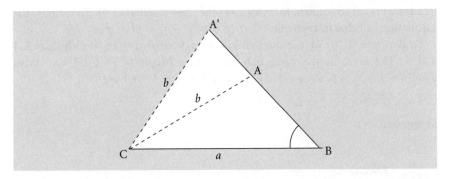

Figure 1.3 If sides *a* and *b* and angle B are given, the triangle is not uniquely defined.

1.1.3 The area of a triangle

The area of a triangle is usually quoted as 'half the base times the height'. As we can see from Figure 1.2, a rectangle on the base AB of height h clearly has twice the area of the triangle. So we can write the area as

$$S = \frac{1}{2}ch = \frac{1}{2}bc\sin A = \frac{1}{2}ac\sin B = \frac{1}{2}ab\sin C \tag{1.7}$$

where, once again, the last entry in (1.7) comes from symmetry considerations.

Since a triangle is defined by its three sides, it must be possible to find the area in terms of the sides. From Figure 1.2,

$$h^2 = a^2 - x^2 = b^2 - (c - x)^2$$

which gives

$$x = \frac{a^2 + c^2 - b^2}{2c}. \tag{1.8}$$

We now find

$$S = \frac{1}{2}ch = \frac{1}{2}c\sqrt{a^2 - x^2} = \frac{1}{2}c\sqrt{a^2 - \frac{(a^2 + c^2 - b^2)^2}{4c^2}},$$

or

$$S = \frac{1}{4}\sqrt{2a^2b^2 + 2b^2c^2 + 2c^2a^2 - a^4 - b^4 - c^4} \tag{1.9}$$

which gives the area in terms of the sides as required. Notice the reassuring symmetry of (1.9); if we had obtained a formula that was not symmetrical in a, b, and c it would surely have been wrong, since the sides have similar roles in defining the area.

There is another form of equation giving the area in terms of sides that is messy to derive but easy to confirm. Given $s = \frac{1}{2}(a + b + c)$, then

$$S = \sqrt{s(s - a)(s - b)(s - c)}. \tag{1.10}$$

Substituting for s in (1.10),

$$S = \frac{1}{4}\sqrt{(a + b + c)(b + c - a)(a + c - b)(a + b - c)}.$$

The reader is left to confirm that this gives the same result as (1.9). To simplify the calculation, it helps to remember that $(p + q)(p - q) = p^2 - q^2$.

To illustrate (1.10) we use the well-known right-angled triangle with sides 3, 4 and 5. The area, in the form half base times height is $\frac{1}{2} \times 4 \times 3 = 6$. With $s = \frac{1}{2}(3 + 4 + 5) = 6$, $s - a = 3$, $s - b = 2$, and $s - c = 1$, we have

$$S = \sqrt{6 \times 3 \times 2 \times 1} = 6,$$

as required.

Exercise 1.3

Find the area of a triangle with $a = 4$, $b = 7$, and $c = 9$.

1.2 **Trigonometric relationships**

In calculations involving trigonometric functions we are often required to express a function of the sum or difference of angles in terms of functions of the individual angles. Examples of this will occur in following chapters. Here we give the most important of the relationships of this kind and show how they are derived.

1.2.1 **The sine of the sum and difference of angles**

Figure 1.4 shows a triangle with the angle C divided into two parts, θ and ϕ, by the line CP that is perpendicular to AB. From (1.4),

$$\frac{\sin(\theta + \phi)}{c} = \frac{\sin A}{a} = \frac{\sin B}{b}.$$

However, since CPB and CPA are right-angled triangles,

$$\sin A = \cos\theta \text{ and } \sin B = \cos\phi$$

so that

$$\frac{\sin(\theta + \phi)}{c} = \frac{\cos\theta}{a} = \frac{\cos\phi}{b}. \tag{1.11}$$

From (1.11),

$$\sin(\theta + \phi) = c\frac{\cos\phi}{b} = x\frac{\cos\phi}{b} + (c - x)\frac{\cos\phi}{b}$$

$$= x\frac{\cos\phi}{b} + (c - x)\frac{\cos\theta}{a}. \tag{1.12}$$

From the right-angled triangles CAP and CPB,

$$\frac{x}{b} = \sin\theta \text{ and } \frac{c - x}{a} = \sin\phi$$

which, inserted in (1.12), gives

$$\sin(\theta + \phi) = \sin\theta\cos\phi + \sin\phi\cos\theta. \tag{1.13}$$

A useful subsidiary result, which comes from making $\phi = \theta$, is

$$\sin 2\theta = 2\sin\theta\cos\theta. \tag{1.14}$$

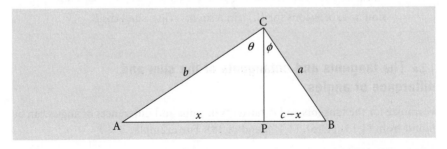

Figure 1.4 A triangle with angle C divided into two parts by CP, which is perpendicular to AB.

From the expansions of $\sin(x)$ and $\operatorname{cosine}(x)$ as power series of x, shown in §4.7.1, it is evident that sine is an **odd function** $\{f(-x)=-f(x)\}$ and cosine an **even function** $\{f(-x)=f(x)\}$ so that

$$\sin(-\theta) = -\sin\theta \ \text{ and } \ \cos(-\theta) = \cos(\theta). \tag{1.15}$$

Substituting $-\phi$ for ϕ in (1.13) gives

$$\sin(\theta - \phi) = \sin\theta\cos\phi - \sin\phi\cos\theta. \tag{1.16}$$

1.2.2 The cosine of the sum and differences of angles

We can obtain the formula for the cosine of a sum of angles from (1.13) as follows:

$$\begin{aligned}\cos(\theta + \phi) &= \sin(\pi/2 - \theta - \phi)\\ &= \sin(\pi/2 - \theta)\cos(-\phi) + \cos(\pi/2 - \theta)\sin(-\phi).\end{aligned}$$

From the symmetry relationships (1.15) this gives

$$\cos(\theta + \phi) = \cos\theta\cos\phi - \sin\theta\sin\phi. \tag{1.17}$$

The corresponding relationship for the difference of angles is

$$\cos(\theta - \phi) = \cos\theta\cos\phi + \sin\theta\sin\phi. \tag{1.18}$$

By making $\phi = \theta$ in (1.17) and using $\sin^2\theta + \cos^2\theta = 1$ we find the useful relationships

$$\cos 2\theta = \cos^2\theta - \sin^2\theta = 2\cos^2\theta - 1 = 1 - 2\sin^2\theta. \tag{1.19}$$

Exercise 1.4

Use the information that $\sin\dfrac{\pi}{4} = \cos\dfrac{\pi}{4} = \dfrac{1}{\sqrt{2}}$, $\sin\dfrac{\pi}{6} = \dfrac{1}{2}$ and $\cos\dfrac{\pi}{6} = \dfrac{\sqrt{3}}{2}$ to find

(i) $\sin\dfrac{5\pi}{12}$;   (ii) $\cos\dfrac{5\pi}{12}$;   (iii) $\sin\dfrac{\pi}{12}$;   (iv) $\cos\dfrac{\pi}{12}$.

Exercise 1.5

Show that (i) $\cos A\cos B = \dfrac{1}{2}\cos(A - B) + \dfrac{1}{2}\cos(A + B)$. Find similar expressions for (ii) $\sin A\sin B$;   (iii) $\sin A\cos B$.

1.2.3 The tangents and cotangents of the sum and difference of angles

Formulae for the tangents and cotangents of sums and differences of angles can be found from (1.13),(1.16), (1.17), and (1.18). For example,

$$\tan(\theta + \phi) = \frac{\sin(\theta + \phi)}{\cos(\theta + \phi)} = \frac{\sin\theta\cos\phi + \sin\phi\cos\theta}{\cos\theta\cos\phi - \sin\theta\sin\phi}.$$

Dividing top and bottom by $\cos\theta\cos\phi$ then gives

$$\tan(\theta + \phi) = \frac{\tan\theta + \tan\phi}{1 - \tan\theta\tan\phi}. \tag{1.20}$$

By similar means it is found that

$$\tan(\theta - \phi) = \frac{\tan\theta - \tan\phi}{1 + \tan\theta\tan\phi}, \tag{1.21}$$

$$\cot(\theta + \phi) = \frac{\cot\theta\cot\phi - 1}{\cot\theta + \cot\phi}, \tag{1.22}$$

$$\cot(\theta - \phi) = \frac{\cot\theta\cot\phi + 1}{\cot\theta - \cot\phi}. \tag{1.23}$$

 Exercise 1.6
Use the information that $\tan\dfrac{\pi}{4} = 1$ and $\tan\dfrac{\pi}{6} = \dfrac{1}{\sqrt{3}}$ to find

(i) $\tan\dfrac{5\pi}{12}$;   (ii) $\cot\dfrac{5\pi}{12}$;   (iii) $\tan\dfrac{\pi}{12}$;   (iv) $\cot\dfrac{\pi}{12}$.

1.2.4 Sum of cosines

Here we are going to find a composite single-term expression for the quantity $\cos A + \cos B$. First we write

$$A = X + Y \text{ and } B = X - Y \tag{1.24}$$

from which we find

$$\cos A = \cos X \cos Y - \sin X \sin Y \tag{1.25a}$$

$$\cos B = \cos X \cos Y + \sin X \sin Y. \tag{1.25b}$$

From (1.25a) and (1.25b),

$$\cos A + \cos B = 2\cos X \cos Y \tag{1.26}$$

From (1.24) we find

$$X = \frac{A + B}{2} \text{ and } Y = \frac{A - B}{2}$$

and substituting this in (1.26),

$$\cos A + \cos B = 2\cos\frac{A + B}{2}\cos\frac{A - B}{2}. \tag{1.27}$$

 Exercise 1.7
Find a composite single-term expression for $\cos A - \cos B$.

1.2.5 Sum of sines

From (1.24) we find

$$\sin A = \sin X \cos Y + \cos X \sin Y, \tag{1.28a}$$

$$\sin B = \sin X \cos Y - \cos X \sin Y. \tag{1.28b}$$

From (1.28a) and (1.28b),

$$\sin A + \sin B = 2 \sin X \cos Y = 2 \sin \frac{A+B}{2} \cos \frac{A-B}{2}. \tag{1.29}$$

Exercise 1.8
Find a composite single-term expression for $\sin A - \sin B$.

1.3 The binomial expansion (theorem)

1.3.1 Integral positive powers

The binomial expansion, or theorem, has to do with the evaluation of quantities of the form $(a+b)^n$, where n can be any real number. A few examples of a binomial expansion, with n as a positive integer, are

$$(a+b)^2 = a^2 + 2ab + b^2,$$

$$(a+b)^3 = a^3 + 3a^2b + 3ab^2 + b^3,$$

$$(a+b)^4 = a^4 + 4a^3b + 6a^2b^2 + 4ab^3 + b^4, \tag{1.30}$$

$$(a+b)^5 = a^5 + 5a^4b + 10a^3b^2 + 10a^2b^3 + 5ab^4 + b^5.$$

A study of these expansions shows that the number of terms in these finite expansions is $n+1$ and also the symmetry of the numerical coefficients.

A general form of the expansion for positive integral n is

$$(a+b)^n = a^n + na^{n-1}b + \frac{n(n-1)}{1 \times 2} a^{n-2}b^2 + \cdots + \frac{n(n-1)\ldots(n-r+1)}{1 \times 2 \times \cdots \times r} a^{n-r}b^r$$

$$+ \cdots + \frac{n(n-1)\ldots 2}{1 \times 2 \times \cdots \times (n-1)} ab^{n-1} + b^n. \tag{1.31}$$

The coefficients of this general expansion seem rather cumbersome in form, but they can be expressed in a more succinct way. The product $1 \times 2 \times 3 \times \cdots \times (n-1) \times n$ is written as $n!$, referred to as **factorial n**. With this notation the coefficient for the $(r+1)$th term in the expansion (1.21) is

$$\frac{n(n-1)\ldots(n-r+1)}{1 \times 2 \times \cdots \times r} = \frac{n!}{r!(n-r)!} \tag{1.32}$$

We shall see in Chapter 12 that the quantity on the right-hand side of (1.32) is rather special in that it gives the number of ways that a selection of r objects can be made from n. For this reason it is represented by a special notation, so we can write

$$\frac{n!}{r!(n-r)!} = {}^nC_r. \tag{1.33}$$

With this notation (1.31) can be written as

$$(a+b)^n = \sum_{r=0}^{n} {}^nC_r a^{n-r} b^r. \tag{1.34}$$

One point must be noted here. The first term of the series is

$${}^nC_0 a^n = \frac{n!}{n!0!} a^n$$

and since we know that the coefficient is unity, the conclusion is that $0! = 1$. That fits in with the idea of making selections, since the number of ways of selecting no objects from n is one – i.e. do nothing!

It has already been pointed out that the coefficients in (1.31) are symmetrical in the sense that the first coefficient equals the last, the second coefficient equals the one from last, and so on. This is because

$${}^nC_r = {}^nC_{n-r} \tag{1.35}$$

which means, as a moment's thought will confirm, the number of ways of selecting r objects from n is the same as the number of ways of selecting $n-r$ objects from n.

Exercise 1.9
Write down the expansion of $(a+b)^6$.

1.3.2 Integral negative powers

We now go on to consider a possible binomial expansion when the power is a negative integer, i.e. we have an expression of the form

$$(a+b)^{-n} = \frac{1}{(a+b)^n}.$$

A simplification we can make is to divide inside the bracket by the larger magnitude of a or b (which can be positive or negative), which we shall take as a in this case to give

$$(a+b)^{-n} = a^{-n}(1+x)^{-n} \tag{1.36}$$

where $x = b/a$ and $|x| < 1$. The expansion of $(1+x)^{-n}$ follows the form of (1.31) with the difference that the number of terms is infinite. Thus

$$(1+x)^{-n} = 1 + (-n)x + \frac{-n(-n-1)}{1 \times 2}x^2 + \cdots + \frac{-n(-n-1)\ldots(-n-r+1)}{1 \times 2 \times \cdots \times r}x^r + \cdots$$

$$= \sum_{r=0}^{\infty}(-1)^r {}^{n+r-1}C_r x^r \qquad (1.37)$$

To see what this means in practice, let us consider the expansion of $(1+0.2)^{-3}$. In the expanded form the first few terms are

$$(1+0.2)^{-3} = 1 - 3 \times 0.2 + \frac{3 \times 4}{1 \times 2}(0.2)^2 - \frac{3 \times 4 \times 5}{1 \times 2 \times 3}(0.2)^3$$

$$+ \frac{3 \times 4 \times 5 \times 6}{1 \times 2 \times 3 \times 4}(0.2)^4 + \cdots.$$

The sums of 1, 2, 3, 4, and 5 terms are respectively 1.0000, 0.4000, 0.6400, 0.5600, and 0.5840, which can be compared with the actual value 0.5787. Note that as the number of terms in the summation increases so the sum oscillates around the true value, alternately greater and less, and becomes closer to the true value. Figure 1.5 shows the values for up to 15 terms.

The oscillation about the true value is evident from the figure, as is also the convergence to the true value. Figure 1.5 shows the effect for a positive value of x; the effect of having a negative value of x is shown in Figure 1.6, which gives the sum for different numbers of terms in the expansion of $(1-0.2)^{-3}$. Once again there is convergence to the true value, 1.953 125, but this time asymptotically from the initial lower value.

We got to the form $(1+x)^n$ by dividing within the bracket by a so the summation we have found has to be multiplied by a^{-n} to give an estimate of the value of $(a+b)^{-n}$. The division within the bracket was made by the term with the largest magnitude in order that $|x| \leq 1$ and the reason for this is that this is a necessary condition for the expansion (1.37) to converge. To illustrate this we show in Figure 1.7 the summation to various numbers of terms of $(1+1.1)^{-3}$, the true value of which is 0.107 980.

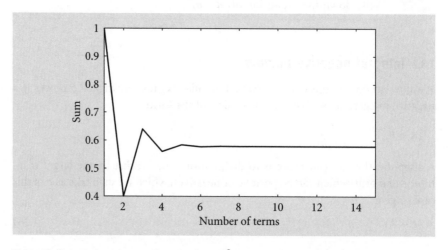

Figure 1.5 Summation of expansion of $(1+0.2)^{-3}$ with increasing number of terms.

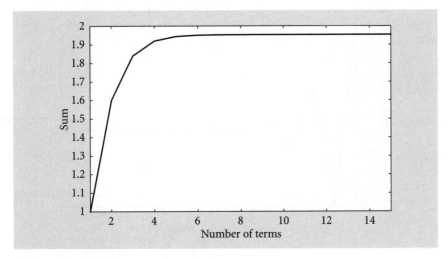

Figure 1.6 Summation of expansion of $(1-0.2)^{-3}$ with increasing number of terms.

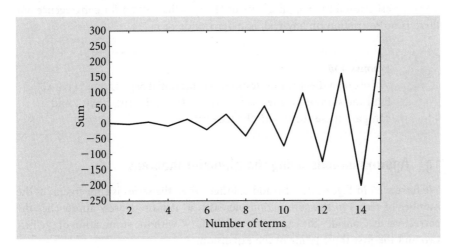

Figure 1.7 Summation of expansion of $(1+1.1)^{-3}$ with increasing number of terms.

In this case it is clear that the sum of the series diverges as more terms are taken. Although $|x| < 1$ ensures convergence of the expansion, if $|x|$ is close to unity the number of terms required to get reasonable accuracy may be very large. On the other hand, as we see in Figures 1.5 and 1.6, with a small value of $|x|$ high accuracy is achieved with only a few terms.

1.3.3 Fractional powers – positive or negative

So far we have taken n to be either a positive or negative integer, but we can drop that restriction and n can be any real number whatsoever. This is illustrated in Figure 1.8, which shows the sum to various numbers of terms in the expansion of $(1+0.3)^{-3.3}$. The summation quickly converges to the true value, 0.420 714.

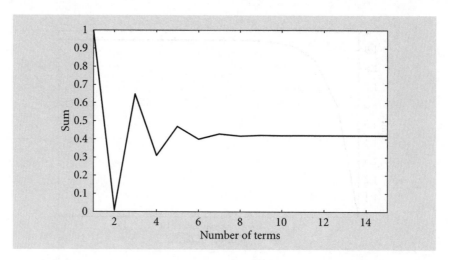

Figure 1.8 Summation of expansion of $(1+0.3)^{-3.3}$ with increasing number of terms.

The only criterion for the applicability of (1.37) is that $|x| < 1$ for convergence but there is no restriction on any real, non-zero value of n.

Exercise 1.10
Write down the first four terms in the binomial expansion of $(1+x)^{0.6}$. Find the sum of four terms for (i) $x = 0.4$ and (ii) $x = 0.1$ and compare these sums with the true values.

1.3.4 Approximations using the binomial theorem

We have seen in Figures 1.5, 1.6 and 1.8 that, when the series is convergent, as the number of terms in the summation increases so it more closely approaches the correct value. Consider the expansion of $(1+0.1)^3$ with the summation of the first two and the first three terms of the expansion:

$(1.0+0.1)^3 = 1.331$; 2 terms give 1.3, 3 terms give 1.33.

In this case, for some applications a two-term summation may be adequate, for many more applications three terms would be good enough. However, if the value corresponding to x is increased these approximations become progressively worse. For example,

$(1.0+0.3)^3 = 2.197$; 2 terms give 1.9, 3 terms give 2.17

$(1.0+0.5)^3 = 3.375$; 2 terms give 2.5, 3 terms give 3.25.

Going in the other direction, if the value of x is made smaller then the two-term summation becomes progressively better as x is reduced. As examples,

$(1.0+0.01)^4 = 1.040\,604\,01$; 2 terms give 1.04

$(1.0+0.001)^4 = 1.004\,006\,004$; 2 terms give 1.004

$(1.0+0.0001)^4 = 1.000\,400\,06$; 2 terms give 1.0004.

It is clear that in the limit of very small ε we can write

$$(1 + \varepsilon)^n \approx 1 + n\varepsilon \tag{1.38}$$

and this can be a useful approximation in many scientific analyses. For example, if the ratio $\frac{1+\varepsilon}{1-\varepsilon}$ appears in an analysis and $\varepsilon \ll 1$, then we can write

$$\frac{1 + \varepsilon}{1 - \varepsilon} = (1 + \varepsilon)(1 - \varepsilon)^{-1} = (1 + \varepsilon)(1 + \varepsilon) = 1 + 2\varepsilon$$

where terms in ε^2 have been neglected.

However, the **linear approximation** (1.38) will sometimes be inadequate, even for very tiny ε, as the following example shows. Consider

$$\left[\frac{(1 + 2\varepsilon)^{-2} - (1 + 4\varepsilon)^{-1}}{\varepsilon^2} \right]_{\varepsilon \to 0}.$$

The linear approximation applied to the numerator gives $1 - 4\varepsilon - 1 + 4\varepsilon = 0$, so that with $\varepsilon \to 0$ the expression becomes 0/0, which is indeterminate. In this particular case we must use the **quadratic approximation**

$$(1 + \varepsilon)^n \approx 1 + n\varepsilon + \frac{n(n - 1)}{2}\varepsilon^2 \tag{1.39}$$

to get a meaningful answer. With this approximation,

$$\left[\frac{(1 + 2\varepsilon)^{-2} - (1 + 4\varepsilon)^{-1}}{\varepsilon^2} \right]_{\varepsilon \to 0} = \frac{1 - 4\varepsilon + 12\varepsilon^2 - 1 + 4\varepsilon - 16\varepsilon^2}{\varepsilon^2} = \frac{-4\varepsilon^2}{\varepsilon^2} = -4.$$

Another example of the use of the binomial approximation is in the solution of equations where the answer is known to be a small quantity. If we take the equation

$$\frac{1}{(1 + x)^4} + x = 1.01$$

then it is clear that $x = 0$ nearly satisfies the equation and that the true value of $x \ll 1$. This being so then, using the linear approximation,

$$1 - 4x + x = 1.01 \quad \text{or} \quad x = -0.003\,33.$$

The precise solution to three significant figures is $x = -0.003\,30$.

Exercise 1.11
Find the binomial linear approximation to $(1 + 0.01)^{-5}$ and compare this with the true value.

1.4 The exponential e

There are many special numbers that occur in mathematics and in everyday life. A well-known example is π, the ratio of the circumference of a circle to its diameter, which crops up in scientific analyses even when circles or circular motion do not seem to be involved. Another such number is $\sqrt{2}$ which, apart from its mathematical occurrences, has properties on which European paper sizes are based. The paper size A0 has an area of $1\,m^2$ and sides in the ratio $\sqrt{2}:1$. Cutting an A0 sheet in half parallel to the short side gives two A1 sheets, again with sides in the ratio $\sqrt{2}:1$. This process continues to give sizes A2, A3, and so on, with A4 being the normal letter size. These useful numbers, π and $\sqrt{2}$, are **irrational numbers**, meaning that it is impossible to represent them *exactly* by the ratio of two integers, no matter how large those integers are. An example of a **rational number** is $0.111\,111\ldots$, where the number of unit digits is infinite. This number is exactly represented by 1/9.

Another irrational number which has a special place in mathematics is the **exponential e**, which can be defined in various ways but is here defined as

$$e = \left[\left(1+\frac{1}{n}\right)^{n}\right]_{n\to\infty}. \tag{1.40}$$

Table 1.1 shows the progressive increase of the right-hand side of (1.40) as n is increased.

The special properties of e that make it important in mathematics will come out in the chapters that follow. It is the basis for describing **exponential growth or decline**, which occurs in many natural situations, such as changes of population or of the radiation from a radioactive source. The output from a radioactive source may be described as

$$N_t = N_0 e^{-\lambda t} \tag{1.41}$$

where N_t and N_0 are the rates of emission at times t and 0 respectively and λ is the **decay constant** for the material. If a particular material emits 10^5 particles per second at a particular time, which we take as $t=0$, and has a decay constant of $0.01\,\text{days}^{-1}$ then after 100 days the emission will be

$$N_{100} = 10^5 e^{-0.01\times100}\,s^{-1} = 10^5\,e^{-1}s^{-1} = 36\,788\,s^{-1}.$$

Table 1.1 Values of (1.40) for different values of n

n	$\left(1+\frac{1}{n}\right)^{n}$	n	$\left(1+\frac{1}{n}\right)^{n}$
1	2.000 000	50	2.691 588
2	2.250 000	100	2.704 814
4	2.370 370	200	2.711 517
6	2.521 626	500	2.715 568
10	2.593 742	10^3	2.716 924
15	2.632 879	10^4	2.718 146
20	2.653 298	10^5	2.718 268
30	2.674 319	∞	2.718 282

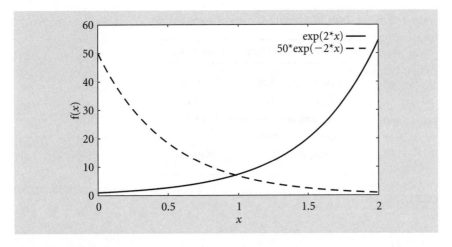

Figure 1.9 The characteristics of exponential growth and decline.

The expression (1.41) gives **exponential decline** since the exponent of e is negative. With a positive exponent the result is **exponential growth**. Figure 1.9 shows the characteristics of exponential growth and decline. It will be seen that exponential growth gives a runaway increase to infinity while exponential decline leads to an asymptotic decline to zero.

Sometimes the power to which e is raised can be a cumbersome expression that would be unclear with the notation given above, so an alternative notation is available. To show the comparison, the expression

$$e^{-\left\{\frac{1}{1+x} + \frac{a}{(1+x)^2}\right\}}$$

is more conveniently written as

$$\exp\left[-\left\{\frac{1}{1+x} + \frac{a}{(1+x)^2}\right\}\right].$$

1.5 Natural logarithms

Before the advent of electronic calculators and computers the only computational aids were slide-rules and logarithms, with the latter the method of choice if precision was required. Such uses are now outmoded, but logarithms still have a role to play in science. The usual logarithms use a base 10 and can be defined as follows:

$$\text{if } y = 10^x \text{ then } x = \log_{10} y. \tag{1.42}$$

When the base is 10 the subscript 10 is usually omitted.

One role of logarithms these days is in representing a large range of numbers in some graphical form. For example, Table 1.2 gives the mean length and mean mass of various kinds of animal.

If the relationship between mass and length is expressed with a linear scale for both quantities, as in Figure 1.10a, the masses for the mouse and rabbit are close to zero and impossible to determine. On the other hand a logarithmic

Table 1.2 Length and mass of some animals

Animal	Length (m)	Mass (kg)
Mouse	0.08	0.10
Rabbit	0.28	2.10
Dog (terrier)	0.50	15.2
Tiger	1.95	150
Elephant	3.30	1510

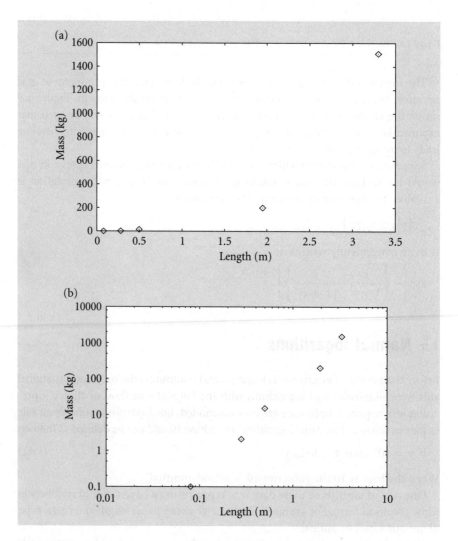

Figure 1.10 The lengths and masses of animals with (a) linear scales and (b) logarithmic scales.

representation, as in Figure 1.10b, spreads the points out and makes the relationship much easier to interpret.

Just as the exponential e plays an important role in mathematics and in the relationships that describe physics and other sciences, so logarithms to the base e are also important. Corresponding to (1.42) we write

$$\text{If } y = e^x \text{ then } x = \log_e y = \ln y. \tag{1.43}$$

Notice the use of 'ln' to represent logarithms to the base e, referred to as **natural logarithms**. As their properties are explored in subsequent chapters, their intrinsic importance will be more greatly appreciated.

Exercise 1.12
A radioactive source is emitting 56 392 alpha-particles per second. After 1 week it is emitting 42 985 particles per second. (i) What will be the emission 3 weeks after the initial observation? (ii) What is the decay constant of the material in units s^{-1}?

Exercise 1.13
(i) If $\ln(1.6) = \log(x)$, what is x?

(ii) If $\log(7.2) = \ln(x)$, what is x?

1.6 Two-dimensional coordinate systems

The simplest type of coordinate system is based on a set of axes that define different directions in space. In Figure 1.11 we show such a pair of axes, x and y, to define a position in a two-dimensional space.

The position of a point is described by its component displacements in the directions of the axes. Such a system is called a **Cartesian coordinate system**. For most purposes it is an advantage to have the defining axes orthogonal (at right angles to each other) and such a system is called a rectangular Cartesian system. Often 'rectangular' is excluded from the description, but Cartesian coordinate systems are usually assumed to be rectangular.

Another common choice of coordinate system in two dimensions is that of **polar coordinates**. In this system, illustrated in Figure 1.12, a point is described with coordinates (r, θ), being equivalent to (x, y) in a rectangular Cartesian system.

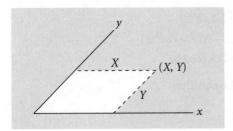

Figure 1.11 A Cartesian system of coordinate axes.

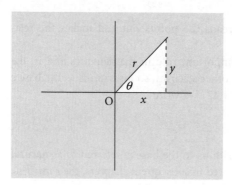

Figure 1.12 The relationship between polar and rectangular Cartesian coordinates.

From simple geometry and trigonometry the transformation from one coordinate system to the other comes from

$$r = \sqrt{x^2 + y^2} \quad \text{and} \quad \theta = \tan^{-1}\left(\frac{y}{x}\right) \tag{1.44a}$$

where the signs of $\sin \theta$ and $\cos \theta$ are given by y and x respectively and

$$x = r \cos \theta \quad \text{and} \quad y = r \sin \theta. \tag{1.44b}$$

Later, in Chapter 13, we introduce coordinate systems for three dimensions.

 Exercise 1.14
Find the polar coordinate equivalent of the rectangular Cartesian coordinates (i) (2, 3); (ii) (3, −2), and the rectangular Cartesian coordinate equivalent of the polar coordinates; (iii) (5, π/3); (iv) (3, 1.42π).

Problems

1.1 A farmer with a quadrilateral field needs to know its area to obtain a farming subsidy. A surveyor measures the following:

Side AB = 100 m, side BC = 90 m, side CD = 56 m, side DA = 60 m and diagonal DB = 95 m. What is the area of the field in hectares?
(1 hectare = 10 000 m²)

1.2 Show that
$$\sin^2(A + B) - \sin^2(A - B) = \sin 2A \sin 2B,$$
$$\sin^2(A + B) + \sin^2(A - B) = 4(1 - \cos^2 A \cos^2 B),$$
$$\cos^2(A - B) - \sin^2(A + B) = \cos 2A \cos 2B.$$

1.3 Find an approximate solution of the equation
$$\frac{1}{(1 - x)^3} + 2x - 1 = 0.001.$$

1.4 Find an approximate solution of the equation

$$\frac{1}{(1+x)^2} + 2x - 1 = 0.001.$$

1.5 The population of India was 550 million in 1972 and 660 million in 1982. Assuming exponential growth, estimate the population in the year 2010.

1.6 Using a calculator find $\ln(1.5)$, $\ln(1.4)$, $\ln(1.3)$, $\ln(1.2)$, $\ln(1.1)$, $\ln(1.01)$, $\ln(1.001)$, and $\ln(1.0001)$. What do you deduce about the value of $\ln(1+x)$?

Dimensions and dimensional analysis

We cannot express time in metres or mass in seconds, since the dimensions of the quantities do not match. The restriction of matching dimensions must be obeyed by plausible equations in physics. In this chapter you will find out about the total range of dimensions that physical quantities can possess, and the units in which these quantities are expressed. You will also discover that in favourable cases it is possible to derive physical equations by consideration of dimensions alone.

2.1 Basic units and dimensions

The concepts of dimensions and units are familiar ones and in everyday use. Cheese is sold by **mass** (although normally we think of **weight** in this context) in kilograms or pounds. Distances between towns are **lengths** expressed in miles or kilometres, and the duration of a sprint race is a **time** measured in seconds. For any particular basic dimension, say length, multiples and subdivisions of units are defined to give convenient units for measuring at different scales. The metre is the basic unit of length in the SI (Système Internationale) set of units used by scientists. A complete list of base SI units is given in Table 2.1.

For distances between towns the kilometre is a more convenient unit and for measuring the diameter of a screw the millimetre is appropriate – the basic goal being to have the numerical value of the distance of a size that can readily be conceptualized. It is not inaccurate to say that the distance from Copenhagen to Rome is 1 247 000 m, but 1247 km is more easily understood.

Combining basic units of mass, length, and time is something else that is commonly done. Speed-limit signs are in the units kilometres per hour ($km\,h^{-1}$) – or in miles per hour in the USA and UK, which still use imperial units for some non-scientific purposes. The output of steel from an ironworks would be expressed in the units tonnes per year, and the density of materials in the units kilograms per cubic metre ($kg\,m^{-3}$). Table 2.1 shows all the units that are necessary to define quantities of various kinds.

Although it is possible, it is not convenient to express all quantities in terms of the base units given in Table 2.1 and many derived units are in common use. One,

Table 2.1 Base SI units. Those marked * are supplementary units

Physical quantity	Unit	Abbreviation
Mass	kilogram	kg
Length	metre	m
Time	second	s
Temperature	kelvin	K
Electric current	ampere	A
Luminous intensity	candela	cd
Amount of substance	mole	mol
*Plane angle	radian	rad
*Solid angle	steradian	sr

Table 2.2 Some derived units for mechanical quantities

Quantity	Name	Symbol	Description	In base units
Force	newton	N		$kg\,m\,s^{-2}$
Pressure	pascal	Pa	$N\,m^{-2}$	$kg\,m^{-1}\,s^{-2}$
Energy	joule	J	$N\,m$	$kg\,m^2\,s^{-2}$
Power	watt	W	$J\,s^{-1}$	$kg\,m^2\,s^{-3}$
Frequency	hertz	Hz		s^{-1}

the unit of charge, the **coulomb (C)**, is the quantity of electric charge that is transported by a current of one ampere flowing for one second. Energy, for example kinetic energy, has the dimensions mass $\times$ length^2 $\times$ time^{-2} and a basic unit of energy would naturally be $kg\,m^2\,s^{-2}$, which indeed it is. However, energy is a commonly occurring quantity in science and this basic unit is given a name of its own, the **joule (J)**. This process can be carried one stage further. The rate of energy output, say of an electric motor, is expressed as energy per unit time and a basic unit is one joule per second ($J\,s^{-1}$). Again, since this is a commonly occurring type of quantity the basic unit is given a distinctive name and $1\,J\,s^{-1}$ is called 1 **watt (W)**. A number of other derived units, not related to electrical or magnetic quantities, are given in Table 2.2.

In describing the luminous intensity of a object it is necessary to introduce another base unit, the **candela**, which is the luminous intensity of $(1/600\,000)\,m^2$ of a black-body surface at a temperature of 2040 K (the melting point of platinum) and a pressure of 1 atmosphere.

The subsidiary angle base units, the **radian** and **steradian**, are really dimensionless but are necessary to indicate the nature of the quantity. Thus an angle in radians can be defined in terms of the length of the arc of a circle subtending that angle at the centre, and is the ratio of the length of the arc to the radius, i.e.

$$\alpha = \frac{s}{r} \tag{2.1a}$$

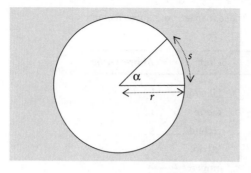

Figure 2.1 The angle α in radians is s/r.

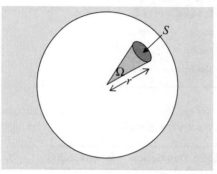

Figure 2.2 The solid angle Ω is given by S/r^2 where S is the area intersected by the angle on the sphere.

– a dimensionless quantity (Figure 2.1). If we say that an angle is 0.6 rad then it is clear that it is not, for example, 0.6° since the degree is another dimensionless unit for measuring angles.

Similarly a solid angle in steradians can be defined as the ratio of the area on the surface of a sphere, S, embracing the solid angle diverging from the centre, divided by the square of the radius (Figure 2.2). Thus we write

$$\Omega = \frac{S}{r^2} \tag{2.1b}$$

and since the surface area of a sphere is $4\pi r^2$ a solid angle taking in all directions in space is therefore 4π sr.

There are some quantities that depend not on the mass of a substance as such but more directly on the number of entities of the substance, where the entities can be atoms, molecules, or ions. Thus the translational energy of the motion of the molecules of a gas at absolute temperature θ, is, on average, $1.5\,k\theta$ per molecule, no matter what the molecular mass, where k is the Boltzmann constant, $1.38 \times 10^{-23}\,\mathrm{J\,K^{-1}}$. One mole of a substance contains a number of entities known as **Avogadro's number** (6.02252×10^{23}) which, in its turn, is defined as the number of atoms in $0.012\,\mathrm{kg}$ of $^{12}\mathrm{C}$, the common isotope of carbon.

 Exercise 2.1
Assuming that the Earth is a perfect sphere of radius 6380 km, find the solid angle subtended at the centre of the Earth by the United States of America, of area $9.373 \times 10^6\,\mathrm{km}^2$.

23

2.2 DIMENSIONAL HOMOGENEITY

2.2 Dimensional homogeneity

Observational and theoretical quantities and physical constants appear together in equations that describe the behaviour of the physical world. A very simple equation, for motion in a straight line, gives the relationship between the final speed of a body, v, related to its initial speed, u, its constant acceleration, a, and the time, t. This is

$$v = u + at \tag{2.2}$$

and this equation illustrates what is known as **dimensional homogeneity**. Using **m**, **l**, and **t** to represent the dimensions of mass, length, and time respectively the dimensions of the terms in (2.2) are

$$v[\mathbf{l\,t^{-1}}], u\;[\mathbf{l\,t^{-1}}] \text{ and } at\;[\mathbf{l\,t^{-2}}] \times [\mathbf{t}] = [\mathbf{l\,t^{-1}}]$$

– that is, they all have the same dimensions. This is a necessary, but not sufficient, property of a correct equation and any equation that does not satisfy the condition of dimensional homogeneity *must* be incorrect. The condition must also be satisfied by purely theoretical equations and when Einstein postulated that $E = mc^2$ the equation was believable because the right-hand side, in the form mass × speed^2, had the dimensions of energy. A postulate $E = mc^3$ would not have been credible.

Many equations involve physical constants and to preserve the concept of dimensional homogeneity these too must have dimensions. The rate of energy output of a black body, Q, of area, A, at a temperature, T, is given by

$$Q = \sigma A T^4 \tag{2.3}$$

where σ is Stefan's constant. The value of Stefan's constant is $5.67 \times 10^{-8}\,\mathrm{J\,s^{-1}\,m^{-2}\,K^{-4}}$ so, using θ as the dimensional symbol for temperature and **E** to represent the dimensions of energy, the right-hand side has the dimensions

$$[\mathbf{E\;t^{-1}l^{-2}}\theta^{-4}] \times [\mathbf{l^2}] \times [\theta^4] = [\mathbf{Et^{-1}}],$$

which has the dimensions of the left-hand side. Note here that we have used **E** as a non-basic composite dimension without breaking it down into its mass, length, and time components.

An important physical constant is the gravitational constant, G, that occurs in the equation that gives the force between two masses, m_1 and m_2, separated by a distance d:

$$F = G\frac{m_1 m_2}{d^2} \quad \text{or} \quad G = \frac{Fd^2}{m_1 m_2}. \tag{2.4a}$$

The dimensions of G are given by

$$[G] = [\mathbf{m\,l\,t^{-2}}][\mathbf{l^2}] \div [\mathbf{m^2}] = [\mathbf{m^{-1}l^3t^{-2}}]. \tag{2.4b}$$

Another aspect of dimensional homogeneity occurs when a combination of several physical quantities appears as the argument of a function such as sin, cos, exp, or log. For example, the displacement for one-dimensional simple harmonic motion varies with the amplitude of the motion, A, the angular frequency of the motion, ω, and the time, t. This is

$$x = a \sin \omega t. \tag{2.5}$$

Another example is the Maxwell–Boltzmann distribution giving the distribution of molecular speeds in a gas. The proportion of molecules with speeds between v and $v + \mathrm{d}v$ is $p(v)\,\mathrm{d}v$ where

$$p(v) = \left(\frac{2\mu^3}{\pi k^3 T^3}\right)^{1/2} v^2 \exp\left(-\frac{\mu v^2}{2kT}\right) \tag{2.6}$$

where μ is the mean molecular mass of the gas, k is the Boltzmann constant, and T is the absolute temperature. Now it is possible to express functions such as sin and exp as infinite power series and it is clear that if the argument of the function has some dimension then each term of the power series will have a different dimension, so dimensional homogeneity will not be preserved. The conclusion is that the combination of quantities in the argument must be dimensionless – just a pure number – for then the power series will also be dimensionless. In (2.5) ω, the angular frequency, has dimension $[\mathbf{t}^{-1}]$ so that ωt is dimensionless, as required. The value of the Boltzmann constant is $1.38 \times 10^{-23}\,\mathrm{J\,K^{-1}}$ so its dimensions are $[\mathbf{E}\theta^{-1}]$. Thus the divisor of the argument of exp in (2.6), kT, clearly has the dimensions of energy – as does the numerator since we know that mass $\times$ speed^2 is the combination giving the kinetic energy of a moving object. Hence the argument of exp is dimensionless and so (2.6) exhibits dimensional homogeneity.

The principle of dimensional homogeneity must be present in any plausible relationship linking physical quantities, and an important consequence of this is that it is possible, in some circumstances, to derive the general form of such relationships without recourse to theoretical or experimental studies.

Exercise 2.2

The force, F, acting on a spherical object of radius a moving with speed v through a liquid with coefficient of viscosity η is given by Stokes's law $F = 6\pi\eta av$. From this find the dimensions of viscosity.

2.3 Dimensional analysis

EXAMPLE 2.1

Consider the problem of finding the period, P, of a simple pendulum in terms of the characteristics of the physical system. Clearly the length of the pendulum, l, is an important factor and, without any prior knowledge, we might suppose that the mass of the pendulum bob, m, is another contributory factor. Finally, since the motion is in the Earth's gravitational field the acceleration of gravity, g, is another likely influence. An obvious form of relationship to investigate is

$$P = Cl^\alpha m^\beta g^\gamma \tag{2.7}$$

where C is a numerical constant and α, β, and γ are powers to be determined. The dimension of period is $[\mathbf{t}]$ and that of g is $[\mathbf{lt}^{-2}]$, so dimensionally we can write

$$[\mathbf{t}] = [\mathbf{l}]^\alpha [\mathbf{m}]^\beta [\mathbf{lt}^{-2}]^\gamma.$$

Equating powers of **m**, **l**, and **t** on the two sides we find $\beta = 0$, $\alpha + \gamma = 0$ and $-2\gamma = 1$, giving $\alpha = \frac{1}{2}$, $\beta = 0$, and $\gamma = -\frac{1}{2}$, or

$$P = C\sqrt{\frac{l}{g}}. \tag{2.8}$$

Dimensional analysis shows that the period is independent of the mass of the pendulum bob and also how it depends on the other two quantities but, of course, it gives no information about the dimensionless number, C. It requires an analysis of the actual physics of a simple pendulum to find that $C = 2\pi$.

EXAMPLE 2.2

A study of the rate of flow of liquid through a pipe illustrates a new facet of dimensional analysis. Figure 2.3 shows the physical arrangement giving rise to this phenomenon. A constant-head device, providing a constant pressure difference ΔP between the ends of a pipe of length l and circular internal cross-section of radius a, forces liquid through the pipe at a rate Q, expressed as volume per unit time.

Obviously the rate of flow depends on ΔP, l, and a, but in addition it must depend on viscosity, η, the property of the fluid that describes its propensity to flow; water has a low viscosity and treacle a high viscosity. Viscosity has the dimensions $[\mathbf{ml^{-1}t^{-1}}]$ and we can see that in applying dimensional analysis there will be the problem that there are four physical quantities involved in defining the rate of flow but that there will be only three equations linking powers of **m**, **l**, and **t**. If we express the relationship as

$$Q = C\eta^\alpha \Delta P^\beta l^\gamma a^\delta \tag{2.9}$$

then, dimensionally,

$$[\mathbf{l^3 t^{-1}}] = [\mathbf{ml^{-1}t^{-1}}]^\alpha [\mathbf{ml^{-1}t^{-2}}]^\beta \, [\mathbf{l}]^\gamma \, [\mathbf{l}]^\delta \tag{2.10}$$

Equating powers of dimensions,

for **m**: $\alpha + \beta = 0$

for **l**: $-\alpha - \beta + \gamma + \delta = 3$

for **t**: $-\alpha - 2\beta = -1$

from which we find $\alpha = -1$, $\beta = 1$, and $\gamma + \delta = 3$. Thus the best relationship that can be found by dimensional analysis alone is

$$Q = C\frac{\Delta P}{\eta} l^\gamma a^{3-\gamma}. \tag{2.11}$$

A simple experiment shows that the rate of flow depends on l^{-1}, giving

$$Q = C\frac{\Delta P a^4}{\eta l}.$$

A full analysis, taking the physics into account, gives **Poiseuille's equation:**

$$Q = \frac{\pi \Delta P a^4}{8\eta l}. \tag{2.12}$$

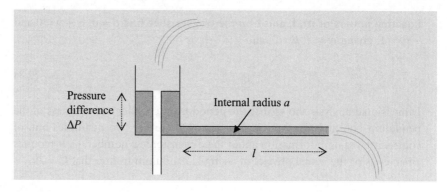

Figure 2.3 Flow of liquid through a pipe under a constant pressure difference.

EXAMPLE 2.3

To illustrate another kind of situation for which dimensional analysis can be used we consider the way in which the density, ρ, of an isothermal atmosphere varies with height given the density at ground level, ρ_0. As with all analyses of this type it is first necessary to identify the variables on which the density depends and these are, together with their dimensions: the mean molecular mass of the atmospheric gas, μ ($[\mathbf{m}]$), the gravitational constant, g ($[\mathbf{lt^{-2}}]$), the height, h ($[\mathbf{l}]$), the temperature of the gas, T ($[\boldsymbol{\theta}]$), and Boltzmann's constant, k ($[\mathbf{E}\theta^{-1}]$) $= ([\mathbf{ml^2t^{-2}\theta^{-1}}])$.

From elementary considerations it is expected that, with all other factors being the same, the density at height h would be proportional to the density at ground level. We may therefore try a relationship of the form

$$\rho = \rho_0 \mu^\alpha g^\beta h^\gamma k^\delta T^\varepsilon$$

and, clearly, the combination $\mu^\alpha g^\beta h^\gamma k^\delta T^\varepsilon$ must be dimensionless. In dimensional terms this means that $[\mathbf{m}]^\alpha \, [\mathbf{lt^{-2}}]^\beta \, [\mathbf{l}]^\gamma \, [\mathbf{ml^2t^{-2}\theta^{-1}}]^\delta [\boldsymbol{\theta}]^\varepsilon$ must be dimensionless. Equating the powers of $\mathbf{m}$, $\mathbf{l}$, $\mathbf{t}$, and $\boldsymbol{\theta}$ to zero:

$$\alpha + \delta = 0,$$

$$\beta + \gamma + 2\delta = 0,$$

$$2\beta + 2\delta = 0,$$

$$-\delta + \varepsilon = 0$$

Since there are fewer equations than unknowns, we can solve for four of the variables in terms of the fifth. Solving in terms of α we find $\beta = \gamma = \alpha$ and $\delta = \varepsilon = -\alpha$, which means we can write

$$\rho = \rho_0 \left(\frac{\mu g h}{kT}\right)^\alpha \tag{2.13}$$

where α is unrestricted from a dimensional point of view. In (2.13), in place of a single term to an unknown power we could have a power series – or essentially any function with argument $\mu g h/kT$. The physics of an isothermal atmosphere gives the relationship as

$$\rho = \rho_0 \exp\left(-\frac{\mu g h}{kT}\right), \tag{2.14}$$

a result consistent with what we have found from dimensional analysis.

These examples have shown both the power and the limitations of dimensional analysis. In favourable cases it can yield a specific relationship, although it can never give the dimensionless constants that only a full physical analysis can provide. In less favourable cases, as for the atmospheric example, it can yield a general form for the relationship but can give no information about the nature of the function – e.g. sin, tan, exp – that may occur.

2.4 Electrical and magnetic units

The concepts of mass, length, time, and temperature are familiar in everyday life but those associated with electricity and magnetism are less so. Here we give a brief resumé of the dimensions and units associated with these quantities, without going into the actual physics associated with them. As will be seen in Table 2.1, the only base electrical unit is the ampere and the derived unit of charge is the coulomb (C). Like the coulomb, all other electrical and magnetic units can be expressed in terms of current, mass, length, and time and a list of such quantities is shown in Table 2.3.

If we use the symbol **c** to represent the dimension of current, then we can check for dimensional homogeneity or use dimensional analysis to derive relationships involving electrical and magnetic quantities. Just as in mechanical systems where it is necessary to bring in physical constants such as the gravitational constant, G, so for electrical and magnetic systems we often need to consider the physical constants μ_0, the **permeability of free space** or **magnetic constant** and ε_0, the **permittivity of free space** or **electric constant**. The value of μ_0 is $4\pi \times 10^{-7}\,\mathrm{N\,A}^{-2}$ and that of ε_0 is $8.854 \times 10^{-12}\,\mathrm{F\,m}^{-1}$. These two quantities are connected by the relationship $\mu_0\varepsilon_0 = 1/c^2$, where c is the speed of light. We can check this dimensionally from the contents of Table 2.3. Using **c** to represent the dimension of current, the dimensions of $\mu_0\varepsilon_0$ are

$$\mathrm{N\,A}^{-2} \times \mathrm{F\,m}^{-1}$$
$$= [\mathbf{mlt}^{-2}][\mathbf{c}^{-2}] \times [\mathbf{m}^{-1}\mathbf{l}^{-2}\mathbf{t}^4\,\mathbf{c}^2][\mathbf{l}^{-1}]$$
$$= [\mathbf{lt}^{-1}]^{-2},$$

which has the dimensions of 1/speed², as is required.

Table 2.3 Derived electrical and magnetic units

Quantity	Name	Symbol	Description	In base units
Electric charge	coulomb	C		$\mathrm{A\,s}$
Electric potential	volt	V	$\mathrm{W\,A}^{-1}$	$\mathrm{kg\,m^2\,s^{-3}\,A^{-1}}$
Capacitance	farad	F	$\mathrm{C\,V}^{-1}$	$\mathrm{kg^{-1}\,m^{-2}\,s^4\,A^2}$
Resistance	ohm	Ω	$\mathrm{V\,A}^{-1}$	$\mathrm{kg\,m^2\,s^{-3}\,A^{-2}}$
Conductance	siemens	S	$\mathrm{A\,V}^{-1}$	$\mathrm{kg^{-1}\,m^{-2}\,s^3\,A^2}$
Magnetic flux	weber	Wb	$\mathrm{V\,s}$	$\mathrm{kg\,m^2\,s^{-2}\,A^{-1}}$
Magnetic flux density	tesla	T	$\mathrm{Wb\,m}^{-2}$	$\mathrm{kg\,s^{-2}\,A^{-1}}$
Inductance	henry	H	$\mathrm{Wb\,A}^{-1}$	$\mathrm{kg\,m^2\,s^{-2}\,A^{-2}}$

Examples of the application of dimensional analysis to electrical and magnetic quantities are given in Problems 2.2 and 2.4.

Exercise 2.3

The rate of change of a potential across a resistor R (ohms), capacitor C (farads), and inductor L (henries) is given by

$$\frac{dV}{dt} = \frac{I}{C} + R\frac{dI}{dt} + L\frac{d^2I}{dt^2}$$

where I is the variable current. Show that this equation is dimensionally sound.

Problems

2.1 Pressure is a measure of energy density, i.e. energy per unit volume. Confirm that this statement is dimensionally sound.

2.2 The magnetic pressure at a point in a magnetic field depends only on the value of the field at that point and the magnetic constant. Find the form of the dependence.

2.3 A long filament of gaseous material of density ρ is gravitationally unstable and breaks up into a string of blobs, each of length λ. The value of λ depends on ρ, on the speed of sound in the gas, c, and the gravitational constant, G. Find the form of the dependency.

2.4 The 'classical radius of an electron', a, can be expressed in terms of the electron charge, e, its mass m_e, the electric constant, ε_0, and the speed of light, c. Find the form of this expression.

Sequences and series

The term series implies that there is a sequence of quantities, in either numerical or functional form, that are systematically related to each other in some way. These sometimes occur in the description of a physical problem, and in this chapter we will find ways of determining the sums of terms of various kinds of series. You will also be introduced to the idea of convergence, which determines whether the sum of an infinite number of terms of a series is finite or goes to infinity.

3.1 Arithmetic series

Consider the series

$$1 \quad 2 \quad 3 \quad 4 \quad 5 \tag{3.1a}$$

or

$$a \quad a+b \quad a+2b \quad a+3b \quad a+4b. \tag{3.1b}$$

These series, in which each term differs from the previous term by a constant, are called **arithmetic series** and the set of numbers or quantities is said to form an **arithmetic progression**.

A series can either be finite, as are those given in (3.1a) and (3.1b), or infinite in extent. Often a quantity of interest is the sum of the terms of a series and in the series given above these sums are readily found to be 15 and $5a + 10b$. However, if the series we wish to sum is

$$S = 1 + 2 + 3 + 4 + \ldots + 999 + 1000 \tag{3.2}$$

then the option of simply adding all the numbers is less attractive. The sum of the first and last terms in (3.2) is 1001, as is the sum of the second term and one from last term, from which it is clear that all the terms can be paired, each pair giving a sum of 1001. This means that the average value of a term is 500.5 and since there are 1000 terms in the series the sum of the terms is 5005. If the series to be summed is of the form

$$S = a + \{a+b\} + \{a+2b\} + \{a+3b\} + \cdots + \{a+(n-2)b\} + \{a+(n-1)b\} \tag{3.3}$$

then the average value of a term is $a + \frac{1}{2}(n-1)b$ and, since there are n terms,

$$S = n\left\{a + \frac{1}{2}(n-1)b\right\}. \tag{3.4}$$

With $a = 1$, $b = 1$, and $n = 1000$ the correct sum is found for the arithmetic series (3.2).

For an arithmetic series with an infinite number of terms ($n=\infty$) the sum will be either $+\infty$ or $-\infty$, and such a series is said to be **divergent**. However, there are other types of series where the sums to an infinite number of terms are finite and such series are said to be **convergent**.

 Exercise 3.1
Find the sum to 40 terms of the arithmetic series $1+3+5+7+\cdots$.

3.2 Geometric series

We now consider the following series of n terms to be summed:

$$S = a + ar + ar^2 + ar^3 + \cdots + ar^{n-2} + ar^{n-1}. \tag{3.5}$$

This series, in which each term is a constant factor times the previous one, is called a **geometric series**. To find the sum we use the equality

$$1 - r^n = (1 - r)(1 + r + r^2 + r^3 + \cdots + r^{n-2} + r^{n-1}), \tag{3.6}$$

which is easily checked. Hence we can find the sum of the geometric series as

$$S = \frac{a(1 - r^n)}{1 - r} \quad \text{or} \quad S = \frac{a(r^n - 1)}{r - 1}. \tag{3.7}$$

We now apply this to the sum

$$S = 1 + 2 + 4 + 8 + \cdots + 128 + 256. \tag{3.8}$$

For this series $a=1$, $r=2$, and $n=9$, giving $S=511$. What is also clear is that this sum increases without limit as more and more terms are taken, and that the series is divergent.

Now we consider another geometrical series:

$$S = 1 + 0.9 + 0.81 + 0.729 + \cdots + (0.9)^8 + (0.9)^9. \tag{3.9}$$

From (3.7),

$$S = \frac{1 - (0.9)^{10}}{1 - 0.9} = 6.5132.$$

Since $r^\infty = 0$, if $r < 1$ then the infinite geometrical series corresponding to (3.9) is convergent with $S = 10$.

As a final example of a geometric series we take

$$S = 1 - 2 + 4 - 8 + 16 - \cdots + (-2)^{n-1} + \cdots \tag{3.10}$$

With each successive term the sum changes sign and becomes larger in magnitude. This is another type of divergent series.

 Exercise 3.2
Using (3.7), sum the geometric series $4+2+1+\frac{1}{2}+\frac{1}{4}+\dots$ (i) to 7 terms and (ii) to 8 terms, and show that the difference equals the eighth term.

3.3 Harmonic series

Another interesting series is the **harmonic series**

$$S = 1 + \frac{1}{2} + \frac{1}{3} + \frac{1}{4} + \cdots + \frac{1}{n} + \cdots. \tag{3.11}$$

This series can be shown to be divergent. We group the terms as follows:

$$S = 1 + \frac{1}{2} + \left(\frac{1}{3} + \frac{1}{4}\right) + \left(\frac{1}{5} + \frac{1}{6} + \frac{1}{7} + \frac{1}{8}\right) + \left(\frac{1}{9} + \frac{1}{10} + \cdots + \frac{1}{15} + \frac{1}{16}\right) + \cdots \tag{3.12}$$

The next parenthesis in the grouping contains 16 terms the first of which is $\frac{1}{17}$ and the last of which is $\frac{1}{32}$. Each of the bracketed terms is greater than $\frac{1}{2}$ and, since the series is an infinite one, there are an infinite number of such bracketed terms – hence the sum is infinity and the series is divergent.

A related series, where the terms have alternate signs, is

$$S = 1 - \frac{1}{2} + \frac{1}{3} - \frac{1}{4} + \cdots + \frac{(-1)^n}{n-1} + \frac{(-1)^{n+1}}{n} + \cdots. \tag{3.13}$$

This series can be shown to be convergent. First we group the terms as

$$S = 1 - \left(\frac{1}{2} - \frac{1}{3}\right) - \left(\frac{1}{4} - \frac{1}{5}\right) - \left(\frac{1}{6} - \frac{1}{7}\right) - \cdots \tag{3.14}$$

and since negative values are repeatedly being taken from the first term it is clear that $S < 1$. Grouping the terms differently,

$$S = 1 - \frac{1}{2} + \left(\frac{1}{3} - \frac{1}{4}\right) + \left(\frac{1}{5} - \frac{1}{6}\right) + \cdots \tag{3.15}$$

from which we find

$$S > \frac{1}{2}.$$

Both the grouped series go smoothly without oscillation to the final sum. We can find this sum in the following way. From the binomial theorem (1.34),

$$(1 + x)^{-1} = 1 - x + x^2 - x^3 + \cdots + (-x)^n + \cdots.$$

Integrating both sides of this expression (see Table 5.1),

$$\ln(1 + x) + C = x - \frac{x^2}{2} + \frac{x^3}{3} - \frac{x^4}{4} + \cdots + (-1)^n \frac{x^{n+1}}{n+1} + \cdots. \tag{3.16}$$

By taking $x = 0$ we find $C = 0$. Now taking $x = 1$,

$$1 - \frac{1}{2} + \frac{1}{3} - \frac{1}{4} + \cdots + \frac{(-1)^{n-1}}{n} + \cdots = \ln 2 = 0.6990. \tag{3.17}$$

Taking the summation, adding one term at a time, gives

$$1.0000, 0.5000, 0.8333, 0.5833, 0.7833, 0.6167, 0.7595, 0.6345, 0.7456, \ldots$$

and the successive values oscillate about the convergence value, getting steadily but very slowly closer.

A number of different series have been described, some convergent and some divergent, and now we consider some tests for determining which series converge and which diverge.

> **?** **Exercise 3.3**
> Show that the series $1 - \dfrac{1}{2} - \dfrac{1}{3} + \dfrac{1}{4} + \dfrac{1}{5} - \dfrac{1}{6} - \dfrac{1}{7} + \cdots$ is convergent, and find some limits for the sum of the infinite series.

3.4 Tests for convergence

A series such as

$$S = a_0 + a_1 + a_2 + a_3 + \cdots + a_{n-1} + a_n + \cdots \qquad (3.18)$$

may contain only positive terms, only negative terms or a mixture of the two. We now consider the series

$$S' = |a_0| + |a_1| + |a_2| + |a_3| + \cdots + |a_{n-1}| + |a_n| + \cdots \qquad (3.19)$$

where the magnitude is taken of each term in (3.18). If (3.19) is convergent then (3.18) must also be convergent and that case (3.18) is said to be **absolutely convergent**. However, if (3.19) is divergent but (3.18) is convergent then (3.18) is said to be **conditionally convergent**. From this we see that (3.13) is conditionally convergent since (3.11), the corresponding series with all terms positive, is divergent.

We now list some tests for convergent series.

3.4.1 Limiting value of terms

Consider a convergent series with a sum S, reached by adding an infinite number of terms. If the last term is removed there are still an infinite number of terms, by the peculiar properties of infinity, so the sum cannot be different from S. The conclusion is that

$$(a_n)_{n \to \infty} = 0. \qquad (3.20)$$

This is a necessary condition for a convergent series but not a sufficient one. The harmonic series (3.11) satisfies the condition (3.20) but it is not convergent.

3.4.2 Comparing series

If two series of positive terms are

$$\sum_{i=0}^{\infty} a_i \quad \text{and} \quad \sum_{i=0}^{\infty} b_i$$

with $a_i \leq b_i$, for all i greater than some finite value, then, if the series $\sum_{i=0}^{\infty} b_i$ is convergent, $\sum_{i=0}^{\infty} a_i$ is also convergent. We notice that the condition $a_i \leq b_i$ needs to apply only after a finite number of terms so that, for example, for the first 20 terms of the series a_i can be larger than b_i as long as they are all smaller thereafter. Any finite value of the first 20 terms affects the sum of an infinite number of terms of the series, but not its convergence property.

3.4.3 d'Alembert's ratio test

This test states that any series will be **convergent** if

$$\left(\frac{a_{n+1}}{a_n}\right)_{n\to\infty} < 1 \tag{3.21a}$$

and **divergent** if

$$\left(\frac{a_{n+1}}{a_n}\right)_{n\to\infty} > 1. \tag{3.21b}$$

To show the convergence property we consider a set of terms from a_p onwards for which

$$\frac{a_{p+1}}{a_p} < s, \quad \frac{a_{p+2}}{a_{p+1}} < s, \quad \frac{a_{p+3}}{a_{p+2}} < s, \text{etc.}, \quad \text{where } s < 1.$$

Then

$$a_{p+1} < sa_p, \quad a_{p+2} < sa_{p+1} < s^2 a_p, \quad a_{p+3} < sa_{p+2} < s^3 a_p \dots,$$

which gives

$$\sum_{i=p}^{\infty} a_i < a_p(1 + s + s^2 + s^3 + \dots) = \frac{a_p}{1-s},$$

since $s < 1$. Since the series is convergent from the pth term, the whole series is convergent.

The divergence property is shown by a similar analysis. If for $i = p$ onwards

$$\frac{a_{i+1}}{a_i} > t \quad \text{where} \quad t > 1$$

then

$$\sum_{i=p}^{\infty} a_i > a_p(1 + t + t^2 + t^3 + \dots) = \frac{a_p(t^\infty - 1)}{t - 1} = \infty$$

and the series is divergent.

If the limiting ratio of successive terms is unity then the series can be either convergent or divergent; some other test must be applied to find out which.

An example of the application of the d'Alembert ratio test is given in §4.7.1.

3.4.4 Series with alternating signs

Any series with alternate positive and negative signs for which the magnitudes of the terms continually decrease must be convergent. The proof of this follows the lines that led to (3.14) and (3.15) where it is shown that successive terms give sums oscillating about the final sum but with a magnitude of difference that steadily decreases.

3.5 Power series

A very commonly occurring type of series is the power series of the form

$$a_0 + a_1 x + a_2 x^2 + a_3 x^3 + \dots + a_n x^n + \dots. \tag{3.22}$$

Such a series may be convergent for all values of x, or convergent only for a range of values of x. In the latter case the d'Alembert ratio test can be used to find the **convergence range**. Consider the series

$$1 + x + \frac{x^2}{2} + \frac{x^3}{3} + \cdots + \frac{x^n}{n} + \cdots. \tag{3.23}$$

Then

$$\left(\left| \frac{a_{n+1}}{a_n} \right| \right)_{n \to \infty} = \left(\left| \frac{nx}{n+1} \right| \right)_{n \to \infty} = |x|. \tag{3.24}$$

Hence, if $|x| < 1$ the series will satisfy the d'Alembert ratio test for convergence. Otherwise, if $|x| > 1$ the series will be divergent. If $x = 1$ the series is the same as the harmonic series (3.11) that was shown to be divergent. However, if $x = -1$ then the series is the same as (3.13) that was shown to be convergent. Hence the convergence range of the series (3.23) is $1 > x \geq -1$.

Problems

3.1 A ball is dropped from a height of 1 m. After each bounce it rises to a height that is 0.8 times the previous height. How far does the ball move before it comes to rest?

3.2 The figure below shows a model of an infinite one-dimensional ionic crystal containing alternate positive and negative ions with charges $\pm e$ a distance r apart. The potential energy of charges q_1 and q_2 separated by a distance r is $\dfrac{q_1 q_2}{4\pi\varepsilon_0 r}$.

(i) Starting from the shaded ion, express the total potential energy of the one-dimensional crystal as an infinite sum.

(ii) Find this sum, expressed in the form $-M \dfrac{e^2}{4\pi\varepsilon_0 r}$. The quantity M is known as the **Madelung constant** for the crystal. Madelung constants can also be found for three-dimensional crystals where the ions are arranged on different kinds of lattices.

3.3 Examine the following series for convergence or divergence:

(i) $\sum\limits_{t=1}^{\infty} \dfrac{t^5}{t!}$; (ii) $\sum\limits_{t=1}^{\infty} \dfrac{e^t}{t!}$; (iii) $\sum\limits_{t=1}^{\infty} \dfrac{1}{t(t+1)}$; (iv) $\sum\limits_{t=1}^{\infty} \dfrac{t^t}{t!}$.

3.4 Find the range of convergence for $\sum\limits_{t=1}^{\infty} \dfrac{(tx)^t}{(t-1)!}$.

Differentiation

In this chapter you meet the most basic idea in calculus, finding the derivative of a function, which is equivalent to finding the gradient of the function. For many simple functions this can be done from first principles, and the technique for doing this is described. More complicated functions can often be described as sums, products, or quotients of simple functions or can be converted to a simpler form by changing the variable in terms of which they are described. In such cases you can apply a number of rules to find the derivatives and these rules are described. Also described are the concepts of the maximum and minimum of a function, of great practical use in finding optimum values of a function satisfying some criterion that has been set. This topic also introduces the idea of higher derivatives. Finally, it is shown how to find expansions for common functions in terms of power series – the Taylor and Maclaurin series that often crop up in mathematical analyses.

4.1 The basic idea of a derivative

The idea of a derivative is most easily introduced by considering a variable y as a function of another variable x in the form

$$y = f(x). \tag{4.1}$$

The relationship between y and x is shown in Figure 4.1.

As the point P′ approaches P, so δx and δy both approach zero and the slope of the line PP′ approaches that of the tangent at P. Thus the slope of the curve at P is given by

$$\frac{\mathrm{d}y}{\mathrm{d}x} = \tan \psi = \left(\frac{\delta y}{\delta x}\right)_{\delta x \to 0} \tag{4.2}$$

where $\mathrm{d}y/\mathrm{d}x$ is the **derivative of y with respect to x**. Basic texts on calculus give the derivatives of various functions, and for convenience we give three common derivatives in Table 4.1.

As a note on notation, where we have a derivative such as $\mathrm{d}y/\mathrm{d}x$ we shall refer to x as the **independent variable** and y as the **dependent variable**. The distinction is somewhat arbitrary, as the two variables are mutually dependent, but the notation

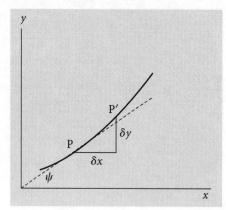

Figure 4.1 As δx and δy tend to zero, fixed at P at one end, so the slope of the line PP′ tends towards dy/dx, the slope of the tangent at P, shown as a dashed line.

Table 4.1 Three simple derivatives

$f(x)$	$\mathrm{d}f(x)/\mathrm{d}x$
$x^n \ (n \neq 0)$	nx^{n-1}
$\sin x$	$\cos x$
$\cos x$	$-\sin x$

will be more meaningful when we come to consider differential equations in Chapters 19 and 20.

As an exercise we now show how, using the results given in Chapter 1, the derivatives of e^x and $\ln(x)$ may be found.

4.1.1 Derivative of e^x

The derivative will be found by the use of (4.2) with

$$y = e^x. \tag{4.3}$$

We then have

$$y + \delta y = e^{x+\delta x} = e^x e^{\delta x}. \tag{4.4}$$

From (1.37)

$$e^{\delta x} = \left[\left(1 + \frac{1}{n} \right)^{n\delta x} \right]_{n \to \infty}$$

which, by use of the binomial theorem, (1.28), gives

$$e^{\delta x} = \left[1 + \frac{n\delta x}{n} + \frac{n\delta x(n\delta x - 1)}{2!n^2} + \frac{n\delta x(n\delta x - 1)(n\delta x - 2)}{3!n^3} + \cdots \right]_{n \to \infty}.$$

With $\delta x \to 0$ and $n \to \infty$ the second term in the square bracket is δx and the third and subsequent terms will tend to zero no matter what assumption is made about the product $n\delta x$. For example, if $n\delta x \ll 1$ then the third term will be of order $\delta x/n$, which will tend to zero. On the other hand if $n\delta x \gg 1$ then the third term will be of order $(\delta x)^2$ which will also tend to zero. The same conclusion can be drawn about subsequent terms in the expansion. Hence we have

$$[e^{\delta x}]_{\delta x \to 0} = 1 + \delta x. \tag{4.5}$$

From (4.2), (4.3), (4.4), and (4.5),

$$\begin{aligned}
\frac{dy}{dx} &= \left[\frac{\delta y}{\delta x}\right]_{\delta x \to 0} \\
&= \left[\frac{e^{x+\delta x} - e^x}{\delta x}\right]_{\delta x \to 0} = \left[\frac{e^x(1 + \delta x - 1)}{\delta x}\right]_{\delta x \to 0} = \left[\frac{e^x \delta x}{\delta x}\right]_{\delta x \to 0} = e^x. \tag{4.6}
\end{aligned}$$

The special property of e^x is that its derivative is also e^x or, in other words, the slope of a graph of e^x is the same as the value of the function itself.

4.1.2 Derivative of ln(x)

We now take

$$y = \ln x,$$

which gives

$$x = e^y. \tag{4.7}$$

From (4.6),

$$\frac{dx}{dy} = e^y$$

or

$$\frac{dy}{dx} = \frac{1}{dx/dy} = e^{-y} = \frac{1}{x}. \tag{4.8}$$

It will be seen from Table 4.1 that the only power of x that cannot be obtained by differentiating a higher power of x is x^{-1}. That deficiency is now removed.

Exercise 4.1
Find, from first principles, the derivative of $\tan x$. Equations (1.20) and (1.38) will be helpful.

4.2 Chain rule

The chain rule describes a simple process by which derivatives can be found when, as an example, the argument or power of a function is itself a function of the independent variable. Examples of situations when the use of the chain rule is appropriate are

$$y = \exp(x^2), \quad y = (1 + x + x^2)^3, \quad y = \exp(\sin x^2).$$

We now illustrate the application of the chain rule with these functions.

EXAMPLE 4.1

The first step is to write

$$u = x^2 \tag{4.9}$$

so that

$$y = \exp(u) \tag{4.10}$$

from which we find

$$\frac{du}{dx} = 2x \quad \text{and} \quad \frac{dy}{du} = \exp(u).$$

The chain rule now states that

$$\frac{dy}{dx} = \frac{dy}{du}\frac{du}{dx}, \tag{4.11}$$

and from this we find that

$$\frac{dy}{dx} = \exp(u) \times 2x = 2x\exp(x^2). \tag{4.12}$$

EXAMPLE 4.2

We write

$$u = 1 + x + x^2$$

so that

$$y = u^3.$$

Hence

$$\frac{dy}{du} = 3u^2 = 3(1 + x + x^2)^2 \quad \text{and} \quad \frac{du}{dx} = 1 + 2x.$$

Applying the chain rule,

$$\frac{dy}{dx} = \frac{dy}{du}\frac{du}{dx} = 3(1 + x + x^2)^2(1 + 2x). \tag{4.13}$$

EXAMPLE 4.3

This requires a three-stage process with the steps

$$u = x^2, \quad v = \sin u, \quad \text{and} \quad y = \exp(v),$$

from which we find

$$\frac{du}{dx} = 2x, \quad \frac{dv}{du} = \cos u = \cos x^2 \quad \text{and} \quad \frac{dy}{dv} = \exp(v) = \exp(\sin u) = \exp(\sin x^2).$$

The chain rule now appears in the form

$$\frac{dy}{dx} = \frac{dy}{dv}\frac{dv}{du}\frac{du}{dx} = 2x\cos x^2 \exp(\sin x^2).$$ (4.14)

The general principles involved in applying the chain rule should be clear from these examples.

 Exercise 4.2
If $y = \sin\{\exp(\tan x)\}$, use the chain rule to find dy/dx.

4.3 Product rule

The product rule is concerned with finding the derivative of a product of two or more functions of the independent variable – for example

$$y = f_1(x)f_2(x).$$ (4.15)

Using the basic method of finding a derivative, we note that when x changes to $x + \delta x$ then y changes to $y + \delta y = (f_1 + \delta f_1)(f_2 + \delta f_2)$, so that

$$
\begin{aligned}
\frac{dy}{dx} &= \left[\frac{(y + \delta y) - y}{\delta x}\right]_{\delta x \to 0} \\
&= \left[\frac{(f_1 + \delta f_1)(f_2 + \delta f_2) - f_1 f_2}{\delta x}\right]_{\delta x \to 0} \quad (4.16) \\
&= \left[f_1\frac{\delta f_2}{\delta x} + f_2\frac{\delta f_1}{\delta x} + \delta f_1\frac{\delta f_2}{\delta x}\right]_{\delta x \to 0}.
\end{aligned}
$$

The final term in the bracket goes to zero as $\delta x \to 0$ since $\delta f_1 \to 0$ and $\delta f_2/\delta x \to df_2/dx$. Hence, from (4.16)

$$\frac{dy}{dx} = f_1\frac{df_2}{dx} + f_2\frac{df_1}{dx}$$ (4.17)

which is the **product rule** for a product of two functions. The use of the product rule is now illustrated by some examples.

EXAMPLE 4.4 $y = x^3 e^x$

We take $f_1 = x^3$ so that $df_1/dx = 3x^2$ and $f_2 = e^x$ so that $df_2/dx = e^x$. Hence, from (4.17),

$$\frac{dy}{dx} = x^3 e^x + e^x \times 3x^2 = x^2 e^x(x + 3).$$

EXAMPLE 4.5 $y = \sin x / x$

This is a function that appears quite frequently in diffraction theory and has the special name sinc(x). To differentiate it we take

$$f_1 = \frac{1}{x} \quad \text{so that} \quad \frac{df_1}{dx} = -\frac{1}{x^2} \quad \text{and} \quad f_2 = \sin x \quad \text{so that} \quad \frac{df_2}{dx} = \cos x.$$

From (4.17),

$$\frac{dy}{dx} = -\frac{\sin x}{x^2} + \frac{\cos x}{x}.$$

EXAMPLE 4.6 $y = x^3 \exp(x^2)$

Taking $f_1 = x^3$ gives $df_1/dx = 3x^2$. We then take $f_2 = \exp(x^2)$ but to find df_2/dx we need to use the chain rule. In this context we take $f_2 = e^v$ and $v = x^2$ so that

$$\frac{df_2}{dv} = e^v = \exp(x^2), \quad \frac{dv}{dx} = 2x, \quad \text{and} \quad \frac{df_2}{dx} = \frac{df_2}{dv}\frac{dv}{dx} = 2x \exp(x^2).$$

We now find

$$\frac{dy}{dx} = f_1 \frac{df_2}{dx} + f_2 \frac{df_1}{dx} = x^3 \times 2x \exp(x^2) + \exp(x^2) \times 3x^2 = x^2 \exp(x^2)(2x^2 + 3).$$

Exercise 4.3

If $y = (x + e^x \sin x)\sin x$ then find dy/dx.

4.3.1 The rule for products of more than two functions

The product rule can be extended to find the derivative of a product of any number of functions. If, for example,

$$y = f_1(x)f_2(x)f_3(x)$$

then, corresponding to (4.16)

$$\frac{dy}{dx} = \left[\frac{(f_1 + \delta f_1)(f_2 + \delta f_2)(f_3 + \delta f_3) - f_1 f_2 f_3}{\delta x} \right]_{\delta x \to 0} \tag{4.18}$$

$$= \left[f_1 f_2 \frac{\delta f_3}{\delta x} + f_1 f_3 \frac{\delta f_2}{\delta x} + f_2 f_3 \frac{\delta f_1}{\delta x} \right]_{\delta x \to 0}$$

where products of two or more δfs are ignored in the expansion of the numerator since they go to zero as $\delta x \to 0$. Applying the limit in (4.18) we obtain

$$\frac{dy}{dx} = f_1 f_2 \frac{df_3}{dx} + f_1 f_3 \frac{df_2}{dx} + f_2 f_3 \frac{df_1}{dx}. \tag{4.19}$$

With $y = x^2 e^x \sin x$ we take

$$f_1 = x^2, \quad \frac{df_1}{dx} = 2x; \quad f_2 = e^x, \quad \frac{df_2}{dx} = e^x \quad \text{and} \quad f_3 = \sin x, \quad \frac{df_3}{dx} = \cos x$$

so that

$$\frac{dy}{dx} = x^2 e^x \cos x + x^2 \sin x \times e^x + e^x \sin x \times 2x$$
$$= xe^x(x \cos x + x \sin x + 2 \sin x).$$

The extension of the product rule to the product of any number of functions should be clear from (4.17) and (4.19). Thus for a product of four functions

$$\frac{dy}{dx} = f_1 f_2 f_3 \frac{df_4}{dx} + f_1 f_2 f_4 \frac{df_3}{dx} + f_1 f_3 f_4 \frac{df_2}{dx} + f_2 f_3 f_4 \frac{df_1}{dx}. \tag{4.20}$$

 Exercise 4.4
If $y = (a + bx)(c + dx)e^x$, find dy/dx.

4.4 Quotient rule

We now consider how to differentiate a quotient of two functions of the form

$$y = \frac{u(x)}{v(x)}. \tag{4.21}$$

From first principles, if x is changed by a small quantity δx, leading to a change in u and v of δu and δv respectively, then y changes to

$$y + \delta y = \frac{u + \delta u}{v + \delta v}. \tag{4.22}$$

Using the binomial theorem approximation (1.38),

$$y + \delta y = \frac{u + \delta u}{v(1 + \delta v/v)} = \frac{u + \delta u}{v}\left(1 - \frac{\delta v}{v}\right) = \frac{uv + v\delta u - u\delta v}{v^2}$$

where products of small quantities have been neglected. This gives

$$\delta y = \frac{v\delta u - u\delta v}{v^2} \quad \text{or} \quad \frac{\delta y}{\delta x} = \frac{v\dfrac{\delta u}{\delta x} - u\dfrac{\delta v}{\delta x}}{v^2}. \tag{4.23}$$

Making $\delta x \to 0$ gives the final result:

$$\frac{dy}{dx} = \frac{v\dfrac{du}{dx} - u\dfrac{dv}{dx}}{v^2}. \tag{4.24}$$

As an example we consider

$$y = \frac{1 + x^2}{\sin x}$$

With $u = 1 + x^2$ and $v = \sin x$ we have $du/dx = 2x$ and $dv/dx = \cos x$, so that

$$\frac{dy}{dx} = \frac{2x \sin x - (1 + x^2) \cos x}{\sin^2 x}. \tag{4.25}$$

A combination of the chain, product, and quotient rules can give the derivatives of quite complex functions.

Exercise 4.5

If $y = \dfrac{x + \sin x}{x + \cos x}$, find $\dfrac{dy}{dx}$.

4.5 Maxima, minima, and higher-order derivatives

The concept of **maximum** and **minimum** in calculus must not be confused with the idea of a maximum value or minimum value of a function. Figure 4.2 shows a function with a maximum at A and a minimum at B.

Point A is a **local maximum** in that it has a greater value than the parts of the curve immediately adjacent to it but it is not the greatest value of y – for example, the point C has a higher value. Similarly B is a **local minimum** and there are lesser values on the curve, such as that for point D. If it happens that a maximum *is* the greatest value of the function then it would be called a **global maximum** and similarly a least value would be a **global minimum**. Sometimes we have to refer to a point as either a maximum or minimum without specifying which it is, and the term **extremum** (plural **extrema**) covers both cases.

The characteristic of an extremum is that the slope of the curve at that point is zero, or that

$$\frac{dy}{dx} = 0. \tag{4.26}$$

The function shown in Figure 4.2 is

$$y = x^3 - 6x^2 + 11x - 6$$

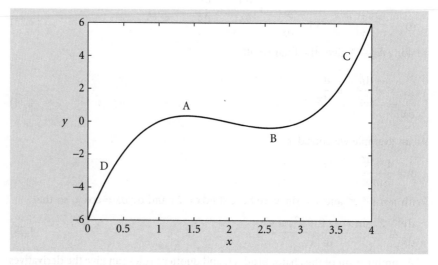

Figure 4.2 The curve $y = x^3 - 6x^2 + 11x - 6$ with a maximum at A and a minimum at B.

giving

$$\frac{dy}{dx} = 3x^2 - 12x + 11$$

and equating dy/dx to zero gives the extrema at $x = 2 \pm 1/\sqrt{3}$. From the graph it is seen that the smaller value corresponds to a maximum and the larger value to a minimum, but it is desirable to have some analytical way of determining which extremum is which.

From Figure 4.2 we can see that the slope of the function, dy/dx, is positive to the left of the point A then, as x increases so dy/dx decreases to zero at the point A and thereafter is negative. The implication of this is that the rate of change of dy/dx with increasing x is negative, or, for the **second derivative**

$$\frac{d^2y}{dx^2} < 0 \tag{4.27a}$$

at the point A, which together with (4.26) is the condition for a maximum. By similar reasoning it is found that, at the point B

$$\frac{d^2y}{dx^2} > 0, \tag{4.27b}$$

which together with (4.26) is the condition for a minimum.

We can now test these conditions on the function that gave rise to Figure 4.2. For this function

$$\frac{d^2y}{dx^2} = 6x - 12.$$

For $x = 2 - 1/\sqrt{3}$, corresponding to A, we find $d^2y/dx^2 = -6/\sqrt{3}$, which is negative so that the point A is at a maximum. Similarly, for the point B, for which $x = 2 + 1/\sqrt{3}$, we find $d^2y/dx^2 = 6/\sqrt{3}$, which is positive so the point B is at a minimum.

We now illustrate the application of finding maxima and minima with some simple examples.

EXAMPLE 4.7

A farmer has available 400 m of fencing with which to enclose an area in the form of a quadrilateral. What is the maximum area he can enclose?

The lengths of the sides are taken as x and $200 - x$, which gives an area

$$A = x(200 - x).$$

From this we find $dA/dx = 200 - 2x$ and equating this to zero to give an extremum gives $x = 100$. We also find $d^2A/dx^2 = -2$, which is negative, independently of the value of x, so the solution we find is a maximum. Thus the area is a square of side 100 m with an area $10^4\,\mathrm{m}^2$.

4 DIFFERENTIATION

EXAMPLE 4.8

An area of $10^4\,\text{m}^3$ is to be enclosed in the form of a quadrilateral. What is the minimum length of fencing required to achieve this?

The sides of the quadrilateral can be taken as x and $10^4/x$. Hence the length of fencing required is

$$l = 2x + \frac{2 \times 10^4}{x}.$$

From this we find

$$\frac{\mathrm{d}l}{\mathrm{d}x} = 2 - \frac{2 \times 10^4}{x^2} \quad \text{and} \quad \frac{\mathrm{d}^2 l}{\mathrm{d}x^2} = \frac{4 \times 10^4}{x^3}.$$

Equating the first derivative to zero gives $x = 100$, and the second derivative is positive for all positive x so that the solution is a minimum. The length of fencing required is 400 m.

Exercise 4.6
Find the (x, y) values for the maximum and minimum of the function $y = 2x^3 + 3x^2 - 36x - 4$.

4.5.1 **Points of inflection**

We consider the function

$$y = 2x^3 - 6x^2 + 6x + 3, \tag{4.28}$$

for which

$$\frac{\mathrm{d}y}{\mathrm{d}x} = 6x^2 - 12x + 6 = 6(x-1)^2 \quad \text{and} \quad \frac{\mathrm{d}^2 y}{\mathrm{d}x^2} = 12(x-1).$$

Equating the first derivative to zero gives two identical solutions, $x = 1$, and this value of x makes the second derivative equal to zero, satisfying neither of the conditions (4.27a) or (4.27b). To see what the condition of the function is at $x = 1$ we plot the curve seen in Figure 4.3.

What we see in Figure 4.3 is a **point of inflection**. As we go from values of x less than that of the point of inflection the slope goes from positive to zero and then, beyond the point of inflection, it increases again. If we plot the gradient against x then we obtain Figure 4.4, from which it is clear that the gradient of $\mathrm{d}y/\mathrm{d}x$ is zero at a point of inflection, or

$$\frac{\mathrm{d}^2 y}{\mathrm{d}x^2} = 0. \tag{4.29}$$

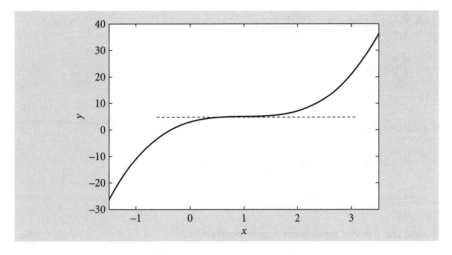

Figure 4.3 The curve $y = 2x^3 - 6x^2 + 6x + 3$ showing a point of inflection at $x = 1$.

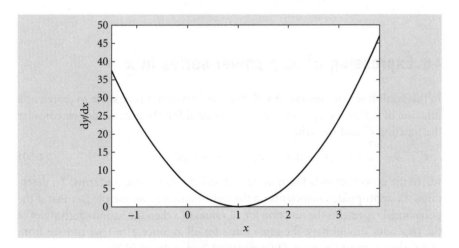

Figure 4.4 The slope of dy/dx is zero at a point of inflection.

4.5.2 **Higher-order derivatives**

In the course of dealing with extrema the second derivative, d^2y/dx^2, was introduced as the result of twice differentiating a function. In a similar way higher-order
derivatives can be defined where $d^n y/dx^n$ is the result of differentiating a function
n times. Here we give two examples.

> **EXAMPLE 4.9**

$$y = 2x^5 + 3x^4 + 4x^3 + 5x^2 + 6x + 7; \quad \frac{dy}{dx} = 10x^4 + 12x^3 + 12x^2 + 10x + 6;$$

$$\frac{d^2y}{dx^2} = 40x^3 + 36x^2 + 24x + 10; \quad \frac{d^3y}{dx^3} = 120x^2 + 72x + 24;$$

$$\frac{d^4 y}{dx^4} = 240x + 72; \quad \frac{d^5 y}{dx^5} = 240;$$

$$\frac{d^6 y}{dx^6} = 0.$$

EXAMPLE 4.10

$$y = \sin x; \quad \frac{dy}{dx} = \cos x; \quad \frac{d^2 y}{dx^2} = -\sin x;$$

$$\frac{d^3 y}{dx^3} = -\cos x; \quad \frac{d^4 y}{dx^4} = \sin x$$

and the pattern repeats for higher derivatives.

To put physical meaning on derivatives, with variables x for position and t for time we have dx/dt representing velocity and $d^2 x/dt^2$ representing acceleration.

4.6 Expressing e^x as a power series in x

In this section we are concerned with the way in which it is possible to represent a function of x, $f(x)$, by a power series in x. To establish the general idea we consider the function e^x and we write

$$e^x = a_0 + a_1 x + a_2 x^2 + a_3 x^3 + a_4 x^4 + \cdots + a_n x^n \cdots \tag{4.30}$$

where the power series is taken as having an infinite number of terms. To determine the appropriate coefficients on the right-hand side we use the fact that if the polynomial represents the function for all values of x then the various derivatives of the two sides should have the same value for all x. Since $e^0 = 1$ we can see from (4.30) that a_0 must be unity. Differentiating both sides of (4.30),

$$e^x = a_1 + 2a_2 x + 3a_3 x^2 + 4a_4 x^3 + \cdots + na_n x^{n-1} + \cdots \tag{4.31}$$

Taking $x = 0$ we find that $a_1 = 1$. Differentiating again,

$$e^x = 2a_2 + 3 \times 2a_3 x + 4 \times 3a_4 x^2 + \cdots + n(n-1)a_n x^{n-2} + \cdots . \tag{4.32}$$

Again taking $x = 0$ we find $a_2 = \frac{1}{2}$. Differentiating once more,

$$e^x = 3 \times 2a_3 + 4 \times 3 \times 2a_4 x + \cdots + n(n-1)(n-2)a_n x^{n-3} + \cdots \tag{4.33}$$

and putting $x = 0$ gives $a_3 = 1/3!$. Continuing this process gives the series expansion of e^x as

$$e^x = 1 + x + \frac{x^2}{2!} + \frac{x^3}{3!} + \frac{x^4}{4!} + \cdots . \tag{4.34}$$

Table 4.2 shows how well this series represents the value of e^x, for various values of x, taking different numbers of terms on the right-hand side.

Various patterns emerge from Table 4.2. First, it is clear that good estimates can be found with very few terms, especially for small values of x. Secondly, it will be seen that as each term is added so the error in the estimate is less. The error is the sum of all the missing terms and is hence greater for larger values of x.

If estimates were required for e^x over the range of x from 0 to 1 then the best estimates are for x values close to 0 and the worst estimates for x close to 1 and the question arises of whether it is possible to spread the error more evenly over the range. It can be done by *centring the series on a value in the middle of the range of proposed application.* This is illustrated by the series

$$e^x = a_0 + a_1(x - 0.5) + a_2(x - 0.5)^2 + \cdots + a_n(x - 0.5)^n + \cdots. \tag{4.35}$$

Taking $x = 0.5$ gives $a_0 = e^{0.5}$. Differentiating both sides of (4.35) gives

$$e^x = a_1 + 2a_2(x - 0.5) + \cdots + na_n(x - 0.5)^{n-1} + \cdots \tag{4.36}$$

and taking $x = 0.5$ then gives $a_1 = e^{0.5}$. Repeating the process of differentiating and then taking $x = 0.5$ enables the rest of the coefficients to be determined. The series appears as

$$e^x = e^{0.5} + e^{0.5}(x - 0.5) + e^{0.5}\frac{(x - 0.5)^2}{2!} + e^{0.5}\frac{(x - 0.5)^3}{3!} + \cdots. \tag{4.37}$$

Table 4.2 Estimates of e^x for various values of x and different numbers of terms. The figures in parentheses are the errors in the estimates in units of the last place of decimals

	Number of terms in series						
x	1	2	3	4	5		True value
0.00	1.000 000	1.000 000	1.000 000	1.000 000	1.000 000	(0)	1.000 000
0.25	1.000 000	1.250 000	1.281 250	1.283 854	1.284 017	(8)	1.284 025
0.50	1.000 000	1.500 000	1.625 000	1.645 833	1.648 438	(283)	1.648 721
0.75	1.000 000	1.750 000	2.031 250	2.101 563	2.114 746	(2254)	2.117 000
1.00	1.000 000	2.000 000	2.500 000	2.666 667	2.708 333	(9949)	2.718 282

Table 4.3 Estimates of e^x for various values of x using (4.37). The figures in parentheses are the errors in the estimates in units of the last place of decimals

	Number of terms in series						
x	1	2	3	4	5		True value
0.00	1.648 721	0.834 361	1.030 451	0.996 102	1.000 396	(396)	1.000 000
0.25	1.648 721	1.236 541	1.288 063	1.283 770	1.284 038	(13)	1.284 025
0.50	1.648 721	1.648 721	1.648 721	1.648 721	1.648 721	(0)	1.648 721
0.75	1.648 721	2.060 902	2.112 424	2.116 718	2.116 986	(14)	2.117 000
1.00	1.648 721	2.473 082	2.679 172	2.713 520	2.717 814	(468)	2.718 282

In Table 4.3 we show the effect of applying (4.37) to the same values of x and with the same maximum power of x as are given in Table 4.2.

It will be seen by comparing Tables 4.2 and 4.3 that (4.37) gives smaller average errors than (4.34) and that those errors are more uniformly distributed over the range of application.

4.7 Taylor's theorem

What we have found for the expression of e^x as a power series can be extended for general functions of x. An important theorem that defines how this can be done is Taylor's theorem. In expressing this theorem we shall make use of the following notation:

$$f'(x) = \frac{df}{dx}, \quad f''(x) = \frac{d^2 f}{dx^2}, \quad \text{and in general} \quad f^n(x) = \frac{d^n f}{dx^n}.$$

Here $f''(a)$ indicates the value of $f''(x)$ with $x = a$. Taylor's theorem says that if within the interval between $\zeta = x$ and $\zeta = a$ a function, $f(\zeta)$, is *single valued*, *continuous* and has *continuous derivatives up to* $f^n(\zeta)$, then one can write

$$f(x) = f(a) + (x-a)f'(a) + \frac{(x-a)^2}{2!}f''(a) + \cdots + \frac{(x-a)^n}{n!}f^n(a) + E_n(x) \tag{4.38}$$

where $E(x)$ is the remainder term given by

$$E_n(x) = \frac{(x-a)^{n+1}}{(n+1)!}f^n(\eta) \tag{4.39}$$

and η has a value between a and x.

The form of (4.38) is exactly the same as that of (4.37), except for the remainder term that enables an estimate of the error when the series is truncated. Thus Table 4.3 is truncated after five terms (corresponding to the fourth derivative) so the remainder is

$$E_5(x) = \frac{(x-0.5)^5}{5!}e^\eta$$

where η is within the interval 0.5 to x. It is clear that if $x < 0.5$ then the sign of $E(x)$ is negative and that its magnitude is somewhere between the values with $\eta = 0$ and $\eta = 0.5$. For $x = 0$ these magnitude limits are

$$\left| \frac{(0-0.5)^5}{5!}e^0 \right| \quad \text{and} \quad \left| \frac{(0-0.5)^5}{5!}e^{0.5} \right|$$

or 0.000 360 and 0.000 429. From Table 4.3 it can be seen that the true remainder term, i.e. what must be added to the table value to get the true value, is -0.000396, which agrees with both the sign and the deduced range of possible values of $E_5(x)$. Conversely, if $x > 0.5$ then $E_5(x)$ is positive, as seen in Table 4.3.

4.7.1 **Taylor and Maclaurin series**

If the conditions for Taylor's theorem are valid for $n \to \infty$ then the function may be represented exactly by an infinite power series as long as the series converges, i.e.

$$E_n(x) = 0. \qquad (4.40)$$
$$\scriptstyle n \to \infty$$

One way of testing for convergence, when all the terms of the series are positive, is the **d'Alembert's ratio test** (§3.4.3). This states that a series of positive terms of the form

$$S = \sum_{r=1}^{\infty} a_r$$

is convergent if $(a_{r+1}/a_r)_{r \to \infty} < 1$ and divergent if $(a_{r+1}/a_r)_{r \to \infty} > 1$. If $(a_{r+1}/a_r)_{r \to \infty} = 1$ then the series could be either convergent or divergent.

When a series has both positive and negative terms then d'Alembert's test may still be able to show convergence if $(|a_{r+1}|/|a_r|)_{r \to \infty} < 1$. Such a series is said to be **absolutely convergent**, and an example is the series

$$f(x) = x - \frac{x^3}{3!} + \frac{x^5}{5!} - \frac{x^7}{7!} + \cdots + (-1)^{(n-1)/2} \frac{x^n}{n!} + \cdots. \qquad (4.41)$$

We find

$$\left(\frac{x^{r+2}/(r+2)!}{x^r/r!} \right)_{r \to \infty} = \left(\frac{x^2}{(r+1)(r+2)} \right)_{r \to \infty} = 0,$$

showing the absolute convergence.

When $E_n(x)_{n \to \infty} = 0$ then $f(x)$ may be represented by the infinite series

$$f(x) = \sum_{r=0}^{\infty} \frac{(x-a)^r}{r!} f^r(a) \qquad (4.42)$$

which is known as a **Taylor series**. For the special case with $a = 0$ then

$$f(x) = \sum_{r=0}^{\infty} \frac{x^r}{r!} f^r(0), \qquad (4.43)$$

which is known as a **Maclaurin series**.

Examples of Maclaurin series for trigonometric functions are

$$\sin x = x - \frac{x^3}{3!} + \frac{x^5}{5!} - \frac{x^7}{7!} + \cdots \qquad (4.44a)$$

$$\cos x = 1 - \frac{x^2}{2!} + \frac{x^4}{4!} - \frac{x^6}{6!} + \cdots \qquad (4.44b)$$

$$\tan x = x + \frac{x^3}{3} + \frac{2x^5}{15} + \frac{17x^7}{315} + \cdots. \qquad (4.44c)$$

A necessary condition for a Maclaurin series expansion to be possible is that the function $f(x)$ and its derivatives are defined and finite for $x = 0$. For this reason it is not possible to have a Maclaurin series for $\log x$ for which the function and its derivatives have infinite magnitudes for $x = 0$. However, it is possible to find a

Taylor series for $\log x$ by expanding around $x=1$. Thus we write

$$\log x = a_0 + a_1(x-1) + a_2(x-1)^2 + a_3(x-1)^3 + a_4(x-1)^4 \cdots. \qquad (4.45a)$$

Taking $x=1$ gives $a_0 = 0$. Differentiating we find

$$\frac{1}{x} = a_1 + 2a_2(x-1) + 3a_3(x-1)^2 + 4a_4(x-1)^3 \cdots \qquad (4.45b)$$

and, again taking $x=1$, this gives $a_1 = 1$. Differentiating once more gives

$$-\frac{1}{x^2} = 2a_2 + 3 \times 2a_3(x-1) + 4 \times 3a_4(x-1)^2 + \cdots \qquad (4.45c)$$

which gives $a_3 = -\frac{1}{2}$ with $x=1$. Continuation of this process gives

$$\log x = (x-1) - \frac{(x-1)^2}{2} + \frac{(x-1)^3}{3} - \frac{(x-1)^4}{4} + \cdots \qquad (4.46a)$$

or, replacing x by $x+1$,

$$\log(1+x) = x - \frac{x^2}{2} + \frac{x^3}{3} - \frac{x^4}{4} + \cdots. \qquad (4.46b)$$

Exercise 4.7
Use the Taylor series to find the expansion formula for $(1+x)^n$ in powers of x.

4.7.2 The Maclaurin series and L'Hôpital's rule

There is sometimes the need to evaluate a ratio of functions when the variable they have in common tends towards some limiting value, i.e. something of the form

$$\left[\frac{f(x)}{g(x)}\right]_{x \to a} = \frac{f(a)}{g(a)}. \qquad (4.47)$$

A problem occurs if both $f(a)=0$ and $g(a)=0$ for then the ratio is indeterminate. An example of this is $[\sin x/x]_{x \to 0}$. In this case the problem can be resolved by using the series expansion of $\sin x$ in the form (4.44a) that gives

$$\left[\frac{\sin x}{x}\right]_{x \to 0} = \left[\frac{x - x^3/3! + x^5/5! - \cdots}{x}\right]_{x \to 0} = \left[1 - \frac{x^2}{3!} + \frac{x^4}{5!} - \cdots\right]_{x \to 0} = 1.$$

In a more general case, using (4.38), the expression to be evaluated can be written in the form

$$\left[\frac{f(x)}{g(x)}\right]_{x \to a} = \left[\frac{f(a) + (x-a)f'(a) + (x-a)^2 f''(a)/2! + \cdots}{g(a) + (x-a)g'(a) + (x-a)^2 g''(a)/2! + \cdots}\right]_{x \to a}. \qquad (4.48)$$

If $f(a)$ and $g(a)$ are both equal to zero then, dividing the rest of the numerator and divisor by $(x-a)$ we have

$$\left[\frac{f(x)}{g(x)}\right]_{x \to a} = \left[\frac{f'(a) + (x-a)f''(a)/2! + \cdots}{g'(a) + (x-a)g''(a)/2! + \cdots}\right]_{x \to a}$$

so that, as long as at least one of f'(a) and g'(a) is not zero, we can write

$$\left[\frac{f(x)}{g(x)}\right]_{x\to a} = \left[\frac{f'(x)}{g'(x)}\right]_{x\to a}. \tag{4.49}$$

It should be noted that (4.49) is only true if the conditions are satisfied that $f(a)$ and $g(a)$ are both zero and that at least one of $f'(a)$ and $g'(a)$ is not zero. In general, L'Hôpital's rule is that

$$\left[\frac{f(x)}{g(x)}\right]_{x\to a} = \left[\frac{f^r(x)}{g^r(x)}\right]_{x\to a} \tag{4.50}$$

where the rth derivative is the first one for which at least one of $f^r(a)$ and $g^r(a)$ is not zero. As an example we consider

$$\left[\frac{x^2}{1-\cos x}\right]_{x\to 0} = \left[\frac{2x}{\sin x}\right]_{x\to 0} = \left[\frac{2}{\cos x}\right]_{x\to 0} = 2.$$

Exercise 4.8

Find the value of $\left(\tan^2 x/x\right)_{x\to 0}$.

Problems

4.1 Differentiate the following functions with respect to x: (i) $\sin(x^3)$;
(ii) $\exp(\sin x)$; (iii) $(1+x^4)^4$; (iv) $x\sin x$; (v) $x\sin(x^2)$; (vi) $\sin(1+x^2)\cos x$;
(vii) $x^2\sin(x^2)\cos(2x)$; (viii) $x^3\sin(1-\cos x)$; (ix) $(1+x^2)/(1+x^3)$;
(x) $x^2/\{(1+x)\cos x\}$

4.2 Show that: (i) $\dfrac{d}{dx}(\tan x) = \sec^2 x$; (ii) $\dfrac{d}{dx}(\cot x) = -\text{cosec}^2 x$;
(iii) $\dfrac{d}{dx}(\sec x) = -\tan x \sec x$; (iv) $\dfrac{d}{dx}(\text{cosec } x) = \cot x \text{ cosec } x$.

4.3 The distribution of speeds in a gas of molecular mass m and temperature T is given by the Maxwell–Boltzmann distribution so that the number of molecules with speeds between v and $v + dv$ is

$$P(v)dv = \left(\frac{2m^3}{\pi k^3 T^3}\right)^{1/2} v^2 \exp\left(-\frac{mv^2}{2kT}\right)dv$$

where k is the Boltzmann constant. Find the value of v for which the distribution function $P(v)$ has a maximum value.

4.4 For a proton of energy E within a star, the probability that it undergoes a nuclear reaction depends on the rate at which it approaches close to another proton, which is

$$P_{\text{app}} = AE^{-1/2}$$

where A is a constant and the probability that if it gets close enough it does react, which is of the form

$$P_{react} = B \exp\left(-\frac{E_G^{1/2}}{E^{1/2}}\right)$$

where B is a constant and E_G is the **Gamov energy**.

The distribution function for the energy of protons is of the form

$$P(E) = CE^{1/2} \exp\left(-\frac{E}{kT}\right)$$

where C is a constant, T the temperature within the star, and k is the Boltzmann constant $(1.38 \times 10^{-23}\,\mathrm{J\,K^{-1}})$. The rate at which reactions are taking place, per unit energy range, is

$$R = P(E)P_{app}(E)P_{react}(E).$$

If the Gamov energy is 493 keV and the temperature is $2 \times 10^7\,\mathrm{K}$ then for what energy of protons, expressed in units of keV, is there the greatest energy production per unit proton-energy range? $(1\,\mathrm{eV} = 1.602 \times 10^{-19}\,\mathrm{J}.)$

4.5 Using the method described in this chapter find the Maclaurin expansion for $f(x) = 1/(1 + x)$. Compare this with expression obtained from the binomial theorem (1.34).

4.6 It is required to use the Maclaurin expansion for $\ln(1 + x)$ to calculate natural logarithms in the interval $x = 0$ to 1 with a precision of 0.001. How many terms are required in the expansion?

4.7 Find a Taylor series expansion for $\sin x$ centred on $a = \pi/4$. How many terms are required to calculate $\sin x$ to an accuracy of 0.001 for x between 0 and $\pi/2$?

4.8 Find the values of the following:

(i) $\left[\dfrac{x^3}{\tan^2 x}\right]_{x \to 0}$; (ii) $\left[\dfrac{1 - x^n}{(1 - x)^n}\right]_{x \to 1}$ $(n > 1)$.

Integration

The operation of differentiation has been described in Chapter 4 and integration may be regarded as the inverse operation. In this chapter you will be shown a variety of methods for integrating various kinds of function. You will also learn about two kinds of integral – indefinite integrals, which may be regarded directly as the inverse operation of differentiation, and definite integrals, which can be thought of as finding 'the area under the curve'.

5.1 Indefinite and definite integrals

If there are two functions $F(x)$ and $f(x)$ such that $dF(x)/dx = f(x)$ then $F(x)$ is said to be the **indefinite integral** of $f(x)$ and is written as

$$F(x) = \int f(x)dx. \tag{5.1}$$

In the expression on the right-hand side of (5.1), $f(x)$ is referred to as the **integrand**.

The **indefinite** part of the description of $F(x)$ comes from the fact that (5.1) does not describe $F(x)$ uniquely. Any constant may be added to $F(x)$, since

$$\frac{d}{dx}\{F(x) + C\} = f(x). \tag{5.2}$$

Hence a full description of the indefinite integral of $f(x)$ is $F(x) + C$.

A function, $f(x)$, is illustrated graphically in Figure 5.1. We now consider the area under the curve, down to the x-axis, between the ordinates at x and $x + \delta x$. This is

$$\delta A \approx f(x)\,\delta x \tag{5.3}$$

where the approximation gets better as δx is made smaller. Hence we have

$$\frac{dA}{dx} = \left(\frac{\delta A}{\delta x}\right)_{\delta x \to 0} = f(x) \tag{5.4}$$

and hence

$$A = \int f(x)\,dx + C. \tag{5.5}$$

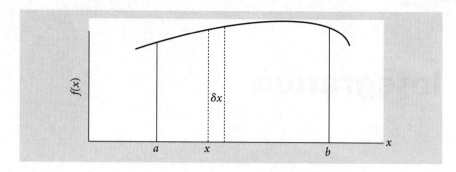

Figure 5.1 The plotted function $f(x)$ with ordinates at $x = a$, x, $x + \delta x$, and b.

The constant C is determined by defining the area under the curve as zero when $x = a$, which means that areas are being measured between the ordinates at $x = a$ and some other value of x, say at $x = b$. This gives

$$0 = \left\{ \int f(x)\mathrm{d}x \right\}_{x=a} + C \quad \text{or} \quad C = -\left\{ \int f(x)\mathrm{d}x \right\}_{x=a}. \tag{5.6}$$

The area under the curve between $x = a$ and $x = b$ is now given by

$$A_a^b = \left\{ \int f(x)\mathrm{d}x \right\}_{x=b} - \left\{ \int f(x)\mathrm{d}x \right\}_{x=a} \tag{5.7}$$

where, when the integrals are evaluated, the constant of integration is made equal to zero. A shorthand way of representing (5.7) is

$$A_a^b = \int_a^b f(x)\mathrm{d}x. \tag{5.8}$$

Integrals of this kind, with upper and lower limits of the variable, are known as **definite integrals**.

5.2 Techniques of evaluating integrals

Since differentiation and integration are inverse operations then a table of standard differentials can also be used as a table of standard integrals. For example, Table 5.1 shows a selection of standard integrals, all of which may be confirmed by checking that differentiating the function in the second column gives the function in the first column.

Tables of standard integrals, such as those distributed in many university departments, are usually quite extensive and may involve 30 or more standard functions. What we shall do is to develop various techniques for finding indefinite integrals and through these develop new members of what can be regarded as a set of standard integrals. These are outlined here and also given in Appendix 1, together with the integrals of functions described in Chapter 15.

Exercise 5.1
Find the indefinite integrals (i) $\int \sec^2 x \, \mathrm{d}x$ and (ii) $\int (1+x)^4 \mathrm{d}x$.

Table 5.1 Some standard integrals

$f(x)$	$\int f(x) dx$
$x^n \ (x \neq -1)$	$\frac{1}{n+1} x^{n+1} + C$
$1/x$	$\ln(x) + C$
e^x	$e^x + C$
$\sin x$	$-\cos x + C$
$\cos x$	$\sin x + C$

5.3 Substitution method

This method is best illustrated by some examples.

EXAMPLE 5.1 $\int \sin ax \ dx$, where a is a constant

We use the substitution $ax = u$, which gives $a \, dx = du$ or $dx = (1/a)du$. From this,

$$\int \sin ax \ dx = \frac{1}{a} \int \sin u \ du = -\frac{1}{a} \cos u + C = -\frac{1}{a} \cos ax + C. \qquad (5.9)$$

EXAMPLE 5.2 $\int \sin(1+x) dx$

Substituting $1 + x = u$ gives $dx = du$. From this,

$$\int \sin(1 + x) dx = \int \sin u \, du = -\cos u + C = -\cos(1 + x) + C. \qquad (5.10)$$

EXAMPLE 5.3 $\int \sin(a + bx) \ dx$

In this more general example, substituting $a + bx = u$ gives $dx = (1/b)du$. From this,

$$\int \sin(a + bx) dx = \frac{1}{b} \int \sin u \, du = -\frac{1}{b} \cos u + C = -\frac{1}{b} \cos(a + bx) + C \qquad (5.11)$$

By a similar process we come to the following general results:

$$\int (a + bx)^n dx = \frac{1}{b(n + 1)} (a + bx)^{n+1} + C, \qquad (5.12)$$

$$\int \frac{1}{(a + bx)} dx = \frac{1}{b} \ln(a + bx) + C, \qquad (5.13)$$

$$\int \exp(a + bx) dx = \frac{1}{b} \exp(a + bx) + C, \qquad (5.14)$$

$$\int \cos(a + bx) dx = \frac{1}{b} \sin(a + bx) + C. \qquad (5.15)$$

The success of the substitution method often depends on the ability to spot what form of substitution can lead to an integration of a standard form.

EXAMPLE 5.4 $\displaystyle\int x\exp(a+bx^2)\mathrm{d}x$

Here the substitution $a + bx^2 = u$ gives $x\,\mathrm{d}x = (1/2b)\mathrm{d}u$, so that

$$\int x\exp(a+bx^2)\,\mathrm{d}x = \frac{1}{2b}\int e^u\,\mathrm{d}u = \frac{1}{2b}e^u + C = \frac{1}{2b}\exp(a+bx^2) + C. \qquad (5.16)$$

EXAMPLE 5.5 $\displaystyle\int \frac{x}{a+bx^2}\,\mathrm{d}x$

Differentiating the divisor gives a numerical factor times the numerator. Again the substitution $a + bx^2 = u$ gives $x\,\mathrm{d}x = (1/2b)\mathrm{d}u$, and we obtain

$$\int \frac{x}{a+bx^2}\,\mathrm{d}x = \frac{1}{2b}\int \frac{1}{u}\,\mathrm{d}u = \frac{1}{2b}\ln u + C = \frac{1}{2b}\ln(a+bx^2) + C. \qquad (5.17)$$

There are a number of integrals, normally provided in lists of standard integrals, which can be derived by the method of substitution. One such is the following example.

EXAMPLE 5.6 $\displaystyle\int \frac{1}{\sqrt{1-x^2}}\,\mathrm{d}x$

Here we make the substitution $x = \sin u$ so that $\mathrm{d}x = \cos u\,\mathrm{d}u$. Now we find

$$\int \frac{1}{\sqrt{1-x^2}}\,\mathrm{d}x = \int \frac{\cos u}{\sqrt{1-\sin^2 u}}\,\mathrm{d}u = \int \mathrm{d}u = u + C = \sin^{-1} x + C. \qquad (5.18a)$$

In this case we could have made the substitution $x = \cos u$, in which case the result is

$$\int \frac{1}{\sqrt{1-x^2}}\,\mathrm{d}x = -\cos^{-1} x + C. \qquad (5.18b)$$

These results are quite consistent, since $\sin^{-1} x = (\pi/2) - \cos^{-1} x$ and the constants in (5.18a) and (5.18b) can be adjusted to give consistency. From (5.18a), again using substitution, we can obtain the more general result

$$\int \frac{1}{\sqrt{a-bx^2}}\,\mathrm{d}x = \sqrt{\frac{1}{b}}\sin^{-1}\left(\sqrt{\frac{b}{a}}x\right) + C. \qquad (5.19)$$

EXAMPLE 5.7 $\displaystyle\int \frac{1}{1+x^2}\,\mathrm{d}x$

This is another integral normally considered as standard. Here the substitution is $x = \tan u$, which gives $\mathrm{d}x = \sec^2 u\,\mathrm{d}u$ (see Problem 4.2(i)). Hence

$$\int \frac{1}{1+x^2}\,\mathrm{d}x = \int \frac{\sec^2 u}{1+\tan^2 u}\,\mathrm{d}u = \int \mathrm{d}u = u + C = \tan^{-1} x + C. \qquad (5.20)$$

A more general result, found by further substitution, is

$$\int \frac{1}{a+bx^2}\,\mathrm{d}x = \sqrt{\frac{1}{ab}}\tan^{-1}\left(\sqrt{\frac{b}{a}}x\right) + C. \qquad (5.21)$$

Sometimes an integral can be converted into the form (5.21) by a process known as **completing the square**. This is also best illustrated by an example.

EXAMPLE 5.8   **Completing the square**

Consider the function

$$\int \frac{1}{x^2 + 2x + 2}\,dx.$$

The first two terms of the divisor are produced by the squared quantity $(x+1)^2$ and we can therefore write

$$\int \frac{1}{x^2 + 2x + 2}\,dx = \int \frac{1}{1 + (x+1)^2}\,dx.$$

With the substitution $x + 1 = u$ giving $dx = du$ we find

$$\int \frac{1}{x^2 + 2x + 2}\,dx = \int \frac{1}{1 + u^2}\,du = \tan^{-1} u + C = \tan^{-1}(x+1) + C. \qquad (5.22)$$

This process will only work if the term accompanying the square is positive, i.e. $+1$ in this case. If we had

$$\int \frac{1}{x^2 + 4x + 3}\,dx = \int \frac{1}{(x+2)^2 - 1}\,dx \qquad (5.23)$$

then the arctan solution would not apply. In general we can write

$$\frac{1}{ax^2 + bx + c} = \frac{1}{\left(\sqrt{a}x + \dfrac{b}{2\sqrt{a}}\right)^2 + \left(c - \dfrac{b^2}{4a}\right)}. \qquad (5.24)$$

With the left-hand side of (5.24) as integrand, the arctan $(= \tan^{-1})$ solution is applicable if $c > b^2/4a$.

We shall meet the substitution method again in the process of describing other techniques of integration.

Exercise 5.2

Find the integrals (i) $\displaystyle\int \tan x \sec^2 x\,dx$; (ii) $\displaystyle\int \frac{\cos x}{\sin x}\,dx$; (iii) $\displaystyle\int \frac{\cos x}{\sin^2 x}\,dx$.

5.3.1 A useful substitution for trigonometric functions

When the integrand contains purely trigonometric functions, the substitution

$$\tan \frac{1}{2}x = t \qquad (5.25)$$

can be very useful. For this substitution

$$\frac{dt}{dx} = \frac{1}{2}\sec^2 \frac{1}{2}x = \frac{1}{2}(1 + t^2)$$

so that in the substitution

$$\mathrm{d}x = \frac{2\mathrm{d}t}{1 + t^2}. \tag{5.26}$$

From (1.20),

$$\tan x = \frac{2\tan\frac{1}{2}x}{1 - \tan^2\frac{1}{2}x} = \frac{2t}{1 - t^2}. \tag{5.27}$$

Now we can write

$$\cos^2 x = \frac{1}{\sec^2 x} = \frac{1}{1 + \tan^2 x} = \frac{(1 - t^2)^2}{(1 + t^2)^2}$$

giving

$$\cos x = \frac{1 - t^2}{1 + t^2}. \tag{5.28}$$

The choice of sign when taking the square root to give (5.28) may be found by considering the value of x as just less than π for which $\cos x$ is almost -1. From (5.25) t is large and positive, so that from (5.28) it is clear that the positive sign must be used in taking the square root.

Finally, we have

$$\sin x = \tan x \cos x = \frac{2t}{1 + t^2}. \tag{5.29}$$

EXAMPLE 5.9 $\qquad I = \displaystyle\int \frac{\cos\theta}{1 + \cos\theta}\,\mathrm{d}\theta.$

With the substitution $\tan\frac{1}{2}\theta = t$

$$I = \int \frac{(1 - t^2)/(1 + t^2)}{1 + (1 - t^2)/(1 + t^2)}\frac{2\mathrm{d}t}{1 + t^2} = \int \frac{1 - t^2}{1 + t^2}\,\mathrm{d}t = \int \frac{2}{1 + t^2}\,\mathrm{d}t - \int \mathrm{d}t$$

$$= 2\tan^{-1} t - t + C = 2\tan^{-1}\left(\tan\frac{1}{2}\theta\right) - \tan\frac{1}{2}\theta + C$$

or

$$I = \theta - \tan\frac{1}{2}\theta + C. \tag{5.30}$$

 Exercise 5.3
Find the integral $\displaystyle\int \frac{\mathrm{d}\theta}{1 + \sin\theta}$.

5.4 Partial fractions

We shall illustrate the principle of the method of partial fractions by taking the simplest case, $\int \frac{1}{1 - x^2}\,\mathrm{d}x$.

EXAMPLE 5.10 $\displaystyle\int \frac{1}{1-x^2}\,dx$

The divisor can be factorized as $(1+x)(1-x)$, so we write

$$\frac{1}{1-x^2} = \frac{A}{1+x} + \frac{B}{1-x} = \frac{A+B+(B-A)x}{1-x^2}.$$

This equality can only be true if $A+B=1$ and $B-A=0$, which gives

$$A = B = \frac{1}{2}.$$

Hence, using (5.13),

$$\int \frac{1}{1-x^2}\,dx = \frac{1}{2}\int \left(\frac{1}{1+x} + \frac{1}{1-x}\right)dx = \frac{1}{2}\{\ln(1+x) - \ln(1-x)\} + C$$

$$= \frac{1}{2}\ln\frac{1+x}{1-x} + C = \ln\sqrt{\frac{1+x}{1-x}} + C.$$

$$(5.31)$$

By substitution we can find the more general result

$$\int \frac{1}{a-bx^2}\,dx = \sqrt{\frac{1}{ab}}\ln\left(\frac{\sqrt{a}+\sqrt{bx}}{\sqrt{a}-\sqrt{bx}}\right)^{1/2}.$$

$$(5.32)$$

If the divisor of the integrand in (5.22) had been $bx^2 - a$ then it still factorizes and the method of partial fractions could be applied. By a suitable substitution this could be used to solve the integral (5.23) or that with (5.24) as integrand if $4ac < b^2$.

We conclude this section with two examples illustrating principles that are involved in the use of partial fractions and also bringing in other techniques that have been described so far.

EXAMPLE 5.11 $\displaystyle I = \int \frac{1}{1-x^3}\,dx$

The divisor factorizes, so we write

$$I = \int \frac{1}{(1-x)(1+x+x^2)}\,dx = \int \frac{a}{1-x}\,dx + \int \frac{bx+c}{1+x+x^2}\,dx \qquad (5.33)$$

To get the correct numerator we require

$$a(1+x+x^2) + (bx+c)(1-x) = 1$$

and equating coefficients of powers of x we find $a = b = \frac{1}{3}$ and $c = \frac{2}{3}$. Notice that in the numerator of the second integral on the right-hand side of (5.33) is a polynomial of power one less than that of the divisor. This gives as many unknowns – a, b, and c – as there were powers of $x - x^2$, x^1, and x^0 – so that the unknowns could be determined.

Now we write

$$I = \frac{1}{3}\int \frac{1}{1-x}\,dx + \frac{1}{3}\int \frac{x+2}{1+x+x^2}\,dx = \frac{1}{3}I_1 + \frac{1}{3}I_2. \qquad (5.34)$$

The differential of the divisor in I_2 is $1 + 2x$ so we now further divide I_2 as

$$I_2 = \frac{1}{2} \int \frac{1 + 2x}{1 + x + x^2} dx + \frac{3}{2} \int \frac{1}{1 + x + x^2} dx = \frac{1}{2} I_3 + \frac{3}{2} I_4. \tag{5.35}$$

Evaluating the integrals, we have

$$I_1 = -\ln(1 - x) \tag{5.36}$$

and, since the numerator is the differential of the divisor,

$$I_3 = \ln(1 + x + x^2). \tag{5.37}$$

For I_4 we complete the square in the divisor so that

$$I_4 = \int \frac{1}{\left(x + \frac{1}{2}\right)^2 + \frac{3}{4}} dx.$$

Substituting $x + \frac{1}{2} = \frac{\sqrt{3}}{2} u$,

$$I_4 = \frac{2}{\sqrt{3}} \int \frac{1}{1 + u^2} du = \frac{2}{\sqrt{3}} \tan^{-1} u = \frac{2}{\sqrt{3}} \tan^{-1} \left\{ \frac{2}{\sqrt{3}} \left(x + \frac{1}{2}\right) \right\}. \tag{5.38}$$

Assembling the components of the solution,

$$I = -\frac{1}{3} \ln(1 - x) + \frac{1}{6} \ln(1 + x + x^2) + \frac{1}{\sqrt{3}} \tan^{-1} \left\{ \frac{2}{\sqrt{3}} \left(x + \frac{1}{2}\right) \right\} + C. \tag{5.39}$$

EXAMPLE 5.12 $\int \dfrac{\tan\theta}{1 + \sin\theta} d\theta.$

Making the substitution $\tan\frac{1}{2}\theta = t$,

$$I = \int \frac{2t/(1 - t^2)}{1 + 2t/(1 + t^2)} \frac{2dt}{1 + t^2} = \int \frac{4t}{(1 - t^2)(1 + t)^2} dt$$

$$= \int \frac{4t}{(1 - t)(1 + t)^3} dt. \tag{5.40}$$

The partial fraction process used in this case is

$$\frac{4t}{(1 - t)(1 + t)^3} = \frac{a}{1 + t} + \frac{b}{(1 + t)^2} + \frac{c}{(1 + t)^3} + \frac{d}{1 - t}$$

which requires

$$a(1 + t)^2(1 - t) + b(1 + t)(1 - t) + c(1 - t) + d(1 + t)^3 = 4t$$

or

$$(a + b + c + d) + (3d + a - c)t + (3d - a - b)t^2 + (d - a)t^3 = 4t.$$

Equating coefficients of the two sides gives $a = \frac{1}{2}, b = 1, c = -2, d = \frac{1}{2}$, so that

$$I = \frac{1}{2} \int \frac{1}{1 + t} dt + \int \frac{1}{(1 + t)^2} dt - 2 \int \frac{1}{(1 + t)^3} dt + \frac{1}{2} \int \frac{1}{1 - t} dt. \tag{5.41}$$

Carrying out the integrations,

$$I = \frac{1}{2}\ln(1 + t) - \frac{1}{1+t} + \frac{1}{(1+t)^2} - \frac{1}{2}\ln(1 - t) + C$$

$$= \frac{1}{2}\ln\left(\frac{1 + \tan\frac{1}{2}\theta}{1 - \tan\frac{1}{2}\theta}\right) - \frac{1}{1 + \tan\frac{1}{2}\theta} + \frac{1}{\left(1 + \tan\frac{1}{2}\theta\right)^2} + C. \tag{5.42}$$

With experience you will learn how to combine the various techniques for integration, as illustrated with this example, to handle quite complex algebraic and trigonometric integrands.

Exercise 5.4
Determine the integral $\displaystyle\int \frac{2 + x + x^2}{1 + x + x^2 + x^3} \, dx$. (Hint: factorize the divisor.)

5.5 **Integration by parts**

Consider the effect of differentiating a product of two functions of x, U and V, giving

$$\frac{d}{dx}(UV) = U\frac{dV}{dx} + V\frac{dU}{dx}.$$

Integrating both sides of this equation gives

$$UV = \int U\left(\frac{dV}{dx}\right) dx + \int V\left(\frac{dU}{dx}\right) dx$$

or

$$\int U\left(\frac{dV}{dx}\right) dx = UV - \int V\left(\frac{dU}{dx}\right) dx. \tag{5.43}$$

This equation is the basis of the **integration by parts** method, the use of which is also best illustrated by examples.

EXAMPLE 5.13 $\displaystyle\int xe^x dx$

We take $U = x$ and $dV/dx = e^x$, giving $dU/dx = 1$ and $V = e^x$. Substituting these values in (5.43) gives

$$\int xe^x dx = xe^x - \int e^x dx = xe^x - e^x + C. \tag{5.44}$$

Notice that if we had taken $dV/dx = x$ then the integrand of the integral on the right-hand side of (5.43) would have been $x^2 e^x$, which could not be directly integrated.

EXAMPLE 5.14 $\int x^2 e^x dx$

We take $U = x^2$ and $dV/dx = e^x$, giving $dU/dx = 2x$ and $V = e^x$. Substituting these values in (5.44) gives

$$\int x^2 e^x dx = x^2 e^x - 2 \int x e^x dx.$$

The right-hand-side integral is given by (5.44) so that

$$\int x^2 e^x dx = x^2 e^x - 2x e^x + 2e^x + C. \tag{5.45}$$

If in the integral x is raised to any integral power, say n, then the integral is solved by a succession of integrations by parts in which the power of x is reduced by one at each step. However, for this to happen U must always be taken as the current power of x.

EXAMPLE 5.15 $\int x \ln x \, dx$

We take $U = \ln x$ and $dV/dx = x$ giving $dU/dx = 1/x$ and $V = \frac{1}{2}x^2$. Inserting in (5.43),

$$\int x \ln x \, dx = \frac{1}{2}x^2 \ln x - \int \frac{1}{2}x^2 \times \frac{1}{x} dx = \frac{1}{2}x^2 \ln x - \frac{1}{4}x^2 + C. \tag{5.46}$$

EXAMPLE 5.16 $\int e^x \cos x \, dx$

In this example we take $U = e^x$ and $dV/dx = \cos x$, giving $dU/dx = e^x$ and $V = \sin x$. Inserting in (5.43) gives

$$\int e^x \cos x \, dx = e^x \sin x - \int e^x \sin x \, dx. \tag{5.47}$$

It seems that we have not made much progress, but in fact we have. In the integral on the right-hand side we take $U = e^x$ and $dV/dx = \sin x$, giving $dU/dx = e^x$ and $V = -\cos x$. Substituting this in (5.47) gives

$$\int e^x \cos x \, dx = e^x \sin x + e^x \cos x - \int e^x \cos x \, dx$$

or

$$\int e^x \cos x \, dx = \frac{1}{2}e^x(\sin x + \cos x) + C, \tag{5.48}$$

where we have included the constant of integration.

These examples illustrate the general principles of using the method of integration by parts. In any particular case you must decide which part of the integrand to allocate to U and which to dV/dx. As a final example we illustrate the use of integration by parts when the integrand contains three distinct components.

EXAMPLE 5.17 $\displaystyle\int xe^x \cos x \, dx$

We take $U = x$ and $dV/dx = e^x \cos x$, giving $dU/dx = 1$ and $V = \frac{1}{2}e^x(\sin x + \cos x)$ (from (5.48)). Hence

$$\int xe^x \cos x \, dx = \frac{1}{2} xe^x(\sin x + \cos x) - \frac{1}{2}\int e^x(\sin x + \cos x)dx \qquad (5.49)$$

From (5.48),

$$\int e^x \cos x \, dx = \frac{1}{2} e^x(\sin x + \cos x)$$

and by a similar process to that giving (5.48),

$$\int e^x \sin x \, dx = \frac{1}{2}e^x(\sin x - \cos x).$$

Substituting in (5.49) and adding the constant of integration,

$$\int xe^x \cos x \, dx = \frac{1}{2}xe^x(\sin x + \cos x) - \frac{1}{2}e^x \sin x + C. \qquad (5.50)$$

Exercise 5.5

Find the following integrals: (i) $\int x \sec^2 x \, dx$; (ii) $\int x^2 \sin x \, dx$.

5.6 Integrating powers of cos x and sin x

As a preliminary to this section we give two results that are used subsequently. These are

$$\int \sin^n x \cos x \, dx = \frac{1}{n+1}\sin^{n+1} x + C \text{ and } \int \cos^n x \sin x \, dx = -\frac{1}{n+1}\cos^{n+1} x + C.$$
$$(5.51)$$

These results may be found by the substitutions $u = \sin x$ and $u = \cos x$ respectively.

(i) $\displaystyle\int \cos^n x \, dx$ and $\displaystyle\int \sin^n x \, dx$

Consider the integral $I_n = \int \cos^n x \, dx$ in which n is a positive integer. We now integrate by parts taking $U = \cos^{n-1} x$ and $dV/dx = \cos x$, which gives $dU/dx = -(n-1)\cos^{n-2} x \sin x$ and $V = \sin x$. This gives

$$I_n = \cos^{n-1} x \sin x + (n-1)\int \cos^{n-2} x \sin^2 x \, dx.$$

Replacing $\sin^2 x$ by $1 - \cos^2 x$ then leads to

$$I_n = \cos^{n-1} x \sin x + (n-1)I_{n-2} - (n-1)I_n$$

or

$$I_n = \frac{1}{n}\cos^{n-1} x \sin x + \frac{n-1}{n}I_{n-2}. \qquad (5.52)$$

This kind of expression, where I_n is given in terms of I_m, where $m < n$, is called a **reduction formula**. How it is used in this case will be illustrated by evaluating

$$I_6 = \int \cos^6 x \, dx.$$

By repeated use of (5.52),

$$I_6 = \frac{1}{6} \cos^5 x \sin x + \frac{5}{6} I_4 = \frac{1}{6} \cos^5 x \sin x + \frac{5}{6} \left\{ \frac{1}{4} \cos^3 x \sin x + \frac{3}{4} I_2 \right\}$$

$$= \frac{1}{6} \cos^5 x \sin x + \frac{5}{24} \cos^3 x \sin x + \frac{5}{8} \left\{ \frac{1}{2} \cos x \sin x + \frac{1}{2} I_0 \right\}. \tag{5.53}$$

Now $I_0 = \int dx = x + C$ so that, with some rearrangement,

$$\int \cos^6 x \, dx = \cos x \sin x \left(\frac{1}{6} \cos^4 x + \frac{5}{24} \cos^2 x + \frac{5}{16} \right) + \frac{5}{16} x + C. \tag{5.54}$$

The same reduction formula (5.52) applies if n is odd, but the final stage of reduction in this case gives $I_1 = \sin x$.

A similar process can be used to evaluate $J_n = \int \sin^n x \, dx$ for any positive integer n. In this case the reduction formula is

$$J_n = -\frac{1}{n} \sin^{n-1} x \cos x + \frac{n-1}{n} J_{n-2}. \tag{5.55}$$

Again, if n is odd the final stage of reduction gives $J_1 = -\cos x$.

(ii) $\int \cos^n x \sin^m x \, dx$

We now consider an integral of the form

$$I_{n,m} = \int \cos^n x \sin^m x \, dx$$

Integration by parts is used with $U = \cos^{n-1} x$ and $dV/dx = \sin^m x \cos x$, which gives $dU/dx = -(n-1) \cos^{n-2} x \sin x$ and $V = \{1/(m+1)\} \sin^{m+1} x$. Then

$$I_{n,m} = \frac{1}{m+1} \cos^{n-1} x \sin^{m+1} x + \frac{n-1}{m+1} \int \cos^{n-2} x \sin^{m+2} x \, dx$$

$$= \frac{1}{m+1} \cos^{n-1} x \sin^{m+1} x + \frac{n-1}{m+1} \int \cos^{n-2} x \sin^m x (1 - \cos^2 x) dx.$$

The integral has components of $I_{n-2,m}$ and $I_{n,m}$ and, when rearranged, gives

$$I_{n,m} = \frac{1}{m+n} \cos^{n-1} x \sin^{m+1} x + \frac{n-1}{m+n} I_{n-2,m}. \tag{5.56}$$

If n is even, then the final stage of reduction gives the integral $I_{0,m}$, i.e. the same as J_m given previously, which can be solved by reduction formula (5.55). If n is odd then the situation is much simpler. The final integral in this case is

$$I_{1,m} = \int \cos x \sin^m x \, dx = \frac{1}{m+1} \sin^{m+1} x + C.$$

By taking $U = \sin^{m-1} x$ and $dV/dx = \cos^n x \sin x$ in the original integral we obtain the resultant reduction formula

$$I_{n,m} = -\frac{1}{m+n}\cos^{n+1} x \sin^{m-1} x + \frac{m-1}{m+n}I_{n,m-2}. \tag{5.57}$$

In this case if m is odd then the final reduction stage gives the simple integral

$$I_{n,1} = \int \cos^n x \sin x \, dx = -\frac{1}{n+1}\cos^n x.$$

Clearly if both n and m are even then two stages of reduction are necessary – either (5.56) followed by (5.55) or (5.57) followed by (5.52). If only n is odd then (5.56) should be used while if only m is odd only (5.57) should be used. If both n and m are odd then the reduction formula should be used that reduces the odd integer to 1 in the smaller number of steps.

EXAMPLE 5.18 $\quad \int \cos^4 x \, \sin^3 x \, dx$

As an illustration we evaluate $\int \cos^4 x \sin^3 x \, dx$. From (5.57),

$$I_{4,3} = -\frac{1}{7}\cos^5 x \sin^2 x + \frac{2}{7}I_{4,1} = -\frac{1}{7}\cos^5 x \sin^2 x + \frac{2}{7}\int \cos^4 x \sin x \, dx$$

$$= -\frac{1}{7}\cos^5 x \sin^2 x - \frac{2}{35}\cos^5 x + C. \tag{5.58}$$

Exercise 5.6

Find (i) $\int \sin^2 x \cos^3 x \, dx$;   (ii) $\int \sin^3 x \cos^2 x \, dx$.

5.7 The definite integral: area under the curve

The way in which the area under a curve is found was given in (5.7). To illustrate this, and the notation we shall be using, we consider the area under the curve

$$f(x) = 1 + 2x$$

between $x = 1$ and $x = 2$. This is indicated by

$$A = \int_1^2 (1 + 2x)dx$$

in which the upper and lower limits, 2 and 1, give the range of x for finding the area. For the next stage the integrand is integrated and we write

$$A = \left| x + x^2 \right|_1^2 = (2 + 2^2) - (1 + 1^2) = 4.$$

EXAMPLE 5.19

As another illustration we find the area under the curve $f(x) = \cos^2 x$ between $x = 0$ and $x = \pi/2$. This is

$$A = \int_0^{\pi/2} \cos^2 x \, dx = \frac{1}{2} \int_0^{\pi/2} (1 + \cos 2x) dx$$

$$= \frac{1}{2} x + \frac{1}{2} \sin 2x \Big|_0^{\pi/2} = \frac{1}{2} \left(\frac{\pi}{2} + \frac{1}{2} \sin \pi \right) - \frac{1}{2} \left(0 + \frac{1}{2} \sin 0 \right) = \frac{\pi}{4}.$$

A definite integral that occurs in many physical situations (see Chapter 22) is of the general form

$$I_n = \int_0^\infty x^n e^{-\alpha x} dx. \tag{5.59}$$

Notice that the upper limit is infinity and the evaluation of this integral requires the following result:

For any finite value of n and α, $\left[x^n e^{-\alpha x} \right]_{x \to \infty} = 0 \tag{5.60}$

The first term, x^n, tends to infinity as x tends to infinity and the second term, $e^{-\alpha x}$, tends to zero, but the second term always dominates.

To evaluate the integral we use integration by parts with $U = x^n$ and $dV/dx = e^{-\alpha x}$, which gives $dU/dx = nx^{n-1}$ and $V = -(1/\alpha)e^{-\alpha x}$. Hence we find

$$I_n = \left| -\frac{x^n}{\alpha} e^{-\alpha x} \right|_0^\infty + \frac{n}{\alpha} \int_0^\infty x^{n-1} e^{-\alpha x} \, dx = \frac{n}{\alpha} I_{n-1}, \tag{5.61}$$

since the first bracketed term disappears at both limits. From this reduction formula

$$I_n = \frac{n!}{\alpha^n} I_0 = \frac{n!}{\alpha^n} \int_0^\infty e^{-\alpha x} dx = \frac{n!}{\alpha^n} \left| -\frac{1}{\alpha} e^{-\alpha x} \right|_0^\infty = \frac{n!}{\alpha^{n+1}}. \tag{5.62}$$

Another definite integral that produces a reduction formula is $I_n = \int_0^{\pi/2} \cos^n x \, dx$. From (5.52),

$$I_n = \left| \frac{1}{n} \cos^{n-1} x \sin x \right|_0^{\pi/2} + \frac{n-1}{n} I_{n-2} = \frac{n-1}{n} I_{n-2} \tag{5.63}$$

since the bracketed term disappears at both limits. As an example, for $n = 10$

$$I_{10} = \frac{9}{10} I_8 = \frac{9 \times 7}{10 \times 8} I_6 = \frac{9 \times 7 \times 5}{10 \times 8 \times 6} I_4 = \frac{9 \times 7 \times 5 \times 3}{10 \times 8 \times 6 \times 4} I_2 = \frac{9 \times 7 \times 5 \times 3 \times 1}{10 \times 8 \times 6 \times 4 \times 2} I_0$$

$$I_0 = \int_0^{\pi/2} dx = |x|_0^{\pi/2} = \frac{\pi}{2}$$

so that

$$\int_0^{\pi/2} \cos^{10} x \, dx = \frac{9 \times 7 \times 5 \times 3 \times 1}{10 \times 8 \times 6 \times 4 \times 2} \times \frac{\pi}{2} = 0.3866.$$

It is important to understand what is meant by the expression **area under the curve**. For example, if we take the definite integral

$$\int_0^\pi \cos x \, dx = |-\sin x|_0^\pi = 0$$

then this implies that the area under the curve is zero. The form of the function cos x between the limits 0 and $\pi/2$ is shown in Figure 5.2. The curve lies both above and below the x-axis. Areas above the x-axis are positive and those below are negative; the resultant of zero in this case is because the positive and negative areas have equal magnitude. There may be situations in which what is required is $\int_b^a |f(x)|dx$ – that is to say that areas both above and below the x-axis are to be taken as positive. It may not be possible to integrate $|f(x)|$ directly and in this case it is necessary to investigate where the curve crosses the x-axis and to evaluate the positive and negative areas separately.

Although scientists are rarely interested in finding areas under a curve in a geometrical sense, the equivalent of that is often of interest. If the speed of a body is a function of time, $v(t)$, then the displacement of the body between times t_1 and t_2 is given by the definite integral

$$d = \int_{t_1}^{t_2} v(t)dt. \tag{5.64}$$

EXAMPLE 5.20

Consider a body projected upwards at an initial speed of $40 \, \mathrm{m\,s^{-1}}$ with an acceleration due to gravity of $-9.8 \, \mathrm{m\,s^{-2}}$, the sign indicating that it is acting in a downward direction. The speed as a function of t is

$$v = (40 - 9.8t) \, \mathrm{m\,s^{-1}}$$

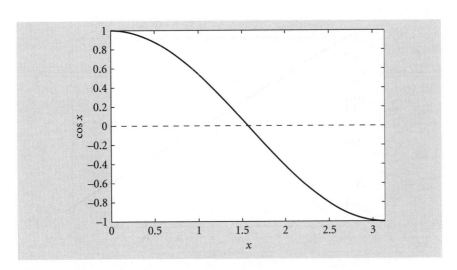

Figure 5.2 The function cos x between $x=0$ and $x=\pi$.

and the displacement from the moment of projection to time τ is

$$d = \int_0^\tau (40 - 9.8t)\,\mathrm{d}t = \left|40t - 4.9t^2\right|_0^\tau = 40\tau - 4.9\tau^2.$$

This expression for d has a maximum value, 81.63 m, at $\tau = 40/9.8$ s and thereafter decreases. At $\tau = 80/9.8$ s the value of d is zero. The first of these values is the maximum height of the projectile, corresponding to the greatest displacement of the projectile from the starting point, and the time for $d = 0$ corresponds to the time to return to the starting point, i.e. zero displacement. The velocity as a function of τ is shown in Figure 5.3.

The distance from the starting point at any time is the area under the curve from the origin to that time. After $40/9.8$ s the area under the curve becomes negative so the projectile is now moving closer to the starting point. If the object of the exercise was to find the total distance *travelled* by the projectile, regardless of direction, then the integral required is

$$d_{\text{trav}} = \int_{t_1}^{t_2} |v|\,\mathrm{d}t.$$

This could be obtained by adding the magnitudes of the areas above and below the curve in Figure 5.3.

Exercise 5.7
Given that $f(x) = x^2 - 6x + 5$, find: (i) $\int_0^2 f(x)\,\mathrm{d}x$; (ii) $\int_0^2 |f(x)|\,\mathrm{d}x$.

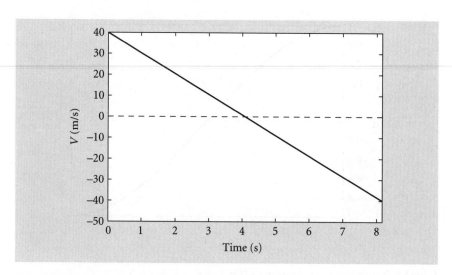

Figure 5.3 Velocity as a function of time.

Problems

5.1 Find the following indefinite integrals: (i) $\int \dfrac{1}{x^2 + 4x + 5} \, dx$; (ii) $\int \cot x \, dx$

(try the substitution $u = \sin x$); (iii) $\int \tan x \, dx$; (iv) $\int \sqrt{\dfrac{1}{4 - x^2}} \, dx$;

(v) $\int \dfrac{1}{\sqrt{3 + 2x - x^2}} \, dx$.

5.2 Find the following indefinite integrals: (i) $\int \dfrac{1}{2x^2 + 3x + 1} \, dx$;

(ii) $\int \dfrac{x + 2}{x^2 + 6x + 8} \, dx$.

5.3 Find the indefinite integral $\int \dfrac{1 - 2\cos\theta}{1 + 2\cos\theta} \, d\theta$.

5.4 Find the following indefinite integrals: (i) $\int x^3 \cos x \, dx$; (ii) $\int \cos^4 x \, dx$;

(iii) $\int \cos^5 x \sin^3 x \, dx$

5.5 Find the area under the curve of $f(x) = x^2 - 5x + 6$ between $x = 0$ and $x = 2$. Then find the area between $x = 0$ and $x = 3$. Why is this smaller than the previous result?

5.6 (i) A rocket motor gives a constant thrust while it is burning but, because the mass of rocket plus fuel is reducing, the rate of acceleration constantly increases. It is found that the upward acceleration, in SI units, can be approximated as $a = 10/(1 - 0.002t)$. If the rocket motor burns for 450 s, what is its velocity when the motor stops?

(ii) Show that $\int \ln x \, dx = x \ln x - x + C$.

(iii) What is the height of the rocket when the motor stops?

6

Complex numbers

The idea of a complex number seems to be remote from reality, and hence at first sight it seems unlikely that these numbers could be useful for the solution of problems in physics. Each complex number consist of two components, one a normal number such as we meet in everyday life and the other related to $\sqrt{-1}$, a quantity that has no meaning in the world of real numbers. In this chapter you will learn about complex numbers and the ways in which they can be represented. You will find, through an application to circuit theory, that they are an extremely powerful mathematical tool that greatly simplifies the analysis of many physical problems.

6.1 Definition of a complex number

The meaning of a complex number is best explained by considering the solutions of quadratic equations. A general quadratic equation of the form

$$ax^2 + bx + c = 0 \tag{6.1}$$

has a pair of solutions

$$x_1 = \frac{-b + \sqrt{b^2 - 4ac}}{2a} \quad \text{and} \quad x_2 = \frac{-b - \sqrt{b^2 - 4ac}}{2a}. \tag{6.2}$$

Thus the solutions of $x^2 - 4x - 3 = 0$ are $x_1 = 2 + \sqrt{7}$ and $x_2 = 2 - \sqrt{7}$. However, applying (6.2) in the same way, the solutions of $x^2 - 2x + 2 = 0$ are

$$x_1 = 1 + \sqrt{-1} \quad \text{and} \quad x_2 = 1 - \sqrt{-1}. \tag{6.3}$$

Since all ordinary numbers, whether positive or negative, have squares that are positive, (6.5) seems to indicate that the original quadratic equation does not have any solutions.

An alternative procedure is to define the quantity $\sqrt{-1}$ in some way and to describe the solution of the quadratic equation in terms of that definition. This is done by writing

$$i = \sqrt{-1} \tag{6.4}$$

and then describing the solutions of the quadratic equation as $x_1 = 1 + i$ and $x_2 = 1 - i$. This seems a rather arbitrary thing to do – the quantity i has no real meaning and is something that comes from the imagination of the mathematician. In fact, in the language of mathematics the solutions of the quadratic equation are **complex numbers**, each with a **real component**, 1, and **imaginary components**, i and −i.

The introduction of the idea of a complex number is not just a mathematical abstraction – it turns out to be a very powerful tool of great utility to scientists in general and to physical scientists in particular.

Let us see what happens when we take powers of i. We have

$$i = \sqrt{-1}, \quad i^2 = -1, \quad i^3 = i^2 \times i = -i, \quad i^4 = i^2 \times i^2 = 1, \quad i^5 = i^4 \times i = i.$$
$$(6.5)$$

EXAMPLE 6.1

To confirm that solutions of quadratic equations in terms of complex numbers are meaningful, we consider the quadratic equation $x^2 - 2x + 3 = 0$, the solutions of which are $x_1 = 1 + \sqrt{2}i$ and $x_2 = 1 - \sqrt{2}i$. Testing the first solution,

$$x_1^2 - 2x_1 + 3 = (1 + \sqrt{2}i)^2 - 2(1 + \sqrt{2}i) + 3$$
$$= 1 + 2\sqrt{2}i + 2i^2 - 2 - 2\sqrt{2}i + 3$$
$$= 1 + 2\sqrt{2}i - 2 - 2 - 2\sqrt{2}i + 3 = 0.$$

You are now invited to check that the solution x_2 is also valid.

Note that for the sum of complex quantities to be zero the real and imaginary parts must each be zero. In general, if complex numbers are added or subtracted then the real and imaginary parts are added separately to give the real and imaginary components of the resultant sum or difference.

EXAMPLE 6.2   **Addition and subtraction of complex numbers**

If

$$z_1 = 3 + 5i, \quad z_2 = -2 + i, \quad z_3 = 4 - 8i$$

then

$$z_1 + z_2 = (3 - 2) + (5 + 1)i = 1 + 6i \quad \text{and}$$
$$z_2 - z_3 = (-2 - 4) + (1 + 8)i = -6 + 9i.$$

Following on from this,

$$z_1 - 2z_2 + z_3 = \{3 - 2 \times (-2) + 4\} + \{5 - 2 \times 1 + (-8)\}i = 11 - 5i.$$

Multiplication also follows the usual pattern, with the exception that real and imaginary components occur.

| EXAMPLE 6.3 | **Multiplication of complex numbers** |

As an example:

$$z_1 z_2 = (3 + 5i)(-2 + i) = -6 - 10i + 3i + 5i^2 = -6 - 10i + 3i - 5$$
$$= -11 - 7i.$$

Division is a little more complicated and is dealt with later.

 Exercise 6.1
Given that $z_1 = 3 - 2i$, $z_2 = 1 + i$, and $z_3 = 2 - 3i$, find:
(i) $z_1 + z_2$; (ii) $z_1 - z_2 + z_3$; (iii) $z_1 z_2$; (iv) $z_1 z_2 z_3$.

6.2 **Argand diagram**

One way of representing ordinary numbers, which stretch from $-\infty$ to $+\infty$, is to place them at the appropriate coordinate on an infinite straight line, say the x axis of a Cartesian coordinate system. For a complex number, which has both real and imaginary components, an obvious corresponding way of representing it is to place it in the x–y plane of a rectangular Cartesian coordinate system with the x coordinate representing the real part and the y coordinate the imaginary part, as shown in Figure 6.1. Such a representation of a complex number is known as an **Argand diagram**.

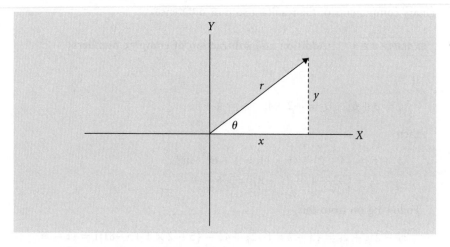

Figure 6.1 The representation of the complex number $x + iy$ on an Argand diagram.

6.3 Ways of describing a complex number

We have already described a complex number in terms of its real and imaginary components in the form

$$z = x + iy. \tag{6.6}$$

However, from Figure 6.1 it can also be described in terms of polar coordinates as

$$z = r\cos\theta + ir\sin\theta = r(\cos\theta + i\sin\theta). \tag{6.7}$$

Referring to the series representations of $\cos x$ and $\sin x$ given by (4.44b) and (4.44a), we can rewrite (6.7) as

$$z = r\left(1 + i\theta - \frac{\theta^2}{2!} - i\frac{\theta^3}{3!} + \frac{\theta^4}{4!} + i\frac{\theta^5}{5!} - \cdots\right). \tag{6.8}$$

We now consider the series expansion of e^x, as given by (4.34), replacing x by $i\theta$. This gives

$$e^{i\theta} = 1 + i\theta + i^2\frac{\theta^2}{2!} + i^3\frac{\theta^3}{3!} + i^4\frac{\theta^4}{4!} + i^5\frac{\theta^5}{5!} + \cdots$$

or, from relationships (6.5),

$$e^{i\theta} = 1 + i\theta - \frac{\theta^2}{2!} - i\frac{\theta^3}{3!} + \frac{\theta^4}{4!} + i\frac{\theta^5}{5!} - \cdots \tag{6.9}$$

Comparing (6.8) and (6.9) we may write

$$z = re^{i\theta} = r(\cos\theta + i\sin\theta). \tag{6.10}$$

In this form of representation of a complex number r is referred to as the **modulus** and θ as the **argument**. As an example we consider changing the complex number $3 - 6i$ to modulus–argument form. From Figure 6.1 we can see that

$$r = \sqrt{x^2 + y^2} = \sqrt{9 + 36} = \sqrt{45}.$$

Similarly we see that

$$\tan\theta = \frac{y}{x} = \frac{r\sin\theta}{r\cos\theta}.$$

We must resolve a possible phase ambiguity by noting that the value of θ is such that x, and hence $\cos\theta$, is positive and that y, and hence $\sin\theta$, is negative. This puts the angle θ between $5\pi/2$ and 2π (or between 0 and $-\pi/2$). Hence we have $\tan\theta = -2$ and $\theta = 5.176$, in radians. This gives

$$3 - 6i = \sqrt{45}\exp(5.176i).$$

Expressing a complex number in modulus–argument form can be very useful in some circumstances. Finally, from (6.10) we have the result

$$\cos\theta + i\sin\theta = e^{i\theta} \tag{6.11a}$$

and replacing θ by $-\theta$ gives

$$\cos\theta - i\sin\theta = e^{-i\theta}. \tag{6.11b}$$

Adding and subtracting these equations gives

$$\cos\theta = \frac{1}{2}\left(e^{i\theta} + e^{-i\theta}\right) \tag{6.12a}$$

and

$$\sin\theta = \frac{1}{2i}\left(e^{i\theta} - e^{-i\theta}\right). \tag{6.12b}$$

Exercise 6.2
Express the following complex numbers in modulus–argument form:
(i) $1 + i$; (ii) $-1 + i$; (iii) $2 - 3i$; (iv) $3 + 2i$.

Exercise 6.3
Express the following complex numbers in the form $x + iy$:
(i) $2e^{2i}$; (ii) $3e^{-2i}$; (iii) e^{-4i}.

6.4 De Moivre's theorem

Representation of complex numbers in the modulus–argument form can greatly simplify many kinds of operations involving complex numbers, as will now be illustrated.

6.4.1 Multiplication and division

Given $z_1 = r_1 e^{i\theta_1}$ and $z_2 = r_2 e^{i\theta_2}$, we have

$$z_1 z_2 = r_1 r_2 \exp\{i(\theta_1 + \theta_2)\} \tag{6.13}$$

and

$$\frac{z_1}{z_2} = \frac{r_1}{r_2} \exp\{i(\theta_1 - \theta_2)\}. \tag{6.14}$$

Thus multiplication and division are particularly simple with the modulus–argument form but, as we have seen previously, addition and subtraction are much simpler with the real-plus-imaginary representation.

If there are many complex numbers in modulus–argument form, the jth of which is $z_j = r_j e^{ir_j}$, then

$$\prod_{j=1}^{N} z_j = \prod_{j=1}^{N} r_j \times \exp\left(i\sum_{j=1}^{N}\theta_j\right), \tag{6.15}^*$$

where the symbol $\prod_{j=1}^{N}$ represents a product of all the quantities with subscript j from $j=1$ to $j=N$.

Exercise 6.4
Given $z_1 = 2e^{2i}$, $z_2 = 3e^{-i}$, and $z_3 = 4e^{-4i}$, find: (i) $z_1 z_2 z_3$; (ii) $z_1 z_2 / z_3$.

6.4.2 Powers of complex numbers – de Moivre's theorem

A special case of taking a product is to form the power z^n where n is some integer. By an extension of (6.15),

$$z^n = r^n (\cos \theta + i \sin \theta)^n = r^n e^{in\theta} = r^n (\cos n\theta + i \sin n\theta). \tag{6.16}$$

Taking $r = 1$ gives

$$(\cos \theta + i \sin \theta)^n = \cos n\theta + i \sin n\theta, \tag{6.17}$$

which is known as **de Moivre's theorem**. This actually applies for negative integer values of n and also for fractional values.

This theorem can be applied in a variety of ways that we now illustrate.

6.4.3 $\cos^n \theta$ and $\sin^n \theta$ in terms of cosines and sines of multiple angles

Taking $z = e^{i\theta} = \cos \theta + i \sin \theta$, we have $z^{-1} = e^{-i\theta} = \cos \theta - i \sin \theta$. Combining these expressions,

$$z + z^{-1} = 2 \cos \theta \tag{6.18a}$$

and

$$z - z^{-1} = 2i \sin \theta. \tag{6.18b}$$

From de Moivre's theorem we have $z^n = \cos n\theta + i \sin n\theta$ and $z^{-n} = \cos n\theta - i \sin n\theta$, giving

$$z^n + z^{-n} = 2 \cos n\theta \tag{6.19a}$$

and

$$z^n - z^{-n} = 2i \sin n\theta. \tag{6.19b}$$

Suppose that we are interested in expressing $\cos^5 \theta$ in terms of multiple angles. We first write, using (1.30),

$$2^5 \cos^5 \theta = (z + z^{-1})^5 = z^5 + 5z^4 z^{-1} + 10z^3 z^{-2} + 10z^2 z^{-3} + 5zz^{-4} + z^{-5}$$
$$= (z^5 + z^{-5}) + 5(z^3 + z^{-3}) + 10(z + z^{-1}). \tag{6.20}$$

Now, using (6.19a),

$$32 \cos^5 \theta = 2 \cos 5\theta + 10 \cos 3\theta + 20 \cos \theta$$

or

$$\cos^5 \theta = \frac{1}{16}(\cos 5\theta + 5 \cos 3\theta + 10 \cos \theta). \tag{6.21}$$

The corresponding calculation for finding an expression for $\sin^5 \theta$ is as follows:

$$2^5 i^5 \sin^5 \theta = (z - z^{-1})^5 = z^5 - 5z^4 z^{-1} + 10z^3 z^{-2} - 10z^2 z^{-3} + 5zz^{-4} - z^{-5}$$

$$= (z^5 - z^{-5}) - 5(z^3 - z^{-3}) + 10(z - z^{-1}). \tag{6.22}$$

Now using (6.19b),

$$32i \sin^5 \theta = 2i \sin 5\theta - 10i \sin 3\theta + 20i \sin \theta$$

or

$$\sin^5 \theta = \frac{1}{16}(\sin 5\theta - 5 \sin 3\theta + 10 \sin \theta). \tag{6.23}$$

Exercise 6.5
Find $\cos^3 \theta$ and $\sin^3 \theta$ in terms of multiple angles.

6.4.4 Sines and cosines of multiple angles in terms of cos θ and sin θ

The converse process to that explained in the previous section is, for example, to express $\cos 4\theta$ in terms of powers and products of $\cos \theta$ and $\sin \theta$. The way to do this is as follows:

$$\cos 4\theta + i \sin 4\theta = (\cos \theta + i \sin \theta)^4$$

$$= \cos^4 \theta + 4 \cos^3 \theta \times i \sin \theta + 6 \cos^2 \theta \times i^2 \sin^2 \theta$$
$$+ 4 \cos \theta \times i^3 \sin^3 \theta + i^4 \sin^4 \theta$$

which may be simplified to

$$\cos 4\theta + i \sin 4\theta = (\cos^4 \theta - 6 \cos^2 \theta \sin^2 \theta + \sin^4 \theta)$$
$$+ i(4 \cos^3 \theta \sin \theta - 4 \cos \theta \sin^3 \theta).$$

If two complex numbers are equal then their real parts and imaginary parts must be separately equal. From this we find expressions for both $\cos 4\theta$ and $\sin 4\theta$ as

$$\cos 4\theta = \cos^4 \theta - 6 \cos^2 \theta \sin^2 \theta + \sin^4 \theta \tag{6.24}$$

and

$$\sin 4\theta = 4 \cos \theta \sin \theta (\cos^2 \theta - \sin^2 \theta). \tag{6.25}$$

Exercise 6.6
Find $\cos 3\theta$ and $\sin 3\theta$ in terms of powers and products of $\cos \theta$ and $\sin \theta$.

6.4.5 Roots of complex numbers

If we are asked to find the square root of 4 or the cube root of 8, or even the cube root of -8, we have no problem. We have also seen that with the definition of imaginary numbers we can even give a value for the square root of -4 as 2i. However, suppose we wish to find the square root of $1 + i$ – how do we do this?

EXAMPLE 6.4   **Finding $\sqrt{1+i}$**

The answer is to express the complex number in modulus–argument form; in this case $r = \sqrt{2}$ and $\theta = \arctan(1/1) = \pi/4$. Hence

$$\sqrt{1+i} = \left(\sqrt{2}\exp(i\pi/4)\right)^{1/2} = 2^{1/4}\exp(i\pi/8). \tag{6.26}$$

This can be expressed in real-plus-imaginary form as

$$\sqrt{1+i} = 2^{1/4}\cos(\pi/8) + i2^{1/4}\sin(\pi/8) = 1.0987 + 0.4551i. \tag{6.27}$$

This solution is now checked:

$$(1.0987 + 0.4551i)^2 = 1.0987^2 - 0.4551^2 + 2 \times 1.0987 \times 0.4551i$$
$$= 1.0000 + 1.0000i.$$

The square root of 4 has two solutions, $+2$ and -2, so you might wonder whether (6.26) is the only solution for $\sqrt{1+i}$. Because of the cyclic nature of angles, where adding or subtracting multiples of 2π gives an equivalent angle, in place of (6.26) we could add 2π to the argument and write

$$\sqrt{1+i} = \left(\sqrt{2}\exp(i9\pi/4)\right)^{1/2} = 2^{1/4}\exp(i9\pi/8) \tag{6.28}$$

which in real-plus-imaginary form gives

$$\sqrt{1+i} = 2^{1/4}\cos(9\pi/8) + i2^{1/4}\sin(9\pi/8) = -1.0987 - 0.4551i. \tag{6.29}$$

This result is just the negative of the previous result and since $(-1)^2 = +1$ it is obviously also a solution.

Just as we find two solutions for the square root of any number, real or complex, so we can find three solutions for a cube root and n solutions for an nth root. We illustrate this by finding the cube root of $1+i$.

EXAMPLE 6.5   **Finding $\sqrt[3]{1+i}$**

First we write $1+i$ in the three equivalent forms:

$$2^{1/2}\exp\left\{i\frac{\pi}{4}\right\}, \quad 2^{1/2}\exp\left(i\frac{9\pi}{4}\right), \quad \text{and} \quad 2^{1/2}\exp\left(i\frac{17\pi}{4}\right). \tag{6.30}$$

We now find the obvious cube root for each form, i.e.

$$2^{1/6}\exp\left\{i\frac{\pi}{12}\right\}, \quad 2^{1/6}\exp\left(i\frac{3\pi}{4}\right), \quad \text{and} \quad 2^{1/6}\exp\left(i\frac{17\pi}{12}\right). \tag{6.31}$$

Expressed in real-plus-imaginary form, these are

$$1.0842 + 0.2905i, \quad -0.7937 + 0.7937i, \quad \text{and} \quad -0.2905 - 1.0842i.$$

There are no solutions other than these three. The final two forms in (6.30) are obtained by adding 2π and 4π to the argument of the first form. If 6π is added then when it is divided by three the result will be equivalent to that given by the first form.

By similar reasoning it is found that there are n nth roots for any number, real or complex.

Exercise 6.7
Find the cube roots of -8.

6.5 Complex conjugate

The complex conjugate of a complex number

$$z = x + iy \qquad (6.32a)$$

is

$$z^* = x - iy \qquad (6.32b)$$

which just involves changing the sign of the imaginary component. A complex number and its complex conjugate are shown in Figure 6.2. From the figure you can see that if we use the modulus–argument form then

$$z = r \exp(i\theta) \qquad (6.33a)$$

and

$$z^* = r \exp(-i\theta). \qquad (6.33b)$$

The sum of a complex number and its complex conjugate is real and the difference of the two is imaginary. This is best seen by expressing the complex

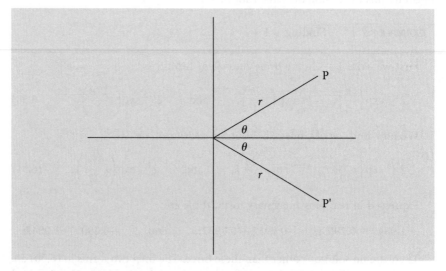

Figure 6.2 The complex number, represented by P, and its complex conjugate, represented by P′.

quantities in real-plus-imaginary components:

$$z + z^* = x + iy + x - iy = 2x \qquad (6.34a)$$

and

$$z - z^* = x + iy - (x - iy) = 2iy. \qquad (6.34b)$$

The product of a complex number and its complex conjugate is real and equals the square of the modulus. This can be shown either by using the modulus–argument form, i.e.

$$zz^* = r\exp(i\theta)r\exp(-i\theta) = r^2 = |z|^2 \qquad (6.35a)$$

or by using the real-plus-imaginary form, i.e.

$$zz^* = (x + iy)(x - iy) = x^2 - i^2y^2 = x^2 + y^2 = r^2 = |z|^2. \qquad (6.35b)$$

Notice the use of the notation $|z| \; (= \sqrt{x^2 + y^2})$, meaning the modulus of z.

Finding the complex conjugate of sums, products, or quotients of complex numbers is particularly straightforward; it just consists of replacing every complex number by its complex conjugate. By expressing the complex numbers in either real plus imaginary or modulus argument form it may be easily verified that

$$(z_1 + z_2)^* = z_1^* + z_2^*, \qquad (6.36a)$$

$$(z_1 z_2)^* = z_1^* z_2^* \qquad (6.36b)$$

$$\left(\frac{z_1}{z_2}\right)^* = \frac{z_1^*}{z_2^*}. \qquad (6.36c)$$

This principle can be extended so that, for example,

$$\left(\frac{z_1 + z_2}{z_3 z_4}\right)^* = \frac{z_1^* + z_2^*}{z_3^* z_4^*}.$$

Exercise 6.8

By evaluating each side separately, show that for $z_1 = 2 + i$ and $z_2 = 3 - i$, $(z_1 z_2)^* = z_1^* z_2^*$.

6.6 Division and reduction to real-plus-imaginary form

We now consider how to convert a quotient of complex numbers, for example

$$z = \frac{1 + 2i}{2 - 3i},$$

to the form $x + iy$. One way of doing this is first to convert the numerator and divisor to modulus–argument form to give

$$z = \frac{\sqrt{5}\exp(1.107i)}{\sqrt{13}\exp(-0.9828i)}$$

which gives

$$z = \sqrt{\frac{5}{13}} \exp(2.0898i) = 0.6202\{\cos(2.0898) + i\sin(2.0898)\}$$
$$= -0.3077 + 0.5385i. \tag{6.37}$$

Another way is to multiply numerator and divisor by the complex conjugate of the divisor. This gives

$$z = \frac{(1+2i)(2+3i)}{(2-3i)(2+3i)} = \frac{-4+7i}{13} = -0.3077 + 0.5385i, \tag{6.38}$$

which is identical to the previous result. This process is less complicated than the former one and uses the fact that the product of a complex number and its complex conjugate is real.

 Exercise 6.9
Express $(1-i)/(2+i)$ in the form $x + iy$.

6.7 Modulus–argument form as an aid to integration

The relationship (6.11a) can often be used to simplify certain forms of integral. Consider, for example, the indefinite integral

$$I = \int e^{2x} \cos x \, dx. \tag{6.39}$$

This can be solved by a two-stage process of integration by parts, as follows. Taking $u = \cos x$ and $dv = e^{2x} \, dx$ we find

$$I = \frac{1}{2}e^{2x} \cos x + \int e^{2x} \sin x \, dx + C,$$

where we always indicate C as the integration constant although it may vary from one integration step to another. Now taking $u = \sin x$ and $dv = e^{2x} \, dx$,

$$I = \frac{1}{2}e^{2x} \cos x + \frac{1}{4}e^{2x} \sin x - \frac{1}{4}\int e^{2x} \cos x \, dx + C$$
$$= \frac{1}{2}e^{2x} \cos x + \frac{1}{4}e^{2x} \sin x - \frac{1}{4}I + C.$$

This gives

$$I = \frac{2}{5}e^{2x} \cos x + \frac{1}{5}e^{2x} \sin x + C.$$

An alternative approach is to write

$$I = R\left[\int e^{2x} e^{ix} dx\right] = R\left[\int \exp\{(2+i)x\}dx\right]$$
$$= R\left[\frac{1}{2+i}e^{2x}(\cos x + i\sin x) + C\right],$$

where R indicates the real part of the expression that follows. Now using the method described in §6.6,

$$I = \frac{1}{5}R\left[e^{2x}(2 - i)(\cos x + i \sin x) + C\right] = \frac{2}{5}e^{2x}\cos x + \frac{1}{5}e^{2x}\sin x + C$$

as previously. The gain in using this process in this simple demonstration case is not great but often, in more complicated cases, taking $\cos x$ as the real part of e^{ix}, or $\sin x$ as its imaginary part ($\sin x = I[e^{ix}]$) can greatly simplify the calculation.

Exercise 6.10

Find the integral $I = \int e^x \sin x \, dx$.

6.8 Circuits with alternating currents and voltages

For direct-current (DC) circuits the relationship between applied potential difference (PD) V, current I, and resistance R, is given by Ohm's law

$$V = IR. \tag{6.40}$$

The resistance, R, is referred to as an **impedance** – the higher its value, the more it impedes the movement of electrons and so the lower is the current. The SI unit of resistance is the ohm (Ω). However, the applied PD can be alternating, that is of the form

$$V = V_0 \sin \omega t \tag{6.41}$$

in which V_0 is the amplitude of the PD, t is time, and ω is the **angular frequency** in radians s^{-1} corresponding to a frequency v normally expressed in hertz (Hz), i.e. cycles per second, which are related by

$$\omega = 2\pi v. \tag{6.42}$$

For an alternating applied PD other kinds of impedance can be present and the resultant current is alternating but may have a **phase difference**, i.e. be out of step with the applied PD. We shall see that the use of complex numbers greatly simplifies the treatment of alternating current circuits.

6.8.1 Expressing an alternating PD or current as a complex quantity

As an alternative to (6.41) it is possible to express an alternating PD in the form

$$V = V_0 \exp i\omega t \tag{6.43a}$$

and, correspondingly, for an alternating current

$$I = I_0 \exp i\omega t. \tag{6.43b}$$

The modulus is the amplitude of the quantity, PD or current, and the argument gives the phase of the quantity compared with that at $t = 0$.

The effect of advancing the phase by δ changes (6.41) to

$$V_0 \sin(\omega t + \delta) = V_0 \sin \omega t \cos \delta + V_0 \cos \omega t \sin \delta \tag{6.44a}$$

and (6.43a) to

$$V_0 \exp i(\omega t + \delta) = V_0 \exp(i\omega t) \exp(i\delta). \tag{6.44b}$$

The change for the complex expression is much simpler, just involving a product with a phase factor, $\exp(i\delta)$, and it is this property that gives the complex notation the advantage.

6.8.2 Inductors

Inductors are devices that depend on two phenomena in physics. The first is that an electric current produces a magnetic field in its vicinity and, in particular, a current flowing through a coil gives a magnetic field within the coil of strength proportional to the current. The second is that a varying magnetic field passing through a coil will induce a PD across it of magnitude proportional to the rate of change of magnetic field. Now imagine that an alternating current passes through a coil of zero resistance, so Ohm's law is not operating and there is no PD across the coil due to resistance. The alternating current produces an alternating, and hence varying, magnetic field within the coil that, in its turn induces an alternating PD across it. The self-induced PD will, at any instant, be proportional to the rate of change of the current. The constant of proportionality depends on the form of the coil and is known as the self-inductance, L. This gives the PD across the coil as

$$V = L\frac{dI}{dt} \tag{6.45a}$$

or

$$V_0 \exp(i\omega t) = L\frac{dI}{dt}. \tag{6.45b}$$

The solution of (6.45b) is

$$I = \frac{V_0}{i\omega L} \exp i\omega t. \tag{6.46}$$

For comparison with Ohm's law, (6.40), we rearrange (6.46) to give

$$V = i\omega L I \tag{6.47}$$

from which it is seen that the inductor is equivalent to a **complex impedance** $i\omega L$. If we replace i by the expression

$$i = \cos\frac{\pi}{2} + i \sin\frac{\pi}{2} = \exp\left(i\frac{\pi}{2}\right) \tag{6.48}$$

we have from (6.46) that

$$I = \frac{V_0}{\omega L} \exp i\left(\omega t - \frac{\pi}{2}\right) \tag{6.49}$$

showing that a PD of form (6.43a) applied across an inductor L gives an alternating current of amplitude $V_0/(\omega L)$ that lags in phase $\pi/2$ behind the applied PD. From Table 2.3 we see that the SI unit of inductance is the **henry** (H).

6.8.3 Capacitors

Capacitors are devices that are, in the most basic form, two conductors separated by an insulating material. In general the charges on the conductors will be equal and opposite, of amount Q on one conductor and $-Q$ on the other. The electric field at any point between the conductors will be proportional to Q, as will be the PD, V, between the conductors. The capacitance of the capacitor, C, is defined as the charge carried by the conductors per unit PD between them, or $C = Q/V$, giving

$$V = \frac{Q}{C}. \tag{6.50}$$

It is convenient to find an expression linking capacitance and PD in terms of current rather than charge, which can be done by differentiating both sides of (6.50) with respect to time. This gives

$$\frac{dV}{dt} = \frac{1}{C}\frac{dQ}{dt} = \frac{I}{C}, \tag{6.51}$$

since the change of charge δQ in an interval δt will be the quantity of charge carried by the current in that interval, i.e. $\delta Q = I\delta t$.

From (6.43a) and (6.51),

$$i\omega V_0 \exp i\omega t = \frac{I}{C},$$

or, using (6.48)

$$V = \frac{1}{i\omega C}I = -\frac{i}{\omega C}I, \tag{6.52}$$

showing that the capacitor is equivalent to a complex impedance $-i/\omega C$.

We also find by rearranging that

$$I = \omega C V_0 \exp i\left(\omega t + \frac{\pi}{2}\right) \tag{6.53}$$

showing that a PD of form (6.43a) applied across a capacitor C gives an alternating current of amplitude $\omega C V_0$ that leads in phase by $\pi/2$ before the applied PD. From Table 2.3 we see that the SI unit of capacitance is the **farad** (F).

6.8.4 A simple circuit

We now consider a simple series circuit containing a resistor, an inductor, and a capacitor driven by an alternating PD as shown in Figure 6.3. Since the impedance elements are in series then, just as for simple resistors, the total impedance, Z, is found by adding the separate impedances. This gives

$$Z = R + i\left(\omega L - \frac{1}{\omega C}\right). \tag{6.54}$$

By analogy with Ohm's law we now write

$$V = IZ$$

or

$$I = \frac{V_0 \exp i\omega t}{R + i\left(\omega L - \dfrac{1}{\omega C}\right)}. \tag{6.55}$$

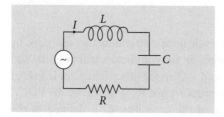

Figure 6.3 An LCR series circuit.

We now use the technique described in §6.6 to rationalize this expression. We find

$$I = \frac{V_0 \exp i\omega t \left\{ R - i\left(\omega L - \dfrac{1}{\omega C} \right) \right\}}{R^2 + \left(\omega L - \dfrac{1}{\omega C} \right)^2}. \tag{6.56}$$

To understand this result more clearly we make the transformation

$$\frac{R}{\sqrt{R^2 + \left(\omega L - \dfrac{1}{\omega C} \right)^2}} = \cos\delta \quad \text{and} \quad \frac{\omega L - \dfrac{1}{\omega C}}{\sqrt{R^2 + \left(\omega L - \dfrac{1}{\omega C} \right)^2}} = \sin\delta. \tag{6.57}$$

Notice that this transformation satisfies the condition $\cos^2\delta + \sin^2\delta = 1$ and defines δ by

$$\tan\delta = \frac{\omega L - (1/\omega C)}{R} \tag{6.58}$$

where $\sin\delta$ and $\cos\delta$ have the signs of the numerator and divisor of (6.58) respectively. In place of (6.56) we can now write

$$I = \frac{V_0 \exp i(\omega t - \delta)}{\sqrt{R^2 + (\omega L - (1/\omega C))^2}} \tag{6.59}$$

from which we see that the impedances in series are equivalent to a single impedance of magnitude $\sqrt{R^2 + (\omega L - (1/\omega C))^2}$ that gives a phase shift of $-\delta$ where δ is defined by (6.58).

The principle that has been used here can be extended to circuits of any complexity. Thus for two impedances z_1 and z_2, complex or otherwise, in parallel the overall impedance, Z, is given by

$$\frac{1}{Z} = \frac{1}{z_1} + \frac{1}{z_2}. \tag{6.60}$$

Exercise 6.11

Find the magnitude of the overall impedance of an inductor of inductance $1\,\text{H}$ in parallel with a capacitor of capacitance $10^{-6}\,\text{F}$ for an applied PD of frequency $50\,\text{Hz}$.

Problems

6.1 (i) Show that if the quadratic equation $ax^2 + bx + c = 0$ has real coefficients and complex solutions then solutions are complex conjugates of each other and their product equals c/a.

(ii) Find the roots of the equation $2x^2 + 3x + 5 = 0$ in modulus–argument form.

6.2 Express $\cos^6 \theta$ as a sum containing terms of the type $\cos n\theta$.

6.3 Express $\cos 4\theta$ and $\sin 4\theta$ in terms of $\cos \theta$ and $\sin \theta$.

6.4 Find the values of $(2 + 3i)^{1/5}$ in the form $x + iy$.

6.5 Express the following quantities in the form $x + iy$:

(i) $\dfrac{2 - i}{3 + 2i}$; (ii) $\dfrac{3 - i}{(1 + i)(2 - i)}$.

6.6 Write down the complex conjugate of $\dfrac{z_1 z_2 + z_3 z_4}{(z_1 + z_2)(z_3 + z_4)}$. Confirm your answer by evaluating the original expression and the complex conjugate for $z_1 = z_2 = 1 + i$ and $z_3 = z_4 = 2 + i$.

6.7 Find the indefinite integral $\int \cos x \sin x \, e^x dx$.

6.8 A source of alternating PD of voltage $100\,\text{V}$ and frequency $50\,\text{Hz}$ is placed across an impedance that consists of a combination of a $0.01\,\text{F}$ capacitor in series with a $5\,\Omega$ resistor that are in parallel with a $0.01\,\text{H}$ inductor. What is the current flowing in the main circuit in both amplitude and phase?

7 Ordinary differential equations

Many physical relationships can be expressed in the form of a differential equation. Solving the differential equation to determine the dependent variable is the main topic of this chapter. Examples are given relating to several variants of problems involving vibrations and alternating-current electrical circuits.

7.1 Types of ordinary differential equation

Many physical relationships can be expressed in the form of a differential equation. For example, the relationship *mass* × *acceleration* = *force* can, in one dimension, be expressed in the form

$$m\frac{d^2x}{dt^2} = F. \tag{7.1}$$

Integrating this relationship in two steps, assuming constant acceleration, gives

$$\frac{dx}{dt} = \frac{F}{m}t + c \tag{7.2a}$$

and

$$x = \frac{F}{2m}t^2 + ct + d \tag{7.2b}$$

where c and d are two constants of integration. These can be determined by the **boundary conditions** associated with the system. Acceleration alone will not tell where a body is in terms of time if we know neither where it starts from nor with what speed it is moving when $t = 0$. From (7.2a) by making $t = 0$ we see that c is the initial velocity of the body. Similarly, with $t = 0$ in (7.2b) we find that d is the initial position of the body.

This example shows the essential characteristics of solving an ordinary differential equation (ODE), which has the characteristic of having one **independent variable,** in this case t. To solve an ODE the processes of integral calculus are used to find an expression for the quantity of interest, the **dependent variable** x and boundary conditions are used to determine the constants of integration. Since (7.1) contains up to second derivatives it is a **second-order ODE**.

An ODE can have more than one dependent variable. Consider a body, P, of negligible mass moving in the *x-y* plane under the gravitational influence of a mass *M* situated at the origin (Figure 7.1). The acceleration of the body is directed towards the origin and is of magnitude

$$a_r = \frac{GM}{r^2} = \frac{GM}{x^2 + y^2} \tag{7.3}$$

where *G* is the gravitational constant, *r* is the distance OP, and (x, y) are the coordinates of P. The force has components (a_x, a_y) where

$$a_x = \frac{d^2 x}{dt^2} = a_r \cos \theta = -\frac{GMx}{(x^2 + y^2)^{3/2}} \tag{7.4a}$$

and

$$a_y = \frac{d^2 y}{dt^2} = a_r \sin \theta = -\frac{GMy}{(x^2 + y^2)^{3/2}}, \tag{7.4b}$$

the negative sign indicating the direction of the acceleration so that, for example, when *x* is positive a_x is in the negative *x* direction. There are two dependent variables in this case, *x* and *y*, and the equations (7.4) are **coupled second-order ODEs**, 'coupled' meaning that some equations of the set contain more than one dependent variable.

In a purely mathematical sense ODEs may occur in an infinite variety of forms, but fortunately the physical world provides us with a small number of frequently occurring types. Here we shall consider some standard techniques for solving fairly simple first-order and second-order ODEs. Later, in Chapter 19, we shall see that even the most complicated ODEs can be solved by numerical methods that enable real physical problems, with non-ideal conditions, to be considered.

We now consider various techniques for solving the **first-order differential equation** of the form

$$\frac{dy}{dx} = f(x, y). \tag{7.5}$$

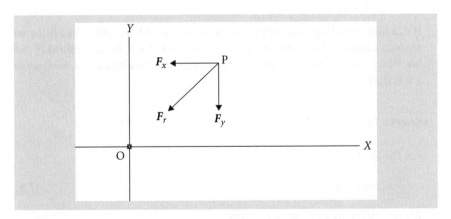

Figure 7.1 A body moving in two dimensions under the gravitational force of a mass at the origin.

This equation is first order because it just involves the first derivative of y with respect to x.

 Exercise 7.1
A body starts from the origin with a speed of $5\,\mathrm{m\,s^{-1}}$ and moves along the x-axis with an acceleration of $6\,\mathrm{m\,s^{-2}}$. How far is it from the origin after 20 s?

7.2 Separation of variables

A simple form of solution for a first-order differential equation occurs when the variables can be separated. To illustrate this, we consider two examples.

EXAMPLE 7.1

For the equation

$$\frac{\mathrm{d}y}{\mathrm{d}x} = \frac{y+2}{x+1}, \tag{7.6}$$

it is possible by rearranging to have quantities involving only x on one side of the equation and quantities only involving y on the other and then integrating both sides, i.e.

$$\int \frac{\mathrm{d}y}{y+2} = \int \frac{\mathrm{d}x}{x+1}. \tag{7.7}$$

Integrating both sides, and retaining just a single constant of integration expressed in the form of a natural logarithm for convenience,

$$\ln(y+2) = \ln(x+1) + \ln C.$$

This is a solution that can be arranged in various ways – for example,

$$y = C(x+1) - 2. \tag{7.8}$$

If this differential equation came from a physical problem, rather than being an abstract mathematical exercise, then the constant C could be determined by the boundary conditions of the problem. For example, if we know that $y=0$ when $x=0$ then $C=2$.

EXAMPLE 7.2

For the ODE

$$\frac{\mathrm{d}y}{\mathrm{d}x} = \sin x \tan y \tag{7.9}$$

the separation of variables now gives

$$\int \cot y \, \mathrm{d}y = \int \sin x \, \mathrm{d}x. \tag{7.10}$$

From the solution to Problem 5.1(ii),

$$\ln(\sin y) = -\cos x + C. \tag{7.11}$$

Taking the exponential of both sides, writing $A = e^C$ and then taking arcsin of both sides,

$$y = \sin^{-1}\{A\exp(-\cos x)\}. \tag{7.12}$$

Exercise 7.2
Solve the differential equation $\dfrac{dy}{dx} = \dfrac{x+1}{y}$.

7.3 Homogeneous equations

Consider the function of x and y

$$f(x, y) = a_1 x^4 + a_2 x^3 y + a_3 x^2 y^2 + a_4 x y^3 + a_5 y^4, \tag{7.13}$$

where a_1 to a_5 are numerical coefficients. Such a function is said to be **homogeneous in x and y of degree 4** since the sum of the powers of x and y are the same and equal to 4 for all the terms. A simplification of a differential equation of the type

$$\frac{dy}{dx} = \frac{f(x, y)}{g(x, y)} \tag{7.14}$$

is possible if $f(x, y)$ and $g(x, y)$ are homogeneous functions of the same degree.

EXAMPLE 7.3

Consider the differential equation

$$\frac{dy}{dx} = \frac{x^2 - y^2}{xy}. \tag{7.15}$$

Dividing top and bottom of the right-hand side by x^2 we have

$$\frac{dy}{dx} = \frac{1 - (y/x)^2}{y/x} \tag{7.16}$$

and we now substitute

$$\frac{y}{x} = u, \tag{7.17}$$

which gives $y = xu$. Differentiating this last equation with respect to x gives

$$\frac{dy}{dx} = x\frac{du}{dx} + u. \tag{7.18}$$

Substituting from (7.17) and (7.18) in (7.16), rearranging, and integrating we find

$$\int \frac{u}{1 - 2u^2}\,du = \int \frac{dx}{x}, \tag{7.19}$$

which gives a separation of the variables u and x. Integrating both sides,

$$-\frac{1}{4}\ln(1 - 2u^2) = \ln x + C$$

or

$$\ln(1 - 2u^2) = -4\ln x + \ln A$$

where $\ln A = -4C$. This is a convenient form for the constant of integration since by taking the exponential of both sides we find

$$1 - 2u^2 = \frac{A}{x^4}.$$

Replacing u from (7.17) we finally find

$$y = \sqrt{\frac{1}{2}\left(x^2 - \frac{A}{x^2}\right)}. \tag{7.20}$$

If the differential equation relates to a real problem with boundary conditions, so that a specific value of y is known for some value of x, then A, the constant of integration, may be determined.

Sometimes a change of variables can convert a non-homogeneous equation into a homogeneous one.

EXAMPLE 7.4

Consider the non-homogeneous differential equation

$$\frac{dy}{dx} = \frac{x + y + 1}{x - y + 3}. \tag{7.21}$$

We now change the variables to

$$x = X + a \quad \text{and} \quad y = Y + b$$

where a and b are numerical constants. This gives in place of (7.21)

$$\frac{dY}{dX} = \frac{X + Y + a + b + 1}{X - Y + a - b + 3}$$

and by selecting a and b such that $a + b + 1 = 0$ and $a - b + 3 = 0$, giving $a = -2$ and $b = 1$, we have turned the differential equation into the form

$$\frac{dY}{dX} = \frac{X + Y}{X - Y} = \frac{1 + Y/X}{1 - Y/X} \tag{7.22}$$

i.e. a standard homogeneous equation.

The substitution $Y = Xu$ gives

$$X\frac{du}{dX} = \frac{1 + u^2}{1 - u}$$

or

$$\int \frac{1 - u}{1 + u^2}\,du = \int \frac{1}{1 + u^2}\,du - \int \frac{u}{1 + u^2}\,du = \int \frac{dX}{X}.$$

Integrating,

$$\tan^{-1} u - \frac{1}{2}\ln(1 + u^2) = \ln(X) + C$$

or

$$\tan^{-1}\left(\frac{Y}{X}\right) - \frac{1}{2}\ln\left\{1 + \left(\frac{Y}{X}\right)^2\right\} = \ln(X) + C. \tag{7.23}$$

In terms of the original variables this becomes

$$\tan^{-1}\left(\frac{y-1}{x+2}\right) - \frac{1}{2}\ln\left\{1 + \left(\frac{y-1}{x+2}\right)^2\right\} = \ln(x + 2) + C. \tag{7.24}$$

This is not a simple relationship in which y can be written as a function of x or x as a function of y. If it is required to find x for a particular value of y, say $y = 2$, and with C fixed by boundary conditions with value 1, say, then the resultant equation that must be solved to give x is

$$\tan^{-1}\left(\frac{1}{x+2}\right) - \frac{1}{2}\ln\left\{1 + \left(\frac{1}{x+2}\right)^2\right\} - \ln(x + 2) - 1 = 0. \tag{7.25}$$

This is called a **transcendental equation**. It is not possible to rearrange the equation with x on the left-hand side and numerical quantities on the right. To find the solution, or solutions, for x requires the use of numerical methods (Chapter 31).

Exercise 7.3

Solve the differential equation $\dfrac{dy}{dx} = \dfrac{x+y}{x-y}$.

7.4 The integrating factor

To illustrate the principle of the **integrating factor** as an aid to the solution of differential equations, we consider another example.

EXAMPLE 7.5

In the equation

$$\frac{dy}{dx} + \frac{y}{x} = x^2, \tag{7.26}$$

the variables cannot be separated and it is not a homogeneous equation, so the methods we have previously considered cannot be used. However, if we multiply both sides by x then we have the equation

$$x\frac{dy}{dx} + y = x^3 \tag{7.27}$$

and we can see that the left-hand is, from (4.17),

$$\frac{d}{dx}(xy) = x\frac{dy}{dx} + y. \tag{7.28}$$

Hence if we integrate both sides of (7.27) we find

$$xy = \int x^3 dx = \frac{1}{4}x^4 + C$$

giving a solution

$$y = \frac{1}{4}x^3 + \frac{C}{x}. \tag{7.29}$$

You should confirm that this is the solution to the original differential equation (7.26).

We now see how to generalize this approach for differential equations of the form

$$\frac{dy}{dx} + P(x)y = Q(x) \tag{7.30}$$

where $P(x)$ and $Q(x)$ are functions of x. We multiply both sides by an integrating factor, which is another function of x, $R(x)$, the form of which we shall determine, to give

$$R(x)\frac{dy}{dx} + P(x)R(x)y = Q(x)R(x). \tag{7.31}$$

The requirement that we make of $R(x)$ is that

$$\frac{d}{dx}\{R(x)y\} = R(x)\frac{dy}{dx} + y\frac{dR(x)}{dx} = R(x)\frac{dy}{dx} + P(x)R(x)y$$

or that

$$\frac{dR(x)}{dx} = P(x)R(x). \tag{7.32}$$

Rearranging (7.32) and integrating,

$$\int \frac{dR(x)}{R(x)} = \int P(x)dx.$$

The left-hand side can be integrated, giving

$$\ln\{R(x)\} = \int P(x)dx$$

or

$$R(x) = \exp\left\{\int P(x)dx\right\}. \tag{7.33}$$

For the example with which we began this section $P(x) = 1/x$, so that

$$R(x) = \exp\left\{\int \frac{1}{x}dx\right\} = \exp(\ln x) = x$$

which is the factor that we used.

We now apply the integrating factor method to

$$\frac{dy}{dx} + y = \cos x. \tag{7.34}$$

The integrating factor is

$$R(x) = \exp\left(\int dx\right) = e^x$$

and the solution of the differential equation is, using (5.48),

$$e^x y = \int e^x \cos x\,dx = \frac{1}{2}e^x(\sin x + \cos x) + C$$

or

$$y = \frac{1}{2}(\sin x + \cos x) + Ce^{-x}. \tag{7.35}$$

The application of the integrating factor method depends on the ability to evaluate the integrals $\int P(x)dx$ and $\int R(x)Q(x)dx$. Clearly, if the functions involved are very complicated, then this may not be possible.

We now illustrate how the integrating factor may be applied to a physical problem.

EXAMPLE 7.6

Consider a mass of 1 kg falling through a fluid under a constant downward force of 5 N, due to gravity partly balanced by liquid upthrust. The liquid is initially cold but heats up to the ambient temperature so that the resisting force due to viscosity is of the form $V\frac{2+t}{1+t}$ N, where the speed V is in m s^{-1} and time, t, is in seconds. We shall find the relationship that gives V as a function of t and the terminal speed for very large t taking the boundary conditions $V = 0$ when $t = 0$.

The differential equation that describes the variation of speed with time is

$$\frac{dV}{dt} = 5 - \frac{2+t}{1+t}V \quad\text{or}\quad \frac{dV}{dt} + \frac{2+t}{1+t}V = 5. \tag{7.36}$$

The integrating factor is

$$\exp\left(\int \frac{2+t}{1+t}dt\right) = \exp\left\{\int\left(1 + \frac{1}{1+t}\right)dt\right\} = \exp\{t + \ln(1+t)\}$$
$$= (1+t)e^t$$

so that

$$(1+t)e^t V = 5\int (1+t)e^t dt = 5te^t + C,$$

where integration by parts was used for the right-hand-side integral. Since $V = 0$ when $t = 0$ we find $C = 0$ and hence

$$V = \frac{5t}{1+t}. \tag{7.37}$$

This gives V as a function of t and when $t \to \infty$ we have the terminal speed as 5 m s^{-1}.

? Exercise 7.4

Solve $\dfrac{dy}{dx} + (y - 1)x = 0$ using the integration-factor approach.

7.5 Linear constant-coefficient equations

The types of first-order differential equations we have dealt with are a special case of a generalized linear equation of order n of the form

$$f_0(x)\frac{d^n y}{dx^n} + f_1(x)\frac{d^{n-1}y}{dx^{n-1}} + \cdots + f_{n-1}(x)\frac{dy}{dx} + f_n(x)y = g(x). \tag{7.38}$$

If all the functions f on the left-hand side are constants and the right-hand side is zero then this is referred to as a **homogeneous constant-coefficient equation**, and these frequently occur in physical contexts, especially as second-order equations. Other kinds of second-order differential equations occur in which the left-hand side coefficients are constant but the right-hand side is a non-zero function of x. Such equations are referred to as **linear inhomogeneous constant-coefficient equations**. We shall illustrate the solution of such second-order equations by physically based examples.

7.6 Simple harmonic motion

We consider a mass hanging on an elastic string that obeys Hooke's law, so that its extension is proportional to the applied stretching force. The mass takes up some equilibrium position and if displaced from that position by a small amount then it experiences a resultant force, proportional to the displacement, directed towards the equilibrium position. Mathematically this is expressed as

$$F_x = m\frac{d^2 x}{dt^2} = -kx \tag{7.39}$$

where k is the force per unit displacement. Writing $\omega_0^2 = k/m$ this becomes

$$\frac{d^2 x}{dt^2} = -\omega_0^2 x, \tag{7.40}$$

a general solution of which is

$$x = C\cos(\omega_0 t + \delta). \tag{7.41}$$

An example of this is shown in Figure 7.2 and shows a displacement–time relationship known as **simple harmonic motion**.

The maximum displacement, C, in this case 5 units, is the **amplitude** of the motion. This has the dimension of length. The quantity ω_0, of value 2, is the **angular frequency** of the motion, with dimension radians time^{-1}, so that the quantity ωt is an angle in radians. The frequency in cycles per second (Hz) is given by

$$v = \omega_0/2\pi. \tag{7.42}$$

The quantity $\delta\,(= -0.7)$ is a **phase angle** that describes a displacement of the whole curve along the t axis. If the phase angle were zero the whole curve would be displaced to the left by 0.7 radians compared with what is shown in the figure.

In Chapter 6 we saw that there is a useful way of representing a wave motion in terms of complex numbers. Consider a point moving round a circular path of radius C, centred on the origin of an x-y plane, at a constant angular speed ω, as illustrated in Figure 7.3. At time $t=0$ the point is at P and after time t it is at Q, having moved round the circle through an angle $\omega_0 t$. The projection of OQ on the x axis, ON, is given by

$$x = C\cos(\omega_0 t + \delta)$$

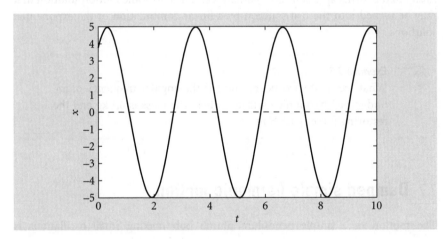

Figure 7.2 A representation of simple harmonic motion.

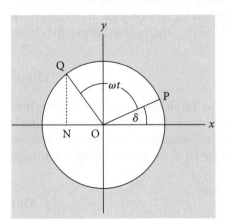

Figure 7.3 Projected motion at a uniform speed around a circle gives simple harmonic motion.

corresponding to (7.41). Considering Figure 7.3 as an Argand diagram, with x and y as the real and imaginary axes, then the position of the point is given by the complex number

$$z = x + iy = C\exp\{i(\omega_0 t + \delta)\}. \tag{7.43}$$

The expression (7.43) is an alternative solution to (7.40) since differentiating it twice with respect to t gives $-\omega_0^2$ times the original expression.

The general solution of (7.40) can be of the form $C\cos(\omega_0 t + \delta)$, $C\sin(\omega_0 t + \delta)$, $C\exp\{\pm i(\omega_0 t + \delta)\}$, or any linear combinations of these expressions. Actually these expressions are themselves linearly related, since

$$\exp\{\pm i(\omega_0 t + \delta)\} = \cos(\omega_0 t + \delta) \pm i\sin(\omega_0 t + \delta) \tag{7.44a}$$

$$2\cos(\omega_0 t + \delta) = \exp\{i(\omega_0 t + \delta)\} + \exp\{-i(\omega_0 t + \delta)\} \tag{7.44b}$$

$$2\sin(\omega_0 t + \delta) = -i[\exp\{i(\omega_0 t + \delta)\} - \exp\{-i(\omega_0 t + \delta)\}]. \tag{7.44c}$$

Any solution of a differential equation for a physical problem, which has to be real, found in the form $\exp(\pm i\omega t)$ is essentially a simple harmonic motion solution that may be turned into the real solution by a linear combination of the exponential solutions.

Exercise 7.5

What are (i) the frequency and (ii) the angular frequency of an undamped harmonic oscillator where the mass is $0.1\,\text{kg}$ and the restoring force is $1.5\,\text{N}\,\text{m}^{-1}$?

7.7 Damped simple harmonic motion

The motion of a simple pendulum plumb bob making small oscillations is approximately linear simple harmonic motion. It is well known that if a simple pendulum is set in motion then the oscillations die down and eventually stop. This is mainly due to the viscosity of air, which gives a force acting on the bob in a direction opposing its motion. For a body moving through a gas at modest speeds the resistance is proportional to the speed and the total force on the plumb bob can be written as

$$F = m\frac{d^2x}{dt^2} = -f\frac{dx}{dt} - kx \tag{7.45a}$$

where f is the constant of proportionality for the air resistance force. The negative sign ensures that, for example, when the bob moves in the positive direction (dx/dt positive) then the force opposes the motion by being in the negative direction. Rearranging (7.45a) gives the second-order ODE for damped simple harmonic motion:

$$m\frac{d^2x}{dt^2} + f\frac{dx}{dt} + kx = 0. \tag{7.45b}$$

A general solution to (7.45b) can be sought in the form

$$x = C \exp(\alpha t)$$

and since $dx/dt = \alpha x$ and $d^2x/dt^2 = \alpha^2 x$ we find from (7.45b)

$$m\alpha^2 + f\alpha + k = 0 \qquad (7.46)$$

which has as a solution for α

$$\alpha = -\frac{f}{2m} \pm \sqrt{\frac{f^2}{4m^2} - \frac{k}{m}}. \qquad (7.47)$$

The corresponding solution of the differential equation (7.45b) is

$$x = \exp\left(-\frac{f}{2m}t\right)\left\{A\exp\left(\sqrt{\frac{f^2}{4m^2} - \frac{k}{m}}\,t\right) + B\exp\left(-\sqrt{\frac{f^2}{4m^2} - \frac{k}{m}}\,t\right)\right\}. \qquad (7.48)$$

The form of this relationship between x and t depends on the value of the quantity under the square-root sign. We now consider three possibilities.

Light damping

$$\frac{f^2}{4m^2} - \frac{k}{m} = -\omega_f^2$$

In this case the quantity is negative and the solution is of the form

$$x = \exp\left(-\frac{f}{2m}t\right)\left\{A\exp\left(i\omega_f t\right) + B\exp\left(-i\omega_f t\right)\right\} \quad .$$

Since the solution must be real then, as was indicated at the end of §7.6, by appropriate choices of A and B, i.e. $A = \frac{1}{2}C\exp(i\delta)$ and $B = \frac{1}{2}C\exp(-i\delta)$ the solution takes the form

$$x = C\exp\left(-\frac{f}{2m}t\right)\cos(\omega_f t + \delta). \qquad (7.49)$$

This situation occurs for smaller values of f and hence corresponds to light damping of the oscillator. It corresponds to a simple harmonic motion of angular frequency ω_f which differs from ω_0, the frequency for an undamped oscillator, where the amplitude falls exponentially with time. Damped and undamped motions are compared in Figure 7.4 for $\frac{k}{m} = 1.0$ and $\frac{f}{2m} = 0.1$.

Critical damping

$$\frac{f^2}{4m^2} - \frac{k}{m} = 0$$

For this case the cosine term in (7.49) is constant and the solution takes on the form

$$x = C\exp\left(-\frac{f}{2m}t\right). \qquad (7.50)$$

This is a simple exponential decline with no oscillatory behaviour.

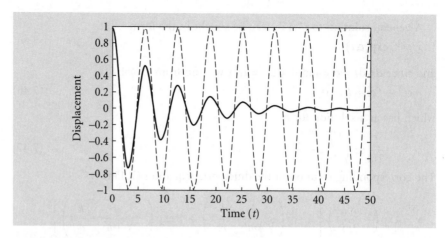

Figure 7.4 Damped (full line) and undamped (dashed line) simple harmonic motion for $k/m = 1$ and $f/2m = 0.1$.

Overcritical damping

$$\frac{f^2}{4m^2} - \frac{k}{m} > 0$$

Since f is now greater than the value required for critical damping, this case is referred to as overcritical damping We now rewrite (7.48) as

$$x = A \exp\left\{-\left(\frac{f}{2m} - \sqrt{\frac{f^2}{4m^2} - \frac{k}{m}}\right)t\right\} + B \exp\left\{-\left(\frac{f}{2m} + \sqrt{\frac{f^2}{4m^2} - \frac{k}{m}}\right)t\right\}.$$

(7.51)

Since

$$\frac{f}{2m} > \sqrt{\frac{f^2}{4m^2} - \frac{k}{m}},$$

it is clear that this expression for x consists of two exponentially-declining terms, one falling off more quickly than the one in (7.50) and the other falling off more slowly. In Figure 7.5 expressions (7.50) and (7.51) are compared with $A = B = \frac{1}{2}C$ so that the two functions have the same initial amplitude.

In this case the critical damping falls off more quickly. The values of the various constants – C, A, B, and δ – that occur in (7.49), (7.50), and (7.51) are fixed by the boundary conditions of the problem. If, for example, in (7.51) we found $A = 0$ then the expression would fall off more quickly than (7.50). However, if $A > 0$, no matter how small it is, then eventually, for some value of t, (7.51) will be larger than (7.50). The proof of this is the subject matter of Problem 7.6.

Exercise 7.6
For a damped simple harmonic motion the mass is 0.1 kg, the restoring force per unit displacement is 1.5 N m^{-1}, and the viscous resisting force is 0.75 N m^{-1} s. What is the angular frequency of the oscillation?

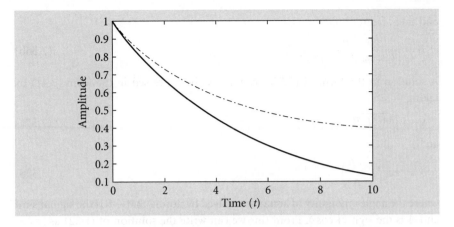

Figure 7.5 Comparison of critical damping (full line) and overcritical damping (broken line).

7.8 Forced vibrations

We now consider a system with both displacement and damping forces that is being driven externally by a periodic force of the form $F\cos\omega t$. The differential equation now appears in the form

$$m\frac{\mathrm{d}^2 x}{\mathrm{d}t^2} + f\frac{\mathrm{d}x}{\mathrm{d}t} + kx = F\cos\omega t. \tag{7.52}$$

We wish to find solution of this equation of the form

$$x = A\cos\omega t + B\sin\omega t, \tag{7.53a}$$

for which

$$\frac{\mathrm{d}x}{\mathrm{d}t} = -A\omega\sin\omega t + B\omega\cos\omega t. \tag{7.53b}$$

and

$$\frac{\mathrm{d}^2 x}{\mathrm{d}t^2} = -A\omega^2\cos\omega t - B\omega^2\sin\omega t. \tag{7.53c}$$

Substituting for x and its derivatives from (7.53) into (7.52) and equating coefficients of $\cos\omega t$ and $\sin\omega t$ on the two sides of the equation, we find

$$(k - m\omega^2)A + f\omega B = F \tag{7.54a}$$

and

$$-f\omega A + (k - m\omega^2)B = 0. \tag{7.54b}$$

From (7.54b),

$$B = \frac{f\omega}{k - m\omega^2}A \tag{7.55}$$

and substituting this value of B in (7.54a) gives an explicit value

$$A = \frac{k - m\omega^2}{(k - m\omega^2)^2 + f^2\omega^2}F \tag{7.56a}$$

and then, from (7.55),

$$B = \frac{f\omega}{(k - m\omega^2)^2 + f^2\omega^2} F.$$

(7.56b)

A solution in the form of (7.53a) can always be expressed in the form (7.41) by taking

$$C = \sqrt{A^2 + B^2}$$

(7.57a)

and

$$\delta = \tan^{-1}\left(\frac{-B}{A}\right)$$

(7.57b)

where the angle ambiguity of arctan is resolved by noting that $-B$ is the sign of $\sin\delta$ and A is the sign of $\cos\delta$. From this we can write the solution of (7.52) as

$$x = \frac{F}{\sqrt{(k - m\omega^2)^2 + f^2\omega^2}} \cos(\omega t + \delta)$$

(7.58a)

where

$$\tan\delta = \frac{-f\omega}{k - m\omega^2}.$$

(7.58b)

However, (7.58a) is not a complete solution to (7.52). The solution of (7.45b), which is similar to (7.52) but with zero on the right-hand side, can be added to (7.58a) and this will still be a solution of (7.52). The added-on solution of (7.45b) can correspond to subcritical damping, (7.49), critical damping, (7.50), or over-critical damping, (7.51), but whichever it is it will contain a factor exponentially declining with time so that after a long enough time this component will disappear, leaving the periodic solution (7.58a). In physical terminology the component that is the solution of (7.45b) is a **transient** – meaning that it has a limited duration. The component (7.58a) is referred to as the **steady-state solution** since, once the transient disappears, this component is unchanging for the lifetime of the system. In mathematical terminology the solution of (7.45b) is a **complementary function**, (7.58a) is a **particular solution**, and the sum of the two is known as the **general solution**.

For a system with particular characteristics of mass m, restoring force per unit displacement k, and resisting force per unit speed f, the amplitude of the steady-state solution will be dependent on the applied angular frequency, ω. This dependence is shown in Figure 7.6 for $F = 10$, $m = 1$, $k = 1$, and $f = 0.2$.

It is clear that there is a resonance effect at an angular frequency around $\omega = 1$. To find the resonance we express the amplitude as

$$C = \frac{F}{\sqrt{(k - m\omega^2)^2 + f^2\omega^2}}.$$

(7.59)

To find a maximum we differentiate and equate $dC/d\omega = 0$. This gives

$$\frac{dC}{d\omega} = -\frac{\omega\{f^2 - 2m(k - m\omega^2)\}}{\{(k - m\omega^2)^2 + f^2\omega^2\}^{3/2}} = 0.$$

(7.60)

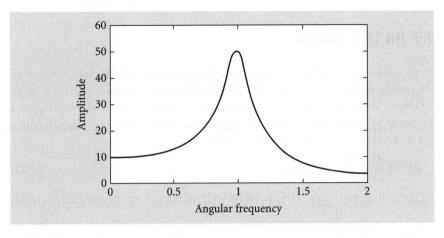

Figure 7.6 Resonance of a forced vibrator.

The solutions to (7.60) are $\omega = 0$, because ω is a factor in the numerator, and $\omega = \infty$, because the denominator of (7.60) involves a higher power of ω than the numerator and the solution of

$$f^2 - 2m(k - m\omega^2) = 0. \tag{7.61}$$

From Figure 7.6 it is clear that $\omega = 0$ and $\omega = \infty$ correspond to minima and the solution of (7.61) gives the maximum. The resonant angular frequency is

$$\omega_{\mathrm{res}}^2 = \frac{k}{m} - \frac{f^2}{2m^2} = \omega_0^2 - \frac{f^2}{2m^2} \tag{7.62}$$

where ω_0 is the natural frequency of the undamped, unforced system. For the parameters used to give Figure 7.6, $\omega_0 = 1.000$ and $\omega_{\mathrm{res}} = 0.990$, very little different.

From (7.62) it is seen that as $f \to 0$ so the resonance frequency approaches ω_0. From (7.58a) it will also be seen that, at resonance, as $f \to 0$ both terms in the divisor of (7.59) tend to zero so that the amplitude C tends to infinity. This extreme condition can never occur in any physical system since the energy of a vibrating system is proportional to the square of the amplitude and infinite energy is not feasible. However, driving a system with small f can build up very large amplitudes and many real-life situations must take account of this. Normally soldiers march in step, but in crossing a bridge (unless it is a very substantial one) they are ordered to break step in order to avoid the possibility that a coordinated periodic force on the bridge would cause violent vibrations that would damage it. A less obvious source of a periodic force on a bridge is the wind. It is well known that blowing on a reed sets it into vibration – a principle employed in many musical instruments. On 7 November 1940 a gusting high wind set the Tacoma Narrows suspension bridge in the USA into vibration. By chance the frequency of the gusts matched the natural bridge frequency, and so the bridge was destroyed.

Exercise 7.7
A mass of 0.1 kg experiences a displacement force of $1.5\,\mathrm{N\,m^{-1}}$ and a viscous force of $0.75\,\mathrm{N\,m^{-1}\,s}$. It is driven by a periodic force of the form $10\cos(4t)\,\mathrm{N}$. What is the amplitude of the steady-state vibration?

7.9 **An LCR circuit**

In Chapter 6 we found the current through a simple series LCR circuit due to an alternating potential difference. The potential difference across a resistor is given by Ohm's law (6.40), that across an inductor is given by (6.45a), and that across a capacitor by (6.50). For an alternating potential difference, $V = V_0 \sin \omega t$, across the three impedances in series we can write

$$RI + L\frac{\mathrm{d}I}{\mathrm{d}t} + \frac{Q}{C} = V_0 \sin \omega t. \tag{7.63}$$

Differentiating both sides of this equation with respect to t, and rearranging the terms,

$$L\frac{\mathrm{d}^2 I}{\mathrm{d}t^2} + R\frac{\mathrm{d}I}{\mathrm{d}t} + \frac{I}{C} = \omega V_0 \cos \omega t \tag{7.64}$$

since $\mathrm{d}Q/\mathrm{d}t = I$. The comparison with (7.52) is clear. In (7.64) I plays the role of x, L plays the role of m, R plays the role of f, $1/C$ plays the role of k, and ωV_0 plays the role of F. Hence from (7.58) the steady-state current is given by

$$I = \frac{\omega V_0}{\sqrt{(1/C - L\omega^2)^2 + R^2\omega^2}} \cos(\omega t + \delta')$$

$$= \frac{V_0}{\sqrt{R^2 + \left(\omega L - \dfrac{1}{C\omega}\right)^2}} \cos(\omega t + \delta') \tag{7.65a}$$

where

$$\tan \delta' = \frac{-R}{(1/C\omega) - L\omega}. \tag{7.65b}$$

Comparing this with (6.59) we see that the amplitude agrees but the time-dependent component is not directly comparable. To show that this part also agrees with (6.59) we must verify that the current lags behind the applied potential difference, $V_0 \sin \omega t$, by δ as given by (6.58), or that

$$\cos(\omega t + \delta') = \sin(\omega t - \delta). \tag{7.66}$$

Expanding (7.66) we must show that

$$\cos \omega t \cos \delta' - \sin \omega t \sin \delta' = \sin \omega t \cos \delta - \cos \omega t \sin \delta.$$

This will be true if $\cos \delta' = -\sin \delta$ and $-\sin \delta' = \cos \delta$.
From (6.58) and (7.65b),

$$\cos \delta' = \frac{(1/C\omega) - L\omega}{\sqrt{R^2 + (L\omega - (1/C\omega))^2}} = -\sin \delta$$

and

$$-\sin \delta' = \frac{R}{\sqrt{R^2 + (L\omega - (1/C\omega))^2}} = \cos \delta.$$

This confirms that the steady-state results from the complex-number approach and the differential-equation approach are identical.

As well as the steady-state current there will be a transient current from when the potential difference is first applied until it dies away. The form of this transient is given by one of (7.49), (7.50), and (7.51) according to the relative values of the electrical components.

Just as for the mechanical case, there can be resonance in a LCR series circuit. The basic SI electrical units were described in Chapter 2, the ones of interest here being amperes (A) for current, volts (V) for potential, ohms (Ω) for resistance, henry (H) for inductance, and farad (F) for capacitance. We now consider a circuit set up as in Figure 6.3 with $R = 250\,\Omega$, $L = 0.2\,H$ and $C = 10^{-6}\,F$. The amplitude of the current divided by the amplitude of the driving potential is

$$\frac{I_0}{V_0} = \frac{1}{\sqrt{R^2 + (\omega L - (1/\omega C))^2}} \tag{7.67}$$

and the resonant angular frequency is given by

$$\omega L - \frac{1}{\omega C} = 0 \text{ or } \omega^2 = \frac{1}{LC}, \tag{7.68}$$

or $\omega = 2236\,s^{-1}$ for this example. The normally expressed resonance frequency of the alternating current is $2236/2\pi$ Hz or 355.9 Hz. The variation of I_0/V_0 is shown in Figure 7.7.

Another quantity of interest is the phase lag of current behind voltage, given by (6.58), which is shown in Figure 7.8. The phase lag goes from $-\pi/2$ at zero frequency to $\pi/2$ for frequencies well above the resonant frequency.

The concepts underlying resonance, illustrated here for a mechanical system and an electrical circuit, also occur in other branches of physics. For example, the electrons in atoms have certain natural frequencies and if an atom is irradiated with X-rays at, or close to, a natural frequency of one of its electrons, then the radiation scattered by the atom in that case is strongly enhanced. This phenomenon, known as **anomalous scattering**, has played an important role in X-ray crystallography, which is at the heart of the modern science of biotechnology.

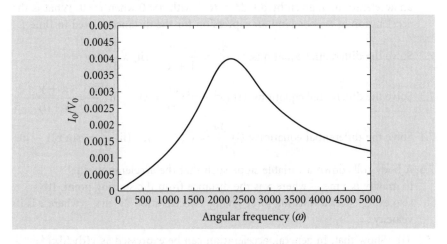

Figure 7.7 Resonance in an LCR series circuit.

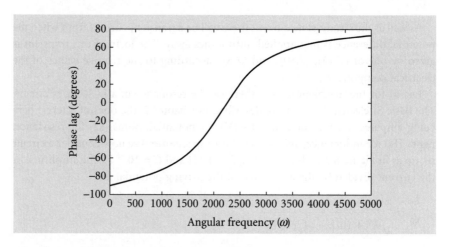

Figure 7.8 Dependence of phase lag on ω for an LCR series circuit.

Exercise 7.8
A circuit has a $10\,\Omega$ resistor, a $0.1\,H$ inductor and a $10^{-5}\,F$ capacitor in series. For what angular frequency, ω_r, of applied potential will the circuit be in resonance? If the applied potential is $100\cos(\omega_r t)$ V, what is the amplitude of the resultant current?

Problems

7.1 A body starts at the origin at rest when $t=0$. It moves along the axis with an acceleration, a, given by $\mathrm{d}a/\mathrm{d}t = te^{-t}$ with $a=0$ when $t=0$. What is the acceleration at $t=\infty$? Find an expression for the distance moved in time t.

7.2 Solve the differential equations: (i) $\dfrac{\mathrm{d}y}{\mathrm{d}x} = \dfrac{1+y^2}{1+x}$; (ii) $x\dfrac{\mathrm{d}y}{\mathrm{d}x} + y = 1$.

7.3 Solve the differential equations: (i) $(x^2 - y^2)\dfrac{\mathrm{d}y}{\mathrm{d}x} = xy$; (ii) $\dfrac{\mathrm{d}y}{\mathrm{d}x} = \dfrac{x+y-2}{x+4y-5}$.

7.4 Solve the differential equations: (i) $\dfrac{\mathrm{d}y}{\mathrm{d}x} = x^2(2-y)$; (ii) $\dfrac{\mathrm{d}y}{\mathrm{d}x} = \sin x(1-y)$.

7.5 A body rolls down a variable slope such that the acceleration due to gravity is $x\,\mathrm{m\,s^{-2}}$ where x is the distance from the starting point. It also experiences a resisting force giving a deceleration $2v^2\,\mathrm{m\,s^{-2}}$ where v is its velocity.

(i) Show that, in general, acceleration can be expressed as $v(\mathrm{d}v/\mathrm{d}x)$.

(ii) Write down a differential equation in terms of v and x that expresses the acceleration of the body.

(iii) Change the variable in the differential equation from v to $K = \frac{1}{2}v^2$.

(iv) If it starts at rest with zero acceleration at $x = 0$ then what is its speed at $x = 1$?

7.6 Show that, if $A \neq 0$, the magnitude of expression (7.51) will be greater than (7.50) for some value of t.

7.7 A mass of 1 kg is suspended from a spring that gives a force per unit displacement of $1\,\mathrm{N\,m^{-1}}$. Air resistance gives a force of $0.01v\,\mathrm{N}$ where v is the velocity of the mass. What is the frequency of oscillation of the mass without air resistance and what is the difference in frequency including air resistance?

7.8 A series AC circuit contains a resistor, $R\,\Omega$, and a capacitor, $C\,\mathrm{F}$, connected in series and being driven by an applied potential difference $V_0 \cos \omega t$. Write down the differential equation that connects the current, I, and t and solve this equation to give I as a function of t.

Matrices I and determinants

In many physical or biological systems there naturally arises a set of quantities that can conveniently be represented as a two-dimensional array, referred to as a matrix. If matrices were simply a way of representing arrays of numbers then they would have only a marginal utility as a means of visualizing data. However, a whole branch of mathematics has evolved, involving the manipulation of matrices, which has become a powerful tool for the solution of physical problems. Some basic operations of matrix algebra are described and then applied to describing the behaviour of lens systems and also to some aspects of special relativity theory. This chapter also deals with another type of array quantity, the determinant, which has special importance in dealing with the solution of sets of linear equations.

8.1 Definition of a matrix

Consider a system containing four particles with coordinates (x_1, y_1, z_1), (x_2, y_2, z_2), (x_3, y_3, z_3), and (x_4, y_4, z_4). These can be expressed as a **matrix:**

$$X = \begin{bmatrix} x_1 & y_1 & z_1 \\ x_2 & y_2 & z_2 \\ x_3 & y_3 & z_3 \\ x_4 & y_4 & z_4 \end{bmatrix}. \tag{8.1}$$

Here we have introduced the notation for expressing a matrix, using a capital letter for the shorthand representation and enclosing the quantities in the matrix within square brackets. In this case we have a matrix of order 4×3 because there are four rows and three columns – notice particularly the order rows-columns. The individual entries in the matrix are referred to as the **elements** of the matrix and we can write an order 6×5 matrix in the general form

$$A = \begin{bmatrix} a_{11} & a_{12} & a_{13} & a_{14} & a_{15} \\ a_{21} & a_{22} & a_{23} & a_{24} & a_{25} \\ a_{31} & a_{32} & a_{33} & a_{34} & a_{35} \\ a_{41} & a_{42} & a_{43} & a_{44} & a_{45} \\ a_{51} & a_{52} & a_{53} & a_{54} & a_{55} \\ a_{61} & a_{62} & a_{63} & a_{64} & a_{65} \end{bmatrix}, \tag{8.2}$$

where a_{ij} is the element for row i and column j.

In many applications the matrices that are obtained are **square matrices**, with the same number of rows and columns, for example

$$S = \begin{bmatrix} s_{11} & s_{12} & s_{13} \\ s_{21} & s_{22} & s_{23} \\ s_{31} & s_{32} & s_{33} \end{bmatrix}. \tag{8.3}$$

For a square matrix the elements with $i = j$ are called **diagonal elements**.

A matrix with only one column, for example

$$s = \begin{bmatrix} s_1 \\ s_2 \\ s_3 \\ s_4 \end{bmatrix} \tag{8.4}$$

is referred to as a **column vector** and a matrix with one row, for example

$$s^T = \begin{bmatrix} s_1 & s_2 & s_3 & s_4 \end{bmatrix}. \tag{8.5}$$

is a **row vector**. There are two things to notice here. The first is that we have used lower-case letters to represent a vector, and the second is that (8.5) is related to (8.4) in that the column of the latter is the row of the former. A matrix produced by interchanging rows and columns is called the **transpose** of the original matrix and, as we see in (8.5), is represented by the superscript ^{T}. From (8.1) we have

$$X^T = \begin{bmatrix} x_1 & x_2 & x_3 & x_4 \\ y_1 & y_2 & y_3 & y_4 \\ z_1 & z_2 & z_3 & z_4 \end{bmatrix}. \tag{8.6}$$

which is an order 3×4 matrix.

Now we describe the basic operations of matrix algebra which make it such a useful tool.

8.2 Operations of matrix algebra

Matrices can be combined by **addition**, but only if they have the same number of rows and columns. Thus if we have

$$X = \begin{bmatrix} x_{11} & x_{12} & x_{13} \\ x_{21} & x_{22} & x_{23} \end{bmatrix} \quad \text{and} \quad P = \begin{bmatrix} p_{11} & p_{12} & p_{13} \\ p_{21} & p_{22} & p_{23} \end{bmatrix}$$

then

$$X + P = \begin{bmatrix} x_{11} + p_{11} & x_{12} + p_{12} & x_{13} + p_{13} \\ x_{21} + p_{21} & x_{22} + p_{22} & x_{23} + p_{23} \end{bmatrix}, \tag{8.7}$$

which is obtained by adding the corresponding elements of the two matrices.

The process of **subtraction** is similar and is obtained by changing all $+$ signs into $-$ signs in (8.7). So far the operations have resembled those with normal numbers, except that they have been carried out with individual elements of the matrices. However, when we come to **matrix multiplication** the process is very different from the ordinary multiplication of numbers. Two matrices can be multiplied together if *the number of columns in the first matrix equals the number of rows in the second*. We illustrate this with the product of a 2×3 matrix and a 3×4 matrix:

$$C = AB = \begin{bmatrix} a_{11} & a_{12} & a_{13} \\ a_{21} & a_{22} & a_{23} \end{bmatrix} \begin{bmatrix} b_{11} & b_{12} & b_{13} & b_{14} \\ b_{21} & b_{22} & b_{23} & b_{24} \\ b_{31} & b_{32} & b_{33} & b_{34} \end{bmatrix}. \tag{8.8}$$

The result of the product of an $m \times n$ matrix with an $n \times p$ matrix is an $m \times p$ matrix, so that C in (8.8) is of order 2×4. The element c_{ij} of C is found by taking the **inner product** of row i of A with column j of B. We illustrate the idea of the inner product by finding c_{11} as the inner product of row 1 of A with column 1 of B:

$$c_{11} = \begin{bmatrix} a_{11} & a_{12} & a_{13} \end{bmatrix} \begin{bmatrix} b_{11} \\ b_{21} \\ b_{31} \end{bmatrix} = a_{11}b_{11} + a_{12}b_{21} + a_{13}b_{31}. \tag{8.9a}$$

Similarly,

$$c_{23} = \begin{bmatrix} a_{21} & a_{22} & a_{23} \end{bmatrix} \begin{bmatrix} b_{13} \\ b_{23} \\ b_{33} \end{bmatrix} = a_{21}b_{13} + a_{22}b_{23} + a_{23}b_{33}. \tag{8.9b}$$

You should confirm your understanding of this process by checking the following numerical example:

$$\begin{bmatrix} 1 & 2 & 1 & 3 \\ -1 & 1 & 0 & 2 \end{bmatrix} \begin{bmatrix} 2 & 1 & 0 \\ 2 & 1 & 3 \\ 1 & -2 & -2 \\ 0 & 1 & 2 \end{bmatrix} = \begin{bmatrix} 7 & 4 & 10 \\ 0 & 2 & 7 \end{bmatrix}. \tag{8.10}$$

Notice that for the matrices defined in (8.8) the matrix product BA does not exist since the number of columns of B (4) does not equal the number of rows of A (2).

Finally, we note that the product of a number and a matrix just involves the multiplication of each element of the matrix by the number. Thus

$$3 \begin{bmatrix} 1 & 2 \\ -1 & 3 \end{bmatrix} = \begin{bmatrix} 3 & 6 \\ -3 & 9 \end{bmatrix}. \tag{8.11}$$

 Exercise 8.1

Evaluate the matrix product $\begin{bmatrix} 1 & 3 & 4 & 1 \\ 2 & 2 & 2 & 6 \\ 3 & 1 & 0 & 2 \end{bmatrix} \begin{bmatrix} 1 & 0 & 2 \\ 2 & 4 & 0 \\ 1 & -1 & 3 \\ 2 & 2 & 1 \end{bmatrix}.$

8.3 Types of matrix

The simple matrix operations can be applied to the solution of some types of physical problems, but before showing this we first describe a number of special kinds of matrix.

8.3.1 Diagonal matrix

A diagonal matrix is a square matrix in which all elements not on the diagonal are zero. A general 4×4 diagonal matrix can be represented as

$$D = \begin{bmatrix} d_1 & 0 & 0 & 0 \\ 0 & d_2 & 0 & 0 \\ 0 & 0 & d_3 & 0 \\ 0 & 0 & 0 & d_4 \end{bmatrix}. \tag{8.12}$$

Pre-multiplying a matrix by a diagonal matrix gives the following result:

$$DA = \begin{bmatrix} d_1 & 0 & 0 \\ 0 & d_2 & 0 \\ 0 & 0 & d_3 \end{bmatrix} \begin{bmatrix} a_{11} & a_{12} & a_{13} & a_{14} \\ a_{21} & a_{22} & a_{23} & a_{24} \\ a_{31} & a_{32} & a_{33} & a_{34} \end{bmatrix} = \begin{bmatrix} d_1 a_{11} & d_1 a_{12} & d_1 a_{13} & d_1 a_{14} \\ d_2 a_{21} & d_2 a_{22} & d_2 a_{23} & d_2 a_{24} \\ d_3 a_{31} & d_3 a_{32} & d_3 a_{33} & d_3 a_{34} \end{bmatrix},$$

$$\tag{8.13a}$$

i.e. the ith row is multiplied by d_i. **Post-multiplying** a matrix by a diagonal matrix gives

$$BD = \begin{bmatrix} b_{11} & b_{12} & b_{13} \\ b_{21} & b_{22} & b_{23} \end{bmatrix} \begin{bmatrix} d_1 & 0 & 0 \\ 0 & d_2 & 0 \\ 0 & 0 & d_3 \end{bmatrix} = \begin{bmatrix} d_1 b_{11} & d_2 b_{12} & d_3 b_{13} \\ d_1 b_{21} & d_2 b_{22} & d_3 b_{23} \end{bmatrix}, \tag{8.13b}$$

i.e. the ith column is multiplied by d_i.

8.3.2 Unit matrix

A unit matrix is a diagonal matrix in which all the diagonal elements are unity. Thus an order-4 unit matrix is

$$I = \begin{bmatrix} 1 & 0 & 0 & 0 \\ 0 & 1 & 0 & 0 \\ 0 & 0 & 1 & 0 \\ 0 & 0 & 0 & 1 \end{bmatrix}. \tag{8.14}$$

From (8.13a) and (8.13b) it is clear that either pre-multiplying or post-multiplying a matrix by a unit matrix leaves the matrix unchanged or, with the matrices in (8.13a) and (8.13b), with all ds equal to unity,

$$IA = A \quad \text{and} \quad BI = B. \tag{8.15}$$

It can be seen that the unit matrix more or less plays the role of the number 1 in normal number theory.

8.3.3 Null matrix

The null matrix is one for which all the elements are zero. Either pre- or post-multiplying a matrix by a null matrix will give a null matrix. In this respect it resembles the number zero in normal number theory. However, we cannot use the number-theory analogy completely. With normal numbers multiplication can only give the result zero if at least one of the numbers being multiplied is zero, but it is possible for the product of two non-null matrices to yield a null matrix. As an example,

$$\begin{bmatrix} 1 & -2 \\ 2 & -4 \end{bmatrix} \begin{bmatrix} 2 & 4 \\ 1 & 2 \end{bmatrix} = \begin{bmatrix} 0 & 0 \\ 0 & 0 \end{bmatrix}. \tag{8.16}$$

8.3.4 Hermitian matrix

Depending on the application, the elements of a matrix can have any character or physical dimension and in some important physical applications they can be complex quantities of the form $z_{ij} = a_{ij} + ib_{ij}$. An important special form of such a matrix is a **Hermitian matrix**, which is a square matrix for which $z_{ji} = z_{ij}^*$, where * indicates **complex conjugate**. Since the diagonal elements have to be their own complex conjugates, they must be real. An example of a Hermitian matrix is

$$H = \begin{bmatrix} 1 & 1+i & 2i \\ 1-i & 3 & 1+3i \\ -2i & 1-3i & 2 \end{bmatrix}. \tag{8.17}$$

The complex conjugate of a matrix with complex elements (which, in general, need not be square) is one for which each element is changed to its complex conjugate so that

$$H^* = \begin{bmatrix} 1 & 1-i & -2i \\ 1+i & 3 & 1-3i \\ 2i & 1+3i & 2 \end{bmatrix} \tag{8.18}$$

From this we see that for a Hermitian matrix

$$H^* = H^T. \tag{8.19}$$

Exercise 8.2

Find the matrix product $\begin{bmatrix} 2 & 1 \\ 1 & 2 \end{bmatrix} \begin{bmatrix} 1 & 1-i \\ 1+i & 3 \end{bmatrix}$.

8.4 Applications to lens systems

A ray of light passing through a lens or a lens system can be defined at any position by its distance from the optical axis, d, and the angle of its travel relative to the optical axis, α, as shown in Figure 8.1.

During its passage through the system the ray is either moving through a uniform optical medium, in which case d changes but α does not, or it is crossing a spherical surface, defined by a lens boundary, in which case α changes but d does not. We shall always be considering **paraxial rays** – ones very close to, and almost parallel to, the optical axis – so that all the angles α are small.

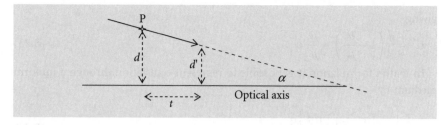

Figure 8.1 A light ray defined at P by d and α.

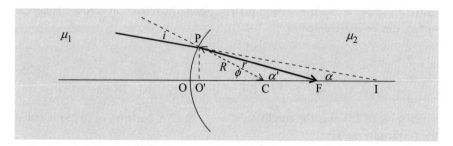

Figure 8.2 The passage of a light ray across a spherical surface.

In Figure 8.1 the effect of a ray moving through a uniform medium a distance t along the optical axis is shown. For small α,

$$d' = d - \alpha t. \tag{8.20}$$

In Figure 8.2 a ray is shown crossing a spherical boundary with radius of curvature R from a medium with refractive index μ_1 to a medium with refractive index μ_2. In passing across the spherical surface the ray is refracted, with angle of incidence i and angle of refraction r, both shown in the figure relative to the normal to the surface, PC. The immediate passage across the surface does not change the distance from the axis, d, but changes the angle relative to the axis from α to α'. From the figure, $\alpha = \phi - i$ and $\alpha' = \phi - r$ so that

$$\frac{i}{r} = \frac{\phi - \alpha}{\phi - \alpha'}. \tag{8.21}$$

Since, for paraxial rays, all angles are small,

$$\frac{i}{r} \approx \frac{\sin i}{\sin r} = \frac{\mu_2}{\mu_1} \quad \text{(Snell's law for refraction)} \tag{8.22}$$

and combining (8.21) with (8.22) we find

$$\alpha' = \phi\left(1 - \frac{\mu_1}{\mu_2}\right) + \frac{\mu_1}{\mu_2}\alpha.$$

Now

$$\phi = \frac{\text{arc PO}}{R} \approx \frac{\text{PO}'}{R} = \frac{d}{R},$$

giving

$$\alpha' = \frac{d}{R}\left(1 - \frac{\mu_1}{\mu_2}\right) + \frac{\mu_1}{\mu_2}\alpha. \tag{8.23}$$

In matrix formulation it is possible to represent translation through a uniform medium by

$$\begin{bmatrix} d' \\ \alpha' \end{bmatrix} = \begin{bmatrix} 1 & -t \\ 0 & 1 \end{bmatrix}\begin{bmatrix} d \\ \alpha \end{bmatrix}, \tag{8.24}$$

which gives (8.20) for the change in d and the result $\alpha' = \alpha$. We refer to the 2×2 matrix in (8.24) as the **translation matrix**. Similarly, refraction across a spherical surface can be represented by

$$\begin{bmatrix} d' \\ \alpha' \end{bmatrix} = \begin{bmatrix} 1 & 0 \\ \frac{1}{R}\left(1 - \frac{\mu_1}{\mu_2}\right) & \frac{\mu_1}{\mu_2} \end{bmatrix}\begin{bmatrix} d \\ \alpha \end{bmatrix}, \tag{8.25}$$

which gives (8.23) plus the condition $d' = d$. The 2×2 matrix in (8.25) is called the **refraction matrix**.

If the above analysis is repeated for a figure similar to Figure 8.2 that shows a spherical surface *concave* towards the oncoming ray, then the equation that results in place of (8.23) is

$$\alpha' = \frac{d}{R}\left(\frac{\mu_1}{\mu_2} - 1\right) + \frac{\mu_1}{\mu_2}\alpha. \tag{8.26}$$

However, if we use the convention that a surface concave to an oncoming ray has a negative radius of curvature then R is replaced by $-R$ in (8.26) and it is then of the same form as (8.23). With this convention we may always use the refraction matrix in the form given in (8.25).

We now have the tools for tackling lens systems of almost any complexity.

8.4.1 Thin lenses

Figure 8.3 shows a thin lens made of material of refractive index μ, whose surfaces have radii of curvature R_1 and R_2. It is situated in air, of refractive index effectively unity. A ray falls on the lens moving parallel to the optical axis so that when it falls on the front surface it may be described by $\{d, 0\}$. Notice the use of curly brackets to indicate a *column* vector – we shall also be using this convention later.

Any real lens must have a finite thickness, but in this case we are assuming that point B is effectively the same as point A. The focal length of the lens, f, is OF and, since α' is small we have

$$\frac{1}{f} = \frac{\alpha''}{d}. \tag{8.27}$$

Following the ray through the lens, the first step is a traverse of the surface of radius R_1. Immediately after crossing this surface the condition of the ray is described as

$$\begin{bmatrix} d' \\ \alpha' \end{bmatrix} = \begin{bmatrix} 1 & 0 \\ \frac{1}{R_1}\left(1 - \frac{1}{\mu}\right) & \frac{1}{\mu} \end{bmatrix}\begin{bmatrix} d \\ 0 \end{bmatrix}$$

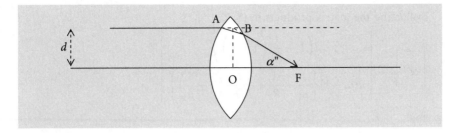

Figure 8.3 A ray passing through a thin lens.

and after crossing the second surface it is

$$\begin{bmatrix} d'' \\ \alpha'' \end{bmatrix} = \begin{bmatrix} 1 & 0 \\ \dfrac{1}{R_2}(1-\mu) & \mu \end{bmatrix} \begin{bmatrix} d' \\ \alpha' \end{bmatrix} = \begin{bmatrix} 1 & 0 \\ \dfrac{1}{R_2}(1-\mu) & \mu \end{bmatrix} \begin{bmatrix} 1 & 0 \\ \dfrac{1}{R_1}\left(1-\dfrac{1}{\mu}\right) & \dfrac{1}{\mu} \end{bmatrix} \begin{bmatrix} d \\ 0 \end{bmatrix}.$$

(8.28)

Evaluating (8.28) we find

$$\begin{bmatrix} d'' \\ \alpha'' \end{bmatrix} = \begin{bmatrix} d \\ d(\mu-1)\left(\dfrac{1}{R_1}-\dfrac{1}{R_2}\right) \end{bmatrix},$$

which, from (8.27), gives

$$\frac{1}{f} = (\mu - 1)\left(\frac{1}{R_1} - \frac{1}{R_2}\right).$$

(8.29)

We illustrate this result with a numerical example taking $R_1 = 1.0\,\text{m}$, $R_2 = -1.0\,\text{m}$ (remember the sign convention) and $\mu = 1.5$. This gives $f = 1.0\,\text{m}$.

The lens illustrated in Figure 8.3 is a converging lens and the focal length is found to be positive. If the surfaces were interchanged so that the first surface was concave to the oncoming ray and the second convex then the sign of f would be negative and this would indicate a diverging lens.

8.4.2 Thick lenses

We now imagine that in Figure 8.3 the two surfaces are moved further apart so that the distance between them is t. Since we are dealing with paraxial rays this would be the distance between the surfaces on the optical axis. Now between the two refractions there is a translation stage, represented by the translation matrix (8.24). Corresponding to (8.28) we now have

$$\begin{bmatrix} d''' \\ \alpha''' \end{bmatrix} = \begin{bmatrix} 1 & 0 \\ \dfrac{1}{R_2}(1-\mu) & \mu \end{bmatrix} \begin{bmatrix} 1 & -t \\ 0 & 1 \end{bmatrix} \begin{bmatrix} 1 & 0 \\ \dfrac{1}{R_1}\left(1-\dfrac{1}{\mu}\right) & \dfrac{1}{\mu} \end{bmatrix} \begin{bmatrix} d \\ 0 \end{bmatrix}.$$

(8.30)

In the case of a thick lens the point O in Figure 8.3 is not at the same distance along the optical axis as the points A and B, but the focal length is still given by

$$\frac{1}{f} = \frac{\alpha'''}{d}.$$

Evaluating the matrix products in (8.30)

$$
\begin{bmatrix} d''' \\ \alpha''' \end{bmatrix} = \begin{bmatrix} d\left(1 - \dfrac{t}{R_1}\left(1 - \dfrac{1}{\mu}\right)\right) \\ d(\mu - 1)\left\{\dfrac{1}{R_1} - \dfrac{1}{R_2} + \dfrac{t}{R_1 R_2}\left(1 - \dfrac{1}{\mu}\right)\right\} \end{bmatrix},
$$

giving

$$
\frac{1}{f} = (\mu - 1)\left\{\frac{1}{R_1} - \frac{1}{R_2} + \frac{t}{R_1 R_2}\left(1 - \frac{1}{\mu}\right)\right\}. \tag{8.31}
$$

This reduces to the thin-lens formula for $t = 0$. Taking the same numerical values as for the thin lens with $t = 0.1\,\mathrm{m}$ we find $f = 1.017\,\mathrm{m}$, a little different from the thin-lens result.

8.4.3 Multiple-lens systems

Evaluating multiple matrix products in symbolic form, as it appears in (8.30), is somewhat tedious. Even for a simple composite lens with two components, as shown in Figure 8.4, it would turn out to be extremely complicated. However, if the problem is expressed directly in numerical form then the evaluation of the focal length is quite straightforward.

For the composite lens in air with $R_1 = -R_2 = R_3 = 1.0\,\mathrm{m}$, $t_1 = t_2 = 0.1\,\mathrm{m}$, $\mu_1 = 1.5$, and $\mu_2 = 1.2$ we find for an original ray parallel to the axis, represented by $\{d, 0\}$,

$$
\begin{bmatrix} d_{\mathrm{final}} \\ \alpha_{\mathrm{final}} \end{bmatrix} = \begin{bmatrix} 1 & 0 \\ \dfrac{1}{R_3}(1 - \mu_2) & \mu_2 \end{bmatrix} \begin{bmatrix} 1 & -t_2 \\ 0 & 1 \end{bmatrix} \begin{bmatrix} 1 & 0 \\ \dfrac{1}{R_2}\left(1 - \dfrac{\mu_1}{\mu_2}\right) & \dfrac{\mu_1}{\mu_2} \end{bmatrix}
$$

$$
\times \begin{bmatrix} 1 & -t_1 \\ 0 & 1 \end{bmatrix} \begin{bmatrix} 1 & 0 \\ \dfrac{1}{R_1}\left(1 - \dfrac{1}{\mu_1}\right) & \dfrac{1}{\mu_1} \end{bmatrix} \begin{bmatrix} d \\ 0 \end{bmatrix}.
$$

With numbers inserted, this appears in the much simpler form

$$
\begin{bmatrix} d_{\mathrm{final}} \\ \alpha_{\mathrm{final}} \end{bmatrix} = \begin{bmatrix} 1 & 0 \\ -0.2 & 1.2 \end{bmatrix} \begin{bmatrix} 1 & -0.1 \\ 0 & 1 \end{bmatrix} \begin{bmatrix} 1 & 0 \\ 0.25 & 1.25 \end{bmatrix} \begin{bmatrix} 1 & -0.1 \\ 0 & 1 \end{bmatrix}
$$

$$
\times \begin{bmatrix} 1 & 0 \\ 0.3333 & 0.6667 \end{bmatrix} \begin{bmatrix} d \\ 0 \end{bmatrix},
$$

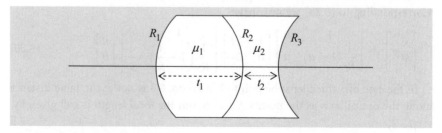

Figure 8.4 A double-lens system.

which is readily evaluated to give

$$\begin{bmatrix} d_{\text{final}} \\ \alpha_{\text{final}} \end{bmatrix} = \begin{bmatrix} 0.900833d \\ 0.609333d \end{bmatrix}. \tag{8.32}$$

As for the other systems we considered,

$$\frac{1}{f} = \frac{\alpha_{\text{final}}}{d}$$

which gives $f = 1.640\,\text{m}$.

In numerical terms very large systems can be dealt with, especially if a simple computer program is available for carrying out the matrix multiplication.

Exercise 8.3

The glass of a planoconvex lens has refractive index 1.5. The curved surface has a radius of 1 m and the thickness of the lens is 0.1 m. What is its focal length?

8.5 Application to special relativity

The theory of special relativity, due to Albert Einstein, is concerned with the way that observers in different inertial, i.e. non-accelerating, frames of reference see the same event. A basic postulate, from which all else follows, is that the speed of light is the same in any frame of reference, a postulate that is in conflict with classical laws. In classical theory the speed of a train relative to a stationary observer is not the same as for an observer moving in a car parallel to the track, and classically the same would be true if a light photon replaced the train. We now consider an event that is seen by an observer, at the origin O of an inertial reference frame, at being at position (x, y, z) and time t. If another observer, at the origin O$'$ of an inertial reference frame moving at velocity v in the x direction relative to O, sees the event at position (x', y', z') and time t', then according to special relativity theory, if the x-axes are collinear,

$$\begin{aligned} x' &= \beta(x - vt) \\ y' &= y \\ z' &= z \\ t' &= \beta\big(t - (vx/c^2)\big) \end{aligned} \tag{8.33}$$

where c is the speed of light and

$$\beta = \frac{1}{\sqrt{1 - (v^2/c^2)}}. \tag{8.34}$$

This transformation from one observer to the other is called the **Lorentz transformation**. Ignoring the trivial transformations in y and z we may write

$$\begin{bmatrix} x' \\ t' \end{bmatrix} = \beta \begin{bmatrix} 1 & -v \\ -v/c^2 & 1 \end{bmatrix} \begin{bmatrix} x \\ t \end{bmatrix} = L(v) \begin{bmatrix} x \\ t \end{bmatrix} \tag{8.35}$$

where the 2×2 matrix is called the **Lorentz transformation matrix**.

We now consider the position shown in Figure 8.5. The frame with origin O' is moving with velocity v relative to O and O'' is moving at velocity v' relative to O'. Classically the velocity of O'' relative to O is $v + v'$ but now we shall find the relative velocity according to special relativity. To do this we consider an event seen at place and time (x, t) by O, (x', t') by O' and (x'', t'') by O''. Applying the Lorentz transformation matrix

$$\begin{bmatrix} x'' \\ t'' \end{bmatrix} = \beta' \begin{bmatrix} 1 & -v' \\ -\dfrac{v'}{c^2} & 1 \end{bmatrix} \begin{bmatrix} x' \\ t' \end{bmatrix} = \beta\beta' \begin{bmatrix} 1 & -v' \\ -\dfrac{v'}{c^2} & 1 \end{bmatrix} \begin{bmatrix} 1 & -v \\ -\dfrac{v}{c^2} & 1 \end{bmatrix} \begin{bmatrix} x \\ t \end{bmatrix}$$

$$= \beta\beta' \begin{bmatrix} 1 + \dfrac{vv'}{c^2} & -(v + v') \\ -\dfrac{v + v'}{c^2} & 1 + \dfrac{vv'}{c^2} \end{bmatrix} \begin{bmatrix} x \\ t \end{bmatrix}$$

$$= \beta\beta' \left(1 + \dfrac{vv'}{c^2} \right) \begin{bmatrix} 1 & -\dfrac{v + v'}{1 + vv'/c^2} \\ -\dfrac{v + v'}{c^2(1 + vv'/c^2)} & 1 \end{bmatrix} \begin{bmatrix} x \\ t \end{bmatrix} \tag{8.36}$$

Comparing (8.35) to the final 2×2 matrix in (8.36) shows that it has the essential pattern for a Lorentz transformation matrix and we deduce that the relative velocity of O'' relative to O is

$$v_{O''O} = \frac{v + v'}{1 + vv'/c^2}, \tag{8.37}$$

which differs from the classical value $v + v'$.

For complete consistency we need to check that

$$\beta\beta' \left(1 + \frac{vv'}{c^2} \right) = \frac{1}{\sqrt{1 - \dfrac{v_{O''O}^2}{c^2}}}.$$

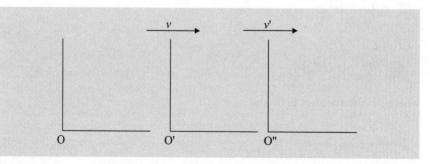

Figure 8.5 Inertial frames of reference in relative motion along x.

Table 8.1 Departure from classical result as velocity increases

v	v'	$v_{O''O}$	Classical
$0.1c$	$0.1c$	$0.198c$	$0.2c$
$0.2c$	$0.2c$	$0.385c$	$0.4c$
$0.5c$	$0.5c$	$0.800c$	$1.0c$
$0.7c$	$0.7c$	$0.940c$	$1.4c$
$1.0c$	$1.0c$	$1.000c$	$2.0c$

We have

$$
\frac{1}{\sqrt{1 + \dfrac{v_{O''O}^2}{c^2}}} = \frac{1}{\sqrt{1 - \dfrac{1}{c^2}\dfrac{(v + v')^2}{(1 + vv'/c^2)^2}}}
$$

$$
= \frac{1 + \dfrac{vv'}{c^2}}{\sqrt{\left(1 + \dfrac{vv'}{c^2}\right)^2 - \dfrac{(v + v')^2}{c^2}}}
$$

$$
= \frac{1 + \dfrac{vv'}{c^2}}{\sqrt{1 + \dfrac{(vv')^2}{c^4} + 2\dfrac{vv'}{c^2} - \dfrac{v^2}{c^2} - \dfrac{v'^2}{c^2} - 2\dfrac{vv'}{c^2}}}
$$

$$
= \frac{1 + \dfrac{vv'}{c^2}}{\sqrt{\left(1 - \dfrac{v^2}{c^2}\right)\left(1 - \dfrac{v'^2}{c^2}\right)}}
$$

$$
= \left(1 + \dfrac{vv'}{c^2}\right)\beta\beta' \tag{8.38}
$$

which completely confirms the consistency of the transformation from O to O''.

Numerical applications of (8.37) show that $v_{O''O}$ cannot exceed c in value and also illustrate the departure from the classical result with increasing velocities (see Table 8.1).

Exercise 8.4
Show that $L(v)L(-v)$ gives a unit matrix.

8.6 **Determinants**

Another type of quantity that involves arrays of elements, which must be square in this case, is the **determinant**. This is illustrated by the simple 2×2 example

$$
D_2 = \begin{vmatrix} a_{11} & a_{12} \\ a_{21} & a_{22} \end{vmatrix} = a_{11}a_{22} - a_{12}a_{21}. \tag{8.39}
$$

Extending to an order-3 determinant,

$$
D_3 = \begin{vmatrix} a_{11} & a_{12} & a_{13} \\ a_{21} & a_{22} & a_{23} \\ a_{31} & a_{32} & a_{33} \end{vmatrix} = a_{11} \begin{vmatrix} a_{22} & a_{23} \\ a_{32} & a_{33} \end{vmatrix}
$$

$$
- a_{12} \begin{vmatrix} a_{21} & a_{23} \\ a_{31} & a_{33} \end{vmatrix} + a_{13} \begin{vmatrix} a_{21} & a_{22} \\ a_{31} & a_{32} \end{vmatrix}, \tag{8.40}
$$

where the individual order-2 determinants are evaluated as in (8.39). Notice in particular the alternating signs of the terms in (8.40). To consolidate the idea we now see an order-4 determinant:

$$
D_4 = \begin{vmatrix} a_{11} & a_{12} & a_{13} & a_{14} \\ a_{21} & a_{22} & a_{23} & a_{24} \\ a_{31} & a_{32} & a_{33} & a_{34} \\ a_{41} & a_{42} & a_{43} & a_{44} \end{vmatrix} = a_{11} \begin{vmatrix} a_{22} & a_{23} & a_{24} \\ a_{32} & a_{33} & a_{34} \\ a_{42} & a_{43} & a_{44} \end{vmatrix} - a_{12} \begin{vmatrix} a_{21} & a_{23} & a_{24} \\ a_{31} & a_{33} & a_{34} \\ a_{41} & a_{43} & a_{44} \end{vmatrix}
$$

$$
+ a_{13} \begin{vmatrix} a_{21} & a_{22} & a_{24} \\ a_{31} & a_{32} & a_{34} \\ a_{41} & a_{42} & a_{44} \end{vmatrix} - a_{14} \begin{vmatrix} a_{21} & a_{22} & a_{23} \\ a_{31} & a_{32} & a_{33} \\ a_{41} & a_{42} & a_{43} \end{vmatrix}.
$$

$$
\tag{8.41}
$$

Again notice the alternating signs of the terms. The individual determinants on the right-hand sides of (8.40) and (8.41) are called the **minors** of their coefficients. Thus the minor of a_{13} in (8.41) is

$$
\alpha_{13} = \begin{vmatrix} a_{21} & a_{22} & a_{24} \\ a_{31} & a_{32} & a_{34} \\ a_{41} & a_{42} & a_{44} \end{vmatrix} \tag{8.42}
$$

and is the determinant formed by removing the row and the column containing a_{13}. We can also see that

$$
D_4 = a_{11}\alpha_{11} - a_{12}\alpha_{12} + a_{13}\alpha_{13} - a_{14}\alpha_{14}. \tag{8.43a}
$$

Illustrating the evaluation of a determinant with a numerical example, we find

$$
D_3 = \begin{vmatrix} 3 & 2 & 1 \\ -1 & 2 & 3 \\ 3 & 2 & 0 \end{vmatrix} = 3 \begin{vmatrix} 2 & 3 \\ 2 & 0 \end{vmatrix} - 2 \begin{vmatrix} -1 & 3 \\ 3 & 0 \end{vmatrix} + 1 \begin{vmatrix} -1 & 2 \\ 3 & 2 \end{vmatrix}. \tag{8.43b}
$$

$$
= 3(0 - 6) - 2(0 - 9) + 1(-2 - 6) = -8
$$

Although we have introduced the minor in terms of the top row of a determinant, it can be associated with any element. The minor of the element a_{ij} is obtained by eliminating the ith row and jth column of the determinant, and the value of the determinant can be expressed as

$$
D_4 = \sum_{j=1}^{4} (-1)^{i+j} a_{ij}\alpha_{ij} \text{ or } D_4 = \sum_{i=1}^{4} (-1)^{i+j} a_{ij}\alpha_{ij}. \tag{8.44}
$$

As an alternative to using the minor it is sometimes convenient instead to use the **cofactor**, defined as

$$A_{ij} = (-1)^{i+j}\alpha_{ij}. \tag{8.45}$$

In terms of the cofactor one may write, for example

$$D_4 = \sum_{j=1}^{4} a_{ij}A_{ij}. \tag{8.46}$$

Exercise 8.5

Find the cofactors A_{13} and A_{23} for the determinant $\begin{vmatrix} 1 & 2 & -1 & 1 \\ 2 & 3 & 2 & -1 \\ 1 & 0 & 0 & 4 \\ 2 & 1 & 3 & 0 \end{vmatrix}$.

8.6.1 Some rules for evaluating determinants

The evaluation of high-order determinants can be complicated if carried out by the process implicit in (8.41). However, it turns out that determinants have certain properties, expressed as a set of rules, which enable us to simplify the process.

Rule 1

Interchanging the rows and columns of a determinant leaves its value unchanged.
 If we interchange the rows and columns for the determinant in (8.43b) we have

$$D_3 = \begin{vmatrix} 3 & -1 & 3 \\ 2 & 2 & 2 \\ 1 & 3 & 0 \end{vmatrix} = 3(0-6) - (-1)(0-2) + 3(6-2) = -8,$$

as found previously.

Rule 2

Interchanging two rows (or columns) of a determinant reverses its value.
 We interchange rows 2 and 3 of the determinant in (8.43b) and find

$$D_3 = \begin{vmatrix} 3 & 2 & 1 \\ 3 & 2 & 0 \\ -1 & 2 & 3 \end{vmatrix} = 3(6-0) - 2(9-0) + 1(6+2) = 8,$$

the reverse of the original value.

Rule 3

The value of a determinant is unchanged if any multiple of one row (or column) is subtracted from another row (or column).
 For the determinant in (8.43b) we subtract the third row from the first row to give

$$D_3 = \begin{vmatrix} 0 & 0 & 1 \\ -1 & 2 & 3 \\ 3 & 2 & 0 \end{vmatrix} = 0(0-6) - 0(0-9) + 1(-2-6) = -8$$

showing that the value is unchanged.

Rule 4

A determinant has value zero if any two rows (or columns) are identical.

This follows from rule 3 since by subtracting one of the rows (or columns) from the other would give a zero row (or column) that automatically makes the whole determinant zero.

Rule 5

Multiplying any one row (or column) of a determinant by λ multiplies the value of the determinant by λ.

8.6.2 Evaluating higher-order determinants

Repeated application of the rules given in the previous section enables higher-order determinants to be evaluated quite readily.

EXAMPLE 8.1

The general principle can be illustrated by the evaluation of

$$D_4 = \begin{vmatrix} 4 & 2 & 1 & 3 \\ 1 & 0 & 1 & 1 \\ 3 & -1 & 2 & 0 \\ 6 & 3 & 1 & 1 \end{vmatrix}. \tag{8.47}$$

First we notice that the second row has the first element equal to unity, so by subtracting appropriate factors of the second row from the other rows we can make the first elements of these other rows all equal to zero. Thus we take four times the second row from the first row, three times the second row from the third row and six times the second row from the fourth row. This gives

$$D_4 = \begin{vmatrix} 0 & 2 & -3 & -1 \\ 1 & 0 & 1 & 1 \\ 0 & -1 & -1 & -3 \\ 0 & 3 & -5 & -5 \end{vmatrix} = -1 \begin{vmatrix} 2 & -3 & -1 \\ -1 & -1 & -3 \\ 3 & -5 & -5 \end{vmatrix} = \begin{vmatrix} -2 & 3 & 1 \\ 1 & 1 & 3 \\ -3 & 5 & 5 \end{vmatrix}. \tag{8.48}$$

For the final order-3 determinant in (8.48) we now add twice the second row to the first row and three times the second row to the third row to obtain

$$D_4 = \begin{vmatrix} 0 & 5 & 7 \\ 1 & 1 & 3 \\ 0 & 8 & 14 \end{vmatrix} = -\begin{vmatrix} 5 & 7 \\ 8 & 14 \end{vmatrix} = -14.$$

Clearly this is a very simple example, easily tackled by hand, but the same principles applied through a computer program can deal with much larger determinants and much more awkward numbers.

Exercise 8.6

Evaluate the determinant $\begin{vmatrix} 15 & 39 & 71 \\ 1 & 3 & 4 \\ 12 & 38 & 51 \end{vmatrix}$.

8.7 **Types of determinant**

A type of determinant that occurs in some problems is the **symmetric determinant**, one for which $a_{ji} = a_{ij}$. An illustration of a symmetric determinant is

$$D_4 = \begin{vmatrix} 1 & 3 & 5 & 2 \\ 3 & 1 & -1 & 4 \\ 5 & -1 & 2 & 0 \\ 2 & 4 & 0 & -3 \end{vmatrix}. \tag{8.49}$$

Notice that the elements below the leading diagonal are just a reflection of those above.

Another type of determinant is one for which $a_{ji} = -a_{ij}$, which is called **skew symmetric**. An example of such a determinant is

$$D_3 = \begin{vmatrix} 0 & -2 & -3 \\ 2 & 0 & 2 \\ 3 & -2 & 0 \end{vmatrix}. \tag{8.50}$$

The terms on the leading diagonal must be zero since $a_{ii} = -a_{ii}$. Actually it is easy to show that a skew diagonal determinant of odd order must be zero. Multiplying each row of a skew-symmetric determinant by -1 transforms the determinant to one that is equivalent to exchanging rows and columns. From rule 1, this leaves the value of the determinant unchanged. However, from rule 5, if the determinant is of order n then multiplying each row by -1 has multiplied the value of the determinant by $(-1)^n$. Hence if n is odd then the value of the determinant satisfies $D = -D$, which means that it has value zero.

Determinants and matrices have in common that they are defined by an array of elements. The determinant formed by taking the elements of a square matrix is called the **determinant of the matrix**. A useful relationship is that

$$\left| \prod_{i=1}^{n} A_i \right| = \prod_{i=1}^{n} |A_i|, \tag{8.51}$$

or, in words, the determinant of a product of matrices is the product of the individual determinants. A simple case will illustrate this point:

$$A_1 = \begin{bmatrix} 1 & 2 \\ 2 & 6 \end{bmatrix} \quad |A_1| = \begin{vmatrix} 1 & 2 \\ 2 & 6 \end{vmatrix} = 2$$

$$A_2 = \begin{bmatrix} 3 & -1 \\ -1 & 4 \end{bmatrix} \quad |A_2| = \begin{vmatrix} 3 & -1 \\ -1 & 4 \end{vmatrix} = 11$$

$$A_3 = \begin{bmatrix} 1 & 2 \\ 7 & 2 \end{bmatrix} \quad |A_3| = \begin{vmatrix} 1 & 2 \\ 7 & 2 \end{vmatrix} = -12$$

$$A_1 A_2 = \begin{bmatrix} 1 & 2 \\ 2 & 6 \end{bmatrix} \begin{bmatrix} 3 & -1 \\ -1 & 4 \end{bmatrix} = \begin{bmatrix} 1 & 7 \\ 0 & 22 \end{bmatrix}$$

$$A_1 A_2 A_3 = \begin{bmatrix} 1 & 7 \\ 0 & 22 \end{bmatrix} \begin{bmatrix} 1 & 2 \\ 7 & 2 \end{bmatrix} = \begin{bmatrix} 50 & 16 \\ 154 & 44 \end{bmatrix}$$

$$|A_1 A_2 A_3| = \begin{vmatrix} 50 & 16 \\ 154 & 44 \end{vmatrix} = -264 = 2 \times 11 \times (-12) = |A_1||A_2||A_3|.$$

? **Exercise 8.7**

If $A_1 = \begin{bmatrix} 2 & -1 \\ 3 & 2 \end{bmatrix}$ and $A_2 = \begin{bmatrix} 3 & -1 \\ -1 & 2 \end{bmatrix}$, show that $|A_1 A_2| = |A_1||A_2|$.

8.8 Inverse matrix

In normal number theory a number multiplied by its own inverse gives unity and now we seek something equivalent for matrices. For a square matrix A we seek another matrix A^{-1}, the **inverse** of A, such that

$$AA^{-1} = A^{-1}A = I. \tag{8.52}$$

We now consider a determinant in the form (8.46), which involves the cofactors

$$D_4 = \sum_{j=1}^{4} a_{ij}A_{ij}.$$

Suppose now that we take the sum

$$S = \sum_{j=1}^{4} a_{ij}A_{kj} = a_{i1}A_{k1} + a_{i2}A_{k2} + a_{i3}A_{k3} + a_{i4}A_{k4}, \tag{8.53}$$

where i and k are different. Expressed in terms of elements of the determinant for $i=1$ and $k=2$, this appears as

$$S = a_{11} \begin{vmatrix} -a_{12} & -a_{13} & -a_{14} \\ -a_{32} & -a_{33} & -a_{34} \\ -a_{42} & -a_{43} & -a_{44} \end{vmatrix} + a_{12} \begin{vmatrix} a_{11} & a_{13} & a_{14} \\ a_{31} & a_{33} & a_{34} \\ a_{41} & a_{43} & a_{44} \end{vmatrix}$$

$$+ a_{13} \begin{vmatrix} -a_{11} & -a_{12} & -a_{14} \\ -a_{31} & -a_{32} & -a_{34} \\ -a_{41} & -a_{42} & -a_{44} \end{vmatrix} + a_{14} \begin{vmatrix} a_{11} & a_{12} & a_{13} \\ a_{31} & a_{32} & a_{33} \\ a_{41} & a_{42} & a_{43} \end{vmatrix}$$

The value of S is the negative of the value of the determinant

$$S = \begin{vmatrix} a_{11} & a_{12} & a_{13} & a_{14} \\ a_{11} & a_{12} & a_{13} & a_{14} \\ a_{31} & a_{32} & a_{33} & a_{34} \\ a_{41} & a_{42} & a_{43} & a_{44} \end{vmatrix} = 0 \tag{8.54}$$

by rule 4, since two rows of the determinant are identical. This leads to the relationship for a general order-n determinant that

$$\sum_{j=1}^{n} a_{ij}A_{kj} = \delta_{ik}D_n \tag{8.55}$$

where D_n is the value of the determinant and δ_{ik} is the **Kronecker delta function** which has value 1 if $i=k$ and zero otherwise.

We now define the **adjoint** of a matrix as the transpose of the matrix of its cofactors. Let us see what this means in practice, using an order-3 square matrix as an example.

For a matrix A we take the cofactors of $|A|$, giving the matrix of cofactors

$$\begin{bmatrix} A_{11} & A_{12} & A_{13} \\ A_{21} & A_{22} & A_{23} \\ A_{31} & A_{32} & A_{33} \end{bmatrix}.$$

The adjoint of A is the transpose of this matrix. Exchanging columns and rows,

$$\text{adj}A = \begin{bmatrix} A_{11} & A_{21} & A_{31} \\ A_{12} & A_{22} & A_{32} \\ A_{13} & A_{23} & A_{33} \end{bmatrix}. \tag{8.56}$$

For an order-4 square matrix we now consider the product

$$A(\text{adj}A) = \begin{bmatrix} a_{11} & a_{12} & a_{13} & a_{14} \\ a_{21} & a_{22} & a_{23} & a_{24} \\ a_{31} & a_{32} & a_{33} & a_{34} \\ a_{41} & a_{42} & a_{43} & a_{44} \end{bmatrix} \begin{bmatrix} A_{11} & A_{21} & A_{31} & A_{41} \\ A_{12} & A_{22} & A_{32} & A_{42} \\ A_{13} & A_{23} & A_{33} & A_{43} \\ A_{14} & A_{24} & A_{34} & A_{44} \end{bmatrix}.$$

If now we take a diagonal term of the product, say $(i, j) = (2, 2)$, this will be

$$a_{21}A_{21} + a_{22}A_{22} + a_{23}A_{23} + a_{24}A_{24} = |A|,$$

from (8.55). However, if we take a non-diagonal term, say $(i, j) = (2, 3)$, this will be

$$a_{21}A_{31} + a_{22}A_{32} + a_{23}A_{33} + a_{24}A_{34} = 0,$$

from (8.55). Hence

$$A(\text{adj}A) = \begin{bmatrix} |A| & 0 & 0 & 0 \\ 0 & |A| & 0 & 0 \\ 0 & 0 & |A| & 0 \\ 0 & 0 & 0 & |A| \end{bmatrix}$$

or

$$\frac{A(\text{adj}A)}{|A|} = \begin{bmatrix} 1 & 0 & 0 & 0 \\ 0 & 1 & 0 & 0 \\ 0 & 0 & 1 & 0 \\ 0 & 0 & 0 & 1 \end{bmatrix}. \tag{8.57}$$

This indicates that the inverse matrix we seek is

$$A^{-1} = \frac{\text{adj } A}{|A|} \tag{8.58}$$

and by reversing the order of the matrices to $(\text{adj } A)A$ it is easily confirmed that (8.52) is valid.

EXAMPLE 8.2

We now illustrate the determination of an inverse matrix for the order-3 square matrix

$$A = \begin{bmatrix} 1 & 3 & 2 \\ 2 & 5 & 1 \\ 4 & 2 & 2 \end{bmatrix}.$$ (8.59)

The cofactors are

$$A_{11} = \begin{vmatrix} 5 & 1 \\ 2 & 2 \end{vmatrix} = 8 \qquad A_{12} = - \begin{vmatrix} 2 & 1 \\ 4 & 2 \end{vmatrix} = 0 \qquad A_{13} = \begin{vmatrix} 2 & 5 \\ 4 & 2 \end{vmatrix} = -16$$

$$A_{21} = - \begin{vmatrix} 3 & 2 \\ 2 & 2 \end{vmatrix} = -2 \qquad A_{22} = \begin{vmatrix} 1 & 2 \\ 4 & 2 \end{vmatrix} = -6 \qquad A_{23} = - \begin{vmatrix} 1 & 3 \\ 4 & 2 \end{vmatrix} = 10$$

$$A_{31} = \begin{vmatrix} 3 & 2 \\ 5 & 1 \end{vmatrix} = -7 \qquad A_{32} = - \begin{vmatrix} 1 & 2 \\ 2 & 1 \end{vmatrix} = 3 \qquad A_{33} = \begin{vmatrix} 1 & 3 \\ 2 & 5 \end{vmatrix} = -1$$

and the transpose of the matrix of the cofactors, the adjoint of A, is

$$\text{adj}A = \begin{bmatrix} 8 & -2 & -7 \\ 0 & -6 & 3 \\ -16 & 10 & -1 \end{bmatrix}.$$

The determinant of the matrix is -24, so

$$A^{-1} = -\frac{1}{24} \begin{bmatrix} 8 & -2 & -7 \\ 0 & -6 & 3 \\ -16 & 10 & -1 \end{bmatrix}.$$

The reader should confirm that this is the inverse matrix of A.

Exercise 8.8

Find the inverse of the matrix $\begin{bmatrix} 3 & 1 \\ 1 & 2 \end{bmatrix}$.

8.9 Linear equations

A commonly occurring requirement is the solution of n simultaneous linear equations for n unknowns. These appear in the form

$$a_{11}x_1 + a_{12}x_2 + \cdots + a_{1n}x_n = b_1$$
$$a_{21}x_1 + a_{22}x_2 + \cdots + a_{2n}x_n = b_2$$
$$\vdots$$
$$a_{n1}x_1 + a_{n2}x_2 + \cdots + a_{nn}x_n = b_n$$ (8.60)

which can be cast in matrix form as

$$Ax = b$$ (8.61a)

where A is an $n \times n$ **coefficient matrix** with elements a_{ij}, x is the **solution vector**, the column vector $\{x_1, x_2, \ldots, x_n\}$, and b is the **right-hand-side vector** $\{b_1, b_2, \ldots, b_n\}$.

Pre-multiplying both sides of (8.61a) by A^{-1}, we find

$$A^{-1}Ax = Ix = x = A^{-1}b. \tag{8.61b}$$

EXAMPLE 8.3

We can illustrate this way of finding a solution by the equations

$$\begin{aligned} x + 3y + 2z &= 2 \\ 2x + 5y + z &= -1 \\ 4x + 2y + 2z &= 6. \end{aligned} \tag{8.62}$$

The coefficient matrix is the matrix given in (8.59), so the solution vector is given by

$$x = A^{-1}b = -\frac{1}{24} \begin{bmatrix} 8 & -2 & -7 \\ 0 & -6 & 3 \\ -16 & 10 & -1 \end{bmatrix} \begin{bmatrix} 2 \\ -1 \\ 6 \end{bmatrix} = \begin{bmatrix} 1 \\ -1 \\ 2 \end{bmatrix}. \tag{8.63}$$

This solution method is only possible if the determinant of A is not zero, technically referred to as the matrix being **non-singular**, for then, from (8.58), the inverse matrix cannot be defined, as in the following example.

EXAMPLE 8.4

The following set of equations cannot be solved:

$$\begin{aligned} x + y + z &= 1 \\ 2x - y + 3z &= 2 \\ x - 2y + 2z &= 4. \end{aligned}$$

Subtracting the first equation from the second gives $x - 2y + 2z = 1$, which has the same coefficients as the third equation but a different right-hand side. The equations are thus **inconsistent**. If the first equation had been $x + y + z = -2$ then subtracting the second equation from the first would have given the third equation. In this case the equations are consistent but **linearly dependent**, and effectively there are only two independent equations to determine the three unknowns.

We shall consider this further, together with various ways of solving linear equations, in Chapter 30. However, one important case we shall consider here is that of **homogeneous equations**.

EXAMPLE 8.5

Consider the homogeneous equations

$$a_{11}x_1 + a_{12}x_2 + a_{13}x_3 = 0$$
$$a_{21}x_1 + a_{22}x_2 + a_{23}x_3 = 0 \qquad (8.64)$$
$$a_{31}x_1 + a_{32}x_2 + a_{33}x_3 = 0.$$

Since the right-hand-side vector is zero then it is clear that applying (8.61) gives only the trivial solution $x_1 = x_2 = x_3 = 0$. If the matrix of coefficients is singular, so that the equations are linearly dependent, then in general there are only two linearly independent equations, say the first two, which may be cast in the form

$$a_{11}\frac{x_1}{x_3} + a_{12}\frac{x_2}{x_3} = -a_{13}$$
$$a_{21}\frac{x_1}{x_3} + a_{22}\frac{x_2}{x_3} = -a_{23}. \qquad (8.65)$$

These may be solved for the ratios x_1/x_3 and x_2/x_3 and hence effectively the ratios of $x_1:x_2:x_3$. You should note that this is *only* possible if the matrix of the coefficients in (8.64) is singular.

Exercise 8.9

Use the result of Exercise 8.8 to solve the simultaneous equations
$$3x + y = a$$
$$x + 2y = b$$
where (a, b) are (i) (4, 3); (ii) (4, 2); and (iii) (2, 2).

Problems

8.1 Find the matrix product

$$\begin{bmatrix} 3 & 1 & 2 & 1 \\ 0 & 2 & 2 & -1 \end{bmatrix} \begin{bmatrix} 3 & 2 \\ 1 & 0 \\ 4 & 2 \\ 1 & 1 \end{bmatrix} \begin{bmatrix} 1 & 1 & 1 \\ 2 & 0 & 2 \end{bmatrix}.$$

8.2 Two biconvex thin lenses (surfaces convex outwards in each case), have all their spherical surfaces with radii of curvature of magnitude 1.0 m. The lenses are separated by 0.1 m along their common optical axis. The refractive index of the material of the lenses is 1.5 and 1.4 respectively. What is the focal length of the combination of lenses situated in air?

8.3 Four inertial frames of reference, with their x-axes collinear, are moving along the x-axes as follows. The frame with origin O' is moving with speed

v relative to the frame with origin O. The frame with origin O″ is moving with speed v' relative to the frame with origin O′. The frame with origin O‴ is moving with speed v'' relative to the frame with origin O″. Use the Lorentz transformation matrix to find the speed of O‴ relative to O. You may find this from the relative values of the elements of the resultant matrix without evaluating the factor $\beta\beta'\beta''$.

Find this relative speed for $v = v' = v'' = 0.5c$.

8.4 Evaluate the determinant

$$\begin{vmatrix} 2 & 1 & 3 & 4 & 1 \\ 1 & 2 & 3 & 2 & 1 \\ 1 & 1 & 5 & 2 & 2 \\ 3 & 0 & 0 & 2 & 1 \\ 1 & -1 & 2 & 0 & -1 \end{vmatrix}.$$

8.5 Find the inverse matrix of $\begin{bmatrix} 3 & 3 & 1 \\ 1 & 0 & 3 \\ 2 & -1 & 1 \end{bmatrix}$ and confirm its validity.

8.6 Use (8.61b) to solve the following simultaneous linear equations:

$$x - y - z = 4$$
$$2x + y - 2z = 3$$
$$x - 2y + z = 5.$$

Vector algebra

Many of the quantities met in physics have not only magnitude but also direction, and both qualities are expressed if the quantity is expressed as a vector. In this chapter the operations of vector algebra will be described. The description of various rotational quantities in terms of vectors will be explained. Applications will be made in geometrical contexts involving lines and planes. Finally, the physical problem of a body moving under a central force, a force directed towards one point, will be dealt with by a vector approach.

9.1 Scalar and vector quantities

Science is concerned with observing or deriving relationships between physical quantities, and the need for dimensional homogeneity in any such relationships was explained in Chapter 2. An example is given in (2.1) that involves the quantities speed, acceleration, and time. This equation is presented in the form

$$v = u + at.$$

Now, when a body moves it not only has a certain speed of movement but also a direction. A speed of $10\,\mathrm{km\,h^{-1}}$ in an easterly direction is quite different from a speed of $10\,\mathrm{km\,h^{-1}}$ in a northerly direction. To specify both the speed and direction of a body, its **velocity**, we have to express its motion as a **vector**, commonly indicated by bold type. In Figure 9.1 we represent the two velocities mentioned above.

Another vector quantity in (2.1) is acceleration, and physical quantities such as displacement, force, and field are also vectors. The remaining quantity in (2.1), time, cannot be associated with direction and is a **scalar** quantity, as are temperature, mass, density, and potential. Operations such as addition, subtraction, multiplication, and division are carried out with scalar quantities just as they are for ordinary numbers.

In (2.1) each individual term is a speed, a scalar quantity, but the equation can also be written in terms of vectors as

$$\mathbf{v} = \mathbf{u} + \mathbf{a}t \tag{9.1}$$

so that the left-hand side is the **vector sum** of the terms on the right-hand side. In Figure 9.2 this equation is represented diagrammatically. If we write (9.1)

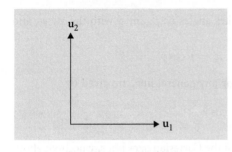

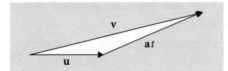

Figure 9.1 Vectors representing $10\,\mathrm{km\,h^{-1}}$ towards the east and $10\,\mathrm{km\,h^{-1}}$ towards the north.

Figure 9.2 The vector diagram corresponding to (9.1).

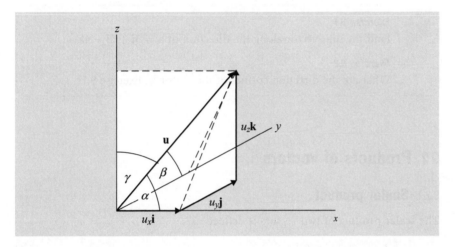

Figure 9.3 Representing a vector in terms of its components along three principal directions. The vector makes angles α, β, and γ with the x-, y-, and z-axes respectively.

as $\mathbf{u} = \mathbf{v} - \mathbf{a}t$ then, taking account of the direction of the vectors, we can interpret Figure 9.2 in terms of the difference of two vectors.

It is useful to represent vectors in terms of their components along the directions of the axes of an orthogonal Cartesian system. Three **unit vectors i, j,** and **k** are defined along the directions x, y, and z respectively. The unit vectors are dimensionless and, obviously, of unit length. Figure 9.3 illustrates the representation of the vector **u** as

$$\mathbf{u} = u_x\mathbf{i} + u_y\mathbf{j} + u_z\mathbf{k}. \tag{9.2}$$

The concept of the unit vector is very useful and can be extended to any direction. For example, if we take a unit vector $\hat{\mathbf{n}}$ in the direction of **u** then we may write

$$\mathbf{u} = u\hat{\mathbf{n}} \tag{9.3a}$$

where u is the magnitude of **u,** or

$$\hat{\mathbf{n}} = \frac{\mathbf{u}}{u}. \tag{9.3b}$$

Finally, we note that if the vector makes angles α, β, and γ with the x-, y-, and z-axes respectively then

$$u_x = u \cos \alpha, \quad u_y = u \cos \beta, \quad \text{and} \quad u_z = u \cos \gamma.$$

The **direction cosines** of the vector, or any general line, are given by

$$l = \cos \alpha, \quad m = \cos \beta, \quad \text{and} \quad n = \cos \gamma \tag{9.4a}$$

and they describe its direction in terms of the Cartesian axes but say nothing about its position, i.e. lateral displacement of the line leaves its direction cosines unchanged. Since $u_x^2 + u_y^2 + u_z^2 = u^2$ it follows that

$$l^2 + m^2 + n^2 = 1. \tag{9.4b}$$

Exercise 9.1
Find the unit vector along the direction of $\mathbf{a} = 3\mathbf{i} + 2\mathbf{j} - 4\mathbf{k}$.

Exercise 9.2
What are the direction cosines of the vector in Exercise 9.1?

9.2 Products of vectors

9.2.1 Scalar product

The **scalar product** of two vectors is defined as

$$\mathbf{a} \cdot \mathbf{b} = \mathbf{b} \cdot \mathbf{a} = ab \cos \theta \tag{9.5}$$

where θ is the angle between the vectors $\mathbf{a}$ and $\mathbf{b}$. Since the angle between a vector and itself is zero, and $\cos 0 = 1$, it follows that

$$\mathbf{a} \cdot \mathbf{a} = a^2 \tag{9.6}$$

and for unit vectors $\mathbf{i}$, $\mathbf{j}$, and $\mathbf{k}$, since they form an orthogonal (i.e. mutually perpendicular) set,

$$\mathbf{i} \cdot \mathbf{i} = \mathbf{j} \cdot \mathbf{j} = \mathbf{k} \cdot \mathbf{k} = 1 \tag{9.7a}$$
$$\mathbf{i} \cdot \mathbf{j} = \mathbf{j} \cdot \mathbf{k} = \mathbf{k} \cdot \mathbf{i} = 0. \tag{9.7b}$$

Expressing *a* and *b* in terms of components, (9.5) appears as

$$\mathbf{a} \cdot \mathbf{b} = (a_x\mathbf{i} + a_y\mathbf{j} + a_z\mathbf{k}) \cdot (b_x\mathbf{i} + b_y\mathbf{j} + b_z\mathbf{k}).$$

Scalar products obey a **distributive law**, i.e.

$$(\mathbf{a} + \mathbf{b}) \cdot (\mathbf{c} + \mathbf{d}) = (\mathbf{a} \cdot \mathbf{c} + \mathbf{a} \cdot \mathbf{d} + \mathbf{b} \cdot \mathbf{c} + \mathbf{b} \cdot \mathbf{d}).$$

Hence, using relationships (9.7a) and (9.7b),

$$\mathbf{a} \cdot \mathbf{b} = a_x b_x + a_y b_y + a_z b_z. \tag{9.8a}$$

From (9.8a) it follows that

$$\mathbf{a} \cdot \mathbf{a} = a^2 = a_x^2 + a_y^2 + a_z^2. \tag{9.8b}$$

To illustrate the scalar product numerically we take the following three vectors with their magnitudes, given by (9.8b), in brackets:

$$\mathbf{a} = 3\mathbf{i} + 4\mathbf{j} - 2\mathbf{k} \qquad (\sqrt{3^2 + 4^2 + 2^2} = \sqrt{29})$$

$$\mathbf{b} = 4\mathbf{i} - 2\mathbf{j} + \mathbf{k} \qquad (\sqrt{21}) \tag{9.9}$$

$$\mathbf{c} = 2\mathbf{i} - \mathbf{j} + \mathbf{k} \qquad (\sqrt{6}).$$

We find the three possible scalar products involving different vectors as follows:

$$\mathbf{a} \cdot \mathbf{b} = 3 \times 4 + 4 \times (-2) + (-2) \times 1 = 2$$

$$\mathbf{b} \cdot \mathbf{c} = 4 \times 2 + (-2) \times (-1) + 1 \times 1 = 11$$

$$\mathbf{c} \cdot \mathbf{a} = 2 \times 3 + (-1) \times 4 + 1 \times (-2) = 0.$$

From (9.5) we find that the angle between the vectors $\mathbf{a}$ and $\mathbf{b}$ is given by

$$\cos \theta_{ab} = \frac{\mathbf{a} \cdot \mathbf{b}}{a \times b} = \frac{2}{\sqrt{29} \times \sqrt{21}} = 0.08104,$$

from which it can be found that $|\theta_{ab}| = 85.35°$. The value of $\mathbf{a} \cdot \mathbf{c}$ clearly indicates that the two vectors involved are orthogonal since neither a nor c is zero.

9.2.2 Vector product

The **vector product**, as its name suggests, is a vector quantity indicated by $\mathbf{a} \times \mathbf{b}$. It has magnitude $ab \sin \theta$ and is perpendicular to the plane defined by $\mathbf{a}$ and $\mathbf{b}$ in the direction indicated in Figure 9.4. The rule for the direction may be remembered in terms of the action of a screwdriver. Turning the screwdriver in the rotational sense of from the first vector in the product to the second through an angle less than π moves the screw in the direction of the vector product.

From the description of the 'screwdriver rule' for defining direction, it follows that

$$\mathbf{a} \times \mathbf{b} = -\mathbf{b} \times \mathbf{a}. \tag{9.10}$$

Because of the orthogonal nature of the unit vectors $\boldsymbol{i}, \boldsymbol{j}$, and $\boldsymbol{k}$ we find

$$\mathbf{i} \times \mathbf{i} = \mathbf{j} \times \mathbf{j} = \mathbf{k} \times \mathbf{k} = \mathbf{0} \tag{9.11a}$$

$$\mathbf{i} \times \mathbf{j} = -\mathbf{j} \times \mathbf{i} = \mathbf{k} \quad \mathbf{j} \times \mathbf{k} = -\mathbf{k} \times \mathbf{j} = \mathbf{i} \quad \mathbf{k} \times \mathbf{i} = -\mathbf{i} \times \mathbf{k} = \mathbf{j}. \tag{9.11b}$$

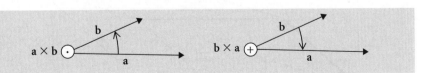

Figure 9.4 The vector $\mathbf{a} \times \mathbf{b}$ is indicated as pointing upwards: the vector $\mathbf{b} \times \mathbf{a}$ points downwards.

Using relationships (9.11) the vector product, which follows a distributive law, can be expressed in terms of the principal components of the vectors – thus

$$\begin{aligned}\mathbf{a} \times \mathbf{b} &= (a_x\mathbf{i} + a_y\mathbf{j} + a_z\mathbf{k}) \times (b_x\mathbf{i} + b_y\mathbf{j} + b_z\mathbf{k}) \\ &= (a_yb_z - a_zb_y)\mathbf{i} + (a_zb_x - a_xb_z)\mathbf{j} + (a_xb_y - a_yb_x)\mathbf{k}.\end{aligned} \tag{9.12a}$$

From (8.40) this can expressed in determinant form as

$$\mathbf{a} \times \mathbf{b} = \begin{vmatrix} \mathbf{i} & \mathbf{j} & \mathbf{k} \\ a_x & a_y & a_z \\ b_x & b_y & b_z \end{vmatrix}. \tag{9.12b}$$

For the vectors **a** and **c** given in (9.9) we find

$$\mathbf{a} \times \mathbf{c} = \begin{vmatrix} \mathbf{i} & \mathbf{j} & \mathbf{k} \\ 3 & 4 & -2 \\ 2 & -1 & 1 \end{vmatrix} = 2\mathbf{i} - 7\mathbf{j} - 11\mathbf{k}.$$

The magnitude of this vector is $\sqrt{2^2 + 7^2 + 11^2} = \sqrt{174}$. Hence from the magnitude of the vector product the angle between the two vectors is given by

$$\sin\theta_{ac} = \frac{|\mathbf{a} \times \mathbf{b}|}{a \times b} = \frac{\sqrt{174}}{\sqrt{29} \times \sqrt{6}} = 1$$

and the conclusion that these two vectors are orthogonal, previously found from the scalar product, is confirmed.

9.2.3 Scalar triple product

Another example of the application of the scalar and vector products is the combination of the two of them in the **scalar triple product**:

$$V = \mathbf{a} \cdot \mathbf{b} \times \mathbf{c}. \tag{9.13}$$

The only logical application of (9.13) is first to determine the vector product of **b** and **c** and then to take the scalar product of the resultant vector with **a**. There is in fact a geometrical interpretation of the scalar triple product. Figure 9.5 shows a parallelepiped constructed from the three vectors **a**, **b**, and **c**. The scalar triple product is the volume of the parallelepiped. The vector product **b** × **c** has

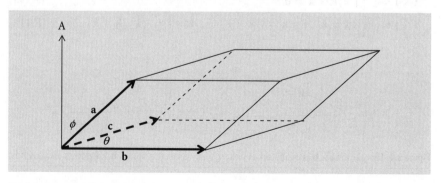

Figure 9.5 A parallelepiped based on the vectors **a**, **b**, and **c**.

magnitude $bc\sin\theta$, which is the area of the base of the parallelepiped as shown. The vector product itself is a **vector area** where the vector is normal to the area it represents. It is shown as **A** in Figure 9.5. Now the volume of the parallelepiped, V, is the area of the base, A, times the height, which is $a\cos\phi$, so

$$V = aA\cos\phi = \mathbf{a}\cdot\mathbf{A} = \mathbf{a}\cdot\mathbf{b}\times\mathbf{c}. \tag{9.14}$$

From (9.12), $\mathbf{b}\times\mathbf{c}$ has components $(b_yc_z - b_zc_y, \; b_zc_x - b_xc_z, \; b_xc_y - b_yc_x)$. Hence

$$\mathbf{a}\cdot\mathbf{b}\times\mathbf{c} = a_x(b_yc_z - b_zc_y) + a_y(b_zc_x - b_xc_z) + a_z(b_xc_y - b_yc_x)$$

$$= \begin{vmatrix} a_x & a_y & a_z \\ b_x & b_y & b_z \\ c_x & c_y & c_z \end{vmatrix}. \tag{9.15}$$

Other kinds of triple product are possible. One is $\mathbf{a}(\mathbf{b}\cdot\mathbf{c})$, which is a vector parallel to $\mathbf{a}$, and another is $\mathbf{a}\times(\mathbf{b}\times\mathbf{c})$. Note the need for the brackets in the final triple product since $(\mathbf{a}\times\mathbf{b})\times\mathbf{c}$ is a completely different vector. Using the result in (9.11) for $\mathbf{b}\times\mathbf{c}$

$$\mathbf{a}\times(\mathbf{b}\times\mathbf{c}) = \begin{vmatrix} \mathbf{i} & \mathbf{j} & \mathbf{k} \\ a_x & a_y & a_z \\ b_yc_z - b_zc_y & b_zc_x - b_xc_z & b_xc_y - b_yc_x \end{vmatrix} \tag{9.16}$$

and the x component is

$$a_y(b_xc_y - b_yc_x) - a_z(b_zc_x - b_xc_z).$$

Similarly, the value of $(\mathbf{a}\times\mathbf{b})\times\mathbf{c}$ is

$$\begin{vmatrix} \mathbf{i} & \mathbf{j} & \mathbf{k} \\ a_yb_z - a_zb_y & a_zb_x - a_xb_z & a_xb_y - a_yb_x \\ c_x & c_y & c_z \end{vmatrix}$$

and the x component is

$$c_z(a_zb_x - a_xb_z) - c_y(a_xb_y - a_yb_x).$$

Since the x components are different, the two vector triple products correspond to different vectors.

It is sometimes useful to think about vector products in terms of the direction. For example, the scalar triple product $\mathbf{a}\cdot\mathbf{a}\times\mathbf{c}$ is zero since $\mathbf{a}\times\mathbf{c}$ is perpendicular to $\mathbf{a}$ (and $\mathbf{c}$) and hence the subsequent scalar product is zero. Geometrically it also corresponds to a parallelepiped with no height – hence no volume.

Exercise 9.3
Find (i) the scalar product and (ii) the vector product of the vectors $\mathbf{a} = \mathbf{i} + \mathbf{j} + \mathbf{k}$ and $\mathbf{b} = 2\mathbf{i} - 2\mathbf{j} + \mathbf{k}$.

Exercise 9.4
Find the angle between the vectors in Exercise 9.3.

Exercise 9.5
A parallelepiped is defined by the two vectors given in Exercise 9.3 plus $\mathbf{c} = 2\mathbf{i} + \mathbf{j} - \mathbf{k}$. What is its volume?

9.3 Vector representations of some rotational quantities

Many physical quantities can be described in terms of the vector product. For example, a force F exerted at vector position r relative to the origin exerts a torque

$$\mathbf{T} = \mathbf{r} \times \mathbf{F} \tag{9.17}$$

about an axis through the origin in the direction indicated by $\mathbf{T}$ (see Figure 9.6).

Another important physical concept that can be represented in vector form is **angular momentum**. The idea that linear velocity is a vector is an easy one to comprehend since a vector implies a direction and the direction in which a body is moving is obvious. In Figure 9.7 we show a body with mass m, at vector position $\mathbf{r}$ relative to an origin O, moving with a speed $\mathbf{v}$. With respect to that origin the body has angular momentum

$$\mathbf{L} = m\mathbf{r} \times \mathbf{v} \tag{9.18a}$$

with the body spinning about an axis parallel to $\mathbf{L}$ through O. If a different origin is chosen then $\mathbf{L}$ changes as does the notional spin axis. The magnitude of the angular momentum is given by

$$L = mrv \sin \theta. \tag{9.18b}$$

The component of the velocity perpendicular to $\mathbf{r}$ is $v\sin\theta$ and hence the angular speed of spin around the axis is

$$\omega = \frac{v \sin \theta}{r}.$$

Combining this result with (9.11b) gives

$$L = mr^2\omega. \tag{9.19a}$$

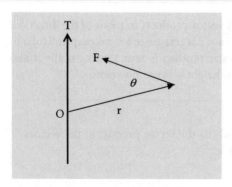

Figure 9.6 The relationship between $\mathbf{r}$, $\mathbf{F}$, and the torque $\mathbf{T}$.

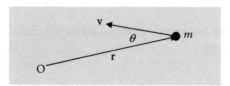

Figure 9.7 A body of mass m moving relative to the origin O.

Defining angular velocity as a vector quantity, $\boldsymbol{\omega}$, parallel to $\mathbf{L}$, we write

$$\mathbf{L} = mr^2\boldsymbol{\omega}. \tag{9.19b}$$

Since the spin axis is along the direction of $\mathbf{r} \times \mathbf{v}$, $\mathbf{r}$ is perpendicular to the axis. The quantity mr^2, where r is the perpendicular distance to the spin axis is the angular momentum, I, of the body around that axis. so that we may write

$$\mathbf{L} = I\boldsymbol{\omega}. \tag{9.19c}$$

If the body is an extended one, and not a point mass as we have been assuming here, then (9.19c) still applies where I is the moment of inertia of that body, found as the sum of the moments of inertia of all the elementary masses that constitute the body. The topic of moments of inertia is dealt with more fully in §13.7.

Exercise 9.6
A force $\mathbf{F} = 2\mathbf{i} + 2\mathbf{j} - 5\mathbf{k}$ acts at a point with vector position $\mathbf{r} = \mathbf{i} + \mathbf{j} + \mathbf{k}$. Find (i) the magnitude of the torque about an axis through the origin and (ii) the unit vector in the direction of the torque.

Exercise 9.7
A body of mass 2 units is situated at vector position $\mathbf{r} = 2\mathbf{i} + \mathbf{j} + \mathbf{k}$ and is moving with velocity $\mathbf{v} = 3\mathbf{i} + 2\mathbf{j}$. What is the angular momentum of its motion about an axis through the origin?

9.4 Linear dependence and independence

The concept of linear dependence was introduced in §8.9 in relationship to sets of linear equations. For example, the following set of three linear equations with three unknowns is linearly dependent because the last of them is obtained by subtracting three times the first one from the second one:

$$\begin{aligned} x - y + z &= 2 \\ 4x + 3y + 2z &= 1 \\ x + 6y - z &= -5. \end{aligned} \tag{9.20}$$

The concept of linear dependence can be extended to types of quantity other than linear equations. If there are n of these quantities, represented by $Q_1, Q_2, \ldots, Q_n$ and if constants $\alpha_1, \alpha_2, \ldots, \alpha_n$, not all zero, exist such that

$$\sum_{i=1}^{n} \alpha_i Q_i = 0, \tag{9.21}$$

where the zero on the right-hand side is the **null quantity** of the Q under consideration, then the quantities are linearly dependent. Otherwise they are linearly independent. To illustrate this if we represent the equations (9.20) by Q_1, Q_2, and Q_3 then $3Q_1 - Q_2 + Q_3$ gives the equation

$$0x + 0y + 0z = 0$$

which is the null version of linear equations of type Q.

The same general concept can be carried over to functions.

EXAMPLE 9.1

We consider the functions

$$f_1 = x^3 + 2x^2 + \sin x, \quad f_2 = -3x^3 + x - 2\sin x, \quad \text{and} \quad f_3 = 6x^2 + x + \sin x.$$

We now find $3f_1 + f_2 - f_3 = 0$, a null function of x. On the other hand if f_3 had been $6x^2 + x + 2\sin x$ then no linear combination of the three functions, with non-zero coefficients, could give the null result and the three functions would then be **linearly independent**.

We now see how these ideas apply to vectors. We first consider the case of two parallel vectors, **u** and **v**, represented in Figure 9.8a. They have magnitudes (lengths) 1 and 2 units respectively so we may write

$$2\mathbf{u} - \mathbf{v} = \mathbf{0} \tag{9.22}$$

where **0** represents the **null vector**, so clearly these vectors are linearly dependent.

Figure 9.8b shows two non-parallel vectors, and it is clear that these must be linearly independent. Although this is obvious we can show it in a systematic way.

EXAMPLE 9.2

Let us suppose that the two vectors are parallel to the x-y plane – we can always choose our axes to make this so. Then we may write

$$\mathbf{u} = a_1\mathbf{i} + b_1\mathbf{j} \quad \text{and} \quad \mathbf{v} = a_2\mathbf{i} + b_2\mathbf{j},$$

and for the combination $\alpha\mathbf{u} + \beta\mathbf{v} = (\alpha a_1 + \beta a_2)\mathbf{i} + (\alpha b_1 + \beta b_2)\mathbf{j}$ to be **0** it requires both $\alpha a_1 + \beta a_2 = 0$ and $\alpha b_1 + \beta b_2 = 0$. This can only be true if $a_1/a_2 = b_1/b_2$, which means that the vectors must be parallel (or antiparallel – i.e. pointing in the opposite direction). Otherwise there is no non-zero solution of the equations for α and β, so the vectors must be linearly independent.

We now consider three vectors, **u**, **v**, and **w** and the combination $\alpha\mathbf{u} + \beta\mathbf{v} + \gamma\mathbf{w}$. If the components of **u** are (u_x, u_y, u_z), with similar notation for the other two

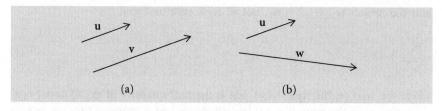

(a)   (b)

Figure 9.8 Examples of (a) linearly dependent vectors and (b) linearly independent vectors.

vectors, then the condition for linear dependence is that the following set of equations should be valid:

$$u_x\alpha + u_y\beta + u_z\gamma = 0$$

$$v_x\alpha + v_y\beta + v_z\gamma = 0$$

$$w_x\alpha + w_y\beta + w_z\gamma = 0.$$

This is a set of linear homogeneous equations similar to (8.64). The condition for a non-zero solution for α, β, and γ is that the matrix of coefficients should be singular – i.e.

$$\begin{vmatrix} u_x & u_y & u_z \\ v_x & v_y & v_z \\ w_x & w_y & w_z \end{vmatrix} = 0, \tag{9.23}$$

which is the condition for linear dependence of the three vectors. If the matrix is non-singular then the vectors are linearly independent.

In §8.1 we introduced the idea of a column vector as a matrix with a single column. For a column vector with three elements we could regard each element as a component of a vector representing some physical quantity. A column vector with two or one elements could then be thought of as representing the components of a physical vector in a one-dimensional (linear) or two-dimensional (planar) space. For some areas of theoretical physics it is an advantage to introduce the idea of a multidimensional vector space – more than the three of the real world. An important property of vectors in a multidimensional space is that $n+1$ vectors defined in a n-dimensional space must be linearly dependent. Indeed, one approach to quantum mechanics uses Hilbert space, a space of infinite dimensions in which the components may be complex quantities – but that is well outside the scope of what we are doing here.

 Exercise 9.8
Show that the vectors
$$\mathbf{a} = \mathbf{i} + \mathbf{j} + \mathbf{k}, \quad \mathbf{b} = 2\mathbf{i} - 2\mathbf{j} + 3\mathbf{k}, \quad \text{and} \quad \mathbf{c} = 3\mathbf{i} + 7\mathbf{j} + 2\mathbf{k} \text{ are}$$
linearly dependent and find the linear relationship that links them.

9.5 **A straight line in vector form**

We are now going to consider the vector representation of lines and planes. However, this is intimately linked with their representation in terms of a rectangular Cartesian coordinate system so this aspect of dealing with lines and planes will also be considered.

A straight line is defined by any two points that lie on it and in Figure 9.9 we show those points as defined by the ends of vectors $\mathbf{u}$ and $\mathbf{v}$ drawn from a chosen origin. The vector $\overrightarrow{\text{UV}}$, which is contained in the line, comes from

$$\mathbf{u} + \overrightarrow{\text{UV}} = \mathbf{v} \quad \text{or} \quad \overrightarrow{\text{UV}} = \mathbf{v} - \mathbf{u}.$$

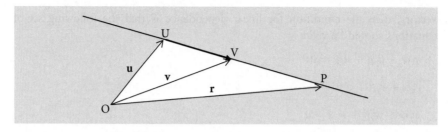

Figure 9.9 A straight line defined by the vectors **u** and **v**.

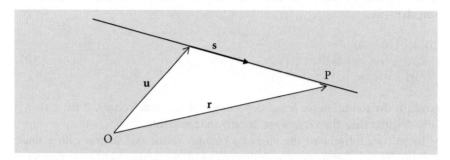

Figure 9.10 A straight line defined by a point and a direction.

We now consider a point P on the line whose position is described by the vector **r** that is given by

$$\mathbf{r} = \mathbf{u} + t(\mathbf{v} - \mathbf{u}) = (1 - t)\mathbf{u} + t\mathbf{v} \tag{9.24}$$

where $t = \mathrm{UP}/\mathrm{UV}$. This is the required equation of the straight line. As t is varied from $-\infty$ to $+\infty$ the point P ranges over the whole infinite extent of the line.

Another way of defining a straight line is by one point in it, given by the end of a vector from the origin, plus the direction of the line, given by another vector. This is shown in Figure 9.10 where the point is given by the vector **u** and the direction by the vector **s**. In this case the equation of the line is

$$\mathbf{r} = \mathbf{u} + t\mathbf{s} \tag{9.25}$$

where, again, as t is varied from $-\infty$ to $+\infty$ the point P ranges over the whole infinite extent of the line.

It is not possible to write a single equation that represents a straight line in a three-dimensional Cartesian system. Let us consider a straight line represented by (9.25) with $\mathbf{u} = \mathbf{i} + \mathbf{j} + \mathbf{k}$ and $\mathbf{s} = 2\mathbf{i} - \mathbf{j} + \mathbf{k}$. This gives

$$\mathbf{r} = x\mathbf{i} + y\mathbf{j} + z\mathbf{k} = (1 + 2t)\mathbf{i} + (1 - t)\mathbf{j} + (1 + t)\mathbf{k}. \tag{9.26}$$

Equating components,

$$x = 1 + 2t, \quad y = 1 - t, \quad \text{and} \quad z = 1 + t.$$

This gives the straight line in terms of the parameter t. Alternatively, finding t in terms of x and then substituting in the other equations gives

$$y = \frac{1}{2}(3 - x) \text{ and } z = \frac{1}{2}(1 + x), \tag{9.27}$$

which also describes the line in a rectangular Cartesian system.

If two lines lie in the same plane then, if they are not parallel, they intersect at some point. However, in three dimensions two non-parallel lines do not necessarily intersect.

EXAMPLE 9.3

Consider the lines $\mathbf{r}_1 = (1 - t)\mathbf{u} + t\mathbf{v}$ and $\mathbf{r}_2 = (1 - s)\mathbf{p} + s\mathbf{q}$ where

$$\mathbf{u} = \mathbf{i} + 2\mathbf{j} + \mathbf{k}, \quad \mathbf{v} = 2\mathbf{i} - \mathbf{j} + 3\mathbf{k}, \quad \mathbf{p} = 2\mathbf{i} + 2\mathbf{j} - 3\mathbf{k}, \quad \mathbf{q} = \mathbf{i} - 3\mathbf{j} + 2\mathbf{k}.$$

For the first equation

$$\mathbf{r}_1 = (1 + t)\mathbf{i} + (2 - 3t)\mathbf{j} + (1 + 2t)\mathbf{k},$$

and for the second

$$\mathbf{r}_2 = (2 - s)\mathbf{i} + (2 - 5s)\mathbf{j} + (5s - 3)\mathbf{k}.$$

If these lines intersect then for some pair of values of t and s the vectors $\mathbf{r}_1$ and $\mathbf{r}_2$ will be the same, which requires

$$1 + t = 2 - s, \quad 2 - 3t = 2 - 5s, \quad \text{and} \quad 1 + 2t = 5s - 3.$$

There are no values of t and s that simultaneously satisfy these three equations, and hence the lines do not intersect.

EXAMPLE 9.4

Now suppose we have

$$\mathbf{u} = -\mathbf{i} - \mathbf{j}, \quad \mathbf{v} = \mathbf{i} + 2\mathbf{j} + 3\mathbf{k}, \quad \mathbf{p} = -\mathbf{j} + 3\mathbf{k}, \quad \mathbf{q} = \mathbf{i} + \mathbf{j} + 4\mathbf{k}.$$

It is readily found that the equations to be satisfied are

$$2t - 1 = s, \quad 3t - 1 = 2s - 1, \quad \text{and} \quad 3t = s + 3.$$

These are simultaneously satisfied by $t = 2$ and $s = 3$, so that the lines intersect and their intersection is at

$$\mathbf{r} = 3\mathbf{i} + 5\mathbf{j} + 6\mathbf{k}.$$

If the lines do not intersect then they do have a distance of closest approach. We shall discover how to find this in §11.6.1.

Exercise 9.9
Find the equations of the straight lines going through the pairs of points (i) $(3, 2, 4)$ and $(-1, 0, -2)$; (ii) $(3, 3, 1)$ and $(-1, -1, 1)$. Show that these lines intersect and find the point of intersection.

9.6 A plane in vector form

A plane is completely defined by the positions of three points within it and we now take those three points as the end of the vectors $\mathbf{u}$, $\mathbf{v}$, and $\mathbf{w}$, as shown in Figure 9.11. Going from the origin to the end of the vector $\mathbf{u}$ gives a point in the plane and then

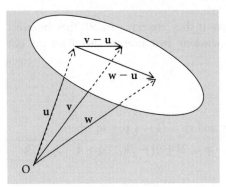

Figure 9.11 A plane defined by three points.

adding $s(\mathbf{v} - \mathbf{u}) + t(\mathbf{w} - \mathbf{u})$ with all possible pairs of values of s and t moves to all points of the plane. Hence we can write

$$\mathbf{r} = \mathbf{u} + s(\mathbf{v} - \mathbf{u}) + t(\mathbf{w} - \mathbf{u}) = (1 - s - t)\mathbf{u} + s\mathbf{v} + t\mathbf{w} \qquad (9.28)$$

as the equation of a plane in vector form.

EXAMPLE 9.5

To illustrate the use of this equation, we consider the problem of showing that the four points with Cartesian coordinates $(x, y, z) = (1, 1, 1), (2, 2, 2), (1, -1, 0)$ and $(2, 0, 1)$ are coplanar.

Taking the first three points as vector positions from the origin, the equation of the plane they define is

$$\mathbf{r} = (1 - s - t)(\mathbf{i} + \mathbf{j} + \mathbf{k}) + s(2\mathbf{i} + 2\mathbf{j} + 2\mathbf{k}) + t(\mathbf{i} - \mathbf{j})$$
$$= (1 + s)\mathbf{i} + (1 + s - 2t)\mathbf{j} + (1 + s - t)\mathbf{k}.$$

If the fourth point is in the plane then for some values of s and t we have $\mathbf{r} = 2\mathbf{i} + \mathbf{k}$, which requires

$$1 + s = 2, \quad 1 + s - 2t = 0, \quad \text{and} \quad 1 + s - t = 1.$$

The values $s = 1$ and $t = 1$ satisfy these equations, so the four points are coplanar.

We will now use a numerical example to illustrate how (9.28) leads to the equation of a plane in a rectangular Cartesian system.

EXAMPLE 9.6

We take

$$\mathbf{u} = \mathbf{i} + \mathbf{j} + \mathbf{k}, \quad \mathbf{v} = 2\mathbf{i} - \mathbf{j} + 2\mathbf{k}, \quad \text{and} \quad \mathbf{w} = \mathbf{i} - 2\mathbf{j} - 2\mathbf{k}.$$

Substituting these in (9.28)

$$\mathbf{r} = x\mathbf{i} + y\mathbf{j} + z\mathbf{k} = (1 + s)\mathbf{i} + (1 - 2s - 3t)\mathbf{j} + (1 + s - 3t)\mathbf{k}$$

and equating components gives

$$x = 1 + s, \quad y = 1 - 2s - 3t, \quad \text{and} \quad z = 1 + s - 3t.$$

First we find s in terms of x from the first equation, then substitute this into the second equation to find t in terms of x and y and then finally substitute for s and t in the third equation to give

$$3x + y - z = 3 \tag{9.29}$$

which is the equation of the plane in Cartesian form.

In general in a Cartesian coordinate system the equation of a plane is of the form

$$ax + by + cz = d. \tag{9.30}$$

It should be noted that while it is possible to deduce the exact form of the Cartesian representation from the vector equation of the plane, it is not possible to determine a unique form of vector representation from a Cartesian form. For any given plane there is an infinite selection of three points in the plane that can be taken as a basis for the vector form.

Exercise 9.10
Find in (i) vector form and (ii) Cartesian form the equation of a plane passing through the points $(1, 1, 1)$, $(2, 1, 2)$, and $(1, -1, 2)$.

9.7 Distance of a point from a plane

Consider a plane in the Cartesian form (9.30). Then the intercepts of this plane on the three axes are:

$$\left(\frac{d}{a}, 0, 0 \right), \left(0, \frac{d}{b}, 0 \right), \text{ and } \left(0, 0, \frac{d}{c} \right)$$

By subtracting pairs of these coordinates we find that the vectors

$$\mathbf{u} = \frac{d}{a}\mathbf{i} - \frac{d}{b}\mathbf{j} \quad \text{and} \quad \mathbf{v} = \frac{d}{a}\mathbf{i} - \frac{d}{c}\mathbf{k}$$

both lie in the plane. We now take the normal from the origin to the plane as

$$\mathbf{p} = x\mathbf{i} + y\mathbf{j} + z\mathbf{k}.$$

This must be normal to the vectors $\mathbf{u}$ and $\mathbf{v}$ so the scalar product of $\mathbf{p}$ with these vectors must be zero. Hence, from (9.8a)

$$x\frac{d}{a} - y\frac{d}{b} = 0 \quad \text{and} \quad x\frac{d}{a} - z\frac{d}{c} = 0$$

from which follows

$$\frac{x}{a} = \frac{y}{b} = \frac{z}{c} = k, \text{ say.} \tag{9.31}$$

Substituting for x, y and z in (9.30) from (9.31) gives

$$k(a^2 + b^2 + c^2) = d \quad \text{or} \quad k = \frac{d}{a^2 + b^2 + c^2}.$$

Now, from (9.31)

$$x = \frac{da}{a^2 + b^2 + c^2}, \quad y = \frac{db}{a^2 + b^2 + c^2}, \quad \text{and} \quad z = \frac{dc}{a^2 + b^2 + c^2}. \tag{9.32}$$

The length of the perpendicular from the origin is now found as

$$p = \sqrt{x^2 + y^2 + z^2} = \frac{d}{\sqrt{a^2 + b^2 + c^2}}. \tag{9.33}$$

The direction cosines (l, m, n) of the perpendicular are proportional to (a, b, c) and are given by

$$l = \frac{a}{\sqrt{a^2 + b^2 + c^2}}, \quad m = \frac{b}{\sqrt{a^2 + b^2 + c^2}}, \quad \text{and} \quad n = \frac{c}{\sqrt{a^2 + b^2 + c^2}}$$

(check $l^2 + m^2 + n^2 = 1$) and the unit vector in the direction of the perpendicular is

$$\hat{\mathbf{p}} = l\mathbf{i} + m\mathbf{j} + n\mathbf{k}. \tag{9.34}$$

We can now find the distance of a general point from the plane. In Figure 9.12 the general point is A with coordinates (x, y, z) and vector position $\mathbf{r}$. OP is the normal from the origin, of length p, and AQ is the normal from the point A, of length s. The line AB is perpendicular to OP and the required distance s is given by

$$s = p - \text{OB} = p - \hat{\mathbf{p}} \cdot \mathbf{r} = \frac{d}{\sqrt{a^2 + b^2 + c^2}} - \frac{xa + yb + zc}{\sqrt{a^2 + b^2 + c^2}}$$

or

$$s = \frac{d - ax - by - cz}{\sqrt{a^2 + b^2 + c^2}}. \tag{9.35}$$

From (9.35) it will be seen that the distance from the origin will be positive if d is positive and negative if d is negative. If d is positive and the distance to the plane, as found from (9.35), is positive then this means that the point is on the same side of the plane as the origin. Conversely, if d is positive and the distance found from (9.35) is negative then the point is on the side of the plane that does not contain the origin. By interchanging 'positive' and 'negative' in the preceding, the rules for negative d are found.

There are various ways of representing a plane in vector form in terms of the distance from the origin, p, and the unit normal to the plane $\hat{\mathbf{p}}$. Figure 9.13a shows

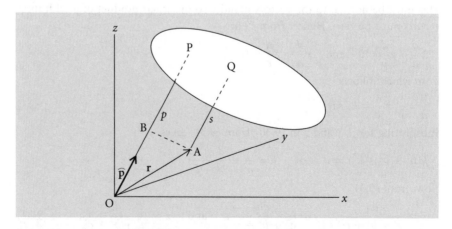

Figure 9.12 Perpendiculars from the origin and a general point A to the plane.

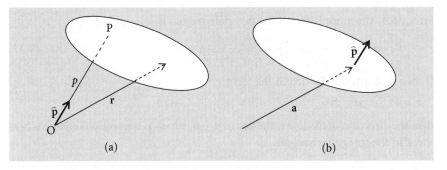

Figure 9.13 A plane: (a) defined by the perpendicular distance from the origin and the unit normal; (b) defined by a point on the plane and the unit normal.

the unit normal, the distance to the plane, and a point on the plane with vector position $\mathbf{r}$. The equation for the plane in this case can be written as

$$\mathbf{r} \cdot \widehat{\mathbf{p}} = p. \tag{9.36}$$

In Figure 9.13b the plane is defined by a point on the plane, with vector position $\mathbf{a}$, and the unit normal. Since $\mathbf{a} \cdot \widehat{\mathbf{p}} = p$ the plane may be described by

$$\mathbf{r} \cdot \widehat{\mathbf{p}} = \mathbf{a} \cdot \widehat{\mathbf{p}}. \tag{9.37}$$

Exercise 9.11

Find the distance of the point $(1, 3, 2)$ to the plane $x + y + z = 2$. Is the point on the same side of the plane as is the origin or on the opposite side?

9.8 Relationships between lines and planes

9.8.1 Angle between lines

As was pointed out in §9.5, in three dimensions non-parallel lines do not normally intersect. When we define an angle between lines we mean the angle between them if one is displaced parallel to itself so as to intersect the other. If the lines are

$$\mathbf{r}_1 = \mathbf{u}_1 + t_1\mathbf{s}_1 \quad \text{and} \quad \mathbf{r}_2 = \mathbf{u}_2 + t_2\mathbf{s}_2$$

then the directions of the lines are given by $\mathbf{s}_1$ and $\mathbf{s}_2$. We write $\mathbf{s}_1$ and $\mathbf{s}_2$ as

$$\mathbf{s}_1 = a_1\mathbf{i} + b_1\mathbf{j} + c_1\mathbf{k} \quad \text{and} \quad \mathbf{s}_2 = a_2\mathbf{i} + b_2\mathbf{j} + c_2\mathbf{k}.$$

The unit vectors in the directions of $\mathbf{s}_1$ and $\mathbf{s}_2$ are

$$\widehat{\mathbf{n}}_1 = \frac{a_1\mathbf{i} + b_1\mathbf{j} + c_1\mathbf{k}}{\sqrt{a_1^2 + b_1^2 + c_1^2}} \tag{9.38a}$$

and

$$\widehat{\mathbf{n}}_2 = \frac{a_2\mathbf{i} + b_2\mathbf{j} + c_2\mathbf{k}}{\sqrt{a_2^2 + b_2^2 + c_2^2}}. \tag{9.38b}$$

From (9.5) the angle between the two unit vectors is

$$\cos \theta = \hat{\mathbf{n}}_1 \cdot \hat{\mathbf{n}}_2 = \frac{a_1 a_2 + b_1 b_2 + c_1 c_2}{\sqrt{a_1^2 + b_1^2 + c_1^2}\sqrt{a_2^2 + b_2^2 + c_2^2}}. \tag{9.39}$$

If the equations are defined in the form

$$\mathbf{r} = (1 - t)\mathbf{u} + t\mathbf{v}$$

then the direction of the line is the vector $\mathbf{v} - \mathbf{u}$, which is treated in the same way as was $\mathbf{s}$ in the foregoing analysis.

Exercise 9.12

Lines are constructed through the pairs of points $(1, 1, 1)$, $(2, 2, 2)$ and $(1, -1, 1)$, $(2, -2, 3)$. What is the angle between these two lines?

9.8.2 Intersection of a line with a plane

In general a line will intersect a plane in one point, unless the line happens to be parallel to the plane. We now consider the intersection of the plane

$$\mathbf{r}_p = (1 - s - t)\mathbf{u} + s\mathbf{v} + t\mathbf{w}$$

and the line

$$\mathbf{r}_l = (1 - m)\mathbf{p} + m\mathbf{q}.$$

The process of finding the point of intersection is best illustrated by a numerical example.

EXAMPLE 9.7

We take

$$\mathbf{u} = \mathbf{i} + \mathbf{j} + \mathbf{k}$$
$$\mathbf{v} = 2\mathbf{i} + 2\mathbf{j} + 2\mathbf{k}$$
$$\mathbf{w} = 2\mathbf{i} - \mathbf{j} + 3\mathbf{k}$$
$$\mathbf{p} = 2\mathbf{i} + \mathbf{k}$$
$$\mathbf{q} = \mathbf{i} + 2\mathbf{k}.$$

From this we find

$$\mathbf{r}_p = (1 + s + t)\mathbf{i} + (1 + s - 2t)\mathbf{j} + (1 + s + 2t)\mathbf{k} \tag{9.40a}$$

and

$$\mathbf{r}_l = (2 - m)\mathbf{i} + (1 + m)\mathbf{k}. \tag{9.40b}$$

At the point of intersection $\mathbf{r}_p = \mathbf{r}_l$ and equating coefficients of the components we find

$$s + t + m = 1, \quad s - 2t = -1, \quad \text{and} \quad s + 2t - m = 0$$

which gives a solution $s = -1/7, t = 3/7$, and $m = 5/7$. Substituting this in (9.40b) we find the point of intersection as the end of the vector $(9/7)\mathbf{i} + (12/7)\mathbf{k}$.

You should notice that in §9.5, in considering the intersection of two lines, there were three equations in two unknowns so that only in special circumstances was there a consistent solution, corresponding to intersecting lines. However, for the intersection of a line and a plane there are three equations for three unknowns so, in general, a solution is expected.

Exercise 9.13

Find the point of intersection of a line passing through the points $(1, 1, 1)$ and $(2, 2, 2)$ with the plane passing through the points $(1, 3, 1), (3, 1, 1),$ and $(1, 1, 3)$.

9.8.3 Line of intersection of two planes

We now consider how to determine the line of intersection of two planes, illustrated in Figure 9.14. This is a problem far better tackled in Cartesian form, and again we illustrate its solution with a numerical example.

EXAMPLE 9.8

Take the two planes

$$3x + 4y + 2z = 5 \quad \text{and} \quad x - y + z = 1. \tag{9.41}$$

The solution can only be found for two of the variables in terms of the third variable, so we rewrite the equations as

$$4y + 2z = 5 - 3x \quad \text{and} \quad -y + z = 1 - x.$$

Solving for y and z in terms of x,

$$y = \frac{1}{2} - \frac{1}{6}x \quad \text{and} \quad z = \frac{3}{2} - \frac{7}{6}x.$$

For each value of x there are corresponding values of y and z and the points so found generate a line.

If there were three planes, then in place of (9.41) there would be a system of three linear equations in the three unknowns. These could give a unique solution (x, y, z) that would correspond to the point of intersection of the three planes.

Exercise 9.14

Find the point of intersection of the planes
$$x + y + z = 4$$
$$x - y + 2z = 4$$
$$2x + 3y - z = 3.$$

9.9 Differentiation of vectors

9.9.1 Differentiating in a Cartesian system

Consider a vector that varies with time, representing, say, the position of a body. In Figure 9.15 we show the positions of the body – $\mathbf{r}(t)$ at time t and $\mathbf{r}(t+\delta t)$ at time $t+\delta t$. In the time δt the vector displacement of the body in moving from P to Q is

$$\delta\mathbf{r} = \mathbf{r}(t + \delta t) - \mathbf{r}(t). \tag{9.42}$$

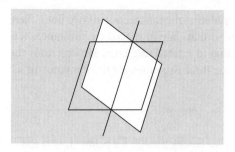

Figure 9.14 The line of intersection of two planes.

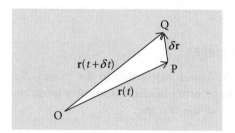

Figure 9.15 A body whose vector position varies with time.

Hence the velocity of the body, a vector quantity, is given by

$$\mathbf{v} = \left(\frac{\delta \mathbf{r}}{\delta t}\right)_{\delta t \to 0} = \frac{d\mathbf{r}}{dt}. \tag{9.43}$$

Thus the differentiation of the vector position with respect to time gives the vector quantity, velocity.

If the vector position varies with time then it is clear that one or more of the components of the position must vary with time. We illustrate this idea with two simple examples.

EXAMPLE 9.9

The equation
$$\mathbf{r} = (a \cos \omega t)\mathbf{i} + (a \sin \omega t)\mathbf{j} \tag{9.44}$$
corresponds to uniform motion around a circle. The velocity at any time is given by
$$\mathbf{v} = (-\omega a \sin \omega t)\mathbf{i} + (\omega a \cos \omega t)\mathbf{j}. \tag{9.45}$$

EXAMPLE 9.10

Consider a projectile launched at speed V at an angle α to the horizontal ground in the Earth's gravitational field, giving it a vertical acceleration $-g$. The position of the projectile is given by
$$\mathbf{r} = (V \cos \alpha \times t)\mathbf{i} + (V \sin \alpha \times t - gt^2/2)\mathbf{j} \tag{9.46}$$
and the corresponding velocity is
$$\mathbf{v} = (V \cos \alpha)\mathbf{i} + (V \sin \alpha - gt)\mathbf{j}. \tag{9.47}$$

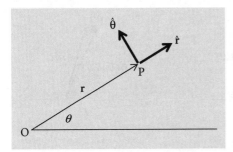

Figure 9.16 A point P at vector position **r**. The unit vectors $\hat{\mathbf{r}}$ and $\hat{\boldsymbol{\theta}}$ are in the directions of increasing r and θ.

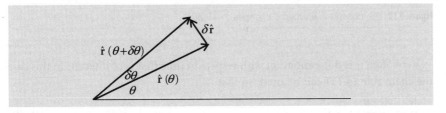

Figure 9.17 The change in $\hat{\mathbf{r}}$ when θ changes to $\theta + \delta\theta$.

Exercise 9.15
A weaving motion is described by the position vector
$\mathbf{r} = 3t\mathbf{i} + 2\cos 2t\mathbf{j}$, where t is time. What are the maximum and minimum speeds involved in this motion?

9.9.2 Differentiating in a polar coordinate system

In describing the vector position of a point in a rectangular Cartesian coordinate system for each component there was a unit vector – **i**, **j** and **k** along the positive x, y, and z directions respectively. Now consider a point described in a two-dimensional polar coordinate system as shown in Figure 9.16.

By an exact analogy with the way that the unit vectors were defined for the Cartesian system, here we show unit vectors $\hat{\mathbf{r}}$ and $\hat{\boldsymbol{\theta}}$ in the directions of increasing r and θ. However, there are important differences between the Cartesian and polar cases. The first difference is that in the polar case the position of the body only involves the distance from the origin and the one unit vector $\hat{\mathbf{r}}$. The second difference is that, while the unit vectors for Cartesian axes do not change with changes in the coordinates, the unit vectors in the polar case do change with changes of coordinates. It can be seen from Figure 9.16 that $\hat{\mathbf{r}}$ and $\hat{\boldsymbol{\theta}}$ are unaffected if r changes while θ remains constant, but vary if θ changes and r remains constant. We now consider the quantities $\hat{\mathbf{r}}$ and $d\hat{\boldsymbol{\theta}}/d\theta$. In Figure 9.17 we show the change in $\hat{\mathbf{r}}$ when θ changes to $\theta + \delta\theta$.

Since the vectors are of unit length then the magnitude of $\delta\hat{\mathbf{r}}$ is $\delta\theta$ and its direction is that of $\hat{\boldsymbol{\theta}}$ in the limit when $\delta\theta \to 0$. Hence we find

$$\frac{d\hat{\mathbf{r}}}{d\theta} = \left(\frac{\delta\hat{\mathbf{r}}}{\delta\theta}\right)_{\delta\theta\to 0} = \hat{\boldsymbol{\theta}}. \tag{9.48}$$

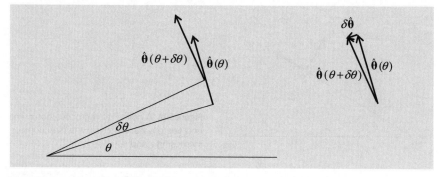

Figure 9.18 The change in $\hat{\boldsymbol{\theta}}$ when θ changes to $\theta + \delta\theta$.

Very often it is differentiation with respect to time that is of interest. In this case the chain rule (4.11) can be used, so that

$$\frac{d\hat{\mathbf{r}}}{dt} = \frac{d\hat{\mathbf{r}}}{d\theta}\frac{d\theta}{dt} = \frac{d\theta}{dt}\hat{\boldsymbol{\theta}}. \tag{9.49}$$

Figure 9.18 shows the change in $\hat{\boldsymbol{\theta}}$ when θ changes to $\theta + \delta\theta$. The unit vectors $\hat{\boldsymbol{\theta}}(\theta)$ and $\hat{\boldsymbol{\theta}}(\theta + \delta\theta)$ are shown with a common origin on the right-hand side of the figure. The angle between them is $\delta\theta$, which is the magnitude of $\delta\hat{\boldsymbol{\theta}}$ that points in the direction of $-\hat{\mathbf{r}}$. From this we find

$$\frac{d\hat{\boldsymbol{\theta}}}{d\theta} = \left(\frac{\delta\hat{\boldsymbol{\theta}}}{\delta\theta}\right)_{\delta\theta \to 0} = -\hat{\mathbf{r}} \tag{9.50}$$

and hence

$$\frac{d\hat{\boldsymbol{\theta}}}{dt} = \frac{d\hat{\boldsymbol{\theta}}}{d\theta}\frac{d\theta}{dt} = -\frac{d\theta}{dt}\hat{\mathbf{r}}. \tag{9.51}$$

EXAMPLE 9.11

To illustrate these principles we consider a particle moving in a circular path of radius a at a constant angular speed ω. The position of the body is given by

$$\mathbf{r} = a\hat{\mathbf{r}} \tag{9.52}$$

and its velocity is

$$\frac{d\mathbf{r}}{dt} = a\frac{d\hat{\mathbf{r}}}{dt} = a\frac{d\theta}{dt}\hat{\boldsymbol{\theta}} = a\omega\,\hat{\boldsymbol{\theta}}. \tag{9.53}$$

This gives the expected result that the instantaneous linear speed at any time is $a\omega$ in the direction of increasing θ, that is, tangential to the circle. Differentiating again with respect to time gives the acceleration of the body:

$$\frac{d^2\mathbf{r}}{dt^2} = a\omega\frac{d\hat{\boldsymbol{\theta}}}{dt} = -a\omega^2\hat{\mathbf{r}}. \tag{9.54}$$

This is the centripetal acceleration directed inwards towards the origin. The only way that this acceleration can be maintained is if there is a centrally directed force that causes it – in physical situations this force is usually either gravitational or electrostatic.

Exercise 9.16

Show that $\dfrac{d^3\hat{\mathbf{r}}}{d\theta^3} + \dfrac{d\hat{\mathbf{r}}}{d\theta} = 0$.

9.10 Motion under a central force

We consider a body moving in some path under the influence of a central force, i.e. one directed towards the origin. Its distance from the origin may vary with time, so we write

$$\mathbf{r} = f(t)\hat{\mathbf{r}} \tag{9.55}$$

where f(t) is some function of time. The speed of the body is

$$\frac{d\mathbf{r}}{dt} = f'(t)\hat{\mathbf{r}} + f(t)\frac{d\hat{\mathbf{r}}}{dt} = f'(t)\hat{\mathbf{r}} + f(t)\theta'\,\hat{\boldsymbol{\theta}}, \tag{9.56}$$

where the prime represents differentiation with respect to time, using the notation of §4.7. This gives the component of velocity along the radius vector as $f'(t) = dr/dt$ and that perpendicular to it as $f(t)\theta' = r\omega$. Differentiating again gives the acceleration

$$\frac{d^2\mathbf{r}}{dt^2} = f''(t)\hat{\mathbf{r}} + f'(t)\theta'\hat{\boldsymbol{\theta}} + f'(t)\theta'\hat{\boldsymbol{\theta}} + f(t)\theta''\hat{\boldsymbol{\theta}} - f(t)\theta'^2\hat{\mathbf{r}}$$
$$= \{f''(t) - f(t)\theta'^2\}\hat{\mathbf{r}} + \{2f'(t)\theta' + f(t)\theta''\}\hat{\boldsymbol{\theta}}. \tag{9.57}$$

The coefficient of $\hat{\mathbf{r}}$ represents the radial acceleration and is due to the central force. If m is the mass of the body then the equation giving the radial acceleration in terms of the central force, F_r, can be written as

$$F_r = m\left\{\frac{d^2 r}{dt^2} - r\left(\frac{d\theta}{dt}\right)^2\right\}, \tag{9.58}$$

where f(t) has been replaced by r, the radial distance from the origin towards which the central force acts.

The form of (9.58) is one that is often poorly understood. If the motion of the body is purely radial, towards and away from the source of the force, then we would only need the first term in the bracket, which is just in terms of the way the distance from the centre changes. However, if the motion has a transverse component then there also has to be acceleration in the direction of the radius vector in order that the distance of the body from the force centre should *not* change. This is illustrated in Figure 9.19. According to Newton's laws of motion, if there were no force acting on the body when it was at the point P it would continue to move in a straight line along the tangent, shown as a dashed line. However, in time δt instead of moving to R it actually moves to Q so it has moved a distance RQ due to a central force. We can also see that, as δt (and $\delta\theta$) tend to zero, so the displacement RQ tends towards the same direction as PO.

The amount by which the body has moved inwards, due to the force acting on it, is

$$s = \frac{r}{\cos\delta\theta} - r \approx \frac{r}{1 - (\delta\theta)^2/2} - r \approx r\left\{1 + \frac{1}{2}(\delta\theta)^2\right\} - r = \frac{1}{2}r(\delta\theta)^2 \tag{9.59}$$

In deriving (9.59) two approximations were used. Firstly (4.44b) was used to approximate for $\cos\delta\theta$ for small $\delta\theta$ and then the binomial approximation (1.38)

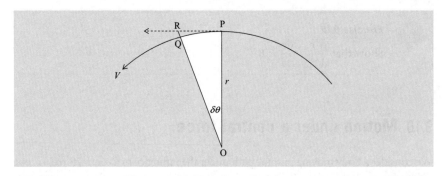

Figure 9.19 Circular motion leading to an inwards acceleration.

was used with $n = -1$. For a body moving under a constant acceleration a for time δt with zero initial speed the distance moved is

$$s = \frac{1}{2}a(\delta t)^2. \tag{9.60}$$

For the body at point P the component of speed along PO is zero, so we may equate the expressions for s in (9.59) and (9.60) to give

$$a = r\left(\frac{\delta\theta}{\delta t}\right)^2. \tag{9.61}$$

This acceleration is inwards and since $(\delta\theta/\delta t)_{\delta t \to 0} = d\theta/dt$ the inwards acceleration required for r *not* to change is identified as that giving the second term in the bracket in (9.58).

The coefficient of $\hat{\theta}$ in (9.57) is quite interesting. Since the force on the body acts along the radius vector it cannot give a transverse acceleration, meaning that the coefficient must be zero. Remembering that $f(t) = r$ we write

$$2r'\theta' + r\theta'' = 0.$$

Multiplying the equation throughout by r gives

$$2rr'\theta' + r^2\theta'' = \frac{d}{dt}(r^2\theta') = \frac{d}{dt}(r^2\omega) = 0. \tag{9.62}$$

This is a very interesting equation, for it tells us that the angular momentum, defined in (9.19), is a constant for a body moving under a central force. In fact in any system of interacting bodies, isolated from any external influence, the angular momentum remains constant. This is one of the fundamental conservation laws on which much of physics is based.

The power of the vector approach is well illustrated by the derivation of (9.58) and (9.62), both involving challenging concepts, in a very straightforward way.

 Exercise 9.17

A body of mass 2 kg is at a distance of 10 m from the source of a central force of strength $10/r^2$ N when r is in metres. It is moving in a plane with a speed of $2\,\mathrm{m\,s^{-1}}$ in a direction making an angle of 45° with the radius vector. Find (i) its radial speed, (ii) its angular velocity around the source point, (iii) $d^2\mathbf{r}/dt^2$, and (iv) d^2r/dt^2.

Problems

9.1 Find the magnitudes of the vectors given below and the scalar products of all pairs of vectors. Hence find the angles between the pairs of vectors:
$$\mathbf{a} = \mathbf{i} + \mathbf{j} + \mathbf{k}; \quad \mathbf{b} = \mathbf{i} + \mathbf{j} - 2\mathbf{k}; \quad \mathbf{c} = -2\mathbf{i} + \mathbf{j} - \mathbf{k}.$$

9.2 Find the vector products $\mathbf{a} \times \mathbf{b}$, $\mathbf{b} \times \mathbf{c}$, and $\mathbf{c} \times \mathbf{a}$ for the vectors given in Problem 9.1. Hence deduce the angles between the vectors and comment on what you find compared with what was found in Problem 9.1.

9.3 Find the scalar and vector products of $\mathbf{a} = \cos \theta \mathbf{i} + \sin \theta \mathbf{j}$ and $\mathbf{b} = \cos \phi \mathbf{i} + \sin \phi \mathbf{j}$. Hence show that $\cos(\theta - \phi) = \cos \theta \cos \phi + \sin \theta \sin \phi$ and $\sin(\theta - \phi) = \sin \theta \cos \phi - \cos \theta \sin \phi$.

9.4 Find the scalar triple product $\mathbf{a} \cdot \mathbf{b} \times \mathbf{c}$ and the vector triple product $\mathbf{a} \times (\mathbf{b} \times \mathbf{c})$ for the three vectors given in Problem 9.1.

9.5 A body is at position $\mathbf{r} = \mathbf{i} - \mathbf{j} + 2\mathbf{k}$ and is moving with a velocity $\mathbf{v} = 2\mathbf{i} + \mathbf{j} + \mathbf{k}$. What is its **intrinsic angular momentum** (angular momentum per unit mass) for rotation about an axis passing through the origin and what is its magnitude? Find the angle that the spin axis makes with the z-axis.

9.6 For all sets of three of the following vectors find which are (i) linearly dependent and (ii) linearly independent:
$$\mathbf{a} = \mathbf{i} - \mathbf{j} + \mathbf{k}; \quad \mathbf{b} = -\mathbf{i} - 5\mathbf{j} + \mathbf{k}; \quad \mathbf{c} = \mathbf{i} + 3\mathbf{j} - 2\mathbf{k}; \quad \mathbf{d} = 2\mathbf{i} + \mathbf{j} + \mathbf{k}.$$

9.7 Find the point of intersection of the line $\mathbf{r}_l = (1 - m)\mathbf{p} + m\mathbf{q}$ with the plane $\mathbf{r}_p = (1 - s - t)\mathbf{u} + s\mathbf{v} + t\mathbf{w}$ for:
$$\mathbf{p} = \mathbf{i} + \mathbf{j} + \mathbf{k}; \quad \mathbf{q} = \mathbf{i} + 2\mathbf{j} + \mathbf{k}; \quad \mathbf{u} = 2\mathbf{i} - 3\mathbf{j} + \mathbf{k}; \quad \mathbf{v} = \mathbf{i} - \mathbf{j} - \mathbf{k};$$
$$\mathbf{w} = 2\mathbf{i} - 3\mathbf{k}.$$

9.8 Find the equation of the plane in Problem 9.7 in Cartesian form. How far is the point (1, 1, 1) from this plane? Is (1, 1, 1) on the same side of the plane as the origin?

9.9 Find the angle between the plane in Problem 9.7 and the plane $\mathbf{r} = (1 - a - b)\mathbf{u}' + a\mathbf{v}' + c\mathbf{w}'$ where $\mathbf{u}' = \mathbf{i} + \mathbf{j} + \mathbf{k}$, $\mathbf{v}' = 2\mathbf{i} + \mathbf{j} + \mathbf{k}$, $\mathbf{w}' = \mathbf{i} + 2\mathbf{j} + \mathbf{k}$.
(Note: the angle between the planes is the angle between their normals.)

9.10 The position of a body is given in terms of time as $x = 3 + t - t^2, y = 2t, z = 3t^2$. Find the velocity and acceleration in both magnitude and direction as a function of time.

10 Conic sections and orbits

In this chapter we describe conic sections, curves that come from the intersection of planes with cones. These occur in physics at a variety of scales, from the motions of planets around the Sun to the motions of atomic particles. We shall show how Newton deduced that the gravitational inverse-square law of attraction leads to the elliptical orbits of the planets and, finally, go through an analysis of alpha-particle scattering, one of the landmark experiments of the twentieth century.

10.1 Kepler and Newton

Early astronomers were very concerned about the movement of bodies in the solar system. The ancient Greeks suggested that the Earth was stationary and that all other bodies, including the Sun, moved around it, an idea that persisted until the sixteenth century. To explain the motions of the planets as observed from the Earth required them to move on quite complicated paths that occasionally made large loops in space. In the sixteenth century Nicolaus Copernicus, a Polish cleric, proposed that the Sun is the central body of the solar system; this required much simpler planetary orbits in which the planet's paths never doubled back on themselves. The exact form of these orbits was deduced from observations by Johannes Kepler in the early years of the seventeenth century. He put forward three laws of planetary motion:

- Planets move in elliptical orbits with the Sun at one focus.
- The radius vector sweeps out equal areas in equal times.
- The square of the period of the orbit is proportional to the square of the mean distance.

In 1687 Isaac Newton published his great scientific work, the *Principia*, which put forward the inverse-square law of gravitation and also showed that this gravitational law led to elliptical orbits. In fact, ellipses are just one kind of motion of one body around another under an inverse-square force but what they all have in common is that they are **conic sections**. Here we shall describe conic sections, show how Newton deduced that an inverse-square law leads to elliptical orbits and finally go through an analysis of alpha-particle scattering, one of the landmark experiments of the twentieth century.

10.2 **Conic sections and the cone**

The term **conic sections** implies, very obviously, that they are curves that can be derived by taking various plane sections through a cone. Figure 10.1 shows a pair of similar coaxial cones with touching apices and various plane sections through them. Some sections give closed curves, the **circle** and **ellipse**, as shown in the figure. Another section gives a **hyperbola**, a conic section that is not closed but goes to infinity. It has two branches, since the section cuts through both cones. The section that marks the interface between ellipses and hyperbolae gives the **parabola**, which is also a non-closed conic section and has only a single branch. We shall now consider these conic sections and give them an analytical form.

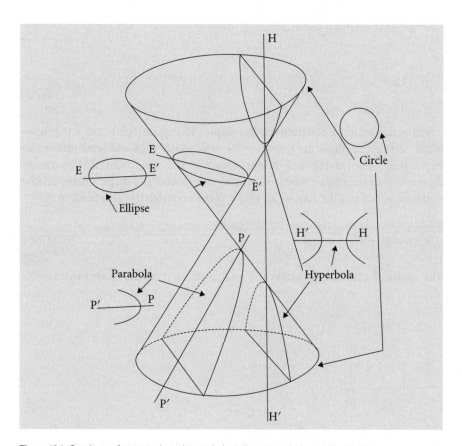

Figure 10.1 Sections of a cone that give a circle, ellipse, parabola, and hyperbola.

10.3 **The circle and the ellipse**

For points on a circle, centred on the origin in the x-y plane, the distance to the origin is a constant, a, so that the equation of the circle is

$$x^2 + y^2 = a^2. \tag{10.1}$$

The relationship of an ellipse to a circle is illustrated in Figure 10.2. The point C is on the circle, shown by the dashed line. The ellipse is produced from the circle by uniformly compressing it by a factor b/a in the y direction, where a is the radius of the circle and $2b$ is the maximum dimension of the ellipse in the y direction. From this construction it is easy to derive the equation of the ellipse in rectangular Cartesian coordinates. We consider the point Q with coordinates (x, y). From our construction of the ellipse,

$$\frac{PC}{PQ} = \frac{PC}{y} = \frac{a}{b}. \tag{10.2}$$

Since C is on the circle,

$$x^2 + (PC)^2 = x^2 + \frac{a^2}{b^2} y^2 = a^2$$

or

$$\frac{x^2}{a^2} + \frac{y^2}{b^2} = 1, \tag{10.3}$$

which is the required equation for the ellipse. The quantities a and b that completely define the ellipse are known as the **semi-major axis** and **semi-minor axis** respectively. The **major** and **minor axes** are the maximum and minimum dimensions of the ellipse. Another quantity that defines the shape, i.e. the relative dimensions but not the size, of an ellipse is the **eccentricity**, e, defined by

$$b^2 = a^2(1 - e^2) \quad \text{or} \quad e = \sqrt{1 - \frac{b^2}{a^2}}. \tag{10.4a}$$

The values of e for ellipses satisfy $0 \leq e < 1$ with $e = 0$ corresponding to a circle.

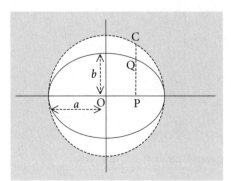

Figure 10.2 The relationship between an ellipse and a circle.

From the way that an ellipse is related to a circle by a uniform compression along y, the area of an ellipse is clearly given by

$$A = \pi a^2 \times \frac{b}{a} = \pi ab. \tag{10.4b}$$

There are two special points associated with an ellipse, known as the **foci**, situated on the major axis at distances $\pm ea$ from the origin. These are shown as F_1 and F_2 in Figure 10.3.

An interesting property of an ellipse is that the sum of the distances from any point on the ellipse to the two foci is a constant. Writing $F_1Q = r_1$ and $F_2Q = r_2$ and Q as the point (x, y),

$$r_1^2 = (x - ae)^2 + y^2. \tag{10.5}$$

From (10.3) and (10.4a)

$$y^2 = (a^2 - x^2)(1 - e^2)$$

so that

$$r_1^2 = (x - ae)^2 + (a^2 - x^2)(1 - e^2) = a^2 - 2eax + e^2x^2 = (a - ex)^2.$$

Since $x \leq a$ and $e < 1$ for a positive value for r_1 we take

$$r_1 = a - ex. \tag{10.6a}$$

Starting from $r_2^2 = (x + ae)^2 + y^2$ we similarly find

$$r_2 = a + ex. \tag{10.6b}$$

Combining (10.6a) and (10.6b),

$$r_1 + r_2 = 2a. \tag{10.7}$$

In Kepler's description of planetary motion the Sun was at one focus of the elliptical orbit. From the point of view of planetary dynamics it is an advantage to describe the motion in terms of distance from the focus and direction from some standard direction – for example, the major axis. The natural choice of coordinate system is then polar coordinates, as described in §1.6.

To find the equation of an ellipse in polar coordinates with a focus as origin we can start with (10.6a) where now r_1 becomes r and x becomes $ea + r \cos \theta$. This gives

$$r = a - e(ea + r \cos \theta)$$

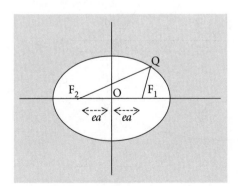

Figure 10.3 The two foci of an ellipse.

or

$$r = \frac{a(1 - e^2)}{1 + e\cos\theta}.$$

(10.8a)

If the other focus had been used as origin we would have found

$$r = \frac{a(1 - e^2)}{1 - e\cos\theta}.$$

(10.8b)

Considering an ellipse in terms of a planetary orbit gives rise to a number of definitions of quantities associated with an orbit. These are illustrated in Figure 10.4. The quantities a, b, r, and θ have already been dealt with. If the Sun were at the focus F then the point A would be the closest approach to the Sun and for that reason is called the **perihelion point** or just **perihelion**, which comes from the Greek meaning 'closest to the Sun'. The term **perihelion** is also used for the closest distance itself, marked q in the figure. Similarly B is at the furthest point from the Sun and the term **aphelion** (most distant from the Sun) is used both for the point and the distance Q. Finally we notice the distance p corresponding to $\theta = \pm\pi/2$. This distance is known as the **semi-latus rectum** and we shall see that it plays an important role in planetary dynamics. From (10.8a) with θ equal to 0, π, and $\pi/2$ respectively we find

$$q = a(1 - e),$$

(10.9a)

$$Q = a(1 + e),$$

(10.9b)

$$p = a(1 - e^2) = q(1 + e).$$

(10.9c)

Finally we note that the angle θ, measured from perihelion, is known as the **true anomaly**.

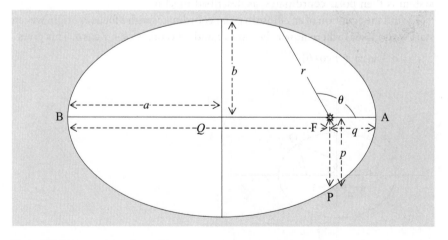

Figure 10.4 The geometry of an ellipse.

Exercise 10.1

A point is at distance 3 units from the focus of an ellipse with semi-major axis 5 units and eccentricity 0.6. What are the possible values of the true anomaly?

Exercise 10.2

What are the possible (x, y) values for the point in Exercise 10.1 if the origin is at the centre of the ellipse and the major axis lies along x?

10.4 **The parabola**

The limits of e for an ellipse exclude the value $e = 1$, and we now consider what happens as e approaches this limiting value. This is best done by considering that the perihelion, q, remains fixed while e varies. From relationships (10.9)

$$Q = q\frac{1 + e}{1 - e} \tag{10.10}$$

so that as $e \to 1$ then $Q \to \infty$. The resultant curve, unlike the ellipse, is an open one: the parabola. From (10.8a) in terms of q and with $e = 1$ its polar coordinate equation is

$$r = \frac{2q}{1 + \cos\theta}. \tag{10.11}$$

Again in terms of q the semi-latus rectum is given by

$$p = a(1 - e^2) = q(1 + e) = 2q. \tag{10.12}$$

For a description in terms of rectangular Cartesian components it is best to use the perihelion point as the origin, as shown in Figure 10.5. The coordinates (x, y) of the point P, with x negative, are given by

$$x = r\cos\theta - q = \frac{2q\cos\theta}{1 + \cos\theta} - q = \frac{q(\cos\theta - 1)}{1 + \cos\theta} \tag{10.13a}$$

and

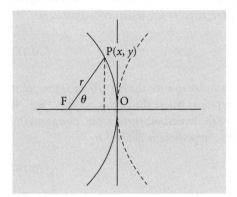

Figure 10.5 Converting a parabola to a Cartesian coordinate system.

$$y = r \sin \theta = \frac{2q \sin \theta}{1 + \cos \theta}. \tag{10.13b}$$

From (10.13a) and (10.13b),

$$\cos \theta = \frac{q + x}{q - x} \quad \text{and} \quad \sin \theta = \frac{y}{q - x}.$$

Now we write

$$\cos^2 \theta + \sin^2 \theta = \left(\frac{q + x}{q - x}\right)^2 + \left(\frac{y}{q - x}\right)^2 = 1$$

which gives

$$y^2 = -4qx \tag{10.14a}$$

The negative sign on the right-hand side comes from the parabola we started with, for which all values of x are negative. By reflecting the curve in Figure 10.5 in the y-axis, to give the dashed curve, the equation of a parabola appears as

$$y^2 = 4qx. \tag{10.14b}$$

Exercise 10.3

A point is at distance 10 units from the focus of a parabola for which $q = 4$ units. If the point is displaced in an anticlockwise sense from the perihelion point then what is its true anomaly?

Exercise 10.4

With the perihelion point as origin, what are the (x, y) coordinates of the point described in Exercise 10.3?

10.5 The hyperbola

Of all the conic sections the hyperbola is by far the most difficult to deal with analytically. What we shall do here is to quote various relationships concerning the hyperbola and pointing out analogies with the results found for the ellipse and the parabola.

A hyperbola is a curve with two branches, the form of which is shown in Figure 10.6. The equation for the hyperbola in rectangular Cartesian coordinates is

$$\frac{x^2}{a^2} - \frac{y^2}{b^2} = 1. \tag{10.15}$$

The origin point O in Figure 10.6 plays a similar role for the hyperbola as the point O in Figure 10.3 plays for the ellipse. The points where the curves cut the x-axis, R and S, are at $x = \pm a$ although in the case of the hyperbola there are no points on the curves with $|x| < a$. The eccentricity of the hyperbola is defined by

$$b^2 = a^2(e^2 - 1) \quad \text{or} \quad e = \sqrt{1 + \frac{b^2}{a^2}}, \tag{10.16}$$

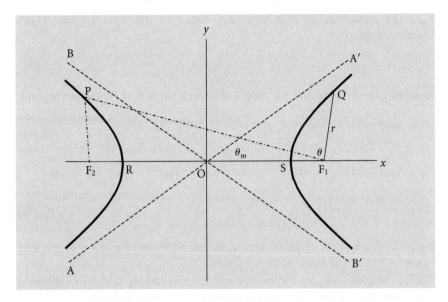

Figure 10.6 The hyperbola and its characteristics.

which is analogous to (10.4a) for the ellipse. There are also two foci, F_1 and F_2 at $x = \pm ea$, exactly the same as for the ellipse. Another characteristic of the hyperbola that has a relationship to one possessed by the ellipse is that the *difference* of the distances of any point from the two foci is a constant, $2a$. It is the *sum* of the distances that equals $2a$ for an ellipse. On Figure 10.6 this means that

$$F_1P - PF_2 = 2a. \tag{10.17}$$

The equation of a hyperbola in polar coordinates is either

$$r = \frac{a(e^2 - 1)}{e \cos \theta + 1} \tag{10.18a}$$

or

$$r = \frac{a(e^2 - 1)}{e \cos \theta - 1}. \tag{10.18b}$$

where r and θ are shown in Figure 10.6. Taking the focus as F_1 in Figure 10.6 the form of equation (10.18a) defines the branch of the hyperbola that encloses F_1. When $\theta = 0$, $r = a(e - 1)$, which is the distance F_1S. The possible values of θ are restricted by the requirement that r must be non-negative, i.e. that $e \cos \theta + 1 \geq 0$ or $\theta \leq \cos^{-1}(-1/e)$. The lines AA' and BB' in Figure 10.6, which make angle θ_m with the x-axis, are asymptotic to the branches of the hyperbola. The restriction of the angle θ is equivalent to

$$\pi + \theta_m \geq \theta \geq \pi - \theta_m. \tag{10.19a}$$

Again with F_1 as focus the form of equation (10.18b) defines the branch of the hyperbola remote from F_1. Now when $\theta = 0$, $r = a(e + 1)$, which is the distance

F_1R. The restriction on θ is now that $e\cos\theta - 1 \geq 0$ or $\theta \leq \cos^{-1}(1/e)$, corresponding to

$$\theta_m \geq \theta \geq -\theta_m. \tag{10.19b}$$

Regarding the hyperbola as a potential orbit with the Sun at F_1, the perihelion is

$$q = \frac{a(e^2 - 1)}{1 + e} = a(e - 1). \tag{10.20a}$$

Also, from (10.18a), the semi-latus rectum, the value of r when $\theta = \pi/2$, is

$$p = \frac{a(e^2 - 1)}{1} = q(1 + e). \tag{10.20b}$$

Notice that the forms of the semi-latus rectum in terms of q and e, as given by (10.9c), (10.12), and (10.20b) are all similar.

This concludes our description of conic sections and we shall now see how it applies to the dynamics of astronomical bodies and alpha-particles.

Exercise 10.5

The points $(2, 2\sqrt{3})$ and $(3, -4\sqrt{2})$ both lie on a hyperbola. What is the equation of the hyperbola (i) in Cartesian coordinates and (ii) in polar coordinates?

10.6 The orbits of planets and Kepler's laws

Consider the motion of a body P, of negligible mass, moving under the gravitational attraction of a body, Q, of mass M as shown in Figure 10.7. According to Newton's law of gravitation the gravitational field at P is in the direction of Q and has magnitude GM/r^2, where r is the distance between P and Q and the gravitational constant, $G = 6.67 \times 10^{-11}\,\mathrm{m^3\,kg^{-1}\,s^{-2}}$. From (9.58), in a frame of reference centred on Q and rotating with P, the radial acceleration is given by

$$\frac{d^2 r}{dt^2} = -\frac{GM}{r^2} + r\left(\frac{d\theta}{dt}\right)^2. \tag{10.21}$$

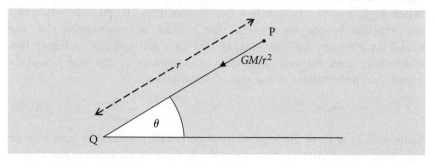

Figure 10.7 The acceleration at P due to a mass M at Q.

From the conservation of angular momentum, derived in (9.62), we have

$$r^2 \frac{d\theta}{dt} = H \tag{10.22}$$

where the constant H is the angular momentum per unit mass (intrinsic angular momentum).

To solve the differential equation (10.21) we change the independent variable from t to θ, using (10.22) in this process. First we write

$$\frac{dr}{dt} = \frac{dr}{d\theta}\frac{d\theta}{dt} = \frac{H}{r^2}\frac{dr}{d\theta}. \tag{10.23}$$

From this we see that the differential operator

$$\frac{d}{dt} = \frac{H}{r^2}\frac{d}{d\theta},$$

so that

$$\frac{d^2r}{dt^2} = \frac{d}{dt}\left(\frac{dr}{dt}\right) = \frac{H}{r^2}\frac{d}{d\theta}\left(\frac{H}{r^2}\frac{dr}{d\theta}\right) = \frac{H}{r^2}\left\{-2\frac{H}{r^3}\left(\frac{dr}{d\theta}\right)^2 + \frac{H}{r^2}\frac{d^2r}{d\theta^2}\right\}. \tag{10.24}$$

Now the dependent variable is changed to $u = 1/r$ giving

$$\frac{dr}{d\theta} = \frac{d}{d\theta}\left(\frac{1}{u}\right) = -\frac{1}{u^2}\frac{du}{d\theta} \tag{10.25}$$

and

$$\frac{d^2r}{d\theta^2} = \frac{2}{u^3}\left(\frac{du}{d\theta}\right)^2 - \frac{1}{u^2}\frac{d^2u}{d\theta^2}. \tag{10.26}$$

Inserting results from (10.24), (10.25) and (10.26) in (10.21) we find

$$\frac{d^2u}{d\theta^2} = -u + \frac{GM}{H^2}. \tag{10.27}$$

The solution of (10.27) is

$$u = A\cos\theta + \frac{GM}{H^2} \tag{10.28}$$

where A is an arbitrary constant. Converting back to r as the dependent variable we find

$$r = \frac{H^2/GM}{1 + (AH^2/GM)\cos\theta}. \tag{10.29}$$

Comparing (10.29) with (10.8a) we can see that this is the equation of an ellipse, and comparing corresponding quantities we find

$$a(1 - e^2) = \frac{H^2}{GM} \quad \text{and} \quad e = \frac{AH^2}{GM}.$$

Since (10.29) represents an ellipse, this shows that the inverse-square-law gravitational field leads to the first of Kepler's laws (§10.1).

Kepler's second law comes directly from (10.22), which is a statement of the conservation of angular momentum. In Figure 10.8 the positions of a planet are shown at the beginning and end of a time interval Δt. The radius vector has rotated through a small angle $\Delta\theta$ and the area swept out is

$$\frac{1}{2}r \times (r + \Delta r) \times \sin(\Delta\theta) \approx \frac{1}{2}r^2\Delta\theta.$$

The rate that area is swept out is thus

$$\left[\frac{1}{2}r^2\frac{\Delta\theta}{\Delta t}\right]_{\Delta t \to 0} = \frac{1}{2}r^2\frac{d\theta}{dt} = \frac{1}{2}H, \tag{10.30}$$

which is a constant. This is equivalent to Kepler's second law (§10.1).

To verify Kepler's third law we equate the numerators of (10.8a) and (10.29) to give

$$H = \{GMa(1 - e^2)\}^{1/2} \tag{10.31}$$

or, from (10.4),

$$H = (GMb^2/a)^{1/2}. \tag{10.32}$$

The period, P, is the total area of the elliptical orbit, given by (10.4b) as πab, divided by the rate at which area is swept out by the radius vector. From (10.30) and (10.32)

$$P = \frac{\pi ab}{H/2} = 2\pi \left(\frac{a^3}{GM}\right)^{1/2}. \tag{10.33}$$

This shows that $P^2 \propto a^3$, which is Kepler's third law.

 Exercise 10.6

A body of mass 1.5×10^{18} kg moves around the Sun (mass 2.0×10^{30} kg) in an orbit with $a = 10^{12}$ m and $e = 0.65$. What are (i) its perihelion distance, (ii) its aphelion distance, (iii) the total angular momentum associated with its orbit, and (iv) the period of the orbit?

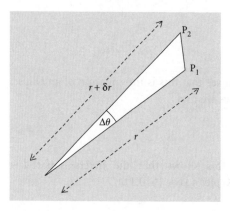

Figure 10.8 The area swept out by the radius vector in a time Δt.

10.7 **The dynamics of orbits**

From (10.9c) and (10.31) the magnitude of the intrinsic angular momentum of an orbiting body can be written as

$$H = \sqrt{GMp}, \tag{10.34}$$

an equation that shows the importance of the semi-latus rectum, p, in the dynamics of the orbit. However, angular momentum is a vector quantity and we should therefore have some means of specifying its direction, which requires the establishment of a coordinate system.

The plane of the Earth's orbit around the Sun is known as the **ecliptic** and this provides a convenient x-y plane for a three-dimensional rectangular Cartesian system. The x-direction is defined by what is called the **first point of Aries**, the direction of the Sun at the time of the **vernal equinox**, when the Sun in its relative motion around the Earth crosses the equator from south to north. The positive z-direction is at right angles to the ecliptic, pointing towards the north. As seen looking towards the ecliptic from the north all planets move in an anti-clockwise direction, and this is known as **direct** or **prograde** motion. Any body moving in an opposite sense is said to have **retrograde** motion.

Prograde motion occurs if the component along z of

$$\mathbf{H} = \mathbf{r} \times \mathbf{v} \tag{10.35}$$

is positive. For example, if with the Sun at the origin the orbiting body is on the positive x-axis so that $\mathbf{r} = (x, 0, 0)$ and it is moving in the positive y-direction so that, with $\dot{y}$ representing dy/dt, $\mathbf{v} = (0, \dot{y}, 0)$ then

$$\mathbf{H} = \begin{vmatrix} \mathbf{i} & \mathbf{j} & \mathbf{k} \\ x & 0 & 0 \\ 0 & \dot{y} & 0 \end{vmatrix} = x\dot{y}\hat{\mathbf{k}}. \tag{10.36}$$

The motion is prograde and the magnitude of the z-component is positive, which agrees with the convention we have established.

For an orbiting body at position $\mathbf{r} = (x, y, z)$ with velocity $\mathbf{v} = (\dot{x}, \dot{y}, \dot{z})$ the angular momentum is

$$\mathbf{H} = H_x\hat{\mathbf{i}} + H_y\hat{\mathbf{j}} + H_z\hat{\mathbf{k}} \quad \text{where} \quad H_x = y\dot{z} - z\dot{y},$$
$$H_y = z\dot{x} - x\dot{z}, \quad \text{and} \quad H_z = x\dot{y} - y\dot{x} \tag{10.37}$$

with the sign of H_z indicating whether the orbit is prograde or retrograde. The angle between the plane of the orbiting body and the ecliptic is called the **inclination**. Since H_z is the component of $\mathbf{H}$ along z, the normal to the ecliptic, the inclination, i, is given by

$$\cos i = H_z/H. \tag{10.38}$$

Another conserved quantity for an orbiting body is the **intrinsic total energy** or the total energy per unit mass. This is the sum of the intrinsic potential and kinetic energies at any point of the orbit, given by

$$E = -\frac{GM}{r} + \frac{1}{2}v^2. \tag{10.39}$$

The perihelion and aphelion are points in the orbit where the motion is orthogonal to the radius vector, so that at these points it is particularly easy to find the speed. At perihelion the distance is $a(1-e)$ and the speed v_P so that the intrinsic angular momentum is

$$H = a(1-e)v_P = \{GMa(1-e^2)\}^{1/2}$$

which gives

$$v_P^2 = \left\{\frac{GM(1+e)}{a(1-e)}\right\}. \tag{10.40}$$

Then from (10.39) it is found that

$$E = -\frac{GM}{a(1-e)} + \frac{1}{2}\left\{\frac{GM(1+e)}{a(1-e)}\right\}$$

which gives

$$E = -\frac{GM}{2a}. \tag{10.41}$$

If both E and H are known then from (10.31) and (10.41) it is possible to find both a and e. From (10.41),

$$a = -\frac{GM}{2E} \tag{10.42}$$

and substituting this in (10.31) and rearranging gives

$$e = \left\{1 + \frac{2EH^2}{G^2M^2}\right\}^{1/2}. \tag{10.43}$$

From (10.43) we can see that for an elliptical (i.e. bound) orbit with $e < 1$ the intrinsic total energy, E, must be negative. If $E = 0$ then $e = 1$ and we have the special case of a parabolic orbit. With $E > 0$ then $e > 1$ and the orbit is hyperbolic.

If the position and velocity of an orbiting body are known then we can find H from (10.37) by

$$H^2 = H_x^2 + H_y^2 + H_z^2. \tag{10.44}$$

The value of E can also be found from (10.39) as

$$E = -\frac{GM}{(x^2 + y^2 + z^2)^{1/2}} + \frac{1}{2}\left\{(\dot{x})^2 + (\dot{y})^2 + (\dot{z})^2\right\}. \tag{10.45}$$

It is evident that both a and e can be found directly from the position and velocity of the body. You should note that the values of a, e, and i do not define the orbit completely. It is also necessary to define some orientation angles, but we shall not deal with that here.

We have been considering orbits in terms of orbits around the Sun, which would apply to planets, asteroids, and comets. However, the same analysis applies to satellites, both natural and artificial, moving around planets.

For the analysis carried out so far we have assumed that the orbiting body has negligible mass. If the mass of the orbiting body cannot be neglected then the formulae we have found need modification. Figure 10.9a illustrates two bodies, B_1 and B_2, of masses M_1 and M_2 respectively, orbiting round each other. The point about which they both move is their **centre of mass**. They follow orbits with the same period, the same shape but different sizes; corresponding points of the two bodies are shown in the figure.

The accelerations of the two bodies relative to the centre of mass are indicated in Figure 10.9b. The acceleration of B_2 relative to B_1, and hence the motion of B_2, relative to B_1 can be found by applying an acceleration to the whole system which cancels that of B_1. This is shown in Figure 10.9c and the effect is equivalent to making B_2 massless and transferring its mass to B_1. To describe the motion of the most massive planet, Jupiter, around the Sun, ignoring the effects of the other planets, involves replacing M_{Sun} by $M_{Sun} + M_{Jup}$ in all the equations describing the motion. Since the mass of Jupiter is 1/1000 that of the Sun, the effect is small but by no means negligible.

Exercise 10.7
A body, distant 1.5×10^{11} m from the Sun, has a speed of $10 \, \mathrm{km \, s^{-1}}$. Find (i) the intrinsic energy associated with its orbit and (ii) the semi-major axis of its orbit.

Exercise 10.8
If the body in Exercise 10.7 is moving at right angles to the radius vector then find (i) the intrinsic angular momentum associated with its orbit and (ii) the eccentricity of its orbit.

10.8 Alpha-particle scattering

When J. J. Thomson discovered the electron in 1897 he proposed what became known as the **plum-pudding model** of an atom. In this model the atom was a large blob of positive charge, within which electrons were embedded like the raisins in a plum pudding. In 1906 two students of Ernest Rutherford, Hans Geiger and Ernest Marsden, carried out an experiment in which they bombarded very thin gold foil with alpha-particles. Rutherford had discovered alpha-particles and had carried out experiments, deflecting them with electric and magnetic fields, so he knew their charge and their mass. The result of the bombardment experiment was a great surprise. About 98% of the alpha-particles went straight through the foil with little deviation. About 2% were scattered through angles that could be readily measured but about 0.01% were scattered *backwards* from the foil. This last outcome was astonishing: Rutherford likened it to the shell of a naval gun bouncing off a sheet of

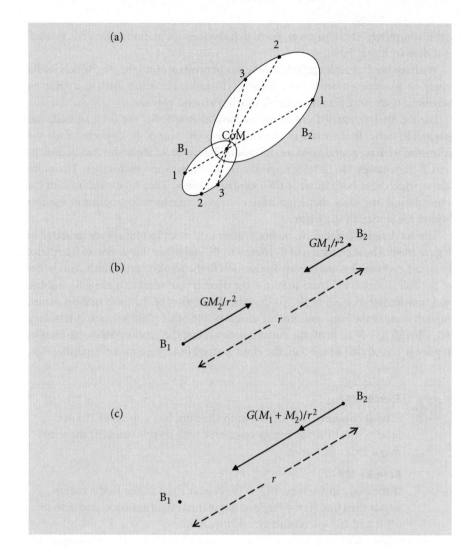

Figure 10.9 (a) Corresponding points on the orbits of two bodies about their centre of mass. (b) The accelerations of the bodies B_1 and B_2. (c) The acceleration of B_2 when an overall acceleration is applied to the system to bring B_1 to rest.

tissue paper. From this experiment there emerged the **nuclear model** of the atom in which all the positive charge was concentrated in a tiny nucleus, much smaller than the atom as a whole. The negatively charged electrons surrounded the nucleus, balancing the charge of the nucleus but contributing very little to its total mass. The alpha-particle scattering experiment could then be understood as being due to the interaction of the alpha-particle with charge zq_e ($z = 2$) with the nucleus of charge Zq_e, where q_e is the electron charge.

The forces between electric charges follow an inverse-square law, as do gravitational forces, but with the important difference that the force between two positive charges is repulsive while gravitational forces are always attractive. Thus for alpha-particle scattering the equation corresponding to (10.21) is

$$\frac{d^2 r}{dt^2} = \frac{zZq_e^2}{4\pi\varepsilon_0 m_\alpha r^2} + r\left(\frac{d\theta}{dt}\right)^2, \tag{10.46}$$

where it is assumed that the mass of the alpha-particle is negligible compared with the mass of the scattering nucleus. The subsequent analysis is similar except that GM is replaced by $-zZq_e^2/4\pi\varepsilon_0 m_\alpha$, which, for brevity, we shall write as $-\Phi$. The alpha-particle approaches the nucleus with high energy E_α (intrinsic energy E_α/m_α) so that, from (10.43),

$$e = \left(1 + \frac{2E_\alpha H^2}{\Phi^2 m_\alpha}\right)^{1/2}, \tag{10.47}$$

which is greater than unity, so the orbit is hyperbolic. The form of the equation of the hyperbola can be found from (10.29) replacing GM with $-\Phi$ as

$$r = \frac{-\dfrac{H^2}{\Phi}}{1 - \dfrac{AH^2}{\Phi}\cos\theta} = \frac{\dfrac{H^2}{\Phi}}{\dfrac{AH^2}{\Phi}\cos\theta - 1}, \tag{10.48}$$

which is of the form (10.18b). Thus the alpha-particle executes a hyperbolic orbit with the atomic nucleus at the *remote* focus. From Figure 10.6 we can see that the deflection of the alpha-particle in going from one end of the hyperbolic path to the other is, using (1.19),

$$\beta = \pi - 2\theta_m \quad \text{or} \quad \cos\beta = -\cos(2\theta_m) = 1 - 2\cos^2\theta_m.$$

Now

$$\cos^2\theta_m = \left(\frac{1}{e}\right)^2 = \left(1 + \frac{2E_\alpha H^2}{\Phi^2 m_\alpha}\right)^{-1} \tag{10.49}$$

which gives

$$\cos\beta = \frac{2E_\alpha H^2 - \Phi^2 m_\alpha}{2E_\alpha H^2 + \Phi^2 m_\alpha}. \tag{10.50}$$

We now have to consider the value of the intrinsic angular momentum of the alpha-particle in its orbit. When it is distant from the nucleus it is travelling in a straight line that would take it to a minimum distance b from the nucleus if it were not deflected. This quantity b is known as the **interaction parameter**, and if the speed of the particle is V_α then, since $E_\alpha = \frac{1}{2}V_\alpha^2$,

$$H^2 = V_\alpha^2 b^2 = \frac{2E_\alpha b^2}{m_\alpha}. \tag{10.51}$$

Inserting from (10.51) in (10.50),

$$\cos\beta = \frac{4E_\alpha^2 b^2 - m_\alpha^2 \Phi^2}{4E_\alpha^2 b^2 + m_\alpha^2 \Phi^2} \tag{10.52}$$

It will be noticed that $m_\alpha \Phi = zZq_e^2/4\pi\varepsilon_0$, which is independent of m_α and inserting this in (10.52) gives

$$\cos\beta = \frac{(8\pi\varepsilon_0 E_\alpha b)^2 - (zZq_e^2)^2}{(8\pi\varepsilon_0 E_\alpha b)^2 + (zZq_e^2)^2}. \tag{10.53}$$

From this result we can immediately see some important characteristics of the scattering experiment. If b is very large, i.e. the path of the alpha-particle keeps it well away from the nucleus, then $\cos\beta \to 1$, or $\beta \to 0$, which means that the particle is not deflected. On the other hand, if $b = 0$, so that the alpha-particle goes straight towards the nucleus, then $\cos\beta \to -1$, or $\beta \to \pi$, which means that the particle is scattered backwards along the original approach path.

To quantify the scattering for any particular value of b requires a knowledge of the physical characteristics of the system.

EXAMPLE 10.1

Consider the following values:

$z = 2$ (α-particle charge)
$Z = 79$ (atomic number of gold)
$q_e = 1.602 \times 10^{-19}$ C (electronic charge)
$\varepsilon_0 = 8.854 \times 10^{-12}$ Fm^{-1}
$E_\alpha = 4$ MeV $= 6.408 \times 10^{-13}$ J.

Figure 10.10 shows the deflection angle for an alpha-particle as a function of the impact parameter. It can be seen that backscatter requires a very small impact parameter, which explains why only a small fraction of a beam of alpha-particles is backscattered.

To compare the Geiger–Marsden results with theory in detail requires the proportion of scattered alpha-particles per unit solid angle to be calculated as a function of scattering angle, which we do not do here. What we have shown is that the phenomenon of alpha-particle scattering can be explained in terms of a hyperbolic orbit under a central repulsive force.

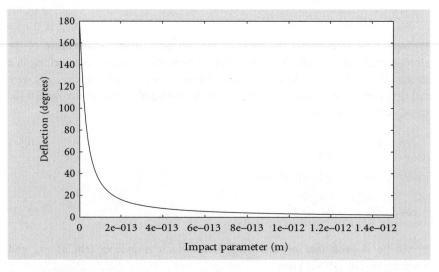

Figure 10.10 Angle of scattering of an alpha-particle by a gold atom as a function of the impact parameter.

Exercise 10.9

An alpha-particle of energy 4 MeV approaches a gold atom with an impact parameter 10^{-13} m. Find:

(i) the speed of the alpha-particle ($1\,eV = 1.60 \times 10^{-19}$ J, $m_\alpha = 6.68 \times 10^{-27}$ kg)

(ii) its intrinsic angular momentum with respect to the gold nucleus

(iii) the value of the constant Φ in (10.47) (electron charge $= 1.60 \times 10^{-19}$ C)

(iv) the eccentricity of its orbit with respect to the gold atom

(v) the value of θ_m (see 10.49)

(vi) the angle of deflection of the alpha-particle.

Problems

10.1 An ellipse is centred on the origin and has its major and minor axes aligned with the x- and y-axes of a Cartesian coordinate system. Two points on it are (1.886, 1) and (1.5, 1.984). Find the lengths of the major and minor axes and the eccentricity.

10.2 A parabola is defined as in (10.11). Show that a point on the parabola is equidistant from the origin and the line $x = 2q$.

10.3 Show that for a point on a hyperbola the difference of the distances to the two foci is $2a$.

10.4 The perihelion distance of an orbiting body is 2.3 AU. When it has travelled an angular distance of 60° from perihelion its distance from the Sun is 3.0 AU. What are its semi-major axis and eccentricity? (Note: 1 astronomical unit (AU) $= 1.496 \times 10^{11}$ m. However, the calculation can be carried out using the AU as the unit of length.)

10.5 An asteroid moving in the ecliptic is seen with coordinates $x = 1.2$ AU, $y = 0.8$ AU moving along a direction making an anticlockwise angle of 55° with the x-axis at a speed of 20 km s^{-1}. What are the semi-major axis and eccentricity of its orbit? (The mass of the Sun is 2.0×10^{30} kg.)

10.6 A Geiger–Marsden type scattering experiment is carried out with protons of energy 2 MeV fired at a target of silver foil. For what impact parameter will the deflection of a proton be 90°? (The atomic number of silver is 47. See Exercise 10.9 for values of physical constants.)

11

Partial differentiation

Chapter 4 was concerned with differentiating functions of one independent variable. In this chapter we shall consider functions of two or more independent variables and find quantities corresponding to rates of change for these. These results are applied to describe relationships between various thermodynamic quantities in respect to different equations of state. Finally we consider the problem of finding maxima and minima of a function defined in a multidimensional space.

11.1 What is partial differentiation?

To take a concrete example of a function of two independent variables, we think of the ground-level temperature within a particular region. It varies with position, and position can be described in terms of a two-dimensional Cartesian system, so we can write the temperature as $T(x, y)$. Now suppose that at a particular point in the region we wish to know the rate of change of temperature in the x-direction with y kept constant. This can be written as

$$\frac{\partial T}{\partial x} = \left\{ \frac{T(x + \delta x, y) - T(x, y)}{\delta x} \right\}_{\delta x \to 0}. \tag{11.1a}$$

Notice the notation for the partial differential of T with respect to x. Similarly,

$$\frac{\partial T}{\partial y} = \left\{ \frac{T(x, y + \delta y) - T(x, y)}{\delta y} \right\}_{\delta y \to 0}. \tag{11.1b}$$

In the case we are considering it is unlikely that the temperature would be known as a function of position in an analytical form, but now let us take an analytical function such as

$$f(x, y) = 4x^2 - 3xy + y^3 - 2. \tag{11.2}$$

To find $\partial f / \partial x$ we simply differentiate $f(x, y)$, treating y as a constant. Doing that we have

$$\frac{\partial f}{\partial x} = 8x - 3y. \tag{11.3a}$$

The term y^3 makes no contribution since it is being treated as a constant. Similarly,

$$\frac{\partial f}{\partial y} = -3x + 3y^2.$$ (11.3b)

To reinforce this idea, we consider another example.

EXAMPLE 11.1

Now we have

$$f(x, y) = xy^2 \exp(x^2 y^2).$$ (11.4)

First we treat y as a constant and differentiate the product of xy^2 with $\exp(x^2 y^2)$. This gives

$$\frac{\partial f}{\partial x} = y^2 \exp(x^2 y^2) + 2x^2 y^4 \exp(x^2 y^2).$$ (11.5a)

Similarly,

$$\frac{\partial f}{\partial y} = 2xy \exp(x^2 y^2) + 2x^3 y^3 \exp(x^2 y^2).$$ (11.5b)

Figure 11.1 shows a geometrical interpretation of $\partial f / \partial x$. The function $f(x, y)$ is plotted along the z-direction of a rectangular Cartesian system and gives the surface indicated. The heavy line is the intersection of the plane $y = y_0$ (represented by ABB$'$A$'$) with the surface. We can now write the slope of the heavy line at the point (x_0, y_0) as $(\partial f / \partial x)_{(x_0, y_0)}$.

Exercise 11.1
Find $\partial f / \partial x$ and $\partial f / \partial y$ for $f(x, y) = 4x^3 + 6x^2 y + 7xy^2 + 3y^3$.

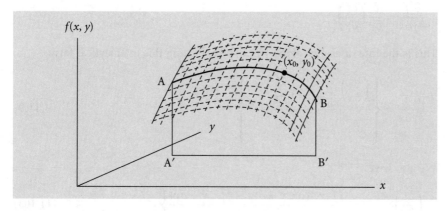

Figure 11.1 A geometrical interpretation of a partial derivative.

11.2 Higher partial derivatives

It is clear from Figure 11.1 that, just as in the case of ordinary differentiation, we can define higher derivatives at points along the heavy line – for example the second derivative $\partial^2 f/\partial x^2$. This can be defined as

$$\frac{\partial^2 f}{\partial x^2} = \left\{ \frac{\left(\frac{\partial f}{\partial x}\right)_{(x+\delta x,\, y)} - \left(\frac{\partial f}{\partial x}\right)_{(x,\, y)}}{\delta x} \right\}_{\delta x \to 0} \tag{11.6}$$

and in practice we simply differentiate $f(x, y)$ twice with respect to x keeping y constant.

EXAMPLE 11.2

With

$$f(x, y) = 6x^3 - 3x^2 y + y^2 x - 4x + y + 6 \tag{11.7}$$

we find

$$\frac{\partial f}{\partial x} = 18x^2 - 6xy + y^2 - 4 \text{ leading to } \frac{\partial^2 f}{\partial x^2} = 36x - 6y \tag{11.8a}$$

and

$$\frac{\partial f}{\partial y} = -3x^2 + 2xy + 1 \text{ leading to } \frac{\partial^2 f}{\partial y^2} = 2x. \tag{11.8b}$$

Next we consider the **mixed double partial derivative**

$$\frac{\partial^2 f}{\partial y \partial x} = \frac{\partial}{\partial y}\left(\frac{\partial f}{\partial x}\right).$$

This is the rate of change along y of $\partial f/\partial x$. Putting this in analytical form,

$$\frac{\partial^2 f}{\partial y \partial x} = \left\{ \frac{\left(\frac{\partial f}{\partial x}\right)_{(x,\, y+\delta y)} - \left(\frac{\partial f}{\partial x}\right)_{(x,\, y)}}{\delta y} \right\}_{\delta y \to 0}. \tag{11.9}$$

We also have

$$\left(\frac{\partial f}{\partial x}\right)_{(x,\, y+\delta y)} = \left\{ \frac{f(x+\delta x, y+\delta y) - f(x, y+\delta y)}{\delta x} \right\}_{\delta x \to 0} \tag{11.10a}$$

and

$$\left(\frac{\partial f}{\partial x}\right)_{(x,y)} = \left\{\frac{f(x+\delta x, y) - f(x,y)}{\delta x}\right\}_{\delta x \to 0} \qquad (11.10b)$$

Inserting (11.10a) and (11.10b) into (11.9) gives

$$\frac{\partial^2 f}{\partial y \partial x} = \left\{\frac{f(x+\delta x, y+\delta y) - f(x,y+\delta y) - f(x+\delta x, y) + f(x,y)}{\delta x \delta y}\right\}_{\delta x, \delta y \to 0}.$$

$$(11.11)$$

The limit in (11.11) is that both δx and δy tend to zero. Since (11.11) is symmetrical in x and y, it follows that $(\partial^2 f / \partial y \partial x) = (\partial^2 f / \partial y \partial x)$. It is interesting, and perhaps not obvious, that the rate of change along y of $\partial f / \partial x$ equals the rate of change along x of $\partial f / \partial y$. We can illustrate this process using the function (11.7):

$$\frac{\partial^2 f}{\partial y \partial x} = \frac{\partial}{\partial y}\left(\frac{\partial f}{\partial x}\right) = \frac{\partial}{\partial y}\left(18x^2 - 6xy + y^2 - 4\right) = -6x + 2y$$

and

$$\frac{\partial^2 f}{\partial x \partial y} = \frac{\partial}{\partial x}\left(\frac{\partial f}{\partial y}\right) = \frac{\partial}{\partial x}\left(-3x^2 + 2xy + 1\right) = -6x + 2y.$$

Higher-order differentials, both mixed and unmixed, can be found if needed.

Exercise 11.2

Find $\dfrac{\partial^2 f}{\partial x^2}$, $\dfrac{\partial^2 f}{\partial y^2}$, and $\dfrac{\partial^2 f}{\partial x \partial y}$ for $f(x,y) = 6x^4 + 5x^2 y^2 + 2y^4$.

11.3 The total derivative

In Equation (4.11) the chain rule was described as a tool for finding derivatives. Consider a function $f(x)$ where $x = \phi(t)$. Then the chain rule states that

$$\frac{df}{dt} = \frac{df}{dx}\frac{dx}{dt}. \qquad (11.12)$$

Now we take an extension of this general situation where we have a function of two variables $f(x, y)$ where $x = \phi_x(t)$ and $y = \phi_y(t)$ and we shall deduce the equivalence of the chain rule in this case. We imagine that t has changed to $t + \delta t$ which has changed x to $x + \delta x$, y to $y + \delta y$ and f to $f + \delta f$. From this we can write

$$\delta f = f(x + \delta x, y + \delta y) - f(x, y). \qquad (11.13)$$

From (11.10a) we have

$$f(x + \delta x, y + \delta y) - f(x, y + \delta y) = \left(\frac{\partial f}{\partial x}\right)_{(x, y+\delta y)} \delta x \qquad (11.14a)$$

and from (11.10b), with y playing the role of x,

$$f(x, y + \delta y) - f(x, y) = \left(\frac{\partial f}{\partial y}\right)_{(x, y)} \delta y \qquad (11.14b)$$

We can modify the right-hand side of (11.14a) using the Taylor theorem taken to two terms in the form

$$\left(\frac{\partial f}{\partial x}\right)_{(x, y + \delta y)} = \left(\frac{\partial f}{\partial x}\right)_{(x, y)} + \left(\frac{\partial^2 f}{\partial y \partial x}\right)\delta y. \qquad (11.14c)$$

which can be understood in words as '$\partial f/\partial x$ at the point $(x, y + \delta y)$ equals $\partial f/\partial x$ at the point (x, y) plus the rate of change of $\partial f/\partial x$ with y times the distance δy'. Adding (11.14a) to (11.14b) and then substituting from (11.14c) gives

$$\delta f = f(x + \delta x, y + \delta y) - f(x, y) = \frac{\partial f}{\partial x}\delta x + \frac{\partial f}{\partial y}\delta y + \frac{\partial^2 f}{\partial x \partial y}\delta x \delta y \qquad (11.15)$$

where the subscripts on the partial derivatives have now been removed since they are all at the point (x, y). Dividing throughout by δt and then making δt tend to zero gives

$$\left(\frac{\delta f}{\delta t}\right)_{\delta t \to 0} = \frac{\partial f}{\partial x}\left(\frac{\delta x}{\delta t}\right)_{\delta t \to 0} + \frac{\partial f}{\partial y}\left(\frac{\delta y}{\delta t}\right)_{\delta t \to 0} + \frac{\partial^2 f}{\partial y \partial x}\left(\frac{\delta x}{\delta t}\right)_{\delta t \to 0}(\delta y)_{\delta t \to 0} \qquad (11.16)$$

The last term has been put into a form that shows clearly that its value is zero since both δx and δy tend to zero as δt tends to zero. The other bracketed terms come to ordinary derivatives, so we have

$$\frac{df}{dt} = \frac{\partial f}{\partial x}\frac{dx}{dt} + \frac{\partial f}{\partial y}\frac{dy}{dt}. \qquad (11.17)$$

This is known as the **total derivative**, and it is the equivalence to the chain rule that we sought.

EXAMPLE 11.3

As an example of the application of the total derivative we take

$$f(x, y) = x^2 + \frac{x}{y} \quad \text{with} \quad x = \frac{\tan t}{t} \quad \text{and} \quad y = e^t.$$

This gives

$$\frac{\partial f}{\partial x} = 2x + \frac{1}{y} = 2\frac{\tan t}{t} + e^{-t}; \quad \frac{\partial f}{\partial y} = -\frac{x}{y^2} = -\frac{\tan t}{t}e^{-2t}$$

$$\frac{dx}{dt} = \frac{\sec^2 t}{t} - \frac{\tan t}{t^2}; \quad \frac{dy}{dt} = e^t.$$

Hence

$$\frac{df}{dt} = \left(2\frac{\tan t}{t} + e^{-t}\right)\left(\frac{\sec^2 t}{t} - \frac{\tan t}{t^2}\right) - \frac{\tan t}{t}e^{-2t} \times e^t$$

$$= \left(2\frac{\tan t}{t} + e^{-t}\right)\left(\frac{\sec^2 t}{t} - \frac{\tan t}{t^2}\right) - \frac{\tan t}{t}e^{-t}.$$

Writing the function f directly in terms of t and differentiating it would give the same result, but the total derivative approach is much more straightforward.

The total derivative idea can be extended. If one can write a function as $f(x_1, x_2, \ldots, x_n)$ where each x is a differentiable function of t, then

$$\frac{df}{dt} = \sum_{i=1}^{n} \frac{\partial f}{\partial x_i} \frac{dx_i}{dt}.$$ (11.18)

Exercise 11.3

Given $f(x, y) = e^x(1 + 2y + 3y^2)$ with $x = \sin t$ and $y = 1/t$, find df/dt using the total derivative approach.

11.3.1 Implicit differentiation

We now consider a situation where we have a function $f(x, y)$ and where y is a function of x. This may be thought of as a special case of a total derivative where $x = t$ so that y is now a function of x and $dx/dt = 1$. Corresponding to (11.17) we now have

$$\frac{df}{dx} = \frac{\partial f}{\partial x} + \frac{\partial f}{\partial y}\frac{dy}{dx}.$$ (11.19)

EXAMPLE 11.4

As an example we take

$$f(x, y) = \cos\left(\frac{x}{y}\right) \quad \text{and} \quad y = \sin x.$$

This gives

$$\frac{\partial f}{\partial x} = -\frac{1}{y}\sin\left(\frac{x}{y}\right) = -\frac{1}{\sin x}\sin\left(\frac{x}{\sin x}\right)$$

$$\frac{\partial f}{\partial y} = \frac{x}{y^2}\sin\left(\frac{x}{y}\right) = \frac{x}{\sin^2 x}\sin\left(\frac{x}{\sin x}\right)$$

and

$$\frac{dy}{dx} = \cos x.$$

Combining these results,

$$\frac{df}{dx} = -\frac{1}{\sin x}\sin\left(\frac{x}{\sin x}\right) + \frac{x}{\sin^2 x}\sin\left(\frac{x}{\sin x}\right)\cos x.$$

In the event that we have an equation of the form

$$f(x, y) = 0$$ (11.20)

then y may be thought of as an **implicit function** of x. Since $f(x, y)$ is zero its total derivative must be zero so that, from (11.19),

$$\frac{dy}{dx} = -\frac{\partial f}{\partial x} \bigg/ \frac{\partial f}{\partial y}. \qquad (11.21)$$

EXAMPLE 11.5

Consider the equation for an ellipse,

$$f(x, y) = \frac{x^2}{a^2} + \frac{y^2}{b^2} - 1 = 0,$$

This gives

$$\frac{\partial f}{\partial x} = \frac{2x}{a^2} \quad \text{and} \quad \frac{\partial f}{\partial y} = \frac{2y}{b^2}$$

and hence

$$\frac{dy}{dx} = -\frac{xb^2}{ya^2},$$

a result that may be confirmed by directly writing y as a function of x.

Exercise 11.4

Given $f(x, y) = \dfrac{x^4}{a^4} + \dfrac{y^4}{b^4} - 1 = 0$, use implicit differentiation to find $\dfrac{dy}{dx}$.

11.4 Partial differentiation and thermodynamics

Thermodynamics is a branch of physics that deals with relationships between quantities such as pressure, density, temperature, internal energy, and entropy, and involves concepts such as reversibility. It has extensive applications in the dynamics of chemical reactions and also impinges on other branches of science. The mathematical tool of the greatest importance in thermodynamics is partial differentiation and relationships between quantities expressed in the form of partial differentials. Here we shall just look at some simple relationships involving the pressure, volume, and temperature of a substance.

Consider a fixed mass of a substance in an experimental situation such that its pressure, P, temperature, T, and volume, V, can be measured. If the volume and temperature are fixed then the pressure is also fixed; in fact knowledge of any two of these quantities gives the value of the third one. The relationship that links these three quantities is known as the **equation of state**. This can be put in one of the forms

$$\text{pressure} = P(V, T), \tag{11.22a}$$

$$\text{volume} = V(P, T), \tag{11.22b}$$

$$\text{temperature} = T(P, V). \tag{11.22c}$$

If pressure P corresponds to a volume V and a temperature T, then changing V to $V+\delta V$ and T to $T+\delta T$ will give a pressure $P+\delta P$. By the process that led to (11.15), eliminating the final term that is negligible compared to the others, we find

$$\delta P = \frac{\partial P}{\partial V}\delta V + \frac{\partial P}{\partial T}\delta T. \tag{11.23a}$$

Many texts on thermodynamics write $(\partial P/\partial V)_T$ instead of $\partial P/\partial V$ to indicate that the partial differential is for constant T. This is necessary when more than three thermodynamic quantities are being considered, but here we are only considering three so the quantity taken as constant is self-evident. Nevertheless, we shall recognize the reality of the additional thermodynamic quantities and use the subscript to give

$$\delta P = \left(\frac{\partial P}{\partial V}\right)_T \delta V + \left(\frac{\partial P}{\partial T}\right)_V \delta T. \tag{11.23b}$$

Similarly we can write

$$\delta V = \left(\frac{\partial V}{\partial P}\right)_T \delta P + \left(\frac{\partial V}{\partial T}\right)_P \delta T. \tag{11.24}$$

Substituting for δV from (11.24) into (11.23b),

$$\delta P = \left(\frac{\partial P}{\partial V}\right)_T \left\{ \left(\frac{\partial V}{\partial P}\right)_T \delta P + \left(\frac{\partial V}{\partial T}\right)_P \delta T \right\} + \left(\frac{\partial P}{\partial T}\right)_V \delta T. \tag{11.25}$$

Now all the increments in the thermodynamic quantities are independently variable so, making $\delta T = 0$, we find

$$\delta P = \left(\frac{\partial P}{\partial V}\right)_T \left(\frac{\partial V}{\partial P}\right)_T \delta P \quad \text{or} \quad \left(\frac{\partial P}{\partial V}\right)_T \left(\frac{\partial V}{\partial P}\right)_T = 1. \tag{11.26}$$

This result is easily visualized. Points corresponding to related values (P, V, T) in the P–V–T space form a surface and the intersection of this surface with the plane $T = $ constant gives the line that is the P–V relationship for that value of T. The two factors on the left-hand side of (11.26) are the rate of change of P with V and the rate of change of V with P along that line. These are clearly reciprocally related.

Another interesting relationship that is derived from (11.25) comes from making $\delta P = 0$, which gives

$$\left\{ \left(\frac{\partial P}{\partial V}\right)_T \left(\frac{\partial V}{\partial T}\right)_P + \left(\frac{\partial P}{\partial T}\right)_V \right\} \delta T = 0.$$

Rearranging and using (11.26) gives

$$\left(\frac{\partial P}{\partial V}\right)_T \left(\frac{\partial V}{\partial T}\right)_P \left(\frac{\partial T}{\partial P}\right)_V = -1, \tag{11.27}$$

a result that is not so easily visualized.

Actually (11.27) is valid for any real equation of state, whether or not it can be expressed in an analytical form.

EXAMPLE 11.6

We can try it out on the van der Waals equation of state for a gas:

$$\left(P + \frac{a}{V^2}\right)(V - b) = RT. \tag{11.28}$$

For 1 mole of gas the constant R is the gas constant, $8.314 \, \mathrm{J\,K^{-1}\,mole^{-1}}$. With $a=0$ and $b=0$ the equation is that for a perfect gas; the constant a takes into account the forces between the gas molecules and b is related to the volume occupied by the molecules.

Writing

$$P = \frac{RT}{V - b} - \frac{a}{V^2}$$

we find

$$\left(\frac{\partial P}{\partial V}\right)_T = \frac{2a}{V^3} - \frac{RT}{(V - b)^2} = \frac{2a}{V^3} - \frac{(P + a/V^2)}{V - b}. \tag{11.29a}$$

From

$$T = \frac{1}{R}\left(P + \frac{a}{V^2}\right)(V - b)$$

we obtain

$$\left(\frac{\partial T}{\partial P}\right)_V = \frac{V - b}{R} \tag{11.29b}$$

and

$$\left(\frac{\partial T}{\partial V}\right)_P = \frac{1}{R}\left(P + \frac{a}{V^2}\right) - \frac{2a}{RV^3}(V - b).$$

Using (11.26), the last result gives

$$\left(\frac{\partial V}{\partial T}\right)_P = R\left\{\left(P + \frac{a}{V^2}\right) - \frac{2a}{V^3}(V - b)\right\}^{-1} \tag{11.29c}$$

From (11.29a), (11.29b), and (11.29c)

$$\left(\frac{\partial P}{\partial V}\right)_T \left(\frac{\partial T}{\partial P}\right)_V \left(\frac{\partial V}{\partial T}\right)_P = -1,$$

which confirms the result (11.27).

As previously mentioned, any physically reasonable equation of state must give this result because it is a statement of a mathematical identity.

Exercise 11.5
The Berthelot equation of state is $P = \dfrac{RT}{V - b} - \dfrac{a}{TV^2}$. Find

$\left(\dfrac{\partial P}{\partial V}\right)_T$ and $\left(\dfrac{\partial P}{\partial T}\right)_V$ and hence $\left(\dfrac{\partial V}{\partial T}\right)_P$.

11.5 Taylor series for a function of two variables

The Taylor series was described in §4.7 applied to a function with a single variable. Here we give the Taylor series for the function of a single variable in a form that is better suited for describing the extension to two variables. This is

$$f(x + h) = f(x) + f'(x)h + f''(x)\frac{h^2}{2!} + f'''(x)\frac{h^3}{3!} + \cdots. \tag{11.30}$$

Now we are going to consider how to derive an expression for $f(x+h, y+k)$ in terms of $f(x, y)$ and various partial derivatives of the function at the point (x, y). We shall restrict ourselves to deriving the first few terms of the series expansion, which will establish the pattern for the whole series.

We start with $f(x+h, y+k)$ and consider that $y+k$ is fixed. We can now use the Taylor series for a single variable but since we are keeping $y+k$ fixed the differentials we use are partial differentials – thus

$$f(x + h, y + k) = f(x, y + k) + \left(\frac{\partial f}{\partial x}\right)_{(x, y+k)} h + \left(\frac{\partial^2 f}{\partial x^2}\right)_{(x, y+k)} \frac{h^2}{2!} + \cdots. \tag{11.31}$$

Now we expand $f(x, y+k)$ around the point (x, y). Since x is kept fixed the differentials are partial differentials with respect to y so that

$$f(x, y + k) = f(x, y) + \frac{\partial f}{\partial y}k + \frac{\partial^2 f}{\partial y^2}\frac{k^2}{2!} + \cdots \tag{11.32}$$

where subscripts are now dropped from the partial differentials since they relate to the central point (x, y). In (11.31) we have two other terms that do not relate to the central point but we can expand them around the central point using the Taylor theorem. Thus, dropping the subscript for differentials at the central point and terminating at second derivatives we find

$$\left(\frac{\partial f}{\partial x}\right)_{(x, y+k)} = \frac{\partial f}{\partial x} + \frac{\partial^2 f}{\partial x\,\partial y}k + \cdots \tag{11.33a}$$

and

$$\left(\frac{\partial^2 f}{\partial x^2}\right)_{(x, y+k)} = \frac{\partial^2 f}{\partial x^2} + \cdots. \tag{11.33b}$$

Inserting (11.32), (11.33a), and (11.33b) into (11.31) gives

$$f(x+h, y+k) = f(x, y) + \left(\frac{\partial f}{\partial x} h + \frac{\partial f}{\partial y} k \right)$$
$$+ \frac{1}{2!} \left(\frac{\partial^2 f}{\partial x^2} h^2 + 2 \frac{\partial^2 f}{\partial x \, \partial y} hk + \frac{\partial^2 f}{\partial y^2} k^2 \right) + \cdots . \tag{11.34}$$

A fuller analysis gives the next term in the series as

$$\frac{1}{3!} \left(\frac{\partial^3 f}{\partial x^3} h^3 + 3 \frac{\partial^3 f}{\partial^2 x \, \partial y} h^2 k + 3 \frac{\partial^3 f}{\partial x \, \partial y^2} hk^2 + \frac{\partial^3 f}{\partial y^3} k^3 \right) . \tag{11.35}$$

The general pattern of the sequence of partial differentials and the powers of h and k associated with them for the $(n+1)$th term in the series should be clear from (11.35), which gives the fourth term. The numerical coefficients within the bracket are the coefficients in the expansion of $(x+y)^n$.

EXAMPLE 11.7

We now give a numerical application of (11.34) truncated at the third term with the function $f(x, y) = \exp(x/y)$. The various partial differentials are:

$$\frac{\partial f}{\partial x} = \frac{1}{y} \exp\left(\frac{x}{y}\right); \quad \frac{\partial f}{\partial y} = -\frac{x}{y^2} \exp\left(\frac{x}{y}\right); \quad \frac{\partial^2 f}{\partial x^2} = \frac{1}{y^2} \exp\left(\frac{x}{y}\right);$$

$$\frac{\partial^2 f}{\partial y^2} = \left(\frac{2x}{y^3} + \frac{x^2}{y^4} \right) \exp\left(\frac{x}{y}\right); \quad \frac{\partial^2 f}{\partial x \, \partial y} = \left(-\frac{1}{y^2} - \frac{x}{y^3} \right) \exp\left(\frac{x}{y}\right).$$

We shall find a value for $f(0.1, 1.1)$ centred on the value at $x=0$ and $y=1$. At the central values $f = 1$,

$$\frac{\partial f}{\partial x} = 1, \quad \frac{\partial f}{\partial y} = 0, \quad \frac{\partial^2 f}{\partial x^2} = 1, \quad \frac{\partial^2 f}{\partial y^2} = 0 \text{ and } \quad \frac{\partial^2 f}{\partial x \, \partial y} = -1.$$

Hence

$$f(0.1, 1.1) = 1 + (1 \times 0.1 + 0 \times 0.1)$$
$$+ 0.5(1 \times 0.1^2 + 2 \times -1 \times 0.1^2 + 0 \times 0.1^2)$$
$$= 1.095.$$

The true value of $\exp(0.1/1.1)$ is $1.095\,17$.

Exercise 11.6

If $f(x, y) = x^2 y^3$ then, using the Taylor series to three terms, find the value of $f(1.1, 0.9)$ centred on the value at $(x, y) = (1, 1)$.

11.6 **Maxima and minima in a multidimensional space**

The conditions for the maxima and minima of one-dimensional functions were dealt with in §4.5. For multidimensional space the configurations of functions can be much more complicated than in the one-dimensional case. Here we shall restrict discussion to a space of two dimensions but then by extrapolation deduce the general conditions for maxima and minima in any number of dimensions.

A well-known example of a two-dimensional function is height as a function of position, often portrayed on maps in the form of contour lines. Figure 11.2 shows some contours relating to different situations where the function shows extremum properties of some kind.

Figure 11.2a shows an isolated maximum. Taking a section through the peak along either x or y will show a profile as shown in Figure 11.2d. From the

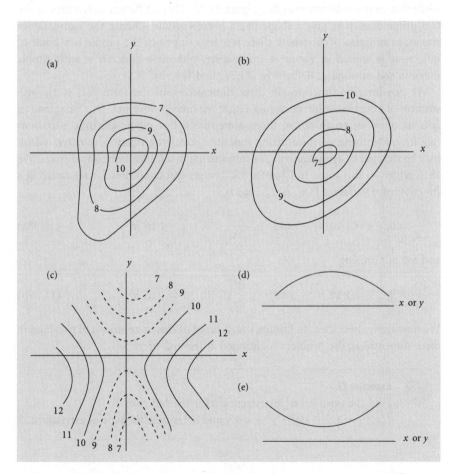

Figure 11.2 Contours corresponding to (a) a maximum, (b) a minimum, and (c) a saddle. Section profile corresponding to (d) a maximum and (e) a minimum.

discussion in §4.5 the condition for a maximum is

$$\frac{\partial f}{\partial x} = 0, \quad \frac{\partial f}{\partial y} = 0, \quad \frac{\partial^2 f}{\partial x^2} < 0, \quad \text{and} \quad \frac{\partial^2 f}{\partial y^2} < 0. \tag{11.36}$$

Figure 11.2b shows an isolated minimum. Taking a section through the peak along either x or y will show a profile as shown in Figure 11.2e. From this we find the condition for a minimum is

$$\frac{\partial f}{\partial x} = 0, \quad \frac{\partial f}{\partial y} = 0, \quad \frac{\partial^2 f}{\partial x^2} > 0, \quad \text{and} \quad \frac{\partial^2 f}{\partial y^2} > 0. \tag{11.37}$$

Other situations exist that give a zero slope at a point in both the x- and y-directions but do not correspond either to a maximum or minimum. Figure 11.2c shows a **saddle configuration**. The profile in the x-direction is like Figure 11.2e while that in the y-direction is like Figure 11.2d. These different curvatures in the two prime directions give a shape like a horse's saddle – hence the name. Other situations can also occur where there is a zero slope but the profile is a point of inflection as shown in Figure 4.3. However, our main concern is with simple maxima and minima as defined in (11.36) and (11.37).

We can imagine functions in three dimensions, of the form $f(x, y, z)$, with surfaces of constant value enclosing either maxima or minima. Our imagination fails us once we go to above three dimensions, but many scientific situations require maximizing or minimizing functions involving many parameters which may be thought of as maximizing or minimizing in a multidimensional space. We shall return to this topic in Chapter 35. The general condition for maximizing a function of n variables $f(x_1, x_2, \ldots, x_n)$ is

$$\frac{\partial f}{\partial x_i} = 0, \quad i = 1 \text{ to } n \quad \text{and} \quad \frac{\partial^2 f}{\partial x_i^2} < 0, \quad i = 1 \text{ to } n. \tag{11.38a}$$

and for minimizing

$$\frac{\partial f}{\partial x_i} = 0, \quad i = 1 \text{ to } n \quad \text{and} \quad \frac{\partial^2 f}{\partial x_i^2} > 0, \quad i = 1 \text{ to } n. \tag{11.38b}$$

We now apply these ideas to finding the distance of closest approach of two lines in three dimensions, the problem mentioned at the end of §9.5.

Exercise 11.7
Find the position of an extremum for the function
$f(x, y) = 3x^2 - 4xy + y^2 + x - y$ and determine the type of extremum.

11.6.1 The closest approach of two lines

The lines we are considering are those through the pairs of points

$$\mathbf{u} = -\mathbf{i} - \mathbf{j} + \mathbf{k}, \quad \mathbf{v} = \mathbf{i} + 2\mathbf{j} + 3\mathbf{k}$$

and

$$\mathbf{p} = -\mathbf{j} + 3\mathbf{k}, \quad \mathbf{q} = \mathbf{i} + \mathbf{j} + 4\mathbf{k}.$$

In parametric form, these lines are

$$\mathbf{r}_1 = (1 - s)\mathbf{u} + s\mathbf{v} = (2s - 1)\mathbf{i} + (3s - 1)\mathbf{j} + (2s + 1)\mathbf{k} \tag{11.39a}$$

and

$$\mathbf{r}_2 = (1 - t)\mathbf{p} + t\mathbf{q} = t\mathbf{i} + (2t - 1)\mathbf{j} + (3 + t)\mathbf{k}. \tag{11.39b}$$

We denote the distance between the two points, defined by s for the first line and t for the second line, as $D(s, t)$. Then

$$D(s, t)^2 = (2s - 1 - t)^2 + (3s - 2t)^2 + (2s - 2 - t)^2. \tag{11.40}$$

For a minimum

$$\frac{\partial(D^2)}{\partial s} = 4(2s - 1 - t) + 6(3s - 2t) + 4(2s - 2 - t) = 0$$

which gives

$$17s - 10t = 6. \tag{11.41a}$$

We must also have

$$\frac{\partial(D^2)}{\partial t} = -2(2s - 1 - t) - 4(3s - 2t) - 2(2s - 2 - t) = 0$$

which gives

$$10s - 6t = 3. \tag{11.41b}$$

The solution of (11.41a) and (11.41b) gives $s = 3$ and $t = 4.5$. The nature of the problem only admits of a minimum, but nevertheless we check to see that these values do give a minimum. Since the values $\partial^2(D^2)/\partial s^2 = 34$ and $\partial^2(D^2)/\partial t^2 = 12$ are both positive, the point is a minimum. Substituting the values of s and t in (11.40) we find

$$D_{\min}^2 = 0.5 \quad \text{or} \quad D_{\min} = 0.707.$$

Problems

11.1 Find $\dfrac{\partial f}{\partial x}, \dfrac{\partial f}{\partial y}, \dfrac{\partial^2 f}{\partial x^2}, \dfrac{\partial^2 f}{\partial y^2}$, and $\dfrac{\partial^2 f}{\partial x \partial y}$ for the following functions:

(i) $f = x^3 + 6x^2y + 2xy^2 + y^3 + 3y + 2x + 6$; (ii) $f = \sin^2(x/y)$;
(iii) $f = \sin^{-1}(x/y)$.

11.2 Given that $f(x, y) = \sin^{-1}(x/y)$, $x = \sin t$, and $y = e^t$, find df/dt.

11.3 If $9x^3 + 8x^2y - 3xy - y^4 - 3 = 0$, find dy/dx at the point $(1, 2)$.

11.4 Show that relationship (11.27) is valid for the perfect gas law $PV = RT$.

11.5 What is the value of $\exp(x^2 + y^2)$ at the point $(1, 1)$? Using this as the centring point use the Taylor expansion (11.34), truncated at second-order terms, to determine the value at the point $(1.1, 0.9)$.

11.6 Lines are defined by the pairs of points

$$\mathbf{u} = 2\mathbf{i} + 2\mathbf{j} - \mathbf{k}, \ \mathbf{v} = \mathbf{i} - \mathbf{j} + \mathbf{k} \quad \text{and} \quad \mathbf{p} = 3\mathbf{i} - \mathbf{j} + \mathbf{k}, \ \mathbf{q} = 2\mathbf{i} - \mathbf{j} - \mathbf{k}.$$

Find the distance of closest approach of the two lines.

Probability and statistics

In this chapter you will be introduced to the concept of probability and how to combine probabilities in particular circumstances. The chapter then moves on to distributions and the various measures by which they may be described. Finally, and of great relevance to experimental scientists, the way of combining different estimates of a quantity, with different measures of reliability, so as to obtain the most reliable combined estimate, is explained.

12.1 What is probability?

The idea of probability is, fortunately, a fairly intuitive one and is certainly understood by all those who indulge in gambling. It is well known that spinning a coin gives equal likelihood of a head or a tail, which mathematically means that the probability of a head is 0.5, as is the probability of a tail. Similarly, in throwing an unbiased die the probability of obtaining a 6 is 1/6, as is the probability of obtaining any other individual number. A roulette wheel contains 37 numbers (0 to 36) but the punter may only place his money on numbers 1 to 36 and is offered odds of 35 to 1 against his number turning up (if successful he will receive 36 times his stake money). If the ball lands on zero then the casino takes all the stakes – that is its profit. Without the zero the punter and the casino would be on equal terms and it would be less a casino than a charity providing free entertainment!

All these examples of *ab initio* probability can be predicted on the basis of the symmetry of the situation and are intellectually based. However, there are other examples of probability that cannot be deduced by thought alone. For example, in 2003 a new virus appeared in Asia, the SARS virus, which carried with it a particular mortality rate. If a victim contracted the virus there was no *ab initio* way of deducing the chance of survival – the only way was to find the proportion of previous victims who had contracted the disease and survived, and assume that it applied to all new cases. This is **empirical probability**, deduced by observation or experiment and, outside the realm of gambling, it is the one that impinges more strongly on everyday activity.

12.2 Combining probabilities

Suppose that there are two events, A and B, that may occur, each with an individual probability, and we ask the question 'What is the combined probability of A and B?'. There is no answer to this question – as it stands it is meaningless – and we have to know more about the relationship of A and B before an answer can be given.

12.2.1 Mutually exclusive events (either – or probability)

We now suppose that A is the probability of throwing a die and obtaining a 1 and B is the probability that on the same throw of the die a 6 is the outcome. Clearly it is impossible for both A and B to occur. They are **mutually exclusive events** – if one of them occurs then the other one cannot occur. A combination of probabilities that is sensible in this case is that of *either* A *or* B occurring – in this case that the throw of the die gives *either* a 1 *or* a 6. The answer in this case is obviously 1/3 since 1 and 6 together constitute 1/3 of the total number of die faces.

To generalize from this particular case, if there are n mutually exclusive events with probabilities $p_1, p_2, \ldots, p_n$ then the probability that *one* of event 1, event 2, up to event n, occurs is

$$P_{\mathrm{me}} = \sum_{i=1}^{n} p_i. \tag{12.1}$$

If, for example, there was a biased die such that the probabilities of obtaining 1, 2, 3, 4, 5, and 6 were 0.15, 0.15, 0.15, 0.18, 0.17, and 0.20 then the probability of throwing a number greater than 3 – i.e *either* a 4 *or* a 5 *or* a 6 – is 0.55, the sum of the last three probabilities.

12.2.2 Independent events (both – and probability)

Events are independent if the probability of one event is not influenced by the occurrence of the other. Spinning a coin gives an equal probability (1/2) of a head or a tail. Throwing an unbiased die gives equal probabilities (1/6) for each of the six possible outcomes. If the coin had been spun before the die was thrown, this would have no influence at all on the outcome of throwing the die. Let us consider the situation that a coin is spun and a die is thrown and we wish to know the probability of obtaining *both* a head *and* a 6. There are 12 possible outcomes from the two actions considered together. They are:

H + 1, H + 2, H + 3, H + 4, H + 5, H + 6, T + 1, T + 2, T + 3, T + 4, T + 5, T + 6

and they are all equally probable. The probability of a head is 1/2 and the probability of a 6 is 1/6, and the joint probability of both together is $1/2 \times 1/6 = 1/12$.

In general if there are n independent outcomes with probabilities $p_1, p_2, \ldots, p_n$ then the probability that all of event 1, event 2 up to event n occur is given by

$$P_i = \prod_{j=1}^{n} p_j. \tag{12.2}$$

As an example we consider the probability that *both* a club is drawn from a pack of cards *and* a 1 is thrown with a die *and* a head comes from spinning a coin. This joint probability is

$$P_i = \frac{1}{4} \times \frac{1}{6} \times \frac{1}{2} = \frac{1}{48}.$$

Strictly speaking the English language does not permit the use of the word *both* in the last example, but we use it to fit with the description of '*both–and*' probability.

> **Exercise 12.1**
> (i) What is the probability of picking either the jack of clubs or the two of diamonds from a pack of cards?
> (ii) What is the probability of both picking the jack of clubs from a pack of cards and getting a head when you spin a coin?

12.3 **Making selections**

Many states run lotteries, which differ in detail but are similar in the way they operate. The UK national lottery is based on selecting 6 numbers from the set 1 to 50. If these 6 numbers are picked at random from a drum containing 50 numbered balls then the lucky lottery player will receive a prize, often some millions of pounds, all for a stake of £1. Clearly there are probability principles involved here, in particular in selecting the 6 numbers from a set of 50.

12.3.1 **Ordered selection**

Suppose we have a bag containing 6 balls, all of different colours, and we wish to know the probability of drawing out of the bag a red ball, a green ball, and a blue ball in that order. The probability of obtaining a red ball on the first selection is clearly 1/6. Once the red ball is removed there are only 5 left, so the probability of next obtaining the green ball is 1/5. This leaves 4 balls in the bag, so that the probability of withdrawing a blue ball on the final selection is 1/4. These three selections are independent events, so that the probability of choosing this ordered arrangement of colours is

$$P_O = \frac{1}{6} \times \frac{1}{5} \times \frac{1}{4} = \frac{1}{120}.$$

Another aspect of this model is to consider the number of ways that three colours can be selected paying regard to order. Now we can say that the number of ways of selecting the first ball is 6, of selecting the second ball is 5, and of selecting the third ball is 4, making the total number of selections $6 \times 5 \times 4 = 120$. Different selections of 3 balls, paying regard to order, are mutually exclusive and of equal probability, so the probability of each of them must be 1/120, as previously found.

The number of selections of 3 objects from 6, paying regard to the order of selection, is

$$N(3,6) = 6 \times 5 \times 4 = \frac{6 \times 5 \times 4 \times 3 \times 2 \times 1}{3 \times 2 \times 1} = \frac{6!}{(6-3)!}$$

where $n!$ (**factorial** n) is the product $1 \times 2 \times 3 \times \ldots \times (n-1) \times n$ (§1.3.1). In general, the number of ways of selecting r objects from n, paying regard to order, is

$$N(r,n) = \frac{n!}{(n-r)!}.\tag{12.3}$$

Exercise 12.2

Ten men, all with different names, are in a room. What is the probability that the first three to leave are Tom, Dick, and Harry in that order?

12.3.2 Selections without order

Suppose that the selection of balls from the bag is made not individually but by taking out 3 balls together (without looking at them – no cheating!). Now no consideration of order arises, but different selections of 3 balls can be made. The probability of picking the red, green, and blue balls is the sum of the probabilities of all the possible ordered selections – RGB, RBG, GRB, GBR, BRG, BGR – which is $6 \times 1/120 = 1/20$. The number of unordered selections is less than the number of ordered selections by the same factor, $6 = 3!$. In general the number of combinations of r objects that can be selected from n *without regard to order* is

$$^nC_r = \frac{n!}{r!(n-r)!}.\tag{12.4}$$

EXAMPLE 12.1

Consider the number of ways of selecting a jury of 12 from 15 individuals. This is

$$^{15}C_{12} = \frac{15!}{12!3!} = \frac{15 \times 14 \times 13}{3 \times 2 \times 1} = 455.$$

It should also be noted that $^nC_r = {}^nC_{n-r}$. The number of ways of selecting 12 jurors from 15 individuals is precisely the same as the number of ways of selecting 3 people who will not be on the jury.

One conclusion we can draw from these selection formulae is that $0! = 1$, a not very obvious result. This definition of $0!$ is necessary if the number of ways of selecting n objects from n, given by (12.4), is to be found as 1.

Exercise 12.3

For the situation described in Exercise 12.2, what is the probability that Tom, Dick, and Harry are the first three to leave without regard to order?

12.4 The birthday problem

There is a well-known problem that neatly illustrates the general ideas of probability associated with selections. The problem is to determine the minimum number of people, chosen at random, that will give a probability over 0.5 that at least 2 of them will have the same birthday. (We assume that a year consists of 365 days, removing the complication of leap years.) First let us consider the number of ways that a group of n people could all have *different* birthdays. The first one can have a birthday in 365 ways. Since all birthdays have to be different the second individual has 364 different possibilities, the third 363, and so on. From this the number of ways that n individuals can have different birthdays is

$$N_{\text{diff}} = \frac{365!}{(365 - n)!}.$$

Next we consider the number of ways that n individuals can have birthdays without any restriction. Since each person can have a birthday in 365 ways the number of different combinations in this case is

$$N_{\text{free}} = 365^n.$$

The probability that n individuals will all have different birthdays is $N_{\text{diff}}/N_{\text{free}}$ and if this is less than 0.5 then the probability that two or more individuals have the same birthday is greater than 0.5. We are thus looking for the smallest value of n that gives

$$\frac{365!}{(365 - n)!365^n} < 0.5. \tag{12.5}$$

The factorials of numbers over 200 or so exceed the capacity of most computers, but this problem can be handled by the use of **Stirling's approximation:**

$$\ln(n!) = n\ln n - n \quad \text{for large } n. \tag{12.6}$$

Taking the natural logarithm of both sides of (12.5) and using Stirling's formula,

$$365\ln 365 - 365 - (365 - n)\ln(365 - n) + (365 - n) - n\ln 365 < \ln 0.5$$

which can be rearranged as

$$(365 - n)\ln\left(\frac{365}{365 - n}\right) - n - \ln(0.5) < 0. \tag{12.7}$$

Finding the minimum value of n to satisfy (12.7) is most straightforwardly done by trial and error, and the process is given in Table 12.1. First n is increased by steps of 10 and this shows that the value required in somewhere between 20 and 30. Then, starting with 21, the value of n is increased by 1 until it is shown that the required value of n is 23. This is much smaller than most intuitive estimates, and shows the importance of making proper probability calculations on matters that are more important than this entertaining problem.

Exercise 12.4
What number of people, chosen at random, will give a greater than 50% chance that at least two of them will have their birthdays in the same month (consider the months as being all equally probable).

Table **12.1** Trial and error determination of n just satisfying (12.7)

n	Expression (12.7)
10	0.555
20	0.135
30	−0.575
21	0.077
22	0.016
23	−0.047

12.5 **Bayes' theorem**

Suppose that a die is thrown. If we could not see the throw we would deduce that the probability of a 6 being the outcome is 1/6, a deduction made on the basis of the symmetry of the problem and an assumption about lack of bias in the die. However, if after the throw we are told that the outcome was larger than 3, then our assessment of the probability of a 6 would be different – again, the symmetry of the problem would indicate that probability in this case is 1/3. In the language of the statistician the probability distribution of outcomes in the first case, with no extra information provided, is the **prior probability distribution** or just **prior**, and the second distribution, in which the condition was applied that the outcome is greater than 3, is known as the **posterior probability distribution** or just **posterior**. For the six possible outcomes of the throw of our die numbers these distributions are

Score	1	2	3	4	5	6
Prior	1/6	1/6	1/6	1/6	1/6	1/6
Posterior	0	0	0	1/3	1/3	1/3.

We now give mathematical form to this discussion. We indicate the **prior probability** of obtaining event A as $P(A)$ and that of obtaining event B as $P(B)$. The **posterior probability** of obtaining A under the condition that B has occurred is $P(A|B)$. The product $P(A|B)P(B)$ is then the probability that both A and B occur but this can also be represented by $P(B|A)P(A)$. This gives

$$P(A|B)P(B) = P(B|A)P(A) \tag{12.8a}$$

or

$$P(A|B) = \frac{P(B|A)P(A)}{P(B)}. \tag{12.8b}$$

This equation, in either of its forms, is known as **Bayes' theorem** and it has many important applications. To translate it into the original die problem:

- $P(A|B)$ is the probability of getting 6 given the outcome is greater than 3, i.e. 1/3.
- $P(B)$ is the prior probability of getting greater than 3, i.e. 1/2.
- $P(B|A)$ is the probability of getting greater than 3 if a 6 is obtained, i.e. 1.
- $P(A)$ is the prior probability of getting a 6, i.e. 1/6.

We now illustrate Bayes' theorem with the following application.

EXAMPLE 12.2

A disease is known to be present in 6% of a population. A test for the disease indicates positive in 95% of those with the disease but also in 1% of those who do not have the disease. If the test on a particular individual is positive, what is the probability that she has the disease?

- The prior probability of an individual with the disease, $P(D) = 0.06$
- The probability of a positive test for someone with the disease, $P(T|D) = 0.95$.

To calculate the prior probability of a positive test we need to take account of positive tests on those who are disease free. Thus

$$P(T) = (0.06 \times 0.95) + (0.94 \times 0.01) = 0.0664.$$

The probability that a positive test indicates someone suffering from the disease is

$$P(D|T) = \frac{P(T|D)P(D)}{P(T)} = \frac{0.95 \times 0.06}{0.0664} = 0.858.$$

Exercise 12.5

A card is drawn from a pack.

(i) What is the prior probability, $P(D)$, that it is a diamond?

(ii) What is the prior probability, $P(NC)$, that it is not a club?

(iii) What is the probability, $P(NC|D)$, that it is not a club if it is known to be a diamond?

(iv) Use Bayes' theorem to find the probability $P(D|NC)$, that the card is a diamond if it is known not to be a club.

12.6 Too much information?

The Ministry of Health wants to find the birthweight of babies born in each year and asks each of its regional offices to provide the information for its area. Each office records the birthweight of all the babies in its area and then sends the list of weights to the ministry. The Minister of Health is now confronted with combined documents containing three-quarters of a million numbers, together about the size of a telephone directory. As she turns the pages she sees the list of weights – 3.124 kg, 3.671 kg, 2.890 kg, etc. – and they convey no useful information other than that most babies weigh about 3 kg at birth, which she knew anyway. She then turns to the list from the previous year and the numbers look no different. She is swamped with information and can learn nothing from it in the way it is presented.

A useful number is the average of the weights. If the regional information was provided in some electronic form suitable for direct computer input then this could quickly be found. A single number, the average, could then be compared with the averages from previous years to see what trends, if any, exist. However, there may be other characteristics of the set of weights that would be of interest, for example the distribution of weights, and this too could be determined from the large number of individual weights once they were in the computer.

Was there any point in the regional offices sending in all the individual weights in the first place? After all, if each office sent in just the number of babies born in its region together with the average weight for the region, then the overall number of babies born nationally and the average weight for all the babies could be ascertained.

The handling of large quantities of data and deriving useful information from them is one of the important roles of **statistics**. Statistical ideas play an important role in science, and in understanding aspects of everyday life, and here we shall concentrate on some common and not-too-complicated aspects of the application of statistics.

12.7 **Mean; variance and standard deviation; median**

12.7.1 **Mean**

If there are N values of a particular quantity, $x_1, x_2, \ldots x_N$, then the average of the values of x is given by

$$\bar{x} = \frac{1}{N} \sum_{i=1}^{N} x_i, \tag{12.9}$$

which is simply the sum of all the quantities x divided by the number of such quantities. The information required for the Minister of Health to know the average birthweight of babies for the whole country is the number of all babies born in each region plus the average weight in each region. The average for the whole country, $\overline{w_T}$, is then given by

$$\overline{w_T} = \frac{\sum_{j=1}^{M} N_j \overline{w_j}}{\sum_{j=1}^{M} N_j}, \tag{12.10}$$

where there are M regions and the total number of babies and average birthweight in the jth region are N_j and $\overline{w_j}$ respectively. This is just the sum of all the birthweights divided by the total number of babies.

12.7.2 **Variance and standard deviation**

The distribution of birthweights has already been mentioned as an important statistical characteristic. A useful measure of the spread of a distribution function, or its **dispersion**, is its **variance**, which is the **mean square deviation from the mean**. For the N quantities that gave (12.9), with mean $\bar{x}$, the variance is

$$V = \frac{1}{N}\sum_{j=1}^{N}\left(x_j - \bar{x}\right)^2, \tag{12.11}$$

Expanding the bracket in the summation,

$$V = \frac{1}{N}\sum_{j=1}^{N}(x_j^2 - 2x_j\bar{x} + \bar{x}^2) = \frac{1}{N}\sum_{j=1}^{N}x_j^2 - 2\bar{x}\frac{1}{N}\sum_{j=1}^{N}x_j + \frac{1}{N}\sum_{j=1}^{N}\bar{x}^2$$

or

$$V = \overline{x^2} - 2\bar{x}\times\bar{x} + \bar{x}^2 = \overline{x^2} - \bar{x}^2. \tag{12.12}$$

Thus the variance is found to be the *average of the square minus the square of the average*. It is important to note the difference between $\overline{x^2}$ and $\bar{x}^2$ in this context.

The variance has the dimensions of the square of the quantities x; a measure of dispersion with the dimensions of x is the **standard deviation**, σ, which is just the square root of the variance, i.e.

$$\sigma = \sqrt{V}. \tag{12.13}$$

To illustrate these quantities we find the mean, variance, and standard deviation of the 15 heights of a sample of male adults given in Table 12.2. The mean height, rounded to a sensible value, is 1.725 m. The standard deviation is given by

$$\sigma = \sqrt{\overline{x^2} - \bar{x}^2} = \sqrt{2.98166 - (1.72467)^2}\ \text{m} = 0.085\ \text{m}. \tag{12.14}$$

For the Ministry of Health to determine the mean and standard deviation of the birthweights for all babies born in the country it is only necessary for each regional office to send in three numbers – the total number of babies born in the region, their mean birthweight, and the standard deviation of their birthweights. If for the jth region the number, mean and standard deviation are N_j, $\bar{w}_j$, and σ_j then, from (12.12) and (12.13),

$$\overline{w_j^2} = \sigma_j^2 + \bar{w}_j^2. \tag{12.15}$$

The mean square birthweight of all the babies from the M regions, $\overline{w_{\text{all}}^2}$, can be found as

$$\overline{w_{\text{all}}^2} = \frac{\displaystyle\sum_{j=1}^{M} N_j\overline{w_j^2}}{\displaystyle\sum_{j=1}^{M} N_j}. \tag{12.16a}$$

Table 12.2 Heights of male adults in a sample

	x (m)	x^2 (m^2)
	1.75	3.0625
	1.68	2.8224
	1.66	2.7556
	1.77	3.1329
	1.59	2.5281
	1.73	2.9929
	1.81	3.2761
	1.79	3.2041
	1.65	2.7225
	1.85	3.4225
	1.60	2.5600
	1.66	2.7556
	1.74	3.0276
	1.70	2.8900
	1.89	3.5721
Sum	25.87	44.7249
Mean	1.72467	2.98166

Similarly the mean birthweight of all the babies is, from (12.9),

$$\overline{w_{\text{all}}} = \frac{\sum\limits_{j=1}^{M} N_j \overline{w_j}}{\sum\limits_{j=1}^{M} N_j}. \tag{12.16b}$$

From (12.16a) and (12.16b) the variance, and hence standard deviation of the birthweight of all the babies is found as

$$V_{\text{all}} = \sigma_{\text{all}}^2 = \overline{w_{\text{all}}^2} - \overline{w_{\text{all}}}^2. \tag{12.17}$$

This calculation is illustrated numerically in Table 12.3 where the first three columns of figures give the input of N_j, $\overline{w_j}$, and σ_j from 10 regions. The fourth column of figures gives $\overline{w_j^2}$ derived from (12.15) and the averages of $\overline{w_j}$ and $\overline{w_j^2}$ are given in the last row. For the whole country 700 641 babies were born with an average weight of 3.060 kg, with standard deviation $\sigma_{\text{all}} = \sqrt{9.424\,592 - 3.059\,87^2}$ kg = 0.249 kg.

When repeated measurements of a particular quantity are made there will normally be a spread of measurements. Intuitively it seems obvious that the mean of many measurements will be more reliable than any single one of them, and that the standard deviation of the measurements will be some measure of the probable

Table 12.3 Data for the birthweights of babies born in different regions

Region	N	$\overline{w}$(kg)	σ (kg)	$\overline{w^2}$(kg^2)
South	66 296	3.062	0.251	9.438 845
South-east	108 515	2.997	0.239	9.039 130
South-west	64 621	3.185	0.267	10.215 514
West	76 225	3.002	0.224	9.062 180
Central	93 496	2.996	0.250	9.038 516
East	41 337	3.099	0.231	9.657 162
North-east	104 212	3.101	0.237	9.672 370
North-west	82 358	3.011	0.226	9.117 197
Far North	37 362	3.167	0.219	10.077 850
Outer Isles	26 219	3.178	0.220	10.148 084
Average (sum)	(700 641)	3.059 87		9.424 592

error. In §12.8 we deal with how to combine different estimates of a quantity to get the best overall estimate.

The concept of variance offers another interpretation of the mean as given in (12.9). The mean can be considered as the value about which the variables have the least mean-square deviation. The mean-square deviation of the variables about a value s is

$$S = \frac{1}{N}\sum_{i=1}^{N}(x_i - s)^2$$

To minimize S we make $dS/ds = 0$, which gives

$$\frac{1}{N}\sum_{i=1}^{N}(x_i - s) = 0 \quad \text{or} \quad s = \frac{1}{N}\sum_{i=1}^{N}x_i = \overline{x}. \tag{12.18}$$

12.7.3 Median

The mean may be thought of as a representative value of a set of quantities, but there are circumstances when some other representative value is more appropriate. The **median** is one such representative value and is the value for which one half of the quantities are greater than the median and one half are less. The median is useful when there are a few extreme quantities that would seriously bias the mean away from being a good representative value. For example, if we take a hypothetical village community of 100 farm workers each earning £15 000 a year with a local manor house occupied by a wealthy individual whose annual income is £1 000 000, then the mean of about £24 750 would be a poor representative value of annual income for the village. The median, £15 000, would be a much more representative figure in this case. The median excludes the effect of **outriders**, values that are either much larger or much smaller than the bulk of the quantities observed or measured. Sometimes in a scientific context there may be several estimates of a quantity and if there are one or more outriders then it is both sensible and legitimate to exclude these from the mean – or, perhaps, to take a median value.

Exercise 12.6

Find (i) the mean, (ii) the median, (iii) the variance, and (iv) the standard deviation of the following set of numbers:

13	20	27	21	9	16	23	26	8	11	17
11	27	26	20	13	24	10	9	24	19	

12.8 Combining different estimates

It can happen that several estimates may be available for the value of a given quantity, say the speed of light, found either by different experimenters or by different kinds of experiment. If each estimate of value is accompanied by an estimate of the probable size of the error associated with it, then a 'best' estimate can be found by combining the individual estimates. It is assumed that the error estimates are in the form of the standard deviations that would be expected if the experiment were repeated many times.

Before we can determine how to find a best estimate we need to establish three simple rules.

Rule 1

If we have a number of estimates, $x_1, x_2, \ldots, x_n$ of a quantity, with weights $w_1, w_2, \ldots, w_n$ then a weighted average of those estimates

$$\bar{x} = \sum_{j=1}^{n} w_j x_j, \qquad (12.19)$$

where

$$\sum_{j=1}^{n} w_j = 1, \qquad (12.20)$$

constitutes an unbiased estimate of the quantity.

Rule 2

If we have a set of quantities $x_1, x_2, \ldots, x_n$ with variance V and standard deviation σ then the set of quantities $kx_1, kx_2, \ldots, kx_n$ has variance $k^2 V$ and standard deviation $k\sigma$. This comes directly from the definitions of variance in (12.11) and (12.12).

Rule 3

The variance of a sum of quantities is the sum of the variances of the individual quantities. If the variance of estimate x_j is V_j then this states that

$$\text{the variance of} \left(\sum_{j=1}^{N} x_j \right) = \sum_{j=1}^{N} V_j \qquad (12.21)$$

However, an important condition for (12.21) to be true is that all the quantities x must be independent, which is to say that the value found for x_i is in no way influenced by the value found for x_j, and vice versa. For *independent* quantities one may write

$$\overline{x_i x_j} = \overline{x_i} \times \overline{x_j} \tag{12.22}$$

or, in words, *the average of the product is the product of the averages.*

From the definition of variance, given in (12.12), and the condition of independence of the quantities being summed, as expressed by (12.22), the validity of (12.21) can be shown. If

$$S = \sum_{j=1}^{N} x_j \tag{12.23}$$

then the variance of S is

$$V_S = \overline{\left(\sum_{j=1}^{N} x_j\right)^2} - \left(\overline{\sum_{j=1}^{N} x_j}\right)^2. \tag{12.24}$$

We have, using (12.22),

$$\overline{\left(\sum_{j=1}^{N} x_j\right)^2} = \sum_{j=1}^{N} \overline{x_j^2} + \sum_{i=1}^{N} \sum_{\substack{i \neq j \\ j=1}}^{N} \overline{x_i x_j} = \sum_{j=1}^{N} \overline{x_j^2} + \sum_{i=1}^{N} \sum_{\substack{i \neq j \\ j=1}}^{N} \overline{x_i x_j}$$

and

$$\left(\overline{\sum_{j=1}^{N} x_j}\right)^2 = \sum_{j=1}^{N} \overline{x_j}^2 + \sum_{i=1}^{N} \sum_{\substack{i \neq j \\ j=1}}^{N} \overline{x_i}\,\overline{x_j}.$$

Hence, since $\overline{x_i x_j} = \overline{x_i}\,\overline{x_j}$ for independent quantities,

$$V_S = \sum_{j=1}^{N} \left(\overline{x_j^2} - \overline{x_j}^2\right) = \sum_{j=1}^{N} V_j, \tag{12.25}$$

which is the required result.

We now seek a weighted sum of the different estimates of the quantity x, in the form (12.19), that will have the least variance because this will be the best estimate that can be made. The individual terms of (12.19) will have variances of the form $w_j^2 V_j$ and from (12.25) the variance of $\overline{x}$ will be

$$V_{\overline{x}} = \sum_{j=1}^{N} w_j^2 V_j \tag{12.26}$$

The requirement is to find a set of weights, w, that will minimize (12.26) subject to the condition (12.20). If V_S is a minimum then we must have the condition that

$$\frac{\partial V_S}{\partial w_j} = 0 \quad \text{for all } j.$$

From (12.26) this gives

$$2 \sum_{i=1}^{N} w_i \frac{\partial w_i}{\partial w_j} V_i = 0 \quad \text{for all } j \tag{12.27}$$

and to maintain condition (12.20) we must have

$$\sum_{j=1}^{N} \frac{\partial w_i}{\partial w_j} = 0 \quad \text{for all } j. \tag{12.28}$$

Whatever j is chosen these equations are both satisfied by the condition that

$$w_i V_i = C, \quad \text{a constant} \quad \text{or} \quad w_i = \frac{C}{V_i}.$$

Substituting these values in (12.20) we find

$$C = \left(\sum_{j=1}^{N} \frac{1}{V_j} \right)^{-1} \tag{12.29}$$

and hence

$$w_i = \frac{1}{V_i} \left(\sum_{j=1}^{N} \frac{1}{V_j} \right)^{-1}. \tag{12.30}$$

This is known as **inverse-variance weighting** and gives a weighted mean with the minimum variance. Substituting from (12.30) into (12.26) the minimum variance is found to be C, as given in (12.29).

The way that this weighting scheme works can be shown by a simple example. Consider three estimates of the speed of light:

$$c_1 = 2.995 \times 10^8 \, \text{m s}^{-1}, \quad \sigma_1 = 0.011 \times 10^8 \, \text{m s}^{-1}$$
$$c_2 = 3.05 \times 10^8 \, \text{m s}^{-1}, \quad \sigma_2 = 0.06 \times 10^8 \, \text{m s}^{-1}$$
$$c_3 = 2.999 \times 10^8 \, \text{m s}^{-1}, \quad \sigma_3 = 0.006 \times 10^8 \, \text{m s}^{-1}.$$

The best estimate, from inverse variance weighting, is, in units of $10^8 \, \text{m s}^{-1}$,

$$\frac{\dfrac{1}{0.011^2} \times 2.995 + \dfrac{1}{0.06^2} \times 3.05 + \dfrac{1}{0.006^2} \times 2.999}{\dfrac{1}{0.011^2} + \dfrac{1}{0.06^2} + \dfrac{1}{0.006^2}} = 2.998.$$

The variance of this estimate, equal to C in (10.28) is, in units of $10^{16} \, \text{m}^2 \, \text{s}^{-2}$,

$$V_{\bar{c}} = \frac{1}{\dfrac{1}{0.011^2} + \dfrac{1}{0.06^2} + \dfrac{1}{0.006^2}} = 2.753 \times 10^{-5}$$

equivalent to a standard deviation of $0.0052 \times 10^8 \, \text{m s}^{-1}$, smaller than the standard deviation of any of the individual estimates.

In any situation where there are several estimates for a particular quantity inverse-variance weighting should be used to find a best estimate. Even if the estimated standard deviations can only be crudely defined, as long as their relative values are reasonably correct an estimate should be found that is more reliable than any individual estimate.

Exercise 12.7
Two estimates of a quantity, with their standard deviations in parentheses, are 15.2 (2.7) and 16.1 (4.2). Find the best estimate of the quantity and its standard deviation.

Problems

12.1 Two dice are thrown and two coins are spun. What is the probability that the sum on the two dice is 7 and that both coins give a head?

12.2 A bridge club has 15 members and is providing a team of 4 for a national tournament. In how many different ways could the club provide its representative team?

12.3 The planet Sumod has a year divided into 100 days. What is the minimum number of Sumodians that would give a more than 50% probability that at least two of them would have birthdays on the same day?

12.4 It is known that 0.01% of the screws made by a particular machine are faulty. A test procedure indicates 99% of faulty screws as faulty although it also indicates some good screws as faulty. If testing a large number of screws gave a fault rate of 0.011% then what is the probability that a screw tested as faulty is actually faulty?

12.5 The six districts of a country send in annual rainfall reports in the form of the mean and standard deviation of the daily rainfall. The submissions for the year are shown in Table 12P.1 Find the mean and standard deviation of the daily rainfall for the whole country.

Table 12P.1 Data for the rainfall in different districts

District	Mean (mm)	σ (mm)
1	2.11	0.85
2	2.50	0.77
3	1.43	0.49
4	1.97	0.52
5	2.24	1.00
6	3.12	1.12

12.6 The acceleration due to gravity, g, is measured at a particular site by three different methods. The results (with estimated standard deviations in parentheses) are:

Simple pendulum   9.841 (0.020) m s^{-2}

Compound pendulum 9.8152 (0.0071) m s^{-2}

Timing a falling sphere 9.8231 (0.0112) m s^{-2}

Determine the best estimate of g with its standard deviation.

Coordinate systems and multiple integration

13

In Chapter 1 we defined two types of coordinate system for describing the position of a point in a plane. This chapter begins by illustrating the processes of multidimensional integration just using the two-dimensional coordinate systems. Then three different coordinate systems are described for three dimensions, together with the processes of integration in each of them. Finally, these ideas are applied to determining the moments of inertia for various one-, two-, and three-dimensional bodies.

13.1 Two-dimensional coordinate systems

Two-dimensional coordinate systems have already been described in §1.6. The most useful coordinate systems in two dimensions are the rectangular Cartesian system and the polar coordinate system. The usefulness of the latter in describing orbital motion became apparent in Chapter 10. Here we shall extend our consideration of these coordinate systems to their applications in integration in a two-dimensional space.

13.2 Integration in a rectangular Cartesian system

First we consider a function $f(x, y)$ in a two-dimensional space defined by a rectangular Cartesian system. If we wish to integrate $f(x, y)$ over a particular area then we can write the integral in the form

$$I = \int_A f(x, y) \, \mathrm{d}A \tag{13.1}$$

where A is the area over which the integral is required and $\mathrm{d}A$ is a small element of area, corresponding to $\mathrm{d}x$ in a one-dimensional integral. Where there are two independent variables then the elemental area is defined by taking a small increment in the x direction, dx, and a small increment in the y direction, dy, to give

$$dA = dx\,dy. \tag{13.2}$$

Now let us suppose that the area of integration is that x goes from x_1 to x_2 and y goes from y_1 to y_2. Then the integral is written as

$$I = \int_{x_1}^{x_2} \int_{y_1}^{y_2} f(x,y) dy\,dx. \tag{13.3}$$

It helps to visualize an integral as a summation over a vast number of tiny elemental areas within each of which the value of the function depends on its position. We now break up the double integral (13.3) as follows:

$$I = \int_{x_1}^{x_2} \left\{ \int_{y_1}^{y_2} f(x,y) dy \right\} dx. \tag{13.4}$$

If the one-dimensional integral within the curly brackets is evaluated, assuming that x is a constant, then the resultant integral is a function of x, $g(x)$, that multiplied by dx gives the contribution to the integral of a strip between x and $x + dx$. Finally the integration over x will complete the double integration process. This may be described as

$$g(x) = \int_{y_1}^{y_2} f(x,y) dy, \tag{13.5a}$$

followed by

$$I = \int_{x_1}^{x_2} g(x) dx. \tag{13.5b}$$

Notice that the order in which the limits are presented in the integral signs is the reverse of the order in which the incremental quantities dx and dy appear. The ordering is dictated by the way in which brackets can be inserted to give (13.4) from (13.3).

The following example shows how this process works.

EXAMPLE 13.1

We consider the double integral

$$I = \int_0^1 \int_0^2 (x^2 + y^2) dy\,dx. \tag{13.6}$$

First we find

$$g(x) = \int_0^2 (x^2 + y^2) dy = \left| x^2 y + \frac{1}{3} y^3 \right|_0^2 = 2x^2 + \frac{8}{3}, \tag{13.7a}$$

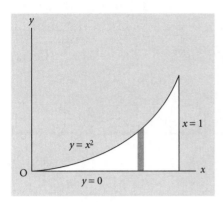

Figure 13.1 The area of integration for (13.8).

followed by

$$I = \int_0^1 \left(2x^2 + \frac{8}{3}\right)dx = \left.\frac{2}{3}x^3 + \frac{8}{3}x\right|_0^1 = \frac{10}{3}. \tag{13.7b}$$

You should confirm that changing the order of the integration to 'first x then y' gives the same result.

You should not think that because rectangular Cartesian coordinates are being used the integration area is restricted to the form of a rectangle. For example, we may be required to integrate for an area bounded by $x=1$, $y=0$, and $y=x^2$, as shown in Figure 13.1. For a particular value of x the limits of y are 0 and x^2. Taking the same integrand as in (13.6), the integral can be written as

$$I = \int_0^1 \int_0^{x^2} (x^2 + y^2)dy\, dx \tag{13.8}$$

and can be split up as

$$g(x) = \int_0^{x^2} (x^2 + y^2)dy = \left.x^2 y + \frac{1}{3}y^3\right|_0^{x^2} = x^4 + \frac{1}{3}x^6, \tag{13.9a}$$

followed by

$$I = \int_0^1 \left(x^4 + \frac{1}{3}x^6\right)dx = \left.\frac{1}{5}x^5 + \frac{1}{21}x^7\right|_0^1 = \frac{1}{5} + \frac{1}{21} = \frac{26}{105}. \tag{13.9b}$$

The quantity $g(x)dx$ gives the integral over the grey strip in Figure 13.1, which has width dx and height x^2. Finally, integral (13.9b) sums over all the strips in the required region.

The order in which the integration is done can sometimes affect the ease with which the calculation is carried out.

EXAMPLE 13.2

By taking horizontal strips instead of vertical strips in Figure 13.1 the limits of x, for a particular value of y, are $\sqrt{y}$ and 1, and the limits of y are 0 and 1. The integral now appears as

$$I = \int_0^1 \left\{\int_{\sqrt{y}}^1 (x^2 + y^2)dx\right\}dy. \tag{13.10}$$

The integral in the curly brackets is

$$g(y) = \left. \frac{1}{3}x^3 + y^2 x \right|_{\sqrt{y}}^{1} = \frac{1}{3} + y^2 - \frac{1}{3}y^{3/2} - y^{5/2}. \tag{13.11a}$$

Finally,

$$I = \int_0^1 \left(\frac{1}{3} + y^2 - \frac{1}{3}y^{3/2} - y^{5/2} \right) dy = \left. \frac{1}{3}y + \frac{1}{3}y^3 - \frac{2}{15}y^{5/2} - \frac{2}{7}y^{7/2} \right|_0^1$$

$$= \frac{1}{3} + \frac{1}{3} - \frac{2}{15} - \frac{2}{7} = \frac{26}{105}, \tag{13.11b}$$

as before.

The method of integrating first over y was somewhat simpler, but the advantage was marginal in this case. Sometimes the order of integration can make a large difference and a prior examination of a double, or multiple, integration to see whether there is an easy order can be worthwhile.

Exercise 13.1
Evaluate (i) $\int_0^1 \int_0^1 xy \, dy \, dx$; (ii) $\int_0^1 \int_0^2 (x + y^2) dy \, dx$.

13.2.1 Separable and inseparable variables

If the function in (13.1) can be separated into a product of two functions, one only involving x and the other involving only y, as in Exercise 13.1(i), then the double integral can be calculated as a product of two one-dimensional integrals:

$$\int_{x_1}^{x_2} \int_{y_1}^{y_2} f(x, y) dy \, dx = \int_{x_1}^{x_2} f_x(x) dx \int_{y_1}^{y_2} f_y(y) dy. \tag{13.12}$$

Actually this was possible for the integral given in (13.6), which could be written as

$$I = \int_0^1 \int_0^2 (x^2 + y^2) dy \, dx = \int_0^1 x^2 dx \int_0^2 dy + \int_0^1 dx \int_0^2 y^2 dy$$

$$= \left(\frac{1}{3} \times 2 \right) + \left(1 \times \frac{8}{3} \right) = \frac{10}{3}.$$

Another example involving **separable variables** is

$$\int_0^1 \int_0^{\pi/2} x^2 \cos y \, dy \, dx = \int_0^1 x^2 dx \int_0^{\pi/2} \cos y \, dy = \left. \frac{1}{3}x^3 \right|_0^1 |\sin y|_0^{\pi/2} = \frac{1}{3} \times 1 = \frac{1}{3}.$$

An example involving **inseparable variables** is

$$\int_0^{\pi/2} \int_0^1 x \cos(xy) dy \, dx = \int_0^{\pi/2} \left\{ \int_0^1 x \cos(xy) dy \right\} dx = \int_0^{\pi/2} |\sin(xy)|_0^1 dx$$

$$= \int_0^{\pi/2} \sin x \, dx = 1.$$

Exercise 13.2

Evaluate $\int_0^1 \int_0^2 x(x+y) dy\ dx$.

13.3 Integration with polar coordinates

For many problems, especially those involving either circular symmetry of the function or of the region of integration, or both, the use of polar coordinates is often required or preferred. For the rectangular Cartesian system the elemental volume was the region defined by the region between x and $x + dx$ and between y and $y + dy$. Similarly, for polar coordinates the elemental volume is defined by the region between r and $r + dr$ and between θ and $\theta + d\theta$. This is shown in Figure 13.2. For small dr and $d\theta$ the elemental area is closely rectangular with sides dr and $rd\theta$, so that

$$dA = r\ dr\ d\theta. \tag{13.13}$$

We now consider a function

$$f(r, \theta) = e^r \cos^2 \theta \tag{13.14}$$

to be integrated over an area bounded by a circle of radius 2 units centred on the origin. Since the variables are separable the integral is

$$I = \int_0^{2\pi} \int_0^2 e^r \cos^2 \theta\ r\ dr\ d\theta = \int_0^2 re^r dr \int_0^{2\pi} \cos^2 \theta\ d\theta. \tag{13.15}$$

For the first integral we use (5.44) to find

$$\int_0^2 re^r dr = |re^r - e^r|_0^2 = e^2 + 1. \tag{13.16a}$$

For the second integral

$$\int_0^{2\pi} \cos^2 \theta\ d\theta = \int_0^{2\pi} \frac{1}{2}(1 + \cos 2\theta) d\theta = \left|\frac{1}{2}\theta + \frac{1}{4}\sin 2\theta\right|_0^{2\pi} = \pi. \tag{13.16b}$$

Combining (13.16a) and (13.16b),

$$I = \pi(e^2 + 1). \tag{13.17}$$

Exercise 13.3

Evaluate $\int_0^{\pi/2} \int_0^1 r^2 \cos \theta\ r\ dr\ d\theta$.

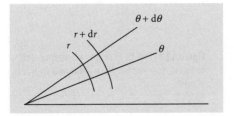

Figure 13.2 The elemental area for polar coordinates.

13.4 Changing coordinate systems

The only two-dimensional coordinate systems we are considering are the rectangular Cartesian and polar coordinate systems, and we have considered them separately using the obvious elemental area representations. However, in principle other two-dimensional coordinate systems are possible and we shall now consider a generalized approach to changing from a rectangular Cartesian system to any other two-dimensional coordinate system.

We consider a change to a coordinate system (u, v) where

$$x = x(u, v) \text{ and } y = y(u, v). \tag{13.18}$$

The area in (u, v) space is delineated between u and $u+\delta u$ and v and $v+\delta v$ so we now find the positions in the (x, y) space of the corners of the area corresponding to the points (u, v), $(u+\delta u, v)$, $(u, v+\delta v)$, and $(u+\delta u, v+\delta v)$. To do this we use the Taylor series expansion in the forms given by (11.32) and (11.34) taken to two terms, recognizing that δu and δv are both very small. These give points in (x, y) space with the following coordinates:

(i) $x(u, v), \quad y(u, v)$

(ii) $x(u, v) + \dfrac{\partial x}{\partial u}\delta u, \quad y(u, v) + \dfrac{\partial y}{\partial u}\delta u$

(iii) $x(u, v) + \dfrac{\partial x}{\partial v}\delta v, \quad y(u, v) + \dfrac{\partial y}{\partial v}\delta v$

(iv) $x(u, v) + \dfrac{\partial x}{\partial u}\delta u + \dfrac{\partial x}{\partial v}\delta v, \quad y(u, v) + \dfrac{\partial y}{\partial u}\delta u + \dfrac{\partial y}{\partial v}\delta v. \tag{13.19}$

These points are shown plotted in Figure 13.3.

From the expressions (13.19), which give the increments in x and y in going from one point to the next, it is evident that the small area delineated is in the form of a parallelogram. This parallelogram is defined by the vectors $\mathbf{p}$ and $\mathbf{q}$ shown in Figure 13.3, which in terms of unit vectors along x and y are

$$\mathbf{p} = \frac{\partial x}{\partial u}\delta u \mathbf{i} + \frac{\partial y}{\partial u}\delta u \mathbf{j} \tag{13.20a}$$

and

$$\mathbf{q} = \frac{\partial x}{\partial v}\delta v \mathbf{i} + \frac{\partial y}{\partial v}\delta v \mathbf{j}. \tag{13.20b}$$

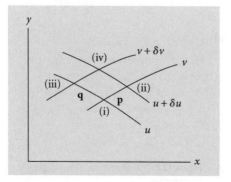

Figure 13.3 The (x, y) points corresponding to (u, v), $(u+\delta u, v)$, $(u, v+\delta v)$ and $(u+\delta u, v+\delta v)$.

From (9.12b) the area of the parallelogram is

$$dA = \text{magnitude of } \mathbf{p} \times \mathbf{q} = \text{magnitude of} \begin{vmatrix} \mathbf{i} & \mathbf{j} & \mathbf{k} \\ \dfrac{\partial x}{\partial u}\delta u & \dfrac{\partial y}{\partial u}\delta u & 0 \\ \dfrac{\partial x}{\partial v}\delta v & \dfrac{\partial y}{\partial v}\delta v & 0 \end{vmatrix}$$

$$= \left(\frac{\partial x}{\partial u}\frac{\partial y}{\partial v} - \frac{\partial y}{\partial u}\frac{\partial x}{\partial v}\right)\delta u \delta v = \begin{vmatrix} \dfrac{\partial x}{\partial u} & \dfrac{\partial x}{\partial v} \\ \dfrac{\partial y}{\partial u} & \dfrac{\partial y}{\partial v} \end{vmatrix}\delta u \delta v. \tag{13.21}$$

The coefficient of $\delta u \delta v$ is known as the **Jacobian of the transformation**, sometimes written as $\frac{\partial(x,y)}{\partial(u,v)}$ or $J\left(\frac{x,y}{u,v}\right)$ (and pronounced with the accent on the second syllable). Note that it is unaffected by interchanging rows and columns.

We can now apply this to the transformation from (x, y) to (r, θ). We start with $x = r\cos\theta$ and $y = r\sin\theta$, which give

$$\frac{\partial x}{\partial r} = \cos\theta, \quad \frac{\partial y}{\partial r} = \sin\theta, \quad \frac{\partial x}{\partial \theta} = -r\sin\theta, \quad \text{and} \quad \frac{\partial y}{\partial \theta} = r\cos\theta.$$

Now, writing $\delta u \delta v$ as $dr\, d\theta$, the elemental volume in (r, θ) space is

$$dA_{r,\theta} = \begin{vmatrix} \cos\theta & -r\sin\theta \\ \sin\theta & r\cos\theta \end{vmatrix} dr\, d\theta = r\, dr\, d\theta, \tag{13.22}$$

which is what was found in (13.13).

> **Exercise 13.4**
> A two-dimensional set of Cartesian axes $X'Y'$ is produced from another set XY by rotation about the origin through an angle α. Coordinates (x, y) are related to coordinates (x', y') by
>
> $$x = x'\cos\alpha - y'\sin\alpha \quad \text{and} \quad y = x'\sin\alpha + y'\cos\alpha.$$
>
> Find the Jacobian of the transformation and hence show that the elemental area in the $X'Y'$ system is $dx'\, dy'$.

13.5 Three-dimensional coordinate systems

The most straightforward way of representing position in three dimensions is by a rectangular Cartesian coordinate system. The set of three mutually orthogonal axes, x, y, and z are taken as a right-handed set that satisfy the condition that if unit vectors $\mathbf{i}$, $\mathbf{j}$, and $\mathbf{k}$ are taken along the positive x-, y-, and z-axes then

$$\mathbf{i} \times \mathbf{j} = \mathbf{k}. \tag{13.23}$$

Such a set is illustrated in Figure 13.4, where the y-axis points into the page.

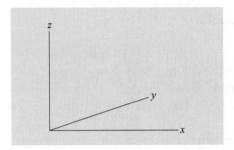

Figure 13.4 A right-handed rectangular Cartesian system.

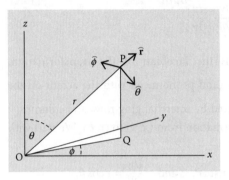

Figure 13.5 The spherical polar coordinate system.

For some purposes where the function under consideration, or the volume within which the function is defined, has a particular symmetry, other choices of coordinate system may simplify calculations.

13.5.1 Spherical polar coordinates

If the function or space being considered has spherical symmetry then it will probably be appropriate to use the spherical polar coordinate system. We show how this defines the coordinates of a point in Figure 13.5 which also shows a set of rectangular Cartesian axes. The coordinate of the point P is given by (r, θ, ϕ) where r is the distance of P from the origin, O, θ is the angle made by OP with the positive z-axis, for which $0 \le \theta \le \pi$, and ϕ is the angle made by OQ with the x-axis, where Q is the projection of P on the x-y plane. The limits of ϕ are $0 \le \phi < 2\pi$.

The relationships between the Cartesian and spherical polar coordinates, which can be deduced from the figure, are as follows:

$$x = r \sin \theta \cos \phi, \quad y = r \sin \theta \sin \phi, \quad z = r \cos \theta \tag{13.24a}$$

and

$$r = \sqrt{x^2 + y^2 + z^2}, \quad \theta = \tan^{-1}\left(\frac{\sqrt{x^2 + y^2}}{z}\right), \quad \phi = \tan^{-1}\left(\frac{y}{x}\right). \tag{13.24b}$$

In determining ϕ and θ the numerator and divisor have the sign of the sine and cosine respectively. In the case of θ the sign of sine is always positive since θ is in the range 0 to π.

The unit vectors along the directions of r, θ, and ϕ increasing are shown in Figure 13.5. They are respectively along OP, perpendicular to OP in the vertical plane, and perpendicular to OQ in the horizontal plane, and they form an orthogonal set. It is possible to work out the elemental volume for three-dimensional integration using spherical polar coordinates from geometrical considerations but we shall do it by means of the Jacobian transformation, which also applies in three dimensions. This is

$$
J\left(\frac{x,y,z}{r,\theta,\phi}\right) = \begin{vmatrix} \dfrac{\partial x}{\partial r} & \dfrac{\partial x}{\partial \theta} & \dfrac{\partial x}{\partial \phi} \\ \dfrac{\partial y}{\partial r} & \dfrac{\partial y}{\partial \theta} & \dfrac{\partial y}{\partial \phi} \\ \dfrac{\partial z}{\partial r} & \dfrac{\partial z}{\partial \theta} & \dfrac{\partial z}{\partial \phi} \end{vmatrix} = \begin{vmatrix} \sin\theta\cos\phi & r\cos\theta\cos\phi & -r\sin\theta\sin\phi \\ \sin\theta\sin\phi & r\cos\theta\sin\phi & r\sin\theta\cos\phi \\ \cos\theta & -r\sin\theta & 0 \end{vmatrix}
$$

$$
= r^2\{\sin^3\theta(\cos^2\phi+\sin^2\phi)+\cos^2\theta\sin\theta(\cos^2\phi+\sin^2\phi)\}
$$

$$
= r^2\sin\theta. \tag{13.25}
$$

Hence the elemental area for spherical polar coordinates is

$$
\mathrm{d}A_{r,\theta,\phi} = r^2\sin\theta\ \mathrm{d}r\ \mathrm{d}\theta\ \mathrm{d}\phi. \tag{13.26}
$$

Exercise 13.5
Find spherical polar coordinates corresponding to the Cartesian coordinates $(3, 2, -1)$.

Exercise 13.6
Find Cartesian coordinates corresponding to the spherical polar coordinates $(4,\ 3\pi/4,\ \pi/3)$.

13.5.2 Cylindrical coordinates

The three-dimensional cylindrical coordinate system is illustrated in Figure 13.6. In this system the point Q is defined by (r, θ) just as in the two-dimensional polar coordinate system. The third coordinate, to give the point P, is obtained just by going a distance z along the z-axis. The transformations to and from the rectangular Cartesian system are

$$
x = r\cos\theta, \quad y = r\sin\theta, \quad z = z \tag{13.27a}
$$

and

$$
r = \sqrt{x^2+y^2}, \quad \theta = \tan^{-1}\left(\frac{y}{x}\right), \quad z = z. \tag{13.27b}
$$

The elemental area for integration is just that for the two-dimensional polar coordinate system, as given by (13.13), multiplied by $\mathrm{d}z$ so that

$$
\mathrm{d}A_{r,\theta,z} = r\ \mathrm{d}r\ \mathrm{d}\theta\ \mathrm{d}z. \tag{13.28}
$$

This can be confirmed by the Jacobian transformation.

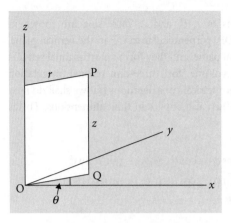

Figure 13.6 The cylindrical coordinate system.

Exercise 13.7
Find cylindrical coordinates corresponding to the spherical polar coordinates $(4, \, 3\pi/4, \, \pi/3)$.

13.6 Integration in three dimensions

13.6.1 Rectangular Cartesian system

The process of integration in three dimensions is just an extension of what was described in §13.2. For the rectangular Cartesian system the integral will appear in the form

$$I = \int_{x_1}^{x_2} \int_{y_1}^{y_2} \int_{z_1}^{z_2} f(x, y, z)dz \, dy \, dx. \tag{13.29}$$

This is carried out in the following stages:

$$g(x, y) = \int_{z_1}^{z_2} f(x, y, z)dz, \tag{13.30a}$$

$$h(x) = \int_{y_1}^{y_2} g(x, y)dy, \tag{13.30b}$$

$$I = \int_{x_1}^{x_2} h(x)dx. \tag{13.30c}$$

All the same considerations apply as for two-dimensional integration. The order of integration may affect the difficulty of the process and the limits of integration may be functions of variables if the integration space is not a rectangular parallelepiped.

As our first example of an integration using the rectangular Cartesian system we evaluate

$$I = \int_0^1 \int_0^1 \int_0^1 z(x + y)dz \, dy \, dx. \tag{13.31}$$

First,

$$g(x,y) = \int_0^1 z(x+y)dz = \left|\frac{1}{2}z^2(x+y)\right|_0^1 = \frac{1}{2}(x+y),$$

next

$$h(x) = \int_0^1 \frac{1}{2}(x+y)dy = \left|\frac{1}{2}xy + \frac{1}{4}y^2\right|_0^1 = \frac{1}{2}x + \frac{1}{4}$$

and finally

$$I = \int_0^1 \left(\frac{1}{2}x + \frac{1}{4}\right)dx = \left|\frac{1}{4}x^2 + \frac{1}{4}x\right|_0^1 = \frac{1}{2}.$$

This integral could have been taken as a sum of two integrals with integrands xz and yz respectively and for each of those integrals the variables could be separated.

We now consider another integral using the rectangular Cartesian system but where the integration space is an octant of a sphere of radius 1 unit. This region is illustrated in Figure 13.7. The octant is bounded by the planes $x=0$, $y=0$, $z=0$, and the part of the spherical surface with apices at A, B, and C. We now consider the limits for the integration. Firstly we may take x between 0 and 1. For any particular value of x the range of y is between 0 and $\sqrt{1-x^2}$. Finally for any particular values of x and y the range of z is 0 to $\sqrt{1-x^2-y^2}$. If the integrand is xyz then the integral is

$$I = \int_0^1\int_0^{\sqrt{1-x^2}}\int_0^{\sqrt{1-x^2-y^2}} xyz\,dz\,dy\,dx. \tag{13.32}$$

The integral is now found in the following steps:

$$g(x,y) = \int_0^{\sqrt{1-x^2-y^2}} xyz\,dz = \left|\frac{1}{2}xyz^2\right|_0^{\sqrt{1-x^2-y^2}} = \frac{1}{2}xy(1-x^2-y^2)$$

$$h(x) = \int_0^{\sqrt{1-x^2}} \frac{1}{2}xy(1-x^2-y^2)dy = \left|\frac{1}{4}xy^2 - \frac{1}{4}x^3y^2 - \frac{1}{8}xy^4\right|_0^{\sqrt{1-x^2}} = \frac{1}{8}x - \frac{1}{4}x^3 + \frac{1}{8}x^5.$$

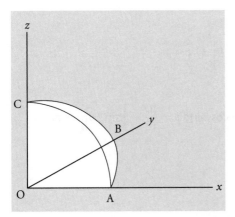

Figure 13.7 An octant of a sphere defined by the region within OABC.

Finally

$$I = \int_0^1 \left(\frac{1}{8}x - \frac{1}{4}x^3 + \frac{1}{8}x^5\right) dx = \left|\frac{1}{16}x^2 - \frac{1}{16}x^4 + \frac{1}{48}x^6\right|_0^1 = \frac{1}{48}.$$

We shall see later that there is an easier way of solving this problem.

 Exercise 13.8
Evaluate the integral $\int_0^1 \int_0^2 \int_0^3 (x + y + z)^2 dz\, dy\, dx$.

13.6.2 Spherical polar coordinates

The stages in evaluating an integral can follow the kind of process described in (13.30). There is, however, a tendency for problems couched in terms of spherical polar coordinates to be in a form where the variables are separable.

EXAMPLE 13.3

In this example the integrand is

$$f(r, \theta, \phi) = r \cos^2 \theta \sin^2 \phi$$

and the integral is within a sphere, centred on the origin, of radius 1. Bearing in mind the form of the elemental volume given in (13.26), the integral is

$$I = \int_0^{2\pi} \int_0^\pi \int_0^1 r \cos^2 \theta \sin^2 \phi \, r^2 \sin \theta \, dr\, d\theta\, d\phi. \qquad (13.33)$$

The variables are separable and I may be written as

$$I = I_r \times I_\theta \times I_\phi$$

where

$$I_r = \int_0^1 r^3 dr = \left|\frac{1}{4}r^4\right|_0^1 = \frac{1}{4},$$

$$I_\theta = \int_0^\pi \cos^2 \theta \sin \theta \, d\theta = \left|-\frac{1}{3}\cos^3 \theta\right|_0^\pi = \frac{2}{3}$$

and

$$I_\phi = \int_0^{2\pi} \sin^2 \phi \, d\phi = \int_0^{2\pi} \frac{1}{2}(1 - \cos 2\phi)d\phi = \left|\frac{\phi}{2} - \frac{1}{4}\sin 2\phi\right|_0^{2\pi} = \pi,$$

giving

$$I = \frac{1}{4} \times \frac{2}{3} \times \pi = \frac{\pi}{6}.$$

EXAMPLE 13.4

As another example of integration using spherical polar coordinates we take the integral (13.32). Using transformations (13.24a) the integrand is

$$f(r, \theta, \phi) = r^3 \sin^2 \theta \cos \theta \sin \phi \cos \phi$$

and an octant of a sphere is defined by the limits r from 0 to 1, θ from 0 to $\pi/2$ and ϕ from 0 to $\pi/2$. Putting in the elemental volume in form (13.26) the variables are separable and $I = I_r \times I_\theta \times I_\phi$, where

$$I_r = \int_0^1 r^5 \, dr = \left| \frac{1}{6} r^6 \right|_0^1 = \frac{1}{6},$$

$$I_\theta = \int_0^{\pi/2} \sin^3 \theta \cos \theta \, d\theta = \left| \frac{1}{4} \sin^4 \theta \right|_0^{\pi/2} = \frac{1}{4}$$

and

$$I_\phi = \int_0^{\pi/2} \sin \phi \cos \phi \, d\phi = \left| \frac{1}{2} \sin^2 \phi \right|_0^{\pi/2} = \frac{1}{2}.$$

This gives $I = \frac{1}{6} \times \frac{1}{4} \times \frac{1}{2} = \frac{1}{48}$ as found previously but, with simpler limits and separable variables, the use of spherical polar coordinates makes the calculation more straightforward.

Exercise 13.9
Integrate the function $r \sin \theta \cos^2 \phi$ over the hemisphere of unit radius above the x-y plane.

13.6.3 Cylindrical coordinates

We now consider an example of three-dimensional integration using cylindrical coordinates.

EXAMPLE 13.5

We consider a cylindrical rod of radius 0.01 m and length 1 m. The density of the rod varies linearly from 2000 kg m^{-3} at one end of the rod to 3000 kg m^{-3} at the other. The problem is to find the total mass of the rod. Taking the ends of the rod as $z = 0$ and $z = 1$ the density may be represented as

$$\rho = 2000 + 1000z. \tag{13.34}$$

The total mass of the rod is

$$M = \int_V \rho \, dV = \int_0^1 \int_0^{2\pi} \int_0^{0.01} (2000 + 1000z) r \, dr \, d\theta \, dz.$$

The variables are separable, so

$$M = I_r \times I_\theta \times I_z$$

where

$$I_r = \int_0^{0.01} r\,dr = \left|\frac{1}{2}r^2\right|_0^{0.01} = 5 \times 10^{-5},$$

$$I_\theta = \int_0^{2\pi} d\theta = |\theta|_0^{2\pi} = 2\pi,$$

$$I_z = \int_0^1 (2000 + 1000z)dz = \left|2000z + 500z^2\right|_0^1 = 2500.$$

Hence the mass of the rod is

$$M = 5 \times 10^{-5} \times 2\pi \times 2500\,\text{kg} = 0.785\,\text{kg}.$$

Exercise 13.10
Find the mass of a cylindrical rod of length 1 m and radius 0.01 m in which the density of the rod varies from 3000 kg m^{-3} along the rod axis to 2000 kg m^{-3} at the cylindrical surface.

13.7 Moments of inertia

Moment of inertia plays a role in rotational motion similar to that of mass in linear motion. The equivalences in the two types of motion are illustrated by some examples in Table 13.1.

For a point mass, m, spinning about some axis the moment of inertia is

$$I_m = mr^2, \tag{13.35}$$

where r is the distance of the point mass from the axis. For a distribution of n point masses the total moment of inertia is the sum of the moments of inertia for the separate masses so that

Table 13.1 Equivalences between linear and rotational motion

Linear motion		Rotational motion	
Scalar mass	m	Moment of inertia	I_m
Vector distance	s	Angle	θ
Vector velocity	v	Angular velocity	ω
Vector acceleration	$\dot{v}$	Angular acceleration	$\dot{\omega}$
For a body with initial speed v_0 at $t=0$	$s = v_0 t + \frac{1}{2}\dot{v}t^2$	For a body with initial angular speed ω_0	$\theta = \omega_0 t + \frac{1}{2}\dot{\omega}t^2$
Linear momentum	mv	Angular momentum	$I_m\omega$
Kinetic energy	$\frac{1}{2}mv^2$	Kinetic energy	$\frac{1}{2}I_m\omega^2$

$$I_m = \sum_{i=1}^{n} m_i r_i^2. \tag{13.36}$$

Where there is an extended continuous body then integration replaces summation and the moment of inertia is of the form

$$I_m = \int_V \rho(\mathbf{r}) r_\perp^2 dV \tag{13.37}$$

where the integral is over the volume V of the body, the density of the body is $\rho(\mathbf{r})$ at vector position $\mathbf{r}$, and $r_\perp$ is the perpendicular distance to the spin axis from position $\mathbf{r}$. We shall now use (13.37) to calculate the moments of inertia of some symmetrical bodies in one, two, and three dimensions.

13.7.1 Moment of inertia of a straight, thin rod

The rod we are considering may be considered as a one-dimensional object with a mass σ per unit length and of length $2l$. The spin axis is taken through the centre of the rod and is at right angles to the rod. This configuration is shown in Figure 13.8. The contribution to the moment of inertia of the small element of length dx at distance x from the spin axis is

$$dI_m = \sigma x^2 dx$$

so that the total moment of inertia is

$$I_m = \int_{-l}^{l} \sigma x^2 dx = \left| \frac{1}{3}\sigma x^3 \right|_{-l}^{l} = \frac{2}{3}\sigma l^3 = \frac{1}{3}Ml^2. \tag{13.38}$$

where $M\ (=2l\sigma)$ is the mass of the rod.

 Exercise 13.11
Find the moment of inertia of a uniform long thin rod of mass M and length l about an axis at one end of the rod and perpendicular to it.

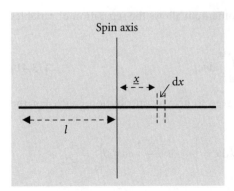

Spin axis

Figure 13.8 A rod spinning about an axis through its centre.

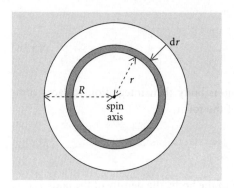

Figure 13.9 A disk spinning about an axis through its centre.

13.7.2 Moment of inertia of a disk

Here we consider a uniform disk of radius R with a mass σ per unit area spinning about an axis through its centre and perpendicular to it (Figure 13.9). All the material in the shaded annulus, of radius r and width $\mathrm{d}r$, is at the same distance from the spin axis so that its contribution to the moment of inertia is

$$\mathrm{d}I_m = 2\pi r\sigma\,\mathrm{d}r \times r^2 = 2\pi r^3\sigma\,\mathrm{d}r.$$

Hence the total moment of inertia is

$$I_m = \int_0^R 2\pi r^3\sigma\,\mathrm{d}r = 2\pi\sigma\left|\frac{1}{4}r^4\right|_0^R = \frac{1}{2}\pi R^4\sigma = \frac{1}{2}MR^2, \tag{13.39}$$

where $M\ (=\pi\sigma R^2)$ is the total mass of the disk.

13.7.3 Moment of inertia of a uniform sphere about a diameter

One way of carrying out this calculation is to imagine that the sphere, of uniform density ρ, consists of a large number of thin disks and use result (13.39). However, a basic approach is more straightforward.

In Figure 13.5 we imagine that we have a small elemental volume $r^2\sin\theta\,\mathrm{d}r\,\mathrm{d}\theta\,\mathrm{d}\phi$ at point P. Its distance from the z-axis, taken as the spin axis, is $r\sin\theta$ so its contribution to the moment of inertia is

$$\mathrm{d}I_m = r^2\sin^2\theta \times r^2\rho\sin\theta\,\mathrm{d}r\,\mathrm{d}\theta\,\mathrm{d}\phi. \tag{13.40}$$

The integral giving the total angular momentum allows the separation of variables so that

$$I_m = \rho\int_0^R r^4\,\mathrm{d}r \times \int_0^\pi \sin^3\theta\,\mathrm{d}\theta \times \int_0^{2\pi}\mathrm{d}\phi. \tag{13.41}$$

The first and third integrals are easily seen to give a contribution $\frac{2}{5}\pi R^5$. For the second integral,

$$\int_0^\pi \sin^3\theta\,\mathrm{d}\theta = \int_0^\pi (1-\cos^2\theta)\sin\theta\,\mathrm{d}\theta = \left|-\cos\theta + \frac{1}{3}\cos^3\theta\right|_0^\pi = \frac{4}{3}.$$

Hence

$$I_m = \frac{8}{15}\pi R^5 \rho = \frac{2}{5}MR^2 = 0.4MR^2, \tag{13.42}$$

where $M(=\frac{4}{3}\pi R^3)$ is the mass of the sphere.

It is interesting to consider the effect of a density distribution in the sphere that is not uniform but is a function of the distance from the centre – for example

$$\rho = \rho_0\left(1 - \frac{r}{R}\right), \tag{13.43}$$

which gives the density varying linearly from a maximum at the centre to zero at the surface. From (13.40), replacing ρ by (13.43), it can be seen that this affects the first integral in (13.41) which becomes

$$\int_0^R r^4\left(1 - \frac{r}{R}\right)dr = \left|\frac{1}{5}r^5 - \frac{1}{6}\frac{r^6}{R}\right|_0^R = \frac{1}{30}R^5. \tag{13.44}$$

This makes the moment of inertia of the sphere

$$I'_m = \frac{4}{45}\pi \rho_0 R^5. \tag{13.45}$$

To put this in the form (13.42) we need to find the mass of the sphere. The mass between spherical shells of radii r and $r + dr$ is

$$dM = 4\pi r^2 dr \times \rho_0\left(1 - \frac{r}{R}\right).$$

making the total mass

$$M = 4\pi\rho_0 \int_0^R r^2\left(1 - \frac{r}{R}\right)dr = 4\pi\rho_0\left|\frac{1}{3}r^3 - \frac{1}{4}\frac{r^4}{R}\right|_0^R = \frac{1}{3}\pi\rho_0 R^3. \tag{13.46}$$

Substituting from (13.46) into (13.45) gives

$$I'_m = \frac{4}{15}MR^2 = 0.267MR^2. \tag{13.47}$$

Many astronomical bodies are close to spheres and have internal structures such that the density is a function of the distance from the centre. Measurements of the motions of satellites, natural or artificial, around astronomical bodies can give estimates of the moment of inertia in the form αMR^2 where α is called the **moment-of-inertia factor**. For a body of uniform density $\alpha = 0.4$ but for a body that is centrally condensed, i.e. has a density that increases towards the centre, the value of α is less than 0.4, as shown by (13.47). The value of α gives information about the internal structure of a body. For example, the value of α for the Moon is 0.392 showing that it is not too far from uniform and limiting the possible size of an iron core. On the other hand for Saturn the value of α is 0.210, showing that it has a high degree of central condensation.

Exercise 13.12
Uniform spheres made of the same material have radii a and b respectively. What is the ratio of their moments of inertia?

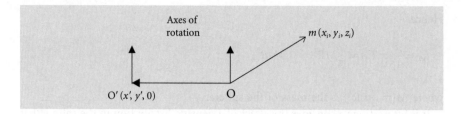

Figure 13.10 A point mass in relation to two parallel rotation axes.

13.8 **Parallel-axis theorem**

Figure 13.10 shows a point mass m_i at position $(x_i,\ y_i,\ z_i)$ relative to an origin O. We consider the moment of inertia of a number of point masses about an axis through O, the centre of mass, in the direction of z. This moment of inertia is

$$I_O = \sum_i m_i(x_i^2 + y_i^2). \tag{13.48}$$

Now consider the moment of inertia of the same system of masses about an axis parallel to the one through O but passing through O' with coordinates $(x',\ y',\ 0)$. This moment of inertia will be

$$\begin{aligned} I_{O'} &= \sum_i m_i \left\{ \left(x_i - x_i'\right)^2 + \left(y_i - y_i'\right)^2 \right\} \\ &= I_O + \left(x_i'^2 + y_i'^2\right) \sum_i m_i - 2x_i' \sum_i m_i x_i - 2y_i' \sum_i m_i y_i. \end{aligned} \tag{13.49}$$

Since O is the centre of mass we have $\sum_i m_i x_i = \sum_i m_i y_i = 0$. In addition $\sum_i m_i = M$, the total mass of the system and $(x_i'^2 + y_i'^2) = a^2$, where a is the distance of the axis through O' from the axis through O. This gives the **parallel-axis theorem** which expresses the moment of inertia about any axis in terms of the moment of inertia about a parallel axis through the centre of mass and the distance between the axes. It is of the form

$$I_{O'} = I_O + Ma^2. \tag{13.50}$$

Although we have derived it for a discrete system of point masses, it also applies to a continuous distribution of material.

EXAMPLE 13.6

As an example of the application of the parallel-axis theorem we find the moment of inertia of a uniform sphere of mass M and radius R about an axis tangential to its surface. Such an axis will be at distance R from an axis through

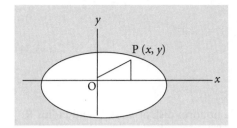

Figure 13.11 A planar object containing orthogonal *x*- and *y*-axes.

the centre of mass for which the moment of inertia is given by (13.42). Hence the moment of inertia we seek is

$$I_{O'} = 0.4MR^2 + MR^2 = 1.4MR^2. \tag{13.51}$$

 Exercise 13.13

Find the moment of inertia of a uniform disk of mass M and radius R about an axis through its edge and perpendicular to the disk.

13.9 Perpendicular-axis theorem

In Figure 13.11 we show a planar object containing two orthogonal axes, x and y, intersecting at the point O. If the density per unit area at P is $\rho(x, y)$ then the moment of inertia of the object about the y-axis is

$$I_y = \int_A \rho(x, y)x^2 \mathrm{d}x\,\mathrm{d}y \tag{13.52a}$$

and about the x-axis it is

$$I_x = \int_A \rho(x, y)y^2 \mathrm{d}x\,\mathrm{d}y, \tag{13.52b}$$

where the integral is over the whole area of the object. If we now consider the moment of inertia about an axis through O perpendicular to the plane then this is

$$I_z = \int_A \rho(x, y)\left(x^2 + y^2\right)\mathrm{d}x\,\mathrm{d}y. \tag{13.52c}$$

From equations (13.52) we have

$$I_z = I_x + I_y, \tag{13.53}$$

which is the **perpendicular-axis theorem.**

EXAMPLE 13.7

As an application of the perpendicular-axis theorem we find the moment of inertia of a uniform disk about a diameter. Taking two perpendicular diameters for each of which the moment of inertia is I_d it is clear that

$$2I_d = I_m$$

where I_m is given by (13.39). Hence

$$I_d = \frac{1}{4}MR^2.$$ (13.54)

 Exercise 13.14

Find the moment of inertia of a uniform disk of mass M and radius R about an axis tangential to the disk in its plane.

Problems

13.1 Evaluate the following integrals: (i) $\int_0^1 \int_0^3 (xy^2 + x^2y)\, dy\, dx$;

(ii) $\int_0^1 \int_0^2 x^2 \exp(-xy)dy\, dx$; (iii) $\int_0^{\pi/2} \int_0^1 \frac{r\cos\theta}{(1 + r\sin\theta)^2}\, dr\, d\theta$

13.2 A rectangular plate made of material with a density $5000\ \mathrm{kg\,m^{-3}}$ has dimensions $X = 0.1\ \mathrm{m}$ and $Y = 0.2\ \mathrm{m}$. With one corner taken as origin the thickness of the plate is $t = (x + 0.1)(y + 0.1)\ \mathrm{m}$ where x and y are in metres. Find the mass of the plate.

13.3 Use the Jacobian transformation to show that the elemental volume for the cylindrical coordinate system is $r\, dr\, d\theta\, dz$.

13.4 A unit sphere centred on the origin has a density that varies as $\rho = \rho_0(1 + x^2)$. Find the mass of the sphere.

13.5 A circular plate of radius R and density ρ has a thickness that varies as $t = t_0\left\{1 - \left(\frac{r}{R}\right)^2\right\}$ where r is the distance from the centre. Find the mass of the plate and the moment of inertia, in the form αMR^2, for rotation about an axis through the centre perpendicular to the plate.

13.6 The Moon can be taken as a sphere of radius $1740\ \mathrm{km}$ with an iron core of uniform density $8000\ \mathrm{kg\,m^{-3}}$ and a mantle of density $3300\ \mathrm{kg\,m^{-3}}$. It is proposed that the core has a radius of $400\ \mathrm{km}$. What is the moment of inertia factor of the Moon for spin about a diameter on the basis of this model? You may assume the result (13.42).

13.7 A pair of equal spheres of uniform density and each of mass M and radius R are connected by a light rod passing along a diameter of each sphere. The moment of inertia of the system about an axis perpendicular to the rod and passing through its midpoint is $5MR^2$. What is the centre-to-centre distance of the two spheres?

13.8 A uniform square plate has mass M and side a. What is its moment of inertia (i) about an axis through its centre of mass perpendicular to the plate and (ii) about an axis through one corner perpendicular to the plate?

Distributions

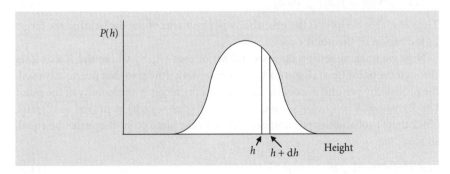

14

This chapter describes the characteristics of three commonly occurring distribution functions. They are all related in some way, but two of them are of particular importance to physicists. The normal distribution occurs frequently in nature and also describes the distribution of results from repeating experimental measurements many times. The Poisson distribution is relevant to the performance of radiation counters, used in many types of physical experiment.

14.1 **Kinds of distribution**

Almost any naturally occurring population will have a variety of characteristics. This is not true for atomic particles so that, for example, all electrons are strictly identical in characteristics – mass and charge – and this plays an important role in their statistical behaviour. On the other hand men, who share the characteristic of being male humans, otherwise have infinite variety in their characteristics. They can vary in height, weight, hair colour, and almost any other characteristic that one can think of. However, if we take height as an example, we find that in any community most men are round about the average height with fewer being either much shorter or taller than the average and a very few being exceptionally short or tall. In fact what we have is a **distribution** of heights, something which can be shown graphically. Figure 14.1 shows such a distribution. The proportion of heights in the range h to $h + \mathrm{d}h$ is $P(h)\mathrm{d}h$ and the distribution is normalized so that

Figure 14.1 A distribution of heights.

$$\int_{h_{\min}}^{h_{\max}} P(x)\,\mathrm{d}x = 1. \tag{14.1}$$

where $h_{\max}$ and $h_{\min}$ give the maximum and minimum heights in the population being represented. Notice that a question such as 'What is the probability that a man has a height of 1.80 m?' is strictly meaningless. It is necessary to specify a range of heights to be able to answer the question. Since the height is specified to the nearest centimetre then it might be interpreted as asking for the probability that a man has height between 1.795 m and 1.805 m – in which case an answer can be given.

Many kinds of distributions occur in nature and in scientific contexts, some continuous and some discrete. An example of a continuous distribution is the one we have considered with heights. An example of a discrete distribution would be the number of leaves on a plant of a particular species – which could be 8, 9, 10, etc. but not 8.4. Such a distribution can be represented graphically as a histogram, an example of which is given in Figure 14.6.

14.2 **Firing at a target**

The goal of firing at a target is to hit the bullseye, but an inexpert marksman will usually spray shots all over the target and may occasionally miss it altogether. However, the outcome of a large number of shots should show a tendency for a greater density of hits closer to the bullseye, with a fall-off of density at increasing distance.

We now assume that the uncertainty in positioning the shot is **isotropic**, that is that the likelihood of missing the bullseye by a certain distance is independent of the direction from the bullseye. Considering the x direction on the target we take the probability of the shot having a deviation between x and $x + \mathrm{d}x$ as $p(x)\mathrm{d}x$. Similarly, for the orthogonal y direction the probability of a deviation between y and $y + \mathrm{d}y$ is $p(y)\mathrm{d}y$. The probabilities of deviations in the orthogonal directions are **independent**, so that the probability that the shot falls in the small area element $\mathrm{d}x\,\mathrm{d}y$ is given by

$$P(x, y)\mathrm{d}x\,\mathrm{d}y = p(x)p(y)\mathrm{d}x\,\mathrm{d}y. \tag{14.2}$$

Thus $P(x, y) = p(x)p(y)$ is the probability per unit area of the shot hitting the target in the region of the point (x, y).

Now we imagine setting up a second set of axes (X, Y) where the X axis goes through the point (x, y) (Figure 14.2). By reasoning similar to that previously used, the probability per unit area of the shot hitting the target in the vicinity of the point (X, Y) where $X = \sqrt{x^2 + y^2}$ and $Y = 0$ is $P(\sqrt{x^2 + y^2}, 0) = p(\sqrt{x^2 + y^2})p(0)$. Since these probabilities per unit area relate to the same point they must be equal, so that

$$p(x)p(y) = p\left(\sqrt{x^2 + y^2}\right)p(0). \tag{14.3}$$

A general solution of (14.3) is

$$p(x) = A \exp(Bx^2) \tag{14.4}$$

since $A \exp(Bx^2) A \exp(By^2) = A \exp\{B(x^2 + y^2)\} A \exp(0)$. To satisfy the requirement that the probability should be positive it is necessary for A to be positive, and to satisfy the requirement that the probability should fall off with increasing distance from the centre, B must be negative. To emphasize the latter condition (14.4) is rewritten as

$$p(x) = A \exp(-\beta^2 x^2). \tag{14.5}$$

The form of this function is shown in Figure 14.3 with $A = 1$ and $\beta = 1$.

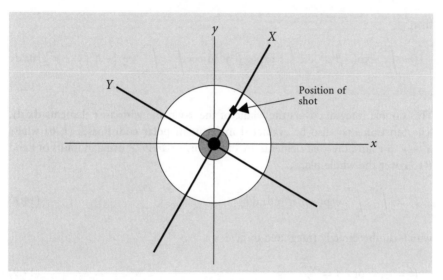

Figure 14.2 On one set of axes the shot is at (x, y), on the other set of axes at $\left(\sqrt{x^2 + y^2}, 0\right)$.

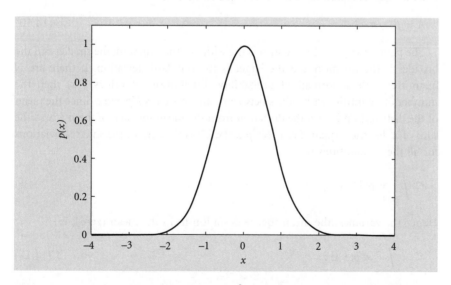

Figure 14.3 The form of the function $p(x) = \exp(-x^2)$.

Since $p(x)dx$ is the probability that the shot lies between x and $x+dx$ in the x direction then for normalization, so that the shot has unit probability of being somewhere between the x limits $+\infty$ and $-\infty$,

$$\int_{-\infty}^{\infty} p(x)dx = A \int_{-\infty}^{\infty} \exp(-\beta^2 x^2)dx = 1. \tag{14.6}$$

The definite integral in (14.6) can be evaluated by a rather interesting substitution. If we write

$$I = \int_{-\infty}^{\infty} \exp(-\beta^2 x^2)dx$$

then

$$I^2 = \int_{-\infty}^{\infty} \exp(-\beta^2 x^2)dx \int_{-\infty}^{\infty} \exp(-\beta^2 y^2)dy = \int_{-\infty}^{\infty}\int_{-\infty}^{\infty} \exp\{-\beta^2(x^2+y^2)\}dxdy. \tag{14.7}$$

This double integral covers the whole of the x-y plane with area elements $dx\,dy$. The function may also be evaluated in terms of polar coordinates (r, θ) where $x^2 + y^2 = r^2$ and the area elements are of the form $r\,dr\,d\theta$. Putting in limits of r and θ to cover the whole plane,

$$I^2 = \int_{0}^{2\pi}\int_{0}^{\infty} \exp(-\beta^2 r^2)r\,dr\,d\theta \tag{14.8}$$

which can be directly integrated to give

$$I^2 = \frac{\pi}{\beta^2} \quad \text{or} \quad I = \frac{\sqrt{\pi}}{\beta}. \tag{14.9}$$

For the normalization of (14.5) it is required that $AI = 1$ or

$$A = \beta/\sqrt{\pi}. \tag{14.10}$$

The parameter β in (14.5) controls the width of the function; the smaller is β the broader is the function and the larger is the standard deviation. If there are N quantities with a normalized probability distribution of values $p(x)$ then the number of quantities with values between x and $x+dx$ is $Np(x)dx$. Since the mean of the distribution is zero the deviation from the mean for each of them is x so the sum of x^2 for these quantities is $Nx^2 p(x)dx$. Thus the sum of the square deviations for all the N quantities is

$$N\int_{-\infty}^{\infty} x^2 p(x)dx.$$

Hence the variance, the mean square deviation from the mean (zero), is

$$V = \int_{-\infty}^{\infty} x^2 p(x)dx. \tag{14.11}$$

Integrating by parts,

$$V = \frac{\beta}{\sqrt{\pi}} \int_{-\infty}^{\infty} x^2 \exp\left(-\beta^2 x^2\right) dx$$

$$= \frac{\beta}{\sqrt{\pi}} \left\{ \left. -\frac{1}{2\beta^2} x \exp(-\beta^2 x^2) \right|_{-\infty}^{\infty} + \frac{1}{2\beta^2} \int_{-\infty}^{\infty} \exp\left(-\beta^2 x^2\right) dx \right\}.$$

The first term in the curly bracket is zero and the value of the second term is indicated by (14.9). We find

$$V = \frac{1}{2\beta\sqrt{\pi}} \times I = \frac{1}{2\beta^2}. \tag{14.12}$$

In terms of σ, the standard deviation, the distribution function (14.5) is thus

$$p(x) = \frac{1}{\sqrt{2\pi\sigma^2}} \exp\left(-\frac{x^2}{2\sigma^2}\right). \tag{14.13}$$

This kind of distribution function occurs frequently in nature and in scientific applications and is called the **Gaussian** or **normal** distribution. In deriving it from the target model the mean of the distribution was zero. A Gaussian distribution with mean $\bar{x}$ and standard deviation σ has the form

$$p(x) = \frac{1}{\sqrt{2\pi\sigma^2}} \exp\left\{ -\frac{(x - \bar{x})^2}{2\sigma^2} \right\}. \tag{14.14}$$

We shall now look at some applications of the normal distribution.

Exercise 14.1
A normal distribution has a mean of 50.0 and a standard deviation 20.0. What is the probability that a value chosen at random from such a distribution will be between 60.95 and 61.05?

14.3 **Normal distribution**

If we take a continuously variable and natural quantity, such as the heights of all male adults in a large community, we find that the distribution of values closely follows a normal distribution. For example, the distribution might be found to have a mean of 1.78 m with a standard deviation of 0.06 m and, from (14.14), this completely defines the distribution. Clearly, a distribution of this kind cannot have a practical range from $+\infty$ to $-\infty$ since there are finite maximum and minimum heights that a man can have. However, a normal distribution is very small beyond four or five standard deviations from the mean and the distribution outside these limits can, for most practical purposes, be ignored. For the example we are using, with a 5σ range from the mean, this will give maximum and minimum heights of 2.08 m and 1.48 m and there will be very few individuals outside these limits.

A question that might be of interest is the proportion of individuals with height greater than some value, say 1.96 m. Since all normal distributions have the same form this is equivalent to finding the area in the tail of the distribution between a value of $s\sigma$ from the mean ($\equiv 1.96$ m) and $+\infty$ (Figure 14.4). This area is given by

$$\Phi(s) = \frac{1}{\sqrt{2\pi\sigma^2}} \int_{s\sigma}^{\infty} \exp\left(-\frac{x^2}{2\sigma^2}\right) dx$$

and by the substitution $t = \sigma x$ this becomes

$$\Phi(s) = \frac{1}{\sqrt{2\pi}} \int_{s}^{\infty} \exp\left(-\frac{1}{2}t^2\right) dt. \tag{14.15}$$

Table 14.1 gives the values of $\Phi(s)$ for s in the range 0 to 5.0 at intervals of 0.1. It is clear from this that the distribution curve effectively terminates beyond about 4–5σ from the mean.

We can now return to the problem of determining the proportion of men with more than a certain height. For the example previously given 1.96 m is three standard deviations from the mean so that, from the table about 0.135% of adult men in our population will be taller than that.

The normal distribution has a wide application in many branches of science and the social sciences. Of particular interest is the way that it impinges on scientific measurements. If an experiment is made many times to measure the speed of light then the results would have a distribution of values and an average of several results would be more reliable than any one of them taken at random. If a sufficient number of results are obtained their values will form a normal distribution with a certain mean and standard deviation.

However, the mean of all the results would not necessarily be an unbiased estimate of the value of the speed of light. There are two kinds of error that can

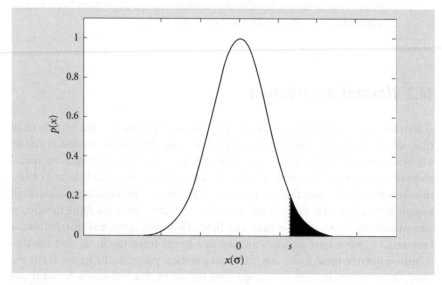

Figure 14.4 The area under the tail of the normal distribution between s standard deviations and $+\infty$.

Table 14.1 Areas under the normal distribution between s standard deviations from the mean and ∞

s	0	1	2	3	4	5
0.0	0.500 00	0.158 66	0.022 75	0.001 350	3.17×10^{-5}	2.98×10^{-7}
0.1	0.460 17	0.135 67	0.017 86	9.68×10^{-4}	2.07×10^{-5}	
0.2	0.420 74	0.115 07	0.013 90	6.87×10^{-4}	1.34×10^{-5}	
0.3	0.382 09	0.096 80	0.010 72	4.83×10^{-4}	8.55×10^{-6}	
0.4	0.344 58	0.080 76	0.008 198	3.37×10^{-4}	5.42×10^{-6}	
0.5	0.308 54	0.066 81	0.006 210	2.33×10^{-4}	3.40×10^{-6}	
0.6	0.274 26	0.054 80	0.004 661	1.59×10^{-4}	2.12×10^{-6}	
0.7	0.242 00	0.044 57	0.003 467	1.08×10^{-4}	1.31×10^{-6}	
0.8	0.211 86	0.035 93	0.002 555	7.24×10^{-5}	8.05×10^{-7}	
0.9	0.184 06	0.028 72	0.001 866	4.81×10^{-5}	4.77×10^{-7}	

occur when making a measurement – random and systematic. **Random errors** are those that contribute to the scatter in the results, shown in the distribution of values. A **systematic error** is one that always pushes the result in one direction and displaces the mean of the distribution. An example of this would be in the measurement of a long distance by the use of a steel tape, just 3 m long. The process of marking off each 3 m length and moving the tape exactly 3 m to measure the next section will undoubtedly introduce random errors. However, if the tape had been stretched by previous use then the result found for the total length would suffer a systematic error. No matter how carefully the measurements were made the length would tend to be underestimated.

Exercise 14.2
Fully grown members of a species of snake have a mean length of 1.2 m and standard deviation 0.2 m. What is the proportion of such snakes that are less than 0.80 m in length?

14.3.1 Central limit theorem

Let us assume that we select a number of variables, $x_1, x_2, \ldots, x_n$, from different distributions. To fix our minds on what this means we can imagine that there are n bags, each containing a set of numbers with different distributions, the ith distribution having mean $\bar{x}_i$ and standard deviation σ_i, and we select one number from each. The sum of the n numbers so selected, X_1, is found, the numbers are returned to the bags, and then the process is repeated many times. For each selection we can write

$$X = \sum_{i=1}^{n} x_i \tag{14.16}$$

and the average of X for many selections is given by

$$\overline{X} = \sum_{i=1}^{n} \overline{x_i}. \tag{14.17}$$

This says that **the average value of the sum of variables is the sum of their individual averages**.

Now we consider what the variance would be of the large number of values of X. First we write

$$\overline{X^2} = \overline{\left(\sum_{i=1}^{n} x_i \right)^2} = \overline{\sum_{i=1}^{n} x_i^2 + \sum_{i=1}^{n} \sum_{\substack{j=1 \\ j \neq i}}^{n} x_i x_j}$$

$$= \sum_{i=1}^{n} \overline{x_i^2} + \sum_{i=1}^{n} \sum_{\substack{j=1 \\ j \neq i}}^{n} \overline{x_i x_j} \tag{14.18a}$$

and

$$\overline{X}^2 = \left(\sum_{i=1}^{n} \overline{x_i} \right)^2 = \sum_{i=1}^{n} \overline{x_i}^2 + \sum_{i=1}^{n} \sum_{\substack{j=1 \\ j \neq i}}^{n} \overline{x_i}\,\overline{x_j}. \tag{14.18b}$$

Since the selections from the different distributions are independent $\overline{x_i x_j} = \overline{x_i}\,\overline{x_j}$ so that the variance of the values of X is given by

$$V = \overline{X^2} - \overline{X}^2 = \sum_{i=1}^{n} \left(\overline{x_i^2} - \overline{x_i}^2 \right) = \sum_{i=1}^{n} \sigma_i^2. \tag{14.19}$$

This says that **the variance of the sum of independent variables is the sum of their individual variances**.

The central limit theorem, which here we just state, is that, no matter what the nature of the individual distributions, if n is large then the distribution of the values of X is a normal distribution with mean $\overline{X}$, given by (14.17), and variance V, given by (14.19). There is no restriction on the nature of the individual distributions and some of them, or indeed all of them, can be the same distribution.

EXAMPLE 14.1

As an example of the application of the central limit theorem we take the selection of 100 random numbers from a distribution consisting of three numbers, 1 with a probability of 1/3, 2 with a probability of 1/3, and 3 with a probability of 1/3. The mean of this distribution is 2 and the variance, the mean square deviation from the mean, is 2/3. The sums of 100 random numbers taken from this distribution should give a normal distribution with mean $100 \times 2 = 200$ and variance $100 \times 2/3 = 200/3$. A simple computer program written to find 100 000 sums of 100 numbers taken from the basic distribution gave the result in Figure 14.5, which shows the comparison of the numerical results with the theoretical normal distribution.

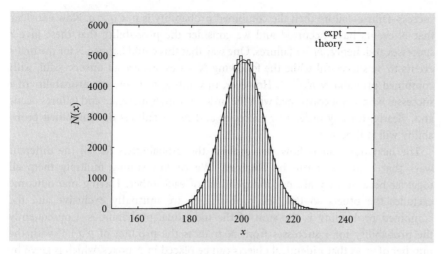

Figure 14.5 The numerical derivation from the central limit theorem (blocks) compared with the theoretical normal distribution (dashed line).

Exercise 14.3
One hundred numbers are chosen from a distribution in which $+1$ and -1 appear with equal probability to give a sum X. What is the mean and standard deviation of a large number of values of X found in this way?

14.4 Binomial distribution

14.4.1 Binomial distribution for 'success-or-failure' events

Suppose that we take a die and throw it six times – or, equivalently, throw six dice all at one time. What is the probability of having a 6 in this case? A simplistic view might be to say that it was unity, obtained by multiplying 1/6, the probability for each throw, by the number of throws, 6, but a moment's thought would soon reject that idea. After all there must be a possibility that there will be no 6s, so the probability cannot be unity. Analysis is necessary to find the probabilities of the various outcomes of throwing a die six times and here we shall generalize that analysis.

Let us consider an event with two outcomes that we will describe arbitrarily as 'success' and 'failure' (i.e. success in obtaining a 6 or failure to do so) with probabilities p and q respectively (1/6 and 5/6 in the case of the die). Since one or other outcome must take place, we must have

$$p + q = 1. \tag{14.20}$$

In §12.2.2 it was stated that if two or more *independent* events occur then the probability of the combined outcome is the product of the probabilities of the separate outcomes. For example, if there are three successful events then the combined probability is $p \times p \times p$ whereas if the outcomes are success–failure–failure then the combined probability is $p \times q \times q$. Now consider that N events have occurred and we consider the probability that these give r successes and, hence, $N - r$ failures. One way that this could happen is for the first r events to be successful while the following $N - r$ events are all unsuccessful, with combined probability $p^r q^{N-r}$. However, in wanting to know the probability of r successes we are not concerned with the order in which successes and failures occur and, clearly, for any order of r successes and $N - r$ failures the combined probability will still be $p^r q^{N-r}$.

The next question is how we combine the probabilities for all the different ways that r successes can be obtained. We cannot simply multiply them all together because they are not independent of each other. Having one outcome excludes the others so that the probabilities are **mutually exclusive** and the combined probability is the sum of the individual probabilities. Consequently the probability for r successes from N trials is the product of $p^r q^{N-r}$ with the number of ways that r identical objects can be placed in N boxes, which is given by (12.4) as

$$^N C_r = \frac{N!}{r!(N - r)!}. \tag{14.21}$$

From this result the probability of obtaining r successes in N trials is

$$P(r) = {}^N C_r p^r q^{N-r}. \tag{14.22}$$

The expression (14.22) is a term in the binomial theorem expansion

$$(p + q)^N = \sum_{r=0}^{N} {}^N C_r p^r q^{N-r} \tag{14.23}$$

and since $p + q = 1$, from (14.22) and (14.23)

$$\sum_{r=0}^{N} P(r) = 1. \tag{14.24}$$

This means that the sum of the probabilities of all the $N + 1$ mutually exclusive outcomes of trials, with from 0 to N successes, is unity, a necessary condition for a plausible probability expression. Because of the relationship of the probabilities to the terms of a binomial expansion the discrete probability function $P(r)$ is called a **binomial distribution**. Its form for $N = 10$ and $p = q = 0.5$ is shown in Figure 14.6.

For $p = q = 0.5$ the distribution is symmetrical about $r = \frac{1}{2} N$ but is asymmetrical if $p \neq q$. As the value of N increases so the outline of the distribution more closely approaches the appearance of a smooth curve. Figure 14.7 shows the distribution for $N = 100$, $p = 0.3$.

Exercise 14.4

What is the probability that if five cards are drawn from a pack, each card being replaced after selection, three of them will be diamonds?

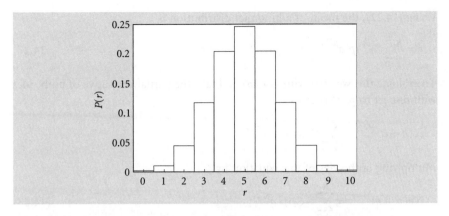

Figure 14.6 The binomial distribution for $N=10$ and $p=q=0.5$ drawn as a histogram.

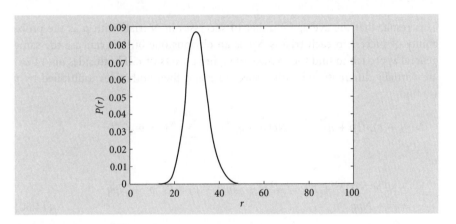

Figure 14.7 The binomial distribution for $N=100$ and $p=0.3$.

14.4.2 **Mean and variance of a binomial distribution**

A continuous distribution for a single variable can, in general, be described by a function $P(x)$ such that the proportion of cases between x and $x+\mathrm{d}x$ is $P(x)\mathrm{d}x$. For such a distribution the average of x^n is given by

$$\overline{x^n} = \int_a^b P(x)x^n\mathrm{d}x \tag{14.25}$$

where a and b are the lower and upper limits of x respectively and the distribution is normalized so that $\int_a^b P(x)\mathrm{d}x = 1$. For a discrete distribution, such as the binomial distribution, the corresponding expression for the average of x^n is

$$\overline{x^n} = \sum_a^b P(x)x^n. \tag{14.26}$$

Using (14.22), the mean of a binomial distribution is

$$\bar{r} = \sum_{r=0}^{\infty} {}^{N}C_{r} r p^{r} q^{N-r}. \tag{14.27}$$

To evaluate this we start with (14.23) and take the partial derivative of both sides with respect to p. This gives

$$N(p+q)^{N-1} = \sum_{r=0}^{N} {}^{N}C_{r} r p^{r-1} q^{N-r}.$$

Multiplying both sides of this equation by p,

$$Np(p+q)^{N-1} = \sum_{r=0}^{N} {}^{N}C_{r} r p^{r} q^{N-r}, \tag{14.28}$$

and, since $p+q=1$,

$$\bar{r} = Np. \tag{14.29}$$

This result, that the average number of successes in N trials with p as the probability of success in each trial is Np, is an obvious one but we can use the same general approach to find the average of higher powers of r. If both sides of (14.28) are partially differentiated with respect to p and then both sides multiplied by p we find

$$N(N-1)p^{2}(p+q)^{N-2} + Np(p+q)^{N-1} = \sum_{r=0}^{N} {}^{N}C_{r} r^{2} p^{r} q^{N-r}$$

or

$$\overline{r^{2}} = N(N-1)p^{2} + Np = N^{2}p^{2} + Np(1-p) = N^{2}p^{2} + Npq$$
$$= \bar{r}^{2} + Npq. \tag{14.30}$$

From this we find the variance of the binomial distribution as

$$\sigma^{2} = \overline{r^{2}} - \bar{r}^{2} = Npq. \tag{14.31}$$

Exercise 14.5

Six dice are thrown all together many times and the number of 6s recorded each time. What is the average number of 6s per throw and the standard deviation of those numbers?

14.4.3 Binomial distribution for large N

You will have noticed that the form of the distribution seen in Figure 14.7 has a superficial resemblance to the shape of a normal distribution. If fact, it can be shown that in the limiting case of $N \to \infty$ the binomial distribution is equivalent to a normal distribution. The approach to demonstrating this is to find the difference between $P(r+1)$ and $P(r)$ and to show that this is the same as the derivative of a normal distribution. This makes the approximation that the smooth function to which the binomial distribution tends when N is very large has a gradient

$$\frac{dP(r)}{dr} = \frac{P(r+1) - P(r)}{r+1-r} = P(r+1) - P(r). \tag{14.32}$$

Normalized curves that always have the same derivatives for the same arguments are identical, so if we can show that (14.32) is the same as the gradient of a normal distribution with mean $\bar{r} = Np$ and variance Npq then we will have shown the equivalence of the binomial and normal distributions.

In the course of this analysis we shall be using various approximations. Since N is very large, then

$$N \gg \sqrt{Npq}(=\sigma) \gg 1 \tag{14.33}$$

and hence any value of r within a few σ of $\bar{r}(= Np)$ will not be very different from $\bar{r}$ in relative terms, so that in the right circumstances we can write

$$r \approx \bar{r}. \tag{14.34}$$

We now have

$$P(r+1) - P(r) = \frac{N!}{(r+1)!(N-r-1)!}p^{r+1}q^{N-r-1} - \frac{N!}{r!(N-r)!}p^r q^{N-r}$$

$$= \frac{N!}{r!(N-r)!}p^r q^{N-r}\left(\frac{N-r}{r+1}\frac{p}{q} - 1\right). \tag{14.35}$$

We now use the approximations (14.33) and (14.34) to give

$$P(r+1) - P(r) = P(r)\left(\frac{Np - r(p+q) - q}{(r+1)q}\right) \approx P(r)\left(\frac{Np - r}{\bar{r}q}\right) \approx P(r)\left(\frac{\bar{r} - r}{Npq}\right)$$

$$= P(r)\frac{\bar{r} - r}{\sigma^2}. \tag{14.36}$$

A normal distribution with mean $\bar{r}$ and variance σ^2 is of the form

$$P_n(r) = \sqrt{\frac{1}{2\pi\sigma^2}}\exp\left\{-\frac{(r - \bar{r})^2}{2\sigma^2}\right\}. \tag{14.37}$$

Differentiating (14.37) gives

$$\frac{dP_n(r)}{dr} = \sqrt{\frac{1}{2\pi\sigma^2}}\frac{\bar{r} - r}{\sigma^2}\exp\left\{-\frac{(r - \bar{r})^2}{2\sigma^2}\right\} = P(r)\frac{\bar{r} - r}{\sigma^2} \tag{14.38}$$

which is identical to (14.36).

The analysis shows that for large N the binomial distribution tends to the normal distribution. This can be a useful approximation in dealing with a binomial distribution for large N, even if it is not so large that the binomial distribution is very close to normal. For example, if there are 100 trials with a probability of success of 0.3 per trial then the probability of less than 25 successes involves calculating 25 terms of the form (14.22), some of which involve the factorials of large numbers. This binomial distribution has a mean of $Np = 30$ and a standard deviation $\sqrt{Npq} = 4.583$. Since we are representing a discrete binomial distribution by a continuous normal distribution we must recognize that the discrete value 25 is equivalent to the range 24.5–25.5 in the continuous distribution. The area of the normal curve we require is that between 24.5 and zero (effectively $-\infty$) and 24.5 is $(30 - 24.5)/4.538\sigma = 0.98\sigma$ from the mean. From Table 14.1, using linear interpolation between 0.9 and 1.0, the area between 0.98σ and ∞ is 0.164, which is the estimated probability of having less than 25 successes. An accurate calculation using the binomial distribution gives the probability as 0.163.

Exercise 14.6

A card is drawn from a complete pack of cards 1 million times. What is the mean and standard deviation of the number of diamonds picked by such a process? What is the probability that more than 251 000 diamonds will be drawn?

14.4.4 Generalized binomial distributions

The binomial distribution has been calculated on the basis of events with two possible outcomes that have been designated as 'success' and 'failure' with probabilities p and q respectively, where $p + q = 1$. The probabilities for N trials have been found as the terms in the expansion of $(p + q)^N$. We now consider N trials with three possible outcomes with probabilities p_1, p_2 and p_3 where $p_1 + p_2 + p_3 = 1$. The probability of having r_1 outcomes of type 1, r_2 outcomes of type 2 and r_3 outcomes of type 3, where $r_1 + r_2 + r_3 = N$ is

$$P(r_1, r_2, r_3) = \frac{N!}{r_1! r_2! r_3!} p_1^{r_1} p_2^{r_2} p_3^{r_3} \tag{14.39}$$

which is a term in the expansion of $(p_1 + p_2 + p_3)^N$.

EXAMPLE 14.2

In a game of football three possible outcomes may be defined as 'home win', 'away win', and 'draw' with probabilities 0.5, 0.3, and 0.2 respectively. We now consider the probability that in 10 games chosen at random there are 6 home wins and 4 away wins. From (14.39), remembering that $0! = 1$, this is

$$P(6, 4, 0) = \frac{10!}{6! 4! 0!} 0.5^6 0.3^4 0.2^0 = 0.0266.$$

The generalization of (14.39) for any number of outcomes is straightforward. If there are m outcomes with probabilities $p_1, p_2, \ldots, p_m$ with $\sum_{i=1}^{m} p_m = 1$ then the probability that in N trials there are r_i occurrences of outcome i for $i = 1$ to m is

$$P(r_1, r_2, \ldots, r_m) = \frac{N!}{r_1! r_2! \ldots! r_m!} p_1^{r_1} p_2^{r_2} \cdots p_m^{r_m}, \tag{14.40}$$

where $\sum_{i=1}^{m} r_i = N$.

Exercise 14.7
A die is thrown six times. What is the probability of obtaining two 2s, two 3s, and two 4s?

14.5 **Poisson distribution**

For a binomial distribution the average number of successes in N trials is Np. We now consider a situation where N is extremely large and p is extremely small, giving a small but finite product Np. Let us suppose that the probability that an asteroid will strike the Earth in any year is 10^{-7} and we want to know the probability that there will be one such strike within the next 10^{6} years. We can regard each of the 10^{6} years as a trial and the probability of one 'success' is

$$P(1) = \frac{10^{6}!}{1!(10^{6} - 1)!}(10^{-7})^{1}(1 - 10^{-7})^{10^{6}-1}.$$

This is an awkward-looking expression, and some care would have to be exercised in computing the final term. However, there is a better way of describing the distribution in a case like this that gives more convenient expressions for the probability.

The probability for r successes in N trials has been given in (14.22) and can be written as

$$P(r) = \frac{N!}{r!(N - r)!}p^{r}q^{N-r} = \frac{N(N - 1)\dots(N - r + 1)}{r!}p^{r}q^{N-r}. \tag{14.41}$$

We now consider that $N \to \infty$ and $p \to 0$ but the average $Np \to a$, a small number. Since p is virtually zero then q is virtually unity so the variance of the distribution is $Npq \approx Np = a$. In this situation the variance of the distribution equals the mean, and since the probability will fall to an infinitesimal value at a few standard deviations from the mean we can only be concerned with values of r that are very much smaller than N. For that reason the quantity $(N - r)!$ in the numerator of the first term in (14.41) is a product of r terms all closely equal to N and so may be replaced by N^{r}. We can thus write

$$P(r) = \frac{(Np)^{r}}{r!}q^{N-r} = \frac{a^{r}}{r!}(1 - p)^{N-r}. \tag{14.42}$$

Since p is very small, and $r \ll N$, we may, without significant error, replace $(1 - p)^{N-r}$ by $(1 - p)^{N}$ or $(1 - a/N)^{N}$. Writing

$$S = \left(1 - \frac{a}{N}\right)^{N} \tag{14.43}$$

we have

$$\ln S = N \ln\left(1 - \frac{a}{N}\right). \tag{14.44}$$

From (4.46b) we have for $\varepsilon \ll 1$

$$\ln(1 + \varepsilon) \approx \varepsilon$$

so that

$$\ln S = N \times \left(-\frac{a}{N}\right) = -a$$

giving

$$S = \left(1 - \frac{a}{N}\right)^N = e^{-a} \quad \text{for} \quad N \to \infty. \tag{14.45}$$

The distribution function (14.42) has now been transformed into

$$P(r) = \frac{a^r}{r!} e^{-a}. \tag{14.46}$$

which is known as the **Poisson distribution**.

EXAMPLE 14.3

We can now apply the Poisson distribution to the asteroid-strike problem. The value of $a = Np = 10^6 \times 10^{-7} = 0.1$ and that quantity alone describes the distribution. The probability of one strike in 10^6 years is therefore

$$P(1) = \frac{(0.1)^1}{1!} e^{-0.1} = 0.090\,48,$$

and the probabilities of no strikes and two strikes are

$$P(0) = \frac{(0.1)^0}{0!} e^{-0.1} = 0.9048 \text{ and } P(2) = \frac{(0.1)^2}{2!} e^{-0.1} = 0.004\,52.$$

The form of a Poisson distribution for $a = 0.7$ is shown in Figure 14.8.

The Poisson distribution occurs in many practical situations.

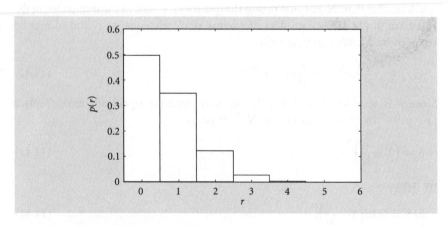

Figure 14.8 The Poisson distribution with average equal to 0.7.

EXAMPLE 14.4

As an example we can take a radiation counter that has a dead-time of 10^{-6} s, which means that any photon arriving within 10^{-6} s of a previous photon is unrecorded. We suppose that 10^6 photons arrive in a period of 1 s and we wish to know how many of these are recorded. The average number of photons arriving per dead-time is 1, and if we consider the dead-time period *before* the arrival of a photon then the probabilities for the number of photons that will have arrived in that period is a Poisson distribution with $a = 1$. The photon will only be recorded if the number arriving in the previous dead time was zero and the probability of this is

$$P(0) = \frac{1^0}{0!}e^{-1} = e^{-1} = 0.368.$$

This is the proportion of the photons that are counted, so the counting process would be very inaccurate. For an average of 10^5 photons arriving per second $a = 0.1$ and the proportion of photons counted is $e^{-0.1} = 0.905$, so less than 10% of the photons are lost. For an arrival rate at 10^4 photons per second the loss would be less than 1%. The proportion of loss depends on the rate of arrival, or intensity, and not on the total number of photons. If the duration of the arrival of a number of photons, say contained in a diffracted beam of X-rays, is unknown, then there is no way of correcting the total count for dead-time loss. However, if the total count and duration of the count are available then, in principle, a correction could be made.

 Exercise 14.8
The number of cars passing a particular spot on a road averages $10\,h^{-1}$. What are the probabilities that in a chosen 10 minute period the number of cars passing will be (i) none, (ii) one, and (iii) two or more?

Problems

14.1 The nth moment of a distribution is defined as $m(n) = \overline{(x - \bar{x})^n}$. For a discrete normalized distribution $m(n) = \sum_r P(r)(r - \bar{r})^n$, and for a continuous distribution, normalized between $x = a$ and $x = b$,

$$m(n) = \int_a^b P(x)(x - \bar{x})^n dx.$$

(i) Find $m(3)$ for binomial distributions with (a) $N = 5$, $p = 0.5$, $q = 0.5$ and (b) $N = 5$, $p = 0.8$, $q = 0.2$.

(ii) Show that for the normal distribution

$$P(x) = \frac{1}{\sqrt{2\pi\sigma^2}} \exp\left(-\frac{x^2}{2\sigma^2}\right) \quad \text{we have} \quad m(n+2) = (n+1)\sigma^2 m(n).$$

Hence find $m(10)$ for the normal distribution.

14.2 In an experiment on thought transference a subject is asked to guess the shape on a card viewed by the experimenter. There are five possible shapes so the likelihood of picking correctly just by chance is 0.2. The subject gives 240 correct guesses in 1000 trials. What is the probability that, just by chance, the subject could have had this number of successful guesses or greater?

14.3 In a throw of 10 dice what is the probability of obtaining three 6s, two 5s, two 4s and one each of 3, 2 and 1?

14.4 In an experiment to measure the distance of an asteroid an intense laser beam is fired towards it and the time is measured to when scattered photons are received by a detector. Calculations indicate that such an experiment, repeated many times, should give an average of 1.2 photons entering the detector. If the detector can only respond to three or more photons then what is the probability that the experiment will succeed?

Hyperbolic functions

Hyperbolic functions are related to the exponential e. They arise naturally in many mathematical and scientific contexts and have properties reminiscent of those of trigonometric functions. Here their properties are described, as are their relationships to trigonometric functions – how they are similar and how they differ.

15.1 Definitions

Hyperbolic functions are defined as follows, their names reflecting the trigonometric functions to which they have some similarity:

$$\sinh x = \frac{1}{2}(e^x - e^{-x}) \tag{15.1}$$

$$\cosh x = \frac{1}{2}(e^x + e^{-x}) \tag{15.2}$$

$$\tanh x = \frac{\sinh x}{\cosh x} = \frac{e^x - e^{-x}}{e^x + e^{-x}} \tag{15.3}$$

$$\operatorname{cosech} x = (\sinh x)^{-1} = \frac{2}{e^x - e^{-x}} \tag{15.4}$$

$$\operatorname{sech} x = (\cosh x)^{-1} = \frac{2}{e^x + e^{-x}} \tag{15.5}$$

$$\coth x = (\tanh x)^{-1} = \frac{e^x + e^{-x}}{e^x - e^{-x}}. \tag{15.6}$$

These are shown graphically in reciprocally related pairs in Figure 15.1. They do not closely resemble the trigonometric functions they are linked to by name. None of the functions is periodic in x. The functions $\sinh x$ and $\cosh x$ are not constrained to have magnitudes less than or equal to unity, unlike $\cos x$ and $\sin x$, and $\tanh x$ is constrained in value to be between -1 and $+1$, unlike $\tan x$. However, the symmetries of the trigonometric and hyperbolic functions with related names are similar in terms of being either **odd** with $f(-x) = -f(x)$, or **even**, with $f(-x) = f(x)$. From Figure 15.1 we can see that $\cosh x$ and $\operatorname{sech} x$ are even and the other functions are odd.

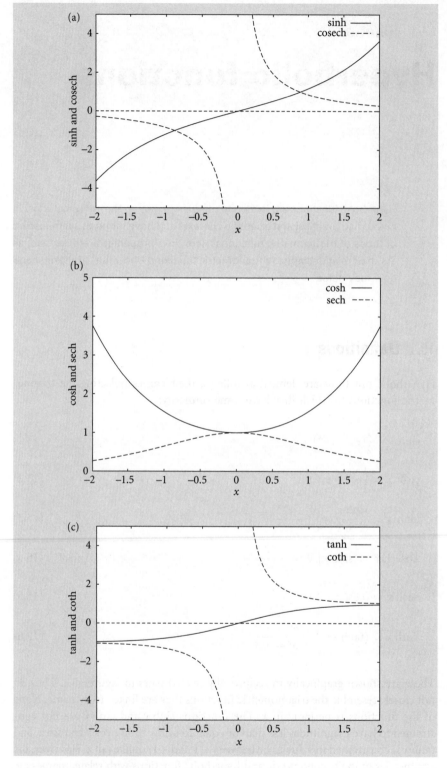

Figure 15.1 (a) sinh (x) and cosech (x); (b) cosh (x) and sech (x); (c) tanh (x) and coth (x).

15.2 **Relationships linking hyperbolic functions**

It has been mentioned that hyperbolic functions are similar to trigonometric functions in some ways, but are very different in their functional forms. The answer to this apparent contradiction is that the hyperbolic functions are linked by relationships that resemble those linking trigonometric functions although sometimes with changes of sign. The first example we take is a relationship linking $\sinh x$ and $\cosh x$. We have

$$\cosh^2 x = \frac{1}{4}(e^{2x} + 2 + e^{-2x}) \quad \text{and} \quad \sinh^2 x = \frac{1}{4}(e^{2x} - 2 + e^{-2x})$$

giving

$$\cosh^2 x - \sinh^2 x = 1. \tag{15.7}$$

The corresponding trigonometric relationship is $\cos^2 x + \sin^2 x = 1$.

Another example is

$$\cosh(x+y) = \frac{1}{2}(e^x e^y + e^{-x} e^{-y})$$
$$= \frac{1}{4}(e^x + e^{-x})(e^y + e^{-y}) + \frac{1}{4}(e^x - e^{-x})(e^y - e^{-y})$$
$$= \cosh x \cosh y + \sinh x \sinh y \tag{15.8}$$

corresponding to the trigonometric relationship $\cos(x+y) = \cos x \cos y - \sin x \sin y$.

A selection of relationships linking hyperbolic functions are given in Table 15.1 with the corresponding trigonometric relationships.

Exercise 15.1
Show that $\cosh^3 x = \frac{1}{4}\cosh 3x + \frac{3}{4}\cosh x$.

Table 15.1 A selection of relationships linking hyperbolic functions with corresponding trigonometric relationships

$\cosh^2 x - \sinh^2 x = 1$	$\cos^2 x + \sin^2 x = 1$
$\cosh(x+y) = \cosh x \cosh y + \sinh x \sinh y$	$\cos(x+y) = \cos x \cos y - \sin x \sin y$
$\cosh^2 x + \sinh^2 x = \cosh 2x$	$\cos^2 x - \sin^2 x = \cos 2x$
$\sinh(x+y) = \sinh x \cosh y + \cosh x \sinh y$	$\sin(x+y) = \sin x \cos y + \cos x \sin y$
$2\sinh x \cosh x = \sinh 2x$	$2\sin x \cos x = \sin 2x$
$1 - \tanh^2 x = \operatorname{sech}^2 x$	$1 + \tan^2 x = \sec^2 x$

Table 15.2 Differentials of hyperbolic functions and corresponding trigonometric functions

$\dfrac{d}{dx}(\sinh x) = \cosh x$	$\dfrac{d}{dx}(\sin x) = \cos x$
$\dfrac{d}{dx}(\cosh x) = \sinh x$	$\dfrac{d}{dx}(\cos x) = -\sin x$
$\dfrac{d}{dx}(\tanh x) = \operatorname{sech}^2 x$	$\dfrac{d}{dx}(\tan x) = \sec^2 x$
$\dfrac{d}{dx}(\operatorname{cosech} x) = -\coth x\,\operatorname{cosech} x$	$\dfrac{d}{dx}(\operatorname{cosec} x) = -\cot x\,\operatorname{cosec} x$
$\dfrac{d}{dx}(\operatorname{sech} x) = -\tanh x\,\operatorname{sech} x$	$\dfrac{d}{dx}(\sec x) = \tan x\,\sec x$
$\dfrac{d}{dx}(\coth x) = -\operatorname{cosech}^2 x$	$\dfrac{d}{dx}(\cot x) = -\operatorname{cosec}^2 x$

15.3 Differentiation of hyperbolic functions

The similarities between hyperbolic and trigonometric functions shown in Table 15.1 also carry over into the field of differentiation. For example, consider the differentiation of $\sinh x$ and $\cosh x$:

$$\frac{d}{dx}(\sinh x) = \frac{d}{dx}\left\{\frac{1}{2}(e^x - e^{-x})\right\} = \frac{1}{2}(e^x + e^{-x}) = \cosh x \tag{15.9a}$$

$$\frac{d}{dx}(\cosh x) = \frac{d}{dx}\left\{\frac{1}{2}(e^x + e^{-x})\right\} = \frac{1}{2}(e^x - e^{-x}) = \sinh x. \tag{15.9b}$$

The results of differentiating the six hyperbolic functions given in §15.1 together with their trigonometric counterparts are listed in Table 15.2.

Exercise 15.2
Find the derivatives of (i) $\cosh^2 x$; (ii) $\sinh^2 x$; (iii) $\tanh^2 x$.

15.4 Taylor expansions of sinh x and cosh x

We now consider $\sinh x$ represented as a power series in the variable x:

$$\sinh x = a_0 + a_1 x + a_2 x^2 + \cdots + a_n x^n \cdots. \tag{15.10a}$$

The characteristics of $\cosh x$ and $\sinh x$ that are needed can be seen in Figures 15.1a and b. These are that, for $x=0$, $\sinh x = 0$ and $\cosh x = 1$. From (15.10a), putting $x=0$, it is clear that $a_0 = 0$. Then, differentiating both sides of (15.10a),

$$\cosh x = a_1 + 2a_2 x + 3a_3 x^2 + \cdots + na_n x^{n-1} \cdots. \tag{15.10b}$$

Now putting $x=0$ gives $a_1 = 1$. Repeatedly differentiating both sides of the equation and then putting $x=0$ at each stage gives the values of the coefficients.

The even coefficients are all zero and the nth coefficient, for n odd, is

$$a_n = \frac{x^n}{n!}.$$ (15.11)

Thus the first few terms in the series expansion of $\sinh x$ are

$$\sinh x = x + \frac{x^3}{3!} + \frac{x^5}{5!} + \frac{x^7}{7!} + \cdots$$ (15.12)

By a similar process we find that

$$\cosh x = 1 + \frac{x^2}{2!} + \frac{x^4}{4!} + \frac{x^6}{6!} + \cdots.$$ (15.13)

The forms of the series expansions of $\sinh x$ and $\cosh x$ are consistent with their properties as odd and even functions respectively. It will also be seen, from (15.12), (15.13) and (3.34), that

$$\sinh x + \cosh x = 1 + x + \frac{x^2}{2!} + \frac{x^3}{3!} + \frac{x^4}{4!} + \cdots = e^x,$$ (15.14)

which also follows from the definitions of the hyperbolic functions (15.1) and (15.2).

The series expansions give an indication of why the hyperbolic functions have trigonometric-like properties. From (4.44a)

$$\sin(ix) = ix - \frac{(ix)^3}{3!} + \frac{(ix)^5}{5!} - \frac{(ix)^7}{7!} + \cdots$$

$$= i\left(x + \frac{x^3}{3!} + \frac{x^5}{5!} + \frac{x^7}{7!} + \cdots\right) = i\sinh x$$ (15.15a)

and from (4.44b)

$$\cos(ix) = 1 - \frac{(ix)^2}{2!} + \frac{(ix)^4}{4!} - \frac{(ix)^6}{6!} + \cdots$$

$$= 1 + \frac{x^2}{2!} + \frac{x^4}{4!} + \frac{x^6}{6!} + \cdots = \cosh x.$$ (15.15b)

Similarly it can be shown that

$$\tan(ix) = i\tanh x.$$ (15.15c)

We can now see how (15.8) arises, starting with (1.17) and using relationships (15.15). We have

$$\cosh(x + y) = \cos\{i(x + y)\} = \cos(ix)\cos(iy) - \sin(ix)\sin(iy)$$
$$= \cosh x \cosh y - i\sinh x \times i\sinh y$$
$$= \cosh x \cosh y + \sinh x \sinh y.$$ (15.16)

The other relationships in Table 15.1 can similarly be derived from the corresponding trigonometric relationships.

? **Exercise 15.3**
From the series expansions (15.12) and (15.13) show that
$\cosh x - \sinh x = e^{-x}$.

15.5 Integration involving hyperbolic functions

From the differentials given in Table 15.2 we find

$$\int \cosh x \, dx = \sinh x + C \tag{15.17a}$$

$$\int \sinh x \, dx = \cosh x + C. \tag{15.17b}$$

Other integrals can be found by standard methods. For example,

$$\int \tanh x \, dx = \int \frac{\sinh x}{\cosh x} dx = \int \frac{d/dx(\cosh x)}{\cosh x} dx = \ln(\cosh x) + C. \tag{15.17c}$$

and similarly

$$\int \coth x \, dx = \ln(\sinh x) + C. \tag{15.17d}$$

An interesting integral, which has previously been solved by the integration-by-parts method, is $\int \frac{1}{1-x^2} dx$, for which the solution is given by (5.31) as $\frac{1}{2}\ln\sqrt{\frac{1+x}{1-x}} + C$. Now we can solve it by making the substitution $x = \tanh u$, which gives $\frac{dx}{du} = \text{sech}^2 u$. Then

$$\int \frac{1}{1-x^2} dx = \int \frac{\text{sech}^2 u}{1 - \tanh^2 u} du = \int du = u + C = \tanh^{-1} x + C. \tag{15.18}$$

It is not obvious that solutions (5.31) and (15.18) are the same. For them to be so requires $\tanh^{-1} x$ and $\ln\sqrt{\frac{1+x}{1-x}}$ either to be equal or to differ from one another by a constant. In fact they are equal. To show this we write

$$y = \ln\sqrt{\frac{1+x}{1-x}} = \frac{1}{2}\ln\left(\frac{1+x}{1-x}\right)$$

or

$$2y = \ln\left(\frac{1+x}{1-x}\right). \tag{15.19}$$

Taking the exponential of both sides of (15.19),

$$e^{2y} = \frac{1+x}{1-x}. \tag{15.20}$$

Solving for x in terms of y,

$$x = \frac{e^{2y} - 1}{e^{2y} + 1} = \frac{e^y - e^{-y}}{e^y + e^{-y}} = \tanh y.$$

Hence $y = \tanh^{-1} x$, which shows the identity.

Other useful integrals involving hyperbolic function solutions are

$$\int \frac{1}{\sqrt{1 + x^2}} \, dx = \sinh^{-1} x + C, \tag{15.21a}$$

found by the substitution $\sinh x = u$, and

$$\int \frac{1}{\sqrt{x^2 - 1}} \, dx = \cosh^{-1} x + C, \tag{15.21b}$$

found by the substitution $\cosh x = u$.

Exercise 15.4

Find the indefinite integrals (i) $\displaystyle\int \frac{1}{\sqrt{1 + 4x^2}} \, dx$; (ii) $\displaystyle\int \frac{1}{\sqrt{4x^2 - 1}}$.

15.6 **Comments about analytical functions**

Take the case of Dr Jones, who is working in a particular area of science that continually throws up solutions to her problems as integrals of the form

$$f(a) = \int_0^{\pi/2} \frac{\sin ax}{(1 + a^x)^{7/2}} \, dx. \tag{15.22}$$

Over and over again she numerically solves these integrals, by the methods described in Chapter 29, and she finds that colleagues working elsewhere are doing the same. Dr Jones decides to calculate $f(a)$ at finely spaced intervals of a and to make these tables available to others. Somewhat immodestly, she decides to call the function of a the Jones function, indicated by $Jn(a)$. After some time this becomes accepted as a standard analytical function and is tabulated in published books of tables. Thereafter, anyone who meets an integral of the form (15.22) will say that he has an analytical solution that is the Jones function $Jn(a)$.

The point being made here is that a so-called analytical function is one that crops up in scientific work often enough for it to appear in published tables and to feature in the standard libraries of computer languages. The functions sine and cosine obviously fall into this category. However, there are some rather less common functions that also come into the category of analytical functions – Bessel functions, for example – as well as much less common ones, such as Hankel functions and Whittaker functions, which most scientists do not meet at all but are important to those who need them.

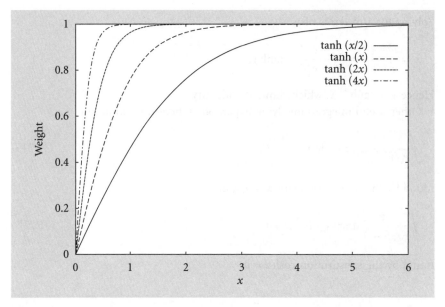

Figure 15.2 The function $\tanh\left(\dfrac{x}{a}\right)$ for various values of a.

Because they are easily obtained in computational work, standard functions can sometimes be used in unorthodox ways for particular purposes. For example, let us suppose that in some application it is desired to use a weighting scheme with observed data such that the weight is linearly dependent on the magnitude of the data for low-magnitude data but saturates to unit weight for high-magnitude data. A weighting scheme of the form

$$w = \tanh\left(\frac{x}{a}\right) \tag{15.23}$$

for datum value x meets the required conditions, and the parameter a can be adjusted to match particular requirements about the detailed shape of the weighting curve, as shown in Figure 15.2.

Knowing the properties of the common analytical functions can be very useful to the working scientist.

Problems

15.1 Show that:

(i) $\sinh A + \sinh B = 2 \sinh\left(\dfrac{A+B}{2}\right) \cosh\left(\dfrac{A-B}{2}\right)$;

(ii) $\cosh A + \cosh B = 2 \cosh\left(\dfrac{A+B}{2}\right) \cosh\left(\dfrac{A-B}{2}\right)$.

15.2 Differentiate the following:

(i) $\dfrac{\sinh x}{1 + x^2}$;

(ii) $\sinh x \cos x$;

(iii) $\tan^{-1}(\sinh x)$.

15.3 Show that the following series are valid:

(i) $\tanh x = x - \dfrac{x^3}{3} + \dfrac{2x^5}{15} + \cdots$;

(ii) $\tanh^{-1} x = x + \dfrac{x^3}{3} + \dfrac{x^5}{5} + \cdots$.

15.4 Evaluate the following integrals:

(i) $\displaystyle\int \dfrac{1}{\sqrt{x^2 + 4x + 3}}\, dx$;

(ii) $\displaystyle\int \dfrac{1}{\sqrt{x^2 + 4x + 5}}\, dx$;

(iii) $\displaystyle\int e^x \cosh^2 x\, dx$;

(iv) $\displaystyle\int_0^1 (\sinh^{-1} x) x \sqrt{1 + x^2}\, dx$.

15.5 A scientist wishes to assign weights to data such that low magnitudes have a weight proportional to the magnitude and for high magnitudes the weight saturates to unity. For magnitude 100 the weight is required to be 0.5. Find a suitable weighting function. (Figure 15.2 may be used for an approximate solution or a more accurate, computer-derived solution may be found.)

16

Vector analysis

In the context of physics we are often interested in a quantity or property which varies in a smooth and continuous way over some one-, two-, or three-dimensional region of space. This constitutes either a scalar field or a vector field, depending on the nature of the property. In this chapter we consider the relationship between a scalar field involving a variable potential and a vector field involving 'field', where this means force per unit mass or charge. The properties of scalar and vector fields are described and how they lead to important concepts, such as that of a conservative field, and the important and useful Gauss and Stokes theorems. Finally, the ideas of vector analysis are applied to develop Maxwell's historic equations that led to the recognition that light was an electromagnetic wave.

16.1 Scalar and vector fields

A newspaper will usually show in its weather section a map showing the variation of atmospheric pressure at ground level. Points where the pressures have a specified value, e.g. 1000 millibars, are connected to show **isobars**, contours of equal pressure. These contour lines give a very graphic picture of the way that the pressure varies over the area in question. Pressure is a scalar quantity and if a particular region has a scalar quantity $\phi(\mathbf{r})$ associated with it, where $\mathbf{r}$ represents a position in the region, then ϕ is said to be a **scalar field** for which, in the case of ground pressure, the region is the surface of the Earth. If, for example, the pressure at different heights were also considered then the associated scalar field would be fully three-dimensional.

Similarly, in some defined region there can be a vector quantity that varies from place to place. The surface currents in the ocean, described by the velocity of the water, $\mathbf{v}(\mathbf{r})$, are one example of this, and we could extend the example into three dimensions by considering currents below the surface. The quantities $\mathbf{v}$ define a **vector field**. Because a vector is defined by two quantities, a magnitude and a direction, vector fields are somewhat more difficult to present visually, even in two dimensions, because one cannot draw contours. Figure 16.1 shows a representation of a two-dimensional vector field as a set of arrows whose length and direction give the vector at each point.

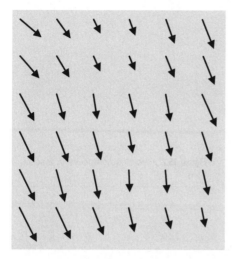

Figure 16.1 Representation of a vector field.

16.2 Gradient (grad) and del operators

We consider a two-dimensional scalar field, $\phi(x,y)$, and two very close contours ϕ and $\phi + \Delta\phi$. In a small region around point O, on contour ϕ, the neighbouring contours will be closely parallel. This is shown in Figure 16.2. The maximum rate of change of ϕ at the point O is in the direction OP, which is perpendicular to the contours. Expressed as a vector this rate of change, called **grad** ϕ, is, in a finite-difference approximation, $(\Delta\phi/\Delta n)\mathbf{n}$, where $\mathbf{n}$ is the unit vector in the direction OP. We can express $\mathbf{n}$ in terms of the unit vectors $\mathbf{i}$ and $\mathbf{j}$ by

$$\mathbf{n} = \frac{\delta x\mathbf{i} + \delta y\mathbf{j}}{\Delta n}.$$

(16.1)

In addition we note from similar triangles that

$$\frac{\delta x}{\Delta n} = \frac{\Delta n}{\Delta x} \quad \text{and} \quad \frac{\delta y}{\Delta n} = \frac{\Delta n}{\Delta y}.$$

(16.2)

From the definition of grad and using (16.1) and (16.2) we find, in a finite-difference approximation,

$$\text{grad}\,\phi = \frac{\Delta\phi}{\Delta n}\mathbf{n} = \frac{\Delta\phi\delta x}{(\Delta n)^2}\mathbf{i} + \frac{\Delta\phi\delta y}{(\Delta n)^2}\mathbf{j} = \frac{\Delta\phi}{\Delta x}\mathbf{i} + \frac{\Delta\phi}{\Delta y}\mathbf{j}.$$

Taking the limit $\Delta n \rightarrow 0$ we find

$$\text{grad}\,\phi = \frac{\partial\phi}{\partial x}\mathbf{i} + \frac{\partial\phi}{\partial y}\mathbf{j}.$$

(16.3)

We may also write (16.3) as

$$\text{grad}\,\phi = \left(\frac{\partial}{\partial x}\mathbf{i} + \frac{\partial}{\partial y}\mathbf{j}\right)\phi.$$

(16.4)

Extending this image to a three-dimensional scalar field, in place of contours we have surfaces of constant value and the gradient at a particular point is perpendicular to the surface going through that point. The equation corresponding

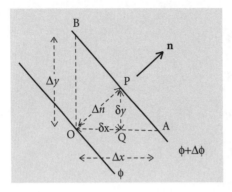

Figure 16.2 Two close contours in a scalar field.

to (16.4) is then

$$\text{grad}\,\phi = \left(\frac{\partial}{\partial x}\mathbf{i} + \frac{\partial}{\partial y}\mathbf{j} + \frac{\partial}{\partial z}\mathbf{k}\right)\phi = \nabla\phi \qquad (16.5)$$

where ∇ (del) is a **vector operator** given by

$$\nabla = \frac{\partial}{\partial x}\mathbf{i} + \frac{\partial}{\partial y}\mathbf{j} + \frac{\partial}{\partial z}\mathbf{k}. \qquad (16.6)$$

In physical situations we may have a potential, ϕ, which varies over a particular region and thus constitutes a scalar field. In one dimension the relationship between potential and **field**, E, is given by

$$E = -\frac{\mathrm{d}\phi}{\mathrm{d}x}$$

and, correspondingly in three dimensions we have

$$\mathbf{E} = -\nabla\phi = -\frac{\partial\phi}{\partial x}\mathbf{i} - \frac{\partial\phi}{\partial y}\mathbf{j} - \frac{\partial\phi}{\partial z}\mathbf{k} = E_x\mathbf{i} + E_y\mathbf{j} + E_z\mathbf{k}. \qquad (16.7)$$

Note the unfortunate use of the word 'field' in two different contexts to mean different things. For electrostatics the potential is the **potential energy per unit charge** at a point and the field is the **force per unit charge**, a vector quantity since force is a vector. For gravitation the potential is the **potential energy per unit mass** at a point and the field is the **force per unit mass**.

EXAMPLE 16.1

As an example we calculate the electric field at point (x, y, z) due to a charge q_1 at $(2, 0, 0)$ and a charge q_2 at $(-2, 0, 0)$ where charges are in coulombs and distances in metres. The potential at the point (x, y, z) is

$$\phi(x, y, z) = \frac{q_1}{4\pi\varepsilon_0\left\{(2-x)^2 + y^2 + z^2\right\}^{1/2}} + \frac{q_2}{4\pi\varepsilon_0\left\{(2+x)^2 + y^2 + z^2\right\}^{1/2}}.$$

The components of the field are $-\dfrac{\partial\phi}{\partial x}$, $-\dfrac{\partial\phi}{\partial y}$, and $-\dfrac{\partial\phi}{\partial z}$ and these are:

$$E_x = -\frac{q_1(2-x)}{4\pi\varepsilon_0\left\{(2-x)^2 + y^2 + z^2\right\}^{3/2}} + \frac{q_2(2+x)}{4\pi\varepsilon_0\left\{(2+x)^2 + y^2 + z^2\right\}^{3/2}}$$

$$E_y = \frac{q_1 y}{4\pi\varepsilon_0 \left\{ (2-x)^2 + y^2 + z^2 \right\}^{3/2}} + \frac{q_2 y}{4\pi\varepsilon_0 \left\{ (2+x)^2 + y^2 + z^2 \right\}^{3/2}}$$

$$E_z = \frac{q_1 z}{4\pi\varepsilon_0 \left\{ (2-x)^2 + y^2 + z^2 \right\}^{3/2}} + \frac{q_2 z}{4\pi\varepsilon_0 \left\{ (2+x)^2 + y^2 + z^2 \right\}^{3/2}} .$$

Exercise 16.1
A scalar field is represented by $\phi = 4x^2 + y^2 + z^2 - 2xy + yz$. Write down an expression for the vector field $\nabla\phi$.

16.3 Conservative fields

The relationship between a field and a potential is one that has been assumed in (16.7) but now this relationship is dealt with in more detail. Consider a particle at vector position $\mathbf{r}$ experiencing a force that varies with $\mathbf{r}$. The force that it experiences, $\mathbf{F}(\mathbf{r})$, constitutes a vector field. Now let us suppose that the body moves from $\mathbf{r}$ to $\mathbf{r} + d\mathbf{r}$. The work done *by the field* in moving the particle is the distance dr times the component of $\mathbf{F}(\mathbf{r})$ along $d\mathbf{r}$, which is

$$dW = \mathbf{F}(\mathbf{r}) \cdot d\mathbf{r} \tag{16.8}$$

If the particle now moves from point A to point B in the field then the work done by the field on the particle is

$$W_{AB} = \int_S \mathbf{F}(\mathbf{r}) \cdot d\mathbf{r} \tag{16.9}$$

where the integral is along the path shown as S in Figure 16.3.

The vector field $\mathbf{F}(\mathbf{r})$ is **conservative** if the work done by the field in going from A to B is independent of the path S. If the particle is then moved further to point C then, from the definition of a conservative field

$$W_{AB} + W_{BC} = W_{AC}. \tag{16.10}$$

In the particular case that point C is the original point A we have

$$W_{AB} + W_{BA} = W_{AA} = 0 \tag{16.11}$$

since the work done by the field must be zero if the particle does not move. Since the path from A to B may be different from the path from B to A we can now write

$$W_{AA} = \oint \mathbf{F}(\mathbf{r}) \cdot d\mathbf{r} = 0 \tag{16.12}$$

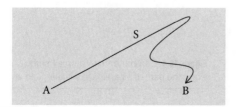

Figure 16.3 Movement of a particle from A to B along path S.

where the integral symbol $\oint$ indicates integration round a closed path.

It can be shown that this is consistent with having a scalar field in the same domain as the vector field such that at point A there is a potential energy ϕ_A. The work done by the field on the particle in going from point A to point B is given by the difference of the two potential energies so that

$$W_{AB} = -(\phi_A - \phi_B). \tag{16.13}$$

Notice the negative sign in (16.13). If the work done by the field *on* the particle is positive, i.e. $W_{AB} > 0$, then one would expect the potential energy of the particle to increase, i.e. $\phi_B > \phi_A$ – which is consistent with (16.13).

The relationship between the vector field $\mathbf{F(r)}$ and the scalar field $\phi(\mathbf{r})$ is as given by (16.7). Using this result we now write

$$dW = \mathbf{F(r)} \cdot d\mathbf{r} = -\nabla\phi(\mathbf{r}) \cdot d\mathbf{r} = -\left(\frac{\partial\phi}{\partial x}\mathbf{i} + \frac{\partial\phi}{\partial y}\mathbf{j} + \frac{\partial\phi}{\partial z}\mathbf{k}\right) \cdot (dx\,\mathbf{i} + dy\,\mathbf{j} + dz\,\mathbf{k})$$

$$= -\left(\frac{\partial\phi}{\partial x}dx + \frac{\partial\phi}{\partial y}dy + \frac{\partial\phi}{\partial z}dz\right). \tag{16.14}$$

Figure 16.4 shows a part of the particle path in going from point A to point B. This has length ds and components (dx, dy, dz).

The first term on the right-hand side of (16.14), with the negative sign, is the negative of the change of potential in going along the path dx and can be written as

$$dW_{dx} = -\{\phi(x + dx, y, z) - \phi(x, y, z)\}.$$

Similarly, along the paths dy and dz we have

$$dW_{dy} = -\{\phi(x + dx, y + dy, z) - \phi(x + dx, y, z)\}$$

and

$$dW_{dz} = -\{\phi(x + dx, y + dy, z + dz) - \phi(x + dx, y + dy, z)\},$$

which gives

$$dW = dW_{dx} + dW_{dy} + dW_{dz}$$
$$= -\{\phi(x + dx, y + dy, z + dz) - \phi(x, y, z)\}. \tag{16.15}$$

From this it can be seen that the work done by the field in going along the complete path is just the negative of the difference between the potential at the end point and the potential at the beginning point.

We can illustrate the basic property of a conservative field by the following example.

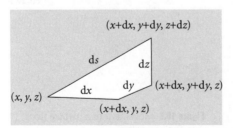

Figure 16.4 Components of a small shift along the path of a particle in a scalar field Φ.

EXAMPLE 16.2

In Figure 16.5 we show two points A and B, separated by a distance a in a conservative field, in line with a point O which is the source of a field of the form $F = K/r^2$ or, in vector form

$$\mathbf{F} = \frac{K}{r^3}\mathbf{r}$$

where $\mathbf{r}$ is the vector position of the point relative to the source point. This form gives both the magnitude and direction of the field at any point. The point A is at distance D from O.

We now consider the work done by the field in moving a particle from A to B by the direct path AB and then this will be compared with the work done by the indirect path ACDB. By the direct path the work done is

$$W_{AB} = \int_D^{D+a} \frac{K}{x^2}\,dx = \left| -\frac{K}{x}\right|_D^{D+a} = K\left(\frac{1}{D} - \frac{1}{D+a}\right). \qquad (16.16)$$

We now consider the path from A to C illustrated in Figure 16.6. When the particle is at E, distant x from A, the force on it is

$$F = \frac{K}{D^2 + x^2}$$

and the work done by the field in moving the particle a distance dx is

$$dW = \frac{K}{D^2 + x^2}\cos\theta\,dx = \frac{Kx}{(D^2 + x^2)^{3/2}}\,dx.$$

Hence

$$W_{AC} = \int_0^b \frac{Kx}{(D^2 + x^2)^{3/2}}\,dx$$

$$= K\left| -\frac{1}{(D^2 + x^2)^{1/2}}\right|_0^b = K\left\{\frac{1}{D} - \frac{1}{(D^2 + b^2)^{1/2}}\right\}. \qquad (16.17)$$

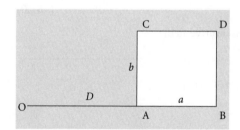

Figure 16.5 Paths from A to B in a conservative field.

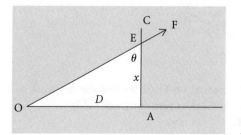

Figure 16.6 The particle path from A to C.

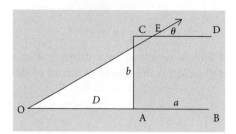

Figure 16.7 The particle path from C to D.

Now we consider the path from C to D as shown in Figure 16.7. The force at the point E, distant x from C is

$$F = \frac{K}{(D+x)^2 + b^2}$$

and the work done in moving a distance dx is

$$dW = \frac{K}{(D+x)^2 + b^2} \cos\theta \, dx = \frac{K(D+x)}{\{(D+x)^2 + b^2\}^{3/2}} dx.$$

Hence

$$W_{CD} = \int_0^a \frac{K(D+x)}{\{(D+x)^2 + b^2\}^{3/2}} dx = K \left| -\frac{1}{\{(D+x)^2 + b^2\}^{1/2}} \right|_0^a$$

$$= K \left\{ \frac{1}{(D^2 + b^2)^{1/2}} - \frac{1}{\{(D+a)^2 + b^2\}^{1/2}} \right\}$$

(16.18)

The work done by the field in the path from D to B may be written down by analogy with (16.17) by replacing D by $D+a$ and reversing the sign of the work since the path is in the opposite direction. This gives

$$W_{DB} = -K \left\{ \frac{1}{D+a} - \frac{1}{\{(D+a)^2 + b^2\}^{1/2}} \right\}.$$

(16.19)

Combining (16.17), (16.18) and (16.19),

$$W_{AC} + W_{CD} + W_{DB} = W_{AB},$$

(16.20)

a result that is consistent with a conservative field.

 Exercise 16.2

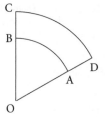

AB and CD are arcs of circles centred on O. Show that the net work done in going along the path ABCDA is zero for force fields
(i) $\mathbf{F} = \frac{K}{r^2} \hat{\mathbf{r}}$ and (ii) $\mathbf{F} = \frac{K}{r} \hat{\boldsymbol{\theta}}$.

16.4 Divergence (div)

Consider a vector field **V** over a small elemental surface of vector area d**A**. If the area is very small then **V** will vary little over the surface and we may take it as constant. Figure 16.8 shows the relationship of **V** to d**A** and we define the flux through the area as

$$dF = \mathbf{V} \cdot d\mathbf{A}, \tag{16.21}$$

which is equivalent to $dA \times$ the component of **V** in the direction of d**A**.

We now consider a small cubical box, with sides of length l parallel to the x, y, and z axes, surrounding a central point O, with position (x, y, z), at which the vector field is **V** as shown in Figure 16.9. We shall now find the total flux *out of the box*. The flux out of the box through the point P, with position $(x + l/2, y, z)$, is

$$F_P = \mathbf{V}(x + \tfrac{1}{2}l, y, z) \cdot l^2 \mathbf{i} = V_x(x + \tfrac{1}{2}l, y, z) l^2$$

$$= \left\{ V_x(x, y, z) + \frac{1}{2}\frac{\partial V_x}{\partial x} l \right\} l^2. \tag{16.22}$$

Similarly the flux out of the box through the point Q, position $(x - l/2, y, z)$ is

$$F_Q = -\left\{ V_x(x, y, z) - \frac{1}{2}\frac{\partial V_x}{\partial x} l \right\} l^2,$$

where the negative sign accounts for the direction of the flux. Thus the total flux out of the box through the x faces is

$$F_x = F_P + F_Q = \frac{\partial V_x}{\partial x} l^3. \tag{16.23}$$

Adding the contributions from the other faces gives the total flux out of the box:

$$F = \left(\frac{\partial V_x}{\partial x} + \frac{\partial V_y}{\partial y} + \frac{\partial V_z}{\partial z} \right) l^3. \tag{16.24}$$

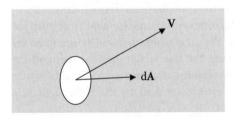

Figure 16.8 The relationship of **V** and the small vector area d**A**.

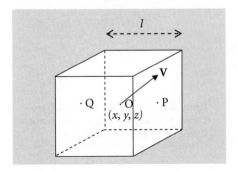

Figure 16.9 The cubical box surrounding the point O.

The outward flux per unit volume at the point O is called the **divergence** of **V** and may be written

$$\text{div } \mathbf{V} = \frac{\partial V_x}{\partial x} + \frac{\partial V_y}{\partial y} + \frac{\partial V_z}{\partial z} = \left(\frac{\partial}{\partial x}\mathbf{i} + \frac{\partial}{\partial y}\mathbf{j} + \frac{\partial}{\partial z}\mathbf{k} \right) \cdot \left(V_x\mathbf{i} + V_y\mathbf{j} + V_z\mathbf{k} \right) = \nabla \cdot \mathbf{V}.$$

(16.25)

> **Exercise 16.3**
> Find an expression for $\nabla \cdot \mathbf{V}$ where
> $\mathbf{V} = (x^2 + y^2)\mathbf{i} + (y^2 + z^2)\mathbf{j} + (z^2 + x^2)\mathbf{k}.$

16.4.1 **Theorems involving divergence**

There is a theorem, called the **divergence theorem**, that relates a volume integral and a surface integral within a vector field. This states that

$$\oiiint_V \nabla \cdot \mathbf{V} \, dv = \oiint_S \mathbf{V} \cdot d\mathbf{A}$$

(16.26)

where the left-hand side is a volume integral, dv being a volume element within the surface, and the right-hand side is a surface integral where $d\mathbf{A}$ is a small vector area on the surface S. Note in particular the integral signs, $\oiiint$ indicating a volume integral and $\oiint$ indicating a surface integral.

Figure 16.10 shows two small neighbouring volume elements within the surface. Since $\nabla \cdot \mathbf{V}$ is the total flux per unit volume at a point in the vector field then $\nabla \cdot \mathbf{V} dv$ is the total flux through the surfaces of a small volume element dv. If we add the flux through the surface elements of two neighbouring volume elements, as shown in Figure 16.10, then it is clear that the contributions through the common face cancel each other since an inward flux for one volume element is an outward flux for the other. If we now imagine that the whole volume is occupied by small abutting volume elements then when the flux is summed for them all the only remaining flux contributions are for volume elements with one side within the surface. The flux through these faces is equivalent to the right-hand side of (16.26).

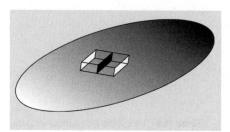

Figure 16.10 A surface within which are shown two neighbouring volume elements with a common face shown black.

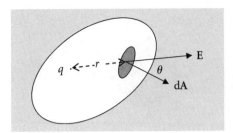

Figure 16.11 A point charge within a closed surface.

A physical situation in which the concept of flux arises is in the electric field flux through a surface surrounding a point charge (Figure 16.11). The flux through the small surface element is

$$dF = \mathbf{E} \cdot d\mathbf{A} = E\, dA \cos\theta = \frac{q\cos\theta\, dA}{4\pi\varepsilon_0 r^2}.$$

The solid angle (§2.1) subtended by the area element at q is

$$d\Omega = \frac{dA\cos\theta}{r^2}$$

so that

$$dF = \frac{q}{4\pi\varepsilon_0}\, d\Omega. \tag{16.27}$$

Integrating (16.27) on both sides over the whole surface,

$$F = \oint \mathbf{E} \cdot d\mathbf{A} = \frac{q}{\varepsilon_0} \tag{16.28}$$

where, again, the symbol $\oint$ represents a **surface integral** and integrating $d\Omega$ over all directions in space gives 4π.

Equation (16.28) is known as **Gauss's law**, which is very useful in deriving fields for some special charge configurations. Since the position of the charge within the surface is not important then q can be replaced by the total charge within the surface, however it is distributed.

EXAMPLE 16.3

As an example of the application of Gauss's law we take the electric field due to a long uniformly charged rod of radius a carrying a linear charge density (charge per unit length) σ. We consider a cylindrical surface of radius r with axis coincident to that of the rod. By symmetry the field is everywhere normal to the cylindrical surface so that the flux through the surface per unit length is

$$F_l = 2\pi r E$$

In the case where $r > a$ (Figure 16.12a) the total included charge per unit length is σ, so from (16.28), $2\pi r E = \sigma/\varepsilon_0$ or

$$E = \frac{\sigma}{2\pi\varepsilon_0 r}. \tag{16.29}$$

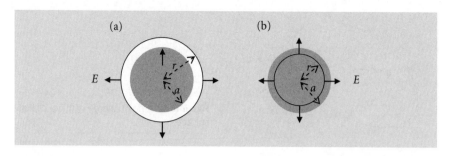

Figure 16.12 The cross-section of the rod (grey) of radius a and the cylindrical surface of radius r.

However, when $r < a$ (Figure 16.12b) the included charge per unit length is $\sigma r^2/a^2$, so that $2\pi r E = \sigma r^2/a^2\varepsilon_0$ or

$$E = \frac{\sigma r}{2\pi a^2\varepsilon_0}.$$ (16.30)

The consequence of Gauss's law is that the total flux through a closed surface contained within an electric field is zero if there is no net charge within the surface. Of course there may be flux through various elements of the surface, inwards and outwards, but the net flux will be zero. For a point within a charge-free vacuum we can write

$$\nabla \cdot \mathbf{E} = 0,$$ (16.31a)

which is one of Maxwell's equations that led to the interpretation of light as an electromagnetic wave motion.

In magnetism it is not possible to have an isolated magnetic pole of some polarity, north or south. Magnetic poles always exist as **dipoles**, a north and a south pole coupled together, so one can always write

$$\nabla \cdot \mathbf{H} = 0,$$ (16.31b)

where $\mathbf{H}$ is the magnetic field. Equation (16.31b) is a second Maxwell equation.

If we consider a fluid flowing with a velocity field $\mathbf{U}$ then, if we consider the vector field $\rho\mathbf{U}$, the net flux through a closed surface will be the rate of mass flow through the surface and $\text{div}(\rho\mathbf{U})$, the net flux divided by the volume, will be the rate of change of density at the point in question. Thus we may write

$$\frac{\partial\rho}{\partial t} = \nabla \cdot (\rho\mathbf{U}) = (\nabla\rho) \cdot \mathbf{U} + \rho\nabla \cdot \mathbf{U}.$$ (16.32)

Note particularly the way that the operator acts on the product $\rho\mathbf{U}$; you can confirm its validity by writing $\mathbf{U}$ in terms of its components and operating on them individually. If the fluid is a gas, which can vary in density both in position and in time, then (16.32) gives the rate of change of density at all points. On the other hand if (16.32) is applied to a liquid, regarded as incompressible, then $\partial\rho/\partial t$ and $\nabla\rho$ are both zero and ρ is a constant, leading to

$$\nabla \cdot \mathbf{U} = 0$$ (16.33)

as a condition of fluid dynamics for an incompressible fluid.

> **Exercise 16.4**
> A sphere of radius a has a uniform volume charge density σ. Find the
> magnitude of the field at distance r from its centre for (i) $r > a$
> and (ii) $r < a$.

16.5 Laplacian operator

A very useful operator that occurs repeatedly in the context of physics is the
Laplacian operator given by

$$\nabla^2 = \nabla \cdot \nabla = \frac{\partial^2}{\partial x^2} + \frac{\partial^2}{\partial y^2} + \frac{\partial^2}{\partial z^2}. \tag{16.34}$$

This operator appears in relation to the wave equation in §20.9. The general three-
dimensional form of the wave equation is

$$\nabla^2 \eta = c^2 \frac{\partial^2 \eta}{\partial t^2}. \tag{16.35}$$

or

$$\frac{\partial^2 \eta}{\partial x^2} + \frac{\partial^2 \eta}{\partial y^2} + \frac{\partial^2 \eta}{\partial z^2} = c^2 \frac{\partial^2 \eta}{\partial t^2} \tag{16.36}$$

The solution to the one-dimensional wave equation, given by (20.31), is a wave in
the x direction. The solution to (16.36) is a wave motion in any arbitrary direction
that we can take as having direction cosines (l, m, n) and is of the form

$$\eta = \sin\left(\frac{2\pi}{\lambda}(lx + my + nz - ct) + \phi\right). \tag{16.37}$$

Given that $l^2 + m^2 + n^2 = 1$, the reader may verify that (16.37) is a solution of (16.36).

Another context in which the Laplacian operator appears is in the Schrödinger
wave equation that is at the heart of quantum mechanics. This will be discussed in
Chapter 21.

The form of ∇ and ∇^2 we have used so far has been for a rectangular Cartesian
system. However, other forms for other coordinate systems and particular symmetries exist and important ones for ∇^2 are as follows.

- **Spherical polar coordinates:**

$$\nabla^2 = \frac{1}{r^2}\frac{\partial}{\partial r}r^2\frac{\partial}{\partial r} + \frac{1}{r^2 \sin\theta}\frac{\partial}{\partial \theta}\left(\sin\theta\frac{\partial}{\partial \theta}\right) + \frac{1}{r^2 \sin^2\theta}\frac{\partial^2}{\partial \phi^2}. \tag{16.38}$$

- **Spherical symmetry.** This is as for spherical polar coordinates but with no
  variation with θ or ϕ:

$$\nabla^2 = \frac{1}{r^2}\frac{\partial}{\partial r}r^2\frac{\partial}{\partial r} = \frac{\partial^2}{\partial r^2} + \frac{2}{r}\frac{\partial}{\partial r}. \tag{16.39}$$

- **Two-dimensional polar coordinates:**

$$\nabla^2 = \frac{1}{r}\frac{\partial}{\partial r}\left(r\frac{\partial}{\partial r}\right) + \frac{1}{r^2}\frac{\partial^2}{\partial \theta^2} = \frac{\partial^2}{\partial r^2} + \frac{1}{r}\frac{\partial}{\partial r} + \frac{1}{r^2}\frac{\partial^2}{\partial \theta^2}.$$
(16.40)

- **Cylindrical coordinates.** This is similar to the equation for two-dimensional polar coordinates with the addition of a z contribution:

$$\nabla^2 = \frac{\partial^2}{\partial r^2} + \frac{1}{r}\frac{\partial}{\partial r} + \frac{1}{r^2}\frac{\partial^2}{\partial \theta^2} + \frac{\partial^2}{\partial z^2}.$$
(16.41)

> **?**
>
> **Exercise 16.5**
> Find $\nabla^2 \psi$ for (i) $\psi = \sin k_x x \sin k_y y \sin k_z z$ and
> (ii) $\psi = r^2 \sin \theta \cos \phi$.

16.6 Curl of a vector field

The curl of a vector field $\mathbf{V}$ is defined as

$$\text{curl } \mathbf{V} = \nabla \times \mathbf{V}$$
(16.42)

or, from (9.12b),

$$\text{curl } \mathbf{V} = \begin{vmatrix} \mathbf{i} & \mathbf{j} & \mathbf{k} \\ \frac{\partial}{\partial x} & \frac{\partial}{\partial y} & \frac{\partial}{\partial z} \\ V_x & V_y & V_z \end{vmatrix} = \left(\frac{\partial V_z}{\partial y} - \frac{\partial V_y}{\partial z}\right)\mathbf{i} + \left(\frac{\partial V_x}{\partial z} - \frac{\partial V_z}{\partial x}\right)\mathbf{j} + \left(\frac{\partial V_y}{\partial x} - \frac{\partial V_x}{\partial y}\right)\mathbf{k}.$$
(16.43)

For a physical interpretation we consider a square contour of side l in the x-y plane with the point O, (x, y, z), at its centre, as shown in Figure 16.13.

We now find an expression for the **circulation**, C_{xy}, which is

$$C_{xy} = \int_C \mathbf{V} \cdot d\mathbf{s},$$
(16.44)

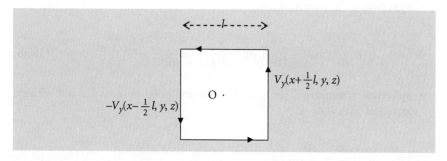

Figure 16.13 The circulation in a square contour in the x-y plane.

where ds is an element of the contour. The contribution from the two sides parallel to y is

$$C_y = l\left\{V_y\left(x+\tfrac{1}{2}l, y, z\right) - V_y\left(x-\tfrac{1}{2}l, y, z\right)\right\} = l\left\{V_y + \tfrac{1}{2}l\,\frac{\partial V_y}{\partial x} - V_y + \tfrac{1}{2}l\,\frac{\partial V_y}{\partial x}\right\}$$

$$= l^2\frac{\partial V_y}{\partial x}, \tag{16.45}$$

where, when no coordinate is specified, it is (x, y, z). By a similar process the contribution to the circulation of the sides parallel to x, taking account of the directions of the components is

$$C_x = -l^2\frac{\partial V_x}{\partial y}$$

so that the total circulation is

$$C_{xy} = l^2\left(\frac{\partial V_y}{\partial x} - \frac{\partial V_x}{\partial y}\right). \tag{16.46}$$

Comparing this expression with (16.43) it is clear that the z component of curl $\mathbf{V}$ is the circulation per unit area in the x-y plane.

The interpretation of this result is as follows. In the immediate vicinity of the point O the vector field $\mathbf{V}$ has a rotatory component in a plane whose normal is in a direction given by the components of curl $\mathbf{V}$. The circulation per unit area in that plane is

$$|\text{curl } \mathbf{V}| = \left\{\left(\frac{\partial V_y}{\partial z} - \frac{\partial V_z}{\partial y}\right)^2 + \left(\frac{\partial V_z}{\partial x} - \frac{\partial V_x}{\partial z}\right)^2 + \left(\frac{\partial V_x}{\partial y} - \frac{\partial V_y}{\partial x}\right)^2\right\}^{1/2}. \tag{16.47}$$

Curl $\mathbf{V}$ is the same at all parts of a body which is in a state of rigid rotation about some axis. If we designate the rotation axis as along the z axis then the motion of all parts of the body is parallel to the x-y plane. Figure 16.14 shows a contour around the point O at $(0, 0, z)$. There is no motion along the radial directions AB and CD and hence no contribution to the circulation. If the angular speed of rotation is ω then the speed at all points along DA is $r\omega$ and the length of the curve DA is $r\theta$. Hence the contribution to the circulation has magnitude $r^2\omega\theta$. Taking an anticlockwise

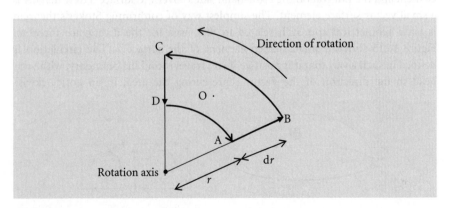

Figure 16.14 A contour for determining curl v for a rigidly rotating body.

circulation as positive it is clear that the total circulation along paths BC and DA is

$$C = (r + dr)^2 \omega\theta - r^2\omega\theta = (2rdr + dr^2)\omega\theta. \tag{16.48a}$$

The area within the contour is

$$A = \left(r + \frac{1}{2}dr\right)\theta \, dr \tag{16.48b}$$

so that

$$\text{curl } \mathbf{V} = \frac{C}{A}\mathbf{k} = 2\omega\mathbf{k}, \tag{16.49}$$

which is independent of position.

We now also show that a radial and spherically symmetric field, $\mathbf{E}$, generated by a point source has $\text{curl }\mathbf{E} = 0$. We may use Figure 16.14 to illustrate this. Now, because of its radial nature, there is no contribution to the circulation along BC and DA. Because of the symmetry of the field the contributions along AB and BC cancel out, so the total circulation is zero. Since fields are additive the curl is zero for the field due to any distribution of point sources, such as electric charges, or indeed a continuous distribution of charges.

 Exercise 16.6
Find $\text{curl }\mathbf{V}$ for $\mathbf{V} = yz\mathbf{i} + zx\mathbf{j} + xy\mathbf{k}$.

16.6.1 Stokes's theorem

We consider a surface S that has a closed non-intersecting boundary, C – the topology of, say, a hammock or one half of a tennis ball. Stokes's theorem states that for a vector field $\mathbf{V}$ within which the surface is situated,

$$\oint_C \mathbf{V} \cdot d\mathbf{r} = \oiint_S (\nabla \times \mathbf{V}) \cdot d\mathbf{A}. \tag{16.50}$$

The integral on the left-hand side is round the boundary within which $d\mathbf{r}$ is a small element and the integral on the right-hand side is over the surface S of which $d\mathbf{A}$ is a small vector surface element. The simplest way of confirming Stokes's theorem is via a geometrical approach related to that used for the divergence theorem. Figure 16.15 shows a small shaded element of the surface S. The circulation is defined in such a way that it is positive if you travel round the boundary, with your head in the direction of the vector representing the area, in an anticlockwise

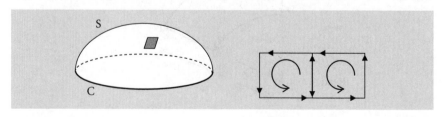

Figure 16.15 The surface S, boundary C, and a small surface element. The circulation around two enlarged surface elements is shown.

direction. The right-hand side of (16.50) corresponds to dividing S into a large (effectively infinite) number of surface elements and adding the circulations for each of them. Figure 16.15 shows two neighbouring elements with the positive sense of the circulation indicated in each of them, and it is clear that the contributions to the circulation along the common boundary cancel each other. If the whole surface is covered with area elements then the only non-cancelling contributions to the sum of the circulations will be along the boundary C, the quantity given by the left-hand side of (16.50).

16.7 **Maxwell's equations and the speed of light**

Here we shall deal with the simplest case where Maxwell's equations are found for free space. Two of the four equations were found in §16.4.1:

$$\nabla \cdot \mathbf{E} = 0 \tag{16.31a}$$

and

$$\nabla \cdot \mathbf{H} = 0. \tag{16.31b}$$

The next equation is a form of Ampère's law that determines the way that magnetic fields are produced by electric currents, which in the present context we present in the form

$$\nabla \times \mathbf{H} = \varepsilon_0 \frac{\partial \mathbf{E}}{\partial t}, \tag{16.51}$$

where ε_0 is the permittivity of free space (§2.4).

The final equation is based on Faraday's law, which determines the way that electric potentials are generated by changes of magnetic flux and can be expressed as

$$\nabla \times \mathbf{E} = -\mu_0 \frac{\partial \mathbf{H}}{\partial t}, \tag{16.52}$$

where μ_0 is the permeability of free space (§2.4).

Taking the curl of both sides of (16.52) and using (16.51),

$$\nabla \times (\nabla \times \mathbf{E}) = -\mu_0 \frac{\partial}{\partial t}(\nabla \times \mathbf{H}) = -\mu_0 \varepsilon_0 \frac{\partial^2 \mathbf{E}}{\partial t^2}. \tag{16.53}$$

The left-hand side of (16.53) can be expanded as follows:

$$\nabla \times (\nabla \times \mathbf{E}) = \nabla(\nabla \cdot \mathbf{E}) - \nabla^2 \mathbf{E} \quad \text{(see Problem 16.2)}$$

and since, from (16.31a) $\nabla \cdot \mathbf{E} = 0$, (16.53) now becomes

$$\nabla^2 \mathbf{E} = \mu_0 \varepsilon_0 \frac{\partial^2 \mathbf{E}}{\partial t^2}. \tag{16.54}$$

Taking the x component of (16.54) we have

$$\frac{\partial^2 E_x}{\partial x^2} + \frac{\partial^2 E_x}{\partial y^2} + \frac{\partial^2 E_x}{\partial z^2} = \mu_0 \varepsilon_0 \frac{\partial^2 E_x}{\partial t^2} \tag{16.55}$$

with similar expressions for the other two components. Comparing this equation with (16.36) it is clear that E_x has a wave-like character with speed $\sqrt{\mu_0 \varepsilon_0}$ and numerically this is found to be equal to the speed of light. The solution to (16.54) is a **vector wave** and it is possible to derive a similar wave equation from Maxwell's equations that involve **H**.

Maxwell's equations have been described as the most important development in physics in the nineteenth century. Certainly they are a very elegant example of the application of vector analysis.

Problems

16.1 A vector field is described by $\mathbf{V} = x\mathbf{i} + y\mathbf{j} + 2z\mathbf{k}$. A box is defined by the six planes $x = \pm 1$, $y = \pm 1$ and $z = \pm 1$. Compare the two sides of (16.26) for this volume and confirm the divergence theorem.

16.2 Verify that a field represented by $\mathbf{F} = (K/r^4)\mathbf{r}$ is conservative by finding the work done by the field in going from A to B directly and by the path ACDB in Figure 16.5.

16.3 By representing the vector field in terms of components, confirm that
$$\nabla \times (\nabla \times \mathbf{V}) = \nabla(\nabla \cdot \mathbf{V}) - \nabla^2 \mathbf{V}.$$

16.4 Verify the following relationships:
$$\nabla \cdot \nabla \times \mathbf{V} = 0 \text{ (div curl V=0)}$$
$$\nabla \times \nabla \phi = 0 \text{ (curl grad } \phi = 0)$$
where $\mathbf{V}$ is a vector field and ϕ a scalar field.

16.5 A square is defined by the boundaries $x = \pm a$ and $y = \pm a$. Confirm Stokes's theorem for this plane surface given the vector field $\mathbf{V} = -y^3\mathbf{i} + x^3\mathbf{j}$.

Fourier analysis

17

Fourier transforms are a powerful approach in many areas of physics and engineering. Some of the areas in which they are applied are:

- optics
- astronomy (e.g. monitoring sunspot activity)
- X-ray crystallography
- quantum mechanics
- Doppler radar
- electrical and electronic engineering
- biomedical engineering

The aim of this chapter is to introduce Fourier methods that are related to the concept of correlation. The approach we use is through a consideration of signals that are of importance in everyday life as well as to the practising scientist.

17.1 Signals

Signals are all around us in everyday life. Signals from mobile phones, radio, television, satellites are omnipresent, although invisible to us. In communications over long distances, these signals are propagated through space in the form of electromagnetic waves. On a smaller scale, within a radio or television set for example, signals are in the form of electrical currents. In medical applications of ultrasound, the signals propagate using ultrasonic vibrations within the medium of interest. The output of many experiments can be in the form that we would conventionally think of as a signal – for example, the information coming in from a satellite sampling the Earth's magnetic environment.

It turns out that we can break down many signals into a set of cosines with varying frequencies. This procedure is called **spectral analysis** and the method that we use is called **Fourier analysis**. If the signal is periodic, then the technique of Fourier series is used. For more general signals, the Fourier transform can be applied.

17.2 The nature of signals

Signals exist in one of two forms, analogue and digital. We will first discuss the differences between these two forms.

17.2.1 Analogue signals

Some signals can be represented as a **continuous** function of time, for example, the cosine signal shown in Figure 17.1. A signal that exists as a continuous function of time is called an **analogue** signal.

17.2.2 Digital signals

Digital signals are obtained from analogue signals by two operations: **sampling** and **binary coding**. We shall discuss just the sampling process here, as this is the most critical stage in the conversion of an analogue signal to digital form. When we sample an analogue signal, we take information of this signal at constant intervals of time, as illustrated in Figure 17.2. This is sampled as shown to obtain the sampled signal, $x_s(t)$. The time between adjacent samples is τ.

The **sampling frequency** is defined as

$$f_s = \frac{1}{\tau}\,\text{Hz} \quad \text{or} \quad \omega_s = \frac{2\pi}{\tau}\,\text{rad s}^{-1}. \tag{17.1}$$

Let us represent the set of samples as $\{x_0, x_1, x_2, \ldots, x_n, \ldots\}$. Now, sampling any signal means that we are ignoring information between samples and hence we are losing this information. The larger the time between samples, or the lower the sampling frequency, then the more of this information is lost. Suppose that we are given a set of samples of a signal as shown in Figure 17.3. The sampling frequency is given by $f_s = 1/0.05 = 20\,\text{Hz}$. Suppose that you are told that it is a cosine signal.

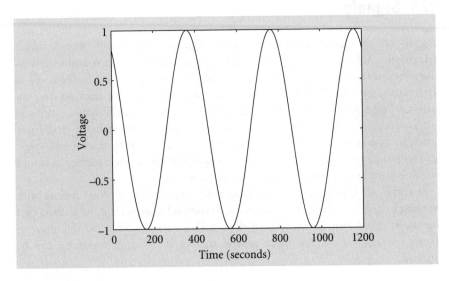

Figure 17.1 Cosine signal.

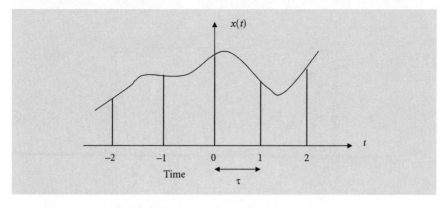

Figure 17.2 Sampled signal.

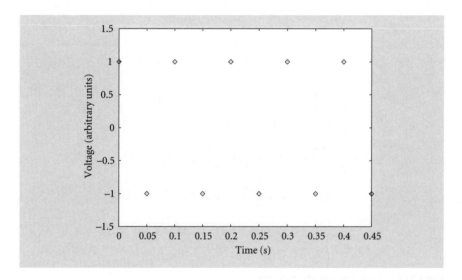

Figure 17.3 Samples of a signal.

Could you reconstruct the original continuous signal? The most obvious answer might be as in Figure 17.4., i.e. a cosine with amplitude $= 1$, phase $= 0$, and frequency $= 10$ Hz. However, another possible reconstruction is shown in Figure 17.5, i.e. a cosine with frequency 30 Hz. Yet another reconstruction, a cosine of frequency 50 Hz, is shown in Figure 17.6.

In fact, considering all possible reconstructions of the continuous signal, the following are possible frequencies: 10, 30, 50,...Hz. So there is an ambiguity here in determining the frequency.

Now suppose that we are given **prior information** about the underlying continuous cosine, that its frequency has a maximum value of 10 Hz . We can then see that the only possible reconstruction is that shown in Figure 17.4; that is, the frequency is 10 Hz. In fact, it can be shown in general, that if a signal has frequency f, then the minimum sampling frequency for us to unambiguously determine the frequency of the underlying signal is $2f$. Or, put another way,

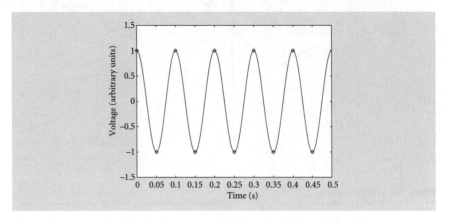

Figure 17.4 A possible reconstruction of the original signal from Figure 17.3.

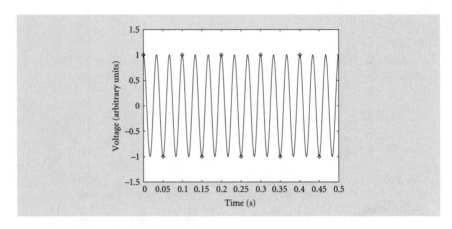

Figure 17.5 Another possible reconstruction.

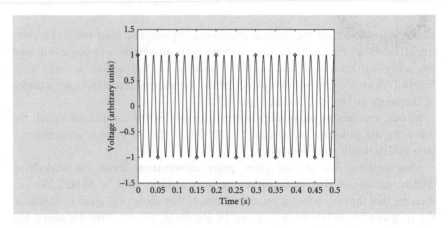

Figure 17.6 Yet another possible reconstruction.

the times between samples T should satisfy the relation

$$T < \frac{1}{2f}. \tag{17.2}$$

This is a basic form of the **sampling theorem** first published by Claude Shannon in 1948; see Shannon and Weaver (1963).

Hence, we can reconstruct the original signal, but we need to know something about the maximum possible value of the underlying frequency.

17.3 **Amplitude–frequency diagrams**

In certain circumstances, it is convenient to represent signals by decomposing them into sums of cosine signals of different amplitudes, phases, and frequencies; this procedure is termed **spectral analysis**. In electronic equipment used in navigation and communication applications, it is desirable to filter off noise, interference, and other effects. It is very difficult to design suitable equipment by analysing these signals as a function of time; however, as we will see later (in Chapter 33) this procedure is greatly simplified if we adopt spectral analysis.

Let us take the analogue signal

$$x(t) = A_1 \cos(2\pi f_1 t + \phi_1). \tag{17.3}$$

This signal has an amplitude A_1 and a phase ϕ_1 (Figure 17.7). Its period is $T_1 = 1/f_1$. Now we plot two graphs. The first is amplitude versus frequency (f). To represent the signal above, we draw a vertical line on the graph at $f = f_1$ and height A_1 to give Figure 17.8. We refer to this graph as the **amplitude spectrum**.

The second graph, Figure 17.9, is phase versus frequency. To represent the signal above, we draw a vertical line on the graph at $f = f_1$ and height ϕ_1. This graph is the **phase spectrum**. The amplitude and phase spectra together are referred to as the **frequency spectrum**. The frequency spectrum defines the signal completely.

So far, the utility of the frequency spectrum is not clear. It is obvious from Figure 17.7 what the amplitude and phase of the signal are. However, consider the following example

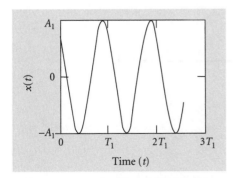

Figure 17.7 Analogue cosine signal.

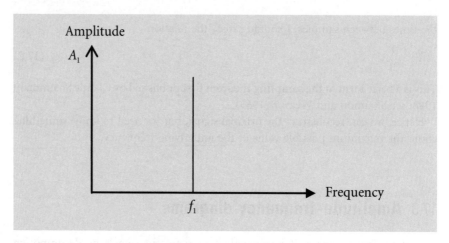

Figure 17.8 Amplitude versus frequency.

Exercise 17.1

Given $s(t) = 3\cos(2\pi(1.5)t + \pi/2)$, sketch the amplitude spectrum and the phase spectrum of the signal.

Exercise 17.2

Derive an expression for the cosine signal $s(t)$, which has the following amplitude and phase spectra.

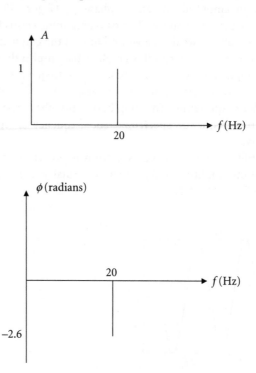

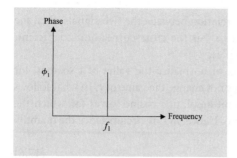

Figure 17.9 Phase versus frequency.

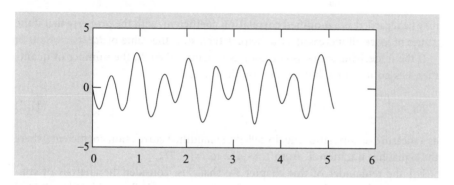

Figure 17.10 Analogue signal consisting of two cosines.

EXAMPLE 17.1

$$x(t) = A_1 \cos(2\pi f_1 t + \phi_1) + A_2 \cos(2\pi f_2 t + \phi_2). \tag{17.4}$$

Suppose we take $A_1 = 1$, $A_2 = 2$, $f_1 = 1\,\text{Hz}$, $f_2 = 1.8\,\text{Hz}$, $\phi_1 = 0.628$ radians, $\phi_2 = 1.885$ radians. This signal is illustrated in Figure 17.10.

It is not very clear from this either how many cosine signals there are (although someone with experience might deduce that there are two) or what the frequencies, amplitudes, and phases are. One method to deduce this is to use **Fourier transforms**, which we will consider next.

17.4 **Fourier transform**

We now look at how correlation methods can be used to determine how many cosines make up a signal, along with their frequencies, amplitudes, and phases.

17.4.1 **Correlation function**

If there are two sets of samples $x = \{x_i\}$ and $y = \{y_i\}$, with i from 1 to N, then the correlation coefficient between the two is

$$r = \frac{\langle xy \rangle - \langle x \rangle \langle y \rangle}{\sigma_x \sigma_y} \tag{17.5}$$

where σ_x and σ_y are the standard deviations of x and y. If the two samples are identical, or one is just a scaled version of the other, then the correlation coefficient is 1. If there is no systematic relation between the two signals, then the correlation coefficient is 0. The nearer to 1 is the cross-correlation coefficient, then the more similar are the two samples.

The effect of the divisors in (17.5) is to normalize the value of r so that, for example, doubling all the values of x does not change the value of r. In what follows the quantities corresponding to y are unit-amplitude cosine waves for which the variance is fixed and the mean, $\langle y \rangle$, is zero. Hence, in this circumstance the quantity

$$R_0 = \langle xy \rangle = \frac{1}{N} \sum_{j=0}^{N-1} x_j y_j \tag{17.6}$$

may be regarded as a **modified correlation coefficient**, with the property that if the values of x are all increased by a factor k then so is the value of R_0.

If the underlying signal is continuous in nature then N, the number of quantities, has no meaning, but we can use the integral

$$R_0 = \int_{-T/2}^{T/2} x(t) y(t) \, dt \tag{17.7}$$

as a measure of correlation, again called the modified correlation coefficient, where the signal has duration T, from $t = -T/2$ to $t = +T/2$.

For the remainder of this chapter we shall just consider the analysis of continuous (analogue) signals. In the next chapter, we shall consider the spectral analysis of digital signals.

17.4.2 Correlating a signal with cosines

We will now describe how correlation methods can be used to determine the amplitude, phase, and frequency of each constituent cosine of a signal. First let us assume that a simple cosine signal under analysis is represented as follows:

$$x(t) = A \cos(2\pi f_0 t + \theta_0). \tag{17.8}$$

In (17.7), let the test cosine be

$$y(t) = \cos(2\pi f t) \tag{17.9}$$

where f is the test frequency. The modified correlation coefficient from (17.7) becomes

$$R_0 = \int_{-T/2}^{T/2} A \cos(2\pi f_0 t + \theta_0) \cos(2\pi f t) \, dt. \tag{17.10}$$

Now we can simplify this integral by using the result from Exercise 1.5(i):

$$\cos(A) \cos(B) = \frac{1}{2} \cos(A + B) + \frac{1}{2} \cos(A - B)$$

in which case (17.10) becomes

$$R_0 = \frac{A}{2} \int_{-T/2}^{T/2} \cos(2\pi(f + f_0)t + \theta_0) dt + \frac{A}{2} \int_{-T/2}^{T/2} \cos(2\pi(f_0 - f)t + \theta_0) \, dt. \tag{17.11}$$

Performing the integrations,

$$R_0 = \frac{A}{4\pi(f + f_0)} \left| \sin(2\pi(f + f_0)t + \theta_0) \right|_{-T/2}^{T/2}$$

$$+ \frac{A}{4\pi(f_0 - f)} \left| \sin(2\pi(f_0 - f)t + \theta_0) \right|_{-T/2}^{T/2}.$$

Putting in the limits,

$$R_0 = \frac{A}{4\pi(f + f_0)} \left[\sin\left(2\pi(f + f_0)\frac{T}{2} + \theta_0 \right) - \sin\left(-2\pi(f + f_0)\frac{T}{2} + \theta_0 \right) \right]$$

$$+ \frac{A}{4\pi(f_0 - f)} \left[\sin\left(2\pi(f_0 - f)\frac{T}{2} + \theta_0 \right) - \sin\left(-2\pi(f_0 - f)\frac{T}{2} + \theta_0 \right) \right].$$

Since sine is an odd function, $\sin(-x) = -\sin(x)$, the second term in each square bracket can be rewritten so that

$$R_0 = \frac{A}{4\pi(f + f_0)} \left[\sin\left(2\pi(f + f_0)\frac{T}{2} + \theta_0 \right) + \sin\left(2\pi(f + f_0)\frac{T}{2} - \theta_0 \right) \right]$$

$$+ \frac{A}{4\pi(f_0 - f)} \left[\sin\left(2\pi(f_0 - f)\frac{T}{2} + \theta_0 \right) + \sin\left(2\pi(f_0 - f)\frac{T}{2} - \theta_0 \right) \right].$$

Using (1.13),

$$R_0 = \frac{A}{2\pi(f + f_0)} \sin\left(2\pi(f + f_0)\frac{T}{2} \right) \cos(\theta_0)$$

$$+ \frac{A}{2\pi(f_0 - f)} \sin\left(2\pi(f_0 - f)\frac{T}{2} \right) \cos(\theta_0). \tag{17.12}$$

We are going to plot R_0 as a function of f. As an example, we take the cosine in (17.8) with the following parameters: $f_0 = 5$ Hz, $A = 10$, and $\theta_0 = 63°$. Suppose that the duration of the data is 2 s. If we plot R_0 in (17.12) versus the test frequency f, we obtain Figure 17.11. There are two peaks. One is at 5 Hz, signifying that the maximum correlation occurs between the underlying signal and the test cosine when the test cosine's frequency equals that of the signal cosine. This is a reassuring result. Hence, we can identify the underlying frequency as 5 Hz. Now the peak at 5 Hz comes mainly from the second term in (17.12). The first term in (17.12) gives us the peak at −5 Hz. This is a strange result: what is meant by a negative

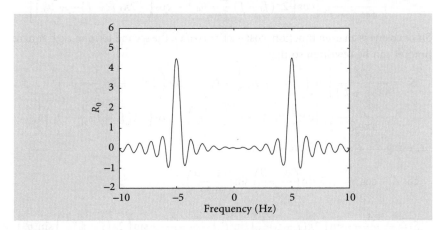

Figure 17.11 Correlation coefficient as a function of frequency.

frequency? In fact, this peak at -5 Hz does not signify anything practical, it is just a consequence of the mathematics of the correlation that we have carried out.

How can we extract the amplitude A and phase θ_0 of the cosine from the results of the correlation calculation? We will look at the second term in (17.12) that gives rise to the positive frequency peak, as this is the one that is physically significant. Using the identity $\sin(-x) = -\sin(x)$ and multiplying both the numerator and denominator by $T/2$, this term can be written as

$$\frac{AT}{2} \frac{\sin(2\pi(f - f_0)(T/2))}{2\pi(f - f_0)(T/2)} \cos(\theta_0).$$

The sinc function, of the form $\sin(x)/x$ (§4.7.2), has a maximum value of unity at $x = 0$ so that the second term of (17.12) has a maximum at $f = f_0$. For this value of f the positive peak has height

$$R_{max} = \frac{AT\cos(\theta_0)}{2}. \tag{17.13}$$

Thus, the maximum of the correlation function is related to the amplitude A and the phase θ_0, but we cannot find each individually.

In order to do this, we can use the following procedure. Instead of correlating the signal with a cosine, we correlate it with a sine instead, giving

$$S_0 = \int_{-T/2}^{T/2} A\cos(2\pi f_0 t + \theta_0) \sin(2\pi f t)\, dt. \tag{17.14a}$$

Using the solution of Exercise 1.5(iii) we can rewrite this as

$$S_0 = \frac{A}{2}\int_{-T/2}^{T/2} \sin(2\pi(f_0 + f)t + \theta_0)\, dt - \frac{A}{2}\int_{-T/2}^{T/2} \sin(2\pi(f_0 - f)t + \theta_0)\, dt. \tag{17.14b}$$

Performing the integrations,

$$S_0 = \frac{-A}{4\pi(f + f_0)}\left|\cos\{2\pi(f + f_0)t + \theta_0\}\right|_{-T/2}^{T/2} + \frac{A}{4\pi(f_0 - f)}\left|\cos\{2\pi(f_0 - f)t + \theta_0\}\right|_{-T/2}^{T/2}.$$

Putting in the limits,

$$S_0 = \frac{-A}{4\pi(f + f_0)}\left[\cos\left(2\pi(f + f_0)\frac{T}{2} + \theta_0\right) - \cos\left(-2\pi(f + f_0)\frac{T}{2} + \theta_0\right)\right]$$
$$+ \frac{A}{4\pi(f_0 - f)}\left[\cos\left(2\pi(f_0 - f)\frac{T}{2} + \theta_0\right) - \cos\left(-2\pi(f_0 - f)\frac{T}{2} + \theta_0\right)\right].$$

Since cosine is an even function, $\cos(-x) = \cos(x)$, the second term in each square bracket can be rewritten so that

$$S_0 = \frac{-A}{4\pi(f + f_0)}\left[\cos\left(2\pi(f + f_0)\frac{T}{2} + \theta_0\right) - \cos\left(2\pi(f + f_0)\frac{T}{2} - \theta_0\right)\right]$$
$$+ \frac{A}{4\pi(f_0 - f)}\left[\cos\left(2\pi(f_0 - f)\frac{T}{2} + \theta_0\right) - \cos\left(2\pi(f_0 - f)\frac{T}{2} - \theta_0\right)\right],$$

or, using the result of Exercise 1.7,

$$\cos A - \cos B = -2\sin\left(\frac{A + B}{2}\right)\sin\left(\frac{A - B}{2}\right),$$

$$S_0 = \frac{A}{2\pi(f + f_0)}\sin\left(2\pi(f + f_0)\frac{T}{2}\right)\sin(\theta_0) - \frac{A}{2\pi(f - f_0)}\sin\left(2\pi(f - f_0)\frac{T}{2}\right)\sin(\theta_0)$$

$$\tag{17.15}$$

where in the second term we have used the identity $\sin(-x) = -\sin(x)$. If we plot S_0 versus f we obtain Figure 17.12.

As for R_0, extrema (maxima or minima) occur at frequencies $\pm f_0$. The extremum at the positive frequency comes from the second term in (17.15). When $f = f_0$, the second term becomes

$$S_{max} = -\frac{AT\sin(\theta_0)}{2}. \tag{17.16}$$

We can now explicitly determine both A and θ_0 from (17.13) and (17.16) as:

$$A = \frac{2}{T}\sqrt{R_{max}^2 + S_{max}^2} \tag{17.17}$$

and

$$\theta_0 = \tan^{-1}\left(-\frac{S_{max}}{R_{max}}\right). \tag{17.18}$$

The ambiguity in the value of θ_0 is resolved by noting that $-S_{max}$ and R_{max} have the signs of $\sin\theta_0$ and $\cos\theta_0$ respectively.

Hence, our general strategy to find the amplitudes and phases of the constituent cosines into which a signal may be resolved is as follows:

Step 1 Calculate the correlation coefficient, $X_R(f)$:

$$X_R(f) = \int_{-T/2}^{T/2} x(t)\cos(2\pi ft)dt \tag{17.19}$$

where (17.10) has been generalized to a general signal $x(t)$.

Step 2 Calculate the correlation coefficient, $X_I(f)$:

$$X_I(f) = \int_{-T/2}^{T/2} x(t)\sin(2\pi ft)dt. \tag{17.20}$$

Step 3 Calculate a function $A(f)$ as follows:

$$A(f) = \sqrt{X_R^2(f) + X_I^2(f)}. \tag{17.21}$$

If there is a component of the signal with frequency f then there will be a maximum of $A(f)$ at that value of f, designated from the function as f_{max}. The cosine with frequency f_{max} has, from (17.17) and (17.18),

$$\text{amplitude} = \frac{2}{T}A(f_{max}) \text{ and phase} = \theta(f_{max}) = \tan^{-1}\left(-\frac{X_I(f_{max})}{X_R(f_{max})}\right). \tag{17.22}$$

For the one-component example illustrated in Figure 17.11, when we correlated with a cosine, the maximum of the correlation function occurs at a frequency of 5 Hz, which is the frequency of the underlying cosine. The maximum of the correlation function in Figure 17.11 is given by

$$X_R(5) = 4.53990. \tag{17.23}$$

When correlating the underlying signal with a sine, the correlation function in Figure 17.12 is close to zero in the positive frequency range, except when the

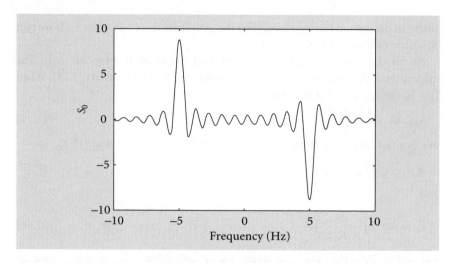

Figure 17.12 Correlation function as a function of frequency.

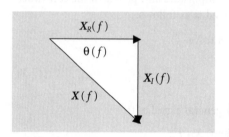

Figure 17.13 The relationship between $x(t)$, $X_R(t)$, and $X_I(t)$.

frequency is 5 Hz when it takes the value of

$$X_I(5) = -8.910\,06. \tag{17.24}$$

Substituting for $X_R(5)$ and $X_I(5)$ into (17.17) and (17.18), with $T = 2$, we find that $A(5) = 10$ and $\theta(5) = 63°$, which are the correct values.

17.4.3 Continuous Fourier transform

There is a geometrical relation between $X_R(f), X_I(f), A(f)$, and $\theta(f)$. Let $X_R(f)$ be the magnitude of a vector $\mathbf{X}_R(f)$ which points in the x-direction and $X_I(f)$ the magnitude of a vector $\mathbf{X}_I(f)$ which points in the negative y-direction. If we add the two vectors, we obtain a third vector given by

$$\mathbf{X}(f) = \mathbf{X}_R(f) + \mathbf{X}_I(f). \tag{17.25}$$

This relation is illustrated vectorially in Figure 17.13. From (17.21) $A(f)$ can be interpreted as the magnitude of the vector $\mathbf{X}(f)$, and from (17.22) $\theta(f_{\max})$ can be interpreted as the phase angle when $f = f_{\max}$, as seen in Figure 17.13.

There is another way of representing the above relations using complex numbers. The complex number i can be represented by a unit vector pointing in the positive y-direction and, similarly, $-i$ represents a vector pointing in the negative y-direction. Hence, we can represent the relation between $\mathbf{X}(f)$, $\mathbf{X}_R(f)$, and $\mathbf{X}_I(f)$ in Figure 17.13, by the following relation:

$$X(f) = X_R(f) - iX_I(f) \tag{17.26}$$

where now $X(f)$ is a complex number. Substituting for $X_R(f)$ from (17.19) and $X_I(f)$ from (17.20), we obtain

$$X(f) = \int_{-T/2}^{T/2} x(t) \cos(2\pi ft) dt - i \int_{-T/2}^{T/2} x(t) \sin(2\pi ft) dt. \tag{17.27}$$

This is equivalent to

$$X(f) = \int_{-T/2}^{T/2} x(t) [\cos(2\pi ft) - i \sin(2\pi ft)] \, dt = \int_{-T/2}^{T/2} x(t) \exp(-2\pi ift) dt. \tag{17.28}$$

We can, quite generally, assume that the signal exists for an infinite time, since even a signal with limited duration can be thought of as being represented by a function of infinite extent that happens to be zero outside the actual duration. Now, taking $T \to \infty$ in (17.28) gives

$$X(f) = \int_{-\infty}^{\infty} x(t) \exp(-2\pi ift) dt. \tag{17.29}$$

The expression on the right-hand side of (17.29) is the **continuous Fourier transform** of $x(t)$. Alternatively, using angular frequency $\omega = 2\pi f$ and changing variables, (17.29) can be rewritten as

$$X(\omega) = \int_{-\infty}^{\infty} x(t) \exp(-i\omega t) dt. \tag{17.30}$$

The amplitude spectrum of the signal, $x(t)$, given in (17.21) is now

$$A(f) = |X(f)| \tag{17.31}$$

and the phase spectrum, $\theta(f)$, given by (17.22) is defined by the phase of the complex function $X(f)$, i.e.

$$\theta(f) = \angle X(f), \tag{17.32}$$

where $\angle$ indicates 'phase of'.

Note that the phase spectrum only has meaning when the amplitude spectrum is non-zero. If $A(f)$ is zero, then no meaning may be attached to the phase spectrum. It may also be shown that the amplitude spectrum is an even function of frequency $A(f) = A(-f)$ and the phase spectrum is an odd function of frequency $\theta(f) = -\theta(-f)$.

Fourier transform theory is of great importance in many areas of physical science apart from signal processing – for example in optical diffraction.

17.4.4 Fourier transform of a cosine signal

We now return to the Fourier transform of a cosine signal with the phase equal to zero, so that $\theta_0 = 0$ in (17.12) and (17.15). In this case, it can be seen that $S_0(f) = 0$ because the integrand in (17.14a) is odd and the integrating range is symmetrical

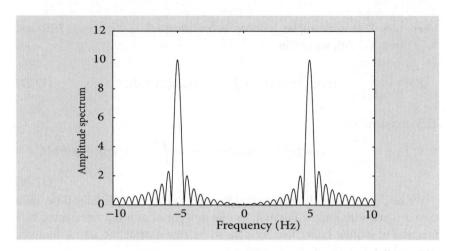

Figure 17.14 Amplitude spectrum of a cosine signal.

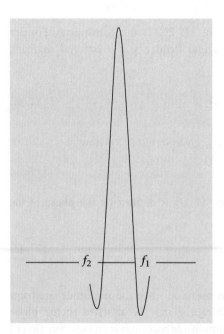

Figure 17.15 Positive frequency peak.

around zero, and $X(f) = R_0(f)$, where R_0 is given by

$$R_0 = \frac{A}{2\pi(f+f_0)} \sin\left(2\pi(f+f_0)\frac{T}{2}\right) + \frac{A}{2\pi(f-f_0)} \sin\left(2\pi(f-f_0)\frac{T}{2}\right).$$

$$(17.33)$$

The second term in (17.33) is the positive frequency peak at frequency $f = f_0$. From (17.13), with phase angle $\theta_0 = 0$, the height of the peak is given by

$$R_0^{\max} = \frac{AT}{2}.$$

$$(17.34)$$

The amplitude spectrum in this particular case, with a single cosine component, is $A(f) = |R_0|$. In Figure 17.14 $A(f)$ is plotted versus f for the cosine signal in (17.8). Although there is a peak at 5 Hz, it can be seen that this representation is not ideal. Ideally, we want there to be a single spike at 5 Hz – that is, a peak of very small width and large height. Now, from (17.34), the height of the peak is given by $AT/2$ so that, as T increases, then so does the height of the peak. The width of the peak is defined as the frequency difference between the points either side of the maximum where the Fourier transform first goes to zero, as illustrated in Figure 17.15.

From the second term in (17.33), the frequency f_1 occurs either when $\sin(2\pi(f - f_0)(T/2)) = 0$ or when $2\pi(f - f_0)(T/2) = \pi$ (remember $\sin\pi = 0$). Solving this equation,

$$f \equiv f_1 = f_0 + \frac{1}{T}.$$

Similarly f_2 occurs when $2\pi(f - f_0)\frac{T}{2} = -\pi$ or

$$f \equiv f_2 = f_0 - \frac{1}{T}.$$

Therefore the width of this peak is given by

$$f_1 - f_2 = \frac{1}{T} + f_0 + \frac{1}{T} - f_0 = \frac{2}{T}. \tag{17.35}$$

Hence, as T increases, the peak becomes *narrower*. One consequence of this can be seen as follows. Suppose that there are two frequencies present in the signal, say at 1 Hz and 1.1 Hz, and the duration of data is 2 s. The positive frequency part of the amplitude spectrum will be as shown in Figure 17.16.

Only one frequency is observed. This is because the peak widths associated with each frequency are so large that they overlap and produce just one peak as shown. We say that the two peaks cannot be **resolved**. If T is now increased to 20 s, then the amplitude spectrum is as shown in Figure 17.17. The two peaks are now clearly resolved.

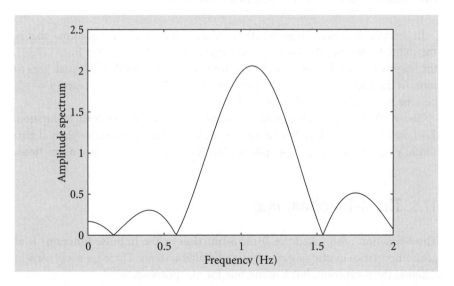

Figure 17.16 Amplitude spectrum for two cosines with frequencies 1 Hz and 1.1 Hz, $T = 2$ s.

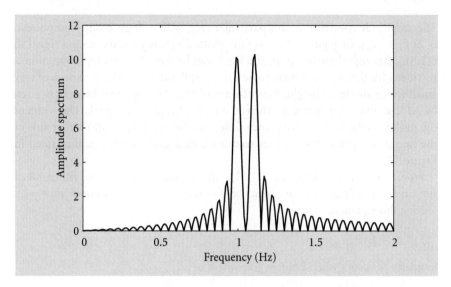

Figure 17.17 Amplitude spectrum for two cosines with frequencies 1 Hz and 1.1 Hz, $T = 20$ s.

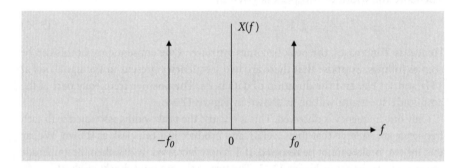

Figure 17.18 An ideal amplitude spectrum for a single cosine.

In general, increasing T improves the resolution of the Fourier transform – that is, the ability to distinguish closely spaced frequencies. As $T \to \infty$, then from (17.34), the height of the peak goes to ∞, whilst from (17.35), the width of the peak goes to zero. In the limit as $T \to \infty$, the peaks in Figure 17.11, say at $f = f_0$ and $f = -f_0$, become spikes as illustrated in Figure 17.18.

The width of the peaks, and hence resolution, is inversely proportional to duration, T, of data analysed. When T is infinite the ideal spikes have infinite height and zero width. When the area under the spike is unity then the spikes are called **δ-functions**.

17.5 The δ-function, $\delta(x)$

The δ-function (also called the **Dirac δ-function** or the **impulse function**) is of great importance in studying analogue and digital systems. There are many ways of defining the δ-function, but a useful one for our purposes is

$$\delta(x) = \frac{1}{\pi} \left(\frac{\sin(xT/2)}{x} \right)_{T \to \infty}. \tag{17.36a}$$

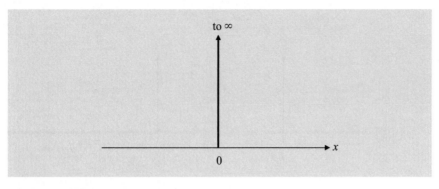

Figure 17.19 A representation of $\delta(x)$.

When $x=0$, $\delta(0) = (T/2\pi)_{T\to\infty}$, which is infinity, and at any other point $|\delta(x)| \leq 1/x$. Hence $\delta(x)$ is drawn as a spike centred at $x=0$ (Figure 17.19).

There is a standard definite integral

$$\int_{-\infty}^{\infty} \frac{\sin(ky)}{y}\mathrm{d}y = \pi \tag{17.36b}$$

which gives the property that

$$\int_{-\infty}^{\infty} \delta(x)\mathrm{d}x = 1 \quad \text{(area under a δ-function is 1)}. \tag{17.37}$$

For any signal $g(t)$,

$$\int_{-\infty}^{\infty} g(t)\delta(t-a)\mathrm{d}t = g(a) \quad \text{(sampling property)}. \tag{17.38}$$

We may understand this by noting that $\delta(t-a)$ is a spike centred at the point $t=a$ and since the δ-function is zero at every other value of t the integral is unchanged by replacing $g(t)$ by $g(a)$. Now $g(a)$ becomes a constant that may be taken outside the integral, and from (17.37) the result (17.38) follows.

From (17.33) and (17.36), the Fourier transform of a cosine signal of amplitude A when $T\to\infty$ can be expressed as

$$X(f) = \pi A\delta(2\pi(f-f_0)) + \pi A\delta(2\pi(f+f_0)). \tag{17.39a}$$

This may be transformed by using a property of δ-functions that

$$\delta(ax) = \frac{1}{a}\delta(x).$$

To confirm this property we consider δ-functions $\delta(x)$ and $\delta(y)$. Then

$$\int_{-\infty}^{\infty} \delta(x)\mathrm{d}x = \int_{-\infty}^{\infty} \delta(y)\mathrm{d}y = 1.$$

If $y=ax$ then

$$\int_{-\infty}^{\infty} \delta(x)\mathrm{d}x = a\int_{-\infty}^{\infty} \delta(ax)\mathrm{d}x = 1,$$

from which the property may be inferred.

If we apply this identity to each of the terms in (17.39a) with $\alpha=2\pi$, we find that we can rewrite (17.39a) as

$$X(f) = \frac{A}{2}\delta(f-f_0) + \frac{A}{2}\delta(f+f_0), \tag{17.39b}$$

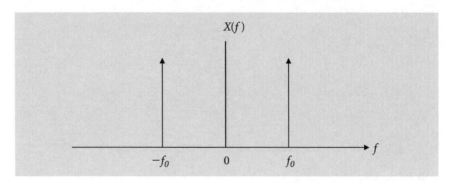

Figure 17.20 Ideal Fourier transform of a cosine.

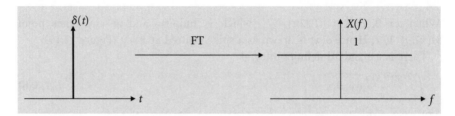

Figure 17.21 Fourier transform of a δ-function.

i.e. a spike at $f = f_0$ and another spike at $f = -f_0$, as illustrated in Figure 17.20.
The Fourier transform of a δ-function is expressed as

$$D(f) = \int_{-\infty}^{\infty} \delta(t) \exp(-2\pi i f t) \mathrm{d}t.$$

From (17.38) with $a = 0$ and $g(t) = \exp(-2\pi i f t)$,

$$D(f) = g(0) = \exp(0) = 1. \tag{17.40}$$

Hence the Fourier transform of a δ-function is 1 – an important result that is
illustrated in Figure 17.21.

17.6 Inverse Fourier transform

It can be shown (Appendix 2) that $x(t)$ can be derived from the Fourier transform,
$X(f)$, from the following relation:

$$x(t) = \int_{-\infty}^{\infty} X(f) \exp(2\pi i f t) \, \mathrm{d}f. \tag{17.41a}$$

Changing the variable to angular frequency, $\omega = 2\pi f$, converts (17.41a) to the
alternative form

$$x(t) = \frac{1}{2\pi} \int_{-\infty}^{\infty} X(\omega) \exp(i\omega t) \, \mathrm{d}\omega. \tag{17.41b}$$

As an example, from the inverse Fourier transform we find

$$\delta(t) = \int_{-\infty}^{\infty} \exp(2\pi i f t) \, \mathrm{d}f = \frac{1}{2\pi} \int_{-\infty}^{\infty} \exp(i\omega t) \, \mathrm{d}\omega. \tag{17.42}$$

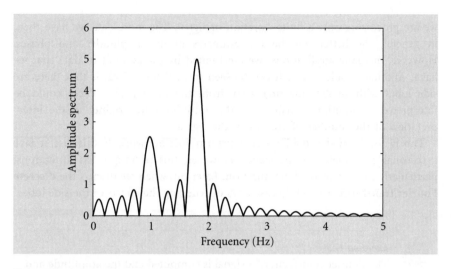

Figure 17.22 Amplitude spectrum of a signal consisting of several cosines.

17.7 **Several cosine signals**

We asked the question how many cosines are there in the signal shown in Figure 17.10 and what are their frequencies, amplitudes, and phases. With $T = 5\,\text{s}$ of data the amplitude spectrum is shown in Figure 17.22. There are two dominant peaks, one at 0.985 Hz with amplitude 2.5237 and the other at 1.8 Hz with amplitude 5. From (17.22), the amplitude of the cosine is related to the amplitude of the corresponding peak in the amplitude spectrum by

$$A_1 = \frac{2}{T} A(f_{\text{max}}). \tag{17.43}$$

Hence, the amplitude of the cosine at 0.985 Hz is given by

$$A_1 = \frac{2(2.5237)}{5} = 1.009\,48$$

and the corresponding amplitude of the 1.8 Hz peak is given by

$$A_{1.8} = \frac{2(5)}{5} = 2.$$

If we compute the phase spectrum, using (17.32), at these two frequencies, then we find that at 0.985 Hz the phase is 38.7° whereas at 1.8 Hz the corresponding value is 108°. Hence, we postulate that the signal in Figure 17.10 is given by

$$s(t) = 1.009\,48\cos(2\pi(0.985)t + 38.70°) \\ + 2.000\,00\cos(2\pi(1.8)t + 108°). \tag{17.44}$$

In fact, the actual signal is given by

$$s(t) = \cos(2\pi(1)t + 36°) + 2\cos(2\pi(1.8)t + 108°). \tag{17.45}$$

Comparing (17.44) and (17.45), we can see that the amplitude, frequency, and phase of the higher-frequency cosine have been found exactly, but approximate values have been found for the lower-frequency cosine. The reason why we have not found exact values for the amplitudes and phases of the lower-frequency cosine follows on from the discussion in §17.4.4; it is because we are processing only a finite duration of signal. The more data we have then, in general, the better will be the estimates of the amplitudes and phases. However, in many applications, we are limited by the amount of data that we have. Another problem, which can be seen from Figure 17.22, is that there are side lobes with smaller maxima away from the main peaks. These could be interpreted as additional cosines, so this introduces uncertainty in the interpretation of the number of cosines in the signal.

This is not to say that the Fourier transform does not work. In this case, it gave a reasonable estimate for the signal. However, this technique has limitations, particularly for signals of short duration. Later on, when we discuss the **discrete Fourier transform**, we will discuss ways of reducing the effects of the side lobes.

? **Exercise 17.3**
The Fourier transform of a signal is computed and the amplitude and phase spectra are as displayed below. The amplitude spectrum is scaled by $2/T$ where T is the length of the data segment. Neglecting the widths of the spectral peaks, the amplitude and phase spectra are as shown below. Derive an expression for the signal.

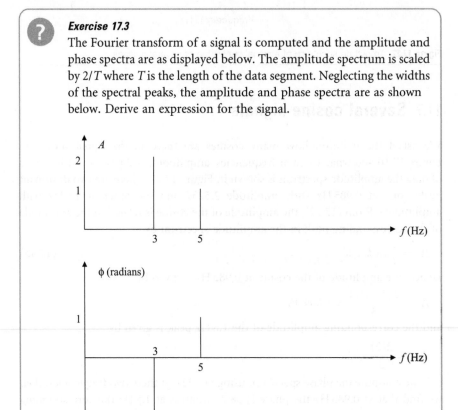

17.8 Parseval's theorem

In our treatment of the Fourier transform thus far we have restricted the quantities involved to time and frequency. However, there are other pairs of variables that can be related by a Fourier transform relationship – examples are given in §17.16.

There is a useful theorem involving Fourier transforms which we now describe in terms of two functions $f(a)$ and $F(b)$ that are related by Fourier transformation so that

$$F(b) = \int_{-\infty}^{\infty} f(a) \exp(-iab) \, da \qquad (17.46a)$$

and

$$f(a) = \frac{1}{2\pi} \int_{-\infty}^{\infty} F(b) \exp(iab) \, db. \qquad (17.46b)$$

Parseval's theorem states that

$$\int_{-\infty}^{\infty} |f(a)|^2 da = \frac{1}{2\pi} \int_{-\infty}^{\infty} |F(b)|^2 db. \qquad (17.47)$$

Taking the complex conjugate of each side of (17.46b) (remember; the complex conjugate of a product is the product of the complex conjugates),

$$f^*(a) = \frac{1}{2\pi} \int_{-\infty}^{\infty} F^*(b') \exp(-iab') \, db'.$$

We take the product $f(a)f^*(a) = |f(a)|^2$ and integrate over the whole range of a:

$$\int_{-\infty}^{\infty} |f(a)|^2 da = \frac{1}{4\pi^2} \int_{b=-\infty}^{\infty} \int_{b'=-\infty}^{\infty} F(b) F^*(b') \left[\int_{-\infty}^{\infty} \exp\{i(b-b')a\} da \right] db' db.$$

From (17.42) this becomes

$$\int_{-\infty}^{\infty} |f(a)|^2 da = \frac{1}{2\pi} \int_{b=-\infty}^{\infty} \int_{b'=-\infty}^{\infty} F(b) [F^*(b') \delta(b - b') db'] db$$

and applying the sampling property (17.38) of the δ-function to the quantity in square brackets we find

$$\int_{-\infty}^{\infty} |f(a)|^2 da = \frac{1}{2\pi} \int_{-\infty}^{\infty} F(b) F^*(b) \, db = \frac{1}{2\pi} \int_{-\infty}^{\infty} |F(b)|^2 db,$$

which is Parseval's theorem.

This theorem relates to energy conservation in physical systems – which could apply, for example, to physical optics or an electric circuit.

EXAMPLE 17.2

To fix our minds on a particular physical application, we consider a $1\,\Omega$ resistor across which there is a time-dependent potential difference $f(t)$, which is a real quantity. Generalizing the duration of the signal as from $t = -\infty$ to $t = \infty$ (it could be zero outside some finite range) then the energy dissipated in the resistor, within which the current is $f(t)$, is

$$E = \int_{-\infty}^{\infty} f(t)^2 dt.$$

We now consider the Fourier transform of $f(t)$, which we write as $F(\omega)$. Parseval's theorem states that

$$\int_{-\infty}^{\infty} f(t)^2 dt = \frac{1}{2\pi} \int_{-\infty}^{\infty} |F(\omega)|^2 d\omega.$$

Since $f(t)$ is real it is found from (17.30), by replacing ω by $-\omega$, that $F(-\omega) = F^*(\omega)$ so that $|F(-\omega)|^2 = |F(\omega)|^2$. Since a negative frequency has no physical meaning, and since $|F(\omega)|$ and hence $|F(\omega)|^2$ is an even function of ω,

Parseval's theorem for energy conservation for an electric current can now be put in the form

$$\int_{-\infty}^{\infty} f(t)^2 \mathrm{d}t = \frac{1}{\pi} \int_0^{\infty} |F(\omega)|^2 \mathrm{d}\omega. \tag{17.48}$$

The quantity on the right-hand side has the dimensions of energy and to conserve energy $|F(\omega)|^2/\pi$ can be identified as the energy spectrum of the current, i.e. the energy in the current per unit angular frequency range. In particular, the energy associated with the frequency band $\omega_1 \leq |\omega| \leq \omega_2$ is given by

$$E(\omega_1, \omega_2) = \frac{1}{\pi} \int_{\omega_1}^{\omega_2} |F(\omega)|^2 \mathrm{d}\omega. \tag{17.49}$$

17.9 Fourier series

Many signals that are encountered in practice are periodic in nature:

- In communications, when transmitting signals using modulation, a switching modulator is used which employs a periodic signal.
- In power electronics, periodic signals are used in power supplies.
- In X-ray crystallography, crystal structures are arrangements of atoms that are periodic in three dimensions and this periodicity affects the way in which X-rays are scattered from the crystal – a process that can be described in terms of the Fourier transform (17.29).

A typical periodic signal is shown in Figure 17.23. This signal has the general property that $s(t + T) = s(t)$, where T is the period. In the above example, $T = 1$ s. It can be seen that this type of signal repeats itself every period T. We define the **fundamental frequency** (in Hz) by

$$f_0 = \frac{1}{T}.$$

The smaller is T the larger is f_0 and hence the more rapidly does the signal repeat itself. In the example in Figure 17.23, $T = 1$ s, hence from the above equation $f_0 = 1$ Hz.

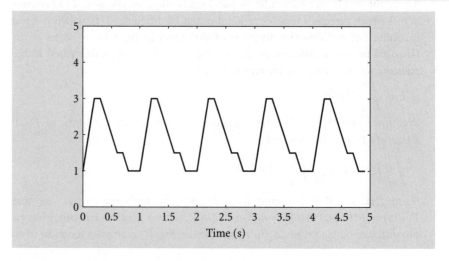

Figure 17.23 A typical periodic signal.

We have seen that it is convenient to express this signal in terms of cosines as described in §17.4. We now look at the Fourier transform of a periodic signal.

17.9.1 **Graphical derivation of the Fourier transform of a periodic signal**

Let us see diagrammatically what happens if we compute the Fourier transform, $X(f)$, of a periodic signal, $x(t)$, given by

$$X(f) = \int_{-\infty}^{\infty} x(t)\exp(-2\pi i f t)\mathrm{d}t. \tag{17.50a}$$

The real part of the Fourier transform is given by

$$X_R(f) = \int_{-\infty}^{\infty} x(t)\cos(2\pi f t)\,\mathrm{d}t. \tag{17.50b}$$

Suppose we choose $f = f_0$. For the example in Figure 17.23 the terms $\cos(2\pi f_0 t)$ and $x(t)$ in the integrand are displayed in Figure 17.24. The product $x(t)\cos(2\pi f_0 t)$ in the integrand of (17.50b) is illustrated in Figure 17.25.

Finally, we could calculate the definite integral in (17.50b) graphically by summing the area above and below $t=0$ (shown dotted in Figure 17.25) using the convention that areas are positive if above $t=0$ and areas below the axis are negative. We are not going to do this quantitatively, but note that the total area under the curve of the integrand has the following properties:

- It is non-zero.
- It is infinite if summed over all time.

The second property is not very useful, but is just a consequence of the fact that the signal $x(t)$ is infinite in extent. We redefine our Fourier transform for periodic signals in (17.50a) by dividing the signal into an infinite number of windows of duration T, where T is the period of the signal. We are going to compute the average value of the Fourier transform, averaged over all windows of duration T. In the $(w+1)$th window, the signal is given by $x(t-wT)$, where $t-wT$ is the equivalent time in the first window. Over $N+1$ windows of the signal, the average value of the Fourier transform, (17.50a), is given by

$$X(f) = \frac{1}{N+1}\sum_{w=-N/2}^{N/2}\int_{-T/2}^{T/2} x(t-wT)\exp(-2\pi i f t)\mathrm{d}t \tag{17.51}$$

where it is assumed, without loss of generality, that N is even and that the total number of windows is odd. Note also that the integration is just being carried out over one period.

In the limit, as $N \to \infty$ this becomes

$$X(f) = \left\{\frac{1}{N+1}\sum_{w=-N/2}^{N/2}\int_{-T/2}^{T/2} x(t-wT)\exp(-2\pi i f t)\,\mathrm{d}t\right\}_{N\to\infty}. \tag{17.52}$$

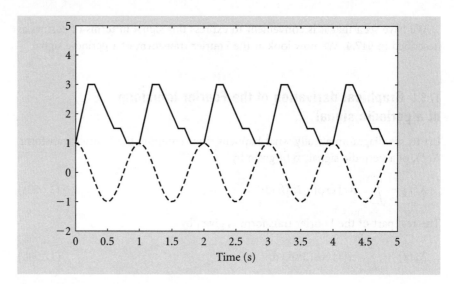

Figure 17.24 Periodic signal and cos($2\pi f_0 t$).

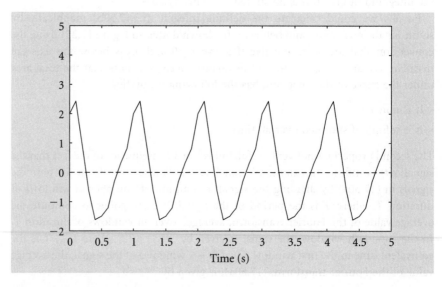

Figure 17.25 Integrand of Equation (17.50b).

The real part, which we are computing, is given by

$$X_R(f) = \left\{ \frac{1}{N+1} \sum_{w=-N/2}^{N/2} \int_{-T/2}^{T/2} x(t - wT) \cos(2\pi ft) \, dt \right\}_{N \to \infty}. \qquad (17.53)$$

Graphically, this is the area under the integrand $x(t - wT) \cos(2\pi ft)$ averaged over all windows.

We look at the particular case when $f = f_0$ in Figure 17.25. It can be seen that the area under the curve is identical for each period of the signal. In this case, $X_R(f)$ in (17.53) becomes equal to the area under the curve for one period of the signal. If we choose this window as $w=0$ then (17.53) becomes

$$X_R(f_0) = \int_{-T/2}^{T/2} x(t) \cos(2\pi f_0 t)\, dt. \tag{17.54}$$

Similarly, if we take the imaginary part of the Fourier transform, (17.50a), and modify it for infinite periodic signals, we obtain the following expression analogous to (17.53) for the real part:

$$X_I(f) = -\left\{ \frac{1}{N+1} \sum_{w=-N/2}^{N/2} \int_{-T/2}^{T/2} x(t - wT) \sin(2\pi f t)\, dt \right\}_{N \to \infty}. \tag{17.55}$$

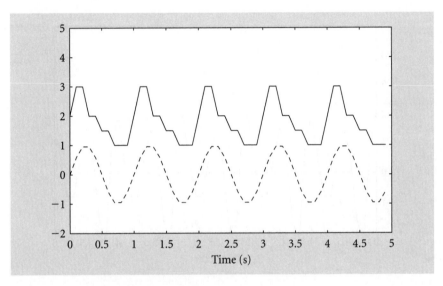

Figure 17.26 Periodic signal and $\sin(2\pi f_0 t)$.

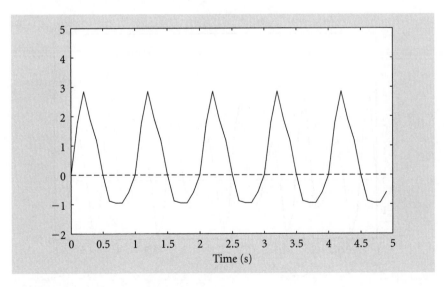

Figure 17.27 $x(t) \sin(2\pi f_0 t)$.

For the example of the periodic signal in Figure 17.23, the original signal and $\sin(2\pi ft)$ are as shown in Figure 17.26, and Figure 17.27 shows $x(t)$ multiplied by $\sin(2\pi f_0 t)$. Again, as for the real part, summing areas above and below the dotted line as for the real part, the imaginary part of the Fourier transform is a non-zero quantity and we can write

$$X_I(f_0) = -\int_{-T/2}^{T/2} x(t) \sin(2\pi f_0 t) \, dt. \tag{17.56}$$

If we carry out the same exercise as above for $f = 2f_0$, we obtain Figures 17.28 and 17.29.

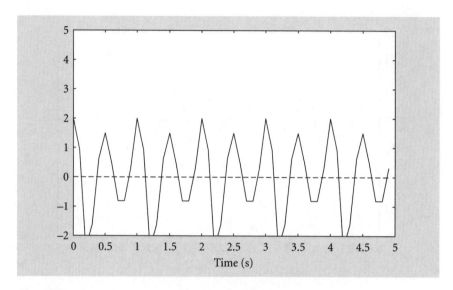

Figure 17.28 Integrand of real part of Fourier transform, $f = 2f_0$.

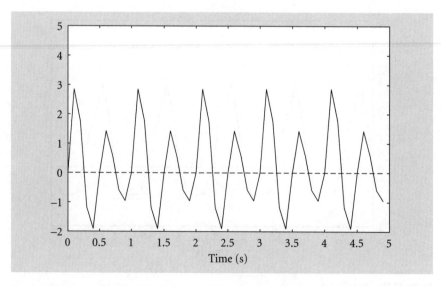

Figure 17.29 Integrand of imaginary part of Fourier transform, $f = 2f_0$.

Summing the areas above and below zero, we can see that these two integrals are non-zero quantities and are equal to the value of the integral over one period:

$$X_R(2f_0) = \int_{-T/2}^{T/2} x(t) \cos(2\pi(2f_0)t)\, dt \tag{17.57}$$

and

$$X_I(2f_0) = -\int_{-T/2}^{T/2} x(t) \sin(2\pi(2f_0)t)\, dt. \tag{17.58}$$

Now let us see what happens if, for example, we put $f = 1.38f_0$ in (17.19) and (17.20). The integrands of (17.19) and (17.20) are shown in Figures 17.30 and 17.31.

Now we can see a different situation to the previous cases. Some of the areas in each period are positive, some negative. In fact, if we carry out this calculation over

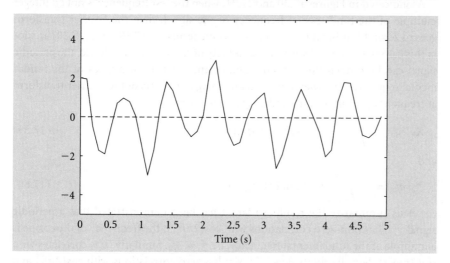

Figure 17.30 Integrand of (17.19), $f = 1.38f_0$.

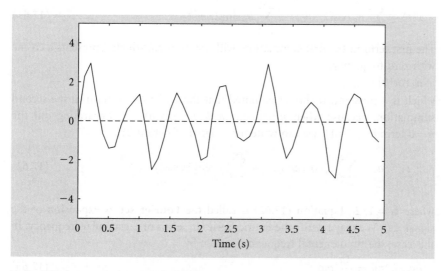

Figure 17.31 Integrand of (17.20), $f = 1.38f_0$.

an infinite number of windows, it can be shown that the total area goes to zero, or put another way,

$$X_R(1.38f_0) = X_I(1.38f_0) = 0.$$

Now why is this situation different from the cases when $f = f_0$ and $f = 2f_0$? Looking at Figures 17.24 and 17.26, we can see that the cosine and sine terms are **synchronized** with $x(t)$. This is because the frequencies of these sines and cosines are the same as the fundamental frequency of the signal. Hence, within each period, the areas are the same and hence the average area is non-zero.

In general, when the frequency of the sine and cosine terms in (17.19) and (17.20) is an integer multiple of the fundamental frequency of the signal then the Fourier transform could be non-zero.

As indicated in Figures 17.30 and 17.31, when the test frequency is not an integer multiple of the fundamental frequency of the signal, then the Fourier transform goes to zero; this is because the sine and cosine terms in (17.19) and (17.20) are not in this case synchronized with the periodicity of the signal $x(t)$. Hence, the periodic signal can be broken up into sines and cosines that are multiples of the fundamental frequency. In general, the real and imaginary parts of the Fourier transform at frequency $f = nf_0$ of a periodic signal are given by

$$X_R(nf_0) = \int_{-T/2}^{T/2} x(t) \cos(2\pi(nf_0)t) \, dt \tag{17.59}$$

and

$$X_I(nf_0) = -\int_{-T/2}^{T/2} x(t) \sin(2\pi(nf_0)t) \, dt. \tag{17.60}$$

Let us summarize the results so far. It has been demonstrated that a periodic signal $x(t)$ has non-zero correlations with $\cos(2\pi ft)$ if the frequency of the cosine is a multiple of the fundamental frequency, i.e. $f = nf_0$. Similarly, $x(t)$ correlates with $\sin(2\pi nf_0 t)$. For all other values of f, $x(t)$ has zero correlations with $\cos(2\pi ft)$ and $\sin(2\pi ft)$. This means that for a signal of period T we can expand it as follows:

$$x(t) = \sum_{n=0}^{\infty} a_n \cos(2\pi nf_0 t) + \sum_{n=1}^{\infty} b_n \sin(2\pi nf_0 t). \tag{17.61}$$

The first term in the first summation, with $n = 0$, is a_0, which represents a cosine with zero frequency,

$$a_0 \cos(0) = a_0,$$

which is a constant. It should be noted that there is no such term in the second summation, because when $n = 0$, $\sin(2\pi nf_0 t) = \sin(0) = 0$. Separating out the $n = 0$ term in (17.61), we can write for a signal of period T:

$$x(t) = a_0 + \sum_{n=1}^{\infty} a_n \cos(2\pi nf_0 t) + \sum_{n=1}^{\infty} b_n \sin(2\pi nf_0 t) \tag{17.62}$$

where $f_0 = 1/T$. Equation (17.62) is called the **Fourier series** expansion of the signal $x(t)$. We may rewrite the Fourier series in terms of the angular frequency. In this case, the fundamental frequency is given by

$$\omega_0 = 2\pi f_0 = \frac{2\pi}{T} \text{ rad s}^{-1} \tag{17.63}$$

and we may write for the Fourier series:

$$x(t) = a_0 + \sum_{n=1}^{\infty} (a_n \cos(n\omega_0 t) + b_n \sin(n\omega_0 t)). \tag{17.64}$$

We shall see that a_0, $\{a_n\}$, and $\{b_n\}$ can be determined from a knowledge of $x(t)$. In (17.64), a_0 is called the **DC term** (coming from the idea of a non-alternating direct current), $a_1 \cos(\omega_0 t) + b_1 \sin(\omega_0 t)$ is called the **fundamental**, and, for $n \geq 2$, $a_n \cos(n\omega_0 t) + b_n \sin(n\omega_0 t)$ is called the **nth harmonic**.

Exercise 17.4

The Fourier series expansion for a given signal with period 0.01 s is as follows:

$$s(t) = 3 + 5\cos(2\pi(100)t) + 3\sin(2\pi(100)t) - 4\sin(2\pi(300)t).$$

State the following: (i) the fundamental frequency; (ii) the DC term; (iii) the fundamental; (iv) the second harmonic; (v) the third harmonic.

17.10 Determination of the Fourier coefficients a_0, $\{a_n\}$, and $\{b_n\}$

17.10.1 Basis functions

The Fourier series for a signal $f(t)$ can be written as

$$f(t) = a_0 1 + \sum_{n=1}^{\infty} \{a_n \cos(n\omega_0 t) + b_n \sin(n\omega_0 t)\}. \tag{17.65}$$

Hence we are expanding $f(t)$ in terms of the following set:

$$B = \{1, \cos(\omega_0 t), \cos(2\omega_0 t), \ldots, \cos(n\omega_0 t), \ldots, \sin(\omega_0 t),$$
$$\sin(2\omega_0 t), \ldots, \sin(n\omega_0 t), \ldots\}. \tag{17.66}$$

B is referred to as the **basis functions** or just **basis**. To determine the Fourier coefficients a_0, $\{a_n\}$, and $\{b_n\}$, we use a special property of the basis functions called **orthogonality**.

17.10.2 Orthogonality

If you multiply two different basis functions in B, given by (17.66), and integrate over a period T, then the integral goes to zero. Consider the integral

$$I = \int_0^T \cos(n\omega_0 t)\cos(m\omega_0 t)\,dt.$$

From the solution to Exercise 1.5(i),

$$I = \frac{1}{2}\int_0^T \cos\{(n+m)\omega_0 t\}dt + \frac{1}{2}\int_0^T \cos\{(n-m)\omega_0 t\}dt$$

$$= \frac{1}{2}\left|\frac{1}{(n+m)\omega_0}\sin\{(n+m)\omega_0 t\}\right|_0^T + \frac{1}{2}\left|\frac{1}{(n-m)\omega_0}\sin\{(n-m)\omega_0 t\}\right|_0^T = 0$$

since $\sin(k\omega_0 t) = 0$ for $t=0$ and $t=T$ when k is an integer. We say that these different basis functions are orthogonal to each other. However, if $n=m$ then

$$I = \int_0^T \cos^2(n\omega_0 t)\, dt = \int_0^T \left\{ \frac{1}{2} + \frac{1}{2}\cos(2n\omega_0 t) \right\} dt = \frac{1}{2}T.$$

In what follows we use the following expressions in which m and n are integers:

$$\int_0^T \cos(n\omega_0 t)\cos(m\omega_0 t)\, dt = 0 \quad (m \neq n) \tag{17.67}$$

$$\int_0^T \sin(n\omega_0 t)\sin(m\omega_0 t)\, dt = 0 \quad (m \neq n) \tag{17.68}$$

$$\int_0^T \sin(n\omega_0 t)\cos(m\omega_0 t)\, dt = 0 \quad (\text{all } m \text{ and } n) \tag{17.69}$$

$$\int_0^T 1\cos(n\omega_0 t)\, dt = 0 \quad (\text{all } n) \tag{17.70}$$

$$\int_0^T 1\sin(n\omega_0 t)\, dt = 0 \quad (\text{all } n) \tag{17.71}$$

$$\int_0^T \cos^2(m\omega_0 t)\, dt = \frac{T}{2} \tag{17.72}$$

$$\int_0^T \sin^2(m\omega_0 t)\, dt = \frac{T}{2}. \tag{17.73}$$

17.10.3 Evaluation of a_0

Integrate both sides of (17.65) between $t=0$ and $t=T$:

$$\int_0^T f(t)dt = \int_0^T a_0 1\, dt + \sum_{n=1}^{\infty} \int_0^T a_n \cos(n\omega_0 t)\, dt + \sum_{n=1}^{\infty} \int_0^T b_n \sin(n\omega_0 t)\, dt$$

$$= a_0 [t]_0^T + \sum_{n=1}^{\infty} a_n \int_0^T \cos(n\omega_0 t)\, dt + \sum_{n=1}^{\infty} b_n \int_0^T \sin(n\omega_0 t)\, dt.$$

The integrals on the right-hand side are 0, from (17.70) and (17.71), and hence

$$\int_0^T f(t)\, dt = a_0 T \quad \text{or} \quad a_0 = \frac{1}{T}\int_0^T f(t)\, dt \tag{17.74}$$

Thus, a_0 is the **mean value** of $f(t)$ over a period T.

17.10.4 Evaluation of a_m

Multiply both sides of (17.65) by $\cos(m\omega_0 t)$, giving

$$f(t)\cos(m\omega_0 t) = a_0 1 \cos(m\omega_0 t) + \cos(m\omega_0 t)\sum_{n=1}^{\infty} a_n \cos(n\omega_0 t)$$

$$+ \cos(m\omega_0 t)\sum_{n=1}^{\infty} b_n \sin(n\omega_0 t).$$

Integrating both sides between $t=0$ and $t=T$ gives

$$\int_0^T f(t)\cos(m\omega_0 t)dt = a_0 \int_0^T \cos(m\omega_0 t)dt + \sum_{n=1}^{\infty} a_n \int_0^T \cos(m\omega_0 t)\cos(n\omega_0 t)dt$$

$$+ \sum_{n=1}^{\infty} b_n \int_0^T \cos(m\omega_0 t)\sin(n\omega_0 t)dt.$$

The first term on the right-hand side is 0 from (17.70) and the third term on the right-hand side is zero from (17.69). In the second term on the right hand side, all terms with $n \neq m$ go to zero, from (17.67). In the second term, the only non-zero term occurs when $n = m$, (17.72), to give, from (17.72)

$$\int_0^T f(t) \cos(m\omega_0 t)\, dt = a_m \int_0^T \cos^2(m\omega_0 t)\, dt = a_m \frac{T}{2}.$$

Hence

$$a_m = \frac{2}{T} \int_0^T f(t) \cos(m\omega_0 t)\, dt. \tag{17.75}$$

17.10.5 Evaluation of b_m

Multiplying both sides of (17.65) by $\sin(m\omega_0 t)$, integrating between $t=0$ and $t=T$, and using (17.69), (17.71), and (17.73), it can be shown that

$$b_m = \frac{2}{T} \int_0^T f(t) \sin(m\omega_0 t)\, dt. \tag{17.76}$$

17.10.6 Alternative expressions

It can be shown that the Fourier coefficients can be found by evaluating (17.74), (17.75), and (17.76) over any interval T. In particular,

$$a_0 = \frac{1}{T} \int_{-T/2}^{T/2} f(t)\, dt \tag{17.77}$$

$$a_m = \frac{2}{T} \int_{-T/2}^{T/2} f(t) \cos(m\omega_0 t)\, dt \tag{17.78}$$

$$b_m = \frac{2}{T} \int_{-T/2}^{T/2} f(t) \sin(m\omega_0 t)\, dt. \tag{17.79}$$

17.10.7 Even and odd symmetry

Odd symmetry
A periodic function, $f(t)$, is **odd** if $f(t) = -f(-t)$ in the interval $-T/2 \leq t \leq T/2$. In this case, it can be shown that $a_0 = a_n = 0$ for all n and

$$f(t) = \sum_{n=1}^{\infty} b_n \sin(n\omega_0 t). \tag{17.80}$$

Note that $\sin(x)$ is an odd function of x. In the special case where $f(t)$ is odd, then the integrand in (17.79) is the product of an odd function with an odd function, which is an even function of time. Hence, the integral in (17.79) can be rewritten as

$$b_m = \frac{4}{T} \int_0^{T/2} f(t) \sin(m\omega_0 t)\, dt. \tag{17.81}$$

Exercise 17.5
Using (17.77) for a_0 and (17.78) for a_n, show that if $f(t)$ has odd symmetry, $a_0 = a_n = 0$. (Hint : look at the symmetry (even or odd) of the integrands in (17.77) and (17.78) when $f(t)$ is odd.)

Even symmetry

A periodic function, $f(t)$, is **even** if $f(t) = f(-t)$ for $-T/2 \leq t \leq T/2$. In this case, it can be shown that $b_n = 0$ for all n so that

$$f(t) = a_0 + \sum_{n=1}^{\infty} a_n \cos(n\omega_0 t) \tag{17.82}$$

with

$$a_0 = \frac{1}{T} \int_{-T/2}^{T/2} f(t)\, dt \tag{17.83}$$

and

$$a_m = \frac{2}{T} \int_{-T/2}^{T/2} f(t) \cos(m\omega_0 t)\, dt. \tag{17.84}$$

In the expression for a_0, (17.83), $f(t)$ is even, hence

$$a_0 = \frac{2}{T} \int_0^{T/2} f(t)\, dt. \tag{17.85}$$

Now $\cos(m\omega_0 t)$ is even, hence so is the integrand in (17.84) for a_m. Hence, for this symmetry of $f(t)$, (17.84) for a_n can be rewritten as

$$a_m = \frac{4}{T} \int_0^{T/2} f(t) \cos(m\omega_0 t)\, dt. \tag{17.86}$$

Exercise 17.6

Using (17.79) for b_n, show that for $f(t)$ even, $b_n = 0$. (Hint: look at the symmetry – even or odd – of the integrand in (17.79).)

17.11 Fourier or waveform synthesis

The Fourier series for a signal has an infinite number of terms but, in practice, only a few are required to represent the main features of the signal. Let us look at an example of a square wave shown in Figure 17.32.

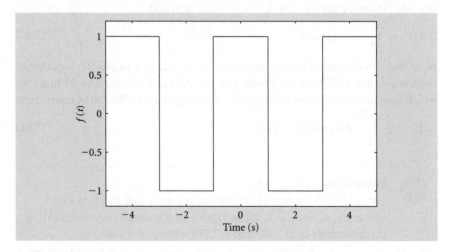

Figure 17.32 Square wave.

EXAMPLE 17.3

297

17.11 FOURIER OR WAVEFORM SYNTHESIS

In evaluating its Fourier coefficients the first point to note is that this signal is *even* between $-T/2 \leq t \leq T/2$ so that $b_n = 0$ for all n and hence has the form of (17.82):

$$\omega_0 = \frac{2\pi}{T} \quad \text{with} \quad T = 4 \tag{17.87}$$

Hence

$$\omega_0 = \frac{\pi}{2} \text{ rad s}^{-1}. \tag{17.88}$$

Also note that $a_0 = 0$. This is because of (17.77) – i.e. the mean value of the signal is zero between $-T/2$ and $T/2$.

Now let us evaluate a_n in (17.86). Substituting for ω_0 and T into (17.86) and putting $m = n$:

$$a_n = \int_0^2 f(t) \cos\left(\frac{n\pi t}{2}\right) dt. \tag{17.89}$$

Now from Figure 17.32, for $0 \leq t \leq 2$,

$$f(t) = \begin{cases} 1 & \text{for } 0 \leq t \leq 1 \\ -1 & \text{for } 1 \leq t \leq 2. \end{cases} \tag{17.90}$$

Splitting up the integral in (17.89) over these two intervals,

$$a_n = \int_0^1 \cos\left(\frac{n\pi t}{2}\right) dt - \int_1^2 \cos\left(\frac{n\pi t}{2}\right) dt. \tag{17.91}$$

Evaluating these integrals,

$$a_n = \frac{2}{n\pi} \left|\sin\left(\frac{n\pi t}{2}\right)\right|_0^1 - \frac{2}{n\pi} \left|\sin\left(\frac{n\pi t}{2}\right)\right|_1^2. \tag{17.92}$$

Putting in the limits,

$$a_n = \frac{2}{n\pi} \sin\left(\frac{n\pi}{2}\right) - \frac{2}{n\pi} \sin(n\pi) + \frac{2}{n\pi} \sin\left(\frac{n\pi}{2}\right). \tag{17.93}$$

or

$$a_n = \frac{4}{n\pi} \sin\left(\frac{n\pi}{2}\right) - \frac{2}{n\pi} \sin(n\pi). \tag{17.94}$$

Now in the second term $\sin(n\pi) = 0$ for all n so that

$$a_n = \frac{4}{n\pi} \sin\left(\frac{n\pi}{2}\right). \tag{17.95}$$

Substituting for a_n from (17.95) into (17.82), and using the fact that $a_0 = 0$ we may write the Fourier series expansion for $f(t)$:

$$f(t) = \sum_{n=1}^{\infty} a_n \cos\left(\frac{n\pi t}{2}\right) \tag{17.96}$$

with a_n given by

$$a_1 = \frac{4}{\pi}, \quad a_2 = 0, \quad a_3 = -\frac{4}{3\pi}, \quad a_4 = 0, \quad a_5 = \frac{4}{5\pi}, \quad a_6 = 0,$$

$$a_7 = -\frac{4}{7\pi}, \text{ etc.} \tag{17.97}$$

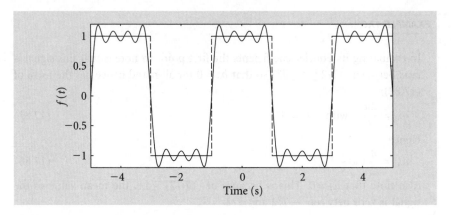

Figure 17.33 Square wave (dashed) and reconstruction from first four terms in Fourier series.

Note that for this particular signal all *even* harmonics are zero. The final form of the Fourier series is

$$f(t) = \frac{4}{\pi} \left[\cos\left(\frac{\pi t}{2}\right) - \frac{1}{3}\cos\left(\frac{3\pi t}{2}\right) + \frac{1}{5}\cos\left(\frac{5\pi t}{2}\right) - \frac{1}{7}\cos\left(\frac{7\pi t}{2}\right) + \cdots \right]. \quad (17.98)$$

How well does this Fourier series expansion represent the original signal? If we take the first four terms in (17.98), up to and including the seventh harmonic, and plot out this signal, we obtain the full curve in Figure 17.33. Comparing this with Figure 17.32, we can seen that, by combining the fundamental, third, fifth, and seventh harmonics, the most significant features of the original signal have been recovered.

You may ask what happens as more harmonics are added; does the reconstructed signal converge to the original signal? Well, the answer is 'yes and no'. Away from the edges, the reconstructed signal converges to the original signal. However, at the edges themselves, the Fourier series will never converge. The reason for this is that the derivatives of the function at the edges are discontinuous and we are trying to approximate them with cosines and sines, which are continuous. The oscillation in the signal close to discontinuous slopes is called the **Gibbs phenomenon**.

17.12 Power in periodic signals

17.12.1 Power dissipated in a $1\,\Omega$ resistor

In §17.8 an expression was given for the total energy dissipated in a $1\,\Omega$ resistor across which there was a potential difference $f(t)$. For a periodic potential the energy in each period is the same and found by integrating between the limits $-T/2$ and $T/2$. This gives the power dissipated in the resistor as

$$P_{\text{TOT}} = \frac{1}{T} \int_{-T/2}^{T/2} f(t)^2 \mathrm{d}t. \quad (17.99)$$

In fact, for a general periodic signal, $f(t)$, whether it is related to an electric current or not, we can refer to expression (17.99) as its **total power**.

17.12.2 Another form of Parseval's theorem

For a periodic signal we may write

$$P_{\text{TOT}} = \frac{1}{T} \int_{-T/2}^{T/2} \left[a_0 + \sum_{n=1}^{\infty} \{ a_n \cos(n\omega_0 t) + b_n \sin(n\omega_0 t) \} \right]^2 dt$$

$$= \frac{1}{T} \int_{-T/2}^{T/2} \left\{ \sum_{n=0}^{\infty} \sum_{m=0}^{\infty} [a_n a_m \cos(n\omega_0 t) \cos(m\omega_0 t) + 2 a_n b_m \cos(n\omega_0 t) \sin(m\omega_0 t) \right.$$

$$\left. + b_n b_m \sin(n\omega_0 t) \sin(m\omega_0 t) \right\} dt \tag{17.100}$$

The application of the orthogonality expressions in §17.10.2 gives the result that

$$P_{\text{TOT}} = a_0^2 + \sum_{n=1}^{\infty} \frac{(a_n^2 + b_n^2)}{2}. \tag{17.101}$$

This is an alternative form of Parseval's theorem.

In (17.101), $(a_n^2 + b_n^2)/2$ is the contribution from the nth harmonic to the total power. If we truncate the Fourier series at the Nth harmonic then we obtain P_N, where

$$P_N = a_0^2 + \sum_{n=1}^{N} \frac{(a_n^2 + b_n^2)}{2}. \tag{17.102}$$

By comparing the value of P_N in (17.102) with P_{TOT} in (17.99), we may assess how well the truncated Fourier series for $f(t)$ approximates the actual signal. Note that one should *not* use the summation form in (17.101) to compute the *total* power. One should use (17.99) instead, which is a lot easier to do!

EXAMPLE 17.4	**Power in a square wave**

Referring to Figure 17.32, the total power $P_{\text{TOT}} = \frac{1}{T} \int_{-T/2}^{T/2} f(t)^2 dt$, with $T = 4$ s. Since $f(t)$ is even then $f(t)^2$ is also even, and hence

$$P_{\text{TOT}} = \frac{2}{T} \int_0^{T/2} f(t)^2 dt. \tag{17.103}$$

Because $f(t)$ is discontinuous the integral is broken up to give

$$P_{\text{TOT}} = \frac{2}{4} \int_0^1 [f(t)]^2 dt + \frac{2}{4} \int_1^2 [f(t)]^2 dt. \tag{17.104}$$

Now

$$f(t) = \begin{cases} 1 & 0 \leq t \leq 1 \\ -1 & 1 \leq t \leq 2, \end{cases}$$

hence

$$P_{\text{TOT}} = \frac{1}{2}\int_0^1 1\,dt + \frac{1}{2}\int_1^2 (-1)^2 dt = \frac{1}{2}\int_0^1 1\,dt + \frac{1}{2}\int_1^2 1\,dt$$

$$= \frac{1}{2}[t]_0^1 + \frac{1}{2}[t]_1^2 = \frac{1}{2} + \frac{1}{2} = 1.$$

We have also seen that

$$f(t) = \sum_{n=1}^{\infty} a_n \cos\left(\frac{n\pi t}{2}\right) \tag{17.105}$$

with

$$a_1 = \frac{4}{\pi}, \quad a_2 = 0, \quad a_3 = -\frac{4}{3\pi}, \quad a_4 = 0, \quad a_5 = \frac{4}{5\pi}, \quad a_6 = 0,$$

$$a_7 = -\frac{4}{7\pi}, \text{ etc.,} \tag{17.106}$$

even harmonics being zero.

Exercise 17.7

The contribution to the total power from the fundamental is

$$\frac{a_1^2}{2} = \frac{1}{2}\left(\frac{4}{\pi}\right)^2 = 0.81 \ (81\%).$$

Determine the percentage contributions to the total power from:
(i) the fundamental and third harmonics combined;
(ii) the fundamental, third, and fifth harmonics combined;
(iii) the fundamental, third, fifth, and seventh harmonics combined.

17.13 Complex form for the Fourier series

17.13.1 Definition

We shall now develop a more compact way of writing the Fourier series. The term representing the nth harmonic is given by

$$a_n \cos(n\omega_0 t) + b_n \sin(n\omega_0 t). \tag{17.107}$$

However, from (6.12a) and (6.12b),

$$\cos(n\omega_0 t) = \frac{1}{2}(\exp(in\omega_0 t) + \exp(-in\omega_0 t)) \tag{17.108}$$

and

$$\sin(n\omega_0 t) = \frac{1}{2i}(\exp(in\omega_0 t) - \exp(-in\omega_0 t)). \tag{17.109}$$

Substituting from (17.108) and (17.109) into (17.107),

$$a_n \cos(n\omega_0 t) + b_n \sin(n\omega_0 t) = \frac{a_n}{2}(\exp(in\omega_0 t) + \exp(-in\omega_0 t))$$

$$+ \frac{b_n}{2i}(\exp(in\omega_0 t) - \exp(-in\omega_0 t))$$

$$= \left(\frac{a_n}{2} + \frac{b_n}{2i}\right)\exp(in\omega_0 t) + \left(\frac{a_n}{2} - \frac{b_n}{2i}\right)\exp(-in\omega_0 t).$$

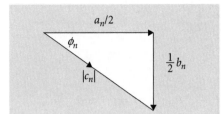

Figure 17.34 The complex quantity c_n.

Noting that $1/i = -i$, the above expression can be written as

$$a_n \cos(n\omega_0 t) + b_n \sin(n\omega_0 t) = \left(\frac{a_n}{2} - i\frac{b_n}{2}\right)\exp(in\omega_0 t) + \left(\frac{a_n}{2} + i\frac{b_n}{2}\right)\exp(-in\omega_0 t).$$

$$(17.110)$$

Let

$$c_n = \frac{1}{2}(a_n - ib_n) \qquad (17.111a)$$

as shown in Figure 17.34, then the complex conjugate

$$c_n^* = \frac{1}{2}(a_n + ib_n). \qquad (17.111b)$$

We define $c_{-n} = c_n^*$ and $c_0 = a_0$. Now, from (17.64), (17.110), and (17.111) the Fourier series (17.65) can be rewritten as

$$f(t) = c_0 + \sum_{n=1}^{\infty}\left(c_n \exp(in\omega_0 t) + c_{-n}\exp(-in\omega_0 t)\right)$$

or in a more compact form

$$f(t) = \sum_{n=-\infty}^{\infty} c_n \exp(in\omega_0 t). \qquad (17.112)$$

17.14 Amplitude and phase spectrum

From (17.111), it can be seen that all the information in the signal is contained in the *complex* Fourier coefficients $\{c_n\}$ and the fundamental frequency ω_0. Now

$$c_n = \frac{1}{2}(a_n - ib_n) = |c_n|\exp(i\phi_n)$$

where, from Figure 17.34,

$$|c_n| = \frac{1}{2}\sqrt{a_n^2 + b_n^2} \quad \text{and} \quad \phi_n = \tan^{-1}\left(\frac{-b_n}{a_n}\right).$$

A useful way to represent the complex Fourier series graphically is to plot $|c_n|$ and ϕ_n versus frequency. If $|c_n|$ is plotted, the **amplitude spectrum** is obtained (Figure 17.35). Since $c_{-n} = c_n^*$ then $|c_{-n}| = |c_n^*| = |c_n|$. This makes the amplitude spectrum symmetric about zero frequency (represented by $n = 0$), i.e. it is an *even* function of frequency.

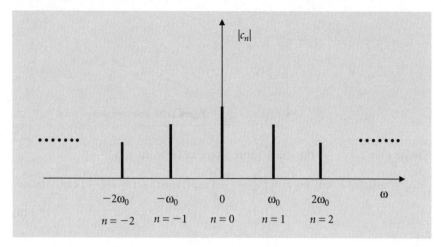

Figure 17.35 Amplitude spectrum.

The line at frequency 0, ($n=0$) term represents the DC term. The larger the amplitude of a line at frequency $n\omega_0$, then the more significant is that harmonic. Note that from (17.110) and (17.111),

$$a_n \cos(n\omega_0 t) + b_n \sin(n\omega_0 t) = c_n \exp(in\omega_0 t) + c_{-n} \exp(-in\omega_0 t).$$

i.e. contributions of equal magnitude but with frequencies $n\omega_0$ and $-n\omega_0$.

The **phase spectrum** is obtained by plotting $\phi_n = \tan^{-1}(-b_n/a_n)$ against frequency. It may be shown that $\phi_n = -\phi_{-n}$ or, in other words, the phase spectrum is an *odd* function of frequency.

Just as for Fourier transforms, negative frequencies ($n<0$) have no physical significance – they just arise from the mathematical properties of Fourier series.

17.15 **Alternative variables for Fourier analysis**

In the discussion so far, our function of interest has been a function of time and the Fourier transform a function of frequency. This pairing of the independent variables occurs, for example, in engineering applications and in quantum mechanics. However, in other applications where Fourier transforms are used, the variables are different. For example, suppose that our variable is position x so that $f(x)$ is the function to be transformed. Analogous to (17.30), the Fourier transform of $f(x)$ can be written as

$$F(k) = \int_{-\infty}^{\infty} f(x) \exp(-ikx) \, dx. \tag{17.113}$$

Similar to (17.41b), the inverse Fourier transform relation can then be written as

$$f(x) = \frac{1}{2\pi} \int_{-\infty}^{\infty} F(k) \exp(ikx) \, dk. \tag{17.114}$$

In this case, k is a called the **wave vector;** this has dimensions of inverse length.

17.16 **Applications in physics**

We now look at two examples of the application of Fourier transforms in physics.

17.16.1 **Scattering of electromagnetic radiation**

Consider a scattering object, O_j, of small size compared to the wavelength of the radiation, situated at vector position $\mathbf{r}$ relative to some origin, O. Figure 17.36 shows parallel oncoming rays scattered in a particular direction both by O_j and also by a hypothetical scatterer at O.

The line OP is perpendicular to the oncoming radiation and O_jP' is perpendicular to the scattered radiation. In the scattered direction the radiation from O_j has phase in advance of that scattered from the origin of

$$\phi_j = \frac{2\pi}{\lambda}(r_j \cos\psi - r_j \cos\psi_j). \tag{17.115}$$

The quantity in brackets is the distance that the scattering from O_j is ahead of that from O; dividing by λ expresses the distance as a fraction of a wavelength and then multiplying by 2π turns it into a phase angle.

Figure 17.36 also shows unit vectors $\mathbf{S}_0$ and $\mathbf{S}$ in the directions of the incident and scattered radiation respectively. From the definition of a scalar product of vectors

$$r_j \cos\psi = \mathbf{r} \cdot \mathbf{S} \text{ and } r_j \cos\psi_j = \mathbf{r} \cdot \mathbf{S}_0$$

and then, writing $\mathbf{s} = (\mathbf{S} - \mathbf{S}_0)/\lambda$, we find

$$\phi_j = 2\pi\mathbf{r} \cdot \mathbf{s}. \tag{17.116}$$

The vector $\mathbf{s}$ is called the **scattering vector**, which is in a space called **reciprocal space** that has a reciprocal relationship to the real space within which the scatterers reside. In some contexts it is found more convenient to use the wave vector, $\mathbf{k} = 2\pi\mathbf{s}$.

If there are a number of scatterers, say N, then, *relative to the scattering from a scatterer at the origin of unit scattering power*, the scattering from the assembly will be

$$\eta(\mathbf{s}) = \sum_{j=1}^{N} f_j \exp(2\pi i \mathbf{r}_j \cdot \mathbf{s}) \tag{17.117}$$

where f_j is the scattering power of the jth scatterer.

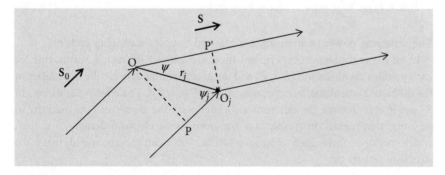

Figure 17.36 Scattering of electromagnetic radiation from a scatterer at O_j and one at the origin.

In most situations, rather than discrete scatterers there is usually a continuous distribution of scattering matter. If the scattering power per unit volume at position $\mathbf{r}$ is $\rho(\mathbf{r})$ then we may write, corresponding to (17.117),

$$\eta(\mathbf{s}) = \int_V \rho(\mathbf{r}) \exp(2\pi i \mathbf{r} \cdot \mathbf{s}) \, d\mathbf{r} \tag{17.118}$$

where $d\mathbf{r}$ is an element of volume within the total scattering volume V. It will be seen that this is the three-dimensional analogue of the *inverse* Fourier transform (17.41a). That being so, we may write the Fourier transform

$$\rho(\mathbf{r}) = C \int_{V_s} \eta(\mathbf{s}) \exp(-2\pi i \mathbf{r} \cdot \mathbf{s}) \, d\mathbf{s} \tag{17.119}$$

where C is a constant that gives the correct relationship of $\rho(\mathbf{r})$ to $\eta(\mathbf{s})$. This equation is important in that it says that if $\eta(\mathbf{s})$ is known throughout reciprocal space, then the distribution of scattering power will be determinable. When an experiment is carried out to form an image with a lens (the lens of the eye, for example), light is scattered by an object and enters the lens. The scattered light that enters the lens from each direction from each region of the object is recombined by the lens to reproduce an image of the object. The scattering process corresponds to an **inverse Fourier transform**, the lens recombination corresponds to a **Fourier transform**, and the combination of the two processes goes back to the original object. Actually the quantities $\eta(\mathbf{s})$ are complex, having both amplitude and phase, but the lens combines all the contributions of the $\eta(\mathbf{s})$ with the correct phases to produce a true image.

An important application is the scattering of X-rays (a form of electromagnetic radiation) by crystals. In this case the object doing the scattering is periodic in three dimensions. A parallelepiped-shaped **unit cell** contains an arrangement of atoms, and the parallelepipeds are packed together to form the whole crystal. Since unit cells are very small compared with crystals, in practice the crystal behaves like an infinite periodic structure. Because of the periodic nature of the crystal the scattering from it can be described by a Fourier series, but in this case a three-dimensional series. The quantity corresponding to $\eta(\mathbf{s})$ is a **structure factor**, $F(h, k, l)$ associated with a set of three integers h, k, and l corresponding to order of diffraction (harmonics) in three directions. The unit cell contains N atoms of which the jth is at the position with **fractional coordinates** (x_j, y_j, z_j); each fractional coordinate is between 0 and 1 and is the position as a fraction of one side of the unit cell relative to an origin at one of the cell corners. The structure factor is now given by

$$F(h, k, l) = \sum_{j=1}^{N} f_j \exp\{2\pi i (hx + ky + lz)\}. \tag{17.120}$$

The scattering power of an atom, f, is referred to as its **scattering factor**.

In an X-ray diffraction experiment there are discrete diffracted beams, one for each possible combination (h, k, l), and the experimental data are the **intensities** of the diffracted beams that are proportional to $|F(h, k, l)|^2$. The effective scatterers of X-rays in a crystal are the electrons associated with the atoms and the quantity to be found that reveals the positions of the atoms is the **electron density**, $\rho(x, y, z)$, which peaks around each atomic position. This can be recovered from the structure factors by

$$\rho(x, y, z) = \frac{1}{V} \sum_h \sum_k \sum_l F(h, k, l) \exp\{-2\pi i (hx + ky + lz)\} \tag{17.121}$$

where V is the volume of the unit cell. The difficulty is that although the **magnitudes** of the structure factors can be found from the measured intensities, the **phases** cannot be measured experimentally and without them the structure cannot be found. This constitutes the **phase problem** in crystallography, and finding ways of solving this problem is a major activity in crystallographic research.

17.16.2 Doppler shift

Applications of the Doppler effect give signals to which Fourier transform techniques can be applied. An example of such an application is given in Problem 17.4. The Doppler effect can be manifested in two ways. The first way is where the moving object is itself a source of a wave motion of frequency f_0 and speed c. If the radial velocity of the object relative to the observer is V then the shift in the frequency of the detected wave motion, f_d, is

$$f_d = \frac{f_0 V}{c}. \tag{17.122a}$$

The frequency shift is positive if the object moves towards the observer, so V is positive for an approaching object and negative if moving away. This kind of Doppler shift can be used to detect the speeds of recession of distant galaxies from the shifts in the wavelengths of known spectral lines they emit.

The second way is where the observer emits the radiation, and the return signal after reflection from the object is recorded – a well-known example being detection of aircraft by radar. Here the frequency shift is doubled because of the two-way process, giving

$$f_d = \frac{2f_0 V}{c}. \tag{17.122b}$$

A very important medical application of the Doppler shift is the use of ultrasound to detect the motion of anatomical structures such as heart walls and valves, where the motions are more or less periodic. The signal that comes from such observations is of the form

$$s(t) = A \cos\{2\pi(f_0 + f_d)t\}. \tag{17.123}$$

This signal is experimentally multiplied by the main frequency to give

$$A\cos\{2\pi(f_0+f_d)t\}\cos(2\pi f_0 t)=\frac{A}{2}[\cos\{2\pi(2f_0+f_d)t\}+\cos(2\pi f_d t)]. \tag{17.124}$$

The first frequency on the right-hand side is filtered out, leaving just $(A/2) \cos(2\pi f_d t)$, which can be analysed by Fourier transform techniques to find the frequency f_d.

17.17 Summary

In this chapter we have looked at how correlation techniques may be applied to determining how many cosines there are in a signal along with the amplitude, frequency, and phase of each cosine. This led us to derive the Fourier transform, which is a technique extensively used in science and engineering. A particular case arises when the underlying signal is periodic, in which case the signal can be expanded as a DC term and a linear combination of sines and cosines that have frequencies that are multiples of the fundamental frequency.

We have also discussed briefly some applications in physics where Fourier transform techniques are used. The techniques that we have looked at have concentrated on continuous signals. In §17.2 we discussed sampled signals. Any signal that is being processed by a computer or any other digital hardware, is going to be in digital form, that is as a set of **samples**. In the next chapter, we shall see how the Fourier transform can be computed for digital signals.

Problems

17.1 The signal shown below has what is known as **half-wave symmetry**. This signal is of importance in power electronics.

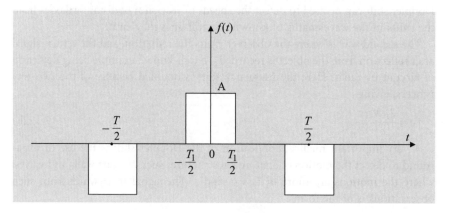

T is the period of the signal. $f(t)$ is given by

$$f(t) = \begin{cases} A & \text{for} \quad |t| \leq \dfrac{T_1}{2} \\[2mm] 0 & \text{for} \quad \dfrac{T_1}{2} \leq |t| \leq \dfrac{T}{2} - \dfrac{T_1}{2} \\[2mm] -A & \text{for} \quad \dfrac{T}{2} - \dfrac{T_1}{2} \leq |t| \leq \dfrac{T}{2}. \end{cases}$$

(i) Show, by inspection of the signal, that $a_0 = 0$ and $b_n = 0$ for all n.

(ii) Show that the Fourier series expansion of $s(t)$ is given by

$$f(t) = \sum_{n=1}^{\infty} a_n \cos(n\omega_0 t), \text{ where}$$
$$a_n = \frac{2A}{n\pi}\left[\sin\left(\frac{n\pi T_1}{T}\right) - \cos(n\pi)\sin\left(\frac{n\pi T_1}{T}\right)\right].$$

(a) Show that all even harmonics are absent.

(b) Evaluate a_n for $n = 1, \ldots, 9$ where $T_1 = T/3$ and comment on the result.

(iii) For $T_1 = T/3$, what percentage of the total power is contained in the fundamental and fifth harmonics combined?

17.2 (i) Let the Fourier transform of the derivative of a signal be given by $F_D(\omega)$, so that

$$F_D(\omega) = \int_{-\infty}^{\infty} \frac{df(t)}{dt} \exp(-i\omega t) dt.$$

Using integration by parts with $u = \exp(-i\omega t)$ and $\dfrac{dv}{dt} = \dfrac{df(t)}{dt}$ and assuming that $f(t) \to 0$ as $t \to \pm\infty$, show that the Fourier transform of $F_D(t)$ is given by $F_D(\omega) = i\omega\, F(\omega)$ where $F(\omega)$ is the Fourier transform of f(t).

(ii) The relation between the voltage across an inductor, $V_L(t)$, and the current through the inductor, $I_L(t)$, is given by $V_L(t) = L\dfrac{dI_L(t)}{dt}$. Using the result in part (i) of this question, derive a relation between the Fourier transforms of the voltage, $_TV_L(\omega)$ and the current, $_TI_L(\omega)$.

(iii) The corresponding voltage : current relation for a capacitor is given by $I_C(t) = C\dfrac{dV_C(t)}{dt}$. Derive a relation between the Fourier transforms of the voltage, $_TV_C(\omega)$ and the current, $_TI_C(\omega)$.

(iv) Compare your answers to parts (ii) and (iii) with the results of §6.8, and comment.

17.3 A signal, $f(t)$, is given by $f(t) = 4u(t)\exp(-t)$ where $u(t)$ is the step function defined by
$$u(t) = \begin{cases} 1 & t \geq 0 \\ 0 & t < 0. \end{cases}$$

(i) With this expression for $f(t)$, determine the total energy.

(ii) Show that the Fourier transform of $f(t)$ is given by $F(\omega) = \dfrac{4}{1 + i\omega}$.

(iii) Using the result for $F(\omega)$ in part (ii) and Parseval's theorem, determine the total energy in the signal and verify that the answer is the same as in part (i).

(iv) The signal in part (ii) is passed through a low-pass filter, with a general cut-off of f_c Hz. This means that the output signal has a Fourier transform that is given by

$$F(\omega) = \begin{cases} \dfrac{4}{1 + i\omega} & : |\omega| \leq 2\pi f_c \\ 0 & : |\omega| > 2\pi f_c. \end{cases}$$

Determine f_c such that the output signal from the filter has 95% of the energy of the input signal.

17.4 In §17.16 we saw that the Doppler shifted signal to be processed using the Fourier transform is given by
$$s_r(t) = A\cos(2\pi f_d t). \tag{P17.1}$$

The Doppler shift for a reflected wave is given by

$$f_d = \frac{2Vf_0}{c}. \tag{P17.2}$$

In the example of target tracking, if the target is travelling *towards* the radar, then $V > 0$ hence $f_d > 0$. However, if the target is travelling *away* from the radar, then $V < 0$ and $f_d < 0$.

One method to determine the frequency f_d is the Fourier transform. The Fourier transform of $\cos(2\pi f_d t)$ is given by

$$S_r(f) = \frac{1}{2}[\delta(f + f_d) + \delta(f - f_d)], \tag{P17.3}$$

where $\delta(x)$ is a δ-function centred at $x = 0$ defined in §17.5.

(i) Sketch the Fourier transform of $s_r(t)$ given by (P17.3).

(ii) Replacing $f_d \rightarrow -f_d$ in (P17.3), determine and sketch the Fourier transform of $s'_r(t) = \cos(-2\pi f_d t)$.

You should have found at this stage that the Fourier transforms of $\cos(2\pi f_d t)$ and $\cos(-2\pi f_d t)$ are identical. Hence, one cannot distinguish between positive and negative frequency shifts when determining the Fourier transform of $s_r(t)$, or, put another way, we cannot tell whether the target is coming towards or away from us. Such information is crucial in many situations! This problem may be solved by using Hilbert transform techniques. Essentially, what happens is that the signal, $s_r(t)$ is shifted in phase by 90° to form the signal $s_i(t)$ given by $s_i(t) = A\sin(2\pi f_d t)$. $s_r(t)$ and $s_r(t)$ can together be thought of as a complex signal:

$$s(t) = s_r(t) + is_i(t) = A\cos(2\pi f_d t) + iA\sin(2\pi f_d t) = A\exp(2\pi if_d t) \tag{P17.4}$$

$s_r(t)$ is referred to as the **in-phase component** and $s_i(t)$ is referred to as the **quadrature component**.

(iii) Given that the Fourier transform of $\sin(2\pi f_d t)$ is $\frac{1}{2}i[\delta(f + f_d) - \delta(f - f_d)]$ and using (P17.3), show that the Fourier transform of $A\exp(2\pi if_d t)$ in (P17.4) has a single peak at f_d only. Show also that the Fourier transform of $A\exp(-2\pi if_d t)$ has a single peak at $-f_d$ only.

Introduction to digital signal processing

In Chapter 17 we introduced Fourier analysis in relation to continuous, or analogue, signals. However, digital methods are widely used in many forms of communication, so now we will see how Fourier analysis can be applied to digital signals. We introduce the discrete Fourier transform (DFT), a sampled Fourier transform at discrete frequencies, which plays an important role in the transmission of signals.

18.1 More on sampling

In §17.2 we dealt with the conversion of an analogue to a digital signal and discussed the information that is lost when sampling a cosine signal. It was shown that we needed to sample the cosine with a sampling frequency that satisfies

$$f_s \geq 2f, \tag{18.1}$$

where f is the frequency of the cosine, in order to be able to reconstruct the original cosine unambiguously from its samples. Now we explore what happens when we sample a general signal that is not a cosine. Figure 18.1 shows the amplitude spectrum of the Fourier transform of a general continuous signal. The original signal can be broken down into an infinite set of cosines of different frequencies and phases. The value of the amplitude spectrum at a particular frequency f_0, indicated above, gives an indication of the importance of the cosine with frequency f_0 that contributes to this signal. To sample this particular cosine would need a sampling frequency that is at least $2f_0$ in order to reconstruct the original continuous cosine from its samples. This would also be sufficient for frequencies f up to and including f_0. However, for cosines with frequencies $f > f_0$ we would be under-sampling the cosines and hence problems will occur. To ensure that all cosines making up the signal are sampled properly the sampling frequency must satisfy the inequality

$$f_s \geq 2f_{max} \tag{18.2}$$

where f_{max} is the maximum significant frequency making up the signal, as shown in Figure 18.1.

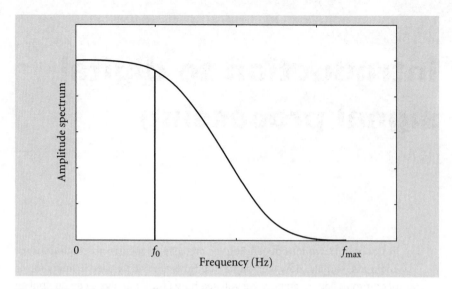

Figure 18.1 Amplitude spectrum of continuous signal.

This process of sampling occurs, for example, in digital television, where it is necessary to reconstruct the original analogue signal at the receiver. The amplitude of each frequency sample is expressed in binary form so the digital signal just consists of a stream of 1s and 0s, which would be unintelligible if viewed on the television screen.

To reconstruct the original continuous signal from the received samples, it is useful to know the Fourier transform of the sampled signal. For the simple case of a sampled cosine of frequency f, if the sampling frequency is f_s then the sampling times are $0, \frac{1}{f_s}, \frac{2}{f_s}, \cdots, \frac{m}{f_s}$, and for a signal of unit amplitude the sample values are

$$1, \quad \cos\left(2\pi\frac{f}{f_s}\right), \quad \cos\left(2\pi\frac{2f}{f_s}\right), \quad \cdots, \quad \cos\left(2\pi\frac{mf}{f_s}\right), \quad \cdots.$$

However, another frequency that would give rise to the same sample values, and many others as well, is $f + nf_s$, where n is an integer, since

$$\cos\left\{2\pi\frac{m(f + nf_s)}{f_s}\right\} = \cos\left(2\pi\frac{mf}{f_s} + 2\pi mn\right) = \cos\left(2\pi\frac{mf}{f_s}\right).$$

In the example given in §17.2 $f_s = 2f$, so a fit could be made to the sampling points with frequencies $f, 3f, 5f$, etc. In the general case there is no restriction on the sign of n so fits to the sampling points also occur for $f - f_s, f - 2f_s, \ldots$

Put another way, the set of samples correlates with cosines with these frequencies and no others. We can thus expand the set of samples as an infinite set of cosines as follows where, for now, the amplitude and phase of each cosine are unspecified:

$$x(t) = \sum_{n=-\infty}^{\infty} a_n \cos(2\pi(nf_s + f)t + \theta_n). \tag{18.3}$$

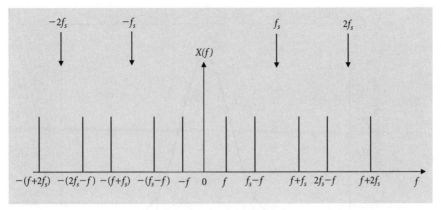

Figure 18.2 Amplitude spectrum of sampled cosine signal.

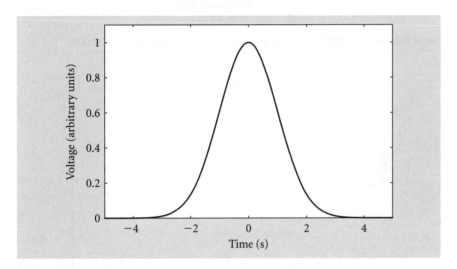

Figure 18.3 Continuous signal.

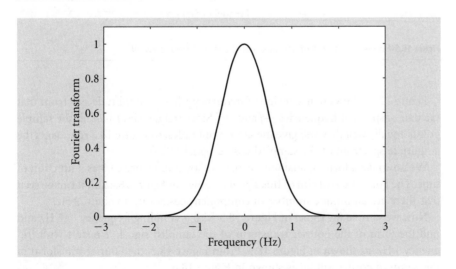

Figure 18.4 Fourier transform of signal in Figure 18.3.

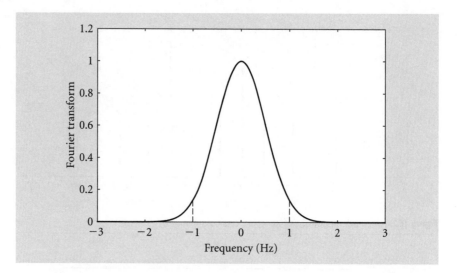

Figure 18.5 Fourier transform of signal with −1 Hz and +1 Hz components indicated.

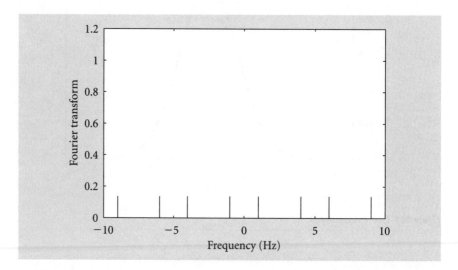

Figure 18.6 Fourier transform of sampled component of frequency 1 Hz.

Figure 17.18 shows that a cosine of frequency f has an amplitude spectrum that is a pair of spikes at frequencies $-f$ and $+f$. Since frequencies $f + nf_s$ fit the sample points equally well they will give the same amplitudes, so Figure 18.2 represents the amplitude spectrum of the sampled cosine signal (18.3).

We now take a form of analogue signal as shown in Figure 18.3 as a function of time. The Fourier transform of this signal is shown in Figure 18.4 and it can be seen that there are an infinite number of component cosines up to about 2 Hz.

Now we sample the signal in Figure 18.3 with a sampling frequency of 5 Hz and find the form of the Fourier transform of the sampled signal. We start with frequency 1 Hz, as shown in Figure 18.5. From Figure 18.2, the Fourier transform of this sampled cosine will be as shown in Figure 18.6.

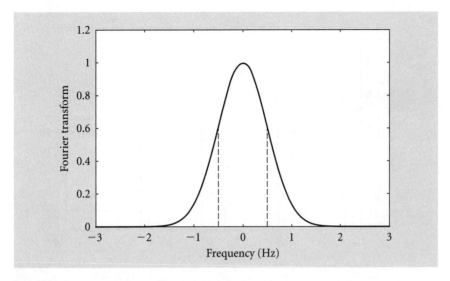

Figure 18.7 Fourier transform of signal with −0.5 Hz and +0.5 Hz components indicated.

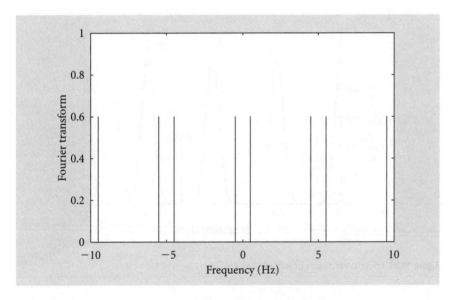

Figure 18.8 Fourier transform of sampled component of frequency 0.5 Hz.

The amplitudes of the spikes in Figure 18.6 are proportional to the corresponding amplitude of the Fourier transform in Figure 18.5.

Next we find the sampled version of the cosine with frequency 0.5 Hz (Figure 18.7). The Fourier transform of the sampled version of the cosine signal with this frequency will look like Figure 18.8 and combining Figures 18.6 and 18.8, the Fourier transforms of the combination of the two sampled cosine signals gives Figure 18.9.

If we continue the above procedure for all frequencies, we find that the Fourier transform of the sampled version of the continuous signal is a set of spikes defining a profile resembling Figure 18.10.

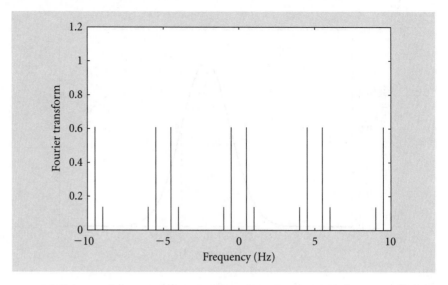

Figure 18.9 Fourier transform of sampled two-cosine signal.

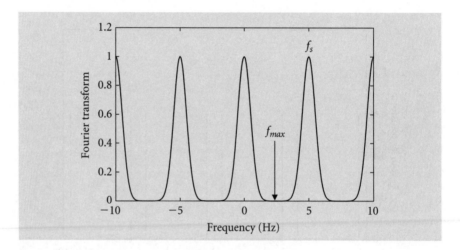

Figure 18.10 Fourier transform of sampled signal.

The Fourier transform of the sampled signal consists of the Fourier transform of the original signal plus sidebands, centred at integer multiples of the sampling frequency, that are replicas of the original spectrum. The positions of f_s and f_{max} in Figure 18.10 give a physical meaning to the condition (18.2). Unless the condition is satisfied the edges of the individual transform profiles in Figure 18.10 will overlap.

This derivation of the Fourier transform of the sampled signal is rather sketchy. In Appendix 3 the same result is derived more rigorously and it is shown that the

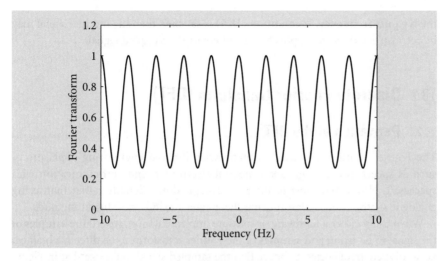

Figure 18.11 Fourier transform of the sampled signal when $f_s = 2\,\text{Hz}$.

Fourier transform of the sampled signal, $X_s(f)$, is related to the Fourier transform of the continuous signal, $X(f)$ by

$$X_s(f) = f_s X(f) + f_s \sum_{n=1}^{\infty} X(f - nf_s) + f_s \sum_{n=1}^{\infty} X(f + nf_s). \tag{18.4}$$

We now go back to the question of how to reconstruct the original analogue signal from its samples. Comparing Figures 18.4 and 18.10, we can see that the Fourier transform of the original continuous signal is represented by the band centred at zero frequency in the Fourier transform of the sampled signal. What is needed is to eliminate the sidebands centred at multiples of the sampling frequency. This can be done by applying what is known as a **filter** to the sampled signal. Filters will be discussed in Chapter 33 – suffice it to say here that it is possible, in theory at least, to eliminate the sidebands and recover the original analogue signal from its samples.

From (18.2), the minimum sampling frequency should be at least $2f_{max}$ where f_{max} is the maximum significant frequency in the amplitude spectrum of the signal. From Figure 18.4, we can see that f_{max} is approximately 2 Hz, so the minimum sampling frequency is 4 Hz. Hence by sampling at 5 Hz it is possible unambiguously to reconstruct the original analogue signal from its samples by using a filter on the sampled signal.

When the sampling frequency is less than $2f_{max}$ the central transform and sidebands overlap giving the Fourier transform of the sampled signal, as shown in Figure 18.11. It is impossible to reconstruct the original analogue signal by filtering, as the band centred at 0 Hz cannot be isolated from the continuous signal forming the sidebands.

To summarize, reconstruction of a continuous signal from its samples is possible only if

$$f_s \geq 2f_{max}, \tag{18.5}$$

which is **Shannon's sampling theorem**. The minimum permitted sampling frequency for reconstruction of the original signal from its samples, $2f_{max}$, is termed the **Nyquist frequency**. If condition (18.5) is satisfied then the original signal may be reconstructed by the application of filters to the sampled signal.

18.2 **Discrete Fourier transform (DFT)**

18.2.1 **Derivation of the DFT**

The Fourier transform is used in many scientific and engineering applications, such as speech processing, radar, medical electronics, and communication with spacecraft. Nowadays, most of the processing is done digitally rather than using analogue methods, one reason being the greater flexibility of digital methods.

When using either a hardware or software implementation, we require **samples** of the signal to be fed in and **samples** of the Fourier transform to be directly obtained, i.e. at discrete frequencies. Suppose that the sampled signal looks like that in Figure 18.12. We will refer to this signal as the **input signal** that is input into the computer or digital hardware. From (18.4), the Fourier transform might resemble Figure 18.13.

Although the input is **discrete**, or **digital**, in nature the Fourier transform is **continuous**, or **analogue**, in nature. Additional sampling of the continuous-output Fourier transform would be required in order to obtain the required **sampled Fourier transform**.

Now let us make the signal in Figure 18.12 periodic, with each period being represented by N samples as indicated in Figure 18.14. Because this extended signal is periodic, it can be expanded as a **Fourier series**. From (17.62), we saw that the frequency spectrum of a periodic signal exists at **discrete** frequencies, which is what we want. We refer to the signal in Figure 18.14 as the **periodically extended signal**. The Fourier transform of this periodically extended signal is called the **discrete Fourier transform (DFT)**, shown in Figure 18.15.

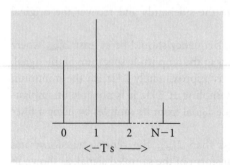

Figure 18.12 Sampled signal.

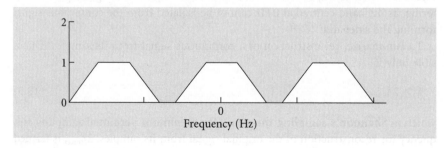

Figure 18.13 Fourier transform of a sampled signal.

To summarize, we have the following procedure to obtain a digital represe-ntation of the Fourier transform if the original analogue signal has a duration T (s):

1 Sample signal

2 Make the sampled signal periodic with period T seconds.

3 Expand periodic sampled signal as a Fourier series.

4 Amplitude and phase spectra now exist at discrete frequencies and together form the discrete Fourier transform.

A potential problem here is step 2, since in general, the signal is not periodic. Thus, the DFT is only going to be an *approximation* to the continuous Fourier transform; this may cause problems, as will be seen later.

A derivation is given in Appendix 4 of how the samples of the Fourier transform are related to the samples of the input signal. It can be shown that the pth har-monic, $X[p]$, of the periodic sampled signal, $\{x[n]\}$, is given by

$$X[p] = \sum_{n=0}^{N-1} x[n] \exp\left(-2\pi i \frac{np}{N}\right) \quad (p = 0, 1, \ldots, N-1) \tag{18.6}$$

where N is the number of samples in the *original* sampled signal in Figure 18.12.

In (18.6), $\{X[p]\}$ is the DFT of $\{x[n]\}$. It should be noted that (18.6) is entirely digital in nature. *Samples* of the Fourier transform, $X[p]$, are related to *samples* of the signal, $x[n]$, so that it would be possible to represent this relation using purely digital hardware.

Compare the DFT in (18.6) with the continuous Fourier transform (expressed in terms of cyclical frequency):

$$X(f) = \int_{-\infty}^{\infty} x(t) \exp(-2\pi i f t) \, dt. \tag{18.7}$$

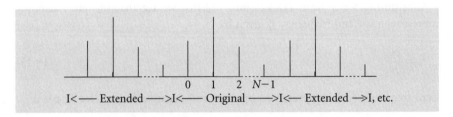

Figure 18.14 Periodic sampled signal.

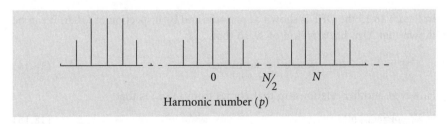

Figure 18.15 Fourier transform of a periodic signal.

Comparing (18.6) and (18.7), we can see that the DFT is effectively a numerical implementation of the continuous Fourier transform. In numerical analysis, definite integrals are often carried out using summations of small areas (Chapter 29).

Given the DFT, $\{X[p]\}$, the sampled signal $\{x[n]\}$, may be obtained using the inverse discrete Fourier transform:

$$x[n] = \frac{1}{N}\sum_{p=0}^{N-1} X[p]\exp\left(2\pi i\frac{pn}{N}\right) \quad (n = 0, 1, \ldots, N-1). \tag{18.8}$$

This result is derived in Appendix 4.

18.2.2 Interpretation of the DFT

We now consider the question of what frequency is represented by the pth harmonic, given in (18.6). We will work with **cyclical frequency** (Hz). Since the extended periodic signal in Figure 18.14 has a period of T, the fundamental has a frequency that is given by

$$F = 1/T \tag{18.9}$$

and the pth harmonic occurs at a frequency of

$$f_p = pF = \frac{p}{T}. \tag{18.10}$$

If the sampling frequency is f_s, then the time between adjacent samples is given by

$$\tau = 1/f_s. \tag{18.11}$$

Thus, if there are N samples in the original signal, Figure 18.12, then T, the duration of the signal, is given by

$$T = N/f_s. \tag{18.12}$$

Substituting for T from (18.12) into (18.10), we see that the pth harmonic represents a spectral frequency f_p given by

$$f_p = \frac{pf_s}{N}. \tag{18.13}$$

Zero frequency (dc) is represented by $p=0$ and negative frequencies by $p<0$.

18.2.3 Periodicity of the DFT

In Figure 18.15 the DFT is shown as periodic and by inspection of (18.6) it can be shown that $X[p]$ has a period of N, so that

$$X[p] = X[p + lN] \quad \text{where } l \text{ is an integer.} \tag{18.14}$$

However, another relationship that comes from (18.6) is that

$$X[-p] = X[p]^*. \tag{18.15}$$

Combining (18.14) and (18.15), it can be seen that it is only necessary to calculate the DFT for $0 \leq p \leq N/2$. For this reason the DFT is usually displayed for $p = 0, 1, \ldots, N/2$. However, there are exceptions to this convention and an example where negative frequencies are displayed is discussed in Problem 18.3.

 Exercise 18.1
Using the definition of a DFT in (18.6), prove (18.14) and (18.15).

18.2.4 **DFT of a cosine signal**

Suppose that the continuous signal $s(t)$ is given by

$$s(t) = A\cos(2\pi f_p t) \tag{18.16}$$

and is sampled with a sampling frequency of f_s.

EXAMPLE 18.1

In (18.16) we take $A = 3$, $f_p = 1.5$ Hz, and we sample with $f_s = 6$ Hz and $N = 32$. We now look at the amplitude spectrum corresponding to this situation. First we note that there are *two* frequency components associated with this signal, one at $+1.5$ Hz, the other at -1.5 Hz. From (18.13), p is related to N, f_p, and f_s by

$$p = \frac{Nf_p}{f_s}. \tag{18.17}$$

Hence, the positive frequency component occurs at a value of p given by

$$p = \frac{32 \times (+1.5)}{6} = 8 \tag{18.18}$$

and the negative frequency component occurs at a value of p given by

$$p = \frac{32 \times (-1.5)}{6} = -8. \tag{18.19}$$

From the periodicity property, (18.14), the peak at $p = -8$ also occurs at

$$p = -8 + 32 = 24. \tag{18.20}$$

If we compute and plot the DFT amplitude spectrum for this signal between $p = 0$ and $p = N/2 = 16$, then the amplitude spectrum looks like Figure 18.16. A peak at $p = 8$ is indeed observed, in agreement with our calculations; note that the shifted negative frequency peak at $p = 24$ is outside the range required to define the DFT. Note also that the DFT can only be evaluated at discrete frequency points. In Figure 18.16 and subsequent graphs these points have been joined up for display purposes, which is often done in practice. The actual DFT consists of two spikes, at $p = 8$ and $p = -8$.

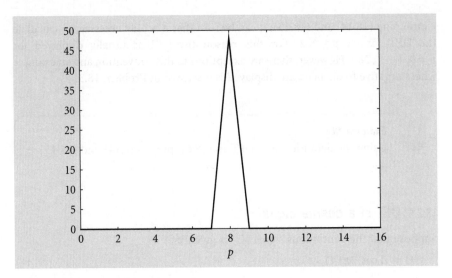

Figure 18.16 Amplitude spectrum for Example 18.1.

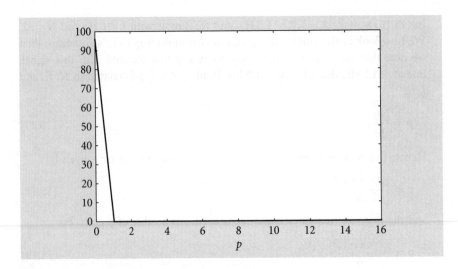

Figure 18.17 Amplitude spectrum for Example 18.2.

EXAMPLE 18.2

Now we take $A=3$, $f_p=1.5\,\text{Hz}$, $f_s=0.75\,\text{Hz}$, and $N=32$. From (18.17) the positive $p=64$ that is equivalent to $p=64-2\times32=0$ is within the defining range for the DFT. If we compute the DFT amplitude spectrum, it looks like Figure 18.17. The peak at $p=0$ corresponds to a frequency of 0 Hz, but the frequency of the input signal is 1.5 Hz! Figure 18.18 shows what has gone wrong. The original continuous signal, duration 8 s, is represented by the full

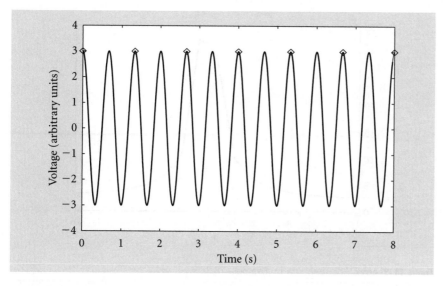

Figure 18.18 Continuous and sampled signal for Example 18.2.

curve and the samples are represented by the dots. The samples occur at every other peak of the signal, all with the same value of 3, and the DFT thinks that it is processing a dc signal and hence the peak occurs at $p=0$.

The frequency of the cosine wave is 1.5 Hz, the Nyquist frequency (18.5), which is the theoretical minimum sampling frequency $2f_p$ which, in this case, is 3 Hz, but *sampling* is only at 0.75 Hz which is less than the Nyquist frequency. Hence the DFT produces erroneous information.

Exercise 18.2

A sine wave has a frequency of 5 Hz, 128 samples are being processed, and the sampling frequency is 16 Hz. At what value of p does the positive frequency peak occur?

Exercise 18.3

A sine wave has a frequency of 5 Hz, 128 samples are being processed, and the sampling frequency is 2.5 Hz.

(i) At what value of p does the positive frequency peak occur?

(ii) The DFT is displayed between 0 and $N/2$ which in this case is 64; using the periodicity property of the DFT where, between $p=0$ and 64, will a peak occur in the DFT amplitude spectrum?

(iii) Comment on the frequency corresponding to this value of p.

In §18.2.1 the assumption was made that the input signal is periodic. This is not usually the case, and can lead to problems in the interpretation of the DFT amplitude spectrum. This is illustrated in the following example.

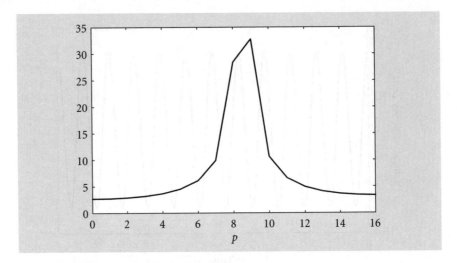

Figure 18.19 Amplitude spectrum for Example 18.3.

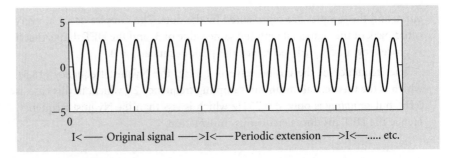

I<—— Original signal ——>I<——Periodic extension——>I<—..... etc.

Figure 18.20 Periodically extended signal for Example 18.1.

EXAMPLE 18.3

Now we take $f_p = 1.6$ Hz, $A = 3$, $f_s = 6$ Hz, and $N = 32$. The DFT amplitude spectrum, calculated from (18.6), looks as in Figure 18.19. In this case, the peak occurs between $p = 8$ and 9. The *positive* frequency peak occurs at

$$p = \frac{32 \times (+1.6)}{6} = 8.53, \tag{18.21}$$

i.e. between $p = 8$ and 9, as observed.

We now have a spectrum of different form from that found in Example 18.1; in Example 18.3, $X[p]$ is *non-zero for all p* and the spectral peak is broader than for Example 18.1. We now consider the reason for this.

To illustrate the problem, let us plot out the signal as a function of time for Example 18.1 and also draw the periodic extension of it which is assumed by the DFT (Figure 18.20). It can be seen that the periodic extensions assumed by the

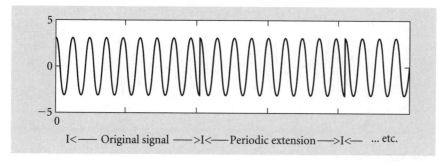

Figure 18.21 Periodically extended signal for Example 18.3.

DFT join smoothly onto the original signal which leads to a sharply defined peak in the amplitude spectrum. In this case, it can be shown that the positions of the peaks predicted by (18.17) occur at *integer* values of p. It can also be shown that the amplitude of the peak is given by $\frac{1}{2}AN$, where A is the amplitude of the cosine and N is the number of samples processed.

Now let us plot the signal in Example 18.3 as a function of time (Figure 18.21). In this case the number of cosine waves, of frequency f_p, within the period of the signal, given by (18.12), is not an integer so that the periodic extensions assumed by the DFT *do not join smoothly on to the original signal*. There are now discontinuities in the periodically extended signal. This causes the peaks in the spectrum to spread out and $X[p]$ takes on non-zero values at all values of p. This spreading is termed **leakage**, which in turn leads to **aliasing** – i.e. overlap of positive and negative peaks in the DFT spectrum. This situation can be recognized by the peak positions, predicted by (18.17), occurring at *non-integer* values of p. It can also be shown that the peak amplitudes are now *less* than $\frac{1}{2}AN$.

18.2.5 Windowing

The effects of the discontinuities mentioned above can be reduced by multiplying the original signal by a window function $w(t)$ which has the property that $w(0) = w(T) = \varepsilon$, where ε is a small number, usually zero. For $\varepsilon = 0$ the periodically extended signal assumed by the DFT now no longer has discontinuities occurring at intervals of T. There are many different window functions that can be used.

EXAMPLE 18.4

If a data window is applied to the input signal in Example 18.3, then the periodically extended signal assumed by the DFT might look like Figure 18.22.

It can now be seen that the discontinuities observed in Figure 18.21 no longer exist, and so leakage effects will be reduced. However, the signal has become distorted between the discontinuities and hence the DFT will be changed. If the

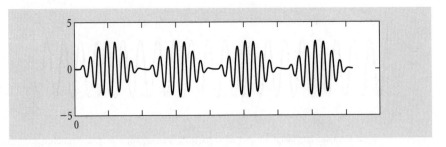

Figure 18.22 Windowed signal.

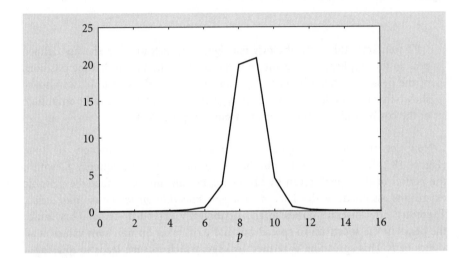

Figure 18.23 Amplitude spectrum for Example 18.4.

von Hann window, which is given by

$$w(t) = \frac{1}{2}\left[1 - \cos\left(\frac{2\pi t}{T}\right)\right], \tag{18.22}$$

is applied to the signal in Example 18.3, then the DFT amplitude spectrum of the windowed signal is as shown in Figure 18.23. There is less leakage of the main peak and the positive and negative frequency peaks no longer overlap. The peak is more prominent but the width of the main peak has increased.

Sometimes this effect can lead to more serious problems than if windowing had not been applied. The following two examples illustrate this.

EXAMPLE 18.5

$$s(t) = 3\cos\{2\pi(1.6)t\} + 3\cos\{2\pi(1.95)t\}. \tag{18.23}$$

If a window is not applied, then peaks should be observed at

$$p = \frac{32 \times 1.6}{6} = 8.53 \quad \text{(for the 1.6 Hz peak)} \tag{18.24a}$$

$$p = \frac{32 \times 1.95}{6} = 10.4 \quad \text{(for the 1.95 Hz peak)}. \tag{18.24b}$$

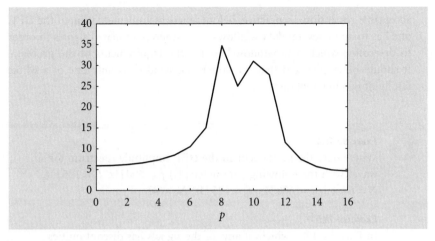

Figure 18.24 Amplitude spectrum for Example 18.5.

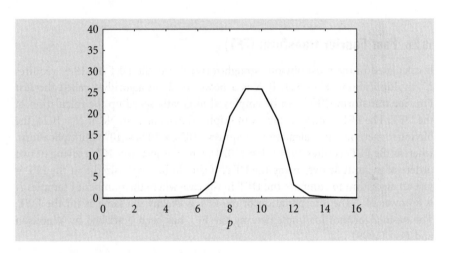

Figure 18.25 Amplitude spectrum for Example 18.6.

The non-integer values of p indicate that there must be discontinuities in the periodically extended signal. The calculated DFT amplitude spectrum is shown in Figure 18.24. The two peaks can be seen, but there is considerable overlap between them due to the leakage effect mentioned earlier.

EXAMPLE 18.6

When a window is applied to the signal (18.23) in Example 18.5, the amplitude spectrum shown in Figure 18.25 is obtained. The leakage has been reduced, but the two peaks have broadened into one! When applying a data window to reduce leakage there is a compromise between reducing leakage and obtaining acceptable resolution. In practice, before practical implementation of the DFT, one has to apply several data windows to computer-simulated signals in order to determine which data window best fits the requirements of the problem. It should be emphasized that the von Hann window is only one of a whole family of window functions.

Exercise 18.4

Where does the peak occur in the DFT amplitude spectrum for sine waves with the following parameters: (i) $f_s = 500\,\text{Hz}$, $f_p = 125\,\text{Hz}$, $N = 64$; (ii) $f_s = 600\,\text{Hz}$, $f_p = 125\,\text{Hz}$, $N = 64$?

Exercise 18.5

In Exercise 18.4, which, if any, of the signals has discontinuities occurring in the periodically extended signal which would lead to leakage in the DFT frequency spectrum?

18.2.6 Fast Fourier transform (FFT)

If calculated in the most obvious straightforward way the DFT in (18.6) requires N^2 multiplications. However, if N is a power of 2, an algorithm called the **fast Fourier transform (FFT)** can be employed to greatly speed up the calculation of the DFT. The FFT requires $N\log_2 N$ multiplications. For example, if $N = 1024$, the obvious method of calculation requires $1024 \times 1024 \approx 10^6$ multiplications, whereas the FFT requires $1024 \times \log_2(1024)$, which is just over 10^4; a saving of two orders of magnitude over using the DFT. It should be emphasized that the FFT is just an algorithm to compute the DFT in the case where the number of samples N is a power of 2; the interpretation of the FFT is exactly the same as for the DFT. The basic algorithm for implementing the FFT has been described by Woolfson and Pert (1999).

Exercise 18.6

Compare the number of multiplications required by a DFT with that required by an FFT for the following number of samples and comment on your answers: (i) 16, (ii) 256, (iii) 8192, (iv) 131 072.

18.3 Some concluding remarks

In this chapter, we have extended the discussion of Fourier analysis in Chapter 17 to sampled signals. These types of signals occur in many applications. The discrete Fourier transform has been discussed; the importance of this technique is that many practical signals of interest are digital in nature, be they signals from the brain or images of planetary satellites transmitted by spacecraft. Some applications are further explored in the problems. MATLAB has its own FFT function, FFT(X), which can implement the FFT of the signal in array X. Many C and FORTRAN program listings for FFTs can be found on the web. One good resource is the Numerical Recipes web site, <*www.numerical-recipes.com*>.

Problems

18.1 The amplitude spectrum of a continuous band-limited signal is as shown below:

    (i) This signal is sampled with a sampling frequency of 22.5 Hz. Sketch the amplitude spectrum of the sampled signal for spectral frequencies up to 30 Hz.

    (ii) For this value of the sampling frequency, is it possible to reconstruct the original continuous signal from the samples?

    (iii) Now suppose that the continuous signal is sampled with a sampling frequency of 10 Hz; note that this is *below* the Nyquist frequency. Sketch the amplitude spectrum for spectral frequencies up to 15 Hz.

    (iv) For this new value of the sampling frequency, is it possible to reconstruct the original signal from its samples? Comment on your answer in the light of the sampling theorem.

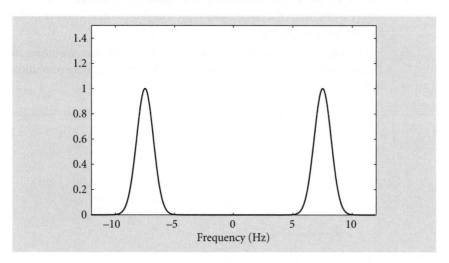

18.2 The signal, $s(t)$, is given by $s(t) = \cos(2\pi(1000)t)$. This signal is sampled with a sampling frequency f_s and a DFT is computed over 32 samples. Sketch the amplitude spectrum for frequency samples $0 \leq p \leq 16$ for the following sampling frequencies: (i) $f_s = 4\,\text{kHz}$; (ii) $f_s = 5\,\text{kHz}$; (iii) $f_s = 600\,\text{Hz}$.

Comment on your answer to part (iii).

18.3 In Problem 17.4 we saw that when measuring the Doppler shift from a moving target, the processed signal is given by $s(t) = \exp(2\pi i f_d t)$ where $f_d = 2V f_0/c$, V is the radial velocity of the target, f_0 is the frequency of the transmitted signal and $c = 3 \times 10^8 \, \mathrm{m\,s^{-1}}$ is the speed of light. The amplitude, A, has been put equal to 1. It was shown in that question that the Fourier transform of $s(t)$ is given by $S(f) = \delta(f - f_d)$. $S(f)$ in this case represents a single peak in the amplitude spectrum at spectral frequency $f = f_d$. In practice, these signals are processed digitally and so one would be using a DFT to determine the Doppler shift of the signal.

Suppose that the DFT of the signal $s(t)$ is being computed using the FFT algorithm. The signals are reflected off an aircraft. The number of samples processed is $N = 1024$. Also assume that the maximum speed of the aircraft is $1000 \, \mathrm{m\,s^{-1}}$ and the radar operating frequency is $1\,\mathrm{GHz}$ ($=10^9\,\mathrm{Hz}$). The aircraft could be moving towards or away from the radar.

(i) What is the range of possible Doppler frequencies?

(ii) From your answer to part (i), suggest an appropriate value for the minimum sampling frequency.

(iii) For the minimum sampling frequency found in part (ii), sketch the DFT amplitude spectrum as a function of frequency point p, when the aircraft is moving towards the radar with a radial velocity of $600 \, \mathrm{m\,s^{-1}}$.

The amplitude spectrum should be sketched for p values between 0 and 1023. Sketch the corresponding amplitude spectrum when the aircraft is moving *away* from the radar with a radial velocity of $600 \, \mathrm{m\,s^{-1}}$.

(iv) One definition of the **resolution** in frequency, δf_d, of the DFT is the difference between the frequencies represented by adjacent points in the DFT amplitude spectrum. Show that $\delta f_d = f_s/N$.

(v) From your answers to parts (ii) and (iv) determine the resolution in frequency of the DFT processor and hence the corresponding resolution in radial velocity.

19 Numerical methods for ordinary differential equations

Some mathematical problems that arise in physics cannot be dealt with by analytical means. However, for such problems numerical methods are available to obtain solutions for explicit values of parameters. In this chapter three basic numerical methods for the solution of ordinary differential equations are described that will deal with the great majority of problems that arise.

19.1 The need for numerical methods

The ordinary differential equations (ODEs) that have been considered so far have been of a kind that can be solved by analytical methods, but not all ODEs can be solved analytically. For example, we consider a modification of (7.45b) in the form

$$m\frac{\mathrm{d}^2x}{\mathrm{d}t^2}+f\left|\frac{\mathrm{d}x}{\mathrm{d}t}\right|^{\alpha-1}\frac{\mathrm{d}x}{\mathrm{d}t}+kx=F\cos\omega t. \tag{19.1}$$

In (19.1) α is an arbitrary constant and the resistance term, as expressed, indicates a force of magnitude $|\mathrm{d}x/\mathrm{d}t|^{\alpha}$ in a direction opposite to that of $\mathrm{d}x/\mathrm{d}t$. To obtain a solution to this equation for some specific problem with a general value of α requires the use of a numerical technique. There are a large number of methods available for the numerical solution of ODEs but here we shall describe three basic methods that will serve the great majority of needs.

19.2 Euler methods

A general form for a first-order ODE is

$$\frac{\mathrm{d}y}{\mathrm{d}x}=f(x,y) \tag{19.2}$$

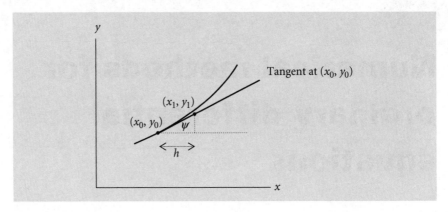

Figure 19.1 Euler method.

with the boundary condition that $x = x_0$ when $y = y_0$. Starting from (x_0, y_0), the process of going a small distance along the curve is approximated by going along the tangent to the curve at (x_0, y_0), as shown in Figure 19.1.

If the integration step in the direction of the independent variable, x, is h then

$$x_1 = x_0 + h \quad \text{and} \quad y_1 = y_0 + h\tan\psi = y_0 + h\left(\frac{dy}{dx}\right)_{x=x_0} = y_0 + hf(x_0, y_0) \quad (19.3).$$

In general the solution is generated by

$$x_{n+1} = x_n + h \quad \text{and} \quad y_{n+1} = y_n + hf(x_n, y_n). \quad (19.4)$$

This is the basis of the **Euler method**, but it is clear from Figure 19.1 that unless h is very small the computed solution will quickly deviate significantly from the true curve.

Exercise 19.1
Apply the Euler method to $dy/dx = \sin x$ starting at $(x, y) = (0, -1)$ and compare the errors of taking two steps with $h = 0.1$ with taking one step with $h = 0.2$.

At the expense of some increase in complexity, the method can be greatly improved by the use of the **Euler predictor–corrector (EPC) method**. The basis of this is illustrated in Figure 19.2.

The desired path from the point (x_0, y_0) is along the chord to (x_1, y_1). To find the slope of this chord we use the Taylor series centred on (x_0, y_0) and truncated after two terms to give

$$y_1 = y_0 + h(dy/dx)_{x=x_0} = y_0 + hf(x_0, y_0) \quad (19.5a)$$

and similarly, centring on (x_1, y_1),

$$y_0 = y_1 - hf(x_1, y_1) \quad (19.5b)$$

Subtracting (19.5b) from (19.5a) we find the slope of the chord as

$$\frac{y_1 - y_0}{h} = \frac{1}{2}\{f(x_0, y_0) + f(x_1, y_1)\}. \quad (19.6)$$

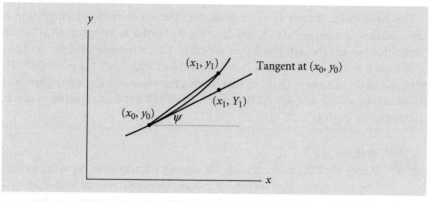

Figure 19.2 Euler predictor–corrector method.

Table 19.1 A comparison of the Euler, EPC and R–K methods for (19.8)

x	e^x-1	Euler		EPC		R–K	
		$h=0.2$	$h=0.1$	$h=0.2$	$h=0.1$	$h=2.0$	$h=1.0$
0.0	0.000 000	0.000 000	0.000 000	0.000 000	0.000 000	0.000 00	0.000 00
0.2	0.221 403	0.200 000	0.210 517	0.222 140	0.221 587		
0.4	0.491 825	0.444 281	0.467 643	0.493 463	0.492 234		
0.6	0.822 119	0.742 646	0.781 698	0.824 857	0.822 804		
0.8	1.225 54	1.107 07	1.165 28	1.229 62	1.226 56		
1.0	1.718 28	1.552 18	1.633 80	1.724 01	1.710 71		1.718 86
1.2	2.320 12	2.095 83	2.206 04	2.327 85	2.322 05		
1.4	3.055 20	2.759 86	2.904 99	3.065 38	3.057 75		
1.6	3.953 03	3.570 90	3.758 67	3.966 20	3.956 33		
1.8	5.049 65	4.561 50	4.801 37	5.066 47	5.053 86		
2.0	6.389 06	5.771 43	6.074 93	6.410 34	6.394 38	6.420 73	6.391 21

To find $f(x_1, y_1)$ requires a knowledge of y_1, the quantity we are trying to find, but an approximation to this is $f(x_1, Y_1)$ where Y_1 is the value found by the simple Euler method. This gives the general form of the EPC method as

$$Y_{n+1} = y_n + hf(x_n, y_n), \quad x_{n+1} = x_n + h$$
$$y_{n+1} = y_n + \frac{1}{2}h\{f(x_n, y_n) + f(x_{n+1}, Y_{n+1})\}. \tag{19.7}$$

To illustrate the application of the simple Euler and EPC methods we consider the solution of

$$\frac{dy}{dx} = e^x \text{ with boundary condition } x = 0; \ y = 0. \tag{19.8}$$

The analytical solution is $y = e^x - 1$, and the numerical solutions are compared with the analytical solutions in Table 19.1.

The error in the simple Euler method over the whole range of integration is approximately proportional to h, which we say is **of order h**, written as $O(h)$, which means that halving the step also halves the error. This is roughly true; for example, with $x = 0.6$ and $h = 0.2$ the error is $-0.079\,473$ but with $h = 0.1$ it is $-0.040\,421$. With the EPC the error is $O(h^2)$ so that halving the value of h gives one quarter of the error. For $x = 0.6$ and $h = 0.2$ the error is $0.002\,738$ but is one quarter as much, $0.000\,685$, for $h = 0.1$.

Exercise 19.2

Apply the EPC method to $dy/dx = \sin x$ starting at $(x, y) = (0, -1)$ and taking one step with $h = 0.2$, and compare the error with that obtained by the Euler method in Exercise 19.1 with a similar step.

19.3 Runge–Kutta method

The four-step Runge–Kutta (R–K) method combines simplicity and accuracy and is a good workhorse for most normal applications. The method is summarized in the following steps:

$$
\begin{aligned}
\delta_1 &= hf(x_0, y_0) \\
\delta_2 &= hf\left(x_0 + \frac{1}{2}h, y_0 + \frac{1}{2}\delta_1\right) \\
\delta_3 &= hf\left(x_0 + \frac{1}{2}h, y_0 + \frac{1}{2}\delta_2\right) \\
\delta_4 &= hf(x_0 + h, y_0 + \delta_3) \\
x_1 &= x_0 + h \\
y_1 &= y_0 + \frac{1}{6}(\delta_1 + 2\delta_2 + 2\delta_3 + \delta_4).
\end{aligned}
\tag{19.9}
$$

In tests with (19.9) applied to differential equation (19.8) for $h = 0.1$ and 0.2 the calculated values (not shown in the table) agree exactly with the true values to the number of significant figures in Table 19.1. The values for $h = 2$ and $h = 1$, which are given in the table, show errors large enough to confirm that the error is approximately $O(h^4)$; the ratio of errors is 14.7 rather than 16, which would be expected from an error proportional to h^4.

The error for R–K with $h = 1.0$ is $0.002\,15$ for $x = 2.0$, and such an error would be obtained with EPC using $h \approx 0.06$. Thus it would require 16 times as many complete steps with EPC although each complete EPC step requires two function evaluations as against four evaluations for R–K. We may therefore conclude that R–K is about eight times as efficient as EPC – at least for this example.

The extension of the R–K method to a set of coupled differential equations is quite straightforward and will be illustrated by the following system with two dependent variables:

$$
\frac{dy}{dx} = f_y(x, y, u)
$$

$$\frac{du}{dx} = f_u(x, y, u) \tag{19.10}$$

with $y = y_0$ and $u = u_0$ when $x = x_0$. The steps to advance the independent variable by h are

$$\delta_{1,y} = hf_y(x_0, y_0, u_0)$$

$$\delta_{1,u} = hf_u(x_0, y_0, u_0)$$

$$\delta_{2,y} = hf_y\left(x_0 + \frac{1}{2}h, y_0 + \frac{1}{2}\delta_{1,y}, u_0 + \frac{1}{2}\delta_{1,u}\right)$$

$$\delta_{2,u} = hf_u\left(x_0 + \frac{1}{2}h, y_0 + \frac{1}{2}\delta_{1,y}, u_0 + \frac{1}{2}\delta_{1,u}\right)$$

$$\delta_{3,y} = hf_y\left(x_0 + \frac{1}{2}h, y_0 + \frac{1}{2}\delta_{2,y}, u_0 + \frac{1}{2}\delta_{2,u}\right)$$

$$\delta_{3,u} = hf_u\left(x_0 + \frac{1}{2}h, y_0 + \frac{1}{2}\delta_{2,y}, u_0 + \frac{1}{2}\delta_{2,u}\right)$$

$$\delta_{4,y} = hf_y(x_0 + h, y_0 + \delta_{3,y}, u_0 + \delta_{3,u})$$

$$\delta_{4,u} = hf_u(x_0 + h, y_0 + \delta_{3,y}, u_0 + \delta_{3,u})$$

$$x_1 = x_0 + h$$

$$y_1 = y_0 + \frac{1}{6}(\delta_{1,y} + 2\delta_{2,y} + 2\delta_{3,y} + \delta_{4,y})$$

$$u_1 = u_0 + \frac{1}{6}(\delta_{1,u} + 2\delta_{2,u} + 2\delta_{3,u} + \delta_{4,u}). \tag{19.11}$$

The same scheme can be used to solve a second-order ODE and will be illustrated by (19.1). Writing

$$\frac{dx}{dt} = u, \tag{19.12a}$$

(19.1) becomes

$$\frac{du}{dt} = \frac{F}{m}\cos\omega t - \frac{f}{m}|u|^{\alpha-1}u - \frac{k}{m}x. \tag{19.12b}$$

It is necessary to fix boundary conditions for x and u for a particular value of t and then the right-hand sides of (19.12) are functions of t, x, and u that play the roles of the functions of x, y, and u on the right-hand side of (19.10).

A program, OSCILLATE, to solve (19.1) in the form presented by (19.12) is provided in outline form in Appendix 5. The program is provided with a set of standard parameter values for m, f, k, F, ω, x_0, and u_0 which may be changed if the user wishes to do so. The value of α is a required input. The program gives amplitude as a function of time for batches of 10 cycles corresponding to the angular frequency ω and up to 6 of these batches, either contiguous or not, can be stored in data files for subsequent graphical output.

The program has been run with the standard values provided:

$$m = 0.001\,\text{kg}, \quad f = 0.003, \quad k = 100\,\text{kg s}^{-2}, \quad F = 0.01\,\text{N}, \quad \omega = 300\,\text{s}^{-1},$$

$$x_0 = 0, \quad u_0 = (dx/dt)_0 = 0,$$

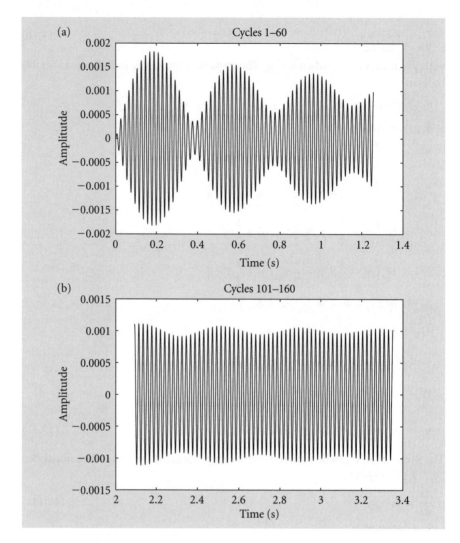

Figure 19.3 A run of the program OSCILLATE with provided parameters and $\alpha = 1.4$ (a) cycles 1–60; (b) cycles 101–160.

and with $\alpha = 1.4$. (The units of f depend on the value of α.) Figure 19.3a shows the initial transients while Figure 19.3b represents the situation some time later where the steady state is being approached but has not quite been reached.

The same general principles may be applied to the solution of ODEs of higher order. For example, the equation

$$\frac{d^3y}{dx^3} + 2\frac{d^2x}{dy^2} - \frac{dy}{dx} + x = 0$$

may be expressed as the coupled equations

$$\frac{dy}{dx} = u, \quad \frac{du}{dx} = v, \quad \frac{dv}{dx} = -2v + u - x. \tag{19.13}$$

Similarly the equations (7.4) can be expressed as four coupled first-order differential equations of the form

$$\frac{dx}{dt} = u, \quad \frac{du}{dt} = -\frac{GMx}{(x^2 + y^2)^{3/2}}, \quad \frac{dy}{dt} = v, \quad \frac{dv}{dt} = -\frac{GMy}{(x^2 + y^2)^{3/2}}. \quad (19.14)$$

Sets of equations similar to (19.11) but extended to deal with three or four dependent variables can be used numerically to solve (19.13) or (19.14), with given boundary conditions.

19.4 Numerov method

A predictor–corrector method that is very efficient is the **Numerov method**. This can be applied to second-order differential equations when no first-order term is present, that is of the form

$$\frac{d^2 y}{dx^2} = f(x, y). \quad (19.15)$$

This method can be explained by the application of the Taylor series to three points (x_0, y_0), (x_1, y_1) and (x_2, y_2) with $x_1 - x_0 = x_2 - x_1 = h$. We write

$$y_2 = y_1 + h y_1^I + \frac{h^2}{2} y_1^{II} + \frac{h^3}{6} y_1^{III} + \frac{h^4}{24} y_1^{IV} + \frac{h^5}{120} y_1^V + \cdots \quad (19.16a)$$

and

$$y_0 = y_1 - h y_1^I + \frac{h^2}{2} y_1^{II} - \frac{h^3}{6} y_1^{III} + \frac{h^4}{24} y_1^{IV} - \frac{h^5}{120} y_1^V + \cdots, \quad (19.16b)$$

where, here, the roman numeral superscripts on the ys indicate the order of differentiation and the subscript the point at which the differential is taken. Truncating each of these equations after four terms and adding gives an estimate for y_2 as

$$Y_2 = 2y_1 - y_0 + h^2 y_1^{II} = 2y_1 - y_0 + h^2 f(x_1, y_1). \quad (19.17)$$

Retaining all the six terms in each of (19.16a) and (19.16b) and adding gives, on rearrangement,

$$y_2 = 2y_1 - y_0 + h^2 y_1^{II} + \frac{h^4}{12} y_1^{IV} = 2y_1 - y_0 + h^2 f(x_1, y_1) + \frac{h^4}{12} y_1^{IV}. \quad (19.18)$$

The final term on the right-hand side of (19.18) can be expressed in terms of second derivatives by another application of the Taylor series in the form

$$y_2^{II} = y_1^{II} + h y_1^{III} + \frac{h^2}{2} y_1^{IV} + \cdots \quad (19.19a)$$

and

$$y_0^{II} = y_1^{II} - h y_1^{III} + \frac{h^2}{2} y_1^{IV} + \cdots. \quad (19.19b)$$

Adding these two equations, rearranging and expressing second derivatives in functional form gives

$$h^2 y_1^{IV} = f(x_2, y_2) - 2f(x_1, y_1) + f(x_0, y_0). \tag{19.20}$$

Inserting from (19.20) into (19.18) and replacing y_2 on the right-hand side by its predicted value, Y_2, we find

$$y_2 = 2y_1 - y_0 + \frac{h^2}{12} \{ f(x_2, Y_2) + 10f(x_1, y_1) + f(x_0, y_0) \}. \tag{19.21}$$

Note that this procedure requires the values of x and y for two previous steps, so it is not self-starting in the sense that the initial conditions alone enable the process to get under way. For this reason, assuming that y_0 and y_0' were known, some other method, such as EPC or R–K, could be used for one step and then Numerov could take over. The Numerov method has the excellent characteristics that it requires only two function evaluations per step and that the error is $O(h^5)$. Where it can be applied, it should be the method of choice.

Exercise 19.3
Use the Numerov method with for a single step with $h = 1$ for the differential equation $d^2 y / dx^2 = y - x - 1$ starting with $(x_0, y_0) = (0, 2)$ and $(x_1, y_1) = (1, 4.718\,282)$.

Problems

19.1 A series circuit containing a resistor of resistance $20\,\Omega$ and a capacitor of capacitance $10^{-4}\,$F is driven by an AC source of $100\,$V at a frequency of $50\,$Hz. Find the resultant steady-state current by expressing potential difference and current in complex form and by writing a simple EPC program to solve the appropriate differential equation.

19.2 Run the program OSCILLATE for $\alpha = 1$ with the standard parameters except for replacing f, by $0.1\,\mathrm{N\,s\,m^{-1}}$ and then by $1.0\,\mathrm{N\,s\,m^{-1}}$. In each case plot the displacement against t for cycles 201–210, by which time the transient has disappeared, and hence estimate the amplitude of the steady-state oscillation. Compare this with the analytical result (7.59).

Applications of partial differential equations

For ordinary differential equations, as dealt with in Chapter 7, there is only one independent variable and any number of dependent variables. For example, a system of gravitationally interacting bodies has a state that is a function of time, the independent variable, and dependent on time are the position and velocity components of all the participating bodies – the dependent variables. However, there are situations in which there can be several independent variables – for example, (x, t), (x, y), (x, y, t) – where one or more additional quantities are dependent both on position and time.

In this chapter we concentrate mostly on situations where there is one dependent variable and two independent variables, which include interesting physical examples such as diffusion, the Poisson equation, the wave equation, and the Schrödinger wave equation. Various computer-based methods will be described appropriate for solving partial differential equations of different types.

20.1 Types of partial differential equation

All the partial differential equations (PDEs) we meet in this chapter are **second-order partial differential equations** of the general form

$$A\frac{\partial^2 \phi}{\partial x^2} + B\frac{\partial^2 \phi}{\partial x\,\partial y} + C\frac{\partial^2 \phi}{\partial y^2} + D\frac{\partial \phi}{\partial x} + E\frac{\partial \phi}{\partial y} + F\phi + G = 0. \tag{20.1}$$

As for ordinary differential equations, the solution of PDEs requires knowledge of boundary conditions and in the case of simpler forms of PDEs – for example, the simple wave equation – it is possible to find analytical solutions. However, in many cases the only way of finding a solution is by a numerical method and here we shall be describing some, but not all, of the methods available for solving various forms of (20.1).

PDEs of the form (20.1) can be divided into three types that lend themselves to different numerical approaches:

- **elliptic equations** for which $B^2 < 4AC$
- **parabolic equations** for which $B^2 = 4AC$
- **hyperbolic equations** for which $B^2 > 4AC$.

Exercise 20.1

Identify the type of second-order PDE for the following:

(i) $\dfrac{\partial^2 \phi}{\partial x^2} + \dfrac{\partial^2 \phi}{\partial y^2} + 3\phi = 0;$ (ii) $\dfrac{\partial^2 \phi}{\partial x} + \dfrac{\partial^2 \phi}{\partial x \partial y} + 4\dfrac{\partial \phi}{\partial x} = 0;$

(iii) $\dfrac{\partial^2 \phi}{\partial x^2} + 2\dfrac{\partial^2 \phi}{\partial x \partial y} + \dfrac{\partial^2 \phi}{\partial y^2} + \dfrac{\partial \phi}{\partial x} + \dfrac{\partial \phi}{\partial y} = 0.$

20.2 Finite differences

To illustrate the principle of finite differences we consider a function $f(x)$ of a single independent variable x and we imagine that we have the function tabulated at a sequence of equispaced values of x with interval h between successive values. Then, by the application of the Taylor series

$$f(X + h) = f(X) + hf'(X) + \frac{h^2}{2}f''(X) + \frac{h^3}{3!}f'''(X) + \cdots. \tag{20.2}$$

This can be rearranged as

$$f'(X) = \frac{f(X + h) - f(X)}{h} - \frac{h}{2}f''(X) - \cdots$$

Thus if for small h the final term, and others following it, can be ignored we have the **forward difference approximation**

$$f'(X) = \frac{f(X + h) - f(X)}{h}. \tag{20.3}$$

Similarly, from the following Taylor series

$$f(X - h) = f(X) - hf'(X) + \frac{h^2}{2}f''(X) - \frac{h^3}{3!}f'''(X) + \cdots \tag{20.4}$$

we can derive a **backwards difference approximation**

$$f'(X) = \frac{f(X) - f(X - h)}{h}. \tag{20.5}$$

These two approximations are illustrated in Figure 20.1.

The better **central difference approximation** is obtained by subtracting (20.4) from (20.2) and rearranging to give

$$f'(X) = \frac{f(X + h) - f(X - h)}{2h} + O(h^2)$$

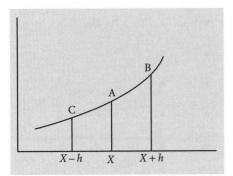

Figure 20.1 The forward difference approximation is the slope of the chord AB; the backward difference approximation is the slope of the chord CA.

where $O(h^2)$ signifies that the extra terms are of the second, or greater power of h. If h is sufficiently small then the extra terms may be neglected so that we can write, with sufficient accuracy,

$$f'(X) = \frac{f(X+h) - f(X-h)}{2h}. \tag{20.6}$$

By adding (20.2) to (20.4), rearranging and neglecting terms of order h^2 or greater power the following finite-difference approximation is found for the second derivative

$$f''(X) = \frac{f(X+h) + f(X-h) - 2f(X)}{h^2}. \tag{20.7}$$

A similar process can be used to express **partial** derivatives for a function, f(x, y), involving two independent variables. Here we imagine that the function is plotted on a two-dimensional rectangular grid with spacing h_x in the x direction and h_y in the y direction (Figure 20.2).

Since for derivatives with respect to one independent variable the other independent variable is kept constant, we have

$$\left(\frac{\partial f}{\partial x}\right)_{(X,Y)} = \frac{f(X+h_x, Y) - f(X-h_x, Y)}{2h_x}, \tag{20.8a}$$

$$\left(\frac{\partial f}{\partial y}\right)_{(X,Y)} = \frac{f(X, Y+h_y) - f(X, Y-h_y)}{2h_y}, \tag{20.8b}$$

$$\left(\frac{\partial^2 f}{\partial x^2}\right)_{(X,Y)} = \frac{f(X+h_x, Y) + f(X-h_x, Y) - 2f(X, Y)}{h_x^2}, \tag{20.9a}$$

$$\left(\frac{\partial^2 f}{\partial y^2}\right)_{(X,Y)} = \frac{f(X, Y+h_y) + f(X, Y-h_y) - 2f(X, Y)}{h_y^2}. \tag{20.9b}$$

An expression for the mixed partial derivative $(\partial^2 f/\partial x\, \partial y)_{(X,Y)}$ is slightly more difficult to derive and since it will not be used in what follows the result is just stated:

$$\left(\frac{\partial^2 f}{\partial x \partial y}\right)_{(X,Y)}$$
$$= \frac{f(X+h_x, Y+h_y) + f(X-h_x, Y-h_y) - f(X+h_x, Y-h_y) - f(X-h_x, Y+h_y)}{4h_x h_y}$$

$$\tag{20.10}$$

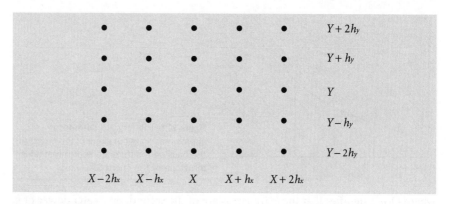

Figure 20.2 The rectangular grid for a function of two independent variables.

We shall now see how to apply these finite-difference expressions to the solution of problems involving one-dimensional diffusion.

 Exercise 20.2

Given $\tan 40° = 0.8391$, $\tan 45° = 1.0000$, and $\tan 50° = 1.1918$, estimate

(i) $\left(\dfrac{\mathrm{d}(\tan x)}{\mathrm{d}x}\right)_{x=45°}$; (ii) $\left(\dfrac{\mathrm{d}^2(\tan x)}{\mathrm{d}x^2}\right)_{x=45°}$.

20.3 Diffusion

Figure 20.3 shows a column of solution of unit cross-section for which the concentration varies along its length. The concentration of the solution, n, at any point can be described as the number of particles of solute (the dissolved substance) per unit volume. There will be a continuous net diffusion of solute from more concentrated to less concentrated regions, so that the concentration will be a function both of position and time.

The net transfer of particles traversing the cross-section A in the positive x direction, into the region between A and B, in a time δt is proportional to the concentration gradient and is given by

$$\delta N_x = -D\left(\frac{\partial n}{\partial x}\right)_x \delta t \tag{20.11a}$$

where D is the coefficient of diffusion. This relationship is known as **Fick's law**.

At the cross-section B, distant δx from A, the transfer of particles in the positive x direction, out of the region between A and B, in the same time interval is

$$\delta N_{x+\delta x} = -D\left(\frac{\partial n}{\partial x}\right)_{x+\delta x} \delta t. \tag{20.11b}$$

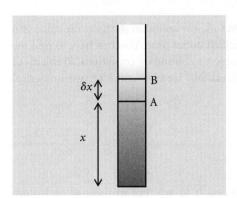

Figure 20.3 A column of solution with a concentration gradient.

The net flow of particles into the volume δx (remember unit cross-section) between A and B gives a change of concentration

$$\delta n = \frac{\delta N_x - \delta N_{x+\delta x}}{\delta x} = -D \left\{ \frac{\left(\frac{\partial n}{\partial x}\right)_x - \left(\frac{\partial n}{\partial x}\right)_{x+\delta x}}{\delta x} \right\} \delta t. \tag{20.12}$$

We now use the two-term Taylor series

$$\left(\frac{\partial n}{\partial x}\right)_{x+\delta x} = \left(\frac{\partial n}{\partial x}\right)_x + \left(\frac{\partial^2 n}{\partial x^2}\right)\delta x$$

together with

$$\frac{\partial n}{\partial t} = \lim_{\delta t \to 0}\left(\frac{\delta n}{\delta t}\right)$$

to give the **diffusion equation**

$$\frac{\partial n}{\partial t} = D\frac{\partial^2 n}{\partial x^2}. \tag{20.13}$$

This equation, which is a parabolic PDE according to the conditions given in §20.1, also occurs in other physical situations. For example, heat flow in a uniform bar well thermally insulated along its length will give a temperature, θ, at cross-sections of the bar varying both in space and time controlled by the relationship

$$\frac{\partial \theta}{\partial t} = \frac{\kappa}{c\rho}\frac{\partial^2 \theta}{\partial x^2} \tag{20.14}$$

where κ is the thermal conductivity of the bar material, c its specific heat capacity, and ρ its density.

Exercise 20.3
A uniform bar has density $\rho = 4000\,\text{kg m}^{-3}$, thermal conductivity $\kappa = 500\,\text{W m}^{-1}\text{K}^{-1}$, and specific heat capacity $c = 1000\,\text{J kg}^{-1}\text{K}^{-1}$. At a particular cross-section of the bar the temperature is 400 K and at cross-sections on either side at distances of 1 cm the temperatures are 383 K and 420 K. Assuming that the bar is well lagged, what is the instantaneous rate of change of temperature with time at the central cross-section?

To solve this type of problem we need to know an initial state, i.e. the value of n (or θ) as a function of position at a specified initial time. We then have to find the distribution of n at subsequent times, subject to boundary conditions. Problems of this kind are called **initial-value problems** and we now consider two numerical approaches for solving such problems.

20.4 **Explicit method**

In solving a diffusion problem the process is to advance in time, at each stage determining the concentration as a function of position. The notation used here is to solve for equally spaced positions x_i, for $i = 0, 1, 2$, etc., such that $x_{i+1} - x_i = \Delta x$ and equally spaced times t_j, for $j = 0, 1, 2$, etc., such that $t_{j+1} - t_j = \Delta t$. The concentration at position x_i and time t_j can then be indicated as $n(i, j)$.

The most straightforward approach to solve (20.13) is to represent the partial derivative with respect to time in a forward-difference form together with a central-difference form for the space derivative. This gives

$$\frac{n(i, j+1) - n(i, j)}{\Delta t} = D \frac{n(i+1, j) + n(i-1, j) - 2n(i, j)}{(\Delta x)^2}$$

or

$$n(i, j+1) = r\{n(i+1, j) + n(i-1, j)\} + (1 - 2r)n(i, j), \qquad (20.15)$$

where

$$r = \frac{D\Delta t}{(\Delta x)^2}. \qquad (20.16)$$

If boundary conditions are known then (20.15) enables a complete set of concentrations at all positions for time t_{j+1}, $n(i, j+1)$, to be determined from the complete set, $n(i, j)$. This calculation is easily programmed and Appendix 6 gives a schematic outline of a program EXPLICIT for following the sequence of temperature profiles in a lagged uniform bar. Boundary conditions are set at both ends of the bar that, in general, give the temperature as a function of time although in this case the two ends are given fixed temperatures of 300 K and 400 K. Initially the central cross-section of the bar has temperature 500 K with a linear variation from the centre to each end. The characteristics of the bar are:

- length 1 m
- thermal conductivity 200 W m^{-1} K^{-1}
- specific heat capacity 1000 J kg^{-1} K^{-1}
- density 2700 kg m^{-3}.

The calculation is run with $\Delta x = 0.05$ m, so requiring 19 internal temperature derivations for each time step. The program allows either Δt or r to be fixed and in this case Δt has been fixed at 10 s. Figure 20.4 gives the output of EXPLICIT for $t = 1000, 2000$, and 3000 s and also shows the initial temperature profile ($t = 0$) and the theoretical final profile at $t = \infty$.

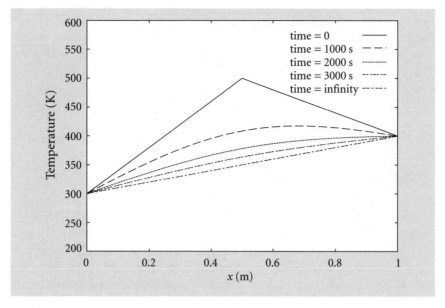

Figure 20.4 Temperature profiles from the program EXPLICIT.

The profile after infinite time is a uniform temperature gradient across the bar. For a uniform gradient $\partial\theta/\partial x = a$ constant or $\partial^2\theta/\partial x^2 = 0$ and hence, from (20.13), $\partial n/\partial t = 0$, corresponding to an equilibrium configuration.

Because of the approximations in the finite-difference approach it will seem obvious that having Δx and Δt both as small as possible is desirable for the greatest accuracy. This is true in general, but there is another overriding requirement in order to have stability in the method; this requirement is that $r \leq \frac{1}{2}$. For the thermal conduction problem

$$r = \frac{\kappa\Delta t}{\rho c(\Delta x)^2} \tag{20.17}$$

and, for the problem giving rise to Figure 20.4, $r = 0.2963$. Changing Δt to 5 s and Δx to 0.025, both smaller than in the previous case, gives $r = 0.5926$, and running the revised problem for 100 s gives the result shown in Figure 20.5.

In calculations of this type there are two criteria to be met – **stability** and **convergence**. The first of these criteria has been illustrated: different finite-difference approaches have different conditions for stability and these can be found by analytical approaches. As previously stated, for the explicit method one must have $r \leq \frac{1}{2}$ for stability, otherwise any small error grows without limit. If the calculation leading to Figure 20.5 is continued to 1000 s the oscillations soon grow to an amplitude of 10^6 K and beyond! Convergence implies that as Δx and Δt both tend to zero, so the computational result tends to that which would be obtained analytically.

Exercise 20.4
A heated-bar problem is set up with the parameters that gave Figure 12.4. Which of the following intervals in the explicit approach give stability of the solution?

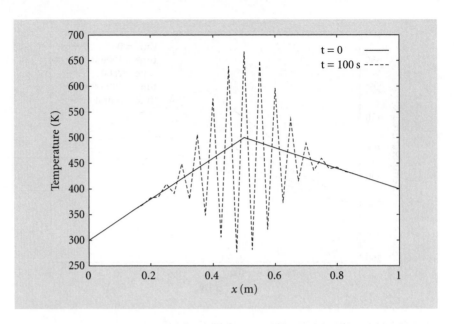

Figure 20.5 The heated bar problem after 100 s with $r = 0.5926$.

> (i) $\Delta t = 50\,\text{s}$, $\Delta x = 0.1\,\text{m}$;   (ii) $\Delta t = 1\,\text{s}$, $\Delta x = 0.01\,\text{m}$;
> (iii) $\Delta t = 1000\,\text{s}$, $\Delta x = 1\,\text{m}$;   (iv) $\Delta t = 0.1\,\text{s}$, $\Delta x = 0.01\,\text{m}$.
>
> Which of these should give the most precise results?

20.5 The Crank–Nicholson method

In the explicit method a forward-difference representation was used for $\partial n/\partial t$ in the form

$$\frac{n(i, j+1) - n(i, j)}{\Delta t}. \tag{20.18}$$

Now, another way of looking at (20.18) is to regard it as a **central-difference** representation of the slope at the point $(i, j + \frac{1}{2})$, but then it would be necessary to represent $\partial^2 n/\partial x^2$ at the same point. In the Crank–Nicholson method this is achieved by averaging of the finite-difference representations of $\partial^2 n/\partial x^2$ at the points (i, j) and $(i, j+1)$. This leads to the equation

$$\frac{n(i, j+1) - n(i, j)}{\Delta t} = \frac{D}{2} \left\{ \frac{n(i+1, j) + n(i-1, j) - 2n(i, j)}{(\Delta x)^2} \right.$$

$$\left. + \frac{n(i+1, j+1) + n(i-1, j+1) - 2n(i, j+1)}{(\Delta x)^2} \right\}$$

which rearranges as

$$-rn(i-1, j+1) + 2(1+r)n(i, j+1) - rn(i+1, j+1)$$
$$= rn(i-1, j) + 2(1-r)n(i, j) + rn(i+1, j),$$

(20.19)

where r is defined in (20.16) and (20.17) for diffusion and heat conduction respectively. This is an **implicit method** in that linear combinations of values at t_{j+1} are given in terms of linear combinations of values at t_j. The system of equations for the unknowns that comes from this method is particularly easy to solve. We shall illustrate this by taking $r = 1$ and using four segments in the space dimension. Writing n_i for $n(i, j+1)$, the resultant set of equations takes the form

$$-n_0 + 4n_1 - n_2 \qquad\qquad\qquad = A$$
$$-n_1 + 4n_2 - n_3 \qquad\qquad\quad = B$$
$$-n_2 + 4n_3 - n_4 \qquad\quad = C$$
$$-n_3 + 4n_4 - n_5 = D$$

Since the boundary-condition values n_0 and n_5 are known, this can be rewritten as

$$4n_1 - n_2 \qquad\qquad\qquad = A + n_0$$ (20.20a)

$$-n_1 + 4n_2 - n_3 \qquad = B$$ (20.20b)

$$-n_2 + 4n_3 - n_4 = C$$ (20.20c)

$$-n_3 + 4n_4 = D + n_5$$ (20.20d)

where everything on the right-hand side is known from the values at time t_j and the boundary conditions. The procedure for solution is as follows:

1 From (20.20a) express n_2 in terms of n_1.
2 Since n_2 is known in terms of n_1 from (20.20b) n_3 is found in terms of n_1.
3 Since n_2 and n_3 are known in terms of n_1 (20.20c) gives n_4 in terms of n_1.
4 Substituting for n_3 and n_4 in terms of n_1 in (20.20d) enables n_1 to be found.
5 Since n_2, n_3, and n_4 have been previously found in terms of n_1 they can now be determined explicitly.

The pattern of coefficients of these equations, which enables the above systematic approach to their solution, is of *tridiagonal form*, which is illustrated as

$$\begin{bmatrix} x & x & & & & & \\ x & x & x & & & & \\ & x & x & x & & & \\ & & x & x & x & & \\ & & & x & x & x & \\ & & & & x & x \end{bmatrix}.$$

The Crank–Nicholson method for solving one-dimensional heat flow problems is available as the computer program HEATCRNI, outlined in Appendix 7. It is run for the parameters that gave rise to Figure 20.4, except that Δt is set at 50 s. The result is shown in Figure 20.6. The results are very similar to those in Figure 20.4 even though, with the value of Δt used, $r = 1.481$, which would make the explicit method highly unstable. In fact, analysis shows that the Crank–Nicholson method is **unconditionally stable**, meaning that it will not go into wild unbounded

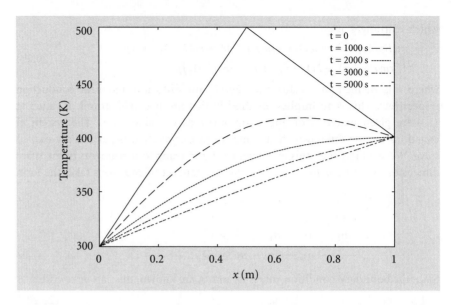

Figure 20.6 Temperature profiles from HEATCRNI with $\Delta t = 50$ s.

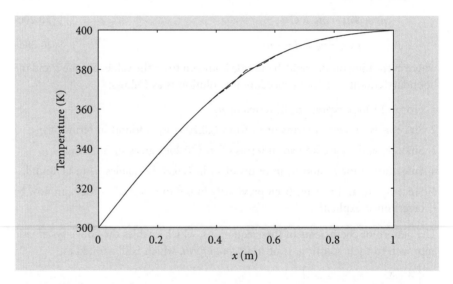

Figure 20.7 Results from HEATCRNI for the same simulated time of 2000 s with $\Delta t = 50$ s (full line) and $\Delta t = 250$ s (broken line).

oscillations no matter how large is the value of r. As an example, Figure 20.7 shows the result of running HEATCRNI for 2000 s with the same parameters that gave Figure 20.6 except that $\Delta t = 250$ s.

The results for the two time intervals are remarkably similar except for a small wiggle in the $\Delta t = 250$ s line at the mid-point – a residual effect of the discontinuity in the initial profile at that point. The value of r for the longer time interval is 7.407, yet it is still stable.

The explicit and Crank–Nicholson methods, particularly the latter, can be used to solve parabolic PDEs derived from a range of scientific sources.

> **Exercise 20.5**
> Solve the set of linear equations
>
> $$\begin{aligned} x_1 + x_2 \quad\quad\quad\quad &= 3 \\ 2x_1 - x_2 + x_3 \quad\quad &= 3 \\ 2x_2 + x_3 - x_4 &= 3 \\ 3x_3 - x_4 &= 5. \end{aligned}$$

20.6 **Poisson's and Laplace's equations**

Consider a thin plate, of thickness z, of material thermally insulated on both surfaces so that heat can only flow parallel to the surface of the plate. This defines a two-dimensional heat-flow problem. In Figure 20.8 a small rectangular element of the plate is shown with sides δx and δy and with a corner at (x, y).

For material with thermal conductivity κ the rate of heat flow *into* the element through the face A, with area $z\delta y$, is

$$\delta Q_x = -\kappa z \delta y \frac{\partial \theta}{\partial x}. \tag{20.21}$$

We now use the relationship that the temperature gradient at $x + \delta x$ is the gradient at x plus the rate of change of the gradient along x times δx – equivalent to the two-term Taylor series

$$\left(\frac{\partial \theta}{\partial x}\right)_{x+\delta x} = \left(\frac{\partial \theta}{\partial x}\right)_x + \left(\frac{\partial^2 \theta}{\partial x^2}\right)_x \delta x.$$

This gives the heat flow *out of* the element through the face B as

$$\delta Q_{x+\delta x} = -\kappa z \delta y \left(\frac{\partial \theta}{\partial x} + \frac{\partial^2 \theta}{\partial x^2} \delta x\right), \tag{20.22}$$

which gives a net rate of flow *into* the element

$$\Delta Q_x = \delta Q_x - \delta Q_{x+\delta x} = \kappa z \delta x \delta y \frac{\partial^2 \theta}{\partial x^2}. \tag{20.23}$$

Taking flow along the y direction into account, the total flow into the element is

$$\Delta Q_x + \Delta Q_y = \kappa z \delta x \delta y \left(\frac{\partial^2 \theta}{\partial x^2} + \frac{\partial^2 \theta}{\partial y^2}\right) = \kappa z \delta x \delta y \nabla^2 \theta, \tag{20.24}$$

where ∇^2 is the **Laplacian operator** for a two-dimensional Cartesian system, similar to (16.34) but without the z component. There could also be a source of

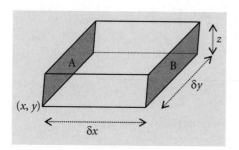

Figure 20.8 An element of a heated plate.

heat within the element that provides heat energy at a rate Q_S per unit volume, where Q_S is a function of position on the plate. If the plate is in a steady state, an equilibrium configuration of temperature, then the net rate of change of heat energy everywhere in the plate must be zero, or

$$\kappa z \delta x \delta y \nabla^2 \theta + z \delta x \delta y\, Q_S = 0.$$

This simplifies to **Poisson's equation**

$$\nabla^2 \theta = -\frac{1}{\kappa} Q_s \tag{20.25}$$

and the special case $Q_s = 0$ gives **Laplace's equation**

$$\nabla^2 \theta = 0. \tag{20.26}$$

From (20.1), for either Poisson's or Laplace's equation A and B are finite and $C = 0$ so we are dealing with elliptic equations.

> **Exercise 20.6**
> In a two-dimensional heat flow problem that gives Laplace's equation, what is the relationship between $\partial^2\theta/\partial x^2$ and $\partial^2\theta/\partial y^2$?

20.7 Numerical solution of a hot-plate problem

We consider the simple case of a square plate with edges kept at constant temperatures (the **Dirichlet problem**) and we consider the internal points situated on a square grid, with temperatures θ_1 to θ_9 and the boundary points at the fixed

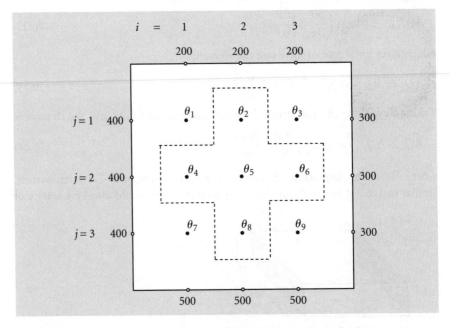

Figure 20.9 A square plate with edges at constant temperatures. A set of points corresponding to the five-star formula is indicated.

temperatures as shown in Figure 20.9. From (20.9a), and given that we are dealing with the Laplace equation

$$\nabla^2\theta = \frac{\partial^2\theta}{\partial x^2} + \frac{\partial^2\theta}{\partial y^2}$$
$$= \frac{\theta(i,j+1) + \theta(i,j-1) - 2\theta(i,j)}{h^2} + \frac{\theta(i+1,j) + \theta(i-1,j) - 2\theta(i,j)}{h^2}$$

or

$$\theta(i,j+1) + \theta(i,j-1) + \theta(i+1,j) + \theta(i-1,j) - 4\theta(i,j) = 0, \qquad (20.27)$$

where h is the spacing of the points in both the x and y directions. The points defined in (20.27) form the shape of a cross and (20.27) is referred to as the **five-star formula**.

The equations corresponding to the points with temperature θ_1 to θ_9 are, from (20.27),

$$
\begin{aligned}
200 + \theta_2 + \theta_4 + 400 - 4\theta_1 &= 0 \\
200 + \theta_3 + \theta_5 + \theta_1 - 4\theta_2 &= 0 \\
200 + 300 + \theta_6 + \theta_2 - 4\theta_3 &= 0 \\
\theta_1 + \theta_5 + \theta_7 + 400 - 4\theta_4 &= 0 \\
\theta_2 + \theta_6 + \theta_8 + \theta_4 - 4\theta_5 &= 0 \qquad (20.28)\\
\theta_3 + 300 + \theta_9 + \theta_5 - 4\theta_6 &= 0 \\
\theta_4 + \theta_8 + 500 + 400 - 4\theta_7 &= 0 \\
\theta_5 + \theta_9 + 500 + \theta_7 - 4\theta_8 &= 0 \\
\theta_6 + 300 + 500 + \theta_8 - 4\theta_9 &= 0.
\end{aligned}
$$

There are nine linear equations and these can be solved to give the temperatures in the body of the plate. Techniques for the solution of sets of linear equations will be dealt with in Chapter 30; for equations of form (20.28), where in the ith equation the coefficient of largest magnitude is that of the term θ_t, a very simple method of solution, the **Gauss–Seidel method**, (§30.4) can be used. The solution found in this case is:

$$
\begin{aligned}
\theta_1 &= 314\,\text{K}, \quad \theta_2 = 286\,\text{K}, \quad \theta_3 = 279\,\text{K} \\
\theta_4 &= 371\,\text{K}, \quad \theta_5 = 350\,\text{K}, \quad \theta_6 = 329\,\text{K} \\
\theta_7 &= 421\,\text{K}, \quad \theta_8 = 414\,\text{K}, \quad \theta_9 = 386\,\text{K}.
\end{aligned}
$$

If the plate has heat sources within it, these must occur at solution points to implement the simple finite-difference method. A source at the point θ_5, for example, modifies the fifth equation of the set (20.28), by the use of (20.25), to

$$\theta_2 + \theta_6 + \theta_8 + \theta_4 - 4\theta_5 = -\frac{h^2 Q_5}{\kappa} \qquad (20.29)$$

with the solution being carried out as previously.

Exercise 20.7

A square plate, insulated on its top and bottom surfaces, has its edges kept at temperatures 300 K, 400 K, 500 K, and 600 K respectively. Write down a five-star equation involving the temperature at the centre of the plate and hence find this temperature.

20.8 Boundary conditions for hot-plate problems

For the problem described in the previous section the boundaries were all at constant temperatures, but other types of boundary condition are possible. One is to have the boundary insulated, which means that the heat flow across it is zero. A zero heat flow implies a zero temperature gradient, and this is simulated by the introduction of **false points** into the calculation. Figure 20.10 shows a part of an insulated plate boundary with false points introduced outside.

The temperatures of points on the boundary are unknowns, but since they are surrounded by other points the temperatures of which are part of the system they can be incorporated into a set of equations such as (20.28) and so determined.

Another form of boundary condition is one that is in radiative equilibrium with its surroundings. The gradient of the temperature perpendicular to the boundary is usually modelled by

$$\kappa \frac{d\theta}{dn} = -K\theta + S \tag{20.30}$$

where K and S are constants. The temperatures of the false points B are determined from the temperatures at points A, the given values of K and S, which give the temperature gradient at the boundary, and the distance AB.

A program SIMPLATE, given in schematic form in Appendix 8, enables problems to be solved for rectangular plates with sides along principal directions with any of the boundary conditions that have been described. The input for this program is illustrated by the example shown in Figure 20.11.

The conditions for each side are shown in the figure and temperatures are to be found at the marked points. The plate is heated throughout with heat energy input $H = 10^7(0.6 - x)(0.6 - y)\,\mathrm{W\,m^{-3}}$, where x and y are the coordinates in metres measured from the lower left-hand corner. The maximum side of the plate is 0.6 m, so that the grid shown has side 0.1 m, and the conductivity is given by $\kappa = 300\,\mathrm{W\,m^{-1}K^{-1}}$. The input for the problem is of the form

150	150	150	150	150	150	200
I	U	U	U	U	U	250
I	U	U	U	U	U	250
I	U	U	U	U	U	250
I	U	U	U	U	U	250
I	U	U	U	U	U	250
I	E	E	E	E	E	250

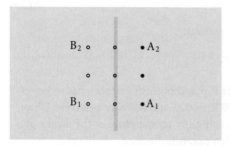

Figure 20.10 The temperatures of the 'false points B$_i$' are set at the same temperatures as internal points A$_i$. Points within the boundary are unknown and determined in the same way as internal points.

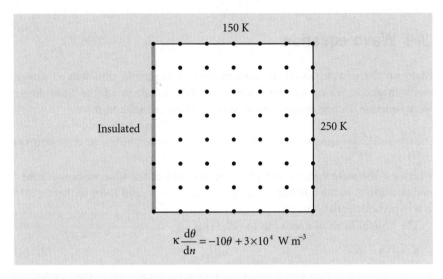

Figure 20.11 A rectangular plate with edges at constant temperature, insulated, and exchanging heat with their surroundings.

THE NUMBER OF CYCLES IS 23						
150	150	150	150	150	150	200
331	322	302	277	253	237	250
433	420	391	355	315	278	250
480	467	436	394	347	298	250
493	481	452	410	361	306	250
489	480	454	414	365	310	250
484	479	455	418	370	314	250

Figure 20.12 The output from SIMPLATE for the problem shown in Figure 20.11.

At the top right-hand corner, where the two constant-temperature sides meet, the average temperature for the two sides, 200 K, is input. The plate is defined in a rectangular region. Insulated sides are marked as I and those exchanging energy with their surroundings are marked as E. The program calls for input of the K and S values for each E point. General points for which the temperatures are required are input as U (for unknown). The output of the program is given in Figure 20.12. From the plotted values it is possible to draw contours that give a clear picture of the temperature distribution in the plate.

20.9 Wave equation

Many problems in physics involve wave motion – for example, vibrations on strings, electromagnetic waves, plasma oscillations, and solutions of the Schrödinger wave equation. In one dimension the wave equation takes the form

$$\frac{\partial^2 \eta}{\partial t^2} = c^2 \frac{\partial^2 \eta}{\partial x^2}, \tag{20.31}$$

where c is the wave velocity and η the displacement of the wave motion at time t and position x. In the notation of (20.1) $A = 1$, $C = -c^2$, and $B = 0$ so that (20.31) is a hyperbolic equation.

The general form of a solution to (20.31) is

$$\eta = f(x \pm ct) \tag{20.32}$$

where f is any function that satisfies the boundary conditions of the problem. If there are several possible solution functions then any weighted linear combination

$$\eta = \sum_{i=1}^{n} w_i f_i(x \pm ct) \tag{20.33}$$

will also be a solution. Very often solutions of the wave equation are expressed as sines, cosines, or in complex exponential form, thus enabling the powerful techniques of the Fourier transform or the Fourier series (§§ 17.4 and 17.9) to be used in solving the problem.

We now imagine that an observer is watching waves moving along an infinite stretched string (Figure 20.13). The wave has speed c, wavelength λ, and frequency v, which are related by

$$c = \lambda v. \tag{20.34}$$

The figure shows the wave, of wavelength λ, frozen in time. The observer sees crests moving at speed c to the right. If the waves are simple sine waves then they can be represented by

$$\eta = A \sin\left\{ 2\pi \left(\frac{x}{\lambda} - vt \right) + \phi \right\}. \tag{20.35}$$

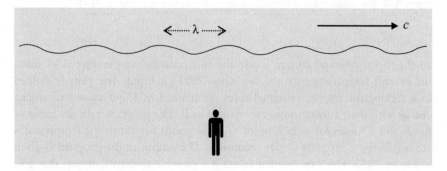

Figure 20.13 Waves moving along an infinite string.

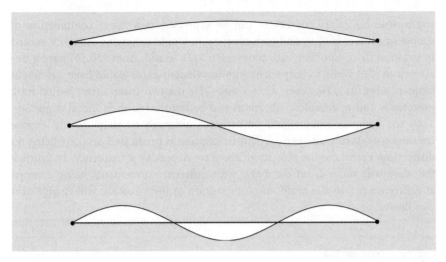

Figure 20.14 Standing waves.

where A is the **amplitude** of the wave and ϕ is a **phase angle**. For a fixed time the argument of sine changes by 2π when x changes by λ, which describes the frozen wave. At a fixed position, directly opposite the observer, the displacement is seen to vary with amplitude A and frequency v.

These waves on an infinite string are **progressive waves** that are not influenced by any boundaries. If there are boundaries that constrain the wave within a restricted region then this gives rise to **standing waves**. A familiar example of such a wave is one on a taut plucked string fixed at both ends. The velocity of such a wave is

$$c = \sqrt{\frac{T}{m}}, \tag{20.36}$$

where T is the tension in the string and m its mass per unit length. If the string is fixed at its ends then these points must represent **nodes** that are stationary points in the vibrating system. For simple sine-wave displacements the types of standing wave that can exist are illustrated in Figure 20.14. These can only exist as a whole number of half-wavelengths, so that the length of the string, L, is given by

$$L = n\frac{\lambda}{2}, \tag{20.37}$$

where n is an integer. A sine-wave standing wave can be constructed by two superimposed wave motions moving in opposite directions, so that

$$\eta_{sw} = \frac{1}{2}A\left[\sin\left\{2\pi\left(\frac{x}{\lambda} + vt\right)\right\} + \sin\left\{2\pi\left(\frac{x}{\lambda} - vt\right)\right\}\right] = A\sin\left(2\pi\frac{x}{\lambda}\right)\cos(2\pi vt). \tag{20.38}$$

At a fixed position the string vibrates with frequency v and amplitude $A\sin(2\pi x/\lambda)$ and at a fixed time the displacements form a wave with amplitude $A\cos(2\pi vt)$.

So far we have only considered simple sine waves, but waves, including those on a string, can be of more complex form. Any well-behaved wave form with the appropriate boundary conditions can be expressed as a linear combination of simple sine waves (§17.9) and each component would behave as though it existed in isolation. A component with wavelength $2L/n$ would, from (20.34) have a frequency $cn/(2L)$ while a component with wavelength $2L/m$ would have a different frequency $cm/(2L)$. However, after a time $2L/c$ the two components would have executed n and m complete vibrations and be restored to their initial condition. Thus the complex wave would be periodic, although its shape would change continuously during one period. This description is predicated on there being no **dispersion**, i.e. where the velocity of the wave depends on frequency. In practice the wave will suffer from damping, with different components being damped at different rates so the harmonic composition of the vibration will change as it dies down.

Exercise 20.8

Show that $f_1(x - c_1 t) + f_2(x - c_2 t)$ is a solution of the wave equation only if $c_1 = c_2$.

20.10 Finite-difference approach for a vibrating string

A finite-difference expression of the basic wave equation (20.31) is

$$\frac{\eta(i, j+1) + \eta(i, j-1) - 2\eta(i, j)}{(\Delta t)^2} = c^2 \frac{\eta(i+1, j) + \eta(i-1, j) - 2\eta(i, j)}{(\Delta x)^2}$$

or

$$\eta(i, j+1) = \left(\frac{c\Delta t}{\Delta x}\right)^2 \{\eta(i+1, j) + \eta(i-1, j) - 2\eta(i, j)\} + 2\eta(i, j) - \eta(i, j-1),$$

$$(20.39)$$

where i and j represent increments in space and time respectively. It will be seen that in order to generate the configuration and time t_{j+1} it is necessary to know the configurations at both times t_j and t_{j-1}, so the process of solution is not self-starting from the initial configuration at time t_0. This limitation has a physical basis because in order to define the subsequent motion of the string it is necessary to know not only the initial displacement of the string but also its initial motion $\dot{\eta}(i, 0)$ for all i. Assuming a hypothetical time point $j = -1$ one may write

$$\dot{\eta}(i, 0) = \frac{\eta(i, 1) - \eta(i, -1)}{2\Delta t}$$

or

$$\eta(i, -1) = \eta(i, 1) - 2\Delta t \dot{\eta}(i, 0), \qquad (20.40)$$

which enables the set of equations (20.39) to be solved. In many applications the string starts at rest that gives, simply, $\eta(i, -1) = \eta(i, 1)$.

An outline of a program STRING1 that solves the vibrating string problem for a string initially at rest, using (20.39) and (20.40), is given in Appendix 9. It has been used to solve the following problem:

length of string : 1.0 m

mass per unit length : 0.001 kg m^{-1}

tension in string : 200 N, $r = \left(\dfrac{c\Delta t}{\Delta x}\right)^2 = 1$

initial configuration :

$$\eta(x,0) = \begin{cases} 4x & 0 \le x < \dfrac{1}{4} \\[2mm] 4(1-x)/3 & \dfrac{1}{4} \le x \le 1 \end{cases}$$

The string was divided into 40 segments for the calculation. Since $r=1$ the time interval for the calculation was $\Delta x/c$ and the period of the vibration, discussed in §20.9, is $2L/c=2.0/c$, since the length of the string, $L=1.0$ m. Since $\Delta x=0.025$ m there are 80 timesteps per period. Figure 20.15 gives the output from STRING1 for this problem at intervals of 20 timesteps, equivalent to one-quarter of a period. The displacements are not specified since the results scale exactly with the initial displacements. It will be seen that after one half of a period the configuration is an antisymmetric form of the original configuration. There is a slight distortion at the point corresponding to the sharp discontinuity in the initial profile of the string, which is not well handled by the finite-difference approach, and this distortion is also evident after one period, which otherwise exactly overlaps the original configuration.

Analysis shows that the finite-difference approach is stable for $r \le 1$, which is known as the **Courant–Friedrich–Lewy condition**, and $r=1$ usually gives a good outcome.

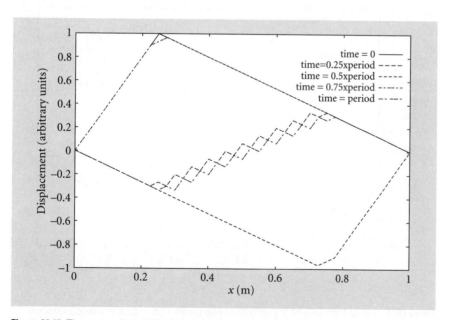

Figure 20.15 The output from STRING1 at quarter-period intervals.

20.11 Two-dimensional vibrations

The general form of (20.31) is

$$\frac{\partial^2 \eta}{\partial t^2} = c^2 \nabla^2 \eta \tag{20.41}$$

and for two space dimensions this is

$$\frac{\partial^2 \eta}{\partial t^2} = c^2 \left(\frac{\partial^2 \eta}{\partial x^2} + \frac{\partial^2 \eta}{\partial y^2} \right). \tag{20.42}$$

However, if the system and its vibrations have circular symmetry then we can use the form (16.40) for ∇^2 without the θ-dependent term and the wave equation becomes

$$\frac{\partial^2 \eta}{\partial t^2} = c^2 \left(\frac{\partial^2 \eta}{\partial s^2} + \frac{1}{s} \frac{\partial \eta}{\partial s} \right) \tag{20.43}$$

where s is the distance from the origin.

In finite-difference form (20.43) becomes

$$\frac{\eta(i, j+1) + \eta(i, j-1) - 2\eta(i, j)}{(\Delta t)^2}$$
$$= c^2 \left(\frac{\eta(i+1, j) + \eta(i-1, j) - 2\eta(i, j)}{(\Delta s)^2} + \frac{\eta(i+1, j) - \eta(i-1, j)}{2s(i)\Delta s} \right) \tag{20.44}$$

where i and j represent increments in the space and time domains respectively.

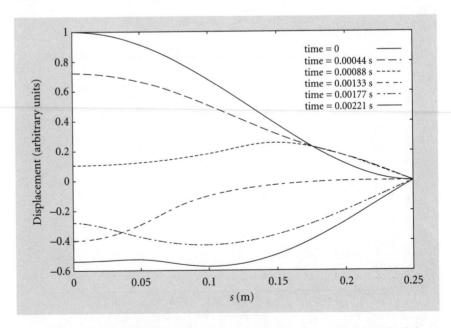

Figure 20.16 Output from DRUM showing the radial profile of a drum skin making circularly symmetric vibrations.

Displacements are updated from

$$\eta(i,j+1) = \left(\frac{c\Delta t}{\Delta s}\right)^2 \left[\eta(i+1,j)\left(1+\frac{\Delta s}{2s}\right) + \eta(i-1,j)\left(1-\frac{\Delta s}{2s}\right)\right]$$
$$+ 2\left\{1 - \left(\frac{c\Delta t}{\Delta s}\right)^2\right\}\eta(i,j) - \eta(i,j-1). \tag{20.45}$$

For a stationary start $\eta(i,-1) = \eta(i,1)$ and the first step becomes

$$\eta(i,1) = \frac{1}{2}\left(\frac{c\Delta t}{\Delta s}\right)^2 \left[\eta(i+1,0)\left(1+\frac{\Delta s}{2s}\right) + \eta(i-1,0)\left(1-\frac{\Delta s}{2s}\right)\right]$$
$$+ \left\{1 - \left(\frac{c\Delta t}{\Delta s}\right)^2\right\}\eta(i,0). \tag{20.46}$$

A system that can give rise to a circularly-symmetric vibration is the skin of a circular drum. The velocity of waves in the drum skin is given by an expression like (20.36) where T is now the tautness of the skin, in units of $N\,m^{-1}$, and m is its mass per unit area. The circular edge of the skin is attached to the body of the drum and so is a node of the vibration. For a circular drum (20.45) and (20.46) can be implemented by the program DRUM outlined in Appendix10. Figure 20.16 shows the result of running DRUM for the following problem: $T = 200\,N\,m^{-1}$, $m = 0.01\,kg\ m^{-2}$, radius $0.25\,m$ with initial displacement $\eta(s,0) = (1 - 16s^2)\cos(2\pi s)$ where s is in metres. For the calculation the radius of the drum was divided into 40 segments and the program was run with $\frac{c\Delta t}{\Delta s} = 1$.

Problems

20.1 The following partial table of $\sin(x)$ is at intervals of $\pi/24$.

x	$\sin(x)$
$\pi/6$	0.500 000
$5\pi/24$	0.608 761
$\pi/4$	0.707 107
$7\pi/24$	0.793 353
$\pi/3$	0.866 025

Estimate the first derivative of $\sin(x)$ for $x = \pi/4$ using (i) the forward-difference formula, (ii) the backward-difference formula, and (iii) the central-difference formula with intervals (a) $h = \pi/12$ and (b) $h = \pi/24$. Show that the errors for (i) and (ii) are approximately dependent on h and for (iii) dependent on h^2.

Use the central-difference formula for the second derivative for (a) and (b) and hence show that the error depends approximately on h^2.

20.2 A uniform metal bar, of length 1 m, is lagged along its length and has an initial temperature profile $\theta(x) = 300 + 200(1 - x^2)$ where x is the distance in metres from one end and temperature is in Kelvin. The temperatures at the ends of the bar are maintained at the initial values of 500 K and 300 K. Using the program EXPLICIT find the temperature profile of the bar after 500, 1000, and 2000 s given thermal conductivity 400 W m^{-1}K^{-1}, density 3400 kg m^{-1}, and specific heat capacity 800 J kg^{-1}K^{-1}. For the purposes of the calculation divide the bar into 20 segments.

20.3 Repeat Problem 20.2 using the HEATCRNI program with a time interval of 50 s.

20.4 A thin plate, insulated on both sides, has boundary conditions as shown below.

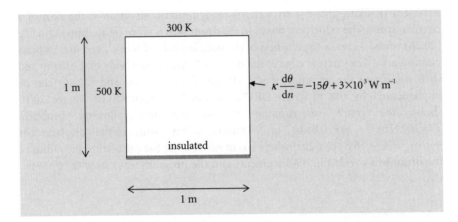

It is heated according to $H = 10^7 xy(1 - x)(1 - y)$ and $\kappa = 300$ W m^{-1}. Find the equilibrium temperature throughout the plate by using the program SIMPLATE and defining the plate within a square 9×9 mesh.

20.5 Run the program STRING1 with the standard parameters provided and 20 distance intervals. Take the initial profile as $\eta = x(1 - x)(0.7 - x)$ and find output for $10n\Delta t$ for $n = 0$ to 4.

20.6 Run the program DRUM with the standard parameters provided and 20 distance intervals. Take the initial profile as $\eta = \cos(s\pi/2S)$, where S is the radius of the drum and s the distance from the centre, and display the output for $10n\Delta t$ for $n = 0$ to 5.

Quantum mechanics I: Schrödinger wave equation and observations

In this chapter you are introduced to one of the most important equations in physics, the Schrödinger wave equation. At the beginning of the twentieth century experiments suggested that material bodies, in particular atomic particles, could display wave-like behaviour. The Schrödinger wave equation achieved this synthesis of the dual particle-and-wave characteristics of matter. The solutions to the equation for various physical systems define what can be observed. They also introduce interesting philosophical questions concerning the relationship between making an observation and the state of the system being observed, and also concerning the limits on how precisely the state of a system can be defined – enshrined as the Heisenberg uncertainty principle.

The analysis involved in solving the Schrödinger wave equation and in interpreting the solutions introduces many mathematical concepts – for example, the idea of eigenvalues and eigenfunctions and the concept of a complete orthonormal set of functions – that can be applied in other physical contexts.

21.1 Transition from classical to modern physics: a brief history

At the end of the nineteenth century the subject of physics was anchored to two well-established theoretical bases – Newtonian mechanics, which described the behaviour of material objects, and Maxwell's equations, which described the behaviour of electromagnetic radiation. At the beginning of the twentieth century both these concepts faced new problems. Newtonian mechanics, which has served, and still serves, so well to describe the macroscopic world, was found to be inadequate to deal with the microscopic world of atomic particles. In addition the laws that described the nature of electromagnetic radiation were found to be unable to deal with the radiation field inside a cavity. A straightforward application of the laws led to the conclusion that the energy density within a cavity would be infinite with a steadily increasing concentration of energy density at shorter wavelengths – a situation dubbed the **ultraviolet catastrophe**.

In 1900 Max Planck showed that the radiation paradox could be solved if light of a particular frequency, v, could only exist in packets, or **photons**, with energy

$$E = hv, \tag{21.1}$$

where h is **Planck's constant**, 6.626×10^{-34} J s. This relationship was reinforced by Einstein's explanation of the photoelectric effect in 1905 – for which he received the Nobel prize. The energy of the electrons knocked off the surface of a metal by light energy was related to the frequency of the light and not its intensity. The light photons acted like a shower of 'bullets' each one of which could bestow its energy, hv, on an electron. The rate of ejection of electrons depended on the intensity of the light, i.e. the number of photons per unit time, but the energy of a particular ejected electron depended on the energy of the individual photons, which was the same for each, less the energy lost in escaping from the surface, which varied from one electron to another.

This particle behaviour of light had been favoured by Newton, but the alternative wave picture of light proposed by Huygens was preferred in the eighteenth century because it readily explained the known interference and diffraction properties of light. The work of Planck and Einstein suggested that light could have both wave and particle properties, depending on the experiment that was being done. In 1924 de Broglie put forward the hypothesis that, just as light could display particle properties, particles may, in the right circumstances, behave like waves. The hypothesis stated that the equivalent wavelength, λ, of a matter wave associated with a particle of momentum p is given by

$$\lambda = \frac{h}{p}. \tag{21.2a}$$

If the particle is a photon then its momentum is given by

$$p = \frac{h}{\lambda} = \frac{hv}{c} = \frac{E}{c} \tag{21.2b}$$

where E is the energy of the photon and c is the speed of light.

In 1925 Davisson and Germer, studying the scattering of electrons from nickel, found that peaks appeared at certain scattering angles indicating that a diffraction process was taking place. The action of an electron microscope depends on the wavelike behaviour of electrons. High-energy electrons impinging on a small object are scattered from it, just as light is scattered from an object in an optical microscope. The lenses in an electron microscope use a combination of magnetic and electric fields and a visible image is produced by focusing the electrons on to a fluorescent screen. The high resolution of an electron microscope is due to the small equivalent wavelength of electrons, as given by (21.2), with energy about

1 keV. For such energies the electrons, with mass m_e and speed v, behave like classical particles with momentum $p = m_e v$ and kinetic energy $E_K = \frac{1}{2} m_e v^2$, from which we find

$$E_K = \frac{p^2}{2m_e} \tag{21.3a}$$

or

$$p = \sqrt{2 m_e E_K}. \tag{21.3b}$$

For $E_K = 1$ keV we find

$$p = \sqrt{2 \times 9.10 \times 10^{-31} \times 1000 \times 1.60 \times 10^{-19}} \text{ kg m s}^{-1} = 1.71 \times 10^{-23} \text{ kg m s}^{-1}.$$

Inserting this in (21.1) gives $\lambda = 3.87 \times 10^{-11}$ m, which can be compared with the shortest blue visible wavelength, 4×10^{-7} m.

In 1897 J. J. Thomson discovered the electron and in 1911, by experiments involving the scattering of alpha-particles from thin metallic foils, Rutherford had established the existence of a very tiny atomic nucleus (§10.8). This led to a model of an atom in which electrons with negative charges moved around a compact nucleus, with a positive charge, in orbits similar to the circular or elliptical orbits of planets around the Sun. However, there was a problem with this model. An orbiting electron is in a constant state of acceleration and according to classical ideas an accelerating charged particle should emit electromagnetic radiation. The energy of this radiation should come from the motion of the electron, which should spiral inwards until it plunges into the nucleus. Since atoms are stable entities, this clearly does not happen. Atoms do emit radiation when they are excited, but they do so with very specific wavelengths that are characteristic of the type of atom.

A model for an atom that satisfied the above requirements was introduced by Niels Bohr in 1913. For a particular atom electrons could exist in stable orbits, with energies E_1, E_2, etc., in which they did not continuously radiate. If the atom was heated, or provided with energy in some other way such as bombardment with particles, then an electron with energy E_n could be moved into another stable orbit with energy E_m. Since physical systems are most stable in configurations of lowest energy the electron will spontaneously jump back into its original state and release energy $E_m - E_n$ which appears as a light photon with frequency given by

$$h\nu = E_m - E_n \quad \text{or} \quad \nu = \frac{E_m - E_n}{h}. \tag{21.4}$$

More complicated transitions of electron energy can take place where the energy jumps can take in intermediate non-radiating orbits, but the essential pattern is as described. Bohr was able to explain the principal spectral lines of hydrogen with his model by assuming that the non-radiating orbits were circular and that the electrons in these orbits had angular momenta that were multiples of

$$\hbar = h/2\pi.$$

It was arbitrary, but it worked for hydrogen although it failed for more complex atoms. There was clearly a need for a new theory to explain the behaviour of matter at an atomic scale and, in particular, the spectral lines emitted by the elements.

The new theory – **quantum, or wave, mechanics**, was highly mathematical in its structure, involved postulates that could not be rigorously derived, and required the interpretation of results in ways that were intuitively reasonable if not formally established. There is also a strong philosophical element underlying the development of quantum mechanics, since it brings into consideration such questions as the relationship between reality and observation. The justification for quantum mechanics is that it is a self-consistent description of the physics of the microscopic world, that it correctly describes what we observe and infer from our observations, and that it has predictive power – it can predict how a system will behave before it has been experimentally investigated.

The aim of this chapter, and Chapter 25, is not to give a complete account of quantum mechanics – such an agenda would require a whole book to itself. Rather it is to introduce the range of mathematical ideas that underpin the subject, thus enabling a further study of quantum mechanics to concentrate on the physical, rather than mathematical, aspects of the subject.

Exercise 21.1
What are (i) the energy and (ii) the momentum of a photon for light of wavelength 500 nm?

Exercise 21.2
An electron jumps from a state with energy $-3.9\,\mathrm{eV}$ to one with energy $-13.6\,\mathrm{eV}$. What is the wavelength of the resultant photon?

21.2 Intuitive derivation of the Schrödinger wave equation

In 1926 Schrödinger introduced the equation that was to become central in dealing with the problem of atomic spectra and many other important phenomena at the atomic scale. Unfortunately there is no rigorous way of deriving the equation, so its validity depends on the fact that it works on a variety of systems, both simple and complex, and always, to date, gives results that are consistent with observation.

The central idea behind the Schrödinger wave equation is that associated with any particle there is a wave-like quantity that gives a full description of its properties and behaviour. The particle is characterized by its mass, momentum, and energy, and the associated wave by wavelength and frequency, so we must find a way of relating the two descriptions. We first assume that the particle moves only in one dimension and the associated wave is similar to (20.35) but expressed in complex form without introducing a phase shift. This wave is

$$\Psi = A\exp\left\{2\pi\mathrm{i}\left(\frac{x}{\lambda} - vt\right)\right\}. \tag{21.5}$$

The particle properties are now brought in by replacing the wavelength and frequency in (21.5) with particle properties using (21.1) and (21.2) to give

$$\Psi = A \exp\left\{2\pi i\left(\frac{p}{h}x - \frac{E}{h}t\right)\right\} = A \exp\left\{i\left(\frac{p}{\hbar}x - \frac{E}{\hbar}t\right)\right\}. \tag{21.6}$$

The total energy of the particle, E, is related to its momentum, p, and potential energy, V, by

$$E = \frac{p^2}{2m} + V, \tag{21.7}$$

where m is the mass of the particle. Partially differentiating (21.6) with respect to t

$$\frac{\partial \Psi}{\partial t} = -\frac{i}{\hbar}E\Psi \tag{21.8a}$$

and partially differentiating twice with respect to x gives

$$\frac{\partial^2 \Psi}{\partial x^2} = -\frac{p^2}{\hbar^2}\Psi. \tag{21.8b}$$

Multiplying both sides of (21.7) by Ψ

$$E\Psi = \frac{p^2}{2m}\Psi + V\Psi$$

which gives, using (21.8a) and (21.8b),

$$-\frac{\hbar^2}{2m}\frac{\partial^2 \Psi}{\partial x^2} + V\Psi = -\frac{\hbar}{i}\frac{\partial \Psi}{\partial t}. \tag{21.9}$$

If the potential energy is a function of both position and time then the **wave function** Ψ will also be a function of time and (21.9) is the **time-dependent Schrödinger wave equation** (TDSWE). We now consider a solution where the wave function is expressed as a product of two functions, one only time dependent and the other only space dependent, thus

$$\Psi(x, t) = \psi(x)f(t). \tag{21.10}$$

Inserting the right-hand side of (21.10) for Ψ in (21.9) gives

$$-\frac{\hbar^2}{2m}f(t)\frac{d^2\psi}{dx^2} + V\psi(x)f(t) = -\frac{\hbar}{i}\psi(x)\frac{df}{dt} \tag{21.11}$$

where we use ordinary differentiation rather than partial differentiation for functions of a single variable. Dividing both sides of the resulting equation by $\psi(x)f(t)$ gives

$$-\frac{\hbar^2}{2m}\frac{1}{\psi(x)}\frac{d^2\psi}{dx^2} + V = -\frac{\hbar}{i}\frac{1}{f(t)}\frac{df}{dt}. \tag{21.12}$$

If the potential energy, V, depends only on position then the left-hand side of (21.12) depends only on position and the right-hand side depends only on time. Since time and position are independent variables, so that one can be changed without changing the other, the only way that (21.12) can be true is if each side equals the same constant, say K. This gives

$$-\frac{\hbar}{i}\frac{1}{f(t)}\frac{df}{dt} = K. \tag{21.13}$$

Multiplying both sides by $\psi(x)f(t)$ $(= \Psi)$ we find

$$-\frac{\hbar}{i}\frac{\partial \Psi}{\partial t} = K\Psi$$

and from (21.8a) we see that $K = E$. Equating the left-hand side of (21.12) to E and rearranging we have

$$-\frac{\hbar^2}{2m}\frac{d^2\psi}{dx^2} + V\psi = E\psi. \tag{21.14}$$

This is the one-dimensional form of the **time-independent Schrödinger wave equation** (TISWE), which is one of the most important equations in modern physics. The quantity ψ completely describes the space-dependent aspect of the state of the system. There is also a time-dependent part given by the solution of (21.13) with $K = E$. This equation can be written as

$$\frac{df}{dt} = -\frac{i}{\hbar}Ef$$

which has the solution

$$f = \exp\left(-\frac{i}{\hbar}Et\right). \tag{21.15}$$

This represents a periodic fluctuation with constant amplitude and we shall see that it has no effect in the interpretation of the state of the system that is totally dependent on the solution of (21.14) – given, of course that the potential energy is time independent.

Now we are going to apply the TISWE to a very simple system through which you will learn about how solutions to the equation may be interpreted and also understand the basic mathematical principles underlying the various ways of applying the results.

21.3 A particle in a one-dimensional box

21.3.1 Solution of the TISWE

We consider a particle of mass m constrained to move in one dimension along the x axis. The potential energy of the particle is defined as

$$V = 0, \quad 0 \leq x \leq L$$
$$V = \infty, \quad x < 0 \quad \text{and} \quad x > L.$$

Since a particle with finite total energy cannot exist with infinite potential energy the particle is constrained within the region $0 \leq x \leq L$ and ψ is zero outside that region. We also impose a boundary condition, applicable to all wave mechanical systems, that the wave function must be continuous and, since it is zero outside the allowed region, the wave function within the allowed region must therefore be zero at $x = 0$ and $x = L$.

Within the allowed region, where $V = 0$, the TISWE is

$$-\frac{\hbar^2}{2m}\frac{d^2\psi}{dx^2} = E\psi$$

or

$$\frac{d^2\psi}{dx^2} = -\frac{2mE}{\hbar^2}\psi. \tag{21.16}$$

The forms of solution of (21.16) are, with $2mE/\hbar^2 = k^2$,

$$\psi = A\sin kx, \quad \psi = A\cos kx, \quad \psi = A\exp(ikx).$$

From the boundary condition that $\psi = 0$ when $x = 0$ we see that only the first solution is possible. In addition the second boundary condition, $\psi = 0$ when $x = L$, requires

$$\sin kL = 0$$

or

$$kL = \sqrt{\frac{2mE}{\hbar^2}}L = n\pi.$$

This shows that only certain energies are possible for the particle given by

$$E_n = \frac{n^2\pi^2\hbar^2}{2mL^2}. \tag{21.17}$$

These energies are the **energy eigenvalues**, the energies associated with the **eigenfunctions**, given by

$$u_n = C\sin\left(\frac{n\pi}{L}x\right) \tag{21.18}$$

that describe the possible states of the system. It should be noted that $n = 0$ is not an allowed eigenfunction as it corresponds to $\psi = 0$, i.e. the non-existence of the particle. You should note that there is no restriction on the constant C which can be real, imaginary, or complex so, in general, the eigenfunctions may be taken as complex quantities. Now we consider what these eigenfunctions look like and what they mean.

Exercise 21.3
An electron is constrained in a one-dimensional box of length 0.1 nm. What is its ground-state energy?

Exercise 21.4
An alpha-particle is constrained in a one-dimensional box of length 10^{-15} m. What is its ground-state energy?

21.3.2 Interpretation of the solution

The appearance of the solutions (21.18) for $n = 1$, 2, and 3 is shown in Figure 21.1. The interpretation of these state functions – not proven but one of the basic

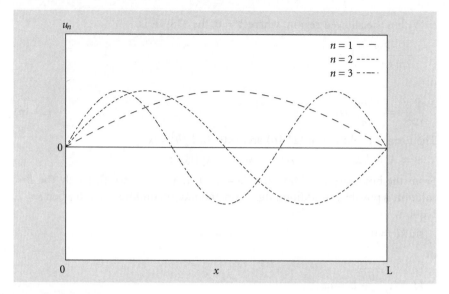

Figure 21.1 The three lowest-energy solutions for the one-dimensional particle in a box.

postulates of quantum mechanics – is that, for a particle in a state represented by $u_n(x)$, the probability density of finding the particle within the box is $|u_n(x)|^2$. This means that if an experiment is done to detect the particle within the box, i.e. to reveal itself in its material aspect, then the probability that it will be found between x and $x + dx$ is $|u_n(x)|^2 dx$. Since the particle is somewhere in the box, the probability of finding it somewhere in the box is unity. From this we find

$$\int_0^L |u_n(x)|^2 dx = 1 \tag{21.19}$$

or

$$|C|^2 \int_0^L \sin^2\left(\frac{n\pi x}{L}\right) dx = 1.$$

Since the average value of 'sine squared' in any whole number of half-sine waves is $\frac{1}{2}$, this gives

$$\frac{1}{2}|C|^2 L = 1 \tag{21.20a}$$

or

$$C = \sqrt{\frac{2}{L}}. \tag{21.20b}$$

A more general solution of (21.20a) is $C = \sqrt{(2/L)}\exp(i\alpha)$, which brings out the point, previously made, that the eigenfunctions can be complex quantities. It also illustrates that when the potential energy is time independent, leading to the TISWE, the time-dependent part of the solution, (21.15), can be discarded as it has unit magnitude and does not influence the probability density for locating the particle. The normalized functions $|u_n(x)|^2$ are shown in Figure 21.2.

It is interesting to compare this interpretation of the quantum-mechanical system with the corresponding classical system. It corresponds to a classical situation where

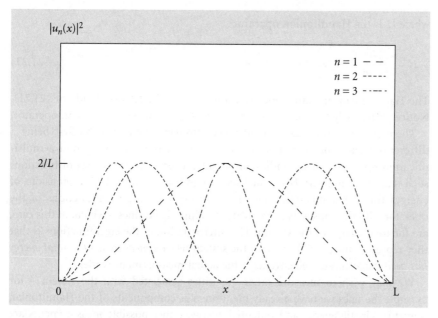

Figure 21.2 The probability density $|u_n(x)|^2$ for the three lowest energy states of a particle in a one-dimensional box.

a particle, constrained to move in one dimension, is bouncing to and fro between two walls with which it makes elastic collisions. In the classical situation there is no constraint on the energy, $\frac{1}{2}mv^2$, whereas the energy can only take discrete value in the quantum-mechanical case. In the classical situation the speed of the particle is constant and the probability density is uniform between the two walls. For the quantum-mechanical case the probability density is far from uniform; for any value of n it is zero at the boundaries of the region and has one or more maxima. For values of n greater than 1 it is zero at other points within the range.

Exercise 21.5
For a particle in a box in the ground state what is the probability of finding it between $x=0$ and $x=L/4$?

21.4 **Observations and operators**

The three-dimensional form of the TISWE, given in (21.14), is

$$-\frac{\hbar^2}{2m}\nabla^2\psi + V\psi = E\psi, \tag{21.21}$$

where ∇^2 is the Laplacian operator described in §16.5. This can be written in the more concise form

$$\hat{H}\psi = E\psi \tag{21.22}$$

where $\hat{H}$ is the **Hamiltonian operator**

$$\hat{H} = -\frac{\hbar^2}{2m}\nabla^2 + V. \qquad (21.23)$$

The Hamiltonian operator, operating on ψ, gives the left-hand side of (21.21). Notice particularly the use of the 'hat' in $\hat{H}$ which indicates that it is an operator.

There are two components of the Hamiltonian operator – the first being a **differential operator** corresponding to kinetic energy and the second a **multiplication operator** corresponding to potential energy. There is a set of functions such that operating with $\hat{H}$ on the function gives a constant (with dimensions of energy) times that function, and these functions are the eigenfunctions of the operator with the constants as the corresponding eigenvalues – which, in this case, are the total energy of the system. The interpretation of the eigenfunctions is that they represent the possible state of the system *when a measurement of total energy has been made* and the eigenvalue is the actual measurement made.

We may wish to measure other quantities associated with the system – for example the ones we have already mentioned as components of the Hamiltonian operator, kinetic energy and potential energy. Other possible measurements are position, momentum, and angular momentum. A postulate of quantum mechanics is that to each possible observable quantity there corresponds an operator. The eigenvalues of the operator are the possible measurements of the quantity that can be made and the associated eigenfunctions give the state of the system *when that measurement has been made*. Thus, if the operator corresponding to the observation is $\hat{A}$ then we write

$$\hat{A}v_n = a_n v_n \qquad (21.24)$$

where v_n is the nth eigenfunction of $\hat{A}$ and a_n the corresponding eigenvalue.

The eigenfunctions of operators corresponding to measurements of different quantities may be different, which means that the states of the system after the measurements have been made are different. Since the system is unaware *a priori* what measurement is about to be made on it, we must accept that the state after the measurement may differ from that before the measurement is made and the original state does not have to be an eigenstate of the measurement that is about to be made. Our model of a quantum-mechanical system is that before a measurement is made it is in some general state. A measurement made on it then changes its state to an eigenstate described by an eigenfunction of the operator corresponding to the measurement that was made.

Later we shall be giving operators corresponding to different kinds of measurement, but now we shall describe the properties that they all share. All quantum-mechanical operators corresponding to measurements are **Hermitian operators**. If ψ_1 and ψ_2 are possible states of the system then the Hermitian operator $\hat{A}$ satisfies the condition

$$\int_S \psi_1^* \hat{A} \psi_2 \, d\tau = \int_S \psi_2 \hat{A}^* \psi_1^* \, d\tau, \qquad (21.25)$$

where S covers the space within which the state function exists and $d\tau$ is an element within that space.

We see in (21.25) that the state function and the operator can be complex quantities. From (21.25) it is possible to show that the eigenvalues of an Hermitian operator are real – a reassuring result, since they correspond to possible measurements.

From (21.25) with $\psi_1 = \psi_2 = v_n$, where v_n is an eigenfunction of $\hat{A}$,

$$\int_S v_n^* \hat{A} v_n \, d\tau = \int_S v_n \hat{A}^* v_n^* \, d\tau.$$

From (21.24) $\hat{A} v_n = a_n v_n$ and so $\hat{A}^* v_n^* = a_n^* v_n^*$ which gives

$$a_n \int_S v_n^* v_n \, d\tau = a_n^* \int_S v_n^* v_n \, d\tau. \tag{21.26}$$

From this we see that $a_n = a_n^*$, which shows that a_n is real since it equals its own complex conjugate.

It is also possible to show that eigenfunctions corresponding to different states are, in general, **orthogonal**, defined by

$$\int_S v_n^* v_m \, d\tau = 0. \tag{21.27}$$

From (21.25) with $\psi_1 = v_n$ and $\psi_2 = v_m$

$$\int_S v_n^* \hat{A} v_m \, d\tau = \int_S v_m \hat{A}^* v_n^* \, d\tau$$

or

$$a_m \int_S v_n^* v_m \, d\tau = a_n \int_S v_n^* v_m \, d\tau. \tag{21.28}$$

If $a_n \neq a_m$ then the result (21.27) follows. From the normalization result (21.19) we have $\int_S v_n^* v_n \, d\tau = \int_S |v_n|^2 \, d\tau = 1$ which, combined with (21.27), gives

$$\int_S v_n^* v_m \, d\tau = \delta_{nm}, \tag{21.29}$$

where δ_{nm} is the **Kronecker delta function**, which is unity if $n = m$ and zero otherwise.

Note that the condition for (21.27) to be valid was given as $a_n \neq a_m$ rather than $n \neq m$, the reason for this being that it is possible to have different states with the same eigenvalue. We shall now consider a simple two-dimensional system that illustrates this point and then discuss the implications of this for defining the states of a system.

21.5 **A square box and degeneracy**

21.5.1 **A particle in a square box**

We now consider a particle in a square box, the potential energy being given by

$V = 0$, both $0 \leq x \leq L$ and $0 \leq y \leq L$

$V = \infty$, otherwise.

The TISWE for this system is

$$-\frac{\hbar^2}{2m}\left(\frac{\partial^2 \psi}{\partial x} + \frac{\partial^2 \psi}{\partial y^2}\right) = E\psi \tag{21.30}$$

and, by arguments similar to those leading to (21.21), a general total energy eigenfunction, satisfying the boundary conditions that $\psi = 0$ for both $x=0$ and L and also $y=0$ and L is of the form

$$u_{n_x n_y} = C \sin\left(\frac{\pi n_x}{L}x\right) \sin\left(\frac{\pi n_y}{L}y\right). \tag{21.31}$$

Inserting $u_{n_x n_y}$ for ψ in (21.30) and evaluating the partial differentials gives

$$E_{n_x n_y} = \frac{\pi^2(n_x^2 + n_y^2)\hbar^2}{2mL^2} \tag{21.32}$$

as the associated total energy eigenvalue.

Normalization of the wave function, to give unit probability of finding the particle somewhere in the box, gives

$$\int_0^L \int_0^L \left|u_{n_x n_y}\right|^2 dx\, dy = C^2 \int_0^L \sin^2\left(\frac{\pi n_x}{L}x\right)dx \int_0^L \sin^2\left(\frac{\pi n_y}{L}y\right)dy = 1. \tag{21.33}$$

A similar argument to the one that gave (21.20a) gives in this case

$$C = \frac{2}{L}.$$

21.5.2 **Degeneracy**

From the expression for the total energy, (21.32), it follows that any states with the same values of $n_x^2 + n_y^2$ will have the same energy. This is so for the pairs of n values (1, 2) and (2, 1) and for the three pairs of n values (1, 7), (7, 1), and (5, 5). States that have accompanying states with the same energy are called **degenerate**. The states corresponding to the n values (1, 2) and (2, 1) are **doubly degenerate**; where there are three states with the same energy then these states are **triply degenerate**. A consequence of degeneracy is that the eigenfunctions for the corresponding energy eigenvalues are not uniquely determined. If we have $\hat{H}u_1 = Eu_1$ and $\hat{H}u_2 = Eu_2$, then

$$\hat{H}(\alpha u_1 + \beta u_2) = E(\alpha u_1 + \beta u_2)$$

so that any linear combination of u_1 and u_2 is also an eigenfunction.

At this stage, to make our notation more concise, we are going to write

$$\int_S u_n^* u_m \, d\tau = \langle u_n | u_m \rangle. \tag{21.34}$$

In fact this notation has a deeper significance than just conciseness and is related to representing eigenstates as unit vectors in a multidimensional vector space, something we shall deal with later.

If u_1 and u_2 are normalized then for the linear combination to be normalized

$$\langle \alpha u_1 + \beta u_2 | \alpha u_1 + \beta u_2 \rangle = |\alpha|^2 \langle u_1 | u_1 \rangle + \alpha^* \beta \langle u_1 | u_2 \rangle + \beta^* \alpha \langle u_2 | u_1 \rangle + |\beta|^2 \langle u_2 | u_2 \rangle = 1.$$

If the orthonormal relationship of u_1 and u_2, as expressed by (21.29) is valid, then

$$|\alpha|^2 + |\beta|^2 = 1. \tag{21.35}$$

In solving the TISWE for an n-fold degenerate state it is possible to find n linearly independent solutions. This means that none of the solutions is equivalent to some linear combination of the others. However, it is possible to form n linear combinations of these solutions that form an orthonormal set by a method known as the **Gram–Schmidt orthogonalization** process. We can illustrate this by forming an orthonormal pair for twofold degeneracy where the non-orthogonal normalized eigenfunctions are u_1 and u_2. One member of the orthogonal pair is taken arbitrarily as u_1 and the other as the linear combination

$$u_2' = \alpha u_1 + \beta u_2.$$

If this is orthogonal to u_1 then

$$\langle u_1 | \alpha u_1 + \beta u_2 \rangle = 0$$

or

$$\alpha \langle u_1 | u_1 \rangle + \beta \langle u_1 | u_2 \rangle = \alpha + \beta \langle u_1 | u_2 \rangle = 0.$$

This gives

$$\alpha = -\beta \langle u_1 | u_2 \rangle$$

and hence

$$u_2' = \beta(u_2 - u_1 \langle u_1 | u_2 \rangle). \tag{21.36}$$

Finally, for normalization β is chosen to make $\langle u_2' | u_2' \rangle = 1$. This process is illustrated in Figure 21.3, which also brings out the idea of representing the state of a system as a vector.

The state of a system represented by the function ψ in the vector representation is $|\psi\rangle$. The complex conjugate of the function, ψ^*, in the vector representation becomes $\langle \psi |$. In Figure 21.3 we see how $|u_2'\rangle$, which is orthogonal to $|u_1\rangle$, can be formed as the sum of two vectors OA and AB that are multiples of $|u_2\rangle$ and $|u_1\rangle$ – which is the Gram–Schmidt orthogonalization process. This process, in an analytical form, can be extended to produce an orthonormal set for any order of degeneracy.

Exercise 21.6

The energy levels in a cubical box of side L are given by

$$E(n_x, n_y, n_z) = \frac{\pi^2(n_x^2 + n_y^2 + n_z^2)\hbar^2}{2mL^2}.$$

What is the degeneracy for a state with energy $99\pi^2\hbar^2/2mL^2$?

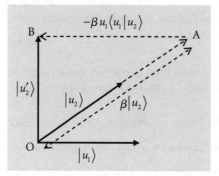

Figure 21.3 A representation of the Gram–Schmidt orthogonalization process.

21.5.3 **Vector space**

The vector representation of states is useful for forming a general picture of the relationship between the states of a system. We imagine that any possible state of the system is described by a unit vector in a multidimensional space – called a **Hilbert space** – which can, in fact, be a space of infinite dimensions. In this space there can exist a large, even infinite, number of orthogonal directions, something that we cannot envisage in our world of three spatial dimensions. The eigenstates corresponding to a particular measurement are a set of orthonormal vectors in this space and any general state function is a linear combination of these eigenvectors. Thus we can write

$$|\psi\rangle = \begin{pmatrix} a_1|u_1\rangle \\ a_2|u_2\rangle \\ a_3|u_3\rangle \\ \vdots \\ a_n|u_n\rangle \end{pmatrix},$$

(21.37)

where we have assumed an n-dimensional space. The complex conjugate state is then represented by the row vector

$$\langle\psi| \equiv (a_1{}^*\langle u_1| \; a_2{}^*\langle u_2| \; a_3{}^*\langle u_3| \; \cdots \; a_n^*\langle u_n|).$$

(21.38)

The square of the magnitude of $|\psi\rangle$ is the **scalar product**

$$\langle\psi|\cdot|\psi\rangle = \langle\psi|\psi\rangle = (a_1{}^*\langle u_1| \; a_2{}^*\langle u_2| \; a_3{}^*\langle u_3| \; \cdots \; a_n^*\langle u_n|) \begin{pmatrix} a_1|u_1\rangle \\ a_2|u_2\rangle \\ a_3|u_3\rangle \\ \vdots \\ a_n|u_n\rangle \end{pmatrix}$$

$$= \sum_{i=1}^{n} |a_i|^2 = 1$$

(21.39)

for normalization.

Measurements of different quantities give different sets of orthonormal eigenvectors but in the *same space* – just as in our three-dimensional world we can set up a set of three orthogonal axes in an infinite number of ways. Any general state of the system can be expressed as a linear combination of the eigenstates corresponding to any measurement and the corresponding eigenfunctions form a **complete orthonormal set** which means that *any function satisfying the required boundary conditions can be represented by a linear combination of these eigenfunctions.*

21.6 **Probabilities of measurements**

We have seen that before any measurement is made on a system it can be in a general, or mixed, state that can be represented as a linear combination of the eigenfunctions of the measurement that is about to be made. We can write

$$\psi = \sum_{i=1}^{n} c_i v_i \tag{21.40}$$

where we have assumed a complete orthonormal set of n of the eigenfunctions, v_i, but where n can be infinity. The coefficients c are constrained by the need to normalize ψ and bearing in mind that $\langle v_i | v_j \rangle = \delta_{ij}$ we have

$$\langle \psi | \psi \rangle = \sum_{i=1}^{n} \sum_{j=1}^{n} \langle c_i v_i | c_j v_j \rangle = \sum_{i=1}^{n} |c_i|^2 = 1. \tag{21.41}$$

There is no way of knowing from a first-principles argument what the probability is that a measurement will yield the value a_i corresponding to the eigenfunction v_i when the system has the state function ψ. It is a postulate of quantum mechanics that this probability is

$$P(i) = |c_i|^2. \tag{21.42}$$

Measurements of the different possible eigenvalues are mutually exclusive events and, since some measurement is made, the sum of the probabilities must be unity or

$$\sum_{i=1}^{n} P(i) = 1, \tag{21.43}$$

which is consistent with (21.42) and (21.41).

Let us now suppose that we have been given the mixed state wave function, ψ, and we know the form of the eigenfunctions. To find the probability of measuring a_k we need to find c_k and this can be done by evaluating the **overlap integral**

$$\langle v_k | \psi \rangle = \sum_{i=1}^{n} c_i \langle v_k | v_i \rangle = c_k. \tag{21.44}$$

This can be illustrated by finding the probability of measuring the ground-state energy (lowest energy state) for a particle in a box when the mixed-state function is

$$\psi = Ax(L - x) \tag{21.45}$$

which satisfies the boundary conditions $\psi = 0$ when $x = 0$ and $x = L$. First we must find the normalizing constant A from

$$A^2 \int_0^L x^2 (L - x)^2 \, dx = \frac{L^5}{30} A^2 = 1$$

or

$$A = \sqrt{\frac{30}{L^5}}. \tag{21.46}$$

To find c_1 we require the overlap integral of ψ with v_1 or

$$c_1 = \sqrt{\frac{30}{L^5}} \sqrt{\frac{2}{L}} \int_0^L x(L - x) \sin\left(\frac{\pi}{L} x\right) dx$$

Evaluating the integral by parts gives

$$c_1 = \sqrt{\frac{30}{L^5}} \sqrt{\frac{2}{L}} \frac{4L^3}{\pi^3} = \frac{4\sqrt{60}}{\pi^3}$$

and hence

$$P(1) = c_1^2 = \frac{960}{\pi^6} = 0.9986.$$

This high probability is due to the fact that the initial mixed-state function is very similar to the ground state function u_1. The two functions are shown in Figure 21.4 and it is clear that the overlap integral had to give close to the maximum possible value.

The probabilities of all the remaining possible measurements must be small, since the sum of all the probabilities is unity. However, without doing any calculations it can be determined that the probabilities $P(i)$ for i even must be zero. The function ψ is symmetrical about $x = L/2$ while all eigenfunctions with i even are anti-symmetric

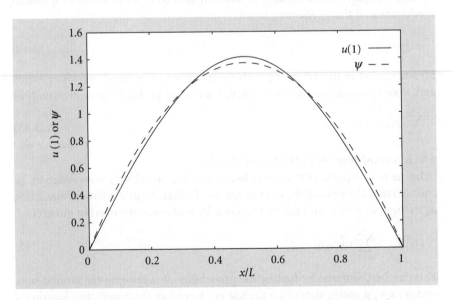

Figure 21.4 The mixed-state function ψ and the ground-state function for a particle in a one-dimensional box.

about $x = L/2$ (see Figure 21.1 for $k = 2$). Hence the overlap integral in the range $L/2$ to L will cancel out the contribution from $x = 0$ to $L/2$ and equal 0.

 Exercise 21.7
What is the probability of a particle in a one-dimensional box of length L being found in the ground state if its wave function is of the form $\psi = x(x - L/2)(x - L)$? (Hint: think of symmetry.)

21.7 Simple harmonic oscillator

21.7.1 Eigenfunctions and eigenvalues

We shall now apply the TISWE to a system that has great relevance to real physical systems, the simple harmonic oscillator. The classical one-dimensional harmonic oscillator was described in §7.6. The oscillator has a natural angular frequency, ω, given by

$$\omega^2 = \frac{k}{m} \tag{21.47}$$

where m is the mass of the oscillating particle and k is the force per unit displacement from the equilibrium point. If the particle starts at the equilibrium point, considered as the origin, and is displaced to position with coordinate X then the total work done in displacing it is equivalent to its potential energy at that point, i.e.

$$V = \int_0^X F(x)\, dx = \int_0^X kx\, dx = \frac{1}{2}kX^2 = \frac{1}{2}m\omega^2 X^2. \tag{21.48}$$

This gives us the form of the TISWE for a one-dimensional oscillator as

$$-\frac{\hbar^2}{2m}\frac{d^2\psi}{dx^2} + \frac{1}{2}m\omega^2 x^2\psi = E\psi. \tag{21.49}$$

A very useful device for dealing with equations with complicated coefficients is to transform the variables so that they become dimensionless – which usually simplifies the form of the equation. In this case if we divide throughout by $\frac{1}{2}\hbar\omega$, which has the dimensions of energy, then (21.49) becomes

$$-\frac{\hbar}{m\omega}\frac{d^2\psi}{dx^2} + \frac{m\omega}{\hbar}x^2\psi = \frac{2E}{\hbar\omega}\psi.$$

If we now write

$$\left(\frac{m\omega}{\hbar}\right)^{1/2} x = y \quad \text{and} \quad \frac{2E}{\hbar\omega} = \mu,$$

making y and μ both dimensionless, the equation takes on the simpler form

$$\frac{d^2\psi}{dy^2} + (\mu - y^2)\psi = 0. \tag{21.50}$$

A solution of this differential equation is of the form

$$\psi = C\exp(-\alpha y^2) \tag{21.51a}$$

giving

$$\frac{d^2\psi}{dy^2} = C(-2\alpha + 4y^2\alpha^2)\exp(-\alpha y^2). \tag{21.51b}$$

Substituting from (21.51a) and (21.51b) in (21.50), and dividing throughout by $C\exp(-\alpha y^2)$ gives

$$(4\alpha^2 - 1)y^2 + \mu - 2\alpha = 0.$$

Since y is an independent variable which can take any value, this relationship requires

$$4\alpha^2 - 1 = 0 \quad \text{or} \quad \alpha = \frac{1}{2}$$

and

$$\mu - 2\alpha = 0 \quad \text{or} \quad \mu = 1.$$

This gives the eigenfunction, which is actually the ground-state function,

$$u_0 = C\exp\left(-\frac{1}{2}y^2\right) = C\exp\left(-\frac{m\omega}{2\hbar}x^2\right), \tag{21.52a}$$

with energy eigenvalue

$$E_0 = \frac{1}{2}\hbar\omega. \tag{21.52b}$$

To find the normalization constant we use

$$\int_{-\infty}^{\infty} |u_0|^2 dx = C^2 \int_{-\infty}^{\infty} \exp\left(-\frac{m\omega}{\hbar}x^2\right) dx = 0. \tag{21.53}$$

This is similar to the normal distribution, normalized between ∞ and $-\infty$, given by (14.13):

$$P(x) = \frac{1}{\sqrt{2\pi\sigma^2}}\exp\left(-\frac{x^2}{2\sigma^2}\right).$$

By comparison we find

$$C = \left(\frac{m\omega}{\pi\hbar}\right)^{1/4}. \tag{21.54}$$

We have anticipated that (21.52a) is the ground state so we now look for higher energy states. These are of the form $u_0\, f(y)$ where $f(y)$ is a polynomial in y. Substituting $\psi = f(y)\exp(-\alpha y^2)$ in (21.50) gives

$$f''(y) - 2yf'(y) + (\mu - 1)f(y) = 0 \tag{21.55}$$

where we have used primes to indicate differentiation with respect to y.

The simplest form of $f(y)$ other than a constant, that gives u_0, is

$$f(y) = ay + b \tag{21.56a}$$

for which

$$f'(y) = a \tag{21.56b}$$

and

$$f''(y) = 0. \tag{21.56c}$$

Substituting from (21.56a), (21.56b), and (21.56c) into (21.55) gives

$$ay(\mu - 3) + (\mu - 1)b = 0. \tag{21.57}$$

There are two possible solutions to (21.57). The first is $a = 0$ and $\mu = 1$, but this gives the solution u_0. The second, and new, solution is $\mu = 3$ and $b = 0$. This gives a solution, which we anticipate as the first excited state of the system,

$$u_1 = Ay\exp(-\alpha y^2) = C_1 x\exp\left(-\frac{m\omega}{2\hbar}x^2\right) \tag{21.58a}$$

with energy eigenvalue

$$E_1 = \frac{\hbar\omega\mu}{2} = \frac{3\hbar\omega}{2}. \tag{21.58b}$$

By taking a quadratic form for $f(y)$ we find in a similar way the second excited state eigenfunction

$$u_2 = C_2\left(\frac{2m\omega}{\hbar}x^2 - 1\right)\exp\left(-\frac{m\omega}{2\hbar}x^2\right) \tag{21.59a}$$

with energy eigenvalue

$$E_2 = \frac{5\hbar\omega}{2}. \tag{21.59b}$$

The pattern is now clear. For the nth excited state function the polynomial $f(y)$ is of degree n and the energy eigenvalue is $(n + \frac{1}{2})\hbar\omega$. In fact the polynomials form a set known as **Hermite polynomials**, usually written as $H(y)$, the first four of which are

$$H_0(y) = 1$$
$$H_1(y) = 2y$$
$$H_2(y) = 4y^2 - 2$$
$$H_3(y) = 8y^3 - 12y.$$

The nth normalized eigenfunction is

$$u_n(x) = \left\{\left(\frac{m\omega}{\pi\hbar}\right)^{1/2}\frac{1}{2^n n!}\right\}^{1/2}\exp\left(-\frac{m\omega}{2\hbar}x^2\right)H_n\left\{\left(\frac{m\omega}{\hbar}\right)^{1/2}x\right\}. \tag{21.60}$$

The form of the first three eigenfunctions is shown in Figure 21.5, together with the associated probability densities. The general pattern, which also occurs for a particle in a box, is that the ground state has no nodes, the first excited state has one node, the second excited state two nodes, and so on.

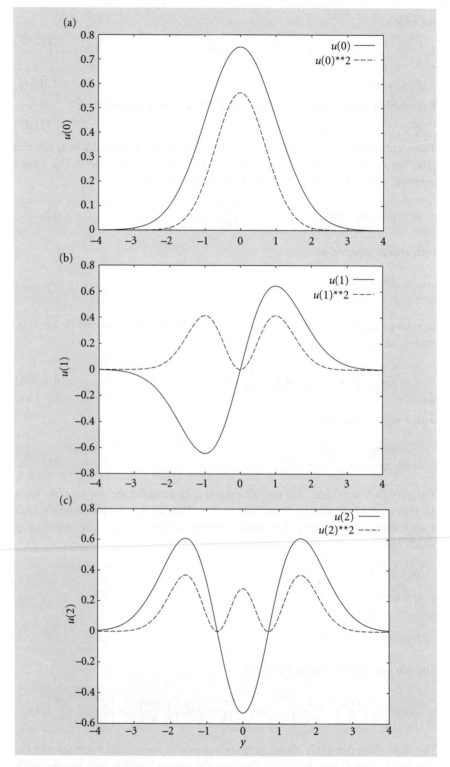

Figure 21.5 The eigenfunctions and probability densities for the three lowest energy states of the one-dimensional harmonic oscillator: (a) u(0), (b) u(1), (c) u(2).

The totality of the eigenfunctions for the one-dimensional oscillator forms a complete orthonormal set for the boundary conditions of this problem. All the eigenfunctions are a product of a finite polynomial with a term of the form $\exp(-\frac{1}{2}y^2)$. For large enough y the exponential term will always be dominant so the boundary condition here is that at $\pm\infty$ the function must behave as $\exp(-\frac{1}{2}y^2)$. A general function with this boundary condition can always be represented by some linear combination of the eigenfunctions for a one-dimensional harmonic oscillator.

21.7.2 Harmonic oscillator in physical systems

There are many situations where atoms are bound together in such a way that when they move from some equilibrium position they experience a restoring force. One such example is an atom bound to another in a diatomic molecule. If the atoms move away from one another from some equilibrium distance then there is an inward restoring force, whereas if they move closer together then they will be pushed apart again. Another situation is where an atom is bound into a rigid framework within a crystal. Any displacement of the atom from some equilibrium position leads to a restoring force. In both these cases, if the displacement is small then the restoring force is proportional to the displacement and a harmonic motion is the outcome. For larger displacements the force will not be proportional to the displacement and an anharmonic vibration will occur.

The average energy of vibration of one of these atomic vibrators will increase with increasing temperature and, classically, would be proportional to the temperature. Thus a conclusion from the classical model is that there will be no energy of vibration at 0 K. However, the quantum-mechanical treatment shows that the minimum possible energy for a one-dimensional harmonic oscillator is $\frac{1}{2}\hbar\omega$. This so-called **zero-point energy** has an important effect on physical systems at very low temperatures.

Exercise 21.8

An atom is constrained to move in one direction under the influence of a restoring force that, for small displacements, causes it to execute simple harmonic motion with angular frequency $10^{15}\,\mathrm{s}^{-1}$. What is its total energy at 0 K?

21.8 Three-dimensional simple harmonic oscillator

An atom bound into a crystal and displaced from its equilibrium position experiences a restoring force proportional to the displacement no matter what the direction of displacement – assuming a displacement that keeps it within the harmonic approximation. However, the magnitude of the force per unit displacement will vary with direction according to the arrangement of neighbouring atoms.

It is always possible to find two orthogonal directions, one of which gives the maximum, and the other a minimum, force per unit displacement. If these directions are taken as x and y axes and the forces per unit displacement along these directions are k_x and k_y, then the z axis can be taken orthogonal to x and y with a force per unit displacement k_z. For a body displaced from the origin to the point (x, y, z) the force on it will have components $(-k_x x, -k_y y, -k_z z)$ and the potential energy at that point will be

$$V = \frac{1}{2}m\left(\omega_x^2 x^2 + \omega_y^2 y^2 + \omega_z^2 z^2\right) \tag{21.61}$$

where the ω are defined as in (21.47). In this case the TISWE for the three-dimensional harmonic oscillator can be written as

$$-\frac{\hbar^2}{2m}\left(\frac{\partial^2 \psi}{\partial x^2} + \frac{\partial^2 \psi}{\partial y^2} + \frac{\partial^2 \psi}{\partial z^2}\right) + \frac{1}{2}m\left(\omega_x^2 x^2 + \omega_y^2 y^2 + \omega_z^2 z^2\right)\psi = E\psi. \tag{21.62}$$

A solution for this equation can be sought in the form of a product of three functions each involving just one of the independent variables, i.e.

$$\psi = \psi_x(x)\psi_y(y)\psi_z(z). \tag{21.63}$$

Substituting this in the partial differentiation factor of the first term of (21.62) gives

$$\frac{\partial^2 \psi}{\partial x^2} + \frac{\partial^2 \psi}{\partial y^2} + \frac{\partial^2 \psi}{\partial z^2} = \left(\frac{1}{\psi_x}\frac{\partial \psi_x}{\partial x} + \frac{1}{\psi_y}\frac{\partial \psi_y}{\partial y} + \frac{1}{\psi_z}\frac{\partial \psi_z}{\partial z}\right)\psi. \tag{21.64}$$

Inserting (21.64) in (21.62), dividing by ψ and rearranging gives

$$-\frac{\hbar^2}{2m}\frac{1}{\psi_x}\frac{\partial \psi_x}{\partial x} + \frac{1}{2}m\omega_x^2 x^2 - \frac{\hbar^2}{2m}\frac{1}{\psi_y}\frac{\partial \psi_y}{\partial y} + \frac{1}{2}m\omega_y^2 y^2 - \frac{\hbar^2}{2m}\frac{1}{\psi_z}\frac{\partial \psi_z}{\partial z} + \frac{1}{2}m\omega_z^2 z^2 = E \tag{21.65}$$

The left-hand side of (21.65) is the sum of three terms, one a function only of x, one a function only of y, and the last a function only of z. Since the variables in these terms can be independently varied and their sum is a constant the only logical conclusion is that each of them is a constant. Hence we can write

$$-\frac{\hbar^2}{2m}\frac{1}{\psi_x}\frac{\partial \psi_x}{\partial x} + \frac{1}{2}m\omega_x^2 x^2 = E_x, \tag{21.66}$$

with similar expressions for the y and z dependent components, where

$$E_x + E_y + E_z = E. \tag{21.67}$$

Multiplying both sides of (21.66) by ψ_x gives

$$-\frac{\hbar^2}{2m}\frac{\partial^2 \psi_x}{\partial x^2} + \frac{1}{2}m\omega_x^2 x^2 \psi_x = E_x \psi_x \tag{21.68}$$

which will be recognized as the TISWE for a one-dimensional simple harmonic oscillator. Thus the eigenfunctions for (21.62) are the products of three one-dimensional eigenfunctions and the eigenvalues are the sum of the three corresponding eigenvalues. Hence a general eigenfunction is

$$u_{n_x n_y n_z}(x, y, z) = u_{n_x}(x)u_{n_y}(y)u_{n_z}(z) \tag{21.69a}$$

where each of the individual functions on the right-hand side are of the form (21.60) and the corresponding eigenvalue is

$$E_{n_x n_y n_z} = (n_x + 1/2)\hbar\omega_x + (n_y + 1/2)\hbar\omega_y + (n_z + 1/2)\hbar\omega_z. \tag{21.69b}$$

For some crystals of high symmetry it is possible that two, or even all three, of the ω can be identical and in such cases degeneracy can occur. For example if $\omega_x = \omega_y$ then $E_{120} = E_{210} = E_{300} = E_{030}$, a fourfold degeneracy.

Exercise 21.9
A three-dimensional isotropic harmonic oscillator has energy $\frac{7}{2}\hbar\omega$. What is its degeneracy?

21.9 The free particle

We now consider a particle moving freely along in a straight line without experiencing any forces, so that the potential energy is always equal to zero. The TISWE for such a particle is

$$-\frac{\hbar^2}{2m}\frac{d^2\psi}{dx^2} = E\psi$$

or

$$\frac{d^2\psi}{dx^2} = -\frac{2mE}{\hbar^2}\psi = -k^2\psi, \tag{21.70}$$

where $k = \sqrt{2mE/\hbar^2}$.

Equation (21.70) is exactly the TISWE for a particle in a box except that now we do not have any boundary conditions – the particle can exist anywhere between $+\infty$ and $-\infty$. In this case the eigenfunctions can be of the form

$$u_k = \exp(ikx) \tag{21.71}$$

and there is no restriction on k and hence on the energy of the particle. We have seen that, in dealing with the TISWE, we normally neglect the time dependent factor given by (21.15). Including this factor gives

$$U_k(x, t) = \exp\left\{i\left(kx - \frac{E}{\hbar}t\right)\right\}. \tag{21.72}$$

Now since the particle only has kinetic energy we can write

$$k^2 = \frac{2mE}{\hbar^2} = \frac{m^2 v^2}{\hbar^2}$$

and using the de Broglie relationship, (21.2), we have

$$k = \frac{mv}{\hbar} = \frac{p}{\hbar} = 2\pi\frac{p}{h} = \frac{2\pi}{\lambda}. \tag{21.73}$$

We see from this that k is a measure of the momentum of the particle.

Using $E = h\nu$, we write

$$\frac{E}{\hbar} = 2\pi \frac{h\nu}{\lambda} = 2\pi\nu. \tag{21.74}$$

Inserting (21.73) and (21.74) in (21.72) gives

$$U_k(x, t) = \exp\left\{2\pi i\left(\frac{x}{\lambda} - \nu t\right)\right\} \tag{21.75}$$

which is the same as (21.5), a simple progressive wave.

One problem of the eigenfunction (21.71) is that it has a constant amplitude over an infinite range and therefore it is not possible to normalize it. Such a wave function corresponds to a physically unrealistic situation where the probability density for finding the particle is the same over the whole infinite range. In any real situation a particle will have its probability density mostly constrained around a fairly narrow range and the corresponding wave function, ψ, would be a mixed state function in the form of a **wave packet**. When there are discrete eigenfunctions, as for a particle in a box or a simple harmonic oscillator, then the mixed-state function is a weighted sum of individual eigenfunctions, as given by (21.40). In this case the eigenfunctions are not discrete and the appropriate form for the mixed state function is

$$\psi(x) = \frac{1}{2\pi} \int_{-\infty}^{\infty} a(k) \exp(ikx)\,dk. \tag{21.76a}$$

The arbitrary factor $1/(2\pi)$, which could be amalgamated with $a(k)$, makes this equation of exactly the same form as (17.41b) showing that $\psi(x)$ is the inverse Fourier transform of $a(k)$ and hence that $a(k)$ is the Fourier transform of $\psi(x)$ given by

$$a(k) = \int_{-\infty}^{\infty} \psi(x) \exp(-ikx)\,dx. \tag{21.76b}$$

As an example we take a Gaussian wave packet moving along the positive x direction with velocity v defined as

$$\psi(x) = C \exp\left(-\frac{(x - vt)^2}{2a^2}\right). \tag{21.77}$$

We are assuming that there is no dispersion, i.e. the wave velocity is independent of wavelength, so that the packet preserves its waveform as it travels. This mixed wave function has the complication that it is a function of both x and t but we can remove this by the transformation $x' = x - vt$, which means that we are measuring the coordinate with respect to an origin at the mean point of the wave packet. The probability density for this wave packet is

$$|\psi(x)|^2 = C^2 \exp\left(-\frac{x'^2}{a^2}\right) \tag{21.78}$$

and, from (14.13), this can be normalized giving

$$C^2 = \frac{1}{\sqrt{\pi a^2}}.$$

For the original space coordinates the k values contributing to the mixed state are distributed around a mean value $\bar{k}$, the average value of k corresponding to the speed of the wave packet, and in the transformed k-space we write $k' = k - \bar{k}$. We now apply (21.76b) with the transformed coordinates so that

$$a(k') = C' \int_{-\infty}^{\infty} \exp\left\{-\frac{x'^2}{2a^2}\right\} \exp(-ik'x')\,dx.$$

The Fourier transform of a Gaussian function is itself a Gaussian function and we find

$$a(k') = D \exp\left(-\frac{a^2 k'^2}{2}\right),$$

or

$$|a(k')|^2 = D^2 \exp(-a^2 k'^2), \tag{21.79}$$

where D is a constant that can be found by the normalization of the distribution $|a(k)|^2$.

The relationship between equations (21.78) and (21.79) will now be discussed.

21.10 Compatible and incompatible measurements

Before a measurement is made on a system it can be regarded as being in some mixed state that is a linear combination of eigenstates. The act of measurement leaves it in the eigenstate corresponding to the eigenvalue that has been measured. Some pairs of measurements have the same set of eigenstates so we now consider what happens when we make a measurement corresponding to operator $\hat{Q}$ with eigenfunctions v_i, $i = 1, 2, 3, \ldots$ and eigenvalues q_i, $i = 1, 2, 3, \ldots$ and *immediately* follow with a measurement corresponding to operator $\hat{R}$ with the same eigenfunctions and eigenvalues equal to r_i, $i = 1, 2, 3, \ldots$. Let us say that the first measurement gives the measurement q_n and leaves the system in the eigenstate represented by v_n. Now, according to the result (21.42) the probability that the following measurement gives r_n is unity since v_n is the only component of the state function before the measurement. Conversely, if the measurements are made in the reverse direction a measurement r_m will inevitably be followed by a measurement q_m. Note particularly the requirement that the second measurement is made immediately after the first, otherwise the system, in response to its environment, could change to some other state.

The measurement of a quantity q on a system with state function ψ can be symbolically represented by $\hat{Q}\psi$ and following this immediately by the measurement of a quantity r can be represented by $\hat{R}\hat{Q}\psi$. Making the measurements in the reverse order can be written as $\hat{Q}\hat{R}\psi$. In a real sense the measurement of quantities q and r are equivalent; if the experiment involved reading the values of the quantities on the dial of an instrument then the dial could be marked in terms of q, of r, or even show both scales simultaneously. Such observations are said to be **compatible**, and compatible pairs can be recognized by the condition that

$$\hat{R}\hat{Q} - \hat{Q}\hat{R} = \hat{0}, \tag{21.80}$$

that is to say that the left-hand side of (21.80) is a **null operator**. This expression is called the **commutator** of the operators and is written in the shorthand form

$$\hat{Q}\hat{R} - \hat{R}\hat{Q} = [\hat{Q}, \hat{R}], \tag{21.81}$$

and operators for which (21.80) is true are said to **commute**. As an example we take the one-dimensional momentum operator, $\hat{p}_x = \frac{\hbar}{i}\frac{\partial}{\partial x}$ and the total energy operator, $\hat{E} = -\frac{\hbar}{i}\frac{\partial}{\partial t}$.

Applying these operators to some arbitrary function F,

$$\hat{E}\hat{p}_x F = \hbar^2 \frac{\partial^2 F}{\partial t\,\partial x} \quad \text{and} \quad \hat{p}_x\hat{E}F = \hbar^2 \frac{\partial^2 F}{\partial x\,\partial t},$$

since $\frac{\partial^2 F}{\partial t\,\partial x} = \frac{\partial^2 F}{\partial x\,\partial t}$ then

$$[\hat{p}_x, \hat{E}] = 0 \tag{21.82}$$

and the operators commute. This means that in a single observation we can simultaneously measure the momentum of a particle and its total energy.

Now we consider another pair of possible measurements, of position and of momentum. The operator corresponding to position is designated by $\hat{x}$ and the operation it represents is that of multiplying the function that follows it by x. So the equation representing the measurement of position is

$$\hat{x}\psi = x\psi \tag{21.83}$$

and this equation imposes no constraints either on x or on ψ. This means that a particle can be measured in any position and that any state function is an eigenfunction for measuring position. Testing for whether or not the operators for position and momentum commute,

$$\hat{p}\hat{x}F = \frac{\hbar}{i}\frac{\partial}{\partial x}(xF) = \frac{\hbar}{i}F + \frac{\hbar}{i}x\frac{\partial F}{\partial x} \tag{21.84a}$$

and

$$\hat{x}\hat{p}F = x\frac{\hbar}{i}\frac{\partial F}{\partial x}. \tag{21.84b}$$

Hence

$$[\hat{p}, \hat{x}]F = \hat{p}\hat{x}F - \hat{x}\hat{p}F = \frac{\hbar}{i}F$$

or

$$[\hat{p}, \hat{x}] = \frac{\hbar}{i}. \tag{21.85}$$

This means that the operators do not commute, so that it is impossible simultaneously to measure the momentum and position of a particle. We can explore the relationship between the measurement of position and momentum by considering the transformed moving wave packet, with probability density $|\psi(x')|^2$ given by (21.78) and the transformed distribution in k-space with probability distribution $|a(k')|^2$. Both the normal space and k-space distributions are Gaussian, and by comparison with (14.13) we find their standard deviations as

$$\sigma_{x'} = \frac{a}{\sqrt{2}} \quad \text{and} \quad \sigma_{k'} = \frac{1}{\sqrt{2}a}. \qquad (21.86)$$

If we now define the uncertainty of the measurement of x' (or x), Δx, as the root mean square deviation from the mean measurement then $\Delta x = \sigma_x$ and similarly $\Delta k = \sigma_{k'}$. This gives

$$\Delta x \Delta k = \frac{1}{2}. \qquad (21.87)$$

From the de Broglie relationship,

$$p = \hbar k \text{ so that } \Delta p = \hbar \Delta k.$$

Combining this with (21.87) gives

$$\Delta x \Delta p = \frac{1}{2}\hbar.$$

It turns out that a Gaussian wave packet gives the minimum possible value for $\Delta x \Delta p$ and more generally we can write

$$\Delta x \Delta p \geq \frac{1}{2}\hbar, \qquad (21.88)$$

an important relationship known as the **Heisenberg uncertainty principle**. It expresses the fundamental limitation in defining the state of a system; if the state function is narrow in normal space so that the uncertainty in measuring the position is small then, correspondingly, the uncertainty in the measurement of momentum will be large. The same incompatibility exists in determining time and measuring the energy of a system, leading to a corresponding expression of the Heisenberg uncertainty principle:

$$\Delta E \Delta t \geq \frac{1}{2}\hbar. \qquad (21.89)$$

Exercise 21.10
An electron, able to move in one dimension, is only known to be somewhere in a box of length 10^{-11} m. What would be the minimum uncertainty in measuring the speed of its motion?

21.11 A potential barrier

We now consider a free particle moving along the x axis in the positive x direction with zero potential energy and total energy (all kinetic) E. When it reaches $x = 0$ its potential energy becomes V_0 ($< E$) and when it reaches $x = a$ its potential energy becomes zero again. The potential barrier given by this description is illustrated in Figure 21.6.

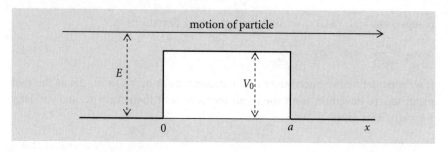

Figure 21.6 A free particle moving across a potential barrier.

The general TISWE for the particle is

$$\frac{d^2\psi}{dx^2} = -\frac{2m(E - V)}{\hbar^2}\psi = -k^2\psi. \tag{21.90}$$

In the region $x < 0$, with $V = 0$ the solution is

$$\psi_1 = A\exp(ik_1 x) + B\exp(-ik_1 x) \tag{21.91a}$$

with $k_1 = \sqrt{2mE/\hbar^2}$. The first term on the right-hand side of (21.91a) represents the oncoming wave moving towards the barrier. The second term represents a wave reflected by the potential energy discontinuity at $x = 0$ moving away from the barrier.

In the region $0 \le x \le a$, with $V = V_0$ the solution is

$$\psi_2 = C\exp(ik_2 x) + D\exp(-ik_2 x) \tag{21.91b}$$

with $k_2 = \sqrt{2m(E - V_0)/\hbar^2}$. Again there is an ongoing and a reflected wave with the latter due to the potential energy discontinuity at $x = a$.

In the final region with $x > a$, with $V = 0$, the solution is

$$\psi_3 = F\exp(ik_1 x) \tag{21.91c}$$

with no reflected wave since there is nothing beyond $x = a$ to produce a reflection.

The intensity of a wave is proportional to the modulus squared of its amplitude so, considering the total effect of the barrier, the incident wave has intensity $|A|^2$, the reflected wave has intensity $|B|^2$ and the transmitted wave has intensity $|F|^2$. What we are interested in finding is the reflection coefficient

$$R = \frac{|B|^2}{|A|^2} \tag{21.92a}$$

and the transmission coefficient

$$T = \frac{|F|^2}{|A|^2}. \tag{21.92b}$$

The physical principle that is used to find these quantities is that at an interface where the change of potential energy is discontinuous but *finite* the wave function *and its derivative* must both be continuous. This means that at $x = 0$

$$\psi_1 = \psi_2 \quad \text{and} \quad \frac{d\psi_1}{dx} = \frac{d\psi_2}{dx}$$

and at $x = a$

$$\psi_2 = \psi_3 \quad \text{and} \quad \frac{d\psi_2}{dx} = \frac{d\psi_3}{dx}.$$

You should note that in §21.3.1, relating to the boundary condition for a particle in a box, the wave function was continuous across the boundary but its derivative was not since the change of potential energy was *infinite*.

Applying the boundary conditions to Equations (21.91),

$$A + B = C + D \tag{21.93a}$$

$$k_1(A - B) = k_2(C - D) \tag{21.93b}$$

$$C \exp(ik_2 a) + D \exp(-ik_2 a) = F \exp(ik_1 a) \tag{21.93c}$$

and

$$k_2(C \exp(ik_2 a) - D \exp(-ik_2 a)) = k_1 F \exp(ik_1 a). \tag{21.93d}$$

Multiplying (21.93a) by k_1 and then adding and subtracting (21.93b) from it gives

$$A = \frac{k_1 + k_2}{2k_1} C + \frac{k_1 - k_2}{2k_1} D \tag{21.94a}$$

and

$$B = \frac{k_1 - k_2}{2k_1} C + \frac{k_1 + k_2}{2k_1} D. \tag{21.94b}$$

Multiplying (21.93c) by k_2 and adding and subtracting (21.93d) from it gives

$$C = \frac{k_1 + k_2}{2k_2} F \exp(i(k_1 - k_2)a) \tag{21.94c}$$

and

$$D = \frac{k_2 - k_1}{2k_2} F \exp(i(k_1 + k_2)a). \tag{21.94d}$$

Substituting from (21.94c) and (21.94d) for C and D in (21.94a) gives

$$A = \frac{F \exp(ik_1 a)}{4k_1 k_2} \left\{ (k_1 + k_2)^2 \exp(-ik_2 a) - (k_1 - k_2)^2 \exp(ik_2 a) \right\}.$$

Using

$$\exp(ik_2 a) + \exp(-ik_2 a) = 2 \cos k_2 a \quad \text{and} \quad \exp(ik_2 a) - \exp(ik_2 a) = 2i \sin k_2 a$$

$$A = \frac{F \exp(ik_1 a)}{2k_1 k_2} \left\{ 2k_1 k_2 \cos k_2 a - i(k_1^2 + k_2^2) \sin k_2 a \right\}.$$

This gives the transmission coefficient

$$T = \left| \frac{F}{A} \right|^2 = \frac{|F|^2}{|A|^2}$$

$$= \frac{4k_1^2 k_2^2}{(k_1^2 + k_2^2)^2 \sin^2 k_2 a + 4k_1^2 k_2^2 \cos^2 k_2 a} = \frac{4k_1^2 k_2^2}{(k_1^2 - k_2^2)^2 \sin^2 k_2 a + 4k_1^2 k_2^2}. \tag{21.95}$$

Substituting from (21.94c) and (21.94d) for C and D in (21.94b) gives

$$B = \frac{F \exp(ik_1 a)}{4k_1 k_2}(k_1^2 - k_2^2)(\exp(-ik_2 a) - \exp(ik_2 a))$$

$$= -i\frac{F \exp(ik_1 a)}{2k_1 k_2}(k_1^2 - k_2^2)\sin k_2 a$$

or

$$\frac{|B|^2}{|F|^2} = \frac{(k_1^2 - k_2^2)^2}{4k_1^2 k_2^2}\sin^2 k_2 a. \tag{21.96}$$

From (21.95) and (21.96) the reflection coefficient is

$$R = \frac{|B|^2}{|A|^2} = \frac{|B|^2}{|F|^2}\frac{|F|^2}{|A|^2} = \frac{(k_1^2 - k_2^2)^2 \sin^2 k_2 a}{(k_1^2 - k_2^2)^2 \sin^2 k_2 a + 4k_1^2 k_2^2}. \tag{21.97}$$

From (21.95) and (21.97)

$$R + T = 1, \tag{21.98}$$

a necessary condition for conservation of energy since none is absorbed in the system. An interesting case is when V_0 is large and negative so that $|E - V_0| \gg |E|$. Since physical systems tend towards states of minimum energy this means that there is a strong attractive force on the particle towards what is now a deep well rather than a barrier. In this case $k_2 \gg k_1$ and from (21.97), as long as $k_2^2 \sin^2 k_2 a \gg 4k_1^2$, we find $R \approx 1$ – that is to say that a strong attractive force *repels* the particle. Although this is a counter-intuitive result it is actually observed in nuclear physics experiments where a low-energy neutron fired at a nucleus, which from a stability point of view it should join, is reflected backwards by the large attractive force. This is not the only unexpected result that is found from quantum mechanics and confirmed by observation or experiment, and we shall now deal with another.

Exercise 21.11
Show that if $k_2 a$ is small and $k_1 \gg k_2$ that the transmission coefficient given by (21.95) is independent of k_2 and find what form it takes.

21.12 Tunnelling

An interesting case arises when $V_0 > E$, a situation which classically gives no transmission and complete reflection of the oncoming particles. From a classical point of view the kinetic energy within the barrier would be negative, which is not permitted.

The TISWE within the barrier is

$$\frac{d^2\psi}{dx^2} = \frac{2m(V_0 - E)}{\hbar^2}\psi = k_3^2\psi \tag{21.99}$$

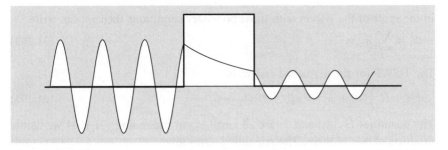

Figure 21.7 The amplitude of the matter wave falls across the barrier to give the tunnelling effect.

giving a solution

$$\psi_2 = C \exp(k_3 x) + D \exp(-k_3 x). \tag{21.100}$$

It is possible to repeat the analysis given in §21.11 with this different expression in place of ψ_2 to obtain the transmission and reflection coefficients but the results can be found from (21.95) and (21.97) by substituting ik_3 in place of k_2. To do this we need the result (15.15a) that $\sin ik_3 = i \sinh k_3$. What we obtain is a **tunnelling transmission coefficient**

$$T_{\text{tun}} = \frac{4k_1^2 k_3^2}{(k_1^2 + k_3^2)^2 \sinh^2 k_3 a + 4k_1^2 k_3^2} \tag{21.101a}$$

and a corresponding reflection coefficient

$$R_{\text{tun}} = 1 - T_{\text{tun}}. \tag{21.101b}$$

A schematic representation of the wave representing the particle passing through the barrier is given in Figure 21.7.

The fact that a particle can tunnel through a potential barrier is completely non-classical behaviour but it does occur in many important physical systems. One such system is the way in which nuclear reactions transform hydrogen into helium in the Sun's interior – a topic dealt with in §22.3.

21.13 Other methods of solving the TISWE

21.13.1 Basic first-order perturbation theory

Consider a system with Hamiltonian operator $\hat{H}$ for which the eigenfunctions u_r and associated eigenvalues E_r are known. The Hamiltonian operator is now changed by a small amount to $\hat{H} + \hat{H}'$. The extra component $\hat{H}'$ represents a small change in the potential energy function, so that the maximum magnitude of $\hat{H}'$ is very small compared to the magnitude of the ground state energy of the system.

A small change in the Hamiltonian operator leads to small changes in the eigenfunctions, from u_r to $u_r + \phi_r$, and in the eigenvalues from E_r to $E_r + \varepsilon_r$. An important condition that must be satisfied is that the boundary conditions for the perturbed eigenfunctions must be the same as those for the unperturbed eigenfunctions. Since the eigenfunctions form a complete orthonormal set for functions in the space of the system with those boundary conditions then we can write

$$\phi_r = \sum_s a_{r,s} u_s. \tag{21.102}$$

The TISWE for the perturbed system is

$$\left(\hat{H} + \hat{H}'\right)(u_r + \phi_r) = (E_r + \varepsilon_r)(u_r + \phi_r). \tag{21.103}$$

The quantities $\hat{H}'$, ϕ_r, and ε_r are all small, so in expanding (21.103) we ignore products of two of them. The remaining terms give

$$\hat{H}u_r + \hat{H}'u_r + \hat{H}\phi_r = E_r u_r + \varepsilon_r u_r + E_r \phi_r$$

or, since $\hat{H}u_r = E_r u_r$,

$$\hat{H}'u_r + \hat{H}\phi_r = \varepsilon_r u_r + E_r \phi_r. \tag{21.104}$$

The next step is to replace ϕ_r by the right-hand side of (21.102), pre-multiply each term on the two sides of (21.104) by u_r^*, and then integrate over the whole of space. This gives

$$\int u_r^* \hat{H}' u_r \, d\tau + \sum_s \int u_r^* \hat{H} a_{s,r} u_s \, d\tau = \varepsilon_r \int u_r^* u_r \, d\tau + E_r \sum_s a_{s,r} \int u_r^* u_s \, d\tau.$$

We now replace $\hat{H}u_s$ by $E_s u_s$ and also introduce a notation based on the vector representation of eigenstates as described in §21.5.2. Thus we write

$$\int \psi_1^* \hat{A} \psi_2 \, d\tau = \langle \psi_1 | \hat{A} | \psi_2 \rangle. \tag{21.105}$$

Taking account of the orthonormal nature of the set of eigenfunctions this gives

$$\left\langle u_r | \hat{H}' | u_r \right\rangle + E_r a_{r,r} = \varepsilon_r + E_r a_{r,r}$$

or

$$\varepsilon_r = \left\langle u_r | \hat{H}' | u_r \right\rangle, \tag{21.106}$$

which enables the perturbation of the energy to be found. This analysis is referred to as **first-order perturbation theory**. There are higher-order perturbation theories that are valid for larger perturbations of the system.

As an example of the application of first-order perturbation theory we take an electron in a one-dimensional box, as described in §21.3, of length 0.1 nm but instead of $V_0 = 0$ inside the box we have

$$V_0 = 0.1 \sin\left(\frac{\pi x}{L}\right) \text{ eV}. \tag{21.107}$$

The ground-state energy for this system, given by (21.17) with $n = 1$, is 37.635 eV so the perturbation is small enough for the analysis leading to (21.106) to be valid. We now calculate the perturbation energy for the ground state. This is, in units eV,

$$\varepsilon_1 = 0.1 \frac{2}{L} \int_0^L \sin^3\left(\frac{\pi x}{L}\right) dx = \frac{0.2}{L} \int_0^L \left(\sin\left(\frac{\pi x}{L}\right) - \sin\left(\frac{\pi x}{L}\right)\cos^2\left(\frac{\pi x}{L}\right)\right) dx$$

$$= \frac{0.2}{L}\left\{\left|-\frac{L}{\pi}\cos\left(\frac{\pi x}{L}\right)\right|_0^L + \left|\frac{L}{3\pi}\cos^3\left(\frac{\pi x}{L}\right)\right|_0^L\right\} = \frac{0.2}{\pi}\left(2 - \frac{2}{3}\right) = 0.085 \text{ eV}.$$

The ground-state energy of the perturbed system is thus 37.720 eV.

This analysis can be carried out for any other state of the original system since the normalized eigenfunctions are known. If the perturbation were too large then the analysis leading to (21.106) would not be valid. For example, if we had

$$V_0 = 20 \sin\left(\frac{\pi x}{L}\right) \text{eV} \tag{21.108}$$

then a straightforward application of (21.106) would give $\varepsilon_1 = 16.977 \text{ eV}$, comparable to the ground state energy so the analysis would not be valid. In this case perturbation theory could not be used but the problem can be solved by a numerical approach.

21.13.2 The shooting method

Any solution of the one-dimensional TISWE (21.14) will represent an eigenstate of the system and the corresponding eigenvalue. A solution is a function that exists within the domain defined by the problem and *satisfies the boundary conditions*. If we take the particle-in-the-box problem then the boundary conditions are that the wave function must be zero at the boundaries of the box.

A characteristic of the TISWE is that it is *linear in ψ*, which means that if ψ is a solution then so is $c\psi$ where c is any constant. If we take a particle in a box then we must have $\psi = 0$ when $L = 0$ but $d\psi/dx$ can take any value since it will be proportional to c. One approach is to start the solution of the differential equation

$$\frac{d^2\psi}{dx^2} - \frac{2m}{\hbar^2}(V_0 - E)\psi = 0$$

at $x = 0$ with $\psi = 0$, an arbitrary value of $d\psi/dx$ and a guessed value of E and then solve the differential equation by some standard method, such as the four-step Runge–Kutta method (§19.3) up to $x = L$. If $\psi \neq 0$ then the value of E is changed, the new value being estimated from the results of previous runs. Eventually a value of E is found that gives $\psi = 0$ when $x = L$ and this is an energy eigenvalue. This is known as the **shooting method** as it has a similarity to the way that gunners change the conditions, in their case the inclination of the gun, until they find the right value to hit the target. We shall illustrate this method by finding the ground-state eigenvalue for

$$\frac{d^2\psi}{dx^2} - \frac{2m}{\hbar^2}\left(V_c \sin\left(\frac{\pi x}{L}\right) - E\right)\psi = 0 \tag{21.109}$$

where V_c is 20 eV, the particle is an electron, and the box has a length of 0.1 nm.

This equation can be simplified in form. Firstly we can use units of electron volts for energy and secondly we can write $x/L = y$. This changes (21.109) to

$$\frac{d^2\psi}{dy^2} = \frac{2m\varepsilon L^2}{\hbar^2}\{V_\varepsilon \sin(\pi y) - E_\varepsilon\}\psi = 0$$

where ε is 1 eV and V_ε and E_ε are expressed in eV. Substituting values for the constants on the right-hand side

$$\frac{d^2\psi}{dy^2} = 0.262\,24\{20 \sin(\pi y) - E_\varepsilon\}, \tag{21.110}$$

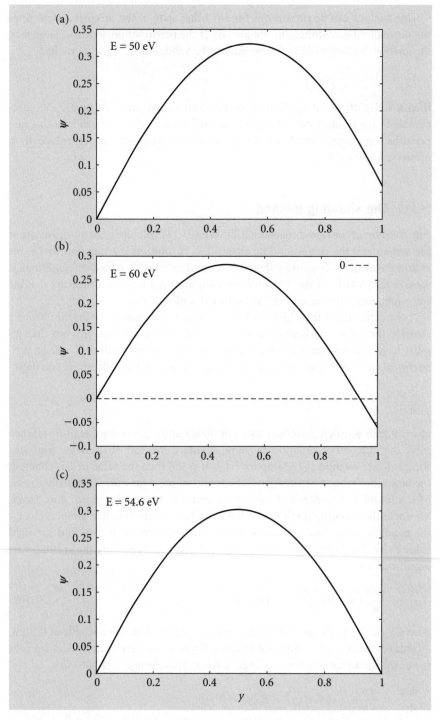

Figure 21.8 Stages in finding the eigenvalue for the TISWE (21.110) showing ψ for (a) 50 eV, which is too low; (b) 60 eV, which is too high; and (c) 54.6 eV, the correct eigenvalue.

and the solution must be found in the range $y = 0$ to 1. The equation is now split up into a pair of coupled first-order differential equations

$$
\frac{d\psi}{dy} = \chi
$$
$$
\frac{d\chi}{dy} = 0.262\,24\{20\sin(\pi y) - E_\varepsilon)\}
$$

(21.111)

and solved as described by (19.11). A program SHOOT for carrying out this calculation is given in Appendix 11. Since the ground state energy with no potential energy within the box is about 38 eV, we first try $E_\varepsilon = 50$ eV. This gives the result shown in Figure 21.8a. The value of ψ for $y = 1$ is positive so $E_\varepsilon = 60$ eV is next tried, giving the result in Figure 21.8b. Since ψ is now negative for $y = 1$ it is clear that the eigenvalue is between 50 and 60 eV. By interpolating between the values at $y = 1$ the next estimate is $E_\varepsilon = 55$ eV, which is a little too large. Eventually the eigenvalue is found as 54.6 eV, which gives the result for the eigenfunction shown in Figure 21.8c. It is interesting to note that first-order perturbation theory gives the result 54.7 eV, very little different from the result found here.

Figure 21.8c gives the form of the ground-state eigenfunction, although this is normally not of much interest as the quantity that can be observed, say in the splitting of a spectral line, depends on the eigenvalue. However, it is possible to find the eigenfunction in the case of first-order perturbation and the way of doing this will now be shown.

21.13.3 The eigenfunction from first-order perturbation theory

The complete eigenfunction for the perturbed system is $u_r + \phi_r$ and ϕ_r is given in terms of the unperturbed eigenfunctions in (21.102). We now replace ϕ_r by the right-hand side of (21.102) in (21.104), pre-multiply by u_t^* $(t \neq r)$ and integrate over the whole of the function space. This gives

$$
\sum_s a_{s,r} \int u_t^* \hat{H} u_s \, d\tau + \int u_t^* \hat{H}' u_r \, d\tau = \varepsilon_r \int u_t^* u_r \, d\tau + E_r \sum_s a_{s,r} u_t^* u_s \, d\tau.
$$

Writing $\hat{H} u_s = E_s u_s$, taking account of the orthonormal relationship of the eigenfunctions and using the notation introduced in (21.108) we find

$$
a_{t,r} E_t + \left\langle u_t | \hat{H}' | u_r \right\rangle = a_{t,r} E_r
$$

or

$$
a_{t,r} = \frac{\left\langle u_t | \hat{H}' | u_r \right\rangle}{E_r - E_t}.
$$

(21.112)

This process enables the coefficients of the expansion of ϕ_r to be determined except $a_{r,r}$. However, in principle the normalization of the total eigenfunction $u_r + \phi_r$ enables the coefficient of u_r to be found from the values of all the other coefficients.

Problems

21.1 An electron constrained to move along the x axis has potential energy $V = 2\,\text{eV}$ $(0 \leq x \leq 0.2\,\text{nm})$, $V = \infty$ otherwise. What are the energies of the ground state and the first excited state?

21.2 From the results (21.18) and (21.31), write down the form of the normalized eigenfunction for a particle in a cubical box of side L. Such a particle has energy $14\pi^2\hbar^2/2mL^2$. What is the degeneracy of this state?

21.3 A particle of mass $10^{-26}\,\text{kg}$ executes one-dimensional simple harmonic motion in its ground state with angular frequency $2 \times 10^{16}\,\text{s}^{-1}$. If it is in a mixed state with un-normalized state function $\psi = \exp(-x^2/a^2)$, with $a = 10^{-12}\,\text{m}$, then what is the probability that a measurement of its energy would give the ground state?

21.4 A stream of electrons, of energy E, falls on a potential barrier of width $0.1\,\text{nm}$ and height $10\,\text{eV}$. Find the proportion of transmitted electrons for (i) $E = 11\,\text{eV}$ and (ii) $E = 9\,\text{eV}$. From the form of Equation (21.95) find the transmission coefficient for $E = 10\,\text{eV}$.

21.5 Use the program SHOOT to find the energy of the first and second excited states of the problem described in §21.13.2. Compare these energies with the particle-in-a-box energies with zero potential energy within the box.

The Maxwell–Boltzmann distribution

22

In a gas at a finite temperature the individual molecules are moving in all directions, occasionally colliding and exchanging energy. The collisions are elastic, meaning that there is no loss of energy of motion in a collision, but any individual molecule is repeatedly losing or gaining energy and changing its direction of motion. The distribution of speeds of the molecules in known as the Maxwell–Boltzmann distribution and is of importance in many areas of physics.

In this chapter the Maxwell–Boltzmann distribution is derived and then applied to two problems in the area of astrophysics. The first is the problem of the retention of a planetary atmosphere. The second is a consideration of energy production in stars, and brings together the Maxwell–Boltzmann distribution with the idea of tunnelling from quantum mechanics.

22.1 Deriving the Maxwell–Boltzmann distribution

The Maxwell–Boltzmann distribution can be theoretically derived by the use of statistical mechanics, but such a derivation is outside the scope of this book. However, on the basis of two assumptions it is possible to derive the distribution in terms of what has been done in previous chapters.

First we must introduce the idea of the **degrees of freedom** of a mechanical system. This is basically the number of modes of motion into which the total motion of the system can be resolved. If the molecules of the gas are single atoms then their only degrees of freedom are in their translational motion and this gives three degrees of freedom associated with motions in three orthogonal directions. A linear molecule, say oxygen (O_2), which consists of two atoms linked together in the x-direction, can have two tumbling modes corresponding to rotations about the y- and z- directions. A three-dimensional molecule, say carbon dioxide (CO_2), can have three tumbling modes. There are also possible modes corresponding to vibrations of various kinds, but they are not relevant here. We are concerned just with the translational motions, giving three degrees of freedom per atom or molecule.

Our two assumptions are as follows:

- The energy associated with each degree of freedom is $\frac{1}{2}kT$ where k is Boltzmann's constant and T the absolute temperature.
- The components of velocity in any specified direction have a normal distribution. The mean velocity has to be zero otherwise there will be bodily motion in one direction.

If the mass of each molecule is μ and the mean square speed of the molecules, say in the x-direction, is $\overline{v_x^2}$, then the average kinetic energy of the molecules is

$$\frac{1}{2}\mu\overline{v_x^2} = \frac{1}{2}kT$$

or

$$\overline{v_x^2} = \frac{kT}{\mu}. \tag{22.1}$$

Since the mean of the normal distribution is zero $\sigma^2 = \overline{v_x^2} = kT/\mu$ and, from (14.13), the distribution of the velocity components, say in the x-direction, is

$$P(v_x) = \left(\frac{\mu}{2\pi kT}\right)^{1/2} \exp\left(-\frac{\mu v_x^2}{2kT}\right). \tag{22.2}$$

The distributions in the y- and z-directions are similar to (22.2) so the probability that the x component of velocity is between v_x and $v_x + dv_x$, the y component between v_y and $v_y + dv_y$ and the z component between v_z and $v_z + dv_z$ is

$$P(v_x)P(v_y)P(v_z)\,dv_x\,dv_y\,dv_z = \left(\frac{\mu}{2\pi kT}\right)^{3/2} \exp\left(-\frac{\mu v^2}{2kT}\right) dv_x\,dv_y\,dv_z, \tag{22.3}$$

where the square of the speed $v^2 = v_x^2 + v_y^2 + v_z^2$. The expression (22.3) shows that the probability density (i.e. probability per unit volume in velocity space) depends only on the speed and not on the components directly. In velocity space the volume between speed v and $v + dv$ is that of a shell of radius v and thickness dv, $4\pi v^2 dv$, so the probability that the speed is in this range is

$$P(v)\,dv = \left(\frac{\mu}{2\pi kT}\right)^{3/2} \exp\left(-\frac{\mu v^2}{2kT}\right)4\pi v^2\,dv = \left(\frac{2\mu^3}{\pi k^3 T^3}\right)^{1/2} v^2 \exp\left(-\frac{\mu v^2}{2kT}\right) dv. \tag{22.4}$$

The function $P(v)$ gives the Maxwell–Boltzmann distribution of molecular speeds in a gas. The peak of distribution is at $v = v_p$ where v_p is the value of v for which $dP(v)/dv = 0$, which gives $v_p = \sqrt{2kT/\mu}$. The distribution is best illustrated in terms of the dimensionless variable $u = v/v_p = v\sqrt{\mu/2kT}$. Substituting u for v in (22.4) gives

$$P(u)\,du = \frac{4}{\sqrt{\pi}}u^2 \exp(-u^2)\,du \tag{22.5}$$

and $P(u)$ is shown in Figure 22.1.

The mean of u is

$$\bar{u} = \int_0^\infty uP(u)\,du = \frac{4}{\sqrt{\pi}}\int_0^\infty u^3 \exp(-u^2)\,du = \frac{2}{\sqrt{\pi}} \tag{22.6}$$

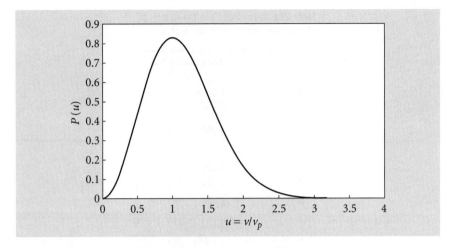

Figure 22.1 Maxwell–Boltzmann distribution.

from which

$$\bar{v} = \sqrt{\frac{2kT}{\mu}}\,\bar{u} = \sqrt{\frac{8kT}{\pi\mu}}. \qquad (22.7)$$

Similarly the value of $\overline{u^2}$ can be calculated from

$$\overline{u^2} = \int_0^\infty u^2 P(u)\,\mathrm{d}u = \frac{4}{\sqrt{\pi}}\int_0^\infty u^4 \exp(-u^2)\,\mathrm{d}u.$$

Two stages of integration by parts and the result (14.9) give $\overline{u^2} = \frac{3}{2}$ and applying the scaling factor v_p that converts u to v,

$$\overline{v^2} = \frac{2kT}{\mu}\,\overline{u^2} = \frac{3kT}{\mu}. \qquad (22.8)$$

This is an expected result, since the average translational kinetic energy of a molecule is $\frac{1}{2}\mu\overline{v^2} = \frac{3}{2}kT$ ($\frac{1}{2}kT$ per degree of freedom), which is consistent with (22.8). Finally we can calculate the variance of the Maxwell–Boltzmann distribution from

$$V_{\mathrm{MB}} = \overline{v^2} - \bar{v}^2 = \frac{3kT}{\mu} - \frac{8kT}{\pi\mu} = 0.454\frac{kT}{\mu}. \qquad (22.9)$$

The Maxwell–Boltzmann distribution is important in many areas of science, with the main interest being in the area in the tail of the distribution. Table 22.1 shows the area from $u = u_c$ to ∞ for integral values of u_c up to 12.

The order of magnitude of the area in the tail changes rapidly with u_c and is better represented in Figure 22.2, which shows log(area) against u_c.

Exercise 22.1
Find $u_{\max}$, the value of u that gives the maximum of $P(u)$ in (22.5), and then calculate the ratio $P(2u_{\max})/P(u_{\max})$.

Table 22.1 Area under the tail of a Maxwell–Boltzmann distribution

u_c	Area under tail
0	1.0000
1	0.5724
2	4.601 (-1)
3	4.398 (-3)
4	5.233 (-7)
5	7.989 (-11)
6	1.592 (-15)
7	4.183 (-21)
8	1.459 (-27)
9	6.784 (-35)
10	4.219 (-43)
11	3.516 (-52)
12	3.933 (-62)

$(-n)$ signifies $\times 10^{-n}$.

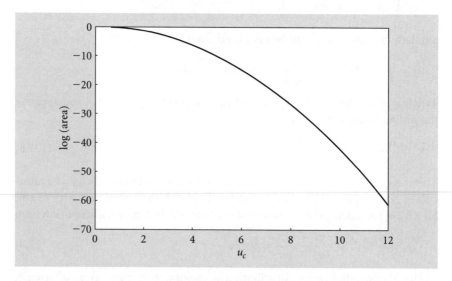

Figure 22.2 log(area in the tail) of a Maxwell–Boltzmann distribution as a function of u_c.

22.2 Retention of a planetary atmosphere

One area of science in which the Maxwell–Boltzmann distribution plays a significant role is the retention of a planetary atmosphere. This is a rather complex topic that has to take into account the detailed density and temperature structure of the atmosphere, but it is possible to illustrate quite simply that the Earth can retain its present atmosphere indefinitely while the Moon, if it ever had a similar atmosphere, would have lost it quite quickly.

The escape speed from a planet of mass M and radius R is

$$v_{esc} = \sqrt{\frac{2GM}{R}} \qquad (22.10)$$

where G is the gravitational constant, $6.67 \times 10^{-11} \, \mathrm{m^3 \, kg^{-1} s^{-2}}$. Any atmospheric molecule with an uninterrupted passage to a region beyond the atmosphere and moving at greater than the escape speed will be lost. Molecules in the lower reaches of the atmosphere are constantly colliding with other molecules and so are inhibited from escaping even if they have the speed to do so. At a high level, in a region known as the **exosphere**, the mean free path between collisions is effectively infinite so that molecules can escape from this region. For the Earth this is at a height of 500 km, at which level the atmospheric temperature is about 1000 K; although near ground level temperature decreases with height, at higher levels the temperature variation is more complex, with some fluctuation, and beyond about 100 km from the surface it steadily increases.

The composition of the Earth's atmosphere is approximately 77% nitrogen, 21% oxygen, and 1% argon, with traces of other gases. For our purpose we shall assume that it is purely nitrogen, the component with the lowest molecular mass and therefore that most likely to escape. The value of v_p for molecular nitrogen at the temperature of the exosphere is

$$v_p = \sqrt{\frac{2kT}{\mu}} = \sqrt{\frac{2 \times 1.38 \times 10^{-23} \times 1000}{28 \times 1.67 \times 10^{-27}}} \, \mathrm{m \, s^{-1}} = 768 \, \mathrm{m \, s^{-1}},$$

where μ is the number of nucleons in a nitrogen molecule (28) times the average nucleon mass, $1.67 \times 10^{-27} \, \mathrm{kg}$.

The mass of the Earth is $5.97 \times 10^{24} \, \mathrm{kg}$ and its radius is 6380 km so the exospheric radius is 6880 km. For these values the escape speed from the exosphere is

$$v_{esc} = \sqrt{\frac{2 \times 6.67 \times 10^{-11} \times 5.97 \times 10^{24}}{6.88 \times 10^6}} \, \mathrm{m \, s^{-1}} = 10.76 \, \mathrm{km \, s^{-1}}.$$

The value of u for escape is

$$u_{esc} = \frac{v_{esc}}{v_p} = 14.0$$

and from Table 22.1 it is clear that the proportion of molecules with escape speed is well below 10^{-61}. The mass of the Earth's atmosphere is just over $5 \times 10^{18} \, \mathrm{kg}$ and hence the number of atmospheric molecules, assumed all to be nitrogen, is about $5 \times 10^{18}/(28 \times 1.67 \times 10^{-27}) \approx 10^{44}$ so it is very unlikely that even one molecule will have escape speed. This illustrates the great stability of the Earth's atmosphere.

The Moon has mass $7.35 \times 10^{22} \, \mathrm{kg}$ and radius 1740 km. We can take the ambient surface temperature as 250 K and consider a nitrogen atmosphere in the vicinity of the surface. The escape speed is

$$v_{esc} = \sqrt{\frac{2 \times 6.67 \times 10^{-11} \times 7.35 \times 10^{22}}{1.74 \times 10^6}} \, \mathrm{m \, s^{-1}} = 2.37 \, \mathrm{km \, s^{-1}}.$$

and

$$v_p = \sqrt{\frac{2 \times 1.38 \times 10^{-23} \times 250}{28 \times 1.67 \times 10^{-27}}} \, \text{m s}^{-1} = 384 \, \text{m s}^{-1}.$$

The value of u_{esc} is 6.2 so, from Table 22.1, a proportion of order 10^{-16} of the molecules will be able to escape at any time. This corresponds to a gradual loss of atmosphere. If instead of a nitrogen atmosphere the Moon had a hydrogen atmosphere then the value of u_{esc} would have been $\sqrt{14}$ times smaller, i.e. 1.66, and 10% or so of the molecules would be able to escape at any time, leading to an almost instantaneous loss of atmosphere.

It must be stressed that the actual physics of atmospheric loss is much more complicated than we have sketched here; this outline is just intended to explain the essential role of the Maxwell–Boltzmann distribution in the loss process.

Exercise 22.2

Find the escape speed from Titan, the largest satellite of Saturn, which has mass 1.36×10^{23} kg and radius 2575 km. It has an exosphere temperature of 100 K and its dense atmosphere consists of molecular nitrogen. Find the root-mean-square speed of the nitrogen molecules in the exosphere.

22.3 Nuclear fusion in stars

Another astronomical process in which the Maxwell–Boltzmann distribution plays an important role is that of energy production in stars. The first step of this process requires two protons to react to produce a positron, a neutrino and a deuterium nucleus. For this to happen the protons must approach to within a proton diameter, about 10^{-15} m. The temperature in the core of the Sun is 1.5×10^7 K so the mean kinetic energy of motion of a proton is

$$\bar{E} = \frac{3}{2}kT = 3.1 \times 10^{-16} \, \text{J}. \tag{22.11}$$

The potential energy of two protons separated by a distance r is

$$E_p = \frac{e^2}{4\pi\varepsilon_0 r} \tag{22.12}$$

which is 2.3×10^{-13} J for $r = 10^{-15}$ m. Even allowing for a spread of proton energies in the tail of the Maxwell distribution it is clear that according to the classical model proton–proton reactions cannot take place in the Sun.

In Figure 22.3 the potential energy is modelled as a function of the separation of two protons. For distances smaller than r_p, the critical distance for a reaction to occur, the potential drops sharply indicating attraction between the two particles.

We can see that the oncoming proton has a barrier to penetrate. The distance it must travel in the non-classical region is

$$d = \frac{e^2}{4\pi\varepsilon_0 E} - r_p \tag{22.13}$$

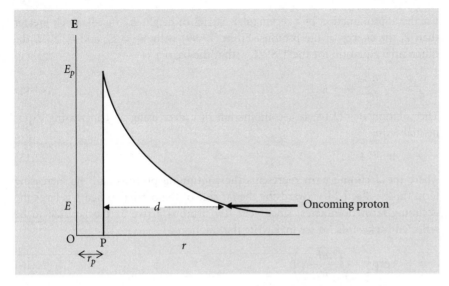

Figure 22.3 Representation of a proton of energy E approaching a Coulomb potential barrier of maximum height E_p.

which, for $E = 3\bar{E}$, equals 2.5×10^{-13} m and the barrier height is, as found previously, 2.3×10^{-13} J. Just to get an estimate of the transmission coefficient we roughly equate this to a rectangular barrier of width $a = 2 \times 10^{-13}$ m and height $V_0 = 1.5 \times 10^{-13}$ J. Allowing for the tail of the Maxwell distribution we take a proton energy $E = 3 \times 10^{-15}$ J – there should be some protons with about this energy. We now substitute these values in (21.101a). The stages in the calculation are:

$$k_1 = \sqrt{\frac{2mE}{\hbar^2}} = 3.00 \times 10^{13} \, \mathrm{m}^{-1}; \; k_3 = \sqrt{\frac{2m(V_0 - E)}{\hbar^2}}$$

$$= 2.10 \times 10^{14} \, \mathrm{m}^{-1}; \; \sinh k_3 a = 8.70 \times 10^{17}$$

which give $T = 1.0 \times 10^{-37}$. This value of the transmission coefficient depends on the rather crude approximation of the Coulomb barrier as a rectangular barrier, so its precise value has no particular significance. It is possible to do a more precise calculation; what we have shown here is just that protons can get together but that this is a difficult process – just one direct approach of protons in 10^{37} leads to them coming close enough to react – but even then some other conditions must be satisfied before a nuclear reaction can take place. The fact that tunnelling takes place enables the Sun, and other stars, to generate energy.

An interesting question to ask is, 'Which part of the Maxwell–Boltzmann distribution of proton energies is most effective in the production of nuclear energy?' The temptation is to say that it will be that part corresponding to the protons of the greatest energy that the Maxwell–Boltzmann distribution can provide – but this is not so. The Maxwell–Boltzmann distribution provides an abundance of protons at lower energies, but these are completely ineffective in penetrating the potential barrier. Those that penetrate the barrier most effectively are those of highest energy – but such energies are in the far tail of the distribution, and there is a vanishingly small number of them. We can consider this question in the following way. First we use the approximation of a rectangular barrier of height E_m that is much greater than E, the energy of the protons. From (21.99) with $V_0 = E_m$ and $E_m \gg E$ the differential equation for the TISWE within the barrier is

$$\frac{d^2\psi}{dx^2} = \frac{2mE_m}{\hbar^2}\psi = k^2\psi. \tag{22.14}$$

The solution of (22.14) is a combination of a decreasing and increasing exponential term:

$$\psi = Ae^{-kx} + Be^{kx} \tag{22.15}$$

where the declining term represents the continuing protons and the increasing term those reflected backwards by the far side of the barrier. When k is large the declining term dominates, since very few protons arrive at the far side to be reflected backwards, so we may write the solution as

$$\psi = A\exp\left(-\sqrt{\frac{2mE_m}{\hbar^2}}x\right). \tag{22.16}$$

Where there are two particles involved the mechanics of the interaction should be described in terms of the *reduced* mass of the two-particle system. For general particles

$$m_{\text{red}} = \frac{m_1 m_2}{m_1 + m_2}$$

and for two protons, each of mass μ_p, $m_{\text{red}} = \frac{1}{2}\mu_p$. This must be used for m in (22.16).

We may take the width of the barrier as d, given in (22.13). For a reasonable maximum energy of proton that would be expected in the Sun's interior, say with $E = 5\mu_p v_p^2 = 4.14 \times 10^{-15}$ J, the value of d is of order 5.6×10^{-14} m so there is little error in neglecting r_p in (22.13). Taking these factors into account the ratio of the amplitude of the wave function when it leaves the barrier to that when it enters is

$$\frac{\psi(d)}{\psi(0)} = \exp\left(-\sqrt{\frac{\mu_p E_m}{\hbar^2}\frac{e^2}{4\pi\varepsilon_0 E}}\right).$$

Since the density of particles is proportional to the square of the amplitude of the wave function we can write the probability of a proton penetrating the barrier (or the proportion of protons that do so) as a function of proton energy as

$$P_{\text{pen}}(E) = \exp\left(-2\sqrt{\frac{\mu_p E_m}{\hbar^2}\frac{e^2}{4\pi\varepsilon_0 E}}\right). \tag{22.17}$$

This indicates that the larger is the value of E the greater is the probability of penetrating the barrier. On the other hand the Maxwell–Boltzmann distribution falls off rapidly for large values of E. The Maxwell–Boltzmann distribution can be found in terms of E by making the substitution $E = \frac{1}{2}\mu v^2$ in (22.4) to give

$$P(E)\,dE = 2\pi\left(\frac{1}{\pi kT}\right)^{3/2} E^{1/2} \exp\left(-\frac{E}{kT}\right) dE. \tag{22.18}$$

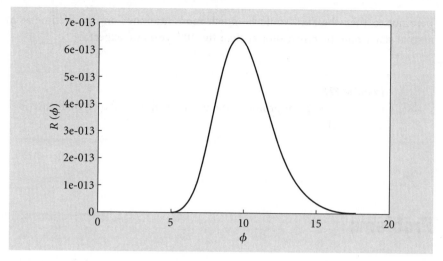

Figure 22.4 Fusion efficiency function.

The efficiency of the barrier penetration process, and hence fusion efficiency, as a function of energy will be of the form

$$R(E)dE = P_{\text{pen}}(E)P(E)\,dE$$

$$= 2\pi\left(\frac{1}{\pi kT}\right)^{3/2} E^{1/2}\exp\left\{-\left(2\sqrt{\frac{\mu_p E_m}{\hbar^2}\frac{e^2}{4\pi\varepsilon_0 E}}+\frac{E}{kT}\right)\right\}dE.$$

As a simplification we can use a non-dimensional form for energy,

$$\Phi = E \div \frac{3}{2}kT,$$

which gives energy as a fraction of the average energy. Finally, substituting for all the physical constants the fusion efficiency function $R(\Psi)$ appears as

$$R(\Phi)\,d\Phi = \left(\frac{27}{2\pi}\right)^{1/2}\exp\left\{-\left(\frac{138}{\Phi}+\frac{3}{2}\Phi\right)\right\}d\Phi. \tag{22.19}$$

The form of $R(\Psi)$ is shown in Figure 22.4 where it will be seen that the maximum, known as the **Gamow peak**, occurs at $\Psi = 9.6$ corresponding to $u = \sqrt{9.6}$, which is to say that the most effective protons are those with speeds somewhat more than three times that corresponding to the peak of the Maxwell–Boltzmann curve.

It must be stressed that this discussion of fusion in stars is far from complete. The rate at which close approaches are made is also important, and this is a function of the speed of the protons and hence their energy. Another factor is the cross-section for an interaction, which also depends on the energy of approach but is a slowly varying function of the energy. What we have done here is just to illustrate the role of the Maxwell–Boltzmann distribution which, together with the quantum-mechanical tunnelling phenomenon, gives the major characteristic of the fusion process as a function of energy.

It will be seen that the fusion process in stars is inefficient in the sense that the proportional rate at which protons are undergoing reactions is very low. If this were not so then the Sun would be much more luminous and its lifetime as a normal star would be much shorter than the 10^{10} years we expect.

 Exercise 22.3
For what energy of proton is the barrier width, as shown by Figure 22.3, equal to $10r_p$?

Problems

22.1 For a Maxwell–Boltzmann distribution, what is the standard deviation of the kinetic energy of a particle? You may assume that $\bar{E} = \frac{3}{2}kT$.

22.2 The ratio of the number of atoms of deuterium (D) to hydrogen (H) is 0.016 for Venus, some 100 times the ratio for the Earth, and 800 times that for the solar system in general. The likely mechanism for producing this ratio is that, early in the life of the planet, water in the atmosphere was dissociated by solar radiation to give

$$H_2O \rightarrow OH + H \text{ and } HDO \rightarrow OH + D$$

and that H was lost preferentially to D. The atomic masses of H and D are 1.67×10^{-27} kg and 3.34×10^{-27} kg respectively. The mass of Venus is 4.87×10^{24} kg, the radius of its exosphere is 6500 km, and the temperature in that region is 400 K. Calculate the ratio of the proportion of H atoms to the proportion of D atoms with greater than escape speed. (Note: You will need to use Figure 22.2 and result will only be approximate.)

22.3 When gas atoms in a container hit a target within it, they give rise to a chemical reaction. Taking all factors into account it is deduced that the probability that a particular gas atom will produce a reaction is proportional to $E^{3/2}$ where E is its energy of motion. For what energy of gas atoms, expressed in terms of the average energy, is the rate of chemical reactions per unit energy range the greatest?

The Monte Carlo method

23

The Monte Carlo method is a technique for solving physical problems that depends on the generation of random numbers to simulate stochastic events, i.e. events related to chance. An example of a system exhibiting such behaviour is a fluid where individual atoms or molecules move about in haphazard and unpredictable ways as they undergo mutual interactions. This chapter first illustrates the general principles of the Monte Carlo method by using a computer-generated set of random numbers, with a uniform distribution between 0 and 1, in the random walk problem, first solved analytically by Einstein. Then you are shown how, using a computer, numbers can be produced with any required distribution – e.g. a normal distribution with given mean and standard deviation or a Maxwell–Boltzmann distribution with a particular mean value. These ideas are then applied to two physical problems – the determination of a liquid equation of state and the operation of an enriched-uranium nuclear reactor.

23.1 Origin of the method

During World War II scientists in the USA were working on the development of the atomic bomb. One problem they faced was to determine how neutrons would penetrate a barrier – a problem that they needed to solve quickly but without any previous experience to guide them. A neutron interacting with an atomic nucleus in the barrier could be unaffected, absorbed, or scattered either elastically or inelastically. The probabilities of these possible interactions, which were dependent on the neutron energy, were known but there was no way of converting this knowledge into an analytical solution of the problem. The problem was eventually solved by two of the scientists working on the project, Stan Ulam and John von Neumann, by a novel numerical procedure. They could estimate how far a neutron would travel before encountering a nucleus, so they simulated the path of a neutron through a barrier. At each interaction they decided what would happen by the generation of a random number that would determine whether the atom was unaffected, absorbed, or scattered and, if scattered, in which direction. By following a large number of such simulated paths they were able to determine the probability that a neutron would penetrate the barrier. The generation of a random number is similar to what happens in various gambling activities – throwing a die or spinning a roulette wheel, for example – so a procedure of this kind is called a **Monte Carlo method**. With the advent of computers that enable large numbers of Monte Carlo trials to be made, the method has been used in many areas, both in science and in the social sciences.

23.2 **Random walk**

A very simple problem that will serve to illustrate the scope and power of Monte Carlo methods is that of the **random walk**, sometimes known as the **drunkard's walk**. Starting at the origin in a two-dimensional space n steps are taken, all of equal length but in random directions. How far is the final point from the origin? Because of the random nature of the process it is clear that different trials will give different outcomes, so a complete description of the solution of this problem will be to find the distribution of the final distances.

Actually this problem is capable of an analytic solution, and was originally solved by Albert Einstein, but let us consider how it may be solved numerically by simply following the path, as Ulam and von Neumann did. For each step a random number is generated that has a uniform probability density in the range 0 to 1. Such a number is said to be a **uniform variate in the range 0 to 1**, which is used so often in what follows that it is convenient to use the shorthand notation uv(0, 1). We take this random number multiplied by 2π as the angle, θ, made by the step with the x-axis. The distance moved along the x-axis by this step is $d\cos\theta$ and along the y-axis it is $d\sin\theta$, where each step is of length d. The final point after n steps has coordinates (X, Y) given by

$$X = \sum_{i=1}^{n} d\cos\theta_i \quad \text{and} \quad Y = \sum_{i=1}^{n} d\sin\theta_i. \tag{23.1}$$

The distance from the origin is then

$$s = \sqrt{X^2 + Y^2}. \tag{23.2}$$

This path simulation is very simply programmed, but requires a random number generator with the required distribution. Generating random numbers that satisfy stringent tests for randomness is not as straightforward as you might imagine. A truly random generator is used for the selection of winning numbers in the Premium Bond lottery run by the UK government. This uses a spark generator, with individual discharges being interpreted as binary bits. However, for scientific purposes it is usually better to use **pseudo-random numbers**, generated by some process that can be repeated if required. A very simple generator, which does not satisfy stringent tests but is adequate for many purposes, generates each random number from the previous one by the algorithm

$$r_{\text{next}} = (\pi + r_{\text{last}})^5, \text{ modulo 1}. \tag{23.3}$$

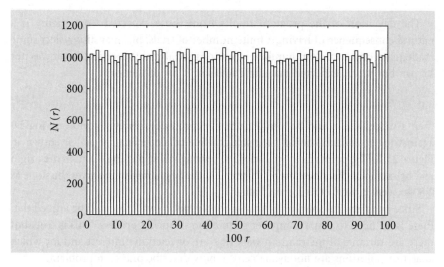

Figure 23.1 Histogram of 100 000 numbers produced by a simple generator.

The notation a, modulo b means the remainder after taking away from a as many entire quantities b as possible. In the context of (23.3) this means that one takes the fractional part of the expression shown. A slightly better generator is obtained by multiplying the outcome of (23.3) by 1000 and taking the resultant of this number modulo 1, but for both forms the outcome is much improved by using double-precision real numbers. To illustrate the result of using this process the result of generating 100 000 numbers is shown as a histogram in Figure 23.1. The unevenness of the histogram is inevitable and the range of variation in the heights of the histogram blocks indicates that the generator may be a reasonable one. There is still the possibility that it is unsatisfactory if there is some correlation between the values of successive numbers generated and further tests would be required to check this.

The detailed outcome of using this generator would depend on the computer that was used, depending on the number of bits representing floating-point numbers. For many purposes it is desirable to have *portable* generators that will give the same result with any computer giving integer numbers represented by some minimum number of bits. A full discussion of the problems in devising such generators is given in texts on numerical computing such as the series by Press *et al.* (see <*www.numerical-recipes.com*>). Some computer languages (e.g. C) have random number generators as part of a standard library of programs.

If instead of allowing steps in any direction the steps are restricted to be just along directions $\pm x$ or $\pm y$ then theory shows that the average distance travelled after n steps is the same as for steps taken in random directions. Table 23.1 shows the result of running a program DRUNKARD, outlined in Appendix 12, that restricts the steps in this way. For each value of n there are 10^6 trials, with unit-length steps, and the table gives the r.m.s. distance from the origin, D, that theoretically equals $\sqrt{n}$.

The deviation of the program results from those expected from theory is a natural consequence of having a finite number of trials, but may also reflect some inadequacy of the random number generator used. If the theoretical result was not known but it was expected that there was some law of the form

$$D = n^{\alpha} \tag{23.4}$$

then a value of α could be found from the computed results by plotting $\log(D)$ against $\log(n)$, which should be a straight line of slope α. The plot is shown in Figure 23.2. The points clearly fall close to a straight line passing through the origin and by methods described later (Chapter 27) the best straight line gives the slope as 0.5006, very close to the correct value.

Since an analytical solution is known for a random walk, it could be argued that there is no need to run a computer simulation of such a process. This is true, but there are variants of the random walk that are of scientific interest and for which analytical solutions are not available. We now describe one such problem.

Table 23.1 Results from program DRUNKARD

n	D (DRUNKARD)	D (theory)
1	1.000 0	1
4	2.027 3	2
16	3.916 4	4
64	7.948 5	8
256	16.149 3	16
1024	32.389 0	32

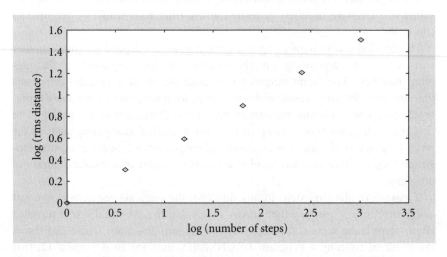

Figure 23.2 Computed results for a random walk.

Exercise 23.1
Table 23.2 lists 100 numbers from a random number generator that gives a distribution uv(0, 1). For these numbers find $\bar{x}, \overline{x^2},$ and $\overline{x^3}$ and check them against the theoretical averages.

23.3 **A simple polymer model**

A polymer chain is a string of identical molecular entities strung together in such a way that the linkages between neighbouring entities are completely flexible. In practice complete flexibility cannot exist, for if the linkage angle between two neighbouring units is very small – thinking of the units as long sausage-like objects – then the atoms of the two units would come too close together. This is part of a more general restriction that a polymer chain cannot cross itself – a restriction that did not apply to the random walk problem. Because of this restriction the distances between the ends of polymer chains with n units have an r.m.s. value greater than $\sqrt{nd}$ where d is the length of each polymer unit.

In the computational approach to the two-dimensional random walk problem each step was restricted to be in one of four directions, along $+x, -x, +y$ or $-y$, and the result is the same as for motions in general directions. A program POLYMER, outlined in Appendix 13, finds the r.m.s. distance from the origin for n steps, again using restricted steps along principal directions, with the constraint that the polymer chain is not allowed to cross itself. The way that this is done is to make a large number of trials but only to record the distances for non-intersecting paths; if during a trial a step leads to an intersection then there are three more attempts to make a non-interacting step. If all four steps lead to an intersection then the trial is immediately abandoned. The r.m.s. distance is calculated for the successful trials. The results are shown in Table 23.3, where polymer chains with up to 64 units have been considered and where the number of trials used was 10 000 and 100 000. For 64 steps the proportion of successful trials is low, about 4%, but a

Table 23.2 100 random numbers

0.086	0.005	0.906	0.236	0.487	0.762	0.244	0.144	0.695	0.770
0.420	0.423	0.934	0.525	0.883	0.817	0.279	0.934	0.183	0.755
0.850	0.966	0.795	0.239	0.765	0.474	0.350	0.858	0.116	0.576
0.165	0.163	0.129	0.618	0.147	0.115	0.061	0.705	0.116	0.032
0.068	0.319	0.970	0.041	0.684	0.584	0.544	0.059	0.162	0.137
0.092	0.579	0.032	0.275	0.474	0.923	0.876	0.287	0.089	0.372
0.701	0.539	0.028	0.663	0.702	0.101	0.453	0.894	0.842	0.299
0.273	0.220	0.772	0.046	0.664	0.271	0.934	0.521	0.765	0.842
0.598	0.063	0.768	0.744	0.875	0.626	0.135	0.466	0.916	0.046
0.663	0.167	0.864	0.818	0.759	0.848	0.547	0.924	0.377	0.413

Table 23.3 Results from the two-dimensional POLYMER program

n	Successful trials	D	Successful trials	D
1	10 000	1.00	100 000	1.00
2	10 000	1.63	100 000	1.63
4	9 781	2.60	97 920	2.61
8	8 878	4.10	88 071	4.12
16	6 349	6.55	63 259	6.54
32	2 844	10.47	27 641	10.48
64	452	17.08	4 165	17.17

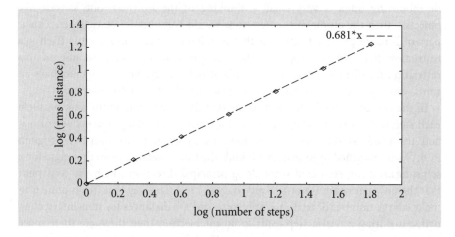

Figure 23.3 Results from the program POLYMER for 100 000 trials.

comparison of the results with 10 000 trials and 100 000 trials shows that the results are probably reliable.

Again if one looks for a relationship between D and n of the form (23.4) then a plot of $\log(D)$ against $\log(n)$ should be a straight line with slope α. This plot is given in Figure 23.3 for the 100 000 trial results together with the best straight line with a slope of 0.681, giving a relationship

$$D = n^{0.681}. \tag{23.5}$$

This run of POLYMER assumed equal probabilities for the three possible steps at each stage – in the direction of the previous step, turn left or turn right. The program has provision for changing these probabilities and in Problem 23.2 at the end of this chapter the effect of making the straight-ahead step twice as likely as each of the other two possible steps is investigated.

The analysis could have been made more realistic as a simulation of polymer conformation by a similar calculation in three dimensions. Since intersections are less common in three dimensions the value of α would be closer to the unconstrained value of 0.5. It is unlikely that this result relating to polymer conformation could be found other than by the Monte Carlo method.

Exercise 23.2
Use sets of four consecutive numbers from the table of random numbers given in Exercise 23.1 to make 25 five-step polymer trials. For random numbers 0–0.333 carry straight on, for 0.334–0.667 turn left, and for larger numbers turn right. Find the r.m.s. distance from starting for the successful trials and compare your answer with (23.5).

23.4 Uniform distribution within a sphere and random directions

A direction in space may be defined in two ways related to a set of rectangular Cartesian axes. One way is by its direction cosines (l, m, n) where l, m, and n are the cosines of the angles made by the direction with the x-, y-, and z-axes respectively (Figure 23.4). The other way is by the angles (θ, ϕ) from the spherical polar coordinate system. The two definitions are linked by

$$
\begin{aligned}
l &= \sin\theta\cos\phi \\
m &= \sin\theta\sin\phi \\
n &= \cos\theta
\end{aligned}
\tag{23.6a}
$$

which may be inverted to give

$$
\begin{aligned}
\tan\phi &= \frac{m}{l} \\
\tan\theta &= \frac{\sqrt{l^2 + m^2}}{n}.
\end{aligned}
\tag{23.6b}
$$

The ambiguity of phase in using arctan to find the actual angles is resolved by taking the signs of the denominator and divisor as those of sine and cosine respectively. Here we shall first describe the generation of random directions that, at the same time, also gives a uniform distribution of points within a sphere.

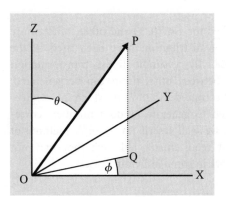

Figure 23.4 The direction is defined by OP. The point Q is the projection of P on the x-y plane.

If we generate three uv(0, 1), r_1, r_2, and r_3, and take

$$x = 2r_1 - 1$$
$$y = 2r_2 - 1 \tag{23.7}$$
$$z = 2r_3 - 1$$

then the points (x, y, z) so generated give a uniform distribution within a cube defined by x, y, and z all between -1 and $+1$. However, if the quantity

$$s = \sqrt{x^2 + y^2 + z^2}$$

is calculated and all (x, y, z) for which $s > 1$ are discarded then the residual points give a uniform distribution within a sphere of unit radius centred on the origin. The directions from the origin to those points then give a uniform distribution of directions in space. Each individual (x, y, z) gives either direction cosines

$$l = x/s, \quad m = y/s, \quad n = z/s$$

or spherical polar coordinate angles

$$\theta = \cos^{-1}(z/s) \quad (0 \le \theta \le \pi) \quad \text{and} \quad \phi = \tan^{-1}(y/x) \tag{23.8}$$

where the ambiguity in taking arctan is resolved as previously described.

Later a more efficient procedure will be described for defining random directions that involves only two uniform variates.

 Exercise 23.3

Find r, the ratio of the volume of a sphere to that of the containing cube. Take sets of three consecutive random numbers from the table in Exercise 23.1 and find the sum of their squares. For the 33 sums of squares so obtained, what proportion are less than unity? Compare this proportion with the value of r.

23.5 Generation of random numbers for non-uniform deviates

A very simple generator has been described for a uv(0, 1) and other, much more sophisticated, generators giving the same distribution have been used in the programs DRUNKARD and POLYMER. By a simple linear transformation random numbers for uniform deviates between other limits can be generated. An application of this idea has already been used in (23.7) in which $2r - 1$ gives uv(−1, 1). However, it is often required to generate random numbers corresponding to other distributions and here we shall describe two simple methods of general application that will meet most requirements although they may not necessarily be the most efficient generators in some individual cases.

23.5.1 Transformation method

The problem here is to generate a series of random numbers that will have a particular distribution defined by some function $p(x)$. Such a normalized function is represented in Figure 23.5. The area under the distribution between the minimum value, $x_{\min}$ and some value X is

$$f(X) = \int_{x_{\min}}^{X} p(x)\,dx. \tag{23.9}$$

If r, a uv(0, 1), is generated and then the value of X is found for which

$$f(X) = r \tag{23.10}$$

then the values of X will have the required distribution. The success of this algorithm depends on being able to integrate $p(x)$ and being able to solve (23.10) for X.

As an example we consider the normalized distribution $p(x) = 3x^2$ between $x = 0$ and $x = 1$. For this distribution

$$f(X) = X^3 = r \quad \text{so} \quad X = r^{1/3}.$$

One million random numbers r were generated by the simple generator (23.3) and the corresponding values of X were found. The distribution found is displayed as a histogram in Figure 23.6, normalized for comparison with the expected distribution $3x^2$. Since 100 histogram blocks are used the average number of entries per histogram block is 10^4. The width of the normalized distribution is unity so its average height is also unity. To show a normalized histogram it is thus necessary to divide the contents of the box by 10^4. The distribution is very close to the theoretical one.

The transformation method can be used to select a random direction in terms of spherical polar coordinate angles θ and ϕ. The angle ϕ has a uniform distribution in the range 0 to 2π and can be taken as $2\pi r_1$ where r_1 is a uv(0, 1). The probability that the direction makes an angle with the z axis between θ and $\theta + d\theta$ is the area on a unit sphere between those angles divided by the area of the sphere (Figure 23.7). This gives $p(\theta) = \frac{1}{2}\sin\theta$. Thus the angles giving the required distribution come from the values of Θ derived from

$$f(\Theta) = \frac{1}{2}\int_{0}^{\Theta} \sin\theta\,d\theta = \frac{1}{2}(1 - \cos\Theta) = r_2$$

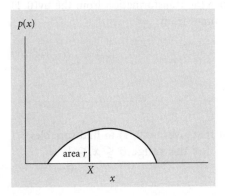

Figure 23.5 Normalized distribution function.

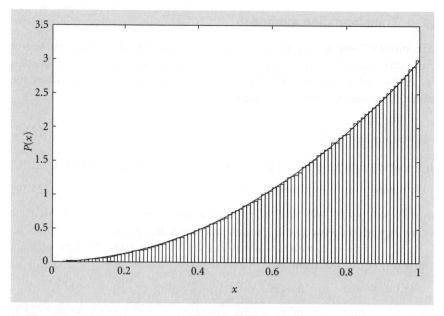

Figure 23.6 The distribution of random numbers for $p(x) = 3x^2$ derived from the transformation method displayed as a histogram. The line $3x^2$ is also shown.

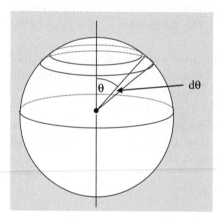

Figure 23.7 The area on the surface of a sphere between the angles θ and $\theta + d\theta$ is $2\pi \sin\theta\, d\theta$.

where r_2 is derived from a uv(0, 1). Hence the transformation from the uv(0, 1) that gives the required distribution of θ comes from

$$\theta = \cos^{-1}(1 - 2r_2). \tag{23.11}$$

Exercise 23.4
Describe how to use a generator that gives uv(0, 1) to find variables with a distribution $P(x) = e^x$ for the range $0 \le x \le 1$.

23.5.2 **Rejection method**

In Figure 23.8 the distribution $p(x)$, restricted between values $x = a$ and $x = b$, is the one for which we require to generate random numbers. Superimposed there is a uniform distribution $u(x)$, between the same limits, which is scaled to have a value of p_m, the maximum value of $p(x)$. The first step of the rejection method is to find a random variable, X, for a uv(a, b) followed by another value, r, from a uv$(0, 1)$. Pairs of values, plotted as $(X, p_m r)$ in Figure 23.8 would uniformly occupy the rectangle defined by $u(x)$. What we wish to do now is to reject those selections which do not fall under the curve $p(x)$ for then what is left will uniformly occupy the area under $p(x)$ and will give the required distribution. This can be done by only retaining the value of X if $p_m r \leq p(X)$, otherwise rejecting it. In this way random numbers are found with the required distribution. However, if the area under the curve is only a small fraction of the total area in the rectangle then the process can be inefficient, in the sense that the rejection rate is high. We shall now explain how, in some circumstances, higher efficiency is possible.

We now consider the situation where we have another normalized distribution, $u(x)$, called a **comparator function** that is similar in shape to $p(x)$ and can be scaled so that $Cu(x) \geq p(x)$ for all x between a and b. This situation is shown in Figure 23.9 in which the scaling constant, C, is adjusted to give coincident peaks. If $u(x)$ is a distribution for which the **transformation method** is appropriate then for each X generated for $u(x)$ the generation of another random number, r, from a uv$(0, 1)$ enables the rejection method to be applied to find variables with distribution $p(x)$.

The plotted number pairs $(X, Cru(X))$ will give a uniform distribution of points under the curve $Cu(x)$ in Figure 23.9. However if all values of X are rejected for which $Cru(X) > p(X)$ then a uniform distribution of points below $p(x)$ will remain and the associated values of X will give the required distribution. Because of the

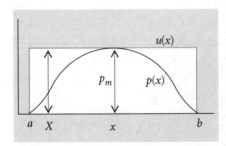

Figure 23.8 The general distribution $p(x)$ and the uniform distribution $u(x)$.

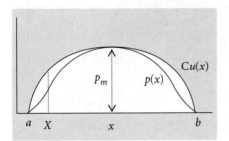

Figure 23.9 The distribution $p(x)$ and the scaled similar distribution $Cu(x)$.

similarity between $p(x)$ and $Cu(x)$ the rejection rate will be low and the process correspondingly more efficient.

An example of the application of the rejection method is in deriving random numbers corresponding to the normalized distribution $p(x) = (2/\pi)/\sin^2 x$ in the range 0 to π. Although this function can be integrated it is not possible to invert (23.10) to obtain X in terms of r. A similar distribution, which can be dealt with by the transformation method, is $u(x) = \frac{1}{2}\sin x$. The transformation relationship, corresponding to (23.10) is that given in (23.11), which gives

$$X = \cos^{-1}(1 - 2r). \tag{23.12}$$

The constant, C, to give the required scaling is $4/\pi$ and the rejection criterion for X is

$$\frac{4}{\pi} \times r \times \frac{1}{2}\sin X > \frac{2}{\pi}\sin^2 X$$

or

$$r > \sin X. \tag{23.13}$$

In a test of the rejection method using these functions 10^6 values of r were found and transformed into values of X by (23.12). Those satisfying (23.13) were rejected and the remaining 784 429 values of X were allocated to 100 histogram boxes in the range 0 to π. The mean value of $p(x)$ over the range is $1/\pi$ (there is unit area under the curve and the width of the function is π) and the histogram contents were scaled to give this mean. The final result is shown in Figure 23.10.

The result is very close to that expected. It should also be noted that the program gave 784 429 accepted values of X, very close to $10^6 \times \pi/4 = 785\,398$, the theoretical value.

The efficiency of a process of generating a distribution may not be of great importance if the generation procedure is part of a much more time-consuming

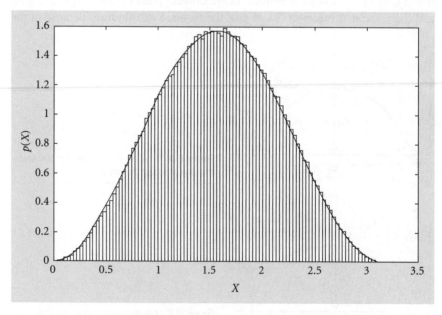

Figure 23.10 Histogram for $2\sin^2 X/\pi$ from the rejection method. The full line is the theoretical distribution.

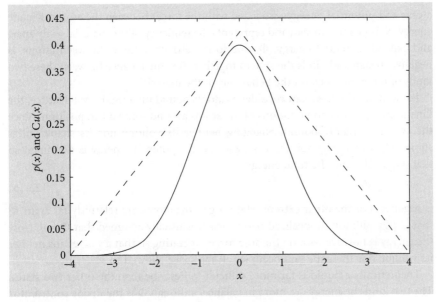

Figure 23.11 The distributions $\exp(-x^2/2)/(2\pi)^{\frac{1}{2}}$ and $0.425(1-0.25|x|)$.

program. The procedures described here will serve for most purposes, although other more efficient generation procedures may be available. As an example, it is possible to generate random numbers corresponding to the normal distribution,

$$p(x) = \frac{1}{\sqrt{2\pi}}\exp\left(-\frac{x^2}{2}\right). \tag{23.14}$$

by use of the rejection method. It is reasonable to assume that $p(x)=0$ for $|x|>4$ so a possible scaled distribution, better than a uniform distribution, that can serve as a template for rejection is

$$Cu(x) = 0.45(1-0.25|x|). \tag{23.15}$$

The two distributions are show in Figure 23.11 and it can be seen that the rejection rate would be of the order of 50%.

There is actually a specific and efficient procedure for generating random numbers with a normal distribution known as the **Box–Müller algorithm** that would be about twice as fast as the rejection method as just described, but what has been given here would work well.

We now consider applications of the Monte Carlo method to two physical problems.

23.6 Equation of state of a liquid

Apart from the ionized state of matter known as a plasma, there are three basic states designated as solid, liquid, and gas. For simplicity, we shall refer to atomic material but the discussion will normally also apply to material consisting of molecular entities. In each of the three states there are two kinds of energy associated with the configuration of atoms – **kinetic energy** associated with the motions of individual atoms and **potential energy** associated with the forces between the atoms. Kinetic energy, K, is positive in sign and represents the tendency of the material to fly apart and expand. Potential energy, Φ, due to the attractive forces between atoms, is negative in sign and binds the system together. It is the balance between these two kinds of force that dictates the behaviour of the material.

For a solid, which we can consider as atoms bound on a rigid crystal lattice, the kinetic energy is due to vibrations of the atoms around some mean position. Since the system is rigidly bound, changing neither its volume nor its shape unless subjected to very large forces, it is clear that the potential energy is dominant so that $|\Phi| \gg |K|$ and the total energy

$$E = K + \Phi \qquad (23.16)$$

is negative. For the other extreme state, a gas, the atoms are relatively far apart so that $|K| \gg |\Phi|$; for an idealized perfect gas it is usually assumed that $\Phi = 0$. Now the energy is positive with no binding forces operating so that a gas retains neither its volume nor its shape and occupies all the space allowed to it.

Theoretically a liquid is far more difficult to describe than the other two states. The two kinds of energy are more in balance, although Φ is the major contributor since, although a fluid readily changes its shape, it is resistant to volume change. The atoms move easily relative to each other but in the environment of any atom at any instant there will be some rough regularity of structure, crudely like that of a solid. However, any semblance of order breaks down at larger distances.

23.6.1 Liquid equation of state

We consider a cubical container, with centre at the origin of a rectangular Cartesian system, of side L containing N atoms (or molecules in general) each of mass m. The ith atom is located at (x_i, y_i, z_i), has a velocity $\mathbf{c}_i$ with components (u_i, v_i, w_i), and experiences a force $\mathbf{F}_i$ with components (F_{xi}, F_{yi}, F_{zi}). The x component of the acceleration of the atom is given by

$$F_{x_i} = m\frac{\mathrm{d}^2 x_i}{\mathrm{d}t^2}, \qquad (23.17)$$

with similar expressions for the other components.

The relationship

$$\frac{\mathrm{d}}{\mathrm{d}t}\left(\frac{\mathrm{d}(x_i^2)}{\mathrm{d}t}\right) = \frac{\mathrm{d}}{\mathrm{d}t}\left(2x_i\frac{\mathrm{d}x_i}{\mathrm{d}t}\right) = 2\left(\frac{\mathrm{d}x_i}{\mathrm{d}t}\right)^2 + 2x_i\frac{\mathrm{d}^2 x_i}{\mathrm{d}t^2} = 2u_i^2 + 2x_i\frac{\mathrm{d}^2 x_i}{\mathrm{d}t^2}$$

together with (23.17) gives

$$F_{xi}x_i = \frac{1}{2}m\frac{\mathrm{d}}{\mathrm{d}t}\left(\frac{\mathrm{d}(x_i^2)}{\mathrm{d}t}\right) - mu_i^2. \qquad (23.18)$$

Adding the components for all three directions and summing over all atoms gives

$$\frac{1}{2}m\sum_{i=1}^{N}\frac{\mathrm{d}}{\mathrm{d}t}\left\{\frac{\mathrm{d}}{\mathrm{d}t}\left(x_i^2 + y_i^2 + z_i^2\right)\right\} = \sum_{i=1}^{N}\left(F_{xi}x_i + F_{yi}y_i + F_{zi}z_i\right) + \sum_{i=1}^{N}mc_i^2$$

$$= \sum_{i=1}^{N}\mathbf{F}_i \cdot \mathbf{r}_i + \sum_{i=1}^{N}mc_i^2, \qquad (23.19)$$

where $\mathbf{r}_i$ is the vector position of the ith atom and $\mathbf{F}_i \cdot \mathbf{r}_i$ is the scalar product of the two vector quantities.

Since the atoms are confined to the cubical box and give a uniform density the mean value of the left-hand side of (23.19) must be zero, although it may fluctuate a little. The final term on the right-hand side is just twice the total kinetic energy for the system of atoms. The average kinetic energy of an atom is $\frac{3}{2}kT$ (§22.1.1) so the final term has value $3NkT$.

The first term on the right-hand side of (23.19) is called the **virial of Clausius** and the force terms it contains are due to two causes – the forces of the atoms on each other and the forces on atoms due to the walls of the container.

We can find the component due to the container walls in the following way. Atoms close to the wall at $x=\frac{1}{2}L$ exert a force on the wall and the wall will exert an equal force on the atoms. The total force on the wall in the x direction is PL^2, where P is the pressure within the container, so this is the force of the wall on the atoms in the *negative* x direction. The component of the virial term due to this wall is thus $-PL^2 \times \frac{1}{2}L = -\frac{1}{2}PV$, where V is the volume of the container. The contribution of all six walls to the virial term is $-3PV$.

We now consider the contribution of the forces between the atoms to the virial term. If the force between two atoms separated by a vector $\mathbf{r}$ is $\mathbf{F}(\mathbf{r})$ then the total force on atom i due to all other atoms is

$$\mathbf{F}_{ai} = \sum_{\substack{j=1 \\ j\neq i}}^{N} \mathbf{F}(\mathbf{r}_{ij})$$

where $\mathbf{r}_{ij}$ is the vector between atom i and atom j and $\mathbf{F}(\mathbf{r}_{ij})$ is the force on atom i due to atom j. This gives the total contribution of all the inter-atomic forces as

$$C = \sum_{i=1}^{N} \sum_{\substack{j=1 \\ j\neq i}}^{N} \mathbf{F}(\mathbf{r}_{ij}) \cdot \mathbf{r}_i. \tag{23.20}$$

Contained in this summation are contributions from all pairs of atoms and the pair i and j contribute

$$\mathbf{F}(\mathbf{r}_{ij}) \cdot \mathbf{r}_i + \mathbf{F}(\mathbf{r}_{ji}) \cdot \mathbf{r}_j = \mathbf{F}(\mathbf{r}_{ij}) \cdot (\mathbf{r}_i - \mathbf{r}_j) = \mathbf{F}(\mathbf{r}_{ij}) \cdot \mathbf{r}_{ij} \tag{23.21}$$

using $\mathbf{F}(\mathbf{r}_{ji}) = -\mathbf{F}(\mathbf{r}_{ij})$. Thus the total contribution to the virial term from all pairs of atoms is

$$C = \sum_{\text{pairs}} \mathbf{F}(\mathbf{r}_{ij}) \cdot \mathbf{r}_{ij} = \sum_{\text{pairs}} F(r_{ij})r_{ij}, \tag{23.22}$$

since $\mathbf{F}(\mathbf{r}_{ij})$ and $\mathbf{r}_{ij}$ are parallel.

Inserting the derived values for the terms in (23.19) including the two contributions to the virial summation we find the liquid equation of state

$$PV = NkT + \frac{1}{3}\sum_{\text{pairs}} F(r_{ij})r_{ij}. \tag{23.23}$$

Excluding the final term in (23.23), which means assuming that atoms do not exert forces on each other, gives the equation of state for a perfect gas. For a liquid the final term is significant since atoms are in close proximity. However, since the atoms are constantly on the move this term will constantly fluctuate in value, although only slightly, and an average of the summation is the appropriate value to use.

23.7 Simulation of a fluid by the Monte Carlo method

23.7.1 The cell model

One cubic millimetre of an average liquid will contain of the order of 10^{19} molecules, and clearly, there is no way of modelling a liquid with anything approaching this number of simulated particles. Another feature that is desirable in a simulation is that it should effectively be considering the liquid well away from boundaries, where its behaviour is modified by physical factors such as the forces due to the container or surface tension effects. These desirable features of a simulation can be obtained by the use of a cell model, illustrated in two dimensions in Figure 23.12.

In three dimensions the liquid is represented by an infinite array of cubical cells each containing the same arrangement of atoms. When an atom moves out of one wall of a cell it re-enters through the opposite wall so the total content of the cell remains constant. The periodicity of the structure will have negligible effect on the behaviour of the system as long as the effective range of the interatomic forces is less than the cell dimension. Forces in fluids tend to be short range and can be ignored beyond a certain distance both because they reduce in magnitude and also because, by symmetry, forces due to distant atoms tend to balance each other.

23.7.2 The form of the equation of state

We have noted that if the virial term in (23.23) were zero it would leave the equation of state of a perfect gas. A useful way of transforming (23.23) is to divide it by NkT throughout to give

$$\frac{PV}{NkT} = 1 + \frac{1}{3NkT} \sum_{\text{pairs}} \mathbf{F}(r_{ij}) r_{ij}, \tag{23.24}$$

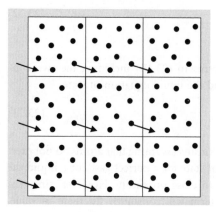

Figure 23.12 A two-dimensional cell model of a liquid.

and the extent to which the left-hand side exceeds unity indicates the degree of departure from the gaseous state.

The virial term depends on the separation of the atoms, and hence on the density of the liquid. It is convenient to express the density in a dimensionless form as

$$V^* = \frac{V}{N\sigma^3}. \tag{23.25}$$

in which a volume V contains N atoms and σ is a notional radius of the atom, defined later. A plot of PV/NkT against V^* gives a good representation of the equation of state of the liquid.

23.7.3 Generating liquid configurations – Metropolis algorithm

From statistical mechanics it is known that the probability of finding a system of interacting particles with total potential energy Φ is proportional to $\exp(-\Phi/kT)$. If all possible arrangements for the system were generated then the average of some quantity Q associated with the system would be

$$\langle Q \rangle = \frac{\sum Q \exp(-\Phi/kT)}{\sum \exp(-\Phi/kT)}. \tag{23.26}$$

In principle it should be possible to find a good estimate for $\langle Q \rangle$ by generating a large number of random configurations, which would be a good sample of all configurations, and evaluating the summations in (23.26) from those although, in practice, it would not work. Virtually all the arrangement would have pairs of particles close together giving a very large Φ for the arrangement and hence an almost zero probability.

The answer to this difficulty was found by Metropolis and his colleagues in 1953. They devised a procedure by which a sequence of configurations could be generated in such a way that the probability of generating a configuration with energy Φ is proportional to $\exp(-\Phi/kT)$. The procedure has the following steps:

1 Generate an initial configuration of N atoms, which can be a uniform distribution, and calculate the value of Φ.

2 Select a random atom. This can be done by generating r from a uv$(0, 1)$ and finding the integral part of $rN + 1$. This integer indicates one of the listed atoms.

3 Choose a random direction by the method described at the end of §23.5.1.

4 The atom is moved a distance δ in the chosen direction where $\delta = r\Delta$, r coming from a uv$(0, 1)$, and Δ is typically 0.05 of the average distance between atoms.

5 Calculate the new potential Φ' and the change of potential $\Delta\Phi = \Phi' - \Phi$.

6 If $\Delta\Phi$ is negative then the new configuration is accepted as one from which the value of $\langle Q \rangle$ will be calculated. Return to step 2 to generate the next configuration.

7 If $\Delta\Phi$ is positive then calculate $\exp(-\Delta\Phi/kT)$, which will be between 0 and 1. Then calculate r from a uv$(0, 1)$ and then
  (a) If $\exp(-\Delta\Phi/kT) \geq r$, the new configuration is accepted and contributes to $\langle Q \rangle$. Return to step 2.
  (b) If $\exp(-\Delta\Phi/kT) < r$, the new configuration is rejected and the original configuration makes an additional contribution to $\langle Q \rangle$. Return to step 2.

This algorithm efficiently generates a set of configurations with the desired distribution. In recent years this process, adapted in various ways, has been used in various areas of physics to solve problems where a number of probable configurations of parameters, which satisfy some optimization condition, are required to be generated. This process is then referred to as **simulated annealing**.

23.7.4 Lennard–Jones potential

The Metropolis generator is incorporated in the program METROPOLIS listed in Appendix 14 which is designed for a Lennard–Jones potential of the form

$$\Phi(r) = \frac{a}{r^{12}} - \frac{b}{r^6}. \tag{23.27}$$

The second term of (23.27) is attractive and has a physical basis. The first term is just a convenient way of representing a short-range repulsion when the atoms approach too closely. The form of the potential, and the corresponding force, as a function of distance are shown in Figure 23.13. The force is given by

$$F(r) = -\frac{d\Phi(r)}{dr} = \frac{12a}{r^{13}} - \frac{6b}{r^7}. \tag{23.28}$$

At distance r_c the force is zero, at distance σ the potential is zero and the depth of the potential well is ε. Equation (23.27) can also be expressed in the alternative form

$$\Phi(r) = 4\varepsilon \left\{ \left(\frac{\sigma}{r}\right)^{12} - \left(\frac{\sigma}{r}\right)^6 \right\}. \tag{23.29}$$

The constants σ and ε (in the form $T_0 = \varepsilon/k$) for some inert gases and molecular nitrogen are given in Table 23.4.

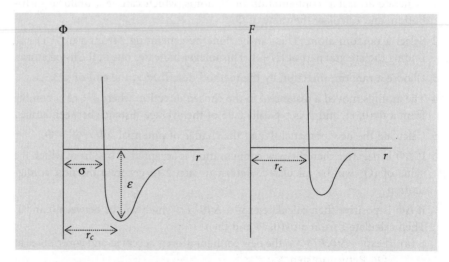

Figure 23.13 The potential and force relationships with distance for a Lennard–Jones potential.

Table 23.4 Lennard–Jones constants for nitrogen and some inert gases

Constant	Nitrogen	Neon	Argon	Krypton	Xenon
σ (nm)	0.370	0.275	0.345	0.360	0.410
$T_0 = \varepsilon/k$ (K)	95	36	120	171	221

Exercise 23.5

Show that the expression for the force between two molecules with a Lennard–Jones potential can be written as

$$F(r) = 4\varepsilon \left(\frac{12\sigma^{12}}{r^{13}} - \frac{6\sigma^6}{r^7} \right).$$

Exercise 23.6

Two argon atoms in a liquid are separated by 0.30 nm. What is their contribution to the virial term?

23.7.5 Simulation results

The METROPOLIS program contains the Lennard–Jones constants for argon and a temperature of 329 K, but these may be changed by the user. The initial configuration of atoms in the cell is on a regular lattice, and 50 000 steps of the Metropolis algorithm are completed before contributions to the virial summation or any other information is collected. This then starts with a configuration of non-negligible probability. Thereafter another 50 000 steps are made, every tenth one of which contribute to the final results. The user input is the value of V^* as defined in (23.25). The output is the value of PV/NkT and also the **radial distribution function**, which is the mean density as a function of the distance from one of the atoms. Although the atoms (or molecules) of a liquid move freely relative to each other, if the configuration in the immediate vicinity of a particular atom is noted at some instant it will be found to resemble the regular arrangement around an atom in a crystal, although somewhat distorted. This shows up by the fact that the distribution of the distances between the central atom and its neighbours shows peaks. For a solid lattice at rest these peaks would be a set of spikes. This regularity gets less obvious the further one goes from the central atom but noticeably persists out to a distance of two or three average atom separations. A typical radial distribution function as found for argon at 329 K with $V^* = 1$ by the METROPOLIS program is shown in Figure 23.14.

Equation of state points found by running the METROPOLIS program for liquid argon (parameters given in the program) are shown in Figure 23.15 compared with the experimental curve. The points calculated from the simulation are all above the curve but follow the general trend of the curve quite well. The theory does have some deficiencies that may explain at least part of the difference

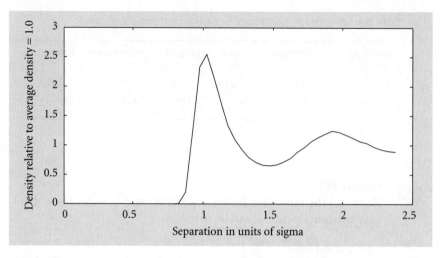

Figure 23.14 A radial distribution function from the program METROPOLIS.

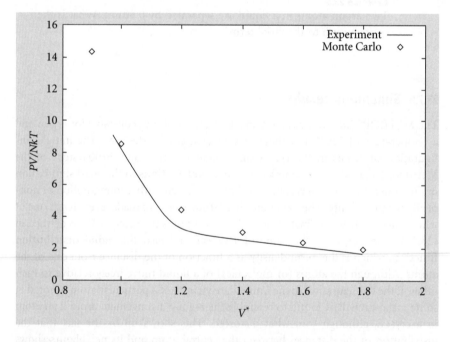

Figure 23.15 A comparison of the results from METROPOLIS with experimental results for argon.

between simulation and experiment. It is assumed that the force between two atoms is independent of the presence of other atoms nearby and this is not true. The attractive component of the Lennard–Jones potential is due to the action of van der Waals dispersion forces. These come from dipole–dipole interactions between atoms that are in turn due to fluctuating dipoles caused by oscillations of the electron clouds surrounding the atomic nuclei. This oscillatory behaviour is affected by the presence of other atoms nearby so the actual potential due to the separation of two atoms deviates from a Lennard–Jones form. More detailed liquid theory takes account of these many-body effects.

23.8 **Modelling a nuclear reactor**

Natural uranium contains two isotopes – ^{238}U, the major component and ^{235}U, which accounts for 0.7194% of the material. When a slow (thermal) neutron, with energy about 0.025 eV, is absorbed by ^{235}U then fission of the nucleus may occur into two major fragments with the production of energy and the release of fast neutrons. The average number of neutrons produced per fission is 2.47 and they are produced with a distribution of energies with a mean of 2 MeV. In order that these fast neutrons can interact with further ^{235}U atoms they must be slowed down by a moderator, a material consisting of light atoms – such as graphite (carbon) or heavy water (containing deuterium) – that is also a poor absorber of neutrons.

In parallel with the moderating process that reduces the neutron energies there is a loss of neutrons, predominantly by absorption in ^{238}U but also in the walls of the containing vessel of the reactor. In order that the reactor should be self-sustaining, on average at least one of the fission-produced neutrons must itself produce fission. The average number of fission producing neutrons that are produced per fission is called the **multiplication factor**. It must be at least unity for sustained energy production but it cannot continuously be greater than unity otherwise a runaway production of energy will take place, leading to an explosion. To prevent this there are feedback-controlled rods, consisting of boron – a heavy absorber of neutrons – which are inserted and withdrawn as required.

If natural uranium is used in a reactor it is formed into slender rods separated by moderator material. This gives a greater chance of neutrons being slowed down before they encounter further atoms of ^{235}U. On the other hand, if the uranium is enriched in ^{235}U then it is possible to have an intimate mixture of uranium and moderator and still have a multiplication factor greater than unity. Here we shall consider a simulation of an enriched fuel reactor, contained in a spherical vessel, making a number of simplifying approximations.

23.8.1 **Interaction cross-sections**

Associated with each type of reaction between a neutron and some other particle there is a **cross-section** which is defined as the effective area in the path of the neutron for the interaction to take place. For nuclear processes cross-sections are normally expressed in terms of the unit **barn** where 1 barn $= 10^{-28} m^2$. If there are n particles per unit volume and the cross-section for a particular type of interaction is σ then the neutron sweeps out a volume $\sigma\, dx$ for possible interactions in moving a distance dx. The number of particles in this volume is, on average, $n\sigma\, dx$ and hence the probability per unit path that an interaction will take place is $n\sigma$. If there are several components in the path of the neutron and several different interactions, say m, may take place then the average number of interactions per unit path, of one sort or another, is

$$\Sigma = \sum_{i=1}^{m} n_i \sigma_i. \qquad (23.30)$$

The **mean free path** of a neutron between collisions is the average distance it will travel between collisions, which is the distance travelled to give an average number of collisions equal to 1. This will be

$$\lambda = \frac{1}{\Sigma}. \qquad (23.31)$$

The probability that a neutron will have a mean free path between l and $l + dl$ is the combined probability that it first travels a distance l without making a collision followed by making a collision within a further distance of travel dl. The first of these probabilities is just that of making no collisions in a distance for which the expected number of collisions is Σl. From the Poisson distribution (§14.5), this is $\exp(-\Sigma l) = \exp(-l/\lambda)$. The second probability is $\Sigma\,dl = dl/\lambda$, so the combined probability of these mutually exclusive events, which is the probability of a mean free path between l and $l + dl$, is

$$p(l)dl = \frac{1}{\lambda}\exp\left(-\frac{l}{\lambda}\right)dl. \qquad (23.32)$$

If the total cross-section has several components, as indicated in (23.30), then when an interaction does take place the probability that it will be of type i is given by the ratio $n_i\sigma_i/\Sigma$.

23.8.2 Scattering

The neutrons are elastically scattered by the various nuclei in the reactor and in the process their energies are changed. It might be thought that the greatest change of energy would be brought about by collision with a heavy nucleus, but this is not so. This can be illustrated by considering an elastic collision between a neutron of mass m, moving at speed V, and a stationary nucleus of mass M as illustrated in Figure 23.16.

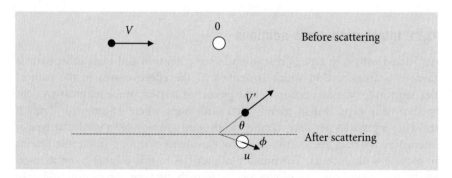

Figure 23.16 Before scattering the neutron has speed V along x and the nucleus is at rest. After scattering the speeds are V' and u in directions making angles θ and ϕ with the x-direction.

Conservation of momentum in the x-direction gives

$$mV = mV' \cos \theta + Mu \cos \phi. \tag{23.33a}$$

For conservation of momentum in the y-direction,

$$mV' \sin \theta = Mu \sin \phi. \tag{23.33b}$$

Eliminating ϕ from (23.33a) and (23.33b), using $\cos^2\phi + \sin^2\phi = 1$, we have

$$M^2 u^2 = m^2 (V^2 + V'^2 - 2VV' \cos \theta). \tag{23.34}$$

Since an elastic scattering process conserves energy,

$$mV^2 = mV'^2 + Mu^2 \tag{23.35}$$

and, eliminating u from (23.34) and (23.35) and writing $V'/V = \psi$, and $M/m = A$,

$$(1 + A)\psi^2 - 2\psi \cos \theta + (1 - A) = 0. \tag{23.36}$$

Equation (23.36) can be solved for ψ, and ψ^2 is the ratio of the neutron energy after the collision to that before. This energy ratio is

$$R_{energy} = \left\{ \frac{\cos \theta + \sqrt{A^2 - \sin^2 \theta}}{1 + A} \right\}^2. \tag{23.37}$$

In Figure 23.17 the value of R_{energy} is shown over the total range of scattering angles, 0 to π, for $A = 12$, corresponding to graphite (carbon) and $A = 238$, corresponding to the dominant uranium nucleus. When the high ratio of the number of moderator to uranium atoms (approximately 1000) is also taken into account it is clear that scattering by moderator atoms is the main agency for slowing

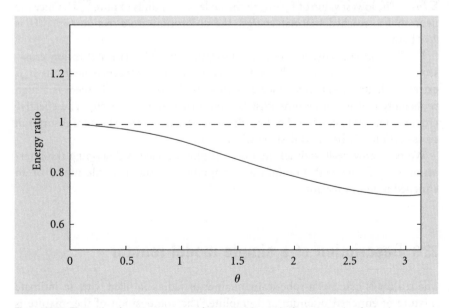

Figure 23.17 The ratio of neutron energy after scattering to before scattering for various scattering angles and for graphite ^{12}C (full line) and ^{238}U (broken line).

Table 23.5 The absorption cross-section, in barns, for ^{238}U as a function of Σ_s/n_{238}

Σ_s/n_{238}	8.3	50	100	300	500	800	1000	2000	∞
$\sigma_{a,238}$	0.601	1.066	1.506	2.493	3.096	3.754	4.138	5.400	15.936

down neutrons to thermal energies. The scattering cross-sections for graphite, ^{235}U, and ^{238}U are

$$\sigma_{s,m} = 4.8 \,\text{barns}, \quad \sigma_{s,235} = 10.0 \,\text{barns}, \quad \text{and} \quad \sigma_{s,238} = 8.3 \,\text{barns}.$$

23.8.3 Absorption and fission

Absorption by moderator atoms is very low, since moderators are chosen to have this property. The cross-section for absorption by graphite is 3.2×10^{-3} barns, but is nevertheless quite important since the amount of moderator in a reactor may exceed that of uranium by a factor of 1000, more or less. By contrast, in the slowing-down region where neutrons are going from fast to thermal, absorption by ^{238}U is quite strong but is complicated by the fact that it is a resonance phenomenon that strongly depends on the neutron energy. However, since at any time there is a whole range of neutron energies present in the reactor it is possible to produce an average value. In doing this it turns out that a quantity that has to be taken into account is the total cross-section for all scattering processes by all types of nucleus per ^{238}U nucleus, or Σ_s/n_{238} where

$$\Sigma_s = n_m \sigma_{s,m} + n_{235}\sigma_{s,235} + n_{238}\sigma_{s,238}.$$

Table 23.5 shows the dependence of the absorption cross-section of ^{238}U on Σ_s/n_{238}. The lowest value of Σ_s/n_{238} in the table corresponds to pure ^{238}U. Once the neutron has reached thermal energy, the absorption cross-section for ^{238}U is 2.73 barns.

Finally, and crucially, we come to absorption by ^{235}U. The absorption cross-section for ^{235}U is very small for fast neutrons but for thermal neutrons it is extremely high – 694 barns. Once a neutron is absorbed by ^{235}U there is a high probability that it will become unstable and that fission will occur. The effective cross-section for fission is 582 barns, which we designate as σ_f. The pure absorption cross-section, 112 barns, is designated as $\sigma_{a,235}$.

We have now dealt with all the essential physical factors describing the interaction of neutrons with uranium and graphite to create a simple model of an enriched uranium reactor.

23.9 Description of a simple model reactor

The reactor is taken as a spherical container of radius R, filled with an intimate mixture of enriched uranium and graphite. The composition of the mixture is defined by the uranium:graphite atomic ratio and the enrichment factor. If the enrichment factor is E then the percentage of uranium atoms that are ^{235}U is

$0.7194 \times E$. From the composition the absorption cross-section for ^{238}U is found by linear interpolation (§28.2) from Table 23.5. For high values of Σ_s/n_{238}, above 2000, a value for the absorption coefficient is found from 15.936 $- 21\ 072 \div (\Sigma_s/n_{238})$, which goes smoothly between the tabulated values of the cross-section for Σ_s/n_{238} between 2000 and ∞.

A detailed analysis shows that the neutron density in a working reactor is not uniform – it tends to rise moving out from the centre but then falls off near the container walls. The form of the neutron density is close to

$$n(r) = \frac{C}{r} \sin\left(\frac{\pi r}{R}\right), \tag{23.38}$$

where r is the distance from the centre and C is some constant. Since the volume between distance r and $r + dr$ from the centre is $4\pi r^2 dr$ the radial density of neutrons varies as

$$p(r)dr = Cr \sin\left(\frac{\pi r}{R}\right) dr \tag{23.39}$$

and the starting points of the fission-produced neutrons are taken with this distribution of distances from the centre. This is done by use of the rejection method (§23.5.2). The neutron then moves off in a random direction, determined as described in §23.5.1. It must now be decided how far the neutron moves before it interacts in some way. The total probability for some interaction or other per unit path is

$$\Sigma = n_m \sigma_{s,\,m} + n_{235}\sigma_{s,\,235} + n_{238}\sigma_{s,\,238} + n_{235}\sigma_{a,\,235} + n_{238}\sigma_{a,\,238} + n_{235}\sigma_f.$$

If the neutron is fast then the interaction cross-sections for pure absorption and fission for ^{235}U are made equal to zero; otherwise they take the values given in §23.8.3. The mean free path for some interaction or other is $\lambda = \Sigma^{-1}$. A random path length with distribution given by (23.32) may be selected using the transformation method (§23.5.1). A value of r, taken from a uv$(0, 1)$, is selected and then the path length is taken as $l = -\lambda \ln(1 - r)$, or, since $1 - r$ is also a uv$(0, 1)$, one may take $l = -\lambda \ln(r)$. If this move takes the neutron outside the containing vessel then it has been absorbed without fission and the next trial can begin. On the other hand, if it is still within the vessel then a choice has to be made of what kind of interaction it underwent.

For simplicity a fixed number of scattering events is taken to slow down a neutron from its first production to thermal energy. We may crudely estimate the mean reduction factor in energy per scattering event for carbon from (23.37) as

$$R_{\text{energy}} = \left(\frac{A}{1+A}\right)^2 = \frac{144}{169}.$$

The number, n, of such factors of reduction to reduce energy from 2 MeV down to 0.025 eV is found from

$$\frac{0.025}{2 \times 10^6} = \left(\frac{144}{169}\right)^n$$

or

$$n = \frac{\log(2 \times 10^6/0.025)}{\log(169/144)} = 114$$

and this is used in the simple model.

If one of two types of interaction, with probabilities p_1 and p_2, can take place then a random choice of which one occurs can be made by generating r from a uv(0, 1). If $r < p_1/(p_1 + p_2)$ then the first interaction is chosen, otherwise the second. A sequence of such processes can select a particular interaction for a neutron. The following outcomes are possible:

- If the neutron is purely absorbed (without fission) then the next Monte Carlo trial is initiated.

- If the neutron is scattered then a new random direction is chosen and the neutron is followed on its next path, as done previously. If the scattering is by a moderator atom then one is added to a counter that was initially set equal to zero. When the counter reaches 114 the neutron is deemed to have become thermalized and absorption and fission for ^{235}U then become activated. Once a neutron is thermalized it stays in that state for all further steps in the process.

- If fission takes place then 2.47 is added to a counter recording fission-generated neutrons and the next Monte Carlo trial is initiated.

After all the Monte Carlo trials have been made, the multiplication factor is found as the ratio of the number of generated neutrons to the number of trials. These ideas are all incorporated in the program REACTOR, outlined in Appendix 15. The multiplication factor as a function of enrichment is shown in Figure 23.18 for a spherical reactor of radius 6 m with a U/C ratio of 0.001 and is found from 10 000 trial neutrons. For the curves marked FORTRAN and C2 it will be seen that a multiplication factor of unity is achieved with an enrichment of about 2.7.

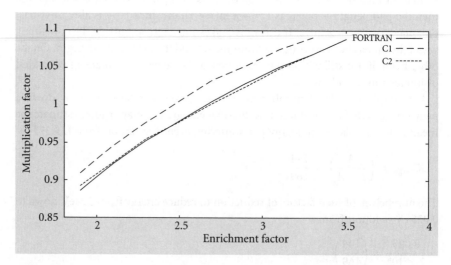

Figure 23.18 Variation of multiplication factor with enrichment for an enriched-uranium reactor.

Much more complex models of nuclear reactors have been designed, including control-rod simulation, and testing a model by computation is a necessary precursor to any construction of a nuclear reactor.

23.10 **A cautionary tale**

With any application of the Monte Carlo method there will be some fluctuation of results due to the stochastic nature of the process, and results may vary somewhat with different random-number generators. Figure 23.18 illustrates the need to be careful in the use of a random-number generator. In the C computer language standard library a random-number generator is provided which, when called in the form rand()%$n + 1$, gives an integer in the range 0 to n with uniform probability. Multiplying this integer by $1/n$ then gives uv(0, 1). The first use of this random-number generator in the REACTOR program, with $n = 10^8$, gave the results C1 in Figure 23.18, which differed markedly from the FORTRAN result, obtained by a FORTRAN subroutine. The FORTRAN and C random-number generators were then tested by generating 100 000 random numbers, r, and finding $\bar{r}, \overline{r^2}$, and $\overline{r^3}$ that, for uv(0, 1), have theoretical values $\frac{1}{2}, \frac{1}{3}$, and $\frac{1}{4}$. For the FORTRAN generator the values were 0.500 596, 0.334 164 and 0.250 882 while for the C generator they were 0.493 947, 0.327 634 and 0.245 264. It was concluded that the FORTRAN generator gave a better distribution. However, when the value of n for the C generator was changed to 10^6 the averages became 0.500 462, 0.333 698, and 0.250 267, much closer to the expected values. Changing the random-number generator in the C version of REACTOR then gave the curve C2 in Figure 23.18, which agrees well with the FORTRAN result. The message is to be wary of random-number generators and to test them if there are any doubts. The moments of the distribution were used here to test the generators but even if they passed that test they may still be faulty if, for example, consecutive numbers are correlated in some way.

Problems

23.1 A random walk, with steps of unit length, allows steps only in the x- and y-directions. A step may not return to the previous point but may go, with equal probabilities, in the other three possible directions. Modify program DRUNKARD to find the rms distance, d, from the starting point after 1, 4, 9, 16, 25, 36, and 49 steps. What is the best relationship of the form $D = n^\alpha$ that describes the results?

23.2 Modify the program POLYMER so that the probability of a forward step is 0.5 and that of steps to the left or right are 0.25. Find the rms distance, d, from the starting point after 1, 4, 9, 16, 25, 36, and 49 steps. What is the best relationship of the form $D = n^\alpha$ that describes the results?

23.3 Write a simple program to use the transformation method to generate 100 000 variates from the normalized distribution $p(x) = 2(1 + x)/3$ in the range $x = 0$ to 1. Show the results on a normalized histogram with 10 blocks together with $p(x)$.

23.4 Write a simple program, using the rejection method with a uniform variate as a comparator function, to generate variates for the normalized distribution $p(x) = 12x^2 (1 - x)$ in the range $x = 0$ to 1. Show the results on a normalized histogram with 10 blocks together with $p(x)$. Compare the proportion of rejected trials found by the program with the theoretical value.

23.5 Modify the parameters in the program METROPOLIS to simulate the behaviour of liquid nitrogen (Table 23.3) at 329 K. Run the program for the following values of V^*: 0.9 by steps of 0.05 to 1.3, then by steps of 0.1 to 1.8 then 2.0, 2.5, and 3.0. Plot the results on a graph.

23.6 (i) Using the REACTOR program find the multiplication factors for a reactor of 6 m radius for both U/C = 0.002 and U/C = 0.005 over a range of enrichment factors from 1.5 to 3.0 in steps of 0.25. Plot the results.

(ii) For a reactor radius of 6 m and an enrichment factor of 2.5 find the multiplication factor for U/C from 0.0015 to 0.004 in steps of 0.00025. Explain the form of the result you obtain.

(For all program runs you are recommended to use 100 000 trial neutrons.)

Matrices II

24

In Chapter 8 you were introduced to the idea of a matrix and how the simple operations of matrix algebra – addition, subtraction, multiplication, and finding an inverse matrix – could be used in various ways to solve problems in physics and mathematics. In this chapter you will become acquainted with some fundamental properties associated with square matrices, referred to as their eigenvalues and eigenvectors. These properties of matrices crop up repeatedly in the context of physics, in particular in the description of vibrating systems, described in this chapter, and in the realm of quantum mechanics, described in the next chapter.

24.1 Population studies

We are now going to describe an aspect of the mathematics of matrices that has great importance in all the physical sciences, not least in physics. However, a useful way of introducing these new ideas happens to be through a mathematical branch of the biological sciences – population studies. As an example we take an aspect of population studies of some commercial interest, the availability of fish stocks as a source of food.

We are accustomed to referring to **generations** in the human context and the same idea can be applied to any other animal species, including fish. For a particular species of fish we arbitrarily divide it into three 'generations' of equal duration that we label infant (I), adult (A), and old (O). If, for example, the maximum lifespan of the species is 9 years then each generation would correspond to a 3 year period.

During the period of one generation the population will change. Surviving infants become adults, surviving adults become old, and the old all die. During the same period there are new births replenishing the population, the parents coming from all three of the generation groups but, usually, mainly from the adults. We now define the following quantities:

- Three **fertility factors**, f_I, f_A, and f_O, which give the number of offspring produced by each group in a generation period as a fraction of the number in the group at the beginning of the period.
- Two **survival factors**, s_I and s_A, which give the number of survivors of the groups as a fraction of the number in the group at the beginning of the period. The survival factor s_O is zero so does not have to be defined.

We can now express the number in each group at the end of a generation period in terms of the numbers at the beginning.

$n'_I = f_I n_I + f_A n_A + f_O n_O$ (sum of infants produced by each group)

$n'_A = s_I n_I$ (surviving infants becoming adults)   (24.1)

$n'_O = s_A n_A$ (surviving adults becoming old).

This can be expressed in matrix form as

$$\begin{bmatrix} n'_I \\ n'_A \\ n'_O \end{bmatrix} = \begin{bmatrix} f_I & f_A & f_O \\ s_I & 0 & 0 \\ 0 & s_A & 0 \end{bmatrix} \begin{bmatrix} n_I \\ n_A \\ n_O \end{bmatrix},$$ (24.2)

where the matrix is the **Leslie matrix** and the vectors are the **population vectors** at the beginning and end of the generation period.

EXAMPLE 24.1

We now take a numerical example with $f_I = 0.2$, $f_A = 0.8$, $f_O = 0.3$, $s_I = 0.8$, and $s_A = 0.7$ with a starting population of one million in each group. After one generation

$$\begin{bmatrix} n'_I \\ n'_A \\ n'_O \end{bmatrix} = \begin{bmatrix} 0.2 & 0.8 & 0.3 \\ 0.8 & 0 & 0 \\ 0 & 0.7 & 0 \end{bmatrix} \begin{bmatrix} 10^6 \\ 10^6 \\ 10^6 \end{bmatrix} = \begin{bmatrix} 1\,300\,000 \\ 800\,000 \\ 700\,000 \end{bmatrix}$$

and after one further generation

$$\begin{bmatrix} n''_I \\ n''_A \\ n''_O \end{bmatrix} = \begin{bmatrix} 0.2 & 0.8 & 0.3 \\ 0.8 & 0 & 0 \\ 0 & 0.7 & 0 \end{bmatrix} \begin{bmatrix} 1\,300\,000 \\ 800\,000 \\ 700\,000 \end{bmatrix} = \begin{bmatrix} 1\,110\,000 \\ 1\,040\,000 \\ 560\,000 \end{bmatrix}.$$ (24.3)

Already the pattern is emerging that the population profile is changing with the infants becoming the dominant group. From (24.3) one can go forward generation by generation finding the way that the population profile changes, but it is a tedious exercise to do by hand. The program LESLIE, provided and described in outline form in Appendix 16, enables jumps of several generations to be made. It not only gives the current number in each generation group but also the ratio of the current number in the group to the number in the previous generation. The following gives the population profile at the end of the 12th generation period together with the ratios giving the change from the previous generation in brackets:

Infant : 1 229 710 (1.004 02)

Adult : 979 825 (1.004 15)

Old : 683 040 (1.003 80).

An interesting feature is the near constancy of the ratio for the different generation groups and going forward shows this more strikingly. At the end of the 22nd generation we have

Infant : 1 280 545 (1.004 06)

Adult : 1 020 296 (1.004 06)   (24.4)

Old : 711 320 (1.004 06).

It might be thought that the final ratio, 1.004 06, depends on the original population profile, but this is not so. Starting with $n_I = 1.5 \times 10^6$, $n_A = n_O = 7.5 \times 10^5$ gives after 22 generations

Infant : 1 377 278 (1.004 06)

Adult : 1 097 369 (1.004 06) (24.5)

Old : 765 054 (1.004 06).

Not only is the ratio of increase from one generation to the next the same, but so is the population profile in relative terms. The relative numbers in the three groups are as 1: 0.796 77:0.555 48 for both (24.4) and (24.5).

This leads to the conclusion that, whatever the starting point, after a sufficient number of generations the change from one generation to the next is described by

$$n' = Ln = \lambda n, \qquad (24.6)$$

where n' is the population vector after the generation period, n the population vector at the beginning, and L the Leslie matrix. In this case $\lambda = 1.004 06$ and there is a steady increase in the fish stock. The survival factors are most strongly affected by predation, of which the human contribution is an important component. Running the program LESLIE with the same fertility factors but with $s_I = 0.78$ and $s_A = 0.65$, corresponding to an increase in fishing activity, gives $\lambda = 0.987 69$, eventually leading to the extinction of the species.

In (24.6) λ is an **eigenvalue** of the matrix L and n is the corresponding **eigenvector**. It should be noted that the eigenvector is defined by the *relative* values of the elements, not the absolute values. If n is an eigenvector of L, satisfying (24.6) then so is cn where c is any constant.

Exercise 24.1
A certain species can be divided into two generations, young (Y) and old (O). The fertility and survival factors are given by $f_Y = 0.6$, $f_O = 0.4$, and $s_Y = 0.8$. Write down the Leslie matrix. Starting with 10 000 in each generation, calculate by hand the numbers after five generations. Estimate the eigenvalue of the Leslie matrix and the eigenvector giving the steady-state ratio of the two generations.

24.2 Eigenvalues and eigenvectors

We now treat the topic of eigenvalues and eigenvectors in more detail in relation to general matrices. For a general square matrix we write

$$Ax = \lambda x \qquad (24.7)$$

and we consider how to determine x, the eigenvector and λ, the eigenvalue. For an order-3 square matrix we have

$$\begin{bmatrix} a_{11} & a_{12} & a_{13} \\ a_{21} & a_{22} & a_{23} \\ a_{31} & a_{32} & a_{33} \end{bmatrix} \begin{bmatrix} \eta_1 \\ \eta_2 \\ \eta_3 \end{bmatrix} = \begin{bmatrix} \lambda\eta_1 \\ \lambda\eta_2 \\ \lambda\eta_3 \end{bmatrix} \qquad (24.8)$$

where x in (24.7) corresponds to $\{\eta_1, \eta_2, \eta_3\}$.

This is equivalent to the set of equations

$$\begin{aligned} (a_{11} - \lambda)\eta_1 + a_{12}\eta_2 + a_{13}\eta_3 &= 0 \\ a_{21}\eta_1 + (a_{22} - \lambda)\eta_2 + a_{23}\eta_3 &= 0 \\ a_{31}\eta_1 + a_{32}\eta_2 + (a_{33} - \lambda)\eta_3 &= 0. \end{aligned} \qquad (24.9)$$

Now we have already seen, at the end of §8.9, that there can only be a non-trivial solution to (24.9) if the matrix of coefficients is singular, or if

$$\begin{vmatrix} a_{11} - \lambda & a_{12} & a_{13} \\ a_{21} & a_{22} - \lambda & a_{23} \\ a_{31} & a_{32} & a_{33} - \lambda \end{vmatrix} = 0. \qquad (24.10)$$

An equation of type (24.10) is known as the **characteristic equation** of the matrix. It is a cubic equation from which three values of $\lambda - \lambda_1$, λ_2, and λ_3 – may be found. Substituting λ_1 in (24.9) gives three homogeneous equations with a singular matrix and any two of the equations gives the relative values of η_1, η_2, and η_3 that define the corresponding eigenvector.

In the fish example taken in §24.1, the process of repeatedly multiplying an initial vector by the Leslie matrix led to a single eigenvalue and its associated eigenvector and the question arises of whether another of the three eigenvalues and eigenvectors could have been found this way. The answer is that, in general, this process leads to the **principal eigenvalue**, the eigenvalue of greatest magnitude, and the corresponding **principal eigenvector**. Finding the complete set of eigenvalues and eigenvectors for a large matrix is complicated, and standard computer programs exist for this purpose.

We now illustrate the numerical determination of eigenvalues and eigenvectors for a simple order-3 square matrix

$$A = \begin{bmatrix} 2 & 3 & 1 \\ 3 & 1 & 2 \\ 1 & 2 & 3 \end{bmatrix}. \qquad (24.11)$$

The characteristic equation corresponding to (24.11) is

$$\begin{vmatrix} 2 - \lambda & 3 & 1 \\ 3 & 1 - \lambda & 2 \\ 1 & 2 & 3 - \lambda \end{vmatrix} = 0, \qquad (24.12)$$

which is equivalent to

$$\lambda^3 - 6\lambda^2 - 3\lambda + 18 = (\lambda^2 - 3)(\lambda - 6) = 0. \qquad (24.13)$$

This gives the three eigenvalues $\lambda_1 = 6$, $\lambda_2 = \sqrt{3}$, and $\lambda_3 = -\sqrt{3}$.

To find the eigenvectors we must substitute the values of the eigenvalues in the equations corresponding to (24.9) and use two of them to find the relative values of the elements of the eigenvector. For λ_1 the first two equations are

$$-4\eta_1 + 3\eta_2 + \eta_3 = 0 \quad \text{and} \quad 3\eta_1 - 5\eta_2 + 2\eta_3 = 0 \tag{24.14}$$

which give the solution $\eta_1 = \eta_2 = \eta_3$ corresponding to the eigenvector $x_1 = \{1, 1, 1\}$. For $\lambda_2 = \sqrt{3}$ the equations to be solved are

$$(2 - \sqrt{3})\eta_1 + 3\eta_2 + \eta_3 = 0 \quad \text{and} \quad 3\eta_1 + (1 - \sqrt{3})\eta_2 + 2\eta_3 = 0. \tag{24.15}$$

The solution of these equations give the eigenvector $x_2 = \{2, \sqrt{3} - 1, -\sqrt{3} - 1\}$. Similarly the final eigenvalue $\lambda_3 = -\sqrt{3}$ gives the eigenvector $x_3 = \{2, -\sqrt{3} - 1, \sqrt{3} - 1\}$.

We have already noted that eigenvectors only give relative values of the elements. Sometimes a **normalized eigenvector** is required for which the sum of the squares of the elements is unity. The normalized eigenvector for λ_1 is

$$\left\{ \frac{1}{\sqrt{3}}, \frac{1}{\sqrt{3}}, \frac{1}{\sqrt{3}} \right\}.$$

Exercise 24.2

Find the eigenvalues and the eigenvectors for the Leslie matrix defined in Exercise 24.1.

24.3 Diagonalization of a matrix

In some applications of matrices it is convenient to **diagonalize** a matrix. If the matrix to be diagonalized is A then this involves finding some matrix Q such that

$$Q^{-1}AQ = \Lambda \tag{24.16}$$

where Λ is a diagonal matrix.

We consider a matrix Q, the columns of which are the eigenvectors of A. For an order-3 square matrix with the jth element of the ith eigenvector indicated by η_{ij} we have

$$AQ = \begin{bmatrix} a_{11} & a_{12} & a_{13} \\ a_{21} & a_{22} & a_{23} \\ a_{31} & a_{32} & a_{33} \end{bmatrix} \begin{bmatrix} \eta_{11} & \eta_{21} & \eta_{31} \\ \eta_{12} & \eta_{22} & \eta_{32} \\ \eta_{13} & \eta_{23} & \eta_{33} \end{bmatrix}.$$

The rth column of the multiplied matrices is given by

$$\begin{bmatrix} a_{11} & a_{12} & a_{13} \\ a_{21} & a_{22} & a_{23} \\ a_{31} & a_{32} & a_{33} \end{bmatrix} \begin{bmatrix} \eta_{r1} \\ \eta_{r2} \\ \eta_{r3} \end{bmatrix} = \begin{bmatrix} \lambda_r \eta_{r1} \\ \lambda_r \eta_{r2} \\ \lambda_r \eta_{r3} \end{bmatrix} \quad \text{from (24.8).}$$

Hence

$$AQ = \begin{bmatrix} \lambda_1 \eta_{11} & \lambda_2 \eta_{21} & \lambda_3 \eta_{31} \\ \lambda_1 \eta_{12} & \lambda_2 \eta_{22} & \lambda_3 \eta_{32} \\ \lambda_1 \eta_{13} & \lambda_2 \eta_{23} & \lambda_3 \eta_{33} \end{bmatrix}. \tag{24.17}$$

From (8.13b) we can now write

$$AQ = \begin{bmatrix} \eta_{11} & \eta_{21} & \eta_{31} \\ \eta_{12} & \eta_{22} & \eta_{32} \\ \eta_{13} & \eta_{23} & \eta_{33} \end{bmatrix} \begin{bmatrix} \lambda_1 & 0 & 0 \\ 0 & \lambda_2 & 0 \\ 0 & 0 & \lambda_3 \end{bmatrix} = Q\Lambda \tag{24.18}$$

where Λ is a diagonal matrix with diagonal elements equal to the eigenvalues. Pre-multiplying both sides of (24.18) by Q^{-1} gives

$$Q^{-1}AQ = Q^{-1}Q\Lambda = I\Lambda = \Lambda. \tag{24.19}$$

Thus the diagonalizing matrix for A is Q and the diagonal matrix that results is Λ.

Exercise 24.3
Write down the diagonalizing matrix, Q, for the Leslie matrix, L, defined in Exercise 24.1. Find its inverse Q^{-1} (see §8.8).

Exercise 24.4
Using the result from Exercise 24.3 show that $Q^{-1}LQ$ gives the required diagonal matrix.

24.4 A vibrating system

We consider a system of three equal masses attached to a taut string, as shown in Figure 24.1.

The system can be imagined as resting on a smooth table so that gravitational forces are not involved. The masses are situated on the string such that the string is divided into four equal sections, each of length a, and the masses are displaced sideways to give very small displacements x_1, x_2, and x_3. The tension, T, in the string gives forces on the masses as shown for the left-hand mass. For very small displacements the restoring force will be dominantly in a direction perpendicular to the un-stretched string. Taking components of the force for small displacements we have, for the left-hand mass

$$m\ddot{x}_1 = -T\left\{\frac{x_1}{a} + \frac{x_1 - x_2}{a}\right\}$$

or

$$\ddot{x}_1 = -\frac{T}{ma}(2x_1 - x_2)$$

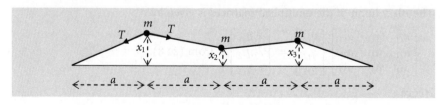

Figure 24.1 Three equal masses equally spaced on a taut string fixed at each end, slightly displaced from their equilibrium positions.

and similarly

$$\ddot{x}_2 = -\frac{T}{ma}(2x_2 - x_1 - x_3)$$

$$\ddot{x}_3 = -\frac{T}{ma}(2x_3 - x_2).$$

These equations can be put in the form

$$\begin{bmatrix} \ddot{x}_1 \\ \ddot{x}_2 \\ \ddot{x}_3 \end{bmatrix} = -\frac{T}{ma} \begin{bmatrix} 2 & -1 & 0 \\ -1 & 2 & -1 \\ 0 & -1 & 2 \end{bmatrix} \begin{bmatrix} x_1 \\ x_2 \\ x_3 \end{bmatrix} \qquad (24.20)$$

or, in symbolic form,

$$\ddot{x} = -\frac{T}{ma}Ax. \qquad (24.21)$$

We now change the vector variable x into a new vector variable X by the transformation

$$x = QX, \text{corresponding to } \ddot{x} = Q\ddot{X} \qquad (24.22)$$

where Q is the diagonalizing matrix for A. From (24.21) we have

$$Q\ddot{X} = -\frac{T}{ma}AQX$$

and pre-multiplying each side by Q^{-1} gives

$$Q^{-1}Q\ddot{X} = -\frac{T}{ma}Q^{-1}AQX.$$

Since $Q^{-1}Q = I$ and $Q^{-1}AQ = \Lambda$ this gives

$$\ddot{X} = -\frac{T}{ma}\Lambda X = -\frac{T}{ma} \begin{bmatrix} \lambda_1 & 0 & 0 \\ 0 & \lambda_2 & 0 \\ 0 & 0 & \lambda_3 \end{bmatrix} \begin{bmatrix} X_1 \\ X_2 \\ X_3 \end{bmatrix}$$

or, in expanded form,

$$\ddot{X}_1 = -\frac{T}{ma}\lambda_1 X_1,$$

$$\ddot{X}_2 = -\frac{T}{ma}\lambda_2 X_2, \qquad (24.23)$$

$$\ddot{X}_3 = -\frac{T}{ma}\lambda_3 X_3.$$

Since the variables are all separated in (24.23) the solutions, corresponding to simple harmonic motions, are found as

$$X_1 = c_1 \exp\left(i\sqrt{\frac{T\lambda_1}{ma}}t\right),$$

$$X_2 = c_2 \exp\left(i\sqrt{\frac{T\lambda_2}{ma}}t\right), \qquad (24.24)$$

$$X_3 = c_3 \exp\left(i\sqrt{\frac{T\lambda_3}{ma}}t\right),$$

where c_1, c_2, and c_3 are arbitrary constants. The complete solution in terms of the original variables is then found from (24.22), which requires knowledge of Q.

We now find the eigenvalues and eigenvectors of A, given in (24.20). The characteristic equation for determining the eigenvalues is

$$\begin{vmatrix} 2-\lambda & -1 & 0 \\ -1 & 2-\lambda & -1 \\ 0 & -1 & 2-\lambda \end{vmatrix} = 0$$

which expands to

$$\lambda^3 - 6\lambda^2 + 10\lambda - 4 = 0.$$

This can be factorized to give

$$(\lambda - 2)(\lambda^2 - 4\lambda + 2) = 0$$

the roots of which are $\lambda_1 = 2$, $\lambda_2 = 2 + \sqrt{2}$, and $\lambda_3 = 2 - \sqrt{2}$.

To find the eigenvectors we substitute the values of λ in the equations corresponding to (24.9). Substituting for λ_1 the equations are

$$-x_2 = 0, \quad -x_1 - x_3 = 0, \quad \text{and} -x_2 = 0.$$

The values of λ are those that give a consistent homogeneous set of equations and from the first two of them we find the non-normalized eigenvector $\{1, 0, -1\}$. From λ_2 we find

$$-\sqrt{2}x_1 - x_2 = 0,$$
$$-x_1 - \sqrt{2}x_2 - x_3 = 0,$$
$$-x_2 - \sqrt{2}x_3 = 0,$$

from the first and third of which the eigenvector is readily found as $\{1, -\sqrt{2}, 1\}$. Similarly, the third eigenvector is $\{1, \sqrt{2}, 1\}$. These eigenvectors are the columns of the normalizing matrix, so

$$Q = \begin{bmatrix} 1 & 1 & 1 \\ 0 & -\sqrt{2} & \sqrt{2} \\ -1 & 1 & 1 \end{bmatrix}. \tag{24.25}$$

Substituting for the eigenvalues in (24.24) we have

$$X_1 = c_1 \exp\left(i\sqrt{\frac{2T}{ma}}t\right),$$

$$X_2 = c_2 \exp\left(i\sqrt{\frac{(2+\sqrt{2})T}{ma}}t\right), \tag{24.26}$$

$$X_3 = c_3 \exp\left(i\sqrt{\frac{(2-\sqrt{2})T}{ma}}t\right).$$

To find the solution in terms of the original variables, x, we use

$$x = \begin{bmatrix} x_1 \\ x_2 \\ x_3 \end{bmatrix} = QX = \begin{bmatrix} 1 & 1 & 1 \\ 0 & -\sqrt{2} & \sqrt{2} \\ -1 & 1 & 1 \end{bmatrix} \begin{bmatrix} X_1 \\ X_2 \\ X_3 \end{bmatrix}.$$

This gives

$$x_1 = c_1 \exp\left(i\sqrt{\frac{2T}{ma}}t\right) + c_2 \exp\left(i\sqrt{\frac{(2+\sqrt{2})T}{ma}}t\right) + c_3 \exp\left(i\sqrt{\frac{(2-\sqrt{2})T}{ma}}t\right),$$

$$x_2 = -\sqrt{2}c_2 \exp\left(i\sqrt{\frac{(2+\sqrt{2})T}{ma}}t\right) + \sqrt{2}c_3 \exp\left(i\sqrt{\frac{(2-\sqrt{2})T}{ma}}t\right),$$

$$x_3 = -c_1 \exp\left(i\sqrt{\frac{2T}{ma}}t\right) + c_2 \exp\left(i\sqrt{\frac{(2+\sqrt{2})T}{ma}}t\right)$$

$$+ c_3 \exp\left(i\sqrt{\frac{(2-\sqrt{2})T}{ma}}t\right). \tag{24.27}$$

The motion represented by (24.27) with general values of c_1, c_2, and c_3 is complicated and non-periodic. However, with special values of c_1, c_2, and c_3 there are simple periodic motions. With $c_2 = c_3 = 0$ only the first terms appear in the equations of (24.27) and this corresponds to a motion with angular frequency $\sqrt{2T/ma}$ and with displacements of the three masses in the ratio $x_1 : x_2 : x_3 = 1 : 0 : -1$. This simple motion is called a **normal mode** of the system and this particular one is represented in Figure 24.2. The masses oscillate between the limits shown, with the centre one always stationary and with the other two oscillating π out of phase with each other.

The second normal mode is found by taking $c_1 = c_3 = 0$. The ratio of displacements is now $x_1 : x_2 : x_3 = 1 : -\sqrt{2} : 1$ and is shown in Figure 24.3. In this case the outer masses oscillate in phase while the central one is π out of phase with the others but with amplitude a factor $\sqrt{2}$ greater.

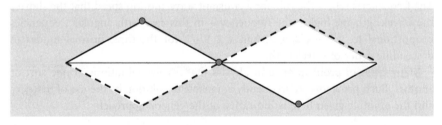

Figure 24.2 The normal mode with angular frequency $\sqrt{2T/ma}$.

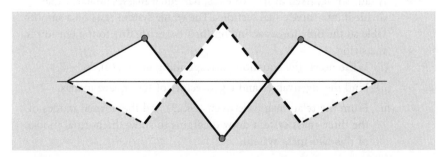

Figure 24.3 The normal mode for angular frequency $\sqrt{(2+\sqrt{2})T/ma}$.

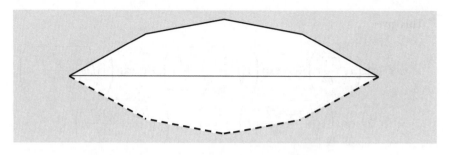

Figure 24.4 The normal mode for angular frequency $\sqrt{(2 - \sqrt{2})T/ma}$.

The final normal mode is found by taking $c_1 = c_2 = 0$. The ratio of displacements is now $x_1 : x_2 : x_3 = 1 : \sqrt{2} : 1$ and is shown in Figure 24.4. All the masses oscillate in phase but with the amplitude of vibration of the central mass $\sqrt{2}$ times as big as that of the other two.

Any general motion of the three masses is some linear combination of the normal modes, the admixture of normal modes being controlled by the coefficients c_1, c_2, and c_3. Another general point to notice is the relationship between the frequency of the vibration and the shape of the normal mode motions as seen in the figures. The profile corresponding to eigenvalue λ_3 has no nodes (i.e. points where the string is stationary) between the fixed ends and it looks something like half a complete wave with an effective wavelength twice the length of the string. Similarly the profile for λ_1 has one node at the centre and is like one complete wave with wavelength equal to the length of the string. Finally, the profile for λ_3 gives two nodes within the string, somewhat resembles three half waves (although they are not of equal length) and a 'wavelength' of two-thirds of the length of the string. The frequencies reflect the rule for a constant wave-motion speed that the shorter the wavelength the higher the frequency – in this case with angular frequencies proportional to $\sqrt{2 - \sqrt{2}}$, $\sqrt{2}$, and $\sqrt{2 + \sqrt{2}}$ for the three normal modes in descending order of wavelength.

Many types of problem arise in physics that involve vibrations of one sort or another. Such problems are frequently amenable to solution by the use of matrices and the example given here is indicative of the general approach.

Exercise 24.5

A taut string, fixed at its two ends, has equal masses attached that divide it into three equal sections. The whole system rests on a smooth table so the only forces acting on the masses are due to the tension in the string.

(i) Write down the equation corresponding to (24.20).

(ii) Find the eigenvalues and eigenvectors of the square matrix.

(iii) From the relationship between (24.25) and the normal modes of the three-mass system draw diagrams to show the normal modes of this two-mass system.

Problems

24.1 Find the eigenvalues and associated eigenvectors for the matrix

$$A = \begin{bmatrix} 1 & 1 & 1 \\ 0 & 2 & 1 \\ 0 & 0 & 3 \end{bmatrix}.$$

24.2 Deer are divided into five generations – infants (I), juniors (J), young adults (Y), adults (A), and elderly (E). The fertility factors for these generations are 0.0, 0.05, 0.35, 0.85, and 0.45 respectively. The survival rates for the first four generations are 0.9, 0.85, 0.8, and 0.7 respectively. Using LESLIE, find the steady-state population profile and determine the factor of increase or decrease per generation.

As a result of instituting a culling policy the survival rates for Y and A are changed to 0.75 and 0.65 respectively. How does this change the situation?

24.3 Determine the angular frequencies and form of the normal modes for a distribution of masses on a taut string, as shown in Figure 24.1, where the outer masses are $2m$ and the central mass m.

Quantum mechanics II: Angular momentum and spin

In §21.4 we introduced the idea that for every kind of observation there exists an operator, the eigenvalues of which represent the quantitative observations it is possible to make and the eigenfunctions of which represent the state function for the system once the observation has been made. The form of the operators can be found intuitively from classical expressions for the observed quantities, and in this chapter we extend this idea into the realm of the rotational quantity, angular momentum. The inclusion of angular momentum as a property of an atomic electron enables us to apply the Schrödinger wave equation to find the allowed states of the electronic structure of the hydrogen atom.

This chapter also deals with the intrinsic spin of an electron, as revealed by experiment. The eigenstates and eigenvalues associated with spin are most efficiently dealt with by matrix operators, and the properties of matrices dealt with in Chapter 24 are used here. The division of particles into two categories, fermions and bosons, dependent on whether the spin is half-integral or integral, is described. Finally, with all the state quantum numbers for an electron now defined, the electronic structure of many-electron atoms is discussed.

25.1 Measurement of angular momentum

25.1.1 Operators for angular momentum

The normal expression for angular momentum is

$$\mathbf{L} = m\mathbf{r} \times \mathbf{v} = \mathbf{r} \times \mathbf{p} \qquad (25.1)$$

which, from (9.12b), can be written as

$$\mathbf{L} = \begin{vmatrix} \mathbf{i} & \mathbf{j} & \mathbf{k} \\ x & y & z \\ p_x & p_y & p_z \end{vmatrix}. \qquad (25.2)$$

Thus the components of angular momentum are given by

$$l_x = yp_z - zp_y$$
$$l_y = zp_x - xp_z \tag{25.3}$$
$$l_z = xp_y - yp_x$$

and the square of the total angular momentum is given by

$$L^2 = l_x^2 + l_y^2 + l_z^2 = x^2(p_y^2 + p_z^2) + y^2(p_z^2 + p_x^2) + z^2(p_x^2 + p_y^2)$$
$$- 2(xyp_xp_y + yzp_yp_z + zxp_zp_x). \tag{25.4}$$

These can be converted into operators for measuring the quantities by writing $\frac{\hbar}{i}\frac{\partial}{\partial x} = \hat{p}_x$ (§21.10) and $x = \hat{x}$ etc., giving

$$\hat{l}_x = \hat{y}\hat{p}_z - \hat{z}\hat{p}_y = \frac{\hbar}{i}\left(y\frac{\partial}{\partial z} - z\frac{\partial}{\partial y}\right)$$

and similarly

$$\hat{l}_y = \frac{\hbar}{i}\left(z\frac{\partial}{\partial x} - x\frac{\partial}{\partial z}\right)$$

$$\hat{l}_z = \frac{\hbar}{i}\left(x\frac{\partial}{\partial y} - y\frac{\partial}{\partial x}\right) \tag{25.5}$$

and, substituting from (25.5) into (25.4),

$$\hat{L}^2 = -\hbar^2\left\{x^2\left(\frac{\partial^2}{\partial y^2} + \frac{\partial^2}{\partial z^2}\right) + y^2\left(\frac{\partial^2}{\partial z^2} + \frac{\partial^2}{\partial x^2}\right) + z^2\left(\frac{\partial^2}{\partial x^2} + \frac{\partial^2}{\partial y^2}\right)\right.$$
$$\left. - 2\left(xy\frac{\partial^2}{\partial x\partial y} + yz\frac{\partial^2}{\partial y\partial z} + zx\frac{\partial^2}{\partial z\partial x}\right)\right\} \tag{25.6}$$

To find the commutator of $\hat{l}_x$ and $\hat{l}_y$ we write

$$\left[\hat{l}_x, \hat{l}_y\right]F = -\hbar^2\left\{\left(y\frac{\partial}{\partial z} - z\frac{\partial}{\partial y}\right)\left(z\frac{\partial F}{\partial x} - x\frac{\partial F}{\partial z}\right) - \left(z\frac{\partial}{\partial x} - x\frac{\partial}{\partial z}\right)\left(y\frac{\partial F}{\partial z} - z\frac{\partial F}{\partial y}\right)\right\}$$

$$= -\hbar^2 \left(\begin{array}{l} y\dfrac{\partial F}{\partial x} + yz\dfrac{\partial^2 F}{\partial x\partial z} - xy\dfrac{\partial^2 F}{\partial z^2} - z^2\dfrac{\partial^2 F}{\partial x\partial y} + xz\dfrac{\partial^2 F}{\partial y\partial z} \\[2ex] - zy\dfrac{\partial^2 F}{\partial z\partial x} + z^2\dfrac{\partial^2 F}{\partial y\partial x} + xy\dfrac{\partial^2 F}{\partial z^2} - x\dfrac{\partial F}{\partial y} - xz\dfrac{\partial^2 F}{\partial y\partial z} \end{array}\right)$$

$$= -\hbar^2\left(y\frac{\partial}{\partial x} - x\frac{\partial}{\partial y}\right)F = i\hbar\hat{l}_z F.$$

From this, and by cyclic permutation of x, y, and z, we find

$$\left[\hat{l}_x, \hat{l}_y\right] = i\hbar\hat{l}_z$$
$$\left[\hat{l}_y, \hat{l}_z\right] = i\hbar\hat{l}_x \tag{25.7}$$
$$\left[\hat{l}_z, \hat{l}_x\right] = i\hbar\hat{l}_y.$$

A similar, but somewhat more complicated, analysis using (25.3) and (25.4) shows that $\hat{L}^2$ commutes with each of $\hat{l}_x$, $\hat{l}_y$, and $\hat{l}_z$ so that

$$[\hat{L}^2, \hat{l}_x] = 0, \quad [\hat{L}^2, \hat{l}_y] = 0, \quad \text{and} \quad [\hat{L}^2, \hat{l}_z] = 0. \tag{25.8}$$

From (21.82) this means that that $\hat{L}^2$ and any one of $\hat{l}_x$, $\hat{l}_y$, or $\hat{l}_z$ are compatible observations, that is to say that $\hat{L}^2$ has some simultaneous eigenfunctions with each of the component operators. However, the component observations are mutually incompatible, so that pairs of them cannot be measured simultaneously nor do they have eigenfunctions in common.

For many purposes it is preferable to have the operators expressed in terms of spherical polar coordinates. The transformation from Cartesian coordinates to spherical polar coordinates, and vice versa, is, from (13.24),

$$x = r \sin\theta \cos\phi \quad y = r \sin\theta \sin\phi \quad z = r \cos\theta \tag{25.9a}$$

and

$$r^2 = x^2 + y^2 + z^2 \quad \tan\phi = \frac{y}{x} \quad \tan\theta = \frac{\sqrt{x^2 + y^2}}{z}. \tag{25.9b}$$

The partial differentials are converted from Cartesian to spherical polar coordinates by relationships of the form

$$\frac{\partial}{\partial z} = \frac{\partial r}{\partial z}\frac{\partial}{\partial r} + \frac{\partial\theta}{\partial z}\frac{\partial}{\partial\theta} + \frac{\partial\phi}{\partial z}\frac{\partial}{\partial\phi}. \tag{25.10}$$

We illustrate the general principle of converting from Cartesian to spherical polar coordinates by transforming the operator $\hat{l}_z$ given in (25.5). Transforming $x\frac{\partial}{\partial y} - y\frac{\partial}{\partial x}$ to spherical polar coordinate form is done by

$$x\frac{\partial}{\partial y} - y\frac{\partial}{\partial x} = x\left(\frac{\partial r}{\partial y}\frac{\partial}{\partial r} + \frac{\partial\theta}{\partial y}\frac{\partial}{\partial\theta} + \frac{\partial\phi}{\partial y}\frac{\partial}{\partial\phi}\right)$$

$$-y\left(\frac{\partial r}{\partial x}\frac{\partial}{\partial r} + \frac{\partial\theta}{\partial x}\frac{\partial}{\partial\theta} + \frac{\partial\phi}{\partial x}\frac{\partial}{\partial\phi}\right). \tag{25.11}$$

The coefficient of $\frac{\partial}{\partial r}$ is $x\frac{\partial r}{\partial y} - y\frac{\partial r}{\partial x}$. From (25.9b) $2r\frac{\partial r}{\partial y} = 2y$ so that $\frac{\partial r}{\partial y} = \frac{y}{r}$ and $\frac{\partial r}{\partial x} = \frac{x}{r}$ and hence the coefficient of $\frac{\partial}{\partial r}$ is zero.

The coefficient of $\frac{\partial}{\partial\theta}$ is $x\frac{\partial\theta}{\partial y} - y\frac{\partial\theta}{\partial x}$. From (25.9b),

$$\sec^2\theta\frac{\partial\theta}{\partial y} = \frac{y}{z\sqrt{x^2 + y^2}} \quad \text{or} \quad \frac{\partial\theta}{\partial y} = \frac{y\cos^2\theta}{z\sqrt{x^2 + y^2}}.$$

Similarly $\frac{\partial\theta}{\partial x} = \frac{x\cos^2\theta}{z\sqrt{x^2+y^2}}$ and hence the coefficient of $\frac{\partial}{\partial\theta}$ is zero.

Finally, the coefficient of $\frac{\partial}{\partial\phi}$ is $x\frac{\partial\phi}{\partial y} - y\frac{\partial\phi}{\partial x}$. From (25.9b),

$$\sec^2\phi\frac{\partial\phi}{\partial y} = \frac{1}{x} \quad \text{and} \quad \sec^2\phi\frac{\partial\phi}{\partial x} = -\frac{y}{x^2}.$$

Hence

$$x\frac{\partial\phi}{\partial y} - y\frac{\partial\phi}{\partial x} = \cos^2\phi\left(1 + \frac{y^2}{x^2}\right) = \cos^2\phi\left(1 + \tan^2\phi\right) = \cos^2\phi\sec^2\phi = 1.$$

From this we find that

$$\hat{l}_z = \frac{\hbar}{i}\frac{\partial}{\partial\phi}. \tag{25.12}$$

By a similar process we find

$$\hat{l}_x = \frac{\hbar}{i}\left(\sin\phi\frac{\partial}{\partial\theta} + \cot\theta\cos\phi\frac{\partial}{\partial\phi}\right),$$

$$\hat{l}_y = \frac{\hbar}{i}\left(\cos\phi\frac{\partial}{\partial\theta} - \cot\theta\sin\phi\frac{\partial}{\partial\phi}\right), \tag{25.13}$$

and

$$\hat{L}^2 = -\hbar^2\left\{\frac{1}{\sin\theta}\frac{\partial}{\partial\theta}\left(\sin\theta\frac{\partial}{\partial\theta}\right) + \frac{1}{\sin^2\theta}\frac{\partial^2}{\partial\phi^2}\right\}. \tag{25.14}$$

It should be noted that these angular-momentum operators are independent of any particular system – for example, they do not involve the definition of V or boundary conditions that distinguish one system from another. Thus their eigenfunctions and eigenvalues are *universally applicable to all systems*.

Given that $\hbar$ is a unit of angular momentum we may write the eigenvalues for $\hat{l}_z$ in the form $m\hbar$ where m is some, as yet undefined or unrestricted, number. The eigenfunctions, Φ, then come from the solutions of

$$\frac{\hbar}{i}\frac{\partial\Phi}{\partial\phi} = m\hbar\,\Phi \tag{25.15}$$

which give eigenfunctions of the form

$$\Phi = e^{im\phi}. \tag{25.16}$$

Since ϕ is an angle, if Φ is to be single-valued then its value for ϕ and $\phi + 2\pi$ must be the same, or

$$e^{im\phi} = e^{im(\phi + 2\pi)}, \tag{25.17}$$

which requires m to be an integer.

Since $\hat{L}^2$ and $\hat{l}_z$ commute they must have common eigenfunctions, so an eigenfunction of the form $P(\theta)e^{im\phi}$ is indicated. Notice that as far as $\hat{l}_z$ is concerned $P(\theta)$ acts like a constant. With this form of the eigenfunction

$$\hat{L}^2\left\{P(\theta)e^{im\phi}\right\} = \mu P(\theta)e^{im\phi}$$

or, from (25.14),

$$-\hbar^2\left\{\frac{1}{\sin\theta}\frac{d}{d\theta}\left(\sin\theta\frac{d}{d\theta}\right) - \frac{m^2}{\sin^2\theta}\right\}P(\theta) = \mu P(\theta). \tag{25.18}$$

We just state here, without proof, that for each value of m there is a family of solutions of (25.18) associated with integers l that satisfy the condition $l \geq m$. These solutions are known as **Legendre polynomials** and are usually expressed in the form $P_l^m(\cos\theta)$. The combination of the Legendre polynomial with $e^{im\phi}$ is called a **spherical harmonic**, indicated by

$$Y_l^m(\theta, \phi) = P_l^m(\cos \theta)e^{im\phi} \tag{25.19}$$

and these are the complete eigenfunctions of $\hat{L}^2$. The associated eigenvalue is

$$\mu_l = l(l+1)\hbar^2 \tag{25.20}$$

A number of normalized spherical harmonic functions are listed in Table 25.1.

Spherical harmonics form a complete orthonormal set of functions, a linear combination of which can represent any well-behaved general function of θ and ϕ. To illustrate that they are eigenfunctions of $\hat{L}^2$ with the stated eigenvalues, we take one example from Table 25.1, $Y_2^2(\theta, \phi)$. Applying the operator to the function

$$\hat{L}^2 Y_2^2(\theta, \phi) = -\hbar^2 \left\{ \frac{1}{\sin \theta} \frac{\partial}{\partial \theta} \left(\sin \theta \frac{\partial}{\partial \theta} \right) + \frac{1}{\sin^2 \theta} \frac{\partial^2}{\partial \phi^2} \right\} \sin^2 \theta e^{2i\phi}$$

$$= -\hbar^2 \left\{ \frac{1}{\sin \theta} \frac{\partial}{\partial \theta} \left(2 \sin^2 \theta \cos \theta \right) \right\} e^{2i\phi} + 4\hbar^2 \frac{1}{\sin^2 \theta} \sin^2 \theta e^{2i\phi}$$

$$= -\hbar^2 e^{2i\phi} \left\{ \frac{2}{\sin \theta} \left(2 \sin \theta \cos^2 \theta - \sin^3 \theta \right) - 4 \right\}$$

$$= -\hbar^2 e^{2i\phi} \left(4 \cos^2 \theta - 4 - 2 \sin^2 \theta \right)$$

$$= 6\hbar^2 \sin^2 \theta e^{2i\phi}$$

$$= 2(2+1)\hbar^2 Y_2^2(\theta, \phi), \tag{25.21}$$

which, since $l = 2$, verifies the form of the eigenfunction and eigenvalue.

An atomic electron possesses angular momentum so that, as we would expect from classical considerations of rotating or orbiting charges, it also has a net

Table 25.1 Some low-order spherical harmonics

l	m	$Y_l^m(\theta, \phi)$
0	0	$\left(\frac{1}{4\pi}\right)^{1/2}$
1	−1	$\left(\frac{3}{8\pi}\right)^{1/2} \sin \theta e^{-i\phi}$
1	0	$\left(\frac{3}{4\pi}\right)^{1/2} \cos \theta$
1	1	$\left(\frac{3}{8\pi}\right)^{1/2} \sin \theta e^{i\phi}$
2	−2	$\left(\frac{15}{32\pi}\right)^{1/2} \sin^2 \theta e^{-2i\phi}$
2	−1	$\left(\frac{15}{8\pi}\right)^{1/2} \sin \theta \cos \theta e^{-i\phi}$
2	0	$\left(\frac{5}{16\pi}\right)^{1/2} (3 \cos^2 \theta - 1)$
2	1	$-\left(\frac{15}{8\pi}\right)^{1/2} \sin \theta \cos \theta e^{i\phi}$
2	2	$\left(\frac{15}{32\pi}\right)^{1/2} \sin^2 \theta e^{2i\phi}$

magnetic dipole moment proportional to the angular momentum. The magnitude of the angular momentum can only be of the form $\sqrt{l(l+1)}\hbar$ and the angular momentum vector will normally point in some random direction in space. However, if the atoms are in a magnetic field then the atom will experience a torque. Classically it would be expected to end up with the dipole axis parallel to the field, just as the needle of a magnet always points towards the north – although it may oscillate around this direction due to thermal effects. However, what actually happens in a quantum-mechanical system is that the angular momentum vector takes up a direction such that it has a component $m\hbar$ in the field direction where $m = 0, \pm 1, \pm 2, \cdots, \pm l$. These directions are shown in Figure 25.1 for the case $l=3$ where the magnitude of the angular momentum is $\sqrt{12}\hbar$.

The spectra from atoms are due to jumps of electrons from a higher energy state to a lower one. For atoms in an electric field the electron magnetic dipoles can align themselves in the way shown in Figure 25.1 and different orientations correspond to small differences in the potential energy and hence the energy levels of the electrons. Thus the change of energy in jumping from one level to another, $\Delta E = E_1 - E_2$, in the environment of a magnetic field becomes

$$\Delta E' = (E_1 + \delta\varepsilon_{m_1}) - (E_2 + \delta\varepsilon_{m_2})$$

where $\delta\varepsilon_{m_1}$ and $\delta\varepsilon_{m_2}$ are the shifts due to the orientation effect in the two energy levels. Thus, when emitting electrons are within a magnetic field, the spectral lines are split into three separate lines – no more, because of selection rules that dictate that

$$\Delta l = \pm 1 \quad \text{and} \quad \Delta m = 0, \pm 1.$$

This is the **normal Zeeman effect**, which illustrates quantization within a magnetic field.

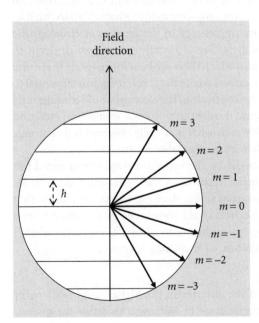

Figure 25.1 Directions of the angular momentum vector in a magnetic field for $l=3$.

Exercise 25.1

What are the eigenvalues for the square of the total angular momentum and for the component in the z-direction for the following eigenfunctions:

(i) $\left(\dfrac{1}{4\pi}\right)^{1/2}$;   (ii) $\left(\dfrac{15}{32\pi}\right)^{1/2} \sin^2\theta e^{-2i\phi}$;   (iii) $\left(\dfrac{15}{32\pi}\right)^{1/2} \sin^2\theta e^{2i\phi}$.

25.2 The hydrogen atom

25.2.1 Bohr model

The classical model of a hydrogen atom consists of an electron orbiting the central proton much as a planet orbits the Sun. Equating the central force and the electron mass times centripetal acceleration for a circular orbit,

$$\frac{mv^2}{r} = \frac{e^2}{4\pi\varepsilon_0 r^2}$$

or

$$mv^2 = \frac{e^2}{4\pi\varepsilon_0 r}. \tag{25.22}$$

The total energy associated with the orbiting electron is the sum of its kinetic and potential energy given by

$$E = \frac{1}{2}mv^2 - \frac{e^2}{4\pi\varepsilon_0 r} = -\frac{e^2}{8\pi\varepsilon_0 r}. \tag{25.23}$$

This model, which allows all possible values of r and hence of E, is untenable. An electron in orbit is in a state of acceleration and, according to Maxwell's laws, such an electron should be continuously emitting energy in the form of electromagnetic radiation. Indeed, this is the basis of the action of a synchrotron in which electrons are sent round a circuit in some parts of which there are bending magnets that cause the electron paths to be along arcs of circles, where they emit radiation tangential to the path. In this classical model of a hydrogen atom the electron would continuously spiral in until it joined the nucleus, and during this process it would emit radiation over a wide range of frequencies. However, what is actually observed is that atomic spectra consist of sharp lines and that atoms are stable entities.

In 1913 Niels Bohr suggested a model for the hydrogen atom that resolved this difficulty and also explained the major features of the hydrogen spectrum. He proposed that there were discrete orbits in which electrons had to exist and within which they would not radiate. An electron could absorb energy and jump from a lower to higher energy state or jump from a higher to a lower energy state, when it would emit a packet of energy, a **photon**, of frequency v, given by

$$hv = E_1 - E_2 \tag{25.24}$$

where E_1 and E_2 are the energies of the electron in the higher and lower energy states respectively. The final hypothesis made by Bohr was to define the discrete orbits by the condition that they had angular momenta that were integral numbers of $\hbar$, or $mvr = n\hbar$. This, with (25.22), gave

$$\frac{mv^2}{r} = \frac{n^2 \hbar^2}{mr^3} = \frac{e^2}{4\pi\varepsilon_0 r^2}$$

and hence

$$r = \frac{4\pi\varepsilon_0 n^2 \hbar^2}{me^2}. \tag{25.25}$$

The value of r for $n = 1$ is the **Bohr radius** of the atom and is given by

$$a_B = \frac{4\pi\varepsilon_0 \hbar^2}{me^2} = 0.53 \times 10^{-10} \text{m} \tag{25.26}$$

Substituting the value of r from (25.25) in (25.23) gives the nth energy level as

$$E_n = -\frac{me^4}{32\pi^2 \varepsilon_0^2 n^2 \hbar^2}. \tag{25.27}$$

Thus the frequency of the emitted radiation in an electron jump from level m to level n is

$$\nu_{mn} = \frac{E_m - E_n}{h} = \frac{me^4}{64\pi^3 \varepsilon_0^2 \hbar^3} \left(\frac{1}{n^2} - \frac{1}{m^2} \right). \tag{25.28}$$

This formula gave excellent agreement with the main observed spectral lines of hydrogen. The frequencies with $n = 1$ are in the ultraviolet part of the spectrum and are called the **Lyman series**. Those with $n = 2$ are in the visible part of the spectrum and form the **Balmer series**.

Despite its success with hydrogen, the Bohr theory cannot explain the spectra of atoms of higher atomic number and the fine structure of spectra where lines are split into two or more components. To explain these features requires the use of the Schrödinger wave equation.

Exercise 25.2
Find the longest and shortest wavelengths of the Balmer line series for hydrogen.

25.2.2 The TISWE applied to hydrogen

The general form of the TISWE is

$$-\frac{\hbar^2}{2m} \nabla^2 \psi + V\psi = E\psi. \tag{25.29}$$

Expressing the Laplacian operator in the form (16.38) and the potential energy as expressed in (25.23), the TISWE is of the form

$$-\frac{\hbar^2}{2m} \left\{ \frac{1}{r^2} \frac{\partial}{\partial r} \left(r^2 \frac{\partial}{\partial r} \right) + \frac{1}{r^2 \sin\theta} \frac{\partial}{\partial\theta} \left(\sin\theta \frac{\partial}{\partial\theta} \right) + \frac{1}{r^2 \sin^2\theta} \frac{\partial^2}{\partial\phi^2} \right\} \psi - \frac{e^2}{4\pi\varepsilon_0 r} \psi$$
$$= E\psi.$$

Comparison of the last two terms in the brackets with (25.14) shows that the TISWE can be written in the form

$$\left\{ -\frac{\hbar^2}{2m}\frac{1}{r^2}\frac{\partial}{\partial r}\left(r^2\frac{\partial}{\partial r}\right) + \frac{\hat{L}^2}{2mr^2} - \frac{e^2}{4\pi\varepsilon_0 r} \right\}\psi = E\psi. \tag{25.30}$$

We now consider solutions of the form

$$u(r,\theta,\phi) = R(r)Y_l^m(\theta,\phi).$$

Applying the operator $\hat{L}^2$ to a solution of this form in (25.30) gives

$$\left\{ -\frac{\hbar^2}{2m}\frac{1}{r^2}\frac{d}{dr}\left(r^2\frac{d}{dr}\right) - \frac{e^2}{4\pi\varepsilon_0 r} + \frac{l(l+1)\hbar^2}{2mr^2} \right\}RY = ERY,$$

and, since to the first term in the bracket Y is a constant, we can divide throughout by Y giving

$$\left\{ -\frac{\hbar^2}{2m}\frac{1}{r^2}\frac{d}{dr}\left(r^2\frac{d}{dr}\right) - \frac{e^2}{4\pi\varepsilon_0 r} + \frac{l(l+1)\hbar^2}{2mr^2} \right\}R = ER. \tag{25.31}$$

This equation can be greatly simplified by the transformation

$$\alpha^2 = -\frac{8mE}{\hbar^2}, \quad \rho = \alpha r, \quad \text{and} \quad n^2 = -\frac{me^4}{32\pi^2\varepsilon_0^2\hbar^2 E}. \tag{25.32}$$

Remember that we are concerned with bound states for which E is negative so that α^2 and n^2 are both positive. Another change we make is to write

$$\frac{1}{r^2}\frac{d}{dr}\left(r^2\frac{d}{dr}\right) = \frac{d^2}{dr^2} + \frac{2}{r}\frac{d}{dr}.$$

By some straightforward, if rather messy, manipulations (25.31) now appears in the form

$$\left\{ \rho^2\frac{d^2}{d\rho^2} + 2\rho\frac{d}{d\rho} + n\rho - l(l+1) - \frac{1}{4}\rho^2 \right\}R = 0. \tag{25.33}$$

To determine the form of solution of (25.33) we first note that when ρ is very large, and only terms in ρ^2 are significant, the equation can be approximated as

$$\frac{d^2R}{d\rho^2} - \frac{1}{4}R = 0$$

with solution $R = e^{-\rho/2}$. The other possible solution, $R = e^{\rho/2}$, is rejected because it diverges for large ρ and cannot be normalized. Knowing the asymptotic behaviour at large ρ we can now try a solution in the form of $R = e^{-\rho/2}f(\rho)$, where $f(\rho)$ is a polynomial in ρ. For a finite polynomial, i.e. one that terminates after a finite number of terms, the exponential term will dominate for large ρ so that the asymptotic behaviour will be as required.

The solution we seek is of the general form

$$R = e^{-\rho/2}\left(a_0 + a_1\rho + a_2\rho^2 + a_3\rho^3 + \cdots + a_s\rho^s + \cdots \right). \tag{25.34}$$

From this we find for the individual terms in (25.33):

$$\begin{aligned} \rho^2\frac{d^2R}{dr^2} = {} & \frac{1}{4}e^{-\rho/2}\left(a_0\rho^2 + a_1\rho^3 + a_2\rho^4 + a_3\rho^5 + \cdots + a_s\rho^{s+2} + \cdots \right) \\ & - e^{-\rho/2}\left(a_1\rho^2 + 2a_2\rho^3 + 3a_3\rho^4 + \cdots + sa_s\rho^{s+1} + \cdots \right) \\ & + e^{-\rho/2}\left(2a_2\rho^2 + 6a_3\rho^3 + \cdots + s(s-1)a_s\rho^s + \cdots \right) \end{aligned}$$

$$2\rho\frac{dR}{d\rho} = -e^{-\rho/2}\left(a_0\rho + a_1\rho^2 + a_2\rho^3 + a_3\rho^4 + \cdots + a_s\rho^{s+1} + \cdots\right)$$

$$+ 2e^{-\rho/2}\left(a_1\rho + 2a_2\rho^2 + 3a_3\rho^3 + \cdots + sa_s\rho^s + \cdots\right)$$

$$n\rho R = e^{-\rho/2}n\left(a_0\rho + a_1\rho^2 + a_2\rho^3 + a_3\rho^4 + \cdots + a_s\rho^{s+1} + \cdots\right)$$

$$- l(l+1)R = -e^{-\rho/2}l(l+1)\left(a_0 + a_1\rho + a_2\rho^2 + a_3\rho^3 + \cdots + a_s\rho^s + \cdots\right)$$

$$- \frac{1}{4}\rho^2 R = -e^{-\rho/2}\frac{1}{4}\left(a_0\rho^2 + a_1\rho^3 + a_2\rho^4 + a_3\rho^5 + \cdots + a_s\rho^{s+2} + \cdots\right).$$

$$(25.35)$$

Using the results (25.35) we can find the coefficient of ρ^{s+1} in (25.33). Since ρ is a variable that can take all possible values and since the right-hand side of (25.33) is zero, this requires all the coefficients of powers of ρ to be zero – which gives for the coefficient of ρ^{s+1}

$$\{(s+1)(s+2) - l(l+1)\}a_{s+1} - (s+1-n)a_s = 0$$

or

$$\frac{a_{s+1}}{a_s} = \frac{s+1-n}{(s+1)(s+2) - l(l+1)}. \qquad (25.36)$$

In fact, (25.36) shows us why it is that, to get a physically sensible solution, the polynomial must be terminated and cannot be allowed to develop to an infinite number of terms. In the limit of large s the ratio of the coefficients of successive terms tends towards $1/(s+2)$. If one considers the expansion

$$e^\rho = 1 + \rho + \frac{\rho^2}{2!} + \frac{\rho^3}{3!} + \cdots + \frac{\rho^s}{s!} + \frac{\rho^{s+1}}{(s+1)!} + \cdots$$

then the ratio of the coefficient of ρ^{s+1} to that of ρ^s is $1/(s+1)$, similar to that found from (25.35). Although this is not formally proved, it indicates that the divergence of an infinite polynomial would overwhelm the convergence of $e^{-\rho/2}$ to give a divergent function that could not be normalized and would be physically unrealistic.

You can see from (25.36) that, since s is an integer, for the series to be truncated n must be some integer. With $n=1$ the series would terminate with $s=0$, i.e. the function $f(\rho)$ would just be a constant. With $n=2$ the series would terminate with $s=1$ and $f(\rho)$ would be linear in ρ. The actual coefficients generated would depend on l although the order of the polynomial depends on n and nothing else. The polynomials generated by (25.36) are known as Laguerre polynomials and here we find the coefficients for the first three of them, un-normalized, for $l=0$:

$n = 1$ and hence $s_{max} = 0$ $f(\rho) = a_0$

$n = 2$ and hence $s_{max} = 1$ $\frac{a_1}{a_0} = -\frac{1}{2}$ and $f(\rho) = a_0(1 - \rho/2)$

$n = 3$ and hence $s_{max} = 2$ $\frac{a_1}{a_0} = -1;\ \frac{a_2}{a_1} = -\frac{1}{6}$ and

$f(\rho) = a_0(1 - \rho + \rho^2/6).$

From the transformations (25.32) we find the energy levels of the electron

$$E_n = -\frac{me^4}{32\pi^2\varepsilon_0^2\hbar^2 n^2}$$

which are precisely the energies found in the Bohr model, in agreement with observations.

Again from the transformations (25.32) and the definition of the Bohr radius (25.26) we find

$$\alpha^2 = -\frac{8mE}{\hbar^2} = \frac{m^2 e^4}{4\pi^2 \varepsilon_0^2 \hbar^4 n^2} = \frac{4}{a_B^2 n^2},$$

which gives

$$\frac{\rho}{2} = \frac{r}{a_B n}. \tag{25.37}$$

It will be noticed that R is a function both of n and l and hence the eigenfunction (25.30) should be designated as

$$u_{n,l,m}(r, \theta, \phi) = R_n^l(r) Y_l^m(\theta, \phi). \tag{25.38}$$

Some low-order normalized eigenfunctions are listed in Table 25.2. All the eigenstates with the same value of n are degenerate and it can be easily found that the order of degeneracy is n^2.

The eigenfunctions for atomic electrons are often called **orbitals**, a word that sounds like 'orbit' but must be distinguished from it. The Bohr model of electrons in orbit around the atomic nucleus could not, for example, give zero angular momentum for the electron; indeed, the whole basis of the way that Bohr's orbits were found was by quantizing the angular momentum with a minimum value $\hbar$. In the quantum mechanical approach the orbitals with $l=0$ have zero angular momentum and wave functions that are spherically symmetric.

The eigenfunctions u_{100}, u_{200}, and u_{300} and the corresponding radial density distributions are illustrated in Figure 25.2. The radial density distributions, $S(r)$ are

Table 25.2 Some normalized eigenfunctions for hydrogen

n	l	m	$u_{nlm}(r, \theta, \Phi)$
1	0	0	$\sqrt{\frac{1}{\pi a_B^3}}\, e^{-r/a_B}$
2	0	0	$\sqrt{\frac{1}{8\pi a_B^3}}\left(1 - \frac{r}{2a_B}\right) e^{-r/2a_B}$
2	1	0	$\sqrt{\frac{1}{8\pi a_B^3}}\frac{r}{2a_B}\, e^{-r/2a_B} \cos\theta$
2	1	± 1	$\sqrt{\frac{1}{16\pi a_B^3}}\frac{r}{2a_B}\, e^{-r/2a_B} \sin\theta e^{\pm i\phi}$
3	0	0	$\sqrt{\frac{1}{27\pi a_B^3}}\left(1 - \frac{2r}{3a_b} + \frac{2r^2}{27a_B^2}\right) e^{-r/3a_B}$
3	1	0	$\sqrt{\frac{2}{81\pi a_B^3}}\frac{r}{3a_B}\left(2 - \frac{r}{3a_B}\right) e^{-r/3a_B} \cos\theta$
3	1	± 1	$\sqrt{\frac{1}{81\pi a_B^3}}\frac{r}{3a_B}\left(2 - \frac{r}{3a_B}\right) e^{-r/3a_B} \sin\theta e^{\pm i\phi}$
3	2	0	$\frac{1}{3}\sqrt{\frac{1}{54\pi a_B^3}}\left(\frac{r}{3a_B}\right)^2 e^{-r/3a_B} (3\cos^2\theta - 1)$
3	2	± 1	$\sqrt{\frac{1}{81\pi a_B^3}}\left(\frac{r}{3a_B}\right)^2 e^{-r/3a_B} \sin\theta \cos\theta e^{\pm i\phi}$
3	2	± 2	$\frac{1}{2}\sqrt{\frac{1}{81\pi a_B^3}}\left(\frac{r}{3a_B}\right)^2 e^{-r/3a_B} \sin^2\theta e^{\pm 2i\phi}$

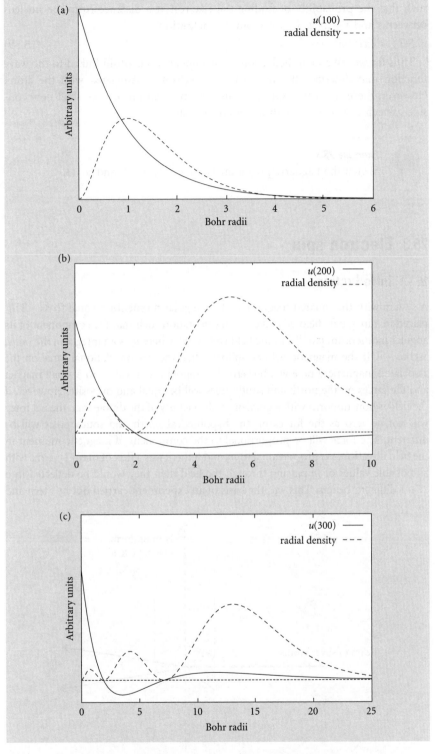

Figure 25.2 Wave function and radial distribution function for (a) u_{100}, (b) u_{200}, (c) u_{300}.

such that the probability of finding the electron at a distance from the nucleus between r and $r + \mathrm{d}r$ is $S(r)\mathrm{d}r$. From this definition,

$$S(r) = 4\pi r^2 |u|^2. \tag{25.39}$$

Thus far we have identified a source of angular momentum linked to the wave function that describes the way that the electron is associated with the atom. However, there is another source of angular momentum that we shall now consider, which is associated with the electron itself.

 Exercise 25.3

Sketch the Laguerre polynomials for $n = 1$, $n = 2$, and $n = 3$.

25.3 Electron spin

25.3.1 Introduction

A system with the squared magnitude of its angular momentum equal to $l(l+1)\hbar^2$ placed in a magnetic field will take up an orientation such that the component of its angular momentum parallel to the field will be $m\hbar$ where m is an integer in the range $+l$ to $-l$. If the magnetic field is uniform then the net translational force on the associated magnetic dipole will be zero. The dipole is equivalent to a small magnet and the forces on the north and south poles will be equal and opposite. However, if the field is non-uniform with a gradient in the vicinity of the dipole then the net force will not be zero as the forces on the hypothetical north and south poles will be different. The force will be proportional to the component of magnetic moment in the field direction, i.e. proportional to m, and if there was a flux of many systems with all possible values of m passing through the field then they would be deflected into $2l + 1$ different beams. This was the basis of an experiment carried out by Stern and

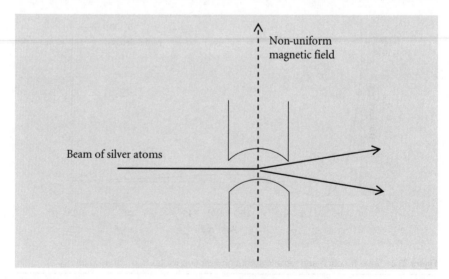

Figure 25.3 Schematic representation of the Stern–Gerlach experiment.

Gerlach in 1922, before the development of quantum mechanics. Their apparatus in schematic form is shown in Figure 25.3. The pole pieces of the magnet are shaped to give a non-uniform field and a beam of atoms is directed perpendicular to the field.

When the experiment was done with silver atoms just two beams were produced, as shown in the figure. A silver atom has 47 electrons and its electronic structure is such that 46 of the electrons form a set of closed shells giving no net angular momentum because the contributions of these electrons cancel each other; this will be explained further in §25.4.1. The final electron, the **valence electron**, has $l = 0$ in the ground state and so $m = 0$, and on this basis there should be an undeflected single beam going through the apparatus.

The interpretation of this result is that the electron has an intrinsic angular momentum that has nothing to do with it being part of an atom. Given that the expected number of beams is $2l + 1$ and that the number observed is 2, we conclude that an electron has half-integral spin and so is of magnitude $\hbar/2$. In the magnetic field the spin angular momentum aligns itself either parallel or anti-parallel to the field so that the quantity that is quantized is the *difference*, $\hbar$, of the component of angular momentum in the direction of the field.

The problem then is how to incorporate electron spin into the formulation of quantum-mechanical theory.

25.3.2 **Mathematical formulation**

In 1925, a year before Schrödinger's development of wave mechanics, an alternative formulation of quantum mechanics, **matrix mechanics**, was put forward by Heisenberg. This was already hinted at in §21.5.3, but here we summarize the relationship between the Schrödinger wave-equation approach and the Heisenberg matrix approach.

- In the wave-equation approach there corresponds to each kind of observation an Hermitian operator. The eigenvalues of these operators, which are real, give the values of the possible observations. In the matrix approach, to each kind of observation there corresponds an Hermitian matrix that, in general, may be of infinite dimensions. The eigenvalues of these matrices, which are real (see Appendix 17), give the values of the possible observations.

- In the wave-equation approach the observable states of a system corresponding to a particular observation are represented by the eigenfunctions of the associated operator. These normalized eigenfunctions form an orthonormal set satisfying (21.29). Where there are degenerate states then orthogonal eigenfunctions for these states can be produced by the Gram–Schmidt orthogonalization process. The eigenfunctions can be complex; only the modulus squared of an eigenfunction, representing a probability density, has any physical significance. Any general state is a linear combination of eigenfunctions.

- For the matrix approach the observable states correspond to the eigenvectors of the associated matrix. These normalized eigenvectors form a complete orthonormal set (see Appendix 17). Where there are degenerate states then orthogonal eigenvectors for these states can be produced by the Gram–Schmidt orthogonalization process. The eigenvectors can have complex components and any general state is a linear combination of eigenvectors. These eigenvectors with complex components exist in a multidimensional space, called a **Hilbert space**. The dimension of the Hilbert space equals the total number of possible eigenstates and could be infinity.

Schrödinger's approach is formally equivalent to Heisenberg's and has generally been preferred for its ease of application. However, including electron spin into quantum mechanics is much more readily done through the Heisenberg approach. There are just two possible states of the system, which we can designate as **spin-up** and **spin-down**, so we just have to deal with 2×2 matrices and two-element vectors.

We now introduce the matrix operators, called **Pauli spin matrices**, which correspond to the measurement of components of electron spin in the x-, y-, and z-directions. These are

$$\hat{s}_x = \frac{1}{2}\hbar \begin{bmatrix} 0 & 1 \\ 1 & 0 \end{bmatrix}, \quad \hat{s}_y = \frac{1}{2}\hbar \begin{bmatrix} 0 & -i \\ i & 0 \end{bmatrix}, \quad \hat{s}_z = \frac{1}{2}\hbar \begin{bmatrix} 1 & 0 \\ 0 & -1 \end{bmatrix}. \tag{25.40}$$

The commutation properties of these operators are similar to those of $\hat{l}_x, \hat{l}_y$ and $\hat{l}_z$ as given in (25.7). Thus

$$[\hat{s}_x, \hat{s}_y] = \frac{1}{4}\hbar^2 \left\{ \begin{bmatrix} 0 & 1 \\ 1 & 0 \end{bmatrix} \begin{bmatrix} 0 & -i \\ i & 0 \end{bmatrix} - \begin{bmatrix} 0 & -i \\ i & 0 \end{bmatrix} \begin{bmatrix} 0 & 1 \\ 1 & 0 \end{bmatrix} \right\}$$

$$= \frac{1}{4}\hbar^2 \left\{ \begin{bmatrix} i & 0 \\ 0 & -i \end{bmatrix} - \begin{bmatrix} -i & 0 \\ 0 & i \end{bmatrix} \right\} = \frac{1}{2}\hbar^2 i \begin{bmatrix} 1 & 0 \\ 0 & -1 \end{bmatrix} = i\hbar\hat{s}_z \tag{25.41a}$$

Similarly

$$[\hat{s}_y, \hat{s}_z] = i\hbar\hat{s}_x \tag{25.41b}$$

$$[\hat{s}_z, \hat{s}_x] = i\hbar\hat{s}_y. \tag{25.41c}$$

The matrix operator for the square of the electron spin angular momentum is

$$\hat{S}^2 = \hat{s}_x^2 + \hat{s}_y^2 + \hat{s}_z^2$$

$$= \frac{1}{4}\hbar^2 \left\{ \begin{bmatrix} 0 & 1 \\ 1 & 0 \end{bmatrix} \begin{bmatrix} 0 & 1 \\ 1 & 0 \end{bmatrix} + \begin{bmatrix} 0 & -i \\ i & 0 \end{bmatrix} \begin{bmatrix} 0 & -i \\ i & 0 \end{bmatrix} + \begin{bmatrix} 1 & 0 \\ 0 & -1 \end{bmatrix} \begin{bmatrix} 1 & 0 \\ 0 & -1 \end{bmatrix} \right\}$$

$$= \frac{3}{4}\hbar^2 \begin{bmatrix} 1 & 0 \\ 0 & 1 \end{bmatrix}. \tag{25.42}$$

Since $\hat{S}^2$ is a constant times a unit matrix it commutes with all matrices of similar order, so that

$$[\hat{S}^2, \hat{s}_x] = [\hat{S}^2, \hat{s}_y] = [\hat{S}^2, \hat{s}_z] = 0, \tag{25.43}$$

which is similar to (25.8).

The eigenvectors and eigenvalues of these matrix operators can be found by the methods described in §24.2. For $\hat{s}_x$ the eigenvalues are found from

$$\begin{vmatrix} -\lambda & \hbar/2 \\ \hbar/2 & -\lambda \end{vmatrix} = 0$$

or

$$\lambda = \pm \hbar/2.$$

This corresponds to the two possible observations of spin-up or spin-down relative to the x-direction. Similar eigenvalues are found for $\hat{l}_y$ and $\hat{l}_z$.

The eigenvectors for $\hat{l}_x$ are found by substituting the values of λ in

$$\frac{1}{2}\hbar\begin{bmatrix} 0 & 1 \\ 1 & 0 \end{bmatrix}\begin{bmatrix} a \\ b \end{bmatrix} = \lambda\begin{bmatrix} a \\ b \end{bmatrix}.$$

For $\lambda = \hbar/2$ we find $a = b$, giving a normalized eigenvector

$$\mathbf{s}_x^+ = \frac{1}{\sqrt{2}}\begin{bmatrix} 1 \\ 1 \end{bmatrix}, \tag{25.44a}$$

while for $\lambda = -\hbar/2$ the normalized eigenvector is

$$\mathbf{s}_x^- = \frac{1}{\sqrt{2}}\begin{bmatrix} 1 \\ -1 \end{bmatrix}. \tag{25.44b}$$

The eigenvalues and associated eigenvectors for $\hat{s}_y$ and $\hat{s}_z$ are

$$\mathbf{s}_y^+ = \frac{1}{\sqrt{2}}\begin{bmatrix} 1 \\ i \end{bmatrix}, \tag{25.45a}$$

$$\mathbf{s}_y^- = \frac{1}{\sqrt{2}}\begin{bmatrix} 1 \\ -i \end{bmatrix}, \tag{25.45b}$$

$$\mathbf{s}_z^+ = \begin{bmatrix} 1 \\ 0 \end{bmatrix}, \tag{25.46a}$$

$$\mathbf{s}_z^- = \begin{bmatrix} 0 \\ 1 \end{bmatrix}. \tag{25.46b}$$

Finally, we find the eigenvalues and eigenvectors for $\hat{S}^2$. The eigenvalues come from

$$\begin{bmatrix} \frac{3}{4}\hbar^2 - \lambda & 0 \\ 0 & \frac{3}{4}\hbar^2 - \lambda \end{bmatrix} = 0$$

so that the eigenvalues are identical and each equal to $\frac{3}{4}\hbar^2$. We note that this is of the form $\frac{1}{2}(\frac{1}{2}+1)\hbar^2$, similar to the $l(l+1)\hbar^2$ we met previously. Inserting this value of λ in

$$\begin{bmatrix} \frac{3}{4}\hbar^2 - \lambda & 0 \\ 0 & \frac{3}{4}\hbar^2 - \lambda \end{bmatrix}\begin{bmatrix} a \\ b \end{bmatrix} = \begin{bmatrix} 0 \\ 0 \end{bmatrix}$$

gives

$$\begin{bmatrix} 0 & 0 \\ 0 & 0 \end{bmatrix}\begin{bmatrix} a \\ b \end{bmatrix} = \begin{bmatrix} 0 \\ 0 \end{bmatrix}$$

which puts no constraints on a and b so that any vector is an eigenvector for this observation.

? **Exercise 25.4**

Derive the commutator relationships $[\hat{s}_y, \hat{s}_z] = i\hbar\hat{s}_x$ and $[\hat{s}_z, \hat{s}_x] = i\hbar\hat{s}_y$.

25.3.3 Observations and experiments with spin

Electron spin, a very simple two-state system, provides us with a good tool for illustrating aspects of the ideas behind making observations and their associated probabilities.

EXAMPLE 25.1

We first consider a two-stage Stern–Gerlach experiment where silver atoms first move along the x-axis through a non-uniform field in the z-direction. This is illustrated schematically in Figure 25.4 and shows the two emitted beams with angular momenta $(\hbar/2)_z$ (spin-up for z-direction) and $-(\hbar/2)_z$ (spin-down for z-direction). (It should be noted that it is not possible to use unattached free electrons for this experiment; the uncertainty principle implies that the two electron beams would spread out and merge into one.)

The beam coming through the first stage corresponding to spin-down is blocked off and then the remaining beam passes through the second stage where the magnetic field is in the y-direction. The atoms coming through the second stage can only be spin-up or spin-down relative to the y-direction and the symmetry of the arrangement would lead us to the conclusion that the beams corresponding to $(\hbar/2)_y$ and $-(\hbar/2)_y$ would be of equal intensity. Now we shall see how to derive this conclusion formally.

The eigenvector corresponding to the atoms with spin-up along z is, from (25.46a), $\mathbf{s}_z^+ = \begin{bmatrix} 1 \\ 0 \end{bmatrix}$. For the second measurement, or observation, where the magnetic field is in the y-direction the eigenvectors are given by (25.45a) and (25.45b) and the entering atoms are in a mixed state. We now express this mixed state as a linear combination of the eigenvectors so that

$$\begin{bmatrix} 1 \\ 0 \end{bmatrix} = c_1 \frac{1}{\sqrt{2}} \begin{bmatrix} 1 \\ i \end{bmatrix} + c_2 \frac{1}{\sqrt{2}} \begin{bmatrix} 1 \\ -i \end{bmatrix}, \tag{25.47}$$

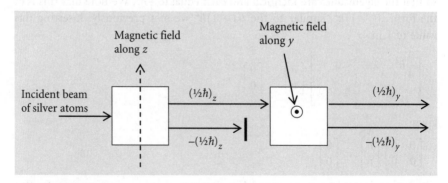

Figure 25.4 Two-stage Stern–Gerlach experiment.

giving the solution

$$c_1 = c_2 = \frac{1}{\sqrt{2}}.$$

Equation (25.47) is analogous to (21.40), which gives a mixed state in terms of the eigenfunctions of the measurement to be made. As indicated by (21.42) the probability of measuring $(\hbar/2)_y$ is $|c_1|^2$, which is 0.5, and similarly the probability of measuring $-(\hbar/2)_y$ is also 0.5.

EXAMPLE 25.2

The next two-stage Stern–Gerlach experiment we can envisage is where the second magnetic field is in the y-z plane but in a direction designated by d at an angle θ to the z-axis, as shown in Figure 25.5.

The operator for the measurement of spin in this direction is

$$\hat{s}_n = \hat{s}_z \cos\theta + \hat{s}_y \sin\theta = \frac{1}{2}\hbar\left\{\cos\theta\begin{bmatrix} 1 & 0 \\ 0 & -1 \end{bmatrix} + \sin\theta\begin{bmatrix} 0 & -i \\ i & 0 \end{bmatrix}\right\}$$
$$= \frac{1}{2}\hbar\begin{bmatrix} \cos\theta & -i\sin\theta \\ i\sin\theta & -\cos\theta \end{bmatrix}. \tag{25.48}$$

We now find the eigenvalues and eigenvectors for this operator. To find the eigenvalues we solve

$$\begin{vmatrix} \frac{1}{2}\hbar\cos\theta - \lambda & -\frac{1}{2}\hbar i\sin\theta \\ \frac{1}{2}\hbar i\sin\theta & -\frac{1}{2}\hbar\cos\theta - \lambda \end{vmatrix} = 0,$$

which gives, as it should, $\lambda = \pm\hbar/2$. The eigenvector for $\lambda = \hbar/2$ comes from

$$\begin{bmatrix} \frac{1}{2}\hbar(\cos\theta - 1) & -\frac{1}{2}\hbar i\sin\theta \\ \frac{1}{2}\hbar i\sin\theta & -\frac{1}{2}\hbar(\cos\theta + 1) \end{bmatrix}\begin{bmatrix} a \\ b \end{bmatrix} = \begin{bmatrix} 0 \\ 0 \end{bmatrix}$$

which gives $a(\cos\theta - 1) - ib\sin\theta = 0$ or, from (1.14) and (1.19),

$$\frac{a}{b} = -i\frac{\cos(\theta/2)}{\sin(\theta/2)}.$$

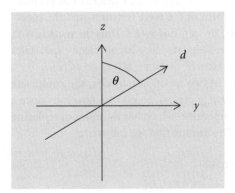

Figure 25.5 A magnetic field along d, in the y-z plane, and making an angle θ with z.

This gives the normalized eigenvector

$$s_d^+ = \begin{bmatrix} \cos(\theta/2) \\ -i\sin(\theta/2) \end{bmatrix}. \tag{25.49a}$$

By a similar process,

$$s_d^- = \begin{bmatrix} \sin(\theta/2) \\ i\cos(\theta/2) \end{bmatrix}. \tag{25.49b}$$

Now we have to represent the state vector coming from the first magnetic field as a linear combination of the eigenvectors for the field along n. We write

$$\begin{bmatrix} 1 \\ 0 \end{bmatrix} = c_1 \begin{bmatrix} \cos(\theta/2) \\ -i\sin(\theta/2) \end{bmatrix} + c_2 \begin{bmatrix} \sin(\theta/2) \\ i\cos(\theta/2) \end{bmatrix},$$

the solution of which is $c_1 = \cos(\theta/2)$ and $c_2 = \sin(\theta/2)$. Hence the probability of a silver atom passing through the second field in the spin-up configuration is $\cos^2(\theta/2)$ and in the spin-down configuration is $\sin^2(\theta/2)$. These probabilities can be regarded either as the probabilities of the outcome of passing through the second magnets for a single atom or as the proportions of a stream of atoms that would leave the second field spin-up and spin-down.

Exercise 25.5

Find the spin state vector corresponding to having a probability of $\frac{3}{4}$ of being observed with a spin $\hbar/2$ along z. What is the probability that the observation will yield a spin $-\hbar/2$ along z?

25.4 Many-electron systems

25.4.1 A two-particle system and particle exchange

The state of a single particle can be defined by some eigenfunction that describes its probability distribution and also conveys information about the total energy, momentum, angular momentum, and many other properties associated with it, including the z-component of angular momentum in the presence of a magnetic field in that direction. To *completely* define the state of the system we must add the description of its spin – either spin-up or spin-down in the prevailing magnetic field. We now consider a particle with wave function $\psi_a^s(\mathbf{r})$ where $\mathbf{r}$ describes its position, the subscript a designates the functional form of the wave function, and the superscript s indicates whether it is spin-up (α) or spin-down (β). Thus the particle with the functional part of its total wave function designated by a has alternative spin states so that its total wave function can be represented by either $\psi_a^\alpha(\mathbf{r})$ or $\psi_a^\beta(\mathbf{r})$.

Next we consider a system consisting of two *identical* particles, the composite wave function of which is represented by $\psi_{comp}(\mathbf{r}_1, s_1, \mathbf{r}_2, s_2)$. We now introduce a new operator, $\hat{P}_e$, the **particle-exchange operator**, which has the effect of replacing particle 1 by particle 2 and vice versa. This means that we can write

$$\hat{P}_e \psi_{comp}(\mathbf{r}_1, s_1, \mathbf{r}_2, s_2) = \psi_{comp}(\mathbf{r}_2, s_2, \mathbf{r}_1, s_1). \tag{25.50}$$

An observer of the system who had been looking away when the particles were exchanged would not have noticed that anything had happened, since the particles are identical and the system would have exactly the same appearance and the same properties. Further insight can be gained into the effect of applying the particle-exchange operator by considering what happens when it is applied twice, for then it restores the system to its original state, or

$$\hat{P}_e\hat{P}_e\psi_{comp}(\mathbf{r}_1, s_1, \mathbf{r}_2, s_2) = \psi_{comp}(\mathbf{r}_1, s_1, \mathbf{r}_2, s_2). \tag{25.51}$$

We infer from this that the eigenvalue corresponding to $\hat{P}_e\hat{P}_e$ is $+1$ and that the eigenvalue for $\hat{P}_e$ is ± 1. From this and (25.50) we have

$$\psi_{comp}(\mathbf{r}_2, s_2, \mathbf{r}_1, s_1) = \pm\psi_{comp}(\mathbf{r}_1, s_1, \mathbf{r}_2, s_2). \tag{25.52}$$

This means that exchanging the particles either has no effect at all on the wave function or multiplies it by -1, and the question is how to decide what actually does happen. The answer is that for systems consisting of half-integral spin particles, such as electrons or protons, the exchange of particles multiplies the wave function by -1, or we may say that the wave function is **antisymmetric**. Such particles are referred to as **fermions** and their energy states are controlled by Fermi–Dirac statistics. Another class of particles, such as photons and π-mesons, have integral spin (spin angular momentum $n\hbar$ where n is zero or a positive integer); they are known as **bosons**, their wave functions are **symmetric** (exchange multiplies wave function by $+1$), and their energies are controlled by Bose–Einstein statistics. A study of the splitting of energy levels in the spectrum of helium confirms that an electronic system, i.e. an atom, consists of fermions, although we shall not deal with the underlying theory here.

From the requirement that the wave function of a two-electron system is antisymmetric it is clear that a composite wave function of the form given in (25.50) cannot be represented by $\psi_a^\alpha(\mathbf{r}_1)\psi_b^\beta(\mathbf{r}_2)$, for example, since interchanging particles 1 and 2 would not just change the function by a factor -1 but would change its form completely, to $\psi_a^\alpha(\mathbf{r}_2)\psi_b^\beta(\mathbf{r}_1)$. However, writing

$$\psi_{comp}(\mathbf{r}_1, s_1, \mathbf{r}_2, s_2) = \frac{1}{\sqrt{2}}\left\{\psi_a^\alpha(\mathbf{r}_1)\psi_b^\beta(\mathbf{r}_2) - \psi_b^\beta(\mathbf{r}_1)\psi_a^\alpha(\mathbf{r}_2)\right\} \tag{25.53}$$

does satisfy the symmetry requirements. Exchanging $\mathbf{r}_1$ and $\mathbf{r}_2$ in (25.53) simply reverses the sign of the right-hand side. The factor $1/\sqrt{2}$ is just a normalization factor. An alternative form of the right-hand side of (25.53) is as a determinant

$$\frac{1}{\sqrt{2}}\begin{vmatrix} \psi_a^\alpha(\mathbf{r}_1) & \psi_b^\beta(\mathbf{r}_1) \\ \psi_a^\alpha(\mathbf{r}_2) & \psi_b^\beta(\mathbf{r}_2) \end{vmatrix}. \tag{25.54}$$

The columns correspond to different states and the rows to different particles. Exchanging the particles is equivalent to exchanging the two rows, and from rule 2 in §8.6.1 we know that interchanging two rows of a determinant reverses its sign.

25.4.2 Pauli exclusion principle and structure of atoms

The arguments that led to (25.54) as an acceptable wave function for a two-electron system suggest that form of a general n-electron system will be

$$\psi_{comp}(\mathbf{r}_1, s_1, \mathbf{r}_2, s_2, \ldots, \mathbf{r}_n, s_n) = \frac{1}{\sqrt{n!}} \begin{vmatrix} \psi_a^{s_1}(\mathbf{r}_1) & \psi_b^{s_2}(\mathbf{r}_1) & \cdots & \psi_n^{s_n}(\mathbf{r}_1) \\ \psi_a^{s_1}(\mathbf{r}_2) & \psi_b^{s_2}(\mathbf{r}_2) & \cdots & \psi_n^{s_n}(\mathbf{r}_2) \\ \vdots & \vdots & \cdots & \vdots \\ \psi_a^{s_1}(\mathbf{r}_n) & \psi_b^{s_2}(\mathbf{r}_n) & \cdots & \psi_n^{s_n}(\mathbf{r}_n) \end{vmatrix}.$$

$$(25.55)$$

This satisfies the antisymmetric rule that exchanging two particles reverses the sign of the composite wave function and it is normalized. The n particles could be a system of n electrons round the nucleus of an atom.

The form of (25.55) leads to an important rule known as the **Pauli exclusion principle**. This says that in a composite system consisting of fermions no two particles can have the same state function. If, for example in determinant (25.55) functional forms a and b were the same and the spin states s_1 and s_2 were the same then the first two columns of the determinant would be identical, and from rule 4 for determinants in §8.6.1 the determinant, and hence the composite wave function, would be zero – which is not a permitted wave function.

Armed with the Pauli exclusion principle we can now consider the electronic structure of the silver atom, used in the Stern–Gerlach experiment. This has 47 electrons and these must all correspond to different state functions. At the same

Table 25.3 Quantum numbers up to $n = 4$

n	l	m	s	Number of states
1	0	0	$\pm\frac{1}{2}$	2
2	0	0	$\pm\frac{1}{2}$	2
2	1	$-1, 0, 1$	$\pm\frac{1}{2}$	6
3	0	0	$\pm\frac{1}{2}$	2
3	1	$-1, 0, 1$	$\pm\frac{1}{2}$	6
3	2	$-2, -1, 0, 1, 2$	$\pm\frac{1}{2}$	10
4	0	0	$\pm\frac{1}{2}$	2
4	1	$-1, 0, 1$	$\pm\frac{1}{2}$	6
4	2	$-2, -1, 0, 1, 2$	$\pm\frac{1}{2}$	10
4	3	$-3, -2, -1, 0, 1, 2, 3$	$\pm\frac{1}{2}$	14

time the energy of the ground state of a silver atom must have the minimum possible energy, which is a general condition for a stable system. In the main the energy of an atomic electron is governed by its total quantum number n; for hydrogen the energy which is always negative is proportional to n^{-2} and hence the lowest energy (negative and of greatest magnitude) is for $n = 1$. For more complex atoms the energy of an electron depends not only on the nuclear charge but also on the presence of the other electrons that partially shield an outer electron from the nuclear charge. However, even for complex atoms the energies of the innermost electrons still increase with n, i.e. they become less negative.

We now consider the silver-atom electrons in terms of their sets of quantum numbers (n, l, m, s). For $n = 1$ there are two possible states, $(1, 0, 0, \frac{1}{2})$ and $(1, 0, 0, -\frac{1}{2})$, and these are both occupied. Table 25.3 gives the possible states up to $n = 4$. In silver the selection of the lowest possible energy levels for 46 electrons gives all the states up to and including $n = 4$ and $l = 2$. Because for each (n, l, m, s) there is a $(n, l, -m, -s)$ the net angular momentum, and hence magnetic moment, is zero for these 46 electrons. However, because of the form of the probability distribution of electron density associated with $n = 4$, $l = 3$, a rather elongated orbital, and the way the shielding by the innermost 46 electrons operates, the energy of this state is higher than that of $(5, 0, 0, \pm\frac{1}{2})$ so that, in the ground state, the valence electron of silver has this latter set of quantum numbers. This means that the net angular momentum of silver is completely dependent on the spin of the 47th electron, which explains its behaviour in the Stern–Gerlach experiment.

Similar considerations can be used to determine the electronic structure of all the atoms in the periodic table.

Problems

25.1 Show that $Y_2^1(\theta, \phi)$ and $Y_1^1(\theta, \phi)$ are eigenfunctions of $\hat{L}^2$ with the expected form of the eigenvalues.

25.2 Show that the spherical harmonics given in Problem 25.1 are orthogonal. This requires the integral $\int_0^{2\pi} \int_0^\pi (Y_l^m)^* Y_{l'}^{m'} \sin\theta \, d\theta \, d\phi = 0$.

25.3 Show that the peak of the radial density function for the ground state of hydrogen is at a distance of one Bohr radius from the nucleus.

25.4 A beam of silver atoms moving in the x-direction passes through six Stern–Gerlach magnetic fields, all parallel to the y-z plane. Each successive field direction makes an angle of $60°$ with the previous one. After each passage through a field the spin-down beam is blocked out. What is the ratio of the final intensity of the beam to the original intensity?

Sampling theory

Sampling theory is a technique for estimating the mean and variance of a large number of quantities from the mean and variance of a random selection of a comparatively small number of them. In this chapter you will be shown how to do this. You will also be able to assess whether the sample matches some particular claim or prediction about the complete set of quantities. An important part of sampling theory is to assess whether or not there is any significant difference between the characteristics of two samples – i.e. are the samples likely to have come from the same source?

26.1 Samples

Let us consider the problem of a company making shoes. Excluding some comparatively rare requirements for extremely large or small shoes there is a range of fairly common shoe sizes for men, perhaps 6 to 13 (UK sizes) in steps of $\frac{1}{2}$, making 15 shoe sizes in all. What proportion of shoes of each size should be made so as best to provide for the customers' needs? Clearly, the proportions of shoes made of each size should match the proportions of men in the population requiring that size, and this is where the technique of sampling comes into its own. Although it is impossible to find out from the whole population of men what the proportions are of each size, it *is* possible to find out from a **sample**, a random selection of a limited number of men, a good estimate of the proportions required. The larger the sample the better the estimate, but large samples can be expensive to collect so the trick is to find the minimum size of sample that will give estimates sufficiently good for the purpose in hand.

26.1.1 Average and standard deviation of a sample

Imagine that we have a large **population** of values x_i with mean μ_p and variance $V_p = \sigma_p^2$. When we use the word 'population' in this sense we do not necessarily mean a large number of human, or other animate, inhabitants of a country or place, but rather the complete set of quantities of the specified type (e.g. shoe sizes for adult men in the UK). If we now take a sample of size n from such a population then the sample mean is given by

$$\mu_s = \frac{1}{n} \sum_{j=1}^{n} x_j \tag{26.1}$$

and the sample variance by

$$V_s = \sigma_s^2 = \frac{1}{n} \sum_{j=1}^{n} (x_j - \mu_s)^2. \tag{26.2}$$

Exercise 26.1
The heights of two samples of men (in cm), are:

A   182   191   167   172   182   195   173   174   169   181
B   184   190   186   179   197   184   191   188

Find the sample mean and sample variance for each sample and for the two samples taken together.

26.1.2 Estimates of the average and variance of the population from the sample

When a sample is taken, it constitutes all the information we have about the total population. The characteristics of the sample, e.g. the mean and variance of the sample values, are known and from these we need to estimate the characteristics of the whole population. We indicate an estimated value of a population quantity by enclosing it in angle brackets, so that $\langle y \rangle$ is the estimated value of y, which is to be contrasted with taking an average or mean value, indicated by $\bar{y}$.

If we were to take samples of n many times then the samples will all have different means and sample variances. To find the mean of the sample means we note that if we take a large (effectively infinite) number of samples of size n then the mean of the sample means is the same as the mean of all the individual values of x in all the samples combined so that

$$\bar{\mu}_s = \mu_p. \tag{26.3}$$

If we only have a single sample then the best estimate we can make of the population mean is

$$\langle \mu_p \rangle = \mu_s. \tag{26.4}$$

To find the variance of the means of many samples, $\sigma_{\mu_s}^2$, we first note that the population variance is σ_p^2, i.e. this is the variance of all the individual xs. Hence if we take the sum of n randomly chosen values of x from the population, $\sum_{i=1}^{n} x_i$, then, from (14.19), the variance of this sum is $n\sigma_p^2$. However, the sample mean, given by (26.1) is

$$\mu_s = \frac{1}{n} \sum_{i=1}^{n} x_i$$

which, using Rule 2 in §12.8, has variance

$$\sigma_{\mu_s}^2 = \frac{1}{n^2} \times n\sigma_p^2 = \frac{\sigma_p^2}{n}. \tag{26.5}$$

We should also note that another way of describing the variance of the sample means is

$$\sigma_{\mu_s}^2 = \frac{1}{n}\sum_{i=1}^{n}(\mu_s - \overline{\mu}_s)^2 = \frac{1}{n}\sum_{i=1}^{n}(\mu_s - \mu_p)^2. \tag{26.6}$$

The assumption that we make is that if we take very many samples of size n then the values of the sample means would have a normal distribution with mean μ_p and variance σ_p^2/n.

We now consider the sample variance as given by (26.2):

$$\sigma_s^2 = \frac{1}{n}\sum_{j=1}^{n}(x_j - \mu_s)^2.$$

If μ_p replaced μ_s in (26.2) then the right-hand side would be the expected value of the mean-square deviation from the population mean, which would be σ_p^2. It might be thought that since the mean value of μ_s, for many samples, is μ_p the corresponding values of σ_s^2 would straddle σ_p^2 and give σ_p^2 as an expected value, but this is not so. We find the actual expected value from

$$\sigma_s^2 = \frac{1}{n}\sum_{j=1}^{n}(x_j - \mu_s)^2 = \frac{1}{n}\sum_{j=1}^{n}\left\{(x_j - \mu_p) + (\mu_p - \mu_s)\right\}^2$$

$$= \frac{1}{n}\sum_{j=1}^{n}(x_j - \mu_p)^2 + \frac{2}{n}\sum_{j=1}^{n}(x_j - \mu_p)(\mu_p - \mu_s) + \frac{1}{n}\sum_{j=1}^{n}(\mu_p - \mu_s)^2. \tag{26.7}$$

Since the second term on the right-hand side is linear in the xs, each can be replaced by its average value, μ_s, leading to

$$\langle\sigma_s^2\rangle = \frac{1}{n}\left\langle\sum_{j=1}^{n}(x_j - \mu_p)^2\right\rangle + \frac{2}{n}\left\langle\sum_{j=1}^{n}(\mu_s - \mu_p)(\mu_p - \mu_s)\right\rangle + \frac{1}{n}\left\langle\sum_{j=1}^{n}(\mu_p - \mu_s)^2\right\rangle$$

$$= \frac{1}{n}\left\langle\sum_{j=1}^{n}(x_j - \mu_p)^2\right\rangle - \frac{1}{n}\left\langle\sum_{j=1}^{n}(\mu_s - \mu_p)^2\right\rangle. \tag{26.8}$$

The first term on the right-hand side is the expected value of the mean square deviation of the individual xs from the population mean, which is σ_p^2. From (26.6) and (26.5) the second term is $\sigma_{\mu_s}^2$ or, equivalently, σ_p^2/n. Hence, substituting in (26.8)

$$\langle\sigma_s^2\rangle = \sigma_p^2 - \frac{\sigma_p^2}{n} = \frac{n-1}{n}\sigma_p^2. \tag{26.9a}$$

The right-hand side of (26.9a) is the average value of a sample variance and when a single sample variance has been found then the best estimate of the variance of the whole population is

$$\langle\sigma_p^2\rangle = \frac{n}{n-1}\sigma_s^2. \tag{26.9b}$$

The factor $n/(n-1)$ which converts the sample variance into the estimate of the population variance is known as the **Bessel correction**. For large samples it is close to unity but for small samples its use is essential to obtain sensible statistical conclusions.

Finally, to estimate the variance of sample means, given by (26.5), the value to use for σ_p^2 is the expected value given by (26.9b) so that

$$\langle \sigma_{\mu_s}^2 \rangle = \frac{\langle \sigma_p^2 \rangle}{n} = \frac{\sigma_s^2}{n-1}. \tag{26.10}$$

To illustrate the application of sample information we take the following example which indicates the general approach to problems of this kind.

EXAMPLE 26.1

A sample of 20 light bulbs gave the following lifetimes in hours:

| 980 | 995 | 1005 | 960 | 972 | 1015 | 1011 | 955 | 971 | 1004 |
| 925 | 1016 | 1080 | 910 | 940 | 990 | 922 | 975 | 946 | 1001 |

The manufacturer claims that the lifetime of the bulbs is greater than 1000 hours. What is the probability that his claim is correct?

The information that we have from the sample is the sample mean, 978.6 hours, and the sample variance, 1532.6 hours^2 corresponding to a standard deviation, 39.2 hours. We now make the assumption that his claim is true, referred to as the **null hypothesis**, and find out on the basis of this hypothesis how likely it is that our sample could have given the result it did. If we took repeated samples of 20 bulbs then the estimate of the standard deviation of the sample means is, from (26.10)

$$\langle \sigma_{\mu_s}^2 \rangle^{1/2} = \frac{\sigma_s}{\sqrt{n-1}} = \frac{39.2}{\sqrt{19}} \text{ hours} = 9.0 \text{ hours}.$$

From the null hypothesis the sample mean is equal to or greater than

$$\frac{1000 - 978.6}{9.0} \langle \sigma_{\mu_s} \rangle = 2.38 \langle \sigma_{\mu_s} \rangle$$

from the population mean. From Table 14.1 the probability of being this far, *or further*, from the mean is 0.0086, which is the probability that the manufacturer's claim is true – i.e. the claim is unlikely to be true.

Exercise 26.2
Find the best estimates of the population mean and the population variance from the sample data in Exercise 26.1 for both sample A and sample B.

26.2 Sampling proportions

A familiar example of taking samples is in sampling opinion in relation to the support for political parties, especially in the periods leading up to elections. The point at issue here is the **proportion** of the electorate supporting each of the parties. To simplify the discussion, we shall consider situations where there are only two choices: a proportion p of the population supports party A and a proportion $q(=1-p)$ supports party B. Some of the results we obtain here are linked to those previously obtained in relation to the binomial distribution dealt with in §14.4.

Let us consider that we select n members (n large) of the population and a number X of them support party A. For many such selections the values of X would form a binomial distribution but since n is large then, from the results in §14.4.3, this would approximate to a normal distribution with mean $\overline{X} = np$ and variance $\sigma_X^2 = npq$.

The best estimate of p from a single sample is

$$\langle p \rangle = \frac{X}{n} \tag{26.11a}$$

and hence

$$\langle q \rangle = 1 - \langle p \rangle \tag{26.11b}$$

and

$$\langle \sigma_p^2 \rangle = \frac{\langle \sigma_X^2 \rangle}{n^2} = \frac{\langle p \rangle \langle q \rangle}{n}$$

or

$$\langle \sigma_p \rangle = \sqrt{\frac{\langle p \rangle \langle q \rangle}{n}}. \tag{26.12}$$

Note that for the large values of n we are assuming in these applications a Bessel correction is unnecessary.

EXAMPLE 26.2

Suppose we poll 1000 randomly chosen electors, of whom 481 support party A and the rest support other parties. What is the probability that for the population at large more than 50% support party A?

The null hypothesis we make is that more than 50% support party A. The standard deviation in the estimates from the sample is given by

$$\langle \sigma_p \rangle = \sqrt{\frac{0.481 \times 0.519}{1000}} = 0.0158.$$

For the null hypothesis to be true, the sample average must be more than

$$\frac{0.500 - 0.481}{0.0158}\langle \sigma_p \rangle = 1.20\langle \sigma_p \rangle$$

from the mean. From Table 14.1 the probability of this is 0.115, which is the probability that party A will receive more than 50% support in the election.

In assessing the significance of the results from sampling, when comparisons or tests of claims are made, it is customary to set a **significance level** – e.g. 1% or 10%. If the probability of the claim being true is lower than the significance level then the assumption is that the claim can be discounted – otherwise the claim *may* have validity. The significance level that is set reflects the importance of the quantity being sampled; tests are more stringent for assessing medical tests than for selling soap powder.

Exercise 26.3

In a large-scale poll of 10 000 people it was found that 48.5% supported the Radical Party. The Radical Party claim to have 50% of the popular vote. Can their claim be supported at the 10% significance level?

26.3 The significance of differences

An important requirement in some circumstances is to compare the means of samples of the same kind of quantity taken from different sources to see whether or not they are significantly different.

As an example we take small samples from the rat populations in two cities and compare their lengths:

London   (cm)   20.5   19.9   21.6   22.3   19.6   18.8   21.1   20.6
Manchester (cm)   19.4   17.6   18.7   19.4   19.2   19.7

Could these be regarded as samples taken from the same population, or is the difference in the mean length sufficiently large as to be able to assert that they are significantly different? To answer this question we first have to set a significance level, say 0.05. The question we are then answering is whether the probability of getting the difference we find, or *some greater difference either positive or negative*, is greater than or less than 0.05, assuming that the samples are from the same population – which is the null hypothesis. If it is greater than 0.05 then we do not regard the difference as significant – otherwise we do. The procedure for carrying this out is done by what is called the **Student *t*-test**, Student being a pseudonym used by the statistician W. S. Gossett.

26.3.1 Student's *t*-test

The basic assumption behind the Student *t*-test is that both samples are drawn from populations that have a normal distribution. The samples drawn from different sources have numbers of members, means and standard deviations as follows:

Sample 1 n_1 μ_1 σ_1
Sample 2 n_2 μ_2 σ_2

If we take the null hypothesis that the samples actually come from the same distribution then theory shows that the expected difference of the sample means is zero and that the expected variance of the **difference** of the sample means is

$$\sigma_{\text{diff}}^2 = \frac{\sigma_1^2 + \sigma_2^2}{n_1 + n_2 - 2}.$$

(26.13)

The quantity in the divisor of the right-hand side of (26.13) is the **degrees of freedom** of the system, usually represented by the symbol ν. It represents the number of variables that can be freely chosen in value, *given* the totals for each of the samples. For each sample the number of degrees of freedom is $n-1$, with the appropriate n, and for the two samples together it is $n_1 - 1 + n_2 - 1 = n_1 + n_2 - 2$.

We now calculate the t value, which is

$$t = \frac{|\mu_1 - \mu_2|}{\sigma_{\mathrm{diff}}} \tag{26.14}$$

and then consult the Student t-test table shown in Table 26.1. For each value of v there is a different probability distribution as a function of t. As $v \to \infty$ so the distribution tends to the normal distribution. In this particular case we are just interested in the significance of the difference of two sample means, and the difference $\mu_1 - \mu_2$ can be either positive or negative in relation to the null hypothesis that the difference is zero. This is what is called a **two-tailed test** and if we have fixed the significance level at 0.05 then the column of Table 26.1 of interest is the one headed $0.05/2 = 0.025$, which is the area under each of the two tails of the distribution.

EXAMPLE 26.3

We now illustrate the application of the Student t-test using the data for the length of rats given on page 471. For this we have, not including dimensions,

$$n_1 = 8, \quad \mu_1 = 20.55, \quad \sigma_1^2 = 1.1075$$

$$n_2 = 6, \quad \mu_1 = 19.00, \quad \sigma_2^2 = 0.4833.$$

This gives $\mu_1 - \mu_2 = 1.55$ and, from (26.13),

$$\sigma_{\mathrm{diff}} = \sqrt{\frac{1.1075 + 0.4833}{12}} = 0.364.$$

Table 26.1 The significance levels of the t-test distribution

v	0.1	0.05	0.025	0.01	0.005	0.001
1	3.078	6.314	12.706	31.821	63.657	318.31
2	1.886	2.920	4.303	6.965	9.925	22.326
3	1.638	2.353	3.182	4.541	5.841	10.213
4	1.533	2.132	2.776	3.747	4.604	7.173
6	1.440	1.943	2.447	3.143	3.707	5.208
8	1.397	1.860	2.306	2.896	3.355	4.501
10	1.372	1.812	2.228	2.764	3.169	4.144
15	1.341	1.753	2.131	2.602	2.947	3.733
20	1.325	1.725	2.086	2.528	2.845	3.552
25	1.316	1.708	2.060	2.485	2.787	3.450
30	1.310	1.697	2.042	2.457	2.750	3.385
40	1.303	1.684	2.021	2.423	2.704	3.307
60	1.296	1.671	2.000	2.390	2.660	3.232
∞	1.282	1.645	1.960	2.326	2.576	3.090

From (26.14),

$$t = \frac{1.55}{0.364} = 4.26.$$

The number of degrees of freedom, v, is 12 and the significance level of 0.05 means that the column 0.025 is relevant in Table 26.1 because it is a two-tailed test. The values of t for a significance level of 0.025 is 2.228 for $v = 10$ and 2.131 for $v = 15$ so it is about 2.2 for $v = 12$. Since t is greater than this, the difference is significant. In fact this value of t would be significant at the 0.002 significance level, corresponding to the column 0.001 in Table 26.1.

As an example of a one-tailed test we consider whether the London sample dis-agrees at the significance level of 0.05 with a null hypothesis that London rats have a mean length of 21.0 cm. In this case the sample mean is less and we only test for the deviation from the null hypothesis in that direction. Since we only have a single sample, the standard deviation of the sample mean is found from (26.10) as

$$\langle \sigma_{\mu_s} \rangle = \sqrt{\frac{1.1075}{7}} = 0.398 \text{ cm} \quad \text{and} \quad t = \frac{21.0 - 20.55}{0.398} = 1.13.$$

The number of degrees of freedom is 7 and the column for the one-tailed test is 0.05 for which the interpolated value in Table 26.1 is 1.90. Since t is less than this the sample mean is *not* significantly different from the null hypothesis.

Exercise 26.4
The two samples in Exercise 26.1 were taken from different regions of a country. Is the difference in mean height of the two samples significant at the 10% level?

Exercise 26.5
It is claimed that the mean height of those from region A is 186 cm. Can this claim be supported at the 10% significance level?

26.3.2 Chi-squared (χ^2) test

The Student t-test is mainly used to assess the significance of the difference of sample means or the difference between a sample mean and some claimed value, cases that give rise to a two-tailed and one-tailed test respectively. The data are assumed to be continuously variable and derived from populations with a normal distribution.

Now we consider a different kind of significance test concerned with the fre-quencies with which a finite range of possible outcomes occur. As an example we consider the number of goals scored in a football match. The relevant statistics for 1000 games are given in Table 26.2.

It is claimed that the 'number of games' in Table 26.2 is related to the 'number of goals' by a Poisson distribution (§14.5) with an average equal to the average number of goals per match, and the question we seek to answer is whether or not there is a significant difference between the observed distribution and a Poisson distribution. The average number of goals per match from the data below is 2.566,

Table 26.2 Numbers of goals scored

Number of goals	Number of games (O)	Poisson (E)
0	84	77
1	190	197
2	261	253
3	202	216
4	150	139
5	62	71
6	33	30
7	9	11
8	4	4
9	5	1

Table 26.3 Values of chi-squared

ν	0.5	0.25	0.1	0.05	0.025	0.01	0.005	0.001
1	0.455	1.323	2.706	3.841	5.024	6.635	7.879	10.827
2	1.386	2.773	4.605	5.991	7.378	9.210	10.597	13.815
3	2.366	4.108	6.251	7.815	9.348	11.345	12.838	16.268
4	3.357	5.385	7.779	9.488	11.143	13.277	14.860	18.465
6	5.348	7.841	10.645	12.592	14.449	16.812	18.548	22.457
8	7.344	10.219	13.362	15.507	17.535	20.090	21.955	26.125
10	9.342	12.549	15.987	18.307	20.483	23.209	25.188	29.588
12	11.340	14.845	18.549	21.026	23.337	26.217	28.300	32.909
14	13.339	17.117	21.064	23.685	26.119	29.141	31.319	36.123
16	15.338	19.369	23.542	26.296	28.845	32.000	34.267	39.252
18	17.338	21.605	25.989	28.869	31.526	34.805	37.156	42.312
20	19.337	23.828	28.412	31.410	34.170	37.566	39.997	45.315

so we can compare the numbers of games with the numbers that result from a Poisson distribution with this average – shown in the final column.

The general form of this kind of problem is that we have several possible outcomes, 1 to n, with actual observed occurrences of number O_1 to O_n and expected number of occurrences E_1 to E_n. The number of degrees of freedom associated with the test is $n-1$; given that there are 1000 football matches in total then only 9 of the numbers in column O of the table can be freely chosen. The value of χ^2 is given by

$$\chi^2 = \sum_{i=1}^{n} \frac{(O_i - E_i)^2}{E_i}. \tag{26.15}$$

Just as for the Student t-test a significance level is chosen and values of χ^2 are tabulated for different significance levels for different degrees of freedom. A range of such values is given in Table 26.3. The null hypothesis is that the observed distribution of goals is a Poisson distribution, and we shall see whether or not this can be supported at the significance level 0.1.

Before calculating the value of χ^2 for the football data we first note in (26.15) that the value will be very sensitive to small changes of small values of E_i. For this reason it is better to lump some data together and in this case we combine the data in the last three categories as the number of matches for which 7 or more goals were scored. This reduces the number of outcomes to 8 and $v = 7$. The value of χ^2 is then given by

$$\frac{(84 - 77)^2}{77} + \frac{(190 - 197)^2}{197} + \frac{(261 - 253)^2}{253} + \frac{(202 - 216)^2}{216}$$
$$+ \frac{(150 - 139)^2}{139} + \frac{(62 - 71)^2}{71} + \frac{(33 - 30)^2}{30} + \frac{(18 - 16)^2}{16} = 4.436.$$

The values of χ^2 for $v = 6$ and $v = 8$ for a significance level 0.1 are 10.645 and 13.362 respectively and since the value of χ^2 is less than this there is no significant difference between the observed data and a Poisson distribution at the significance level 0.1.

Another situation in which the chi-squared test can be used is in judging the efficacy of a medical treatment.

EXAMPLE 26.4

In a particular test, of 50 people not given the treatment for a potentially fatal disease, 11 died. During the same period, of 28 people receiving the treatment, 2 died. We want to know whether the treatment is effective at the significance level 0.01. Our null hypothesis in this case is that the treatment is *ineffective*. If the value of χ^2 is greater than that in the table for a significance level of 0.01 then we can claim with some confidence that the treatment is effective. If the value of χ^2 is less then it may still be effective but there would be less confidence in that conclusion.

We show the observed data in tabular form (Table 26.4*a*). At the edges of the table we add totals of rows and columns and, given these totals, there is one degree of freedom. Once a single entry is put into the table then, given the totals, the other entries are fixed.

Accepting the null hypothesis that there is no difference between treated and untreated patients, we see that $13/78 = 1/6$ of them die. That being so the number of untreated patients expected to survive is $50 \times 5/6$ and the number expected to die is $50/6$. The equivalent numbers are $28 \times 5/6$ and $28/6$ for the treated patients. This gives a table of expected outcomes for the two categories of patient (Table 26.4*b*).

Although fractions of a patient do not make physical sense, they are included in the table for statistical purposes. We now have observed values in the first table and expected values in the second table, which is what is required to evaluate the value of χ^2. However, before we do so there is one more feature of the chi-squared test that we have to take into account. For only one degree of freedom, the theory behind the chi-squared test is not really valid. One way of correcting this, called the **Yates correction**, involves subtracting 0.5 from the magnitude of $O - E$, but statisticians are not agreed about whether it is better to do this or not. Here we shall not apply the Yates correction, but the reader should note its existence.

Table 26.4 Outcomes of treatment

Outcome	No treatment	Treatment	Total
a. Observed			
Survived	39	26	65
Died	11	2	13
Total	50	28	
b. Expected			
Survived	41.67	23.33	65
Died	8.33	4.67	13
Total	50	28	

The value of χ^2 for the present problem is

$$\chi^2 = \frac{(39 - 41.67)^2}{41.67} + \frac{(26 - 23.33)^2}{23.33} + \frac{(11 - 8.33)^2}{8.33} + \frac{(2 - 4.67)^2}{4.67} = 2.86.$$

This value of χ^2 falls far short of the value for a significance level of 0.01 (6.635), so one cannot be sure of the efficacy of the treatment at the significance level 0.01.

The value of χ^2 depends on the scale of the tests being made. If the number of patients had been larger, say by a factor of 4, and all the numbers in the tables were larger by a factor of 4 then the value of χ^2 would be greater by a factor of 4 and would be highly indicative of the positive value of the treatment.

Exercise 26.6
Two methods of teaching children to read are tested. With method A, 81 out of 100 children passed the final reading test. With method B, only 59 out of 100 passed the test. Can it be claimed at the 0.1 level of significance that method A is superior?

Problems

26.1 The following 10 measurements were made of the density of a mineral grain (in units $kg\,m^{-3}$):

2891   2924   2886   2911   2927   2909   2892   2930   2887   2891.

Are these readings consistent with a mineral density of $2914\,kg\,m^{-3}$ at the significance level 0.1?

26.2 Geologists are trying to locate an underground pocket of dense material by measurements of the acceleration due to gravity. Ten measurements are made at each of two fairly close points. These are, in units $\mathrm{m\,s^{-2}}$,

9.821   9.813   9.830   9.822   9.817   9.823   9.819   9.804   9.817   9.823

and

9.814   9.832   9.826   9.809   9.826   9.831   9.830   9.831   9.809   9.824

Is the difference of the means significant at the significance level 0.1?

26.3 A random digit generator produces the following output of 1000 digits:

0	1	2	3	4	5	6	7	8	9
115	92	87	111	104	109	91	88	116	87

Are these results consistent with a properly working random-digit generator at the significance level 0.1?

27 Straight-line relationships and the linear correlation coefficient

It is often known from theory, by experiment, or from general observation that pairs of quantities are connected by a straight-line relationship. When pairs of values (x, y) are available from an experiment, the equation of the 'best' straight line linking them depends on the relative precision of the different variables. This chapter explains how to determine the best straight line when one of the variables has no error and also when both variables have expected errors of known magnitude. It also describes a measure of how well points define a straight-line relationship through the linear correlation coefficient, first mentioned in Chapter 17 (where it was just called the 'correlation coefficient', as it often is). Finally we will show how some non-linear equations can be converted into a linear form to enable us to exploit the simplicity of linear-equation theory to determine unknown quantities from experiment.

27.1 General considerations

When a mass is suspended from a spring the extension of the spring is found to be proportional to the mass (Figure 27.1). This has been deduced by experimental measurements, and the relationship is referred to as **Hooke's law**.

Another example is in the observation of Newton's rings (Figure 27.2). Monochromatic light incident normally on to a planoconvex lens resting on a plane mirror produces a series of dark rings when viewed along the axis of the system, such that the square of the radius of the nth ring from the centre is proportional to n. The linear relationship between the pairs of quantities (mass, extension) and (number, radius-squared) is revealed by a simple linear plot.

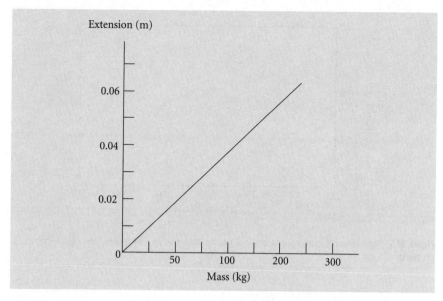

Figure 27.1 Relationship between the extension of a spring and the suspended mass.

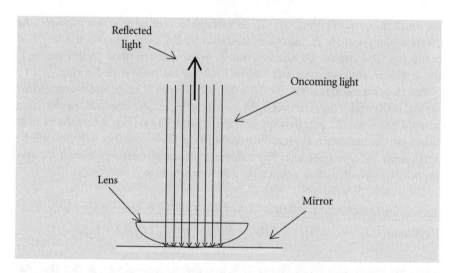

Figure 27.2 Formation of Newton's rings. The light reflected from the bottom surface of the lens and from the mirror interferes to give a set of dark rings corresponding to destructive interference. Due to a phase change of π when light is reflected going from a less optically dense region to a denser one, the central point (corresponding to $n = 0$ and no path difference) is dark.

In both these cases the quantities were connected by **proportionality** – that is, a relationship of the form

$$y = kx \tag{27.1}$$

where k is the constant of proportionality. The points (x, y) lie on a straight line of slope k that passes through the origin. However, a more general linear relationship

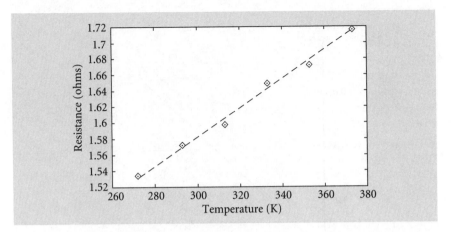

Figure 27.3 Plotted (temperature, resistance) points with the line of regression of resistance on temperature.

is of the form

$$y = mx + c \tag{27.2}$$

for which the pairs of points (x, y) lie on a straight line of slope m that intersects the y-axis at the point $(0, c)$. Such a relationship would come about if the *length* of the spring were considered as a function of the suspended mass. In this case m is the extension per unit mass and c is the length of the unstretched spring.

For the two examples we have considered, Newton's rings and the stretched spring, with careful measurements the pairs of values (x, y) would all be close to a straight line – but not precisely on it. There are other situations where the pairs of values would indicate a straight-line relationship fairly convincingly, but still lie appreciably off a straight line. The following are some measurements of the resistance of a coil of wire as a function of its temperature:

Temperature (K)	273	293	313	333	353	373
Resistance (Ω)	1.533	1.571	1.598	1.649	1.673	1.718.

These are shown plotted in Figure 27.3, and everyone would agree that the relationship between the two quantities is almost certainly linear, even though no straight line goes through all the points. There would be less agreement about how to draw the 'best' straight line, although the deviations between the different straight lines would not be great.

 Exercise 27.1
Three points – (1, 1), (3, 4), and (6, 7) – come from measurements (with errors) of quantities connected by a straight-line relationship. Calculate the m and c for straight lines joining each pair of points, and draw the line corresponding to the average m and c.

27.2 **Lines of regression**

Depending on the nature of the data, one can set up mathematical criteria that enable a 'best' straight line to be found. The simplest such line is a **line of regression**, which is appropriate when one of the variables is error-free. For example, the number of a Newton's ring is an integer not subject to error and all the error in a plot would be associated with the measurement of the radius of the ring. By contrast, in the resistance example the measurement of both resistance and temperature would be subject to error. However, even in this case, if temperature could be measured with high accuracy compared with measurements of resistance then a line of regression would still be appropriate.

We now consider a general case where the ith pair of observations is (x_i, y_i) for which x_i has no associated error, but with the same **standard error**, ε_y, for all y measurements. This means that if a particular y_i were to be measured many times the values would have a normal distribution with mean $_t y_i$, the true value of y_i, and standard deviation ε_i. Thus the probability of the ith measurement of y being within an interval δ with y_i at its centre is

$$P_i(y_i)\delta = \frac{1}{\sqrt{2\pi\varepsilon_y^2}} \exp\left(-\frac{(y_i - _t y_i)^2}{2\varepsilon_y^2}\right)\delta. \tag{27.3}$$

The straight line that we seek, of the form (27.2), is such that all the points $(x_i, _t y_i)$ fall on it and the relative probability that an individual measurement y_i would be made is

$$p_i(y_i) = C\exp\left\{-\frac{(y_i - _t y_i)^2}{2\varepsilon_y^2}\right\}, \tag{27.4}$$

where C is some proportionality constant. The combined relative probability of making the complete set of independent measurements of y is the product of the separate probabilities (§12.2.2) or

$$P(\{y\}) = \prod_{i=1}^{n} C\exp\left(-\frac{(y_i - _t y_i)^2}{2\varepsilon_y^2}\right) = C^n \exp\left(-\frac{1}{2\varepsilon_y^2}\sum_{i=1}^{n}(y_i - _t y_i)^2\right). \tag{27.5}$$

The 'best' line is now defined as that which makes the combined probability of the set of observations of y a maximum or that which makes the quantity

$$S = \sum_{i=1}^{n}(y_i - _t y_i)^2 \tag{27.6}$$

a minimum. Since the point $(x_i, _t y_i)$ lies on the line $y = mx + c$ we may replace $_t y_i$ by $mx_i + c$ so that the quantity to be minimized is

$$S = \sum_{i=1}^{n}(y_i - mx_i - c)^2. \tag{27.7}$$

Figure 27.4 shows the points (x_i, y_i) and $(x_i, _t y_i)$, from which it is clear that the minimization condition corresponds to finding the line that minimizes the sum of

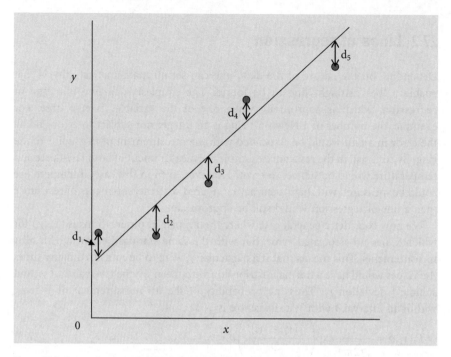

Figure 27.4 A set of points and the line of regression of y on x.

the squares of the vertical distances, d_i, from the observed points to the line. This is known as the **line of regression of y on x.**

The minimum of S will occur when $\partial S/\partial m = 0$ and $\partial S/\partial c = 0$. Partial differentiation with respect to m and the condition for minimization gives

$$\frac{\partial S}{\partial m} = -2 \sum_{i=1}^{n} x_i(y_i - mx_i - c) = 0$$

or

$$\sum_{i=1}^{n} x_i y_i - m \sum_{i=1}^{n} x_i^2 - c \sum_{i=1}^{n} x_i = 0. \tag{27.8}$$

Dividing (27.8) throughout by n gives

$$\langle xy \rangle - m\langle x^2 \rangle - c\langle x \rangle = 0 \tag{27.9}$$

where $\langle xy \rangle$ in this chapter means the average value of the products xy.

Similarly, by partial differentiation with respect to c and the minimization condition we find

$$\langle y \rangle - m\langle x \rangle - c = 0. \tag{27.10}$$

Solving (27.9) and (27.10) for m and c

$$m = \frac{\langle xy \rangle - \langle x \rangle \langle y \rangle}{\langle x^2 \rangle - \langle x \rangle^2} = \frac{\langle xy \rangle - \langle x \rangle \langle y \rangle}{\sigma_x^2} \quad \text{and} \quad c = \frac{\langle y \rangle \langle x^2 \rangle - \langle x \rangle \langle xy \rangle}{\sigma_x^2} \tag{27.11}$$

Table 27.1 Derivation of quantities for the line of regression of resistance on temperature

	Temperature x	Resistance y	x^2	xy
	273	1.533	74 529	418.51
	293	1.571	85 849	460.30
	313	1.598	97 969	500.17
	333	1.649	110 889	572.09
	353	1.673	124 609	590.57
	373	1.718	139 129	640.81
Σ	1938	9.742	632 974	3182.45
$\langle\rangle$	323	1.6237	105 496	530.40

where σ_x^2 is the standard deviation of the set of values of x_i. It should be noted that the value of ε_y is not directly involved in deriving the characteristics of the line of regression. It is necessary, however, that the standard error for y should be the same for all observations and, of course, for a line of regression of y on x the values of x must be error-free.

A line of regression of y on x can be used to predict values of y from values of x taking all the data into account. By interchanging the roles of x and y it is possible to find a line of regression of x on y if this is required.

27.3 A numerical application

We now apply this result (27.11) to the resistance data given on page 480; the necessary quantities are shown in Table 27.1.

From (27.11), $m = 1.8314 \times 10^{-3}$ and $c = 1.03212$. Figure 27.3 shows the line of regression that would enable estimates to be made of the resistance within the temperature range and, by extrapolation, outside the range. However, extrapolation is not a safe procedure and should not be taken too far outside the range of observations.

Exercise 27.2
Find the line of regression of y on x for the three points given in Exercise 27.1.

27.4 The linear correlation coefficient

A property of interest for a set of measured (x, y) data is how well the points fit a straight line. Figure 27.5 shows three sets of points, together with lines that have the property that the sum of the squares of the **perpendicular distance** of the

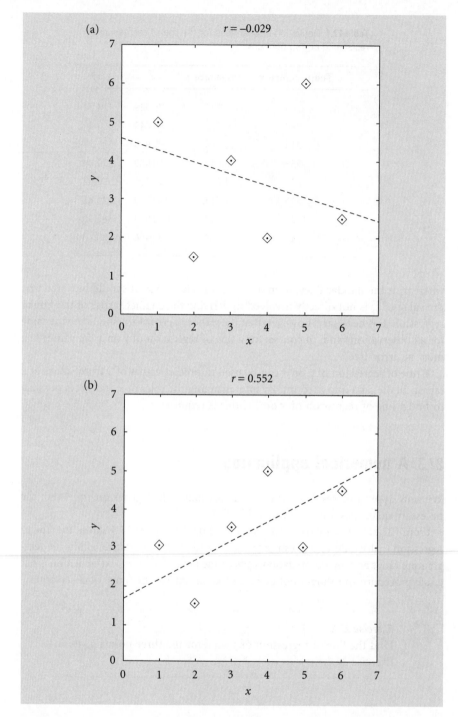

Figure 27.5 Sets of points with linear correlation coefficients (a) −0.029, (b) 0.552, and (c) −0.919.

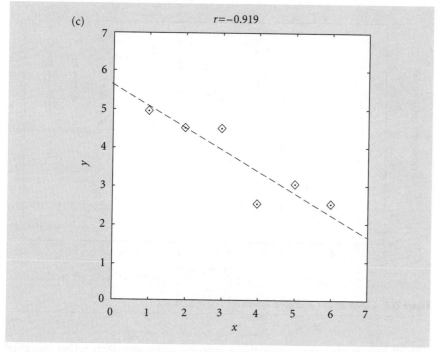

Figure 27.5 (*Contd.*)

points to the line is a minimum (see §27.5). It is clear that the points of set (a) fit the line so badly that the line could never be found just by visual judgement. The points of set (b) are well off the line but, nevertheless, define the general position of the line moderately well, while set (c) fits quite well and a visually judged line would not be too different from that shown.

A quantity that describes the quality of the collinearity of a set of points is the **linear correlation coefficient**, to which reference has already been made in (17.5):

$$r = \frac{\langle xy \rangle - \langle x \rangle \langle y \rangle}{\sigma_x \sigma_y} \tag{27.12}$$

where $\sigma_x = (\langle x^2 \rangle - \langle x \rangle^2)^{1/2}$, the standard deviation of the values of x, with a corresponding expression for σ_y. By the nature of the expression (27.12) the value of r is restricted to the range $1 \le r \le -1$. If $r = 1$ this means that the points all fall *exactly* on a straight line with a positive slope; for $r = -1$ the points again fall *exactly* on a line but this time with a negative slope. The values of r are given in Figure 27.5 for the three distributions of points.

There are many situations where correlation is sought – for example, to measure the relationship between the yield of a particular crop and the quantity of fertilizer used. However, we shall use an example from the field of education.

EXAMPLE 27.1

In the passage from school to university in the UK, results gained in the final school examinations (A-levels) are used as a criterion for making offers. It is

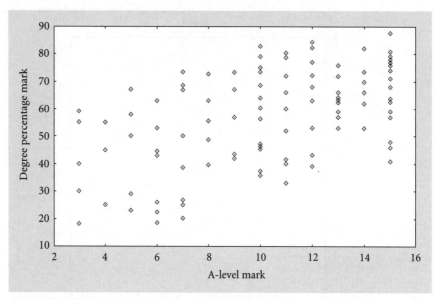

Figure 27.6 A-level and final degree marks for 100 individuals.

well known that very often those entering university with rather marginal A-level results get higher-grade degrees and, conversely, those entering with good A-level results occasionally get lower-grade degrees or perhaps no degree at all. In the development of educational policy it is useful to determine a linear correlation coefficient linking A-level results and degree results, but to do this it is necessary numerically to quantify the two kinds of result. One convenient way of doing this for A-level results is to assign numbers such as A = 5, B = 4, C = 3, D = 2, and E = 1 to the passing grades and to take the marks for the best three subjects if more than three are taken. This gives a scale of numbers from 0 to 15. Degree classifications are a rather coarse measure of performance, so the percentage performance in the final examination would be an appropriate quantity to take here – a scale going from 0 to 100. Figure 27.6 shows the A-level and degree results for 100 individuals. It is clear that the correlation is far from perfect but it can be quantified by the linear correlation coefficient (27.12), which is 0.529, for this example.

Exercise 27.3
Find the linear correlation coefficient for the three points given in Exercise 27.1.

27.5 A general least-squares straight line

When both the quantities (x, y) are subject to measurement errors, as could be the situation with the resistance example given above, then the definition of a 'best' straight line is still possible but a little more complicated. The least-squares straight

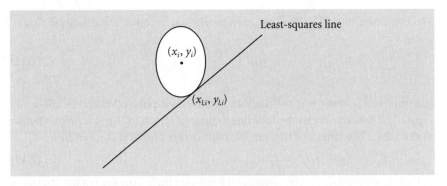

Figure 27.7 The elliptical contour shown around the measured point (x_i, y_i) is that tangential to the least-squares line. The point of contact of ellipse with line is at $(x_{l,i}, y_{l,i})$.

line (LSSL), which is used in this case, then depends on the standard (root-mean-square) error of measuring each of the quantities – ε_x and ε_y, respectively. For probability distributions of type (27.3) for both x and y, the probability density for the true values around a measured point (x_i, y_i) is of the general form

$$P(x, y) = D \exp\left\{-\left[\frac{(x - x_i)^2}{2\varepsilon_x^2} + \frac{(y - y_i)^2}{2\varepsilon_y^2}\right]\right\},\qquad(27.13)$$

where D is some constant. Contours of constant probability are given by

$$\frac{(x - x_i)^2}{2\varepsilon_x^2} + \frac{(y - y_i)^2}{2\varepsilon_y^2} = \phi\qquad(27.14)$$

for which ϕ increases for decreasing probability, and these describe a family of ellipses all centred on the point (x_i, y_i). The least squares situation is now indicated by Figure 27.7, which shows the measured point (x_i, y_i) and the LSSL.

The probability associated with the point $(x_{l,i}, y_{l,i})$ is of the form

$$P(x_{l,i}, y_{l,i}) = D \exp\left\{-\left[\frac{(x_i - x_{l,i})^2}{2\varepsilon_x^2} + \frac{(y_i - y_{l,i})^2}{2\varepsilon_y^2}\right]\right\},\qquad(27.15)$$

and the LSSL is that which makes the product of the probabilities for all i a maximum.

Finding the LSSL directly by this approach is not straightforward, but a better way is readily available. The numerical values of both ε_x and ε_y depend on the system of units being used so their values are not immutable. By suitably scaling one or other of x or y the standard errors can be made equal in magnitude and in that case contours of equal probability are not general ellipses but are circles. If we change x to $x' = kx$ where

$$k = \frac{\varepsilon_y}{\varepsilon_x}\qquad(27.16)$$

then the standard error of x' will equal ε_y. It should be noted that ε_x and ε_y are dimensioned quantities with the dimensions of x and y respectively.

Replacing x_i by x'_i in (27.15) gives

$$P(x'_{l,i}, y_{l,i}) = D \exp\left\{-\frac{(x'_i - x'_{l,i})^2 + (y_i - y_{l,i})^2}{2\varepsilon_y^2}\right\}.\qquad(27.17)$$

The condition that $\prod_{i=1}^{n} P(x'_{l,i}, y_{l,i})$ is a maximum becomes the condition that

$$S = \sum_{i=1}^{n} \left\{ (x'_i - x'_{l,i})^2 + (y_i - y_{l,i})^2 \right\} \tag{27.18}$$

is a minimum, where n is the number of measured points. When the ellipses in Figure 27.7 become circles then the line joining (x'_i, y_i) to $(x'_{l,i}, y_{l,i})$ is perpendicular to the LSSL. The length of the line joining the two points is d_i given by

$$(x'_i - x'_{l,i})^2 + (y_i - y_{l,i})^2 = d_i^2, \tag{27.19}$$

so that the straight line being sought is the one that makes the sum of the squares of the perpendicular distances of the points to the line a minimum. The procedure for finding the LSSL in terms of the unscaled values of x is thus as follows:

1 Transform x to x' by $x' = kx$.
2 Find the line $y = m'x' + c'$ that makes S a minimum.
3 Transform x' to x to give the required LSSL, $y = mx + c$ where $m = km'$ and $c = c'$.

We must now consider how to achieve step 2.

27.5.1 LSSL for equal standard errors

We use a standard result, derived in Appendix 18, that the perpendicular distance of the point (x'_i, y_i) to the line $y = m'x' + c$ is

$$d_i = \frac{y_i - m'x'_i - c}{\sqrt{1 + m'^2}}. \tag{27.20}$$

Hence the quantity to be minimized with respect to m' and c is

$$S = \sum_{i=1}^{n} \frac{(y_i - m'x'_i - c)^2}{1 + m'^2}. \tag{27.21}$$

The condition $\partial S / \partial c = 0$ for minimization gives

$$-\frac{2}{1 + m'^2} \sum_{i=1}^{n} (y_i - m'x'_i - c) = 0.$$

Ignoring the factor preceding the summation and dividing by n leads to

$$\langle y \rangle - m' \langle x' \rangle - c = 0.$$

or

$$c = \langle y \rangle - m' \langle x' \rangle. \tag{27.22}$$

Substituting this value of c in (27.21) gives

$$S = \sum_{i=1}^{n} \frac{\left\{ (y_i - \langle y \rangle) - m' \left(x'_i - \langle x' \rangle \right) \right\}^2}{1 + m'^2} \tag{27.23a}$$

and writing $Y_i = y_i - \langle y \rangle$ and $X_i = x'_i - \langle x' \rangle$ gives the function to be minimized:

$$S = \sum_{i=1}^{n} \frac{(Y_i - m' X_i)^2}{1 + m'^2}. \tag{27.23b}$$

Now the other condition for minimization is

$$\frac{\partial S}{\partial m'} = -\frac{2m'}{(1+m'^2)^2}\sum_{i=1}^{n}(Y_i - m'X_i)^2 - \frac{2}{1+m'^2}\sum_{i=1}^{n}X_i(Y_i - m'X_i) = 0. \qquad (27.24)$$

Expanding and simplifying this gives a quadratic equation for m' of the form

$$\langle XY \rangle m'^2 + \{\langle X^2 \rangle - \langle Y^2 \rangle\}m' - \langle XY \rangle = 0.$$

Now

$$\langle X^2 \rangle = \langle (x' - \langle x' \rangle)^2 \rangle = \sigma_{x'}^2$$

and similarly

$$\langle Y^2 \rangle = \sigma_{y'}^2.$$

Also

$$\langle XY \rangle = \langle (x' - \langle x' \rangle)(y - \langle y \rangle) \rangle = \langle x'y \rangle - \langle x' \rangle \langle y \rangle = A', \text{ say.} \qquad (27.25)$$

The equation for m' now appears as

$$A'm'^2 + (\sigma_{x'}^2 - \sigma_y^2)m' - A' = 0. \qquad (27.26)$$

The solutions of this quadratic equation are

$$m' = \frac{\sigma_y^2 - \sigma_{x'}^2 \pm \sqrt{\left(\sigma_{x'}^2 - \sigma_y^2\right)^2 + 4A'^2}}{2A'}. \qquad (27.27)$$

and we must decide which of the two possible solutions is appropriate in this case. The square-root quantity in the numerator of (27.27) is larger in magnitude than the sum of the two terms that precede it. Hence, if the positive sign is taken the numerator is positive and m' has the same sign as A'. Conversely, if the negative sign is taken then the numerator is negative and the sign of m' is opposite to that of A'. We now examine (27.12), the numerator of which is

$$\langle xy \rangle - \langle x \rangle \langle y \rangle = A = A'/k \qquad (27.28)$$

and it is clear that the sign of the linear correlation coefficient, r, which is also the sign of the slope of the line, is the same as that of A. This indicates that it is the positive sign that must be taken in (27.27). The value of m, the slope of the LSSL for the unscaled data, is thus

$$m = km' = \frac{\sigma_y^2 - k^2\sigma_x^2 + \sqrt{(k^2\sigma_x^2 - \sigma_y^2)^2 + 4k^2A^2}}{2A} \qquad (27.29)$$

when all the transformation from primed to unprimed quantities is carried out. The value of c can then be obtained from (27.22) replacing $m'\langle x' \rangle$ by the equivalent quantity $m\langle x \rangle$.

Table 27.2 Variation of the volume of a liquid with temperature

	Temperature (K)	Volume (m³)
	291	1.059 0
	307	1.064 0
	334	1.064 2
	344	1.069 4
	360	1.070 8
	371	1.076 1
σ	28.052	$5.524\,1 \times 10^{-3}$

The regression line of y on x is a special case of an LSSL when $\varepsilon_x \to 0$ or $k \to \infty$. When this is so we may neglect σ_y^2 compared with $k^2\sigma_x^2$ and (27.29) may be written as

$$m = \frac{-k^2\sigma_x^2 + k^2\sigma_x^2\left(1 + \dfrac{4A^2}{k^2\sigma_x^4}\right)^{1/2}}{2A}. \qquad (27.30)$$

The square-root term in (27.30) is of the form $(1 + \alpha)^{1/2}$, where $\alpha \ll 1$ and so, from (1.38) can be approximated by $1 + \frac{1}{2}\alpha$. This gives, from (27.30),

$$m = \frac{A}{\sigma_x^2} = \frac{\langle xy \rangle - \langle x \rangle \langle y \rangle}{\sigma_x^2},$$

which is the result found for the regression line of y on x in (27.11).

27.5.2 Numerical determination of a LSSL

In the determination of the line of regression in §27.3 it was assumed that temperature could be measured without error. The assumption that the standard error of a temperature measurement is 0.5 K ($\equiv \varepsilon_x$) and that of measuring resistance is 0.005 Ω ($\equiv \varepsilon_y$) gives a value of $k = 0.01$. From the values in Table 27.1 and (27.29) and (27.22), the LSSL is found as

$$y = 1.83177x + 1.03200$$

which is hardly distinguishable from the line of regression shown in Figure 27.3.

One situation in which a line of regression and a LSSL will be similar is if the measured points are very closely collinear. However, even if the points are far from collinear the LSSL may still be similar to a line of regression. Table 27.2 gives data for a not very precise measurement of the volume of a liquid as a function of its temperature. These are plotted in Figure 27.8, together with the regression lines of volume on temperature and temperature on volume. The two lines are similar but quite distinctive. That of volume on temperature is

$$V = 1.871819 \times 10^{-4}T + 1.00464 \qquad (27.31a)$$

whereas that of temperature on volume is

$$V = 2.071699 \times 10^{-4}T + 0.99795. \qquad (27.31b)$$

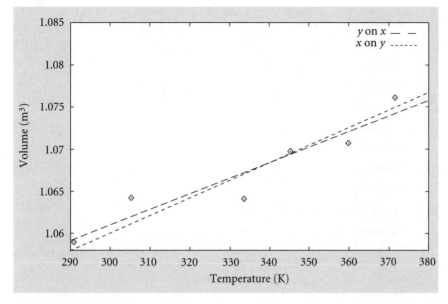

Figure 27.8 Experimental points for the variation of the volume of a liquid with temperature together with the two lines of regression.

The values of ε_x and ε_y do not give a true picture of the relative accuracy with which the quantities x and y have been measured. For example, if $\varepsilon_x = 0.1$ then this would be a large error if the values of x were of the order 0.5 but insignificant if the values of x were closer to 10^5. However, of more importance is the spread of the values of x. If all the measured values of x were in the range 100 000 to 100 001 then $\varepsilon_x = 0.1$ would represent a significant error. For this reason exactly where a LSSL falls between the two regression lines (27.31) depends on the value of

$$k' = \frac{\sigma_y}{\sigma_x}, \tag{27.32}$$

which is $5.5241 \times 10^{-3}/28.0520 = 1.9692 \times 10^{-4}$ in this example. If $k \gg k'$, the LSSL resembles the line of regression of volume on temperature; conversely, if $k \ll k'$, the LSSL resembles the line of regression of temperature on volume. The variation of m and c for values of k spanning k' is shown in Figure 27.9.

The LSSL depends on the ratio of standard errors and not on their individual values. As long as some reasonable estimate of the *relative* accuracy of estimating the quantities involved can be made, then the LSSL will give a good representation of their relationship. If it turns out that k is either much less or much greater than k' then one or other of the lines of regression would be appropriate.

Exercise 27.4

If the standard errors of the values of x and y in Exercise 27.1 are equal then:

  (i) What is k as defined in (27.16)?

  (ii) What is A as defined in (27.28)?

  (iii) Find m from (27.29).

  (iv) Find c from (27.22).

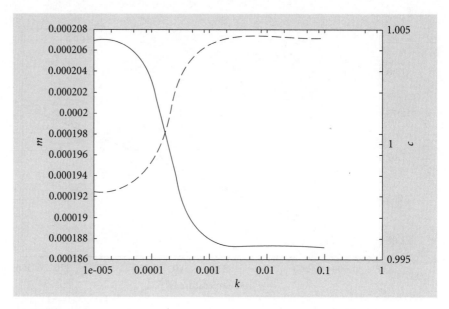

Figure 27.9 Variation of m (full line) and c (broken line) for values of k spanning k'.

27.6 **Linearization of other forms of relationship**

The number of atoms of a radioisotope changes with time according to the differential equation

$$\frac{\mathrm{d}N}{\mathrm{d}t} = -\lambda N \tag{27.33}$$

where N is the number of atoms present and λ is the **decay constant**. The solution of this equation is

$$N = N_0 e^{-\lambda t} \tag{27.34}$$

where the number of atoms at time $t = 0$ is N_0.

The **half-life** of the radioisotope is the time for one-half of the atoms to decay and is given by

$$\tau_{1/2} = \frac{\ln(2)}{\lambda} = \frac{0.6931}{\lambda}. \tag{27.35}$$

For a radioactive source of short half-life – of the order of days or weeks – the decay constant can be determined by measuring the activity, as the number of disintegrations per unit time, at intervals over a period of time. From (27.33) and (27.34) the relationship is

$$-\frac{\mathrm{d}N}{\mathrm{d}t} = \lambda N_0 e^{-\lambda t} = r \tag{27.36}$$

where r is the activity of the source after time t. The relationship between r and t is not linear but by taking natural logarithms

$$\ln(r) = -\lambda t + \ln(\lambda N_0) \tag{27.37}$$

and $\ln(r)$ is linearly related to t. A plot such as $\ln(r)$ against t is called a **log-linear plot**. An application of (27.37) is involved in Problem 27.5.

Another kind of non-linear relationship that can be linearized is where powers are involved, for example

$$y = Ax^n. \qquad (27.38)$$

This time we take normal logarithms to base 10 to give

$$\log(y) = n\log(x) + \log(A) \qquad (27.39)$$

and the **log-log plot**, of $\log(y)$ against $\log(x)$, is linear.

EXAMPLE 27.2

A log-log relationship describes the distance travelled by an object falling in the Earth's gravitational field as a function of time. This is

$$s = \tfrac{1}{2}gt^2 \qquad (27.40)$$

or

$$\log(s) = 2\log(t) + \log(\tfrac{1}{2}g). \qquad (27.41)$$

We now estimate g from the following data measured in a dropping experiment:

s (m)	10	20	30	40	50	60	70
t (s)	1.4	2.0	2.5	2.8	3.2	3.5	3.8.

In this case $\log(s)$ plays the role of y, $\log(t)$ the role of x and $\log(\tfrac{1}{2}g)$ the role of c in (27.2). We know that $m = 2$ and does not have to be determined. Thus the function to be minimized, equivalent to (27.7), is

$$S = \sum_{i=1}^{n} (y_i - 2x_i - c)^2 \qquad (27.42)$$

and only c has to be determined. The minimization condition $\partial S/\partial c = 0$ leads to

$$c = \langle y \rangle - 2\langle x \rangle$$

which for this example gives

$$\log(\tfrac{1}{2}g) = \langle \log(s) \rangle - 2\langle \log(t) \rangle. \qquad (27.43)$$

Table 27.3 gives the required calculations. From these results,

$$\log(\tfrac{1}{2}g) = 1.529 - 2 \times 0.417 = 0.695$$

and, putting in the units,

$$g = 2 \times 10^{0.695} = 9.91 \text{ m s}^{-2}.$$

In this example the slope of the line was fixed and the quantity to be determined was the intercept. In Problem 27.2 it is the intercept that is known, since the straight line goes through the origin, and it is the slope that must be determined.

Table 27.3 Results of the dropping experiment

s	$\log(s)$	t	$\log(t)$
10	1.000	1.4	0.146
20	1.301	2.0	0.301
30	1.477	2.5	0.398
40	1.602	2.8	0.447
50	1.699	3.2	0.505
60	1.778	3.5	0.544
70	1.845	3.8	0.580
$\langle\ \rangle$	1.529		0.417

Exercise 27.5

The relationship between the length, l, of an animal and its weight, w, is of the form $w = al^3$. Three animals have the following lengths and weights:

l (cm)	w (kg wt)
10	0.20
45	21.0
150	650.0

Find the best value of a for predicting values of w from values of l. Use your value of a to obtain predicted values of w from the given values of l.

Problems

27.1 A sphere is dropped in a vacuum tube and its passage is timed electronically at intervals of 0.1 m. Timing starts when it has fallen 0.3 m and the times recorded are:

s (m)	0.3	0.4	0.5	0.6	0.7	0.8
t (s)	0.000	0.037	0.073	0.109	0.128	0.161

where s is the distance from the starting point. Show analytically that t and s are related by

$$t = \sqrt{\frac{2}{g}}\sqrt{s} - t_0$$

where g is the acceleration due to gravity. By finding the line of regression of t on $\sqrt{s}$ estimate the value of g.

An important conclusion is drawn from this problem. Read the solution before passing on to other problems.

27.2 For a Newton's rings experiment, the radius of the nth ring is given by $r_n^2 = 2R\lambda n$ where R is the radius of curvature of the convex lens surface and λ is the wavelength of the light. For a lens surface of radius 0.5 m the measured radii of rings were as follows:

n	5	10	15	20	25	30
r_n (mm)	1.61	2.28	2.81	3.12	3.44	3.93

Find the line of regression of r_n^2 on n and hence estimate the wavelength of the light being used. (Note: The line must pass through the origin and be of the form $y = mx$.) Derive an expression for the best value of m based on the analysis given in §27.2.

27.3 The integrated power received in 1 s periods from a weak astronomical radio source is recorded. To test for possible periodicity the linear correlation coefficient is calculated for pairs of readings at time t and $t+\tau$ where τ is a possible period to be tested. Corresponding signals, s, for $\tau = 0$, 5 s and 8 s are:

$s(t)$	1	4	5	7	7	6	8	2	0	4	6	10	11	7	5
$s(t+5)$	7	5	6	2	1	1	4	6	8	2	0	4	6	10	11
$s(t+8)$	0	4	6	10	11	7	5	6	2	4	5	7	7	6	8

Calculate the linear correlation coefficients for pairs of times $(t, t+5)$ and $(t, t+8)$ and comment on your results.

27.4 A cord is attached to a point in a very high church tower that cannot easily be accessed. To estimate the length of the cord a mass is attached to it to produce a simple pendulum, the period of which is measured. The period, P, when the mass is attached at different distances, x, from the bottom of the chord are as follows:

x	0.0	0.5	1.0	1.5	2.0
P	8.95	8.79	8.70	8.57	8.48

If the length of the cord is l then show that

$$P^2 = -\frac{4\pi^2}{g}x + \frac{4\pi^2}{g}l,$$

where g is the acceleration due to gravity.

It is estimated that the standard error in measuring x is 0.05 m and that of measuring P^2 is $0.2\,\text{s}^2$ (note: this corresponds to an error in measuring P itself of about 0.01 s). Calculate an LSSL for the data and hence estimate l and g.

27.5 A substance has a slight contamination of an unknown α-emitting radioactive material and it is required to determine the nature of the contaminant without chemical analysis. The unknown substance can be identified if its half-life can be determined. A series of measurements are made at one-day intervals of the number of alpha-particles emitted per minute with the following result:

t (days)	0	1	2	3	4	5
number	995	361	141	54	18	8

From this data, estimate the half-life of the contaminant.

Interpolation

When numerical information is provided in tabular form, the process of interpolation can be used to find values between the tabulated points. This chapter describes interpolation techniques at various levels of complexity, from linear interpolation, where a linear relationship is assumed between tabulated points, to cubic spline interpolation, which requires the interpolation function to be continuous in both slope and second derivative at tabulated points. The process of multidimensional interpolation is also described.

28.1 Applications of interpolation

In any situation in which one variable, y, is a continuous function of another variable, x, the relationship can be expressed by giving tabulated values of x and y. Table 28.1 expresses the height of a plant at intervals of 2 weeks during the growing season.

If we have to estimate the height at 7 weeks then, clearly, it is somewhere between 9.9 cm and 14.3 cm. Since the height is required at the midpoint of the interval an average of these two values, 12.1 cm, would obviously be a sensible estimate. In taking this estimate we have instinctively used the process of **linear interpolation**, which is the topic of §28.2, but other kinds of interpolation are possible.

Another kind of application comes from the field of computation. We can envisage a computational exercise, generated by the solution of a physical problem, where it is required to calculate 10^8 times the value of $\sin^{-1}(1 - 2e^{-x})$ for values of x between 0 and ∞. This could be the most time-consuming part of the computation, each calculation involving two intrinsic functions, so any time saved in calculating the expression would be important. From the nature of the function the upper limit of x can be truncated; for $x = 100$ the value of the function equals that for $x = \infty$ ($\pi/2$) to better than 1 part in 10^{10}. If the function is then tabulated for 10 001 values between $x = 0$ and $x = 100$ at intervals of x equal to 0.01, and interpolation used between the tabulated values, quite precise values can be found over the range with a tiny fraction of the computational effort required for individual function calculations. The number of tabulated values used and the type of interpolation employed depend on the precision required in the overall calculation.

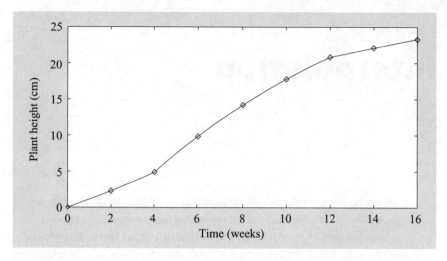

Figure 28.1 Linear interpolation graph corresponding to Table 28.1.

Table 28.1 The height of a plant at 2 week intervals from sowing

Week	Height (cm)
0	0.0
2	2.2
4	4.9
6	9.9
8	14.3
10	17.8
12	21.0
14	22.2
16	23.4

28.2 **Linear interpolation**

Figure 28.1 shows the plotted data points given in Table 28.1 linked by straight lines and this represents the continuous function of plant height against time as given by linear interpolation.

To find the height h corresponding to some time t it is first necessary to find the times $t(i)$ and $t(i+1)$ that bracket the value of t. The value of h is then given by

$$h = \frac{t(i+1) - t}{t(i+1) - t(i)} h(i) + \frac{t - t(i)}{t(i+1) - t(i)} h(i+1). \tag{28.1}$$

To check that (28.1) represents the straight line linking point i and point $i+1$ we first note that it only involves t to power 1 and hence represents a straight line. If $t = t(i)$

Table 28.2 Interpolation from data in Table 28.1

Time (weeks)	Linear interpolation	Quadratic interpolation
7.0	12.10	12.21
9.4	16.75	16.78
12.2	21.12	21.21
14.319	22.39	22.39

then the second term vanishes and the first term gives $h(i)$; if $t = t(i+1)$ then the first term vanishes and the second term gives $h(i+1)$. A straight line going through the correct end-points is the line required. A simple linear interpolation program gives the interpolated heights for different times as shown in Table 28.2.

Interpolation is faster if the data points are equally spaced in the x, or equivalent, variable, for then the flanking values i *and* $i+1$ are easily found. For example, taking the first row of the tabulated values in Table 28.1 as corresponding to $i = 0$ the lower flanking i value for $t = 7$ weeks is $\text{int}(7/2) = 3$, where int means 'the integral part of'.

Exercise 28.1

Find the estimated height of the plant after 6.3 weeks from Table 28.1 by linear interpolation.

28.2.1 Errors of linear interpolation

Most functions or data sets, unless they represent straight lines, have finite curvature, so representing them by a series of straight lines must introduce error. This error will be greatest if few divisions are taken within the range of the function and where the function has its greatest curvature. This varies from one function to another, but a general approach is possible to ascertain the maximum error for any particular function.

Consider a function $\phi(x)$, which is known for $n+1$ equally spaced values of x – $x_0, x_1, \ldots, x_n$ where the interval between neighbouring values of x is δx. We now compare the value of the function at $x_i + f\delta x$ $(0 < f \leq 1)$ with that found by linear interpolation from the known values. The error will be

$$\Delta = f\phi(x_{i+1}) + (1-f)\phi(x_i) - \phi(x_i + f\delta x). \tag{28.2}$$

Assuming that in the interval the function is not sinuous and does not cross the straight line, the pattern is that the error is zero for $f = 0$ and $f = 1$ and has a maximum magnitude somewhere between. To find this extremum we find the value of f for which $d\Delta/df = 0$ or

$$\phi(x_{i+1}) - \phi(x_i) - \frac{d}{df}\phi(x_i + f\delta x) = 0. \tag{28.3}$$

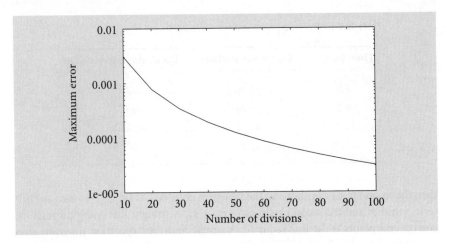

Figure 28.2 Maximum error in the range 0 to π/2 for sin(x) tabulated with different numbers of divisions.

EXAMPLE 28.1

To illustrate this procedure we consider the function $\sin(x)$ in the interval 0 to $\pi/2$, tabulated at intervals of $\pi/2n$ where the integer n is the number of intervals into which the range of x is divided. For this function (28.3) becomes

$$\sin\left\{\frac{\pi}{2n}(i+1)\right\} - \sin\left\{\frac{\pi i}{2n}\right\} - \frac{\pi}{2n}\cos\left\{\frac{\pi}{2n}(i+f)\right\} = 0$$

which gives

$$f = \frac{2n}{\pi}\cos^{-1}\left[\frac{2n}{\pi}\left\{\sin\left(\frac{\pi(i+1)}{2n}\right) - \sin\left(\frac{\pi i}{2n}\right)\right\}\right] - i. \qquad (28.4)$$

This value of f gives the maximum magnitude of error in the interval i to $i+1$ from

$$|\Delta| = \left|f\sin\left\{\frac{\pi}{2n}(i+1)\right\} + (1-f)\sin\left\{\frac{\pi i}{2n}\right\} - \sin\left\{\frac{\pi}{2n}(i+f)\right\}\right|. \qquad (28.5)$$

Figure 28.2 gives the maximum error in the whole range from 0 to $\pi/2$ for 10 to 100 divisions of the range in steps of 10 divisions. If, for example, a maximum error of 0.0001 was acceptable in the application being made then 60 intervals (each 1.5°) in the range would be adequate. A table of 2001 entries would give a maximum error of 10^{-7}, which would be adequate for all but the most demanding applications.

Exercise 28.2
A function $f(x) = x^2$ is tabulated at intervals $\Delta x = 0.1$ for the range $x = 0$ to $x = 1$. What is the maximum interpolation error in the intervals (i) $x = 0$ to $x = 0.1$ and (ii) $x = 0.9$ to $x = 1.0$?

28.3 Parabolic interpolation

Linear interpolation assumes that a curve can be approximated by a series of straight lines, and the closer are the data points the better is this approximation. The next level of approximation is to take account of the curvature of the function and to fit a parabola to three neighbouring points. The values defining the parabola are (x_{i-1}, y_{i-1}), (x_i, y_i) and (x_{i+1}, y_{i+1}) with $x_{i+1} - x_i = x_i - x_{i-1} = h$. To find the parabola we transform the x coordinate to $X = x - x_i$ so that the three points, in the form (X, y) are $(-h, y_{i-1})$, $(0, y_i)$, and (h, y_{i+1}). We now connect these points by a parabolic equation

$$y = aX^2 + bX + c \tag{28.6}$$

and find the values of a, b, and c from

$$y_{i-1} = ah^2 - bh + c \tag{28.7a}$$

$$y_i = c \tag{28.7b}$$

$$y_{i+1} = ah^2 + bh + c. \tag{28.7c}$$

These give

$$a = \frac{y_{i-1} + y_{i+1} - 2y_i}{2h^2}, \tag{28.8a}$$

$$b = \frac{y_{i+1} - y_{i-1}}{2h}, \tag{28.8b}$$

$$c = y_i \tag{28.8c}$$

with the form of the parabola, in terms of x, as

$$y = a(x - x_i)^2 + b(x - x_i) + c. \tag{28.9}$$

The procedure for applying parabolic interpolation is first to identify the value of x_i closest to the x for which y is required. Equations (28.8) are used to determine a, b, and c and then (28.9) to find the value of y. Applying parabolic interpolation to the data in Table 28.1 for the x values chosen in Table 28.2 gives results that are, in general, different from those obtained from linear interpolation. The results for $x = 14.319$ are the same because the final three points are precisely linear, making the parabolic and linear interpolation processes identical in outcome.

The value $x = 7$ weeks in Table 28.2 falls exactly halfway between two tabulated values. In the program used to find the interpolated plant height the larger tabulated value was taken as the closest. If the lower tabulated value had been taken instead then the interpolated height would have been 12.175 cm, some 0.035 cm different from that given in Table 28.2. This underlines a characteristic of parabolic interpolation, that any particular value of x falls within two local parabolic representations of the function and these are not necessarily the same. In general this does not matter; it is accepted that an interpolation process involves error and the user must ensure that the error is acceptable for the application being made.

? **Exercise 28.3**
Find the estimated height of the plant after 6.3 weeks from Table 28.1 by parabolic interpolation.

28.4 **Gauss interpolation formula**

After fitting a straight line to two points for linear interpolation and a parabola to three points for parabolic interpolation, the next logical step is to fit a cubic to four points. There is a general procedure for fitting a polynomial of degree n to $n+1$ points that gives what is called **Gauss interpolation**.

To illustrate the principle of Gauss interpolation, consider a polynomial of degree 4 passing through the five points (x_i, y_i) for $i = 0$ to 4:

$$
\begin{aligned}
y = {} & \frac{(x - x_1)(x - x_2)(x - x_3)(x - x_4)}{(x_0 - x_1)(x_0 - x_2)(x_0 - x_3)(x_0 - x_4)} y_0 \\
& + \frac{(x - x_0)(x - x_2)(x - x_3)(x - x_4)}{(x_1 - x_0)(x_1 - x_2)(x_1 - x_3)(x_1 - x_4)} y_1 \\
& + \frac{(x - x_0)(x - x_1)(x - x_3)(x - x_4)}{(x_2 - x_0)(x_2 - x_1)(x_2 - x_3)(x_2 - x_4)} y_2 \\
& + \frac{(x - x_0)(x - x_1)(x - x_2)(x - x_4)}{(x_3 - x_0)(x_3 - x_1)(x_3 - x_2)(x_3 - x_4)} y_3 \\
& + \frac{(x - x_0)(x - x_1)(x - x_2)(x - x_3)}{(x_4 - x_0)(x_4 - x_1)(x_4 - x_2)(x_4 - x_3)} y_4.
\end{aligned}
\tag{28.10}
$$

It will readily be seen that (28.10) is a quartic equation, involving powers of x^4 and no higher. When $x = x_0$ all terms other than the first term disappear and $y = y_0$. Similarly when $x = x_1$ then only the second term is non-zero and $y = y_1$, and so on. The equation clearly satisfies the condition of being a quartic going through the specified five points.

In general it is unwise to use interpolation of higher order than cubic, or perhaps quartic, since the process can become unstable and give rather sinuous and unrealistic interpolation curves, especially if the errors or uncertainties give a non-smooth progression of the data points.

28.5 **Cubic spline interpolation**

Interpolation can be regarded as finding a function linking a number of data points (x, y) such that values of y can be found anywhere within the range of the values of x. Linear interpolation gives the function as a set of linked straight lines, with the obvious fault that the slope is discontinuous at each data point. Quadratic interpolation has the fault that the function is not uniquely defined over the whole range of x – the function in the range i to $i+1$ is different for the quadratics fitted in the ranges $i-1$ to $i+1$ and i to $i+2$. Thus the interpolated values may make a discontinuous jump in moving from one side to the other of the x value represented by $i+\frac{1}{2}$. Higher-order interpolation, as can be carried out by the Gauss interpolation formula, has the same fault as the quadratic method.

A procedure which fits all the data points, does not stray into rather risky high-order interpolation, and preserves continuity of first and second derivatives over the whole range, is fitting a **cubic spline**. This fits a cubic between all neighbouring pairs of points with the coefficients of the cubic function determined so as to give the required continuity conditions. The description is simplest when the data points have a uniform interval in x, and that will be assumed here. The $n+1$ data points are (x_i, y_i) for $i=0$ to n and the function between point i and $i+1$ is

$$_iy = a_i + b_if + c_if^2 + d_if^3, \tag{28.11}$$

where f is the **fractional distance** between points i and $i+1$ so that $0 \le f < 1$. Matching the values of y and dy/df (y') at the two ends of the range gives

$$a_i = y_i \ (f = 0)$$
$$a_i + b_i + c_i + d_i = y_{i+1} \ (f = 1)$$
$$b_i = y'_i \ (f = 0)$$
$$b_i + 2c_i + 3d_i = y'_{i+1}(f = 1). \tag{28.12}$$

Equations (28.12) can be solved for a_i, b_i, c_i, and d_i as

$$a_i = y_i$$
$$b_i = y'_i$$
$$c_i = 3(y_{i+1} - y_i) - 2y'_i - y'_{i+1} \tag{28.13}$$
$$d_i = 2(y_i - y_{i+1}) + y'_i + y'_{i+1}.$$

From (28.13) it can be seen that a knowledge of the values of y and y' at points i and $i+1$ would give the individual coefficients and hence the spline function $_iy$.

To determine the unknown first derivatives we now introduce the condition that the second derivatives are continuous, that is that $_iy''(1) = {}_{i+1}y''(0)$. This gives

$$2c_i + 6d_i = 2c_{i+1}$$

or

$$6(y_{i+1} - y_i) - 4y'_i - 2y'_{i+1} + 12(y_i - y_{i+1}) + 6y'_i + 6y'_{i+1}$$
$$= 6(y_{i+2} - y_{i+1}) - 4y'_{i+1} - 2y'_{i+2},$$

giving

$$y'_i + 4y'_{i+1} + y'_{i+2} = 3(y_{i+2} - y_i). \tag{28.14}$$

This treatment does not apply for the end sections, $i=0$ and $i=n$. These sections do not have continuity conditions at both extremes of the x range, and it is necessary to apply boundary conditions. For some physically based problems this may be in the form of fixed slopes but, in the absence of any other condition, it is common to use what is called the **natural cubic spline** which makes the second derivatives zero at both boundaries.

We now consider $_0y$ with the imposed boundary condition. This gives

$$a_0 = y_0$$
$$c_0 = 0 \text{ (zero second derivative for } f = 0)$$
$$a_0 + b_0 + c_0 + d_0 = y_1$$
$$b_0 + 2c_0 + 3d_0 = y_1'. \tag{28.15}$$

These equations give

$$a_0 = y_0$$
$$b_0 = \frac{3}{2}(y_1 - y_0) - \frac{1}{2}y_1'$$
$$c_0 = 0 \tag{28.16}$$
$$d_0 = \frac{1}{2}(y_1' - y_1 + y_0).$$

We now introduce the unknown slope at the boundary:

$$y_0' = b_0 = \frac{3}{2}(y_1 - y_0) - \frac{1}{2}y_1'$$

which gives

$$2y_0' + y_1' = 3(y_1 - y_0). \tag{28.17a}$$

A similar treatment at the other boundary gives

$$y_{n-1}' + 2y_n' = 3(y_n - y_{n-1}). \tag{28.17b}$$

The $n+1$ equations given by (28.14) and (28.17) enable the $n+1$ values of y' to be determined in terms of the $n+1$ values of y. The a, b, c, and d values can then be evaluated which then give the individual spline sections. The form of the equations giving y' is:

$$2y_0' + y_1' = 3(y_1 - y_0)$$
$$y_0' + 4y_1' + y_2' = 3(y_2 - y_0)$$
$$y_1' + 4y_2' + y_3' = 3(y_3 - y_1)$$
$$\cdots$$
$$\cdots$$
$$y_{n-2}' + 4y_{n-1}' + y_n' = 3(y_n - y_{n-2})$$
$$y_{n-1}' + 2y_n' = 3(y_n - y_{n-1}). \tag{28.18}$$

Systems of linear equations can be expressed in matrix notation and the coefficient matrix on the left-hand side of (28.18) is of **tridiagonal form** (§20.5). Simultaneous linear equations in this form are straightforward to solve, no matter what the size of the system. The first equation gives y_1' in terms of y_0'; the second equation then gives y_2' in terms of y_0', and so on to the penultimate equation that gives y_n' in terms of y_0'. Substituting for y_{n-1}' and y_n' in the final equation in terms of y_0' enables y_0' to be explicitly found. Since all the other y' were previously found in terms of y_0', they too can now be evaluated.

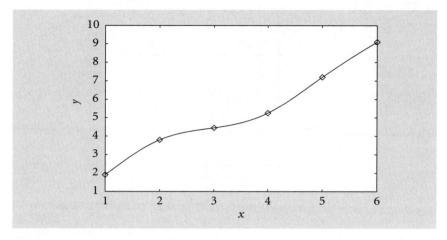

Figure 28.3 Natural cubic spline fitted to given data points.

An example of a cubic spline is shown in Figure 28.3 fitted to the data set

x	1	2	3	4	5	6
y	1.90	3.80	4.45	5.25	7.20	9.10

In accordance with the 'natural' form of the cubic spline the curve has zero curvature at the ends, $x=1$ and $x=6$.

It should be noted that the given analysis depends on the data being given at uniform intervals of x. If that is not so, then the boundary condition

$$_iy'(1) = {_{i+1}}y'(0)$$

would not be true since y' is dy/df and not dy/dx. It is still possible to find a cubic spline for unequal x intervals but the analysis then introduces factors corresponding to the ratios of neighbouring intervals.

Computer programs exist for interpolation by the cubic spline process (see for example *Numerical Recipes* by Press *et al.* – details on p. 407).

28.6 **Multidimensional interpolation**

Suppose that we have a two dimensional table of a function $f(x, y)$ tabulated at points (x_i, y_j). To use two-dimensional linear interpolation to find $f(x, y)$ we first need to identify the **cell** within which the point (x, y) exists, i.e. the values of i and j such that

$$x_i \le x < x_{i+1} \quad \text{and} \quad y_j \le y < y_{j+1}. \tag{28.19}$$

The cell containing the point (x, y) and the surrounding cells are shown in Figure 28.4.

The linearly interpolated value at the point (x, y) is given by

$$f(x, y) = w_{i,j}f(i, j) + w_{i+1,j}f(i+1, j) + w_{i,j+1}f(i, j+1)$$
$$+ w_{i+1,j+1}f(i+1, j+1), \tag{28.20}$$

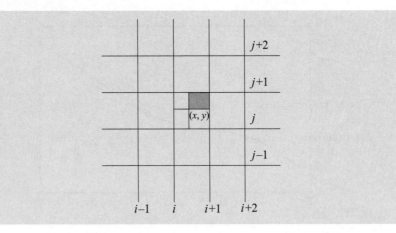

Figure 28.4 The point (x, y) within the cell with origin at (i, j).

where the weights are proportional to areas within the cell, one of which is shown shaded in Figure 28.4. The weight for each of the four points is the area of the 'opposite' rectangle as a fraction of the total area of the cell, i.e.

$$w_{i,j} = \frac{\text{shaded area}}{(x_{i+1} - x_i)(y_{j+1} - y_j)}. \tag{28.21}$$

Writing $A = (x_{i+1} - x_i)(y_{j+1} - y_j)$, we have

$$w_{i,j} = \frac{1}{A}(x_{i+1} - x)(y_{j+1} - y) \quad w_{i+1,j} = \frac{1}{A}(x - x_i)(y_{j+1} - y)$$

$$w_{i,j+1} = \frac{1}{A}(x_{i+1} - x)(y - y_j) \quad w_{i+1,j+1} = \frac{1}{A}(x - x_i)(y - y_j), \tag{28.22}$$

which, inserted into (28.20), give the desired interpolated value.

Two-dimensional interpolation can be very valuable to reduce computing times when a function of two variables needs to be calculated many times. The same principle can be extended to higher-dimensional interpolation for functions of many variables – for example, $f(x, y, z)$ tabulated at points (x_i, y_j, z_k). The values of (i, j, k) at the origin of the rectangular parallelepiped cell containing the point (x, y, z) must first be identified. Then

$$f(x, y, z) = w_{i,j,k} f(i, j, k) + w_{i+1,j,k} f(i + 1, j, k)$$

$$+ w_{i,j+1,k} f(i, j + 1, k) + w_{i,j,k+1} f(i, j, k + 1)$$

$$+ w_{i+1,j+1,k} f(i + 1, j + 1, k) + w_{i+1,j,k+1} f(i + 1, j, k + 1)$$

$$+ w_{i,j+1,k+1} f(i, j + 1, k + 1) + w_{i+1,j+1,k+1} f(i + 1, j + 1, k + 1). \tag{28.23}$$

For $V = (x_{i+1} - x_i)(y_{j+1} - y_j)(z_{k+1} - z_k)$ we have

$$w_{a,b,c} = \frac{1}{V} \phi_a \phi_b \phi_c,$$

where

for $a = i$ $\phi_a = (x_{i+1} - x)$ and for $a = i + 1$ $\phi_a = (x - x_i)$

for $b = j$ $\phi_b = (y_{j+1} - y)$ and for $b = j + 1$ $\phi_b = (y - y_j)$ (28.24)

for $c = k$ $\phi_c = (z_{k+1} - z)$ and for $c = k + 1$ $\phi_c = (z - z_k)$.

This general pattern can be repeated for interpolation in more than three dimensions.

Higher-order interpolation (e.g. parabolic) is also possible for multidimensional tables but, in general, linear interpolation is sufficient for most applications.

28.7 Extrapolation

Extrapolation is the process by which estimated values for y are found outside the given range of the x values. It is a very uncertain procedure and assumes that the function behaves in a well-mannered and predictable way both inside and outside the interval of x available. A possible procedure is to find the *interpolation* equation in the neighbouring section within the range of x and to use this to estimate the extrapolated value. As long as the extrapolation is not too far outside the range of x, a useful estimate may be found.

Although there are algorithms for extrapolation that can be applied in special circumstances the general advice is 'do not extrapolate unless you really have to'.

Problems

28.1 For the data set

x	0	1	2	3
y	0	1.5	2.0	3.5

find interpolated values of y for $x = 2/3$ and $x = 1.7$ for (i) linear interpolation, (ii) parabolic interpolation, and (iii) natural cubic spline interpolation.

28.2 The following is part of a two-dimensional table of $f(x, y)$:

			x		
		0.1	0.2	0.3	0.4
	0.1	0.000316	0.000894	0.001643	0.002530
	0.2	0.001265	0.003578	0.006573	0.010119
y	0.3	0.002848	0.008050	0.014789	0.022768
	0.4	0.005060	0.014311	0.026291	0.040477

By linear interpolation, find the value of the function at (2.1, 2.7).

Quadrature

The mathematics of definite integrals has already been dealt with, in Chapters 5 and 13. In all the examples dealt with in those chapters we were able to integrate the integrands and find the integrals by pure analysis. However, there are many situations where analytical integration cannot be carried out and then we have to use numerical integration or quadrature to obtain a solution. The remainder of this chapter will describe various methods of quadrature, with different degrees of complexity.

29.1 Definite integrals

The simplest form of a definite integral, involving a single variable, is

$$A(a, b) = \int_a^b f(x)\, dx. \tag{29.1}$$

Definite integrals involving more than one independent variable can also occur. The simplest case is

$$B(a, b, c, d) = \int_{y=c}^d \int_{x=a}^b f(x, y)\, dx\, dy. \tag{29.2}$$

This is the volume under the surface $f(x, y)$ within the rectangular area defined by $x = a$ to b and $y = c$ to d. Integrals with two independent variables, x and y, can be of a more complex form, for example

$$C(a, b) = \int_a^b \int_{g_1(x)}^{g_2(x)} f(x, y)\, dy\, dx. \tag{29.3}$$

This is the integral of the function $f(x, y)$ within the shaded area shown in Figure 29.1.

Now we describe a number of methods of numerically evaluating definite integrals.

29.2 Trapezium method

The trapezium method (often called the **trapezium rule**) consists of approximating the area under the curve by the sum of the areas of the trapezia shown in Figure 29.2. The range of x, between a and b is divided into n equal segments, each

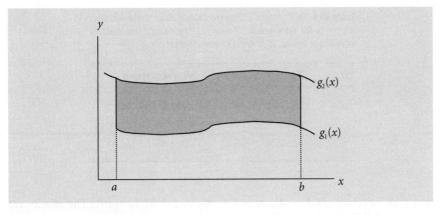

Figure 29.1 Integration area for (29.3).

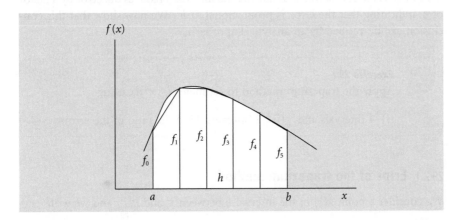

Figure 29.2 The area under the curve is approximated by the sum of the areas of the trapezia.

of width h so that

$$h = \frac{b-a}{n}.$$ (29.4)

The area of each trapezium is the product of the width and the average height, so that

$$I = \int_a^b f(x)\,dx \approx \frac{1}{2}(f_0 + f_1)h + \frac{1}{2}(f_1 + f_2)h + \frac{1}{2}(f_2 + f_3)h + \frac{1}{2}(f_3 + f_4)h$$
$$+ \frac{1}{2}(f_4 + f_5)h.$$

For the general case of n segments this becomes

$$I = \frac{1}{2}h(f_0 + 2f_1 + 2f_2 + \cdots + 2f_{n-2} + 2f_{n-1} + f_n).$$ (29.5)

It is clear that the greater the number of segments, n, the closer will be the estimate of the area. As an example we take the definite integral

$$I = \int_0^{\pi/2} \sin x\,dx$$ (29.6)

Table 29.1 Estimate of (29.6) by the trapezium and Simpson methods for different h. The error, in terms of the least significant figure, is shown in parenthesis.

h	Trapezium result	Simpson result
$\pi/8$	0.987 12 (1288)	1.000 134 59 (13 459)
$\pi/16$	0.996 79 (321)	1.000 008 30 (830)
$\pi/32$	0.999 20 (80)	1.000 000 52 (52)

and take intervals $\pi/8$ $(n=4)$, $\pi/16$ $(n=8)$, and $\pi/32$ $(n=16)$. The results are shown in Table 29.1 and are compared with the correct result, $I=1$. An examination of Table 29.1 shows that halving the interval h reduces the error by a factor of 4, indicating that the error is proportional to h^2. We now show that this conclusion can be verified by an analytical approach.

Exercise 29.1
Apply the trapezium method to estimate $\int_0^1 x^2 \, dx$ using
(i) 4 intervals and (ii) 8 intervals. Find the ratio of the errors.

29.2.1 Error of the trapezium method

We consider a point $x(s)$ in the interval h between x_i and x_{i+1} and we write

$$x(s) = x_i + s$$

where $0 \le s < h$. By the use of Maclaurin's theorem (4.43),

$$f(x(s)) = f(x_i) + sf'(x_i) + \frac{s^2}{2!}f''(x_i) + \frac{s^3}{3!}f'''(x_i) + \cdots. \tag{29.7}$$

Hence the true area under the curve between points i and $i+1$ is

$$A = \int_0^h f(x(s)) \, ds = hf(x_i) + \frac{h^2}{2}f'(x_i) + \frac{h^3}{3!}f''(x_i) + \frac{h^4}{4!}f'''(x_i) + \cdots. \tag{29.8}$$

From (29.7), with $s=h$

$$f(x_{i+1}) = f(x_i) + hf'(x_i) + \frac{h^2}{2!}f''(x_i) + \frac{h^3}{3!}f'''(x_i) + \cdots. \tag{29.9}$$

The area of the trapezium in this interval is

$$A_{tr} = \frac{1}{2}h\{f(x_i) + f(x_{i+1})\}$$
$$= hf(x_i) + \frac{h^2}{2}f'(x_i) + \frac{h^3}{4}f''(x_i) + \frac{h^4}{12}f'''(x_i) + \cdots. \tag{29.10}$$

The contribution of this interval to the total error in the estimate, $A - A_{tr}$, is dominated by the term with the lowest power of h, since h is assumed small. This is

$$\varepsilon_i = A - A_{tr} \approx -\frac{h^3}{12}f''(x_i). \tag{29.11}$$

The total error for the integral is

$$E = \sum_{i=1}^{n} \varepsilon_i = -\frac{h^3}{12} \sum_{i=1}^{n} f''(x_i) = -\frac{h^3}{12} n \langle f''(x) \rangle, \tag{29.12}$$

where $\langle f''(x) \rangle$ is the average value of the second derivative of $f(x)$ in the range $x = a$ to b. Using (29.4) to eliminate n gives

$$E = -\frac{(b-a)h^2}{12} \langle f''(x) \rangle \tag{29.13}$$

which shows the dependence of E on h^2.

For $f(x) = \sin x$, $f''(x) = -\sin x$ and the average of $f''(x)$ in the range $x = 0$ to $\pi/2$ is $2/\pi$. With $b - a = \pi/2$ and $h = \pi/8$ (29.13) gives $E = 0.012\,85$, in satisfactory agreement with the entry in Table 29.1. The quantitative agreement between theory and calculation could be easily checked in this particular case but the general result, that the error depends on h^2, is always valid.

29.3 Simpson's method (rule)

Instead of joining two neighbouring points on the curve by a straight line we now consider fitting a parabola to the three points $i-1$, i, and $i+1$. To simplify the discussion we take the x coordinates as $-h$, 0, and h and the corresponding ordinates of the curve as f_{i-1}, f_i, and f_{i+1}. This is illustrated in Figure 29.3. The parabola fitted to the three points is

$$f(x) = ax^2 + bx + c \tag{29.14}$$

and from (28.8)

$$a = \frac{f_{i-1} + f_{i+1} - 2f_i}{2h^2},$$

$$b = \frac{f_{i+1} - f_{i-1}}{2h},$$

$$c = f_i.$$

The area under the parabola is

$$I_i = \int_{-h}^{h} (ax^2 + bx + c)\,dx = \frac{1}{3}h(2ah^2 + 6c) = \frac{h}{3}(f_{i-1} + 4f_i + f_{i+1}). \tag{29.15}$$

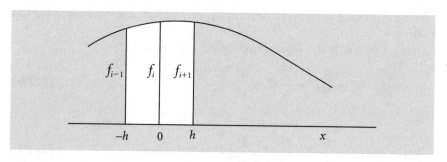

Figure 29.3 Two intervals for the application of Simpson's method.

To apply the Simpson method the range from a to b must be divided into an *even* number of intervals and a parabola is fitted to pairs of intervals. The estimate of the total area is then

$$
\begin{aligned}
I_{\text{Sim}} &= \frac{h}{3}\left(f_0 + 4f_1 + f_2 + f_2 + 4f_3 + f_4 + \cdots + f_{n-4} + 4f_{n-3} + f_{n-2} + f_{n-2} \right. \\
&\quad \left. + 4f_{n-1} + f_n \right) \\
&= \frac{h}{3}\left(f_0 + 4f_1 + 2f_2 + 4f_3 + 2f_4 + \cdots + 2f_{n-4} + 4f_{n-3} + 2f_{n-2} + 4f_{n-1} + f_n \right).
\end{aligned}
$$
(29.16)

The result of applying this formula to the integral (29.6) is given in Table 29.1. It will be seen that, for a given interval, it gives much greater precision than does the trapezium method. In addition halving the value of h decreases the error by a factor of 16 showing that the error depends on h^4, a result that will now be shown by analysis.

Exercise 29.2
Apply Simpson's method to estimate $\int_0^1 x^2 \, dx$ using   (i) 4 intervals and (ii) 8 intervals. Comment on the errors.

29.3.1 Error of Simpson's method

We consider a point $x(s)$ in the interval h between x_{i-1} and x_{i+1} and we write

$$
x(s) = x_i + s
$$

where $-h \le s < h$. By the use of Maclaurin's theorem (4.43), using the present notation,

$$
f(x(s)) = f_i + sf_i' + \frac{s^2}{2!}f_i'' + \frac{s^3}{3!}f_i''' + \frac{s^4}{4!}f_i^{IV} + \cdots .
$$
(29.17)

Hence the true area under the curve between points $i-1$ and $i+1$ is

$$
A = \int_{-h}^{h} f(x(s)) \, ds = 2hf_i + \frac{h^3}{3}f_i'' + \frac{h^5}{60}f_i^{IV} + \cdots
$$
(29.18)

where the number of terms retained in the summation is just sufficient for this analysis.

To find the Simpson's rule estimate of this area we note that

$$
f_{i-1} = f_i - hf_i' + \frac{h^2}{2!}f_i'' - \frac{h^3}{3!}f_i''' + \frac{h^4}{4!}f_i^{IV} - \cdots
$$
(29.19a)

and

$$
f_{i+1} = f_i + hf_i' + \frac{h^2}{2!}f_i'' + \frac{h^3}{3!}f_i''' + \frac{h^4}{4!}f_i^{IV} + \cdots .
$$
(29.19b)

The estimate is

$$A_{Sim} = \frac{h}{3}(f_{i-1} + 4f_i + f_{i+1}) = 2hf_i + \frac{h^3}{3}f_i'' + \frac{h^5}{36}f_i^{IV} + \cdots. \quad (29.20)$$

Thus the contribution to the total error of the estimate from this pair of intervals is, taking only the lowest power of h in the error term,

$$\varepsilon_i = A - A_{Sim} = \frac{h^5}{60}f_i^{IV} - \frac{h^5}{36}f_i^{IV} = -\frac{h^5}{90}f_i^{IV}. \quad (29.21)$$

Now we sum over the $\frac{1}{2}n$ pairs of intervals that give the total estimate as

$$E = \sum_{i=1}^{n/2}\varepsilon_i = -\frac{h^5}{90}\sum_{i=1}^{n/2}f_i^{IV} = -\frac{nh^5}{180}\langle f^{IV}\rangle = -\frac{(b-a)h^4}{180}\langle f^{IV}\rangle \quad (29.22)$$

where $\langle f^{IV}\rangle$ is the average value of the fourth derivative over the range $x = a$ to b. This clearly demonstrates the dependence of the error on h^4. For integral (29.6) the fourth derivative of $\sin x$ is $\sin x$ and has an average value of $2/\pi$ over the range. For $h = \pi/8$ the error is

$$E = -\frac{\pi/2 \times (\pi/8)^4}{180}\frac{2}{\pi} = 0.000\,132\,12,$$

which is in satisfactory agreement with the figure given in Table 29.1.

One result that comes from (29.22) is that Simpson's method gives an exact answer for any function that is a polynomial of cubic power or less, for then the fourth derivative is zero. Since a parabola is being fitted to sets of three points it might seem surprising that an exact result can be obtained for a cubic function. The reason is that when a quadratic function is fitted to a cubic the positive and negative errors exactly cancel each other.

29.4 Romberg method

In applying the trapezium rule in Table 29.1 the range was divided into 4, 8, and 16 intervals. The error for four intervals can be written as

$$\varepsilon_4 = I - T_4 \quad (29.23a)$$

where I is the exact value of the integral and T_4 is the estimate found with four intervals. Similarly we can write

$$\varepsilon_8 = I - T_8. \quad (29.23b)$$

From (29.13) we found that the error for the trapezium rule is proportional to h^2. Consequently, since h is twice as big for four intervals as it is for eight intervals, we have $\varepsilon_4 = 4\varepsilon_8$ so that, from (29.23a) and (29.23b),

$$I - T_4 = 4(I - T_8)$$

or

$$I = \frac{4T_8 - T_4}{3}. \quad (29.24)$$

Substituting the values of T_4 and T_8 from Table 29.1 we find $I = 1.000\,008\,3$, which happens to be the result using Simpson's rule with eight intervals. It can be demonstrated that (29.24) is equivalent to using Simpson's rule. If three successive values of the function, separated by an interval h, are f_0, f_1, and f_2 then using one interval over the $2h$ range

$$T_1 = \frac{1}{2} \times 2h \times (f_0 + f_2).$$

Using two intervals gives

$$T_2 = h \times (f_0 + 2f_1 + f_2).$$

We then find

$$\frac{4T_2 - T_1}{3} = \frac{h}{3}(f_0 + 4f_1 + f_2),$$

which is the Simpson's rule equation.

Since we know that error varies as h^4 for Simpson's rule, we can find a new estimate for I from two Simpson rule values. If these are S_2 for two intervals and S_4 for four intervals then

$$I - S_2 = 16(I - S_4) \quad \text{or} \quad I = \frac{16S_4 - S_2}{15}. \tag{29.25}$$

This estimate, which we shall call R_4, is equivalent to the result of fitting a quartic to the five ordinates delineating the four intervals. Fitting two quartics to two neighbouring sets of four intervals would give R_8, and so on, and the error in the estimate would vary as h^6. This successive combination of estimates to give estimates equivalent to fitting higher-order polynomials is called the **Romberg method** and it is illustrated in Table 29.2.

The way in which each column is derived from the previous column is as follows:

$$S_{2n} = \frac{4T_{2n} - T_n}{3}, \quad R_{4n} = \frac{16S_{4n} - S_{2n}}{15}, \quad Q_{8n} = \frac{64R_{8n} - R_{4n}}{63}$$

$$P_{16n} = \frac{256Q_{16n} - Q_{8n}}{255}, \quad O_{32n} = \frac{1024P_{32n} - P_{16n}}{1023}.$$

Table 29.2 Structure of the Romberg method

No. of intervals						
1	T_1					
		S_2				
2	T_2		R_4			
		S_4		Q_8		
4	T_4		R_8		P_{16}	
		S_8		Q_{16}		O_{32}
8	T_8		R_{16}		P_{32}	
		S_{16}		Q_{32}		
16	T_{16}		R_{32}			
		S_{32}				
32	T_{32}					

Table 29.3 Romberg method for integral (29.6) with eight segments

$T_1 = 0.785\,398\,164$

$\qquad S_2 = 1.002\,279\,877$

$T_2 = 0.948\,059\,449$ $\qquad\qquad\qquad R_4 = 0.999\,991\,567$

$\qquad S_4 = 1.000\,134\,586$ $\qquad\qquad\qquad\qquad Q_8 = 1.000\,000\,007$

$T_4 = 0.987\,115\,802$ $\qquad\qquad\qquad R_8 = 0.999\,999\,875$

$\qquad S_8 = 1.000\,008\,295$

$T_8 = 0.996\,785\,172$

The application of this process to integral (29.6) with eight segments is shown in Table 29.3.

The final column of Table 29.3 is equivalent to fitting a polynomial of degree eight to nine ordinates. In practice, in applying the Romberg method, there is usually a tolerance limit appropriate to the problem in hand. As soon as all the figures in a column agree to within that tolerance the process can be terminated and the last figure in the column taken as the result. Thus, if agreement is required to four places of decimals then R_8 could be taken as the solution.

The Romberg method involves a whole series of arithmetic operations at each stage of which round-off errors will occur that affect the least significant digits of the numbers being processed. It is therefore prudent to use double-precision numbers in applying the Romberg method if high precision is being sought.

Exercise 29.3

Use the Romberg method with four intervals to evaluate $\int_0^1 \dfrac{1}{1+x^2}\,\mathrm{d}x$.

Give your answer to four significant figures and compare it with the analytical solution.

29.5 Gauss quadrature

In all the methods of quadrature described so far, the intervals at which the function has been evaluated have been equally spaced. There is a different approach, **Gauss quadrature**, in which the number of evaluations of the function is fixed but the points are not equally spaced. The situation is illustrated in Figure 29.4. The integration range is $b - a = \Delta$ and the problem is to find the best estimate of integral (29.6) in the form

$$I = \Delta\left\{ w_1 f(a + \alpha_1\Delta) + w_2 f(a + \alpha_2\Delta) \right\}. \tag{29.26}$$

Since Δ is the range of the integral then the expression in curly brackets in (29.26) is an estimate of the average value of the function in the range. It is not possible to find values of w_1, w_2, α_1, and α_2 that will give a precise solution for every possible function $f(x)$ but what we can do is to find values that give the correct answer for

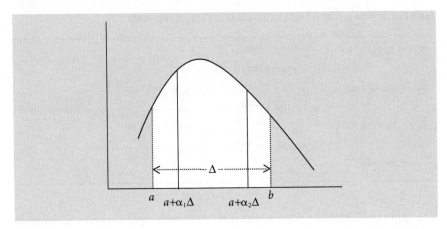

Figure 29.4 The ordinates, $a + \alpha_1\Delta$ and $a + \alpha_2\Delta$, for Gauss two-point quadrature.

$f(x) = 1$, $f(x) = x$, and $f(x) = x^2$, and hence for any linear combinations giving a quadratic function.

The first condition gives

$$\int_a^{a+\Delta} 1 \, dx = \Delta = \Delta(w_1 + w_2) \quad \text{or} \quad w_1 + w_2 = 1 \qquad (29.27)$$

The second condition, using (29.27), gives

$$\int_a^{a+\Delta} x \, dx = \frac{1}{2}\{(a + \Delta)^2 - a^2\} = \frac{1}{2}\Delta(2a + \Delta)$$

$$= \Delta\Big\{ w_1(a + \alpha_1\Delta) + w_2(a + \alpha_2\Delta) \Big\} = \Delta\Big\{ a + \Delta(w_1\alpha_1 + w_2\alpha_2) \Big\}$$

or

$$2(w_1\alpha_1 + w_2\alpha_2) = 1. \qquad (29.28)$$

The final condition is

$$\int_a^{a+\Delta} x^2 \, dx = \Delta\Big(a^2 + a\Delta + \frac{1}{3}\Delta^2 \Big) = \Delta\Big\{ w_1(a + \alpha_1\Delta)^2 + w_2(a + \alpha_2\Delta)^2 \Big\}$$

which, using (29.27) and (29.28), gives

$$3(w_1\alpha_1^2 + w_2\alpha_2^2) = 1. \qquad (29.29)$$

The three relationships (29.27), (29.28), and (29.29) are not sufficient to determine the four quantities α_1, α_2, w_1, and w_2, so the condition is imposed that the points are symmetrically placed around the centre of the range, which implies

$$\alpha_1 + \alpha_2 = 1. \qquad (29.30)$$

Substituting from (29.27) and (29.30) in (29.28)

$$2\Big\{ w_1\alpha_1 + (1 - w_1)(1 - \alpha_1) \Big\} = 1$$

which gives

$$(2w_1 - 1)(2\alpha_1 - 1) = 0. \qquad (29.31)$$

One solution of (29.31), that α_1, and hence also α_2, equals $\frac{1}{2}$, is unacceptable since it gives a single point and not two as specified. Hence we accept the other possible solution:

$$w_1 = w_2 = \frac{1}{2}. \tag{29.32}$$

With the weights known and with relationship (29.30) giving α_2 in terms of α_1, we find from (29.29)

$$\frac{3}{2}\left\{\alpha_1^2 + (1 - \alpha_1)^2\right\} = 1$$

giving

$$2\alpha_1^2 - 2\alpha_1 + \frac{1}{3} = 0.$$

The solutions of this equation give α_1 and α_2 as

$$\alpha_1 = \frac{1}{2} - \frac{1}{2\sqrt{3}} = 0.211\ 324\ 9 \quad \text{and} \quad \alpha_2 = \frac{1}{2} + \frac{1}{2\sqrt{3}} = 0.788\ 675\ 1.$$

Applying these values of w and α the two-point Gauss gives for integral (29.6)

$$I \approx \frac{\pi}{2}\left\{\frac{1}{2}\sin(0.211\ 324\ 9\frac{\pi}{2}) + \frac{1}{2}\sin(0.788\ 675\ 1\frac{\pi}{2})\right\} = 0.998\ 472\ 6.$$

This is much better than the trapezium method with two points (0.7854) and slightly better than Simpson's method with three points (1.002 279 9).

The Gauss two-point formula gives a precise solution even for a cubic function. You should confirm that

$$\int_a^b x^3\,dx = \frac{\Delta}{4}(4a^3 + 6a^2\Delta + 4a\Delta^2 + \Delta^3)$$

$$= \Delta\left[\frac{1}{2}\left\{a + \left(\frac{1}{2} - \frac{1}{2\sqrt{3}}\right)\Delta\right\}^3 + \frac{1}{2}\left\{a + \left(\frac{1}{2} + \frac{1}{2\sqrt{3}}\right)\Delta\right\}^3\right]$$

$$= \Delta\left\{w_1(a + \alpha_1)^3 + w_2(a + \alpha_2)^3\right\}.$$

Similar, but more complicated, analyses can give the weights and abscissae (values of α) for Gauss n-point formulae with values of $n > 2$. Tables of weights and abscissae can be found in mathematical tables (e.g. Abramowitz and Stegun 1964), and values for a number of n are given in Table 29.4.

A Gauss n-point formula gives a precise result for polynomials up to degree $2n - 1$. It is an efficient process and, since it is usually applied in a computer program, the fact that it evaluates functions at arbitrary points is no disadvantage.

Exercise 29.4

Use the Gauss 3-point method to evaluate $\int_0^1 \frac{1}{1 + x^2}\,dx$. Give your answer to four significant figures and compare it with the analytical solution.

Table 29.4 Abscissae and weights for Gauss quadrature

n	Abscissae	Weights
3	0.112 701 7	0.277 777 8
	0.5	0.444 444 4
	0.887 298 3	0.277 777 8
4	0.069 431 8	0.173 927 4
	0.330 009 5	0.326 072 6
	0.669 990 5	0.326 072 6
	0.930 568 2	0.173 927 4
7	0.025 446 0	0.064 742 5
	0.129 234 4	0.139 852 7
	0.297 077 4	0.190 915 0
	0.5	0.208 979 6
	0.702 922 6	0.190 915 0
	0.870 765 6	0.139 852 7
	0.974 554 0	0.064 742 5
10	0.013 046 7	0.033 335 7
	0.067 468 3	0.074 725 7
	0.160 295 2	0.109 543 2
	0.283 302 3	0.134 633 4
	0.425 562 8	0.147 762 1
	0.574 437 2	0.147 762 1
	0.716 697 7	0.134 633 4
	0.839 704 8	0.109 543 2
	0.932 531 7	0.074 725 7
	0.986 953 3	0.033 335 7

29.6 **Multidimensional quadrature**

In the methods of quadrature we have considered so far we have dealt with definite integrals of the form (29.6) in which the integration is over a single variable. The general principle of numerical quadrature can be extended to any number of dimensions and this can be illustrated by an integral of the form

$$I = \int_{y=c}^{d} \int_{x=a}^{b} f(x, y) \, \mathrm{d}x \, \mathrm{d}y. \tag{29.33}$$

The x range is divided into n intervals each of width h and the y range into m intervals each of width k so that $nh = b - a$ and $mk = d - c$. If Simpson's rule were being applied then

$$I = \frac{hk}{9} \sum_{j=1}^{m} \sum_{i=1}^{n} {}_n w_{im} w_j f(a + ih, c + jk) \tag{29.34}$$

where $_nw_i$ and $_mw_j$ follow the pattern 1, 4, 2, 4, ..., 4, 2, 4, 1 with $n+1$ and $m+1$ members respectively. Similarly it is possible to apply Gauss integration with n points in the x-direction and m points in the y-direction in the form

$$I = (b-a)(d-c) \sum_{j=1}^{m} \sum_{i=1}^{n} {}_nw_i \, {}_mw_j f\left\{a + {}_n\alpha_i(b-a), c + {}_m\alpha_j(d-c)\right\} \qquad (29.35)$$

where the αs and ws are the abscissae and weights for the appropriate Gauss points.

To illustrate the application of (29.35) we consider the integral

$$I = \int_0^{\pi/2} \int_0^{\pi/2} \sin(x+y) \, dx \, dy. \qquad (29.36)$$

A simple program, given in Box 29.1, evaluates this integral with arbitrarily chosen 3-point Gauss integration in the x-direction and 4-point Gauss integration in the y-direction. The abscissae, α, and weights, w, are derived from Table 29.4.

Box 29.1 Program GAUSS2

This program evaluates integral (29.36) using Gauss 3-point integration in the x-direction and 4-point integration in the y-direction.

Set coefficients for Gauss 3-point $- \alpha_{3,1}, \alpha_{3,2}, \alpha_{3,3}, w_{3,1}, w_{3,2}, w_{3,3}$
Set coefficients for Gauss 4-point $- \alpha_{4,1}, \alpha_{4,2}, \alpha_{4,3}, \alpha_{4,4}, w_{4,1}, w_{4,2}, w_{4,3}, w_{4,4}$
sum = 0
Loop from i = 1 to 3
Loop from j = 1 to 4

   $xpy = 0.5\pi(\alpha_{3,i} + a_{4,j})$

   $sum = sum + w_{3,i}w_{4,j}\sin(xpy)$

End loop
End loop
Output sum $\times \pi^2/4$
stop

The estimate of this integral found by a version of the GAUSS2 program is 2.000 02, compared with the correct value of 2.0 precisely.

Exercise 29.5

(i) Evaluate the integral $\int_0^1 \int_0^1 xy^2x \, dy$ using the trapezium rule with two intervals in each direction. Do not factorize the integral, but treat it as a two-dimensional problem. Compare your solution with the analytical solution.

(ii) Use Simpson's rule for this integral and comment on the result.

29.7 Monte Carlo integration

There are some physical problems in which multidimensional integrals occur. For example, for a system of n particles with pair potential energy $\phi(r_{ij})$, where r_{ij} is $|\mathbf{r}_i - \mathbf{r}_j|$, the distance between particles i and j, the classical partition function, is

$$Z = \int \int \cdots \int \int \exp\left\{-\frac{\phi(r_{ij})}{kT}\right\} d^3r_1 \ldots d^3r_n \tag{29.37}$$

where d^3r_i represents a volume element occupied by particle i so that (29.37) is an integral of dimension $3n$. With $n = 20$, a not very large system of particles, the dimension of the integral is 60 and even with Gauss 2-point integration in each dimension the number of function evaluations would be 10^{18}, which would take 30 years to compute at 10^9 function evaluations per second.

An alternative approach, based on Monte Carlo ideas (Chapter 23), considers (29.37) in the form

$$Z = \overline{\exp\left\{-\sum_{\text{pairs}} \frac{\phi(r_{ij})}{kT}\right\}} V^n, \tag{29.38}$$

where V is the multidimensional 'volume' of the system. An approximation to the average of the integrand is found by generating a large number of sets of random values for the variables r_i, for $i = 1$ to n. In the case $n = 20$ it might be decided to generate 10^{10} function evaluations, which would take less than 1 hour even at 10^7 function evaluations per second.

All the numerical values of the integrands found in the Monte Carlo method form a sample from a population with some mean value $\bar{x}$ (for which an estimate is sought) and some unknown variance σ^2. The standard deviation of the mean of the sample can be estimated from the standard deviation of the individual members of the sample, σ_s, from $\langle\sigma\rangle = \sigma_s/\sqrt{n}$, given by (26.10) with n replacing $n-1$ for very large n, and multiplying this by the multidimensional 'volume' of the integration space gives the standard deviation of the derived value of the integral.

As an illustration of this principle we consider the evaluation of

$$I = \int_0^{\pi/2}\int_0^{\pi/2}\int_0^{\pi/2}\int_0^{\pi/2}\int_0^{\pi/2}\int_0^{\pi/2}\int_0^{\pi/2}\int_0^{\pi/2} \sin\left\{\frac{1}{8}\sum_{i=1}^{8} x_i\right\} dx_1\, dx_2\, dx_3\, dx_4\, dx_5\, dx_6\, dx_7\, dx_8.$$

$$\tag{29.39}$$

The analytical solution of this integral gives

$$I = \left\{16\sin\left(\frac{\pi}{32}\right)\right\}^8 \sin\left(\frac{\pi}{4}\right) = 25.87385.$$

The program MULGAUSS, outlined in Appendix 19, applies 3-point Gauss integration in eight dimensions and gives $I \approx 25.873\,85$.

The program MCINT, outlined in Appendix 20, finds the estimated value of integral (29.39), and its standard deviation, by the Monte Carlo method with numbers of trials input by the user. The results of running MCINT, with errors in units of the least significant decimal point in parentheses and calculated standard deviations, are as follows:

10 000 trials	25.88758 (1373)	$\sigma = 0.04237$
100 000 trials	25.86876 (509)	$\sigma = 0.01331$
1 000 000 trials	25.86954 (431)	$\sigma = 0.00420$

The superiority of the Gauss method is clear in this case, but this is not always so. The values of the integrand found by the Monte Carlo method are random selections from a population with an average equal to the true mean value of the integrand in the integration space with a standard deviation that depends on the variability of the function. Regardless of the number of dimensions of the problem the standard deviation in the estimate of the mean is $O(n^{-\frac{1}{2}})$. If the dimension of the problem is m and we make n integrand evaluations, the Simpson or Gauss methods would use n/m points in each dimension. For all such methods the error will depend on some power, say α, of the integration interval and hence will be $O\{(m/n)^{\alpha}\}$, which will also be the order of error of the whole m-dimensional integral. For a fixed n at some value of m the balance of favour will switch from the conventional factorized method to the Monte Carlo method. As an example we take the integral

$$I = \int_0^1 \int_0^1 \int_0^1 \int_0^1 \int_0^1 \int_0^1 \int_0^1 \int_0^1 \int_0^1 \int_0^1 2^{10} \prod_{i=1}^{10} \{\sin^3(\pi x_i)\, dx_i\}. \qquad (29.40)$$

This can be found analytically by factorizing it into 10 similar integrals, and its value is 0.1942. A Monte Carlo estimate based on 3^{10} (59 049) points gives an estimated value of 0.1943 with a standard deviation of 0.0117, whereas a 10-dimensional Gauss integration with three points in each direction gives the rather poor value 0.5117.

For multidimensional integration, with n greater than 4 or 5, Monte Carlo integration is often preferred for its simplicity of programming even if it cannot be justified on grounds of precision.

Problems

29.1 Estimate the value of the integral $\int_0^{\pi/2} \dfrac{1}{1+x^2}\, dx$ using the trapezium rule with 1, 2, 4, 8, and 16 intervals. Given that the true value of the integral is $\pi/4$, confirm that the error varies as h^2.

29.2 Apply the Romberg method to the results from Problem 29.1 to find the quadrature results equivalent to Simpson's rule for 2, 4, 8, and 16 intervals.

29.3 Modify the programs MULGAUSS and MCINT to find the integral (29.39) but with the limits $\pi/4$ and 0 in each dimension. For MCINT use 1000, 10 000, and 100 000 points and also find the standard deviation of the result in each case.

29.4 Find the integral $\int_0^1 \int_0^1 \int_0^1 \dfrac{1}{1+x^2+y^2+z^2}\, dx\, dy\, dz$ using Gauss 3-point quadrature in each dimension.

30 Linear equations

Simultaneous linear equations involving many variables frequently crop up in the numerical solution of scientific problems. This chapter describes a number of methods of solving such sets of equations. Equations coming from the result of scientific measurements, which are prone to error, sometimes yield sets of equations that are inconsistent with each other in the sense that no set of values of the variables can satisfy all of them. In such cases one must seek a 'best' solution; what this is, and how it is found, are important parts of the contents of this chapter.

30.1 Interpretation of linearly dependent and incompatible equations

In §8.9 it was shown that a set of n linear equations linking n variables can be expressed in matrix notation as

$$Ax = b \tag{30.1}$$

where A is an $n \times n$ coefficient matrix, x the solution vector, and b the right-hand-side vector. The solution can be expressed as

$$x = A^{-1}b \tag{30.2}$$

where A^{-1} is the inverse of A. The requirement for obtaining a solution is that the matrix A should be **non-singular**, i.e. that it should not have a zero determinant.

To understand the process of solving linear equations it helps to use graphical illustrations, or mental imagery, for systems of two or three equations. A simple system of two linear equations

$$\begin{aligned} a_{11}x_1 + a_{12}x_2 &= b_1 \\ a_{21}x_1 + a_{22}x_2 &= b_2 \end{aligned} \tag{30.3}$$

is represented in Figure 30.1. The intersection of the two straight lines gives the solution.

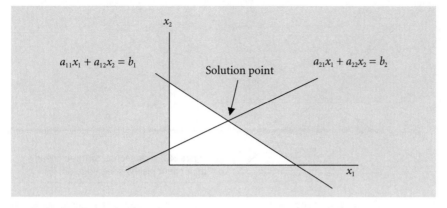

Figure 30.1 Graphical representation of a system of two linear equations.

Now we may mentally extend the representation to three dimensions corresponding to the solution of

$$a_{11}x_1 + a_{12}x_2 + a_{13}x_3 = b_1$$
$$a_{21}x_1 + a_{22}x_2 + a_{23}x_3 = b_2 \qquad\qquad (30.4)$$
$$a_{31}x_1 + a_{32}x_2 + a_{33}x_3 = b_3.$$

Each of the three equations represents a plane and the values of (x_1, x_2, x_3) satisfying the first two equations fall on a straight line that is the intersection of the two planes. The plane represented by the third equation is intercepted by the straight line at the point that corresponds to the solution of the equations.

We now consider the possibility that the final plane is not intercepted by the straight line at a point but actually contains the line or, in other words, the three planes all contain a common line, as illustrated in Figure 30.2.

In this case the three planes are **linearly dependent**, meaning that the equation for any one of them can be expressed as a linear combination of the other two. As an example we consider

$$3x_1 - 2x_2 + x_3 - 4 = 0$$
$$x_1 + x_2 - x_3 - 2 = 0 \qquad\qquad (30.5a)$$
$$5x_1 - 5x_2 + 3x_3 - 6 = 0$$

for which

$$3x_1 - 2x_2 + x_3 - 4 = \frac{1}{2}\{(x_1 + x_2 - x_3 - 2) + (5x_1 - 5x_2 + 3x_3 - 6)\}. \qquad (30.5b)$$

For linearly dependent equations no unique solution is possible, but any two of the equations can be used to solve for two of the variables in terms of the third one.

Another situation that can occur is where the line of intersection of the first two planes does not cut the third plane because it is parallel to it – which is equivalent to the condition that the intersections of pairs of the planes give three parallel lines. This is shown in Figure 30.3.

In this case the equations are **incompatible**, that is to say that no solution of any kind is possible – not even expressing two of the variables in terms of the third as is possible for linearly dependent equations. The presence of incompatible equations is indicated when a linear combination of some of the equations gives the

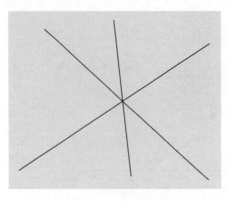

Figure 30.2 Intersection of three linearly-dependent planes seen edge-on.

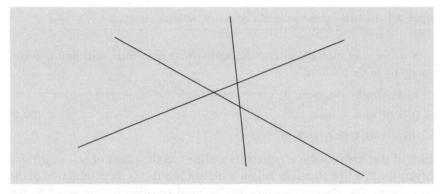

Figure 30.3 Three planes, seen edge-on, corresponding to incompatible equations.

coefficients for one of the other equations but with a different constant. Thus the three equations

$$x_1 + x_2 + x_3 - 1 = 0$$
$$2x_1 + 2x_2 - x_3 - 2 = 0$$
$$4x_1 + 4x_2 + x_3 - 3 = 0$$

are incompatible, since

$$2(x_1 + x_2 + x_3 - 1) + (2x_1 + 2x_2 - x_3 - 2) = 4x_1 + 4x_2 + x_3 - 4,$$

which gives the coefficients of the third equation but a different constant.

The conditions that give linearly dependent or incompatible equations can be extended to systems involving more than three variables although it is not possible to use graphical representations for such cases.

Exercise 30.1
Identify each of the following sets of equations as **independent, linearly dependent**, or **incompatible**:

(i) $x + y - z = 2$, $2x - y - 2z = 3$, $x - 2y - z = 1$;

(ii) $2x + 3y - 2z = 3$, $x + y + z = 3$, $4x - y - 2z = 3$;

(iii) $x - y - z = 2$, $2x + 3y - 4z = 1$, $x + 4y - 3z = 1$.

30.2 **Gauss elimination method**

With modern computers it is possible to invert large matrices very quickly, so there is little motivation to use faster methods. However, there may be situations where many sets of linear simultaneous equations, all with different coefficients, must be solved and then a method that is faster than inverting a matrix could be useful. In most applications it is not worth using a faster method to reduce computing time from four seconds to two, but it certainly is worthwhile to reduce four hours to two.

One method for solving a set of linear equations is the Gauss elimination method that is here illustrated by the solution of a set of three equations:

$$4x_1 - x_2 + x_3 = 5 \tag{30.6a}$$

$$x_1 + 2x_2 - 2x_3 = -1 \tag{30.6b}$$

$$2x_1 - 2x_2 + x_3 = 1. \tag{30.6c}$$

Equation (30.6a) is used to express x_1 in terms of the other variables giving

$$x_1 = \frac{5}{4} + \frac{x_2}{4} - \frac{x_3}{4} \tag{30.7a}$$

and this value of x_1 is substituted in (30.6b) and (30.6c) to give

$$\frac{9}{4}x_2 - \frac{9}{4}x_3 = -\frac{9}{4} \tag{30.7b}$$

and

$$-\frac{3}{2}x_2 + \frac{1}{2}x_3 = -\frac{3}{2}. \tag{30.7c}$$

Next (30.7b) is used to express x_2 in terms of x_3 and this expression for x_2 is inserted in (30.7c) giving

$$x_2 = x_3 - 1 \tag{30.8a}$$

and

$$x_3 = 3. \tag{30.8b}$$

With $x_3 = 3$ explicitly given by (30.8b) the value of $x_2 = 2$ can be found from (30.8a) and then the value of $x_1 = 1$ from (30.7a).

This process, known as **Gauss elimination**, can be extended to any size of system and is usually carried out by a computer program. The variable chosen as the one to be expressed in terms of the remaining ones at each stage – x_1 in (30.6a) and x_2 in (30.7b) – is known as the **pivot** and for the highest precision it is best to take the coefficient of the pivot with as large a magnitude as possible. Actually the values of the pivots are arbitrary since, for example, multiplying (30.6b) throughout by 2 would give the same solution for the set of equations but double the coefficient of x_2, the pivot, in (30.7b). In principle, the choice of the maximum magnitude of a pivot can be made on an absolute basis by initially multiplying each equation throughout by a factor that made the mean square coefficient equal to unity. In practice, in normal circumstances it is sufficient to make sure that the pivot is not so small that the number of significant figures given by the computer would be inadequate to give the accuracy required.

Exercise 30.2
Solve the following set of equations by the Gauss elimination method:
$7x - y - z = 7$, $x - y - z = 1$, $4x + 3y + 2z = 5$.

30.3 **Conditioning of a set of equations**

The solution point is very clearly indicated in Figure 30.1, since the angle between the lines is large and the coefficients of the two equations are very different. In Figure 30.4 we show a less favourable situation. The two lines in Figure 30.4 cross at a very small angle and the coefficients of the two lines are clearly very similar. As an example we can take

$$1.99x_1 + 1.51x_2 = 4 \tag{30.9a}$$

and

$$2.01x_1 + 1.49x_2 = 6. \tag{30.9b}$$

There is no difficulty in finding a solution to any degree of precision required, and for these equations $x_1 = 44.2857$ and $x_2 = -55.7143$. The problem arises if the coefficients and the values of the right-hand-side vector are derived from experiment and therefore subject to experimental error. Small changes in the coefficients and right-hand sides, for example to

$$1.98x_1 + 1.52x_2 = 4.05 \tag{30.10a}$$

and

$$2.02x_1 + 1.48x_2 = 5.95 \tag{30.10b}$$

give the solution $x_1 = 21.7857$, $x_2 = -25.7143$, very different from the previous one.

This sensitivity of the solution to small changes in the coefficients of the equations is referred to as the **conditioning** of the equations. It is known that if the matrix of coefficients is singular then a solution of the equations is not

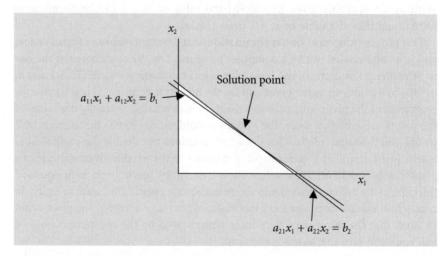

Figure 30.4 Simultaneous linear equations with similar coefficients.

possible since the equations are linked by linear dependencies in some way. However, if the determinant of the coefficient matrix is very small in value, as is the case for (30.9a) and (30.9b), then a solution can be found but will be sensitive to small changes in the coefficients. The problem is to be able to judge what is, or is not, a small value of the determinant. Multiplying each of a set of n simultaneous linear equations for n variables by λ multiplies the determinant by λ^n without changing either the solution or the conditioning of the equations. In order to be able to judge conditioning each equation should be multiplied by a factor that makes the mean-square coefficient equal to unity. The maximum magnitude of the determinant is then equal to $n^{n/2}$ (a diagonal determinant with diagonal elements each $\pm\sqrt{n}$). What constitutes good or bad conditioning depends on the source of the coefficients – for example, if it was known that the coefficients and right-hand sides of (30.9a) and (30.9b) were correct to 1 part in 10^8 then the solution would be reliable to four or five significant figures. In normal circumstances, as a rule of thumb, a magnitude of determinant less than $0.001n^{n/2}$ would probably indicate a conditioning problem.

Exercise 30.3
Solve the equations (i) $8x + 9y = 1$, $7x + 8y = 1$; (ii) $8x + 9y = 1.01$; $7x + 8y = 0.99$.
Comment on the results.

30.4 **Gauss–Seidel method**

This method of solving simultaneous linear equations is applicable when the coefficient matrix has diagonal elements of large magnitude. As an example we take the set of equations

$$3x_1 + x_2 - x_3 + 2x_4 = 12$$
$$x_1 + 4x_2 - 2x_3 + x_4 = 14$$
$$2x_1 - x_2 + 5x_3 - x_4 = -8 \qquad\qquad (30.11)$$
$$x_1 + x_2 + x_3 + 3x_4 = 11.$$

These are now recast in a form where for each equation the variable corresponding to the coefficient of largest magnitude is expressed in terms of the other variables, thus

$$x_1 = \frac{12}{3} - \frac{x_2}{3} + \frac{x_3}{3} - \frac{2x_4}{3} \qquad\qquad (30.12a)$$

$$x_2 = \frac{14}{4} - \frac{x_1}{4} + \frac{2x_3}{4} - \frac{x_4}{4} \qquad\qquad (30.12b)$$

$$x_3 = -\frac{8}{5} - \frac{2x_1}{5} + \frac{x_2}{5} + \frac{x_4}{5} \qquad\qquad (30.12c)$$

$$x_4 = \frac{11}{3} - \frac{x_1}{3} - \frac{x_2}{3} - \frac{x_3}{3}. \qquad\qquad (30.12d)$$

One begins with some estimates of the values of the variables, and here we will start with initial estimates $_0x_1 = {_0x_2} = {_0x_3} = {_0x_4} = 0$. The first step is to use (30.12a) with the current estimates to find a new estimate for x_1 which gives $_1x_1 = 4$. Now, using the current estimates $_1x_1$, $_0x_3$, and $_0x_4$ in (30.12b) a new estimate for x_2 is determined, $_1x_2 = 2.5$. From (30.12c) with estimates $_1x_1$, $_1x_2$, and $_0x_4$ one finds $_1x_3 = -2.7$ and then from (30.12d) with estimates $_1x_1$, $_1x_2$ and $_1x_3$ one finds $_1x_4 = 2.4$. The process moves in cyclic fashion through the set of equations, always using the latest estimates of other variables to update the estimate of the left-hand-side variable and eventually the values converge to the solution. Doing this process by hand is very tedious and computer programs exist to apply the **Gauss–Seidel method**. For small systems it is possible to write 'quick fix' programs to carry out the calculations and Appendix 21 gives an example of an outline program to give the solution of Equations (30.11). The condition for terminating the calculation is designed safely to give a solution accurate to 0.0001. The output from a version of GS is

4.0000	2.5000	−2.7000	2.4000
0.6667	1.3833	−1.1100	3.3533
0.9333	1.8733	−0.9280	3.0404
1.0393	2.0161	−1.0044	2.9830
1.0045	2.0009	−1.0050	2.9999
0.9981	1.9980	−0.9997	3.0012
1.0000	1.9999	−0.9998	3.0000
1.0001	2.0001	−1.0000	2.9999
1.0000	2.0000	−1.0000	3.0000
1.0000	2.0000	−1.0000	3.0000

and the final row gives the solution $x_1 = 1$, $x_2 = 2$, $x_3 = -1$, and $x_4 = 3$. When the coefficient matrix has a strong diagonal the Gauss–Seidel method *usually* gives a solution and it is a very efficient process and easy to program. Under the condition that for each equation, indicated by subscript i,

$$|a_{ii}| \geq \sum_{\substack{j=1 \\ j \neq i}}^{n} |a_{ij}| \qquad (30.13)$$

the method will *always* converge to the correct solution but even when (30.13) is not satisfied the method is often successful.

Exercise 30.4

Use a hand application of the Gauss–Seidel method to solve $10x + y = 1.392$, $2x + 9y = 2.682$.

30.5 Homogeneous equations

Linear homogeneous equations are linear equations with a zero constant term, such as the set

$$x_1 + 2x_2 - x_3 = 0$$
$$2x_1 - x_2 + x_3 = 0 \qquad (30.14)$$
$$x_1 + 3x_2 - 2x_3 = 0.$$

The coefficient matrix is not singular and the only solution is $x_1 = x_2 = x_3 = 0$. However, in finding eigenvalues and eigenvectors, as described in §24.2, the process automatically generates a set of homogeneous linear equations for which the coefficient matrix *is* singular. In the example of that section each of the values of λ found from the solution of the cubic equation (24.10), when substituted into (24.9) gives a linearly dependent set of equations from any two of which two of the variables may be expressed in terms of the third. As an example, the equations might be of the form

$$x_1 + x_2 - 3x_3 = 0 \quad \text{and} \quad 3x_1 - x_2 - x_3 = 0$$

giving the solution $x_1 = x_3$ and $x_2 = 2x_3$. The special case of this solution, with all variables equal to zero, is usually not of interest where the equations are derived from a physical problem. If x_1, x_2, and x_3 are the components of an eigenvector then only their relative values are of interest and in the example we have just given the eigenvector can be written as $\{1, 2, 1\}$ or normalized as $(1/\sqrt{6})\{1, 2, 1\}$.

30.6 Least-squares solutions

Suppose that we are given three linear equations

$$
\begin{aligned}
x_1 - x_2 &= 1 \\
2x_1 + x_2 &= 5 \\
x_1 - 2x_2 &= 0.
\end{aligned}
\tag{30.15}
$$

The equations are compatible, and any pair of them will give the solution $x_1 = 2$, $x_2 = 1$. These equations are represented graphically in Figure 30.5. Now we suppose that the final equation is changed to $x_1 - 2x_2 = 1$. The equations are incompatible, and a different solution is found by taking different pairs of equations. This set of equations is illustrated graphically in Figure 30.6. The three equations represented in Figure 30.6 might come about as a result of three different experiments, the purpose of which is to determine the pair of values (x_1, x_2). The 'best' value is one that falls as close as possible to the three lines and the point P shown in the figure is roughly equidistant from, and close to, each of the lines.

The 'best' point that is usually chosen in such cases is the point corresponding to the **least-squares solution**, which will be illustrated for the set of equations

$$
\begin{aligned}
a_{11}x_1 + a_{12}x_2 + a_{13}x_3 &= b_1 \\
a_{21}x_1 + a_{22}x_2 + a_{23}x_3 &= b_2 \\
a_{31}x_1 + a_{32}x_2 + a_{33}x_3 &= b_3 \\
a_{41}x_1 + a_{42}x_2 + a_{43}x_3 &= b_4.
\end{aligned}
\tag{30.16}
$$

The least-squares solution, (X_1, X_2, X_3), is the one that makes the sum of the squares of the differences between the left-hand and right-hand sides of the equations a minimum, or

$$S = \sum_{i=1}^{4} (a_{i1}X_1 + a_{i2}X_2 + a_{i3}X_3 - b_i)^2 \text{ is a minimum.} \tag{30.17}$$

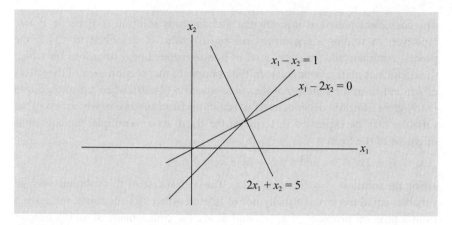

Figure 30.5 Representation of three consistent linear equations for two variables.

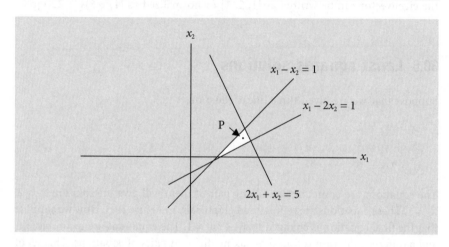

Figure 30.6 Representation of inconsistent equations. A 'best' solution is point P.

The condition for (30.17) to be satisfied is

$$\frac{\partial S}{\partial X_1} = \frac{\partial S}{\partial X_2} = \frac{\partial S}{\partial X_3} = 0. \tag{30.18}$$

Applying (30.18) to (30.17) gives the following three equations that give the least-squares solution

$$\left(\sum_{i=1}^{4} a_{i1}^2 \right) X_1 + \left(\sum_{i=1}^{4} a_{i1} a_{i2} \right) X_2 + \left(\sum_{i=1}^{4} a_{i1} a_{i3} \right) X_3 = \sum_{i=1}^{4} a_{i1} b_i$$

$$\left(\sum_{i=1}^{4} a_{i2} a_{i1} \right) X_1 + \left(\sum_{i=1}^{4} a_{i2}^2 \right) X_2 + \left(\sum_{i=1}^{4} a_{i2} a_{i3} \right) X_3 = \sum_{i=1}^{4} a_{i2} b_i \tag{30.19}$$

$$\left(\sum_{i=1}^{4} a_{i3} a_{i1} \right) X_1 + \left(\sum_{i=1}^{4} a_{i3} a_{i2} \right) X_2 + \left(\sum_{i=1}^{4} a_{i3}^2 \right) X_3 = \sum_{i=1}^{4} a_{i3} b_i.$$

These are called the **normal equations,** from which the solution we want is obtained. It is possible to express them in a much more succinct way using matrix notation. The original equations (30.16) can be written as

$$
\begin{bmatrix} a_{11} & a_{12} & a_{13} \\ a_{21} & a_{22} & a_{23} \\ a_{31} & a_{32} & a_{33} \\ a_{41} & a_{42} & a_{43} \end{bmatrix} \begin{bmatrix} x_1 \\ x_2 \\ x_3 \end{bmatrix} = \begin{bmatrix} b_1 \\ b_2 \\ b_3 \\ b_4 \end{bmatrix}. \tag{30.20}
$$

Pre-multiplying both sides by the transpose of the coefficient matrix gives

$$
\begin{bmatrix} a_{11}\ a_{21}\ a_{31}\ a_{41} \\ a_{12}\ a_{22}\ a_{32}\ a_{42} \\ a_{13}\ a_{23}\ a_{33}\ a_{43} \end{bmatrix} \begin{bmatrix} a_{11}\ a_{12}\ a_{13} \\ a_{21}\ a_{22}\ a_{23} \\ a_{31}\ a_{32}\ a_{33} \\ a_{41}\ a_{42}\ a_{43} \end{bmatrix} \begin{bmatrix} x_1 \\ x_2 \\ x_3 \end{bmatrix} = \begin{bmatrix} a_{11}\ a_{21}\ a_{31}\ a_{41} \\ a_{12}\ a_{22}\ a_{32}\ a_{42} \\ a_{13}\ a_{23}\ a_{33}\ a_{43} \end{bmatrix} \begin{bmatrix} b_1 \\ b_2 \\ b_3 \\ b_4 \end{bmatrix} \tag{30.21a}
$$

or

$$
\begin{bmatrix} a_{11} & a_{21} & a_{31} & a_{41} \\ a_{12} & a_{22} & a_{32} & a_{42} \\ a_{13} & a_{23} & a_{33} & a_{43} \end{bmatrix} \begin{bmatrix} a_{11}x_1 + a_{12}x_2 + a_{13}x_3 \\ a_{21}x_1 + a_{22}x_2 + a_{23}x_3 \\ a_{31}x_1 + a_{32}x_2 + a_{33}x_3 \\ a_{41}x_1 + a_{42}x_2 + a_{43}x_3 \end{bmatrix}
$$

$$
= \begin{bmatrix} a_{11}b_1 + a_{21}b_2 + a_{31}b_3 + a_{41}b_4 \\ a_{12}b_1 + a_{22}b_2 + a_{32}b_3 + a_{42}b_4 \\ a_{13}b_1 + a_{23}b_2 + a_{33}b_3 + a_{43}b_4 \end{bmatrix} \tag{30.21b}
$$

Multiplying the matrices on the left-hand-side of (30.21b) gives a three-element vector and equating these elements to those of the right-hand-side vector gives the three equations of (30.19), showing that the solution will be $(x_1, x_2, x_3) = (X_1, X_2, X_3)$.

Converting this process into matrix notation, (30.20) is

$$
Ax = b, \tag{30.22a}
$$

(30.21a) is

$$
A^T Ax = A^T b, \tag{30.22b}
$$

and from (30.2),

$$
x = (A^T A)^{-1} A^T b, \tag{30.22c}
$$

which is the least-squares solution of the set of equations.

This process is now applied to the set of equations illustrated in Figure 30.6. The equations are

$$
\begin{aligned} x_1 - x_2 &= 1 \\ x_1 - 2x_2 &= 1 \\ 2x_1 + x_2 &= 5, \end{aligned} \tag{30.23}
$$

for which

$$
A = \begin{bmatrix} 1 & -1 \\ 1 & -2 \\ 2 & 1 \end{bmatrix} \quad \text{and} \quad b = \begin{bmatrix} 1 \\ 1 \\ 5 \end{bmatrix}.
$$

We find

$$
A^T A = \begin{bmatrix} 1 & 1 & 2 \\ -1 & -2 & 1 \end{bmatrix} \begin{bmatrix} 1 & -1 \\ 1 & -2 \\ 2 & 1 \end{bmatrix} = \begin{bmatrix} 6 & -1 \\ -1 & 6 \end{bmatrix}, \quad (A^T A)^{-1} = \frac{\text{adj}A}{|A|} = \frac{1}{35} \begin{bmatrix} 6 & 1 \\ 1 & 6 \end{bmatrix}
$$

and

$$A^T b = \begin{bmatrix} 1 & 1 & 2 \\ -1 & -2 & 1 \end{bmatrix} \begin{bmatrix} 1 \\ 1 \\ 5 \end{bmatrix} = \begin{bmatrix} 12 \\ 2 \end{bmatrix},$$

giving the least-squares solution

$$\begin{bmatrix} x_1 \\ x_2 \end{bmatrix} = \left(A^T A \right)^{-1} A^T b = \frac{1}{35} \begin{bmatrix} 6 & 1 \\ 1 & 6 \end{bmatrix} \begin{bmatrix} 12 \\ 2 \end{bmatrix} = \begin{bmatrix} 74/35 \\ 24/35 \end{bmatrix}.$$

Standard computer programs are available for handling larger systems. One feature that sometimes arises is that the equations are of unequal reliability, so that different weights are applied to them. For the present example, if the three equations in the order presented in (30.23) have associated weights 1, 2, and 3 respectively then the equations are multiplied throughout by their associated weight and would appear as

$$x_1 - x_2 = 1$$
$$2x_1 - 4x_2 = 2 \tag{30.24}$$
$$6x_1 + 3x_2 = 15.$$

Now we find

$$A^T A = \begin{bmatrix} 1 & 2 & 6 \\ -1 & -4 & 3 \end{bmatrix} \begin{bmatrix} 1 & -1 \\ 2 & -4 \\ 6 & 3 \end{bmatrix} = \begin{bmatrix} 41 & 9 \\ 9 & 26 \end{bmatrix},$$

$$\left(A^T A \right)^{-1} = \frac{\text{adj} A}{|A|} = \frac{1}{985} \begin{bmatrix} 26 & -9 \\ -9 & 41 \end{bmatrix}$$

and

$$A^T b = \begin{bmatrix} 1 & 2 & 6 \\ -1 & -4 & 3 \end{bmatrix} \begin{bmatrix} 1 \\ 2 \\ 15 \end{bmatrix} = \begin{bmatrix} 95 \\ 36 \end{bmatrix},$$

giving the least-squares solution

$$\begin{bmatrix} x_1 \\ x_2 \end{bmatrix} = \left(A^T A \right)^{-1} A^T b = \frac{1}{985} \begin{bmatrix} 26 & -9 \\ -9 & 41 \end{bmatrix} \begin{bmatrix} 95 \\ 36 \end{bmatrix} = \begin{bmatrix} 2146/985 \\ 621/985 \end{bmatrix}.$$

This is somewhat different from the previous solution, and weighting schemes are often important to get the best from observed data when least-squares solutions are found.

Exercise 30.5
Find a least-squares solution of the following equations: $4x = 10$, $x = 2$, $3x = 7$.

30.7 Refinement procedures using least squares

There are many areas of science where the problem is to find the best values of some quantities from **overdetermined** systems, i.e. those where there are many more observations than unknowns. One such case is in crystallography, where the problem is to find the positions of atoms within the repeat unit of the crystal and where the observations are the intensities of X-ray beams diffracted from the crystal, as described in §17.14. Normally the number of X-ray data exceeds the number of unknown coordinates of the atoms by a considerable factor, but the equations linking the coordinates with the X-ray intensities are complicated so that finding the positions of the atoms is a major preoccupation of crystallographers. They have devised various techniques for solving crystal structures but these often give only rough coordinates of the atoms and some refinement procedure is necessary. Here we illustrate this process by a simple example of a hypothetical one-dimensional crystal containing three atoms, at positions x_1, x_2, and x_3, which should satisfy the relationships

$$\begin{aligned}
F_1 &= \cos 2\pi x_1 + \cos 2\pi x_2 + \cos 2\pi x_3 = -1.0 \\
F_2 &= \cos 4\pi x_1 + \cos 4\pi x_2 + \cos 4\pi x_3 = -0.9 \\
F_3 &= \cos 6\pi x_1 + \cos 6\pi x_2 + \cos 6\pi x_3 = 0.5 \\
F_4 &= \cos 8\pi x_1 + \cos 8\pi x_2 + \cos 8\pi x_3 = -1.5
\end{aligned} \tag{30.25}$$

with an approximate solution $_0x_1 = 0.2$, $_0x_2 = 0.35$ and $_0x_3 = 0.6$. We wish to obtain better values for the three variables. Substituting the approximate values, we find for the four expressions $F_1' = -1.087$, $F_2' = -0.809$, $F_3' = 0.450$, $F_4' = -1.309$. If, for example, the variable $_0x_1$ is changed by a small amount Δx_1 then the first term in the first equation is changed to

$$\cos 2\pi(_0x_1 + \Delta x_1) = \cos 2\pi_0 x_1 \cos 2\pi \Delta x_1 - \sin 2\pi_0 x_1 \sin 2\pi \Delta x_1.$$

If $(\Delta x_1)^2 \ll 1$ then $\cos 2\pi \Delta x_1 \approx 1$ and $\sin 2\pi \Delta x_1 \approx 2\pi \Delta x_1$, so that

$$\cos 2\pi(_0x_1 + \Delta x_1) \approx \cos 2\pi_0 x_1 + 2\pi \sin 2\pi_0 x_1 \Delta x_1.$$

Now we assume that each of the variables is changed by a small amount that gives the correct values for the Fs; then for the first equation

$$F_1 - F_1' = -2\pi \sin 2\pi_0 x_1 \Delta x_1 - 2\pi \sin 2\pi_0 x_2 \Delta x_2 - 2\pi \sin 2\pi_0 x_3 \Delta x_3 = 0.087.$$

Substituting the approximate values of the variables and rearranging, this becomes

$$0.9511\Delta x_1 + 0.8090\Delta x_2 - 0.5878\Delta x_3 = -0.0138$$

and similarly the other equations give

$$\begin{aligned}
0.5878\Delta x_1 - 0.9511\Delta x_2 + 0.9511\Delta x_3 &= 0.0072 \\
-0.5878\Delta x_1 + 0.3090\Delta x_2 - 0.9511\Delta x_3 &= -0.0027 \\
-0.9511\Delta x_1 + 0.5878\Delta x_2 + 0.5878\Delta x_3 &= 0.0076.
\end{aligned} \tag{30.26}$$

The first step of obtaining a least-squares solution, finding $A^T A x = A^T b$, gives

$$\begin{bmatrix} 2.5002 & -0.5303 & 0 \\ -0.5303 & 2.0000 & -1.3285 \\ 0 & -1.3285 & 2.5002 \end{bmatrix} \begin{bmatrix} \Delta x_1 \\ \Delta x_2 \\ \Delta x_3 \end{bmatrix} = \begin{bmatrix} -0.0145 \\ -0.0144 \\ 0.0220 \end{bmatrix}. \tag{30.27}$$

You will notice that the coefficient matrix in (30.27) has a dominant diagonal, a common feature of the normal equations since diagonal elements are the sums of squares of the original coefficients. This being so, the Gauss–Seidel method can be used for solution with the equations in the form

$$\Delta x_1 = 0.2121\Delta x_2 - 0.0058$$
$$\Delta x_2 = 0.2652\Delta x_1 + 0.6643\Delta x_3 - 0.0072 \qquad (30.28)$$
$$\Delta x_3 = 0.5314\Delta x_2 + 0.0088.$$

As shown in §30.4, we can write a 'quick fix' program to find a solution. Such a program gives $\Delta x_1 = -0.0068$, $\Delta x_2 = -0.0049$, $\Delta x_3 = 0.0062$ which, added to the initial estimates, give the new estimates $_1x_1 = 0.1932$, $_1x_2 = 0.3451$, $_1x_3 = 0.6062$. From (30.25) we find $F_1' = -0.999$, $F_2' = -0.889$, $F_3' = 0.516$, $F_4' = -1.478$. This first cycle of refinement has given new estimates that reduce the differences between F' and F from $\{-0.087, 0.091, -0.050, 0.191\}$ to $\{0.001, 0.011, 0.016, 0.022\}$. The process can be repeated until the estimates change by less than whatever tolerance is required.

Problems

30.1 Use Gauss elimination to solve the following equations:

$$2x + 4y - 3z = 8$$
$$x + y - z = 3$$
$$2x - 2y + 3z = -1.$$

30.2 Use the Gauss–Seidel method to solve the following equations:

$$8x_1 - x_2 + x_3 - x_4 + 2x_5 = 19$$
$$x_1 - 4x_2 + x_3 + x_4 + x_5 = 8$$
$$x_1 + 2x_2 + 7x_3 - 2x_4 - x_5 = 14$$
$$2x_1 - x_2 - x_3 + 9x_4 + 2x_5 = -11$$
$$x_1 + x_2 + x_3 + x_4 - 5x_5 = -15.$$

You should write a simple program similar to GS (Appendix 21).

30.3 Find the least-squares solution of

$$x + 2y = 5$$
$$2x - y = 0$$
$$5x + y = 6.$$

30.4 The results of an experiment show that x_1 and x_2 are related by

$$\cos x_1 + \cos x_2 = -0.621$$
$$\cos 2x_1 + \cos 2x_2 = -0.112$$
$$\cos 3x_1 + \cos 3x_2 = -1.237.$$

Given an initial estimate $x_1 = 1.25$ and $x_2 = 2.75$ (in radians), carry out one stage of a refinement process. Show that the process you use decreases only two of the residuals of the equations, i.e. the magnitudes of the differences between the left- and right-hand sides, but still improves the overall agreement.

Numerical solution
of equations

31

A problem that sometimes arises is to find values of x that satisfy an equation of the form $f(x) = 0$ where the function is so complicated that no analytical solution can be found. In this chapter two methods for finding numerical solutions in such cases are described, and the conditions required for them to be successful are derived.

31.1 The nature of equations

If we have an equation of the form $f(x) = 0$ then solving the equation means finding the value, or values, of x that satisfy it. If $f(x)$ is a linear, quadratic, or cubic expression then routine analytical methods exist to find the solutions, or **roots**, of the equation. If, on the other hand, $f(x)$ involves higher powers of x, either explicitly or implicitly by including functions such as $\tan(x)$ which can be represented by an infinite series of powers of x, then the solutions can only be found by numerical means. In this chapter we present two practical methods that will deal with the great majority of cases.

Normally, a necessary precondition for finding a solution is to start with an approximation that is a good enough starting point to enable the numerical method to converge to the accurate solution. One obvious way of doing this is to plot the function and to find approximately where it cuts the x-axis. If the function is not too sinuous then this will usually reveal the approximate solution value, or values, very easily. However, if the function is very sinuous and complicated then there is no alternative but to calculate it at fine intervals to find where it crosses the x-axis.

  Exercise 31.1

Plot the function $1 + \sin x - x = 0$ to find the approximate solutions.

31.2 Fixed-point iteration method

We shall illustrate this method by an example.

EXAMPLE 31.1

Consider the equation

$$e^x + x - 2 = 0. \tag{31.1}$$

A simple plot confirms that this has a single solution somewhere around $x = 0.45$.

Now we rewrite the equation in the form

$$x = \ln(2 - x). \tag{31.2}$$

The process on which we are now going to embark is to put the current estimate of x in the expression on the right-hand side of (31.2) to get a new estimate. The new estimate is then put on the right-hand side to get another estimate, and so on. If the estimate after the nth iteration is x_n then the process can be summarized as

$$x_{n+1} = \ln(2 - x_n) \tag{31.3}$$

and we start with $x_0 = 0.45$. Working to three-figure accuracy, the successive estimates are:

0.438   0.446   0.441   0.444   0.442   0.443   0.443

and it is clear that we have found the solution. Actually the solution process is extremely robust, and even starting with $x = 0$ or $x = 1$ gets to the solution in 15 or 16 iterations.

The procedure outlined here is as follows. The original equation of the form

$$f(x) = 0 \tag{31.4}$$

is transformed to appear as

$$x = g(x) \tag{31.5a}$$

and the solution is found by the iterative process

$$x_{n+1} = g(x_n) \tag{31.5b}$$

starting with an initial estimate x_0.

It is clear that the form (31.5a) can be produced from (31.1) in another, and more obvious, way, i.e. as

$$x = 2 - e^x \tag{31.6}$$

and we now seek to use this to obtain a solution. The first few steps are:

0.432   0.460   0.416   0.484   0.377   0.542   0.281

We can see that successive estimates are oscillating with an ever-increasing amplitude about the initial estimate, and it is evident that no solution will be found.

We now have to find out why it is that a solution could be found from (31.2) but not from (31.6).

Exercise 31.2

Use the fixed-point iteration method to find a solution near $x=1$ of $x \sin x - 1 = 0$.

31.2.1 Convergence for fixed-point iteration

Let the error in the nth estimate x_n be e_n so that

$$x_n = x_s + e_n \tag{31.7}$$

where x_s is the solution value. Equation (31.5b) can now be written as

$$x_s + e_{n+1} = g(x_n + e_n). \tag{31.8}$$

If the estimate is not too far from the solution, so that e_n is small, then using the Taylor series (4.42) to two terms

$$x_{n+1} = x_s + e_{n+1} = g(x_s) + e_n g'(x_s). \tag{31.9}$$

However, for the solution

$$x_s = g(x_s)$$

so that

$$e_{n+1} = e_n g'(x_s). \tag{31.10}$$

This gives the condition for convergence. If

$$|g'(x_s)| < 1 \tag{31.11}$$

the magnitude of the error will progressively lessen as the process continues and the correct solution will be found. Under this condition, if $g'(x_s)$ is positive the solution will be steadily approached from one direction. If $g'(x_s)$ is negative then the successive approximations will oscillate about the solution, but will get closer and closer. The latter condition was present in getting the solution from (31.3). We can check this. With

$$g(x) = \ln(2 - x)$$

$$g'(x_s) = -\frac{1}{2 - x_s} = -0.642,$$

which satisfies the convergence condition.

However, if

$$|g'(x_s)| \geq 1 \tag{31.12}$$

then there will be no convergence. For (31.6), $g(x) = 2 - e^x$ and $g'(x_s) = -e^{x_s} = -1.557$. The behaviour of the attempted solution in this case is now understandable: because $g'(x_s)$ is negative the successive approximations oscillate about the starting estimate, and because $|g'(x_s)| > 1$ the oscillations have an ever-increasing amplitude.

Exercise 31.3

Examine the convergence properties of the formulation of the fixed-point iteration formula used in Exercise 31.2.

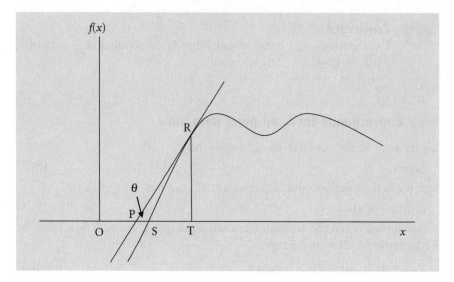

Figure 31.1 A function $f(x)$.

31.3 Newton–Raphson method

The principle of the Newton–Raphson method can best be understood by examination of a diagram. In Figure 31.1 a function $f(x)$ is shown over a range of x and what is required is a solution of the equation

$$f(x) = 0 \tag{31.13}$$

represented by the point S.

We start with an approximate value, x_n, represented by the point T, and RP is a tangent to the curve. The point P is the next approximation, x_{n+1}, and repeating the process eventually gives a solution if certain conditions are satisfied. In particular, if the first trial point is taken in the undulating region to the right of T the process is unlikely to work, although it may do so if one of the tangents ends up close to S.

From Figure 31.1

$$TP = \frac{RT}{\tan\theta} = \frac{f(x_n)}{f'(x_n)}$$

so that

$$x_n - x_{n+1} = \frac{f(x_n)}{f'(x_n)}$$

or

$$x_{n+1} = x_n - \frac{f(x_n)}{f'(x_n)}, \tag{31.14}$$

which is the basic Newton–Raphson refinement procedure.

This is now applied to (31.1). With $f(x) = e^x + x - 2$ we have $f'(x) = e^x + 1$ and

$$x_{n+1} = x_n - \frac{e^{x_n} + x_n - 2}{e^{x_n} + 1}. \tag{31.15}$$

Starting with $x_0 = 0.45$ we find successive values, expressed to six significant figures, as

0.442 870   0.442 854   0.442 854

and it is evident that, to four significant figures, the solution is reached in a single step.

It is a characteristic of the Newton–Raphson process that it is extremely efficient, especially if the first trial is fairly close to the solution.

Exercise 31.4

Use the Newton–Raphson method to find a solution near $x=1$ of $x \sin x - 1 = 0$. You may find a simple 'quick and dirty' computer program useful.

31.3.1 Convergence for the Newton–Raphson procedure

The error, e_n, in the nth estimate, x_n, is defined as in (31.7) and the solution value is taken as x_s. Then from (31.14)

$$x_s + e_{n+1} = x_s + e_n - \frac{f(x_s + e_n)}{f'(x_s + e_n)}. \tag{31.16}$$

The Taylor expansion is used both on $f(x_s + e_n)$ and $f'(x_s + e_n)$ up to the second derivative term, giving

$$e_{n+1} = e_n - \frac{f(x_s) + e_n f'(x_s) + \frac{1}{2}e_n^2 f''(x_s)}{f'(x_s) + e_n f''(x_s)}. \tag{31.17}$$

Since x_s is the solution,

$$f(x_s) = 0 \tag{31.18a}$$

and we also use the approximation (1.38) to write

$$\frac{1}{f'(x_s) + e_n f''(x_s)} = \frac{1}{f'(x_s)}\left\{1 - e_n\frac{f''(x_s)}{f'(x_s)}\right\}. \tag{31.18b}$$

Substituting from (31.18a) and (31.18b) in (31.17) and ignoring terms in e_n^3 we find

$$e_{n+1} = \frac{f''(x_s)}{2f'(x_s)} e_n^2. \tag{31.19}$$

The method will converge if

$$\left|\frac{e_n}{e_{n+1}}\right| = \left|\frac{2f'(x_s)}{e_n f''(x_s)}\right| > 1$$

or

$$|e_n| < 2\left|\frac{f'(x_s)}{f''(x_s)}\right|. \tag{31.20}$$

In this analysis the derivatives that appear are those at the solution value and in the refinement process the derivatives are at different values – those corresponding to the current approximation. However, condition (31.20) does represent a *necessary* condition for convergence.

Exercise 31.5
Examine the convergence properties of the Newton–Raphson method for the equation used in Exercise 31.4.

Problems

Note: Simple computer programs should be written for both these problems.

31.1 Use the fixed-point iteration method to find the solution of $x^3 e^x + 3(1 - x^2) = 0$. There is a solution close to $x = -1$. Check your fixed-point iteration equation for its convergence properties.

31.2 Use the Newton–Raphson method to find the solution of the equation given in Problem 31.1.

Signals and noise

As we saw in Chapter 17, we encounter signals constantly in everyday life. Unfortunately, we also encounter noise. Noise exists in electronic equipment because electrons move even in the absence of voltage sources. The thermal energy of the electrons gets converted to kinetic energy and random motions of the electrons, which, in turn, leads to random currents; this is heard as background hiss in an analogue loudspeaker. In this chapter you will learn how to reduce the effects of noise and retrieve the information contained in the signal.

32.1 Introduction

Noise corrupts signals. In many applications it is important to reduce the effects of noise and retrieve the information contained in the signal. One example is the global positioning system (GPS) that enables you to determine your position, for example if you are driving in an area you don't know, or walking in the hills. The GPS system operates on the principle illustrated in Figure 32.1. Various satellites S_1 to S_4 send digital transmissions to the receiver A (in practice there are normally eight satellites visible at any time). The receiver produces the code that is expected to be received from each satellite at a particular time. This code is digital in nature consisting of 1's and -1's. The transmission from satellite S_1 to receiver might look like the one illustrated in Figure 32.2.

The receiver produces an identical code at the same time that the satellite transmits this code. Because of the time delay for the code to travel from the satellite to the receiver, these two codes will be shifted in time with respect to each other by an amount T. Hence the distance between the satellite and receiver can be estimated by

$$R = cT, \tag{32.1}$$

where c is the speed of light. By measuring the time delays from three or more satellites, the receiver can compute its position with respect to the satellites. This sounds easy, but there is a significant problem. A typical signal coming from the satellite S_1 in Figure 32.1 might look more like that in Figure 32.3 than Figure 32.2. It is very difficult to see the digital transmission from the satellite – it is buried underneath a lot of noise. By the end of this chapter, we will have come up with a method that would allow you to detect the underlying signal. We will be asking the

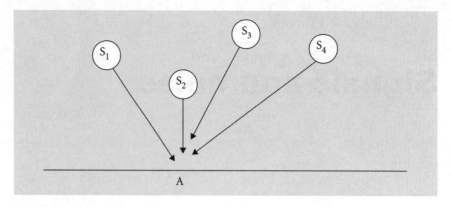

Figure 32.1 Simplified version of GPS.

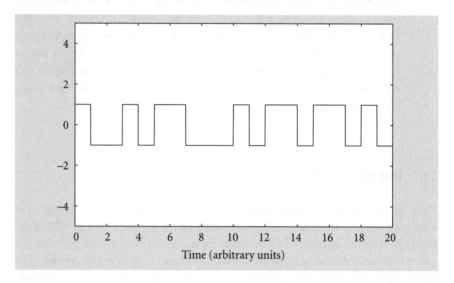

Figure 32.2 Example of a clean GPS signal.

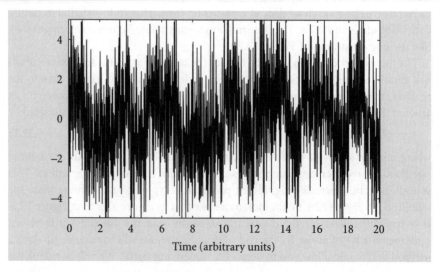

Figure 32.3 Example of a noisy GPS signal.

question: how can we distinguish noise from a signal *mathematically*? When we know how to do this, we can then devise methods to detect the presence of signals in noise. One important method on which we will concentrate in this chapter is **correlation**.

32.2 **Signals, noise, and noisy signals**

First, we consider the basic difference between signals and noise. One possible example of a signal is shown in Figure 32.4. This is a cosine signal, of amplitude 2, phase 0, and frequency 0.2 Hz. Hence, we can represent it as

$$s(t) = 2\cos(2\pi(0.2)t). \tag{32.2}$$

Although the signal is only displayed for times $-10 \leq t \leq 10$ s, we can predict what its value will be at later times – all we do is put the appropriate value of t in (32.2). The signal in Figure 32.4 is hence **predictable**. To illustrate this further, the signal at time $t = 2$ s has a value of -1.618 to three significant figures. We can predict with certainty that, at $t = 5$ s, this signal will have a value of 2.

Another example of a signal was shown in Figure 32.2. As we mentioned earlier, the series of 1s and -1s is random to the outside world. However, as far as the satellite and receiver are concerned this signal is known; it is a signal to the satellite and receiver although it is noise to everyone else. The definition of 'signal' and 'noise' in this case depends on who you are! This signal is more complicated than a cosine, but is nevertheless predictable, in a certain sense. The signal is positive between times 0 and 1, negative between times 1 and 3 and positive between times 3 and 4. Hence there is some predictability in this signal in the sense that we know that it is either $+1$ or -1.

Now, let us look at an example of noise (Figure 32.5). In contrast to the signals illustrated above, noise looks like a mess: it is completely unpredictable. Let us use

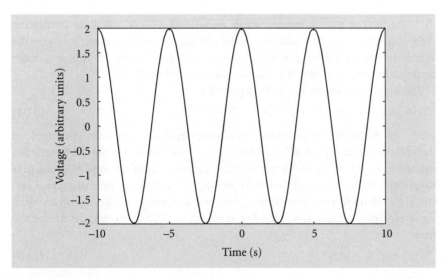

Figure 32.4 Cosine signal.

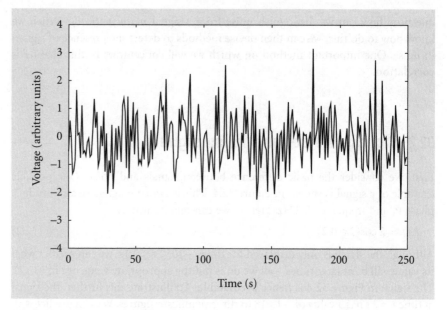

Figure 32.5 Noise.

the same analysis as for the cosine. In Figure 32.5, at $t = 200$ s, the noise signal has a value of -0.186. What is its value at $t = 500$ s? We can say with some confidence that the value will have a high probability of being between -3 and 3; we will discuss this point later. However, the value at $t = 500$ s could be almost anything. Noise is unpredictable or **random**.

Most signals we encounter in practice are corrupted by noise. When a signal is amplified, the amplifier introduces noise. Transmitted images of planetary surfaces are initially corrupted with considerable noise – we only see the image after the original noisy image has been cleaned up. Hence, if we are analysing analogue signals, we normally have to deal with noisy signals $x(t)$ given by

$$x(t) = s(t) + n(t) \tag{32.3}$$

in which $s(t)$ is the underlying clean signal and $n(t)$ is the noise corrupting the signal. This is a simple **additive** model of noise, where noise directly adds to the signal. In **multiplicative noise** the noise multiplies the signal; this is a more complicated problem, which is outside the scope of this book.

The digital equivalent to (32.3) is given by

$$x[p] = s[p] + n[p] \tag{32.4}$$

where $s[p]$ is the pth sample of the underlying signal and $n[p]$ is the pth sample of the additive noise. An example of a noisy signal was shown in Figure 32.3; there is a signal part of this signal and a noise part. A common signal-processing problem is to estimate the unknown $s(t)$ in (32.3) (or $s[p]$ in (32.4)) given the noisy signal $x(t)$ (or $x[p]$). If noise were predictable, then the estimation of the unknown clean signal $s(t)$ (or $s[p]$) would be easy. For example, in (32.3), we could determine $s(t)$ from

$$s(t) = x(t) - n(t). \tag{32.5}$$

However, as we have seen, noise is unpredictable and we do not have a convenient mathematical description of it that would allow us to evaluate it at any time point. We need to be able to characterize noise in some way that distinguishes it from a signal, in order to be able to reduce its effects on a signal. Note that it is normally impossible to eliminate noise completely, because of its unpredictable behaviour.

First, we will look at some statistical properties of noise that will be useful in detecting signals in noise.

32.2.1 Mean and standard deviation

In the previous section, we saw that noise is unpredictable and we cannot write down an equation for the noise voltage as a function of time. Two particular noise signals are shown in Figure 32.6. They both look completely random, but there is one striking difference: noise signal B has a larger variation about 0 than noise signal A. If we calculate the averages, then the mean for noise signal A is found to be 0.0096 and the corresponding value for noise signal B is −0.0868. These values

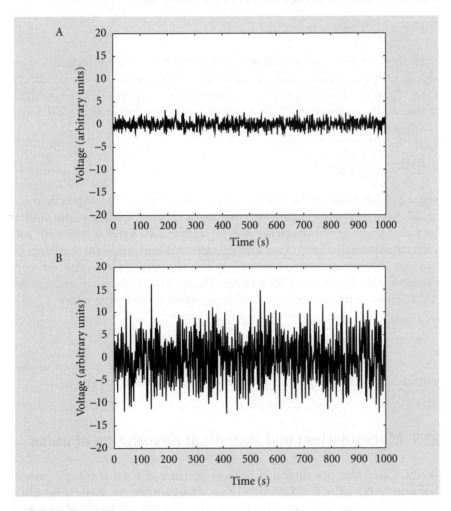

Figure 32.6 Two noise signals.

are very small compared with the deviations of the noise about zero. It can be shown that if we take larger and larger segments of any noise signal, then the average value will tend to 0 for both noise signals.

Mathematically, we can write for any digital signal, $\{x[i]\}$, the average value over N samples as

$$\bar{x}[N] = \frac{1}{N} \sum_{i=1}^{N} x[i]. \tag{32.6}$$

For the particular case of noise, the average value is given by

$$\bar{n}[N] = \frac{1}{N} \sum_{i=1}^{N} n[i] \tag{32.7}$$

and in the limit as $N \to \infty$:

$$\lim_{N \to \infty} \bar{n}[N] = \lim_{N \to \infty} \left[\frac{1}{N} \sum_{i=1}^{N} n[i] \right] = 0. \tag{32.8}$$

Equation (32.8) is only valid for infinite segments of noise, but for large enough segments of noise we can rewrite it as an approximation:

$$\bar{n}[N] = \frac{1}{N} \sum_{i=1}^{N} n[i] \approx 0. \tag{32.9}$$

Hence the mean is not a good way to distinguish between two different noise signals. Of greater distinguishing power is the variance, or the standard deviation (§12.7.2), of the noise signal. Since the mean is effectively zero then the standard deviation, the square root of the variance, equals the root-mean-square value (*RMS*) given by

$$RMS = \sqrt{\frac{1}{N} \sum_{i=1}^{N} (n[i])^2} \tag{32.10}$$

where $n[i]$ is the value of the ith noise sample. The computed *RMS* values for noise signals A and B in Figure 32.6 are 0.999 for A and 4.772 for B. The larger value of *RMS* for noise signal B signifies that it has more power than noise signal A. However, these values need to be calculated over a large enough number of samples to be reliable. In addition, different segments of noise of the same length will give slightly different values for *RMS*. If we increase the number of samples to 10^6 and compute the mean and *RMS* values for noise signals A and B, we obtain the following values:

A : mean $= -0.000718$, *RMS* $= 1.001$;

B : mean $= 0.0015$, *RMS* $= 4.999$.

The *RMS* values of A and B are actually 1 and 5, hence the estimates of this parameter over 10^6 samples are accurate to 0.1% and 0.02% respectively.

32.3 Mathematical and statistical description of noise

We have seen that one difference between signals and noise is that we cannot characterize the latter as a function of time. However, we can characterize noise using parameters related to its statistical nature, and this will be described next.

32.3.1 Probability distribution functions

Standard deviations are one way to distinguish different noise signals. Look at the two noise signals in Figure 32.7, where the noise voltages in each case have been plotted as points rather than joining lines between adjacent samples. The two noise signals look different, although their means (0) and standard deviations (1) are identical. However, visually it is apparent that the noise voltages for C have a greater tendency to have smaller rather than larger magnitudes. In the case of noise signal D, the noise values appear to have any magnitudes with equal probability up to a limiting value. Beyond this limit there is no noise.

Although these two noise signals have the same mean and standard deviation, they look, and are, different, and we need another way to distinguish them. What we can say that the **distributions** of the voltages for the two noise signals are different.

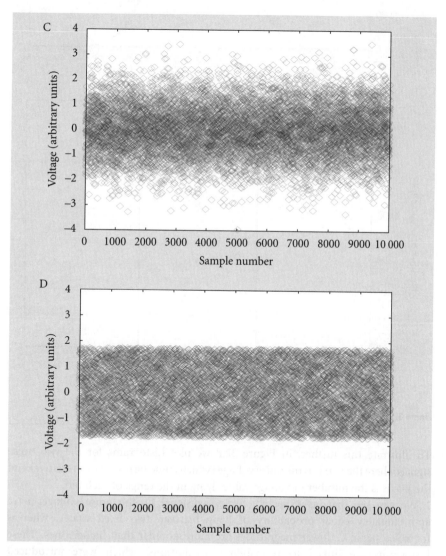

Figure 32.7 Another two noise signals.

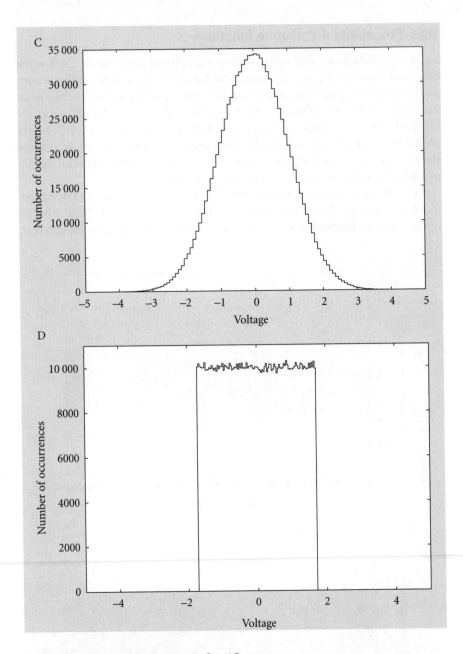

Figure 32.8 Histograms for noise signals C and D.

To illustrate this further, in Figure 32.8 we plot histograms for the two noise signals where the x-axis is the noise voltage (which could be positive or negative) and the y-axis is the number of voltage values lying in the range of each bin.

We can see that, as expected, for noise signal D, the noise voltages have approximately equal probability of lying between two fixed values, whereas for noise signal C, lower magnitudes are more popular than higher ones. These histograms are linked to probability distributions, which were introduced in Chapter 14. The distribution for noise signal C is a **Gaussian** or **normal**

distribution whereas that for noise signal D is called a **uniform distribution**. These distributions are normalized so that the area under the curves is 1. The Gaussian distribution is

$$p(x) = \frac{1}{\sqrt{2\pi}\sigma} \exp\left(-\frac{x^2}{2\sigma^2}\right) \tag{32.11}$$

where σ is the standard deviation and the variable x is the voltage. The normalized probability distribution function for the uniform distribution between voltages $-A$ and $+A$ is given by

$$p(x) = \begin{cases} \dfrac{1}{2A} & \text{for } -A \le x \le A \\ 0 & \text{otherwise.} \end{cases} \tag{32.12}$$

We interpret the probability distribution function, $p(x)$, as follows. The probability that the noise voltage is between values a and b volts is given by

$$P(a \le x \le b) = \int_a^b p(x) \, \mathrm{d}x. \tag{32.13}$$

Note that, by definition, noise has a probability of 1 of having a value between $-\infty$ and $+\infty$, in other words it must have some value between these limits, so that

$$\int_{-\infty}^{\infty} p(x) \, \mathrm{d}x = 1. \tag{32.14}$$

Thus noise can be characterized by its probability distribution function. One way to characterize noise with a particular probability distribution is through its **moments**.

32.3.2 Moments

The mean is an example of a moment of a distribution and the variance is a function of two moments. The nth moment of a distribution is defined through its probability distribution function as follows:

$$\bar{x}_n = \int_{-\infty}^{\infty} x^n p(x) \, \mathrm{d}x. \tag{32.15}$$

For $n = 1$ we obtain the **mean** of the noise:

$$\bar{x}_1 = \int_{-\infty}^{\infty} x p(x) \, \mathrm{d}x. \tag{32.16}$$

If the probability distribution is an even function of voltage x then the integrand in (32.16) is an odd function of x. When an odd function is integrated between equal but opposite limits the result is zero. This is the case for all noise distributions, that the mean value when taken over an infinite amount of time is zero:

$$\bar{x}_1 = 0. \tag{32.17}$$

If we put $n = 2$, we then obtain the **second moment** or mean-square value of the noise:

$$\bar{x}_2 = \int_{-\infty}^{\infty} x^2 p(x) \, \mathrm{d}x. \tag{32.18}$$

For the uniform distribution this is

$$\bar{x}_2 = \frac{A^2}{3} \tag{32.19}$$

while for the Gaussian distribution it is

$$\bar{x}_2 = \sigma^2. \tag{32.20}$$

Results (32.19) and (32.20) are derived in Appendix 22.

To summarize, the variance of any function is defined by

$$V = \bar{x}_2 - (\bar{x}_1)^2 \tag{32.21}$$

and since, for noise, the mean value, $\bar{x}_1$, tends to zero as the number of samples tends to infinity,

$$V = \bar{x}_2. \tag{32.22}$$

For a finite segment of data, (32.22) is an approximation as, from (32.9), the mean value $\bar{x}_1$ is then only approximately zero.

In general, it can be seen from (32.15), that as the probability distribution of noise is an even function of x then, if n is odd, the integrand will be an odd function of x and hence from (32.15), all odd moments will be zero.

32.4 Auto- and cross-correlation functions

We have seen that the clean signal (Figure 32.4) is a predictable function of time. The value of the signal at one time point is strongly related to its value at a nearby subsequent time point. For example, when $t = 3$ s the cosine has a value of -1.618; 0.02 s later its value is -1.588, which is close in value. If we take any two points on the signal this amount of time apart, then we will always get two values that are close to each other when compared to the total variation in amplitude of the signal. Another way of describing this relationship is to say that the value of the signal at time 3.02 s is **correlated** with its value at time 3 s. However, in Figure 32.5 the noise signal at time 50 s is in no way related to its value at the subsequent sample. We can say that the values of the noise signal at any two adjacent samples are **uncorrelated**. These correlation properties can be used to distinguish between signal and noise.

We introduced the idea of correlation in §17.4.1. If there are two sets of samples $x = \{x_i\}$ and $y = \{y_i\}$, then the correlation coefficient between the two is

$$r = \frac{\langle xy \rangle - \langle x \rangle \langle y \rangle}{\sigma_x \sigma_y} \tag{32.23}$$

where σ_x and σ_y are the standard deviations of x and y. For identical signals the **cross-correlation coefficient** is 1; for unrelated signals it is 0. The nearer to 1 is the cross-correlation coefficient, then the more similar are the two signals. In the following discussion, we will use the following modified definition of correlation:

$$r_{xy} = \langle xy \rangle. \tag{32.24}$$

In this case, the two signals are multiplied sample by sample without normalization or subtraction of the mean.

To illustrate what is happening when computing the cross-correlation coefficient between two signals, we first align the signals:

$$
\begin{array}{ccccc}
x[-2] & x[-1] & x[0] & x[1] & x[2] \\
y[-2] & y[-1] & y[0] & y[1] & y[2]
\end{array}
$$

Here, each 'block' represents a sample of a signal and we assume for now that each signal is infinite in extent. We now use the notation that $x[m]$ represents the mth sample of signal x, with the same convention used for signal y. The cross-correlation coefficient (32.24) is then calculated as follows:

$$r[0] = \sum_{m=-\infty}^{\infty} x[m]y[m]. \tag{32.25}$$

The argument '[0]' on r will be explained later.

Now, suppose that we move signal y one sample to the left, keeping signal x in the same position:

$$\begin{array}{ccccc} x[-2] & x[-1] & x[0] & x[1] & x[2] \\ y[-1] & y[0] & y[1] & y[2] & y[3] \end{array}$$

The correlation coefficient between signal x and the shifted y signal is given by

$$r_{xy}[1] = \sum_{m=-\infty}^{\infty} x[m]y[m+1]. \tag{32.26}$$

The argument '[1]' on r_{xy} signifies that y has been shifted to the left by one sample whereas in (32.25), the argument '[0]' on r indicates that there is no shift between signals x and y when calculating the correlation coefficient.

One could also shift signal y one sample to the right:

$$\begin{array}{ccccc} x[-2] & x[-1] & x[0] & x[1] & x[2] \\ y[-3] & y[-2] & y[-1] & y[0] & y[1] \end{array}$$

to give

$$r_{xy}[-1] = \sum_{m=-\infty}^{\infty} x[m]y[m-1] \tag{32.27}$$

where the argument '[-1]'on r signifies that the signal y has been shifted one sample to the right before calculating the correlation coefficient. We can repeat the above procedure moving signal y to the left and right by an increasing number of samples to work out the correlation coefficients $\{r_{xy}[p]\}$, where $r_{xy}[p]$ is the correlation coefficient between signal x and signal y shifted p samples to the left. Note, $p < 0$ indicates a right shift of the signal y.

In general we may write

$$r_{xy}[p] = \sum_{m=-\infty}^{\infty} x[m]y[m+p], \tag{32.28}$$

where $p = \ldots, -2, -1, 0, 1, 2, \ldots$. The set of correlation coefficients $\{r_{xy}[p]\}$ is referred to as the **cross-correlation function** and $r_{xy}[p]$ is termed the **cross-correlation function** at lag p. If $y = x$ then we are comparing a signal with itself, shifted by 0, ±1, ±2, etc. samples to give

$$r_{xx}[p] = \sum_{m=-\infty}^{\infty} x[m]x[m+p] \tag{32.29}$$

where the subscript xx on r signifies that we are comparing signal x with itself. In this case, $r_{xx}[p]$ is called the **autocorrelation function** of x at lag p.

We will see later that the cross-correlation function is a powerful method of distinguishing signals from noise. However, before doing this, we should note that (32.28) and (32.29) are unrealistic, because it is assumed that we are computing the autocorrelation and cross-correlation lags over an infinite number of samples of x and y. In practice, we would only be doing this over a finite number of samples. In many applications, the following situation arises:

$$x[0] \quad x[1] \quad x[2] \quad x[3] \quad x[4]$$
$$y[0] \quad y[1] \quad y[2] \quad y[3] \quad y[4]$$

We are now calculating the cross-correlation function between a signal x, which is of finite duration, and a stream of data y, which could be of finite or infinite duration. In the satellite positioning example, y represents the stream of data coming from a satellite and x represents a 'template' stored in the receiver. We shall refer to this type of correlation as **sliding-window correlation**.

In the above example, the sliding-window cross-correlation coefficient is given by

$$r_{xy}[0] = \sum_{m=0}^{4} x[m]y[m]. \tag{32.30}$$

The choice of time origin of y where $m=0$ is arbitrary here, depending on the particular application.

If we shift y one sample to the left, we obtain

$$x[0] \quad x[1] \quad x[2] \quad x[3] \quad x[4]$$
$$y[1] \quad y[2] \quad y[3] \quad y[4] \quad y[5]$$

In this case, the cross-correlation coefficient is given by

$$r_{xy}[1] = \sum_{m=0}^{4} x[m]y[m+1]. \tag{32.31}$$

Similarly, if we shift signal y one sample to the right, we obtain the cross-correlation function at lag -1 as:

$$r_{xy}[-1] = \sum_{m=0}^{4} x[m]y[m-1]. \tag{32.32}$$

In general, for the pth lag,

$$r_{xy}[p] = \sum_{m=0}^{4} x[m]y[m+p]. \tag{32.33}$$

Generalizing, for a template x with M samples, the cross-correlation function is

$$r_{xy}[p] = \sum_{m=0}^{M-1} x[m]y[m+p]. \tag{32.34}$$

In many applications, the set of samples $\{y[i]\}$ represents data that we are analysing and $\{x[i]\}$ represents a 'template'. We are trying to detect the presence of the template in the data $\{y[i]\}$. In the satellite positioning example, x represents the code transmitted by a particular satellite. The sequence y is part of an infinite stream of data coming from all satellites. Embedded somewhere in $\{y[i]\}$ is the same sequence as the template x that is stored in the receiver and we want to detect when this match occurs; however added on to this are the codes from the other satellites and noise. We return to this problem later.

The sliding-window cross-correlation function between two identical signals is called the **autocorrelation function** and this is given by putting $y = x$ in (32.34) and is

$$r_{xx}[p] = \sum_{m=0}^{M-1} x[m]x[m+p]$$

(32.35)

where $x[m] = 0$ for $m < 0$ and $m > M - 1$.

As a simple example, we suppose that signal $\{x[i]\}$ is given by the set of samples

$$\{1, 1, 1, 1, 1, 1\}$$

(32.36)

illustrated in Figure 32.9. We are trying to detect the presence of $\{x[i]\}$ in the incoming data stream $\{y[i]\}$ given by the set of samples

$$\{0, 0, 0, 0, 0, 1, 1, 1, 1, 1, 1, 0, 0, 0, 0, 0\}.$$

(32.37)

This looks like Figure 32.10.

Aligning the template in Figure 32.9 with the beginning part of the data stream in Figure 32.10, we have

```
1  1  1  1  1  1
0  0  0  0  0  1  1  1  1  1  1  0  0  0  0  0
```

Using (32.34), with $M = 6$, the cross-correlation coefficient at zero shift is:

$$r_{xy}[0] = (1 \times 0) + (1 \times 0) + (1 \times 0) + (1 \times 0) + (1 \times 0) + (1 \times 1) = 1.$$

(32.38)

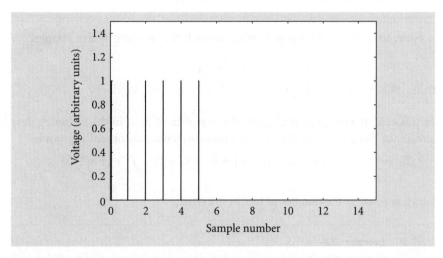

Figure 32.9 Signal x.

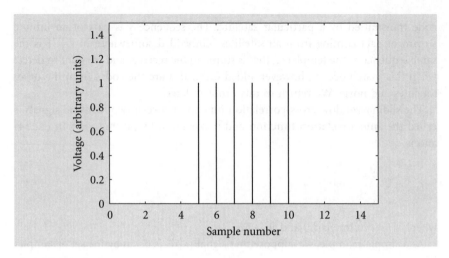

Figure 32.10 Signal y.

Shifting the template one sample to the right and working out the cross-correlation coefficient for a shift of one sample gives

```
    1   1   1   1   1   1
0   0   0   0   0   1   1   1   1   1   1   0   0   0   0   0
```

and

$$r_{xy}[1] = (1 \times 0) + (1 \times 0) + (1 \times 0) + (1 \times 0) + (1 \times 1) + (1 \times 1)$$
$$= 2. \tag{32.39}$$

We can repeat this procedure for shifts of two or more samples.

 Exercise 32.1
Show that $r_{xy}[2] = 3$, $r_{xy}[3] = 4$, and $r_{xy}[4] = 5$.

Now, let us look what happens when we shift the template by five samples:

```
            1   1   1   1   1   1
0   0   0   0   0   1   1   1   1   1   1   0   0   0   0   0
```

In this case, the template is aligned with a replica of itself in the incoming data stream. In this case, from (32.34), the cross-correlation coefficient is given by

$$r_{xy}[5] = (1 \times 1) + (1 \times 1) + (1 \times 1) + (1 \times 1) + (1 \times 1) + (1 \times 1)$$
$$= 6 \tag{32.40}$$

which is equal to the number of bits in the signal.

 Exercise 32.2
Show that $r_{xy}[6] = 5$, $r_{xy}[7] = 4$, $r_{xy}[8] = 3$, $r_{xy}[9] = 2$, and $r_{xy}[10] = 1$.

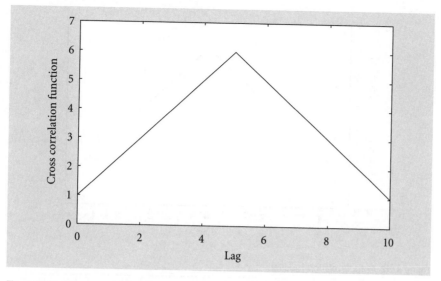

Figure 32.11 Cross-correlation function between signals y and x.

The cross-correlation function as a function of lag gives Figure 32.11. We can see that the maximum value of the cross-correlation function is given by 6 and it occurs at a lag of 5. However, how do we know that signal x occurs in signal y and not some similar signal? Well, we can find this from knowledge of the auto-correlation function of x. Suppose that in Figure (32.9) x is aligned with itself giving

$$
\begin{array}{cccccc}
1 & 1 & 1 & 1 & 1 & 1 \\
1 & 1 & 1 & 1 & 1 & 1
\end{array}
$$

The autocorrelation function, for zero lag, is given by (32.35) with $M = 6$:

$$r_{xx}[0] = (1 \times 1) + (1 \times 1) + (1 \times 1) + (1 \times 1) + (1 \times 1) + (1 \times 1) = 6.$$

$$(32.41)$$

Shifting one version by one sample to the right gives

$$
\begin{array}{ccccccc}
1 & 1 & 1 & 1 & 1 & 1 & \\
& 1 & 1 & 1 & 1 & 1 & 1
\end{array}
$$

and the autocorrelation function at lag 1 is given by

$$r_{xx}[1] = (1 \times 0) + (1 \times 1) + (1 \times 1) + (1 \times 1) + (1 \times 1) + (1 \times 1) + (0 \times 1) = 5.$$

$$(32.42)$$

where we have used the convention that if a 1 aligns with nothing, then the contribution to the correlation function is zero.

If we now shift one version of x to the left by one sample we have

$$
\begin{array}{ccccccc}
& 1 & 1 & 1 & 1 & 1 & 1 \\
1 & 1 & 1 & 1 & 1 & 1 &
\end{array}
$$

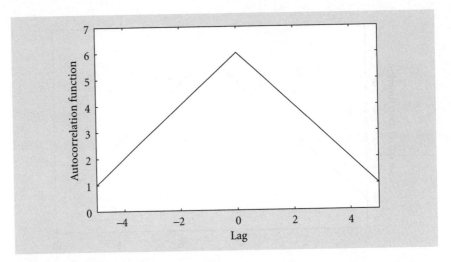

Figure 32.12 Autocorrelation function of *x*.

and the autocorrelation function at lag -1 is:

$$r_{xx}[-1] = (0 \times 1) + (1 \times 1) + (1 \times 1) + (1 \times 1) + (1 \times 1) + (1 \times 1) + (1 \times 0) = 5.$$

$$(32.43)$$

 Exercise 32.3
Compute the autocorrelation function at lags ± 2, ± 3, ± 4, and ± 5.

If we plot out the autocorrelation function, we arrive at the graph shown in Figure 32.12. Comparing Figures 32.11 and 32.12, we can see that the cross-correlation function between *x* and *y* is equal to the autocorrelation function of *x* shifted by five samples. In particular, the maximum values of the cross- and autocorrelation functions in this case are the same (6). Hence, signal *x* is embedded in *y* and is shifted five samples with respect to the template stored in the receiver.

In general, for signals, the autocorrelation function has the following three properties:

1 It is an even function of *p*, so that $r_{xx}[p] = r_{xx}[-p]$.

2 It has a maximum value at $p = 0$.

3 It is non-zero for a certain range of shifts *p*.

Almost all signals that you will meet in practice have these three properties. In particular, property 3 tells you that the value of a signal at a certain time is connected or correlated with its value at a later time. Shifted versions of the signal have similarities with the original signal itself. In the case of some signals, for large enough shifts, this similarity disappears.

Autocorrelation functions are useful when detecting signals in noise because signals and noise have very different types of autocorrelation functions. To see this, let us look at the autocorrelation function for a segment of 100 samples of noise, displayed in Figure 32.13. This obeys properties 1 and 2 above but not 3, since the autocorrelation function is only significantly non-zero for lag zero. For zero lag, $p=0$, in (32.29),

$$r_{xx}[0] = \sum_{m=-\infty}^{\infty} x[m]x[m] = \sum_{m=-\infty}^{\infty} \{x[m]\}^2 = \sum_{m=0}^{99} \{x[m]\}^2. \quad (32.44)$$

In the last step, we are using the fact that the segment of noise contains only 100 samples; we are assuming the noise to be zero outside this segment. Now the mean square (MS) value of any function x, whether it represents signal or noise, is given by

$$MS(x) = \frac{1}{100} \sum_{m=0}^{99} \{x[m]\}^2. \quad (32.45)$$

Comparing (32.44) with (32.45), we can see that

$$r_{xx}[0] = 100 \times MS(x). \quad (32.46)$$

In general, for a signal, or noise, consisting of N samples,

$$r_{xx}[0] = N \times MS(x). \quad (32.47)$$

The spike at zero lag in Figure 32.13 is essentially a measure of the energy in the segment of noise. Now we consider non-zero lags starting with lag 1:

$$r_{xx}[1] = \sum x[m]x[m+1] \quad (32.48)$$

where the summation is over all non-zero values of the noise samples. For 'pure' noise the value at sample n bears no relation to its value at the next sample $n+1$. For noise, $x[n+1]$ has an equal probability of being positive or negative. Hence the product $x[n] \times x[n+1]$ has an equal probability of being positive or negative. Figure 32.14 shows each of the terms $x[n] \times x[n+1]$ in (32.48) as a function of n.

The sum of these terms will tend to cancel each other out, and hence $r_{xx}[1]$ in (32.48) will tend to a small value. In the limit as $n \to \infty$, this value will go to zero,

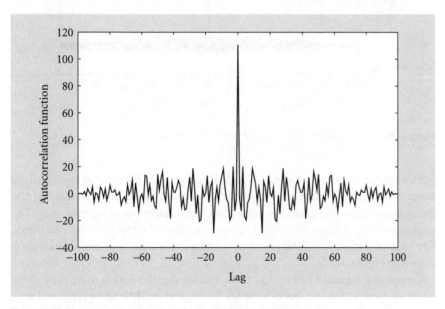

Figure 32.13 Autocorrelation function of noise with 100 samples.

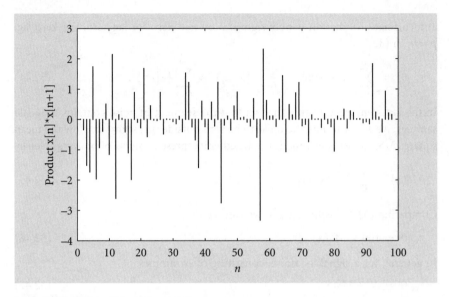

Figure 32.14 $x[n]x[n + 1]$ for noise signal.

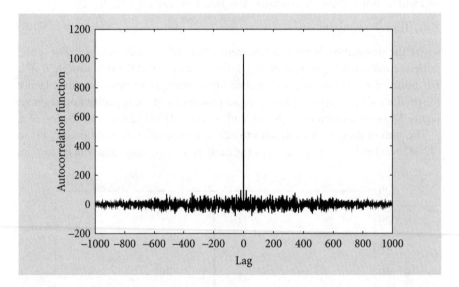

Figure 32.15 Autocorrelation of noise signal with 1000 samples.

but for finite number of samples $r[1] \approx 0$. This argument holds for all lags since the noise at sample m is not related to any of its values at future or past times. Hence, the autocorrelation function for noise has significant values at zero lag only. The peak at zero lag becomes more marked as the number of samples of noise increases. For example, if we have 1000 samples of noise, the autocorrelation function is shown in Figure 32.15.

Comparing Figures 32.13 and 32.15 we can see that the peak is more prominent when we increase the number of samples. The reason for this can be explained with reference to (32.47), where it can be seen that $r_{xx}[0] \propto N$. On the other hand, for a

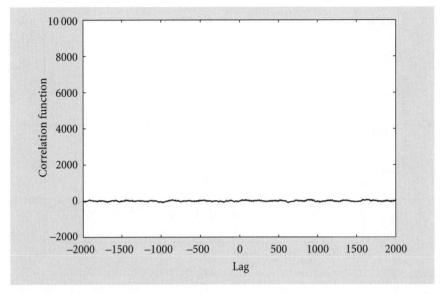

Figure 32.16 Cross-correlation function between signal and noise.

lag other than zero the autocorrelation value will come from a sum of N random numbers equally likely to be $+1$ or -1 in this particular case. Hence $r_{xx}[p] \approx 0$ with *RMS* value for all $r_{xx}[p]$ with $p \neq 0$ proportional to $\sqrt{N}$ (§23.2).

Property 3 of the autocorrelation function does not hold for noise, and we will use this property later to detect the presence of a signal in noisy data.

So far we have looked at the differences between the autocorrelation functions of signals and noise. A final correlation property that is important when detecting signals in noise is the cross-correlation function between a signal and noise. To illustrate this, let us look at the sliding window cross-correlation function between the signal in Figure 32.2 and noise (Figure 32.16). Compare this to the auto-correlation functions of the signal shown in Figure 32.17 and of the noise shown in Figure 32.18.

In Figure 32.16, it can be seen that cross-correlation function between signal and noise is negligible compared to the two other correlation functions. This is to be expected, as there is no relation between the signal samples and the noise samples. For example, suppose that we evaluate the sliding window correlation function at zero lag; in this case, we align the signal and noise and calculate the sum of the products of corresponding samples. We show this in Figure 32.19 for a few samples of signal (impulses) and noise (points) from which it is clear that the cross-correlation will tend to be small, approaching zero. Hence, to a good approximation:

$$r_{sn}[p] = 0 \text{ for all lags } p, \tag{32.49}$$

where the subscripts s and n denotes signal and noise respectively. This identity only holds for an infinite segment of noise correlated with an infinite segment of signal.

To explain (32.49) further, note that the cross-correlation function between samples of a signal, $s[m]$ and those of noise, $n[m]$ is given by

$$r_{sn}[p] = \sum s[m+p]n[m] \tag{32.50}$$

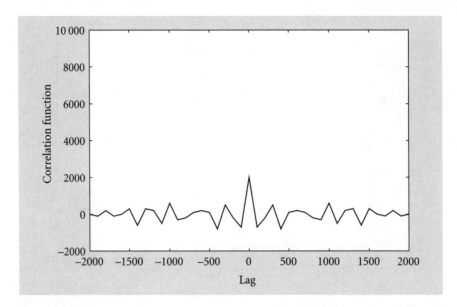

Figure 32.17 Autocorrelation function of signal.

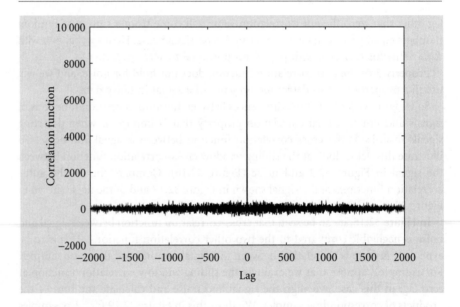

Figure 32.18 Autocorrelation function of noise.

where the summation index has been omitted. For a GPS signal, $s[i]$ is either -1 or $+1$, hence we can write

$$r_{sn}[p] = \sum n'[m] \tag{32.51}$$

where

$$n'[m] = \begin{cases} n[m] & \text{if } s[m+p] = 1 \\ -n[m] & \text{if } s[m+p] = -1. \end{cases} \tag{32.52}$$

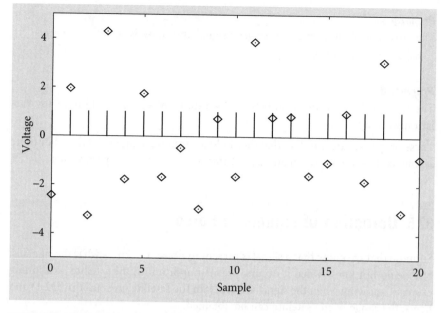

Figure 32.19 Signal samples (impulses) aligned with noise samples (points).

Since $n'[m]$ is just another noise signal, (32.51) shows that $r_{sn}[p]$ is proportional to the mean of the noise signal $\{n'[m]\}$. Comparing (32.51) with (32.8), we can see that for finite N:

$$r_{sn}[p] \approx 0. \tag{32.53}$$

The cross-correlation function of one signal with another could be anything. If the two signals are the same, then the cross-correlation function will be identical to the autocorrelation function, because we are then effectively correlating a signal with itself. On the other hand, it is possible to find two signals that have a cross-correlation function which is almost zero for all lags; we shall see that this is an essential property to be able to use two or more satellites for positioning.

32.4.1 **Summary of correlation properties of signals and noise**

The following is a summary of the ideal correlation properties of signals and noise.

Property 1
The autocorrelation function of a signal is non-zero for some values of lag p:

$$r_{ss}[p] \neq 0 \qquad \text{for} \qquad -M \leq p \leq M. \tag{32.54}$$

Property 2
The autocorrelation function of white noise is zero for all lags except zero lag:

$$\text{for } N > 0, \, r_{nn}[p] = \begin{cases} N & \text{if} \quad p = 0 \\ 0 & \text{otherwise.} \end{cases} \tag{32.55}$$

Property 3

The cross-correlation function between signal and noise is zero for all lags:

$$r_{sn}[p] = 0 \quad \text{for all } p. \tag{32.56}$$

Property 4

The cross-correlation function between two different signals could have any variation with lag.

These properties are 'ideal' in that they hold for infinite segments of data. For finite segments they are approximations, but they suffice for most applications.

32.5 **Detection of signals in noise**

Suppose that we know that a signal is present in noise. We know what the signal is, but we do not know when it occurs. This is important in the satellite positioning exercise; knowing when the signal arrives from the satellite gives us T in (32.1) and hence the range of the satellite can be obtained.

Let the mth sample of the underlying clean signal be $s[m]$ and the corresponding sample of noise be $n[m]$. The noisy signal entering the receiver is then given by

$$y[m] = s[m] + n[m]. \tag{32.57}$$

We can now correlate the incoming noisy signal $y[m]$ with a 'template' $s[m]$ stored in the receiver – we know what $s[m]$ looks like, we just do not know when it occurs. Performing the sliding-window cross-correlation between the template $s[m]$ and the noisy signal coming in, $y[m]$, where the template consists of N samples, gives the cross-correlation function at lag p:

$$r_{sy}[p] = \sum_{m=0}^{N-1} s[m]y[m+p]. \tag{32.58}$$

From (32.57),

$$y[m+p] = s[m+p] + n[m+p]. \tag{32.59}$$

Substituting for $y[m+p]$ from (32.59) into (32.58) and using the definitions in (32.34) and (32.35),

$$r_{sy}[p] = \sum_{m=0}^{N-1} s[m](s[m+p] + n[m+p]) = \sum_{m=0}^{N-1} s[m]s[m+p] + \sum_{m=0}^{N-1} s[m]n[m+p]$$

$$= r_{ss}[p] + r_{sn}[p].$$

Now using the correlation property 3 in §32.4, $r_{sn}[p] = 0$ for all p and hence

$$r_{sy}[p] = r_{ss}[p]. \tag{32.61}$$

This is a significant result. It suggests that the result of correlating the noisy signal with the template is the autocorrelation of the signal *independent of how much noise there is*! However, we must be cautious here. Property 3 is only true for processing *infinite* segments of data, but in practice, we are using *finite* segments of data; hence, in practice $r_{sn}[p] \approx 0$ and (32.61) becomes an approximation.

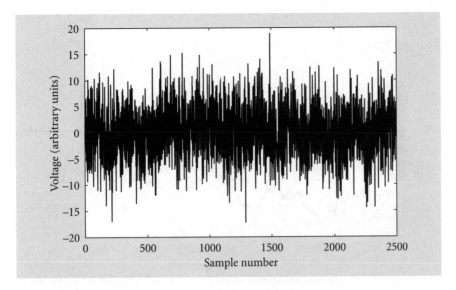

Figure 32.20 Signal buried in noise.

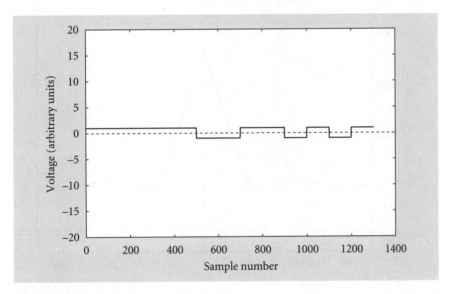

Figure 32.21 Clean signal.

To see how good this approximation is in practice we take a noisy signal as shown in Figure 32.20. Buried somewhere within this noisy signal there is the clean signal shown in Figure 32.21. If we substitute for $s[n]$ and $y[n]$ into (32.58), then the cross-correlation function is found to look like Figure 32.22. Compare this with the autocorrelation function of the underlying signal plotted in Figure 32.23 between lags -350 and $+350$.

Comparing Figures 32.20 and 32.21, it is impossible to see in Figure 32.20 whether the clean signal is there or what it looks like. However, comparing Figures 32.22 and 32.23, we can see that the cross-correlation function between the template and the noisy signal is very similar to that for the clean signal, except for

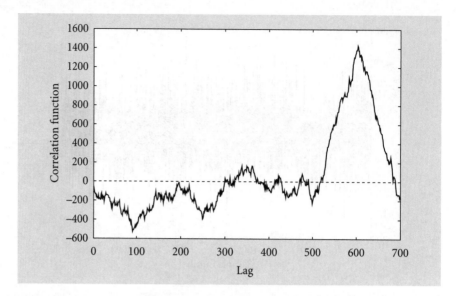

Figure 32.22 Cross-correlation function between template and noisy signal.

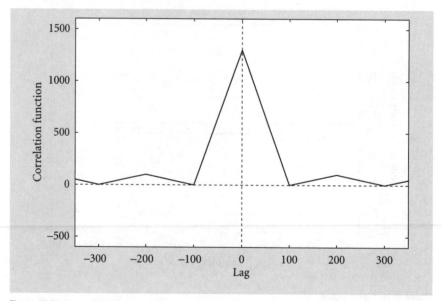

Figure 32.23 Autocorrelation function of template.

the residual effects of noise and the cross-correlation function is shifted by 600 samples with respect to the autocorrelation function. Hence the underlying signal Figure 32.21 is in the noisy signal, Figure 32.20, starting at 600 samples.

This problem of detecting the presence of known signals in noise occurs in many applications. Detecting fetal electrocardiogram signals in the maternal abdominal signal is one example. Another is the detection of edges in an image; in this application, the image is represented by a two-dimensional signal as a function of *position* rather than time and the template is what we expect an edge to look like. In both cases, the signal to noise ratio can be so small that sometimes the signal of interest is not visible in the raw noisy signal. In effect, what we are doing when correlating the template with the noisy signal is to enhance the signal above noise. Or put another way, we are filtering off noise from the noisy signal $y[n]$. The cross-correlation function, (32.58), is sometimes called a **matched filter**.

Now let us return to the satellite positioning problem described in §32.1. A binary signal is transmitted by the satellite and is also produced in the receiver at the same time. Because of the time it takes for the satellite signal to reach the receiver, the two codes will be out of synchronization. The result for this is that the cross-correlation between the template and the satellite signal will have a maximum at a lag equal to the delay. Knowing this delay, the distance R from the satellite to the receiver can be found from (32.1).

However, there may be a problem. Does the template from satellite 1 correlate strongly with the transmissions from satellites 2, 3, and 4? This question is addressed in Problem 32.2.

32.6 **White noise**

In order to design filters to reduce the effects of noise from signals, it is useful to have a model of noise in terms of cosines. What happens if we take the Fourier transform of a segment of noise, and plot the amplitude and phase spectra? If we take Gaussian noise, for example, then typically we obtain amplitude and phase spectra, as shown in Figures 32.24a and 32.24b. Similar spectra are obtained for uniform noise.

Two observations arise from these spectra. First, the amplitude spectrum deviates randomly about an average value for all frequencies. Second, the phase spectrum varies randomly with frequency between $-180°$ and $+180°$. In fact, if we take longer and longer segments of data, then the amplitude spectrum tends to a constant and the phase spectrum stays random.

We can formalize this result by looking at the discrete Fourier transform (DFT) of a segment of noise. Let the noise voltage at the mth sample be $n[m]$ with M the number of noise samples present. The DFT of the segment of noise is, from (18.6), given by

$$N[p] = \sum_{m=0}^{M-1} n[m] \exp\left(-\frac{2\pi imp}{M}\right) \qquad (p = 0, 1, \ldots, M-1) \tag{32.62}$$

where m is the time sample index.

The **energy spectrum**, $E[p]$ is defined as follows:

$$E[p] = |N[p]|^2 \tag{32.63}$$

where

$$|N[p]|^2 = N[p] N^*[p]. \tag{32.64}$$

Taking complex conjugates of both sides of (32.62), remembering that $n[m]$ is real:

$$N^*[p] = \sum_{k=0}^{M-1} n[k] \exp\left(\frac{2\pi ikp}{M}\right) \tag{32.65}$$

where now k is the time sample variable.

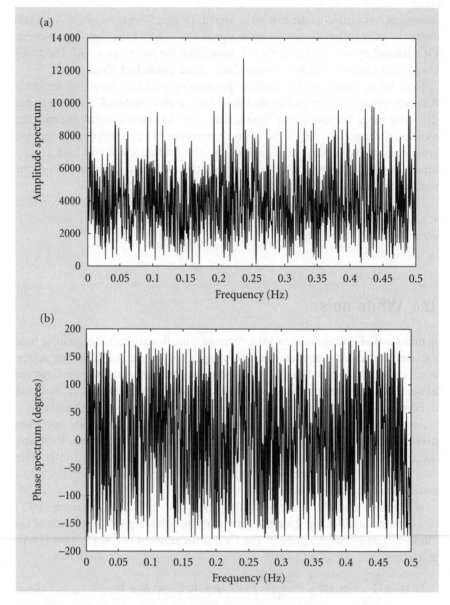

Figure 32.24 (a) Amplitude spectrum and (b) phase spectrum of noise.

Substitute for $N[p]$ from (32.62) and $N^*[p]$ from (32.65) into (32.64):

$$|N[p]|^2 = \sum_{m=0}^{M-1} \sum_{k=0}^{M-1} n[m]n[k] \exp\left(-\frac{2\pi i m p}{M}\right) \exp\left(\frac{2\pi i k p}{M}\right). \tag{32.66}$$

Multiplying the two exponentials together we obtain the following expression:

$$|N[p]|^2 = \sum_{m=0}^{M-1} \sum_{k=0}^{M-1} n[m]n[k] \exp\left(\frac{2\pi i (k-m) p}{M}\right). \tag{32.67}$$

We can now separate this double summation into two summations, one where

$k = m$ and one where $k \neq m$:

$$|N[p]|^2 = \sum_{m=0}^{M-1} (n[m])^2 + \sum_{m=0}^{M-1} \sum_{\substack{k=0 \\ m \neq k}}^{M-1} n[m]n[k] \exp\left(\frac{2\pi i(k-m)p}{M}\right) \quad (32.68)$$

since $\exp(i0) = 1$. Referring to (32.35), the first term in (32.68) is the auto-correlation function of noise at zero lag; from (32.55) this value is non-zero. In the second term in (32.68) we are taking products of samples of noise *taken at different times*. With reference to Figure 32.14 we have already seen that the sum of products of noise taken at different samples tends to zero. Hence, for large enough M we can make the approximation that the second term on the right-hand side of (32.68) is zero and hence

$$|N[p]|^2 \approx \sum_{m=0}^{M-1} (n[m])^2. \quad (32.69)$$

Hence the amplitude spectrum of noise is given by

$$|N[p]| = \sqrt{|N[p]|^2} \approx \sqrt{\sum_{m=0}^{M-1} (n[m])^2}. \quad (32.70)$$

The interesting result here is that the right hand sides of (32.69) and (32.70) are *independent of the frequency index p*. Or, put another way, the amplitude spectrum of noise is a *constant*. It can be seen that this is only approximately the case in Figure 32.24a; however, if we were to average several such spectra of this noise then the average would tend to a constant.

The constant amplitude spectrum indicates that we model the noise as an infinite set of cosines with *all frequencies being given equal weight*. This is implied by (32.70) in that all frequencies in the noise signal have equal power. As indicated in Figure 32.24b, the phases of these cosines are random within the range -180 to $+180°$. This is known as the **white noise model** of the noise. (The term is analogous to white light, which consists of all frequencies in the visible electromagnetic spectrum.)

This is a useful model of noise, because in the next chapter we are going to use filters to reduce the effects of noise. Noise contributes a fairly constant background to the amplitude spectrum of a noisy signal, while signals contribute a 'peaky' structure. In the absence of any other information about the signal, we need to be able to estimate the range of spectral frequencies in the signal. This could be taken as when the peaky structure from the spectrum of the signal merges into the constant background from the noise. Then, if we design a network or algorithm to extract just the peaky frequency components and reduce the other components, then we can enhance the signal above noise. Some methods for doing this will be discussed in the following chapter. At this stage, we are just illustrating by example why the white noise model is a useful one in estimating the parameters of a signal in noise.

32.7 Concluding remarks

As stated in the introduction, most signals that are dealt with by experimental physicists are accompanied by noise, interference, and other effects. We have already seen that, if we know what the signal looks like, then we may apply matched filtering to detect the presence of this signal in noise. As an example, we looked at detecting known sequences of 1s and −1s in a signal transmitted by a satellite. There are many other applications, for example the detection of edges in an image.

The limitation of the matched filter is that we need to know something about the shape of the signal that we want to detect. Suppose that we do not know exactly what the signal looks like, but we do know approximately what its bandwidth is.

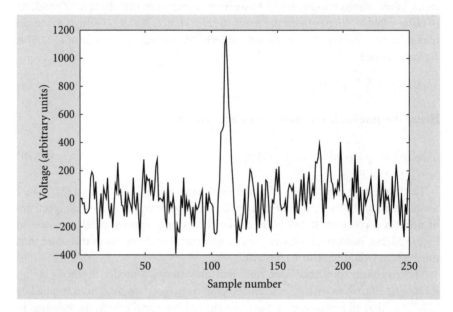

Figure 32.25 ECG signal in noise.

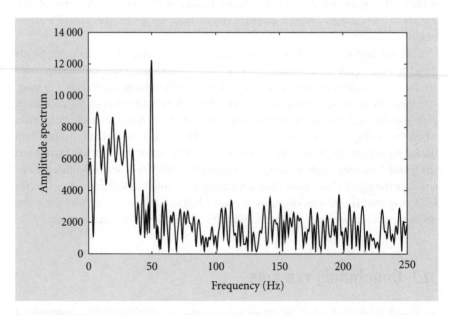

Figure 32.26 DFT amplitude spectrum of signal in Figure 32.25.

For example, look at the signal shown in Figure 32.25, which represents an ECG signal corrupted by noise and mains interference. It looks quite messy. If we take the DFT of this signal, the amplitude spectrum is illustrated in Figure 32.26. From our discussion of the amplitude spectrum of white noise, we can identify the constant background as coming from noise. The spike at 50 Hz comes from the mains current. We may guess that the spectral content between 0 Hz and about 100 Hz (excluding the 50 Hz spike) comes from the ECG signal that we want to extract. Can we still use this spectral information in Figure 32.26 to clean up the signal in Figure 32.25? The answer is yes, and there are many filtering methods that can be used, two of which are described in the next chapter.

Problems

32.1 The sliding-window cross-correlation function between a template $\{x[n]\}$ and an incoming signal is defined from (32.34) as

$$r_{xy}[p] = \sum_{m=0}^{M-1} x[m]y[m+p]$$

where M is the number of samples in the template. Suppose that the y is made up of lots of sub-signals with the mth sample of y being given by

$$y[m] = s_1[m] + s_2[m] + s_3[m] + \cdots + s_N[m]$$

where $s_i[m]$ is the mth sample of sub-signal i. Show that

$$r_{xy}[p] = r_{xs_1}[p] + r_{xs_2}[p] + r_{xs_3}[p] + \ldots + r_{xs_N}[p].$$

This result will be of use in the next problem where we consider GPS with more than one satellite.

32.2 Suppose that in GPS there are N satellites transmitting binary codes, with each code being particular to each individual satellite. Let $s_i[m]$ be the mth sample of the transmission from satellite i. Ideally the composite code entering a GPS receiver is given by

$$y[m] = s_1[m] + s_2[m] + s_3[m] + \cdots + s_N[m].$$

However, in practice this signal is corrupted by white noise and the signal entering the receiver is given by

$$y[m] = s_1[m] + s_2[m] + s_3[m] + \cdots + s_N[m] + n[m],$$

where $n[m]$ is the mth sample of noise.

The set of samples $\{y[m]\}$ is correlated with the templates $\{x_i[m]\}$ stored in the receiver, where the ith template is identical to the signal transmitted by satellite i so that $x_i[n] = s_i[n]$.

Using the result of Problem 32.1 and the correlation properties of noise, derive an expression for the cross-correlation function between template

$\{x_1[m]\}$ stored in the receiver and the incoming signal $\{y[m]\}$ in terms of the cross-correlation functions between the template and each of the satellite signals.

It is required that the template $\{x_1[m]\}$ just picks up the component $\{s_1[m]\}$ of the incoming signal $\{y[m]\}$. Determine the conditions on the cross-correlation functions $r_{s_i s_j}[p]$ ($i \neq j$) between the codes transmitted by different satellites for this requirement to be achieved.

32.3 In some applications, for example Doppler ultrasound measurements of a beating heart, we may be interested in monitoring a periodic signal in noise. We may not know exactly what the signal looks like, but the underlying period may be required. In fact, correlation methods may be used to determine this period. To see this, suppose that the underlying clean signal is $\{s[m]\}$ and the underlying (white) noise is $\{n[m]\}$. The noisy signal is thus given by $y[m] = s[m] + n[m]$.

(i) Show that the pth lag of the autocorrelation function of $\{y[m]\}$ is given by

$r_{yy}[p] = r_{ss}[p]$ for $p \neq 0$

$r_{yy}[0] = r_{ss}[0] + r_{nn}[0]$ for zero lag.

(ii) Now suppose that the underlying signal s has a period of N samples, so that $s[m + N] = s[m]$. Show that in this case the autocorrelation function for s satisfies the condition $r_{ss}[p] = r_{ss}[p + N]$.

(iii) Using your answers to (i) and (ii), show how you can use correlation methods to determine the period of a periodic signal in noise.

Digital filters

Theories in physics are tested by comparison with experimental data. However, in practice, experimental data are contaminated by noise coming from the instrumentation and other effects and these contaminations need to be eliminated, or at least reduced, before the data can be handed to the theoretician. In this chapter we describe two approaches for the elimination of noise from digital signals, one based on the use of Fourier transforms and the other on a windowing technique. The advantages and disadvantages of the two types of method are discussed.

33.1 Introduction

An important example of contaminated data from the field of medical physics is the modelling of the electrical signals in the human heart that produce an electrocardiogram signal (ECG). This is a complicated problem that involves modelling the electrical properties of tissue and fluids inside the body. The electrical behaviour of the heart is approximated by modelling it as a dipole. If there are mechanical defects in the heart or problems with the production of the electrical signal, then this affects the measured ECG and it is of interest to see what this effect is from the model. In order to verify that the model is a good one, we need to compare it with experimental data.

Suppose that, because of poor shielding of the equipment from the mains and inadequate filters in the equipment, you are given the data shown in Figure 33.1. The main peak of the ECG is visible, but the rest of the signal is contaminated with noise and other interferences. Hence, before comparing this data with theory, you need to process the signal to reduce the effects of noise and interference and, hopefully, obtain a signal that is as close to the actual signal as possible. The process of reducing the effects of noise and getting rid of unwanted signals is called **filtering**.

We shall assume that the signal that you have to process has already been digitized. Hence, any processing that you carry out will be on the samples of the signal. For this reason, these filters are referred to as **digital filters**. **Analogue filters** are applied before digitization of the analogue signal, which is a continuous function of time.

There are many mathematical techniques for the design of digital filters. We shall discuss two methods of digital filtering that are based on applying the Fourier

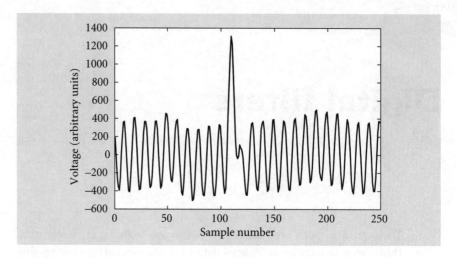

Figure 33.1 ECG signal contaminated by noise and interference.

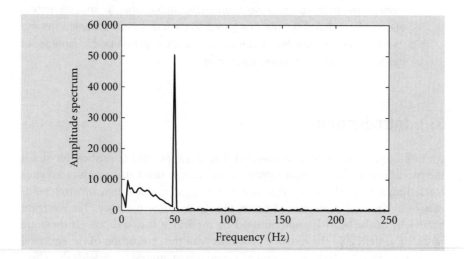

Figure 33.2 Amplitude spectrum of the signal in Figure 33.1.

transform and the discrete Fourier transform (DFT), which were introduced in Chapters 17 and 18.

33.2 Fourier transform methods

If we take the DFT of the signal in Figure 33.1 and plot the amplitude spectrum for *positive frequencies only*, we obtain Figure 33.2. The negative frequencies (not shown) will be a reflection of this spectrum in the *y*-axis.

We need to identify which parts of the amplitude spectrum represent the underlying clean signal and which parts represent noise and interference. Such assessments cannot be made blindly; we need to have some prior knowledge of the bandwidths of typical ECG signals. Such bandwidths can vary from one ECG to another, but typically most adult ECGs have bandwidths less than 80 Hz, so we can take this as a typical bandwidth of this signal.

It can be seen that the amplitude spectrum of the signal is non-zero for frequencies greater than 80 Hz. The amplitude spectrum fluctuates around a constant value so, from our discussion in §32.6, we can associate this part of the spectrum with white noise which contains no useful information for us.

Another striking feature of the amplitude spectrum in Figure 33.2 that deserves comment is a prominent peak at 50 Hz. We know that this peak is absent in the amplitude spectra of clean ECGs. From §17.4.4, we saw that a cosine signal has an amplitude spectrum that, ideally, consists of two spikes at frequencies $-f_0$ and $+f_0$. In Figure 33.2, we are just plotting positive frequencies, so we can associate the peak at $+50$ Hz as coming from a cosine signal with frequency 50 Hz. This spike represents mains interference – in the USA it would occur at 60 Hz. We need to get rid of this mains interference as it is masking the underlying signal.

Now the amplitude spectrum in Figure 33.2 was computed using the DFT. The x-axis was scaled to frequency by using the relation in (18.13):

$$f = \frac{pf_s}{N} \tag{33.1}$$

where f_s is the sampling frequency and N is the total number of samples. For the data in Figure 33.1, $N=250$ and $f_s=500$ Hz. From (33.1), we can determine p from f using

$$p = \frac{Nf}{f_s} \tag{33.2}$$

We can in fact filter off the mains interference and reduce the effects of noise by carrying out the following steps:

1 Let $\{y[m]\}$ be the set of samples representing the noisy signal, with m representing the mth sample index. Compute the DFT $\{Y[p]\}$, with p representing the pth harmonic. Plot the amplitude spectrum for positive frequencies to $p = N/2$. This is shown in Figure 33.2. Equation (33.1) has been used to scale the x-axis of the spectrum to frequency. Note, from §18.2.3, that $N/2 \le p \le N-1$ are frequency points corresponding to negative frequencies; we will need to consider these later, but for now we will deal just with positive frequencies.

2 Identify those frequencies coming from noise and interference. In our example, Figure 33.2, these frequencies are 50 Hz (coming from the mains) and $f \ge 80$ Hz coming from the noise.

3 Convert these frequency ranges to p values using (33.2). From (33.2), the 50 Hz mains frequency corresponds to a p value:

$$p = \frac{250 \times 50}{500} = 25$$

and, similarly the 80 Hz limit corresponds to $p=40$ and the maximum frequency displayed, 250 Hz, corresponds to $p=125$. Hence the displayed range identified as white noise corresponds to $40 \le p \le 125$.

4 To explain how to use the DFT to filter off unwanted frequency components of the signal, we write out the expression for the DFT of the signal in terms of the sample values. From (18.6), this is given by

$$Y[p] = \sum_{n=0}^{N-1} y[n] \exp\left(-\frac{2\pi i n p}{N}\right) \tag{33.3}$$

where y and Y in (33.3) replace x and X in (18.6). What we are now going to do is to put to zero those values of $Y[p]$ that occur at those values of p corresponding to noise and interference frequencies. In our ECG example, we put $Y[p] = 0$ for $p = 25$ corresponding to mains interference and $40 \leq p \leq 125$ corresponding to high-frequency noise.

Suppose that the modified DFT, with unwanted frequency components of the original DFT put to zero, is denoted by $Y'[p]$. What we are going to do is to find the inverse DFT (18.8) of $Y'[p]$ by

$$y'[m] = \frac{1}{N} \sum_{p=0}^{N-1} Y'[p] \exp\left(\frac{2\pi i m p}{N}\right) \tag{33.4}$$

so that $y'[m]$ is thus the original signal with the unwanted components filtered out. Note that in the summation (33.4) the values of p include the range $p > N/2$ corresponding to negative frequencies – remember that these are an essential part of interpreting signals in terms of cosine components. The frequency -50 Hz, which must also be removed to eliminate the 50 Hz cosine signal, corresponds to $p = -25$ giving, from the periodicity property (18.14),

$$Y[-25] = Y[-25 + 250] = Y[225].$$

This value of p corresponds to frequency 450 Hz. Similarly, the white noise range $-40 \leq p \leq -125$ is equivalent to $210 \geq p \geq 125$ (frequency range of 420 Hz $\geq f \geq 250$ Hz) so that $Y'[p]$ is made equal to zero for these values of p. The final transformed amplitude spectrum with the 50 Hz component and white noise removed is shown in Figure 33.3 for the frequency range 0–250 Hz.

If we now take the inverse DFT we obtain the filtered signal (full curve) compared with the clean signal (dashed curve) as shown in Figure 33.4. Comparing the full curve in Figure 33.4 with that in Figure 33.1, we can see that the signal looks a lot clearer and more like an ECG signal. In fact, this is an example of a simulated ECG signal where noise and mains have been simulated on a computer and added on to an actual ECG signal. The clean, underlying signal is displayed by a dashed line in Figure 33.4, and we can see that there are only minor differences between the two curves.

The filtered signal still has small oscillations that do not correspond to the original signal. This is because, although we have filtered off noise outside the bandwidth of the signal, we have not been able to filter off noise within the bandwidth of the signal. This is a problem with all constant-coefficient digital filters; noise can be reduced but, in general, not completely eliminated. This unfiltered noise will have lower frequency components than the noise that has already been filtered off. In addition, we put to zero the Fourier component at 50 Hz, which is within the bandwidth of the signal, so that those parts of the underlying signal that have frequency components around 50 Hz will be filtered out of the signal.

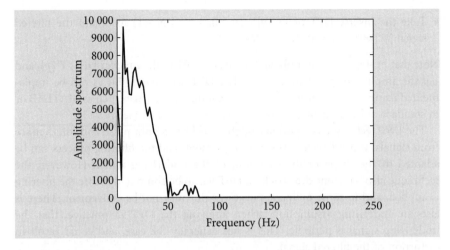

Figure 33.3 Amplitude spectrum of ECG signal with 50 Hz component and high-frequency noise removed. The frequency range 250–500 Hz is equivalent to negative frequencies.

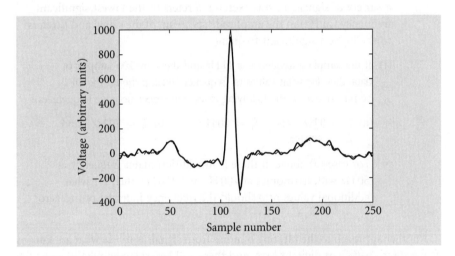

Figure 33.4 Filtered ECG signal. Filtered signal (full curve); clean signal (dashed curve).

To summarize, the generalized algorithm for the DFT filtering method is as follows:

1 Let $\{y[m]\}$ be the set of samples representing the noisy signal, with m representing the mth sample index. Compute the DFT $\{Y[p]\}$, with p representing the pth harmonic. Plot the amplitude spectrum just for positive frequencies to $p = N/2$.

2 Identify those frequency components coming from the signal and those from noise and other interference.

3 For all values of p that are judged to correspond to noise and interference, put the DFT equal to zero. Put the corresponding negative frequency component $N-p$ also equal to zero. Let the DFT with unwanted components zeroed be $\{Y'[p]\}$.

4 Take the inverse DFT of $\{Y'[p]\}$, $(p=0, 1,\ldots, N-1)$ to obtain the filtered signal $\{y'[m]\}$, $(m = 0, 1, \ldots, N-1)$.

Note that in step 4 we are applying the inverse DFT to the *complex* DFT, $Y'[p]$, and *not* the amplitude spectrum, $|Y'[p]|$. The DFT and inverse DFTs can be implemented using standard procedures that are found in packages such as MATLAB or in standard software listings in books such as Press *et al.* (see p. 715).

The DFT technique is a popular method of filtering out noise and interference from signals. It is relatively simple to implement, and the filtering process can be adapted to the shape of the spectrum of the underlying signal. However, the technique also has some drawbacks: a DFT has to be computed before the filtering is applied, and in some online applications this may not be convenient. There is also an underlying assumption, when applying the DFT in practice, that the underlying signal is periodic. This is not generally the case, and could result in distortion of the filtered signal.

Exercise 33.1

It is required to use the DFT method to reduce the effects of noise from a variety of signals. In this exercise, f_l refers to the lowest significant frequency (in Hz) in the amplitude spectrum of the signal and f_h refers to the highest significant frequency.

(i) If the sampling frequency is 1 kHz and there are 200 samples of data, then for what values of frequency index p should the DFT be put to zero in the following cases (remember negative frequencies)?

    (a) $f_l = 0\,\text{Hz}$ and $f_h = 20\,\text{Hz}$;   (b) $f_l = 20\,\text{Hz}$ and $f_h = 200\,\text{Hz}$.

(ii) For case (a) above, if there is in addition mains frequency at 50 Hz with harmonics at 100 Hz and 150 Hz, then for what additional values of p should the frequency index be put to zero?

Another approach to the filtering of noise from signals is to use what are known as constant-coefficient digital filters, and these will be explained next.

33.3 Constant-coefficient digital filters

33.3.1 Introduction

Another approach to designing digital filters is to carry out simple mathematical operations directly on the samples of the noisy signal to obtain a signal with the noise and interference effects reduced. Suppose that we have a noisy signal as shown in Figure 33.5. Now suppose that we take a segment of N samples of the signal and form an average:

$$y[n] = \frac{1}{N}(x[n] + x[n-1] + \cdots + x[n-N+1]). \tag{33.5}$$

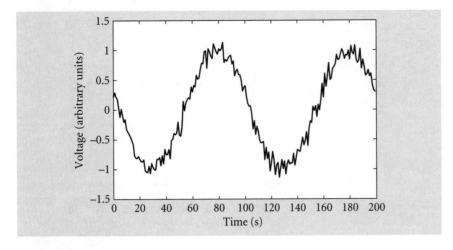

Figure 33.5 Noisy signal.

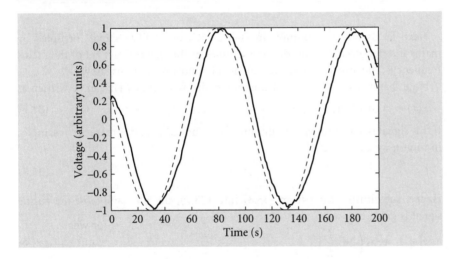

Figure 33.6 Filtered signal (full curve) and underlying clean signal (dashed curve).

We can move the window along the signal one sample at a time (by increasing n by 1) and work out the average for each window. We refer to this process as taking a **moving average**. For $N = 10$, we can write this as:

$$y[n] = \frac{1}{10}(x[n] + x[n-1] + \cdots + x[n-9]) \tag{33.6}$$

where $n = 9, 10, 11, \ldots$ If we apply this moving average to the samples of the signal, then we obtain the signal shown by the full curve in Figure 33.6.

The filtered signal is now much closer to the underlying signal, shown as a dashed curve in Figure 33.6, than is the noisy signal in Figure 33.5. The filtered signal has a delay with respect to the underlying signal, but this is usually of no importance. The signal is smoother and more resembles the underlying signal, which is a cosine.

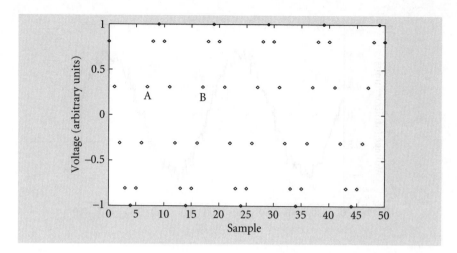

Figure 33.7 Samples of 50 Hz mains signal.

Next, let us take the example in Figure 33.1, of an ECG signal corrupted by mains interference. Can we combine samples of the signal to get rid of the mains frequency? The answer is that we can, as the following analysis shows.

Note that a cosine signal modelling the mains frequency f Hz can be written as

$$x(t) = A\cos(2\pi f t).\tag{33.7}$$

If this signal is sampled with a sampling frequency of f_s, and if there are n samples in time t, we can write

$$t = \frac{n}{f_s}.\tag{33.8}$$

Hence, substituting for t from (33.8) into (33.7), the nth sample of the cosine signal is given by

$$x[n] = A\cos\left(2\pi n\frac{f}{f_s}\right).\tag{33.9}$$

For the previous example of the contaminated ECG signal, we have $f_s = 500$ Hz and the mains frequency is $f = 50$ Hz. Substituting for f and f_s into (33.9),

$$x[n] = A\cos\left(2\pi n\frac{50}{500}\right) = A\cos\left(2\pi\frac{n}{10}\right).\tag{33.10}$$

The samples are plotted in Figure 33.7.

Now look at the signal at samples A and B. The values of the mains voltage at these sample points are identical. To see this, suppose that at B, the voltage value is given by (33.10). At A, which is 10 samples behind B, the voltage is given by

$$x_m[n-10] = A\cos\left(2\pi\frac{(n-10)}{10}\right)$$
$$= A\cos\left(2\pi\frac{n}{10} - 2\pi\right) = \cos\left(2\pi\frac{n}{10}\right) = x_m[n],\tag{33.11}$$

where x_m is the mains contribution to the signal. Therefore

$$x_m[n] - x_m[n-10] = 0.$$

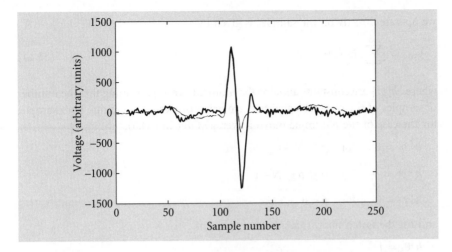

Figure 33.8 Filtered signal (full curve) and underlying signal (dashed curve).

This provides a method to get rid of mains frequency. If we subtract samples of the ECG signal that are 10 samples apart, i.e. take

$$y[n] = x[n] - x[n-10] \tag{33.12}$$

with $n = 10, 11, 12, \ldots$, then the mains contribution to the signal will be eliminated. If we carry out this operation on the signal in Figure 33.1, we obtain the full curve in Figure 33.8.

The underlying clean signal is shown as a dashed curve, and we can see that the mains frequency has been eliminated. However, the peaks of the signal have been distorted because filtering has also been carried out on the signal's components around 50 Hz. The background white noise is still present in the filtered curve because the filter in (33.12) is designed just to filter off the mains frequency and not the background noise; we will return to this problem later.

Exercise 33.2

(i) If the sampling frequency is changed to 120 Hz, can the digital filter, (33.12) still get rid of 50 Hz mains frequency?

(ii) With this new value for the sampling frequency, would this filter work in the United States where the mains frequency is 60 Hz?

The formulae (33.6) and (33.12) are two examples of a digital filter. In both cases, we are combining samples of the noisy signal using the following operations:

1 addition/subtraction of samples

2 multiplication of samples by a constant   (33.13)

3 storing past samples.

Formula (33.5) is an example of what is known as a **low-pass filter**, which filters off high-frequency noise, and (33.12) is an example of a **notch filter**, which eliminates specific frequencies from the signal. A general expression for a **linear filter** where we operate directly on the samples is given by:

$$y[n] = \sum_{p=-\infty}^{\infty} h[p]x[n-p],$$ (33.14)

where $\{h[p]\}$ are constants and, for the moment, we are assuming that the number of samples of the original signal that are being processed is infinite. For example, for the case of the N-sample moving average filter in (33.5):

$$h[p] = 0 \quad \text{for} \quad p > N-1$$

$$h[p] = \frac{1}{N} \quad \text{for} \quad 0 \le p \le N-1$$

$$h[p] = 0 \quad \text{for} \quad p < 0$$

and for the notch filter, (33.12),

$$h[0] = 1$$

$$h[10] = -1$$

$$h[p] = 0 \quad \text{otherwise.}$$

Formula (33.14) represents what is known as a **non-recursive filter**, where each filtered sample depends on the input noisy samples, not on any previous filtered values. This type of filter is also referred to as a **finite impulse response (FIR) filter**.

In general, (33.14) is referred to as the **convolution** between the sequence of coefficients $\{h[p]\}$ and the discrete signal samples $\{x[n]\}$. We can think of convolution as replacing the sample $\{x[n]\}$ by a linear combination of $x[n]$ and some samples around $x[n]$. It is possible to construct a filter where the output from the filter at time sample n, $y[n]$, depends on previous **outputs** $\{y[n-1], y[n-2], \ldots\}$ as well as on present and previous **inputs** $\{x[n], x[n-1], x[n-2], \ldots\}$. This is known as an **infinite impulse response (IIR) filter**. In this chapter we will concentrate on FIR filters as these are more widely used, for reasons given later.

The task of filter design is to choose the coefficients $\{h[n]\}$ in (33.14) in order to filter out unwanted components of the signal of interest. It turns out that this is best achieved by working with the Fourier transforms of the input and output signals. The **convolution theorem** helps us here, and this is the subject of the next section.

33.3.2 Convolution theorem

First, we will generalize (33.14) to continuous signals. Although we are interested in designing digital filters, it is more convenient at this stage to work with continuous (analogue) signals; we will specialize to digital signals later. To go to the continuous version of (33.14), the summation becomes an integral, p is replaced by a continuous variable τ and n is replaced by a continuous variable t:

$$y(t) = \int_{-\infty}^{\infty} h(\tau)x(t-\tau)\, d\tau$$ (33.15)

The right-hand side of (33.15) is known as the **convolution integral**.

For the next step, we will determine the Fourier transform, $Y(f)$, of $y(t)$ in terms of the Fourier transforms $H(f)$ and $X(f)$ of $h(t)$ and $x(t)$ respectively. This relation is derived in Appendix 23, where it is shown that:

$$Y(f) = H(f)X(f). \tag{33.16}$$

The combination (33.15) and (33.16) together constitute the **convolution theorem**. Function $H(f)$ is known as the **frequency response** of the filter. In the case where only the three operations indicated in (33.13) are carried out, it can be shown that $H(f)$ is *independent* of the input and output signals; this function of frequency depends only on the operations (addition, multiplication by a constant, delay) making up the filter. Now $H(f)$ is *complex*. How do we interpret this function in practice?

First, using (6.10), we write $H(f)$ in terms of its amplitude and phase as

$$H(f) = \alpha(f)\exp\{i\phi(f)\}. \tag{33.17}$$

If f is varied, then we can plot $\alpha(f)$ and $\phi(f)$ as a function of f as illustrated in Figure 33.9. The graph of $\alpha(f)$ versus f is termed the **amplitude response** of the filter, and $\phi(f)$ versus f is called the corresponding **phase response**. The amplitude response is an even function of frequency:

$$\alpha(f) = \alpha(-f) \tag{33.18}$$

and the phase response is an odd function of frequency:

$$\phi(f) = -\phi(-f). \tag{33.19}$$

In Appendix 24 it is shown that if the signal

$$x(t) = A\cos(2\pi f_0 t)$$

is input to the filter, then the output is given by:

$$y(t) = \alpha(f_0)\,A\cos(2\pi f_0 t + \phi(f_0))$$

i.e. a change in amplitude and a shift in phase. Note, however, that the **frequency**, f_0, of the output signal is the same as that of the input signal; this is a general property of **linear filters**.

In the example shown in Figure 33.9, the filter passes low frequencies and attenuates high frequencies – this is an example of a **low-pass filter**. This type of filter is good for reducing noise on a signal.

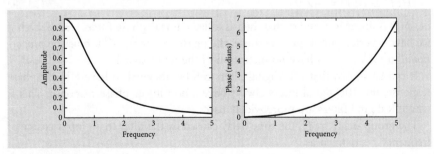

Figure 33.9 Examples of the amplitude and phase responses of a filter.

33.3.3 Condition for an undistorted signal

Although filters can reduce the effects of noise on unwanted signals, they can also distort the signal itself. Thus, when we design digital filters, we need to know the conditions on the amplitude and phase responses such that the underlying signal is undistorted.

A signal is said to be **undistorted** by a filter if the output, $y(t)$, is related to the input, $x(t)$, by the following relation:

$$y(t) = K \times x(t - \tau), \tag{33.20}$$

i.e. the filter *delays* signal by τ and *scales* it by K. We now ask the question: what is the frequency response of an ideal filter that does not distort the underlying signal? To answer this, we take Fourier transforms of both sides of (33.20): for the left-hand side,

$$y(t) \rightarrow Y(f).$$

To find the Fourier transform of the right-hand side, we write

$$FT\{x(t - \tau)\} = \int_{-\infty}^{\infty} x(t - \tau) \exp(-2\pi i f t) \, dt.$$

Making the substitution $t' = t - \tau$,

$$FT\{x(t - \tau)\} = \int_{-\infty}^{\infty} x(t') \exp(-2\pi i f t') \exp(-2\pi i f \tau) \, dt'$$

$$= \exp(-2\pi i f \tau) X(f), \tag{33.21}$$

hence

$$Y(f) = K \exp(-2\pi i f \tau) X(f). \tag{33.22}$$

The frequency response, $H(f)$, is given, from (33.16), by

$$H(f) = \frac{Y(f)}{X(f)} = K \exp(-2\pi i f \tau). \tag{33.23}$$

We can identify the required amplitude response as $|H(f)| = K$, which is an **all-pass filter** that passes all frequencies equally. In practice, this requirement states that the amplitude response should be flat over the bandwidth of the signal; this requirement can be met only approximately.

The required phase response is

$$\angle H(f) \equiv \phi(f) = -2\pi f \tau \tag{33.24}$$

i.e. $\phi(f)$ is linear in f, and such a filter is called a **linear phase filter**. Hence, even if we have a perfect amplitude response, the phase response, $\phi(f)$, must be *proportional* to f if we are to have no distortion of the input signal.

It can be shown that FIR digital filters can be designed to have a linear phase response, but IIR digital filters always have a non-linear phase response. This is why FIR digital filters are more widely used.

The **group delay**, $\tau(f)$, the time delay caused by the filter to an input cosine of frequency f is defined as

$$\tau(f) = -\frac{1}{2\pi} \frac{d\phi(f)}{df}. \tag{33.25}$$

Suppose that the filter has a perfect amplitude response. If $\phi(f) = -2\pi f \tau$, then from (33.25), $\tau(f) = \tau = $ a constant; the whole signal is delayed by an amount τ seconds and, according to (33.20), the signal is undistorted by the filter. However, if $\phi(f)$ is a non-linear function of f, then $\tau(f)$ is a function of f and different frequency components of the input signal are delayed by different amounts. This non-linearity in the phase response of $H(f)$ can introduce unacceptable distortion of the original signal at the output of the filter, irrespective of how good the amplitude response is.

33.3.4 Designing digital filters from specified H(f)

From (33.17) we can see that all the characteristics of a filter are defined in terms of the amplitude and phase responses. In particular, the amplitude response, $|H(f)|$, tells us what frequencies are selected by the filter and which frequencies are filtered out. As an example, suppose that we wish to design a filter subject to the following conditions:

1 All frequencies $|f| \le f_c$ are kept, where f_c is the **cut-off frequency** of the filter.

2 All frequencies $|f| > f_c$ are filtered out.

3 The underlying signal should not be distorted; this means that the filter should be linear phase and the phase response should be proportional to frequency, We can, if we wish, put $\tau = 0$ in (33.24), meaning that there is no delay between input and output and, from (33.23), $H(f)$ is a constant.

In this case the frequency response, $H(f)$, is identical to the amplitude response and is displayed in Figure 33.10.

Mathematically, the frequency response can be written as:

$$H(f) = \begin{cases} 0 & \text{for} \quad |f| > f_c \\ 1 & \text{for} \quad |f| \le f_c. \end{cases} \tag{33.26}$$

Now, when we are implementing the filter, we require the coefficients $\{h[p]\}$ in (33.14). To find these we first work out $h(t)$, which is the inverse Fourier transform of $H(f)$ from

$$h(t) = \int_{-\infty}^{\infty} H(f) \exp(2\pi i f t) \, df. \tag{33.27}$$

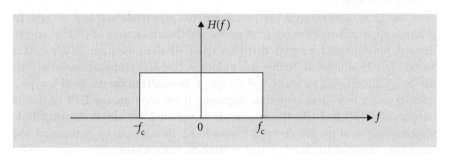

Figure 33.10 Ideal frequency response for low-pass filter.

The quantity $h(t)$ is also called the **impulse response** of the filter.

Because $H(f)$ is zero for $f > f_c$,

$$h(t) = \int_{-f_c}^{f_c} \exp(2\pi i f t) df + 0. \tag{33.28}$$

Hence

$$h(t) = \frac{1}{i2\pi t} [\exp(i2\pi f t)]_{-f_c}^{f_c} = \frac{1}{i2\pi t} [\exp(i2\pi f_c t) - \exp(-i2\pi f_c t)]$$

$$= \frac{\sin 2\pi f_c t}{\pi t}. \tag{33.29}$$

The right-hand side in formula (33.29) is found from using the identity

$$\sin(x) = \frac{1}{2i} [\exp(ix) - \exp(-ix)],$$

with $x = 2\pi f_c t$.

To find the coefficients $\{h[p]\}$ in (33.14), we need to sample $h(t)$ by replacing the continuous time variable t by p/f_s, where f_s is the sampling frequency. This gives

$$h[p] = \frac{f_s}{\pi p} \sin\left(2\pi f_c \frac{p}{f_s}\right). \tag{33.30}$$

Substituting for $h[p]$ from (33.30) into (33.14) the implementation of the digital low-pass filter is given by:

$$y[n] = \frac{f_s}{\pi} \sum_{p=-\infty}^{\infty} \left\{ \frac{\sin(2\pi f_c p / f_s)}{p} x[n-p] \right\}. \tag{33.31}$$

The quantity $h[p]$ in (33.30) is straightforward to evaluate on a computer, except for $p = 0$ where

$$h[0] = \frac{f_s}{\pi} \lim_{p \to 0} \left[\frac{\sin(2\pi f_c(p/f_s))}{p} \right] = 2f_c$$

since $[(\sin Ax)/x]_{x \to 0} = A$ (see §4.7.2).

We now look at an example where the required cut-off frequency $f_c = 80$ Hz and the sampling frequency is $f_s = 500$ Hz in (33.30). If we plot $\{h[p]\}$ versus p, for $-25 \le p \le 25$ we obtain the graph shown in Figure 33.11. Here, to keep numbers small, $h[p]$ has been normalized so that $\sum_{p=-25}^{25} h[p] = 1$. We now have the problem that we are required to sum from sample numbers $-\infty$ to $+\infty$ in (33.31). We cannot do this so, in practice, we have to make do with summing over a finite number of samples.

From (33.30) and Figure 33.11 we can see that $h[p]$ tends to 0 as $p \to \pm\infty$. In practical applications, there is a limit on the number of samples of $\{h[p]\}$ that can be used; for example for signals that have relatively short duration. Suppose, that we take just 25 samples of h either side of $h[0]$ so that $h[p]$ is approximated to zero for $|p| \ge 26$. Ignoring values of $h[p]$ for $|p| \ge 26$ results in the designed low-pass filter having a non-ideal amplitude response. If we compute the DFT of the 51 samples of $\{h[p]\}$ and take the magnitude of the DFT, we obtain the amplitude response shown as the full curve in Figure 33.12. In this case, by convention, the amplitude response has been plotted in decibels (dB), where

$$A_{dB} = 20 \log_{10} A.$$

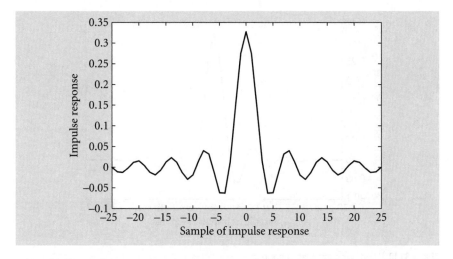

Figure 33.11 Impulse response for designed digital filter.

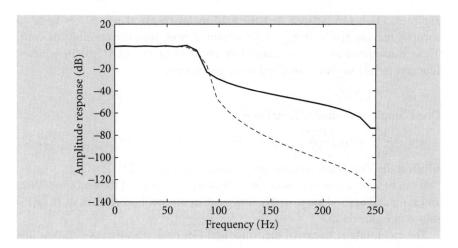

Figure 33.12 Amplitude response (dB) for designed digital filter. No windowing (full curve); windowing (dashed curve).

We can see that there is not an ideal sharp cut-off at 80 Hz; the amplitude response is non-zero, albeit relatively small, for frequencies larger than the cut-off.

Let now assume, without loss of generality, that the number of retained coefficients of $\{h[p]\}$ is odd. We can thus adapt (33.14) for a finite number of h-coefficients as

$$y[n] = \sum_{p=-N/2}^{N/2} h[p]x[n-p] \tag{33.32}$$

where the total number of coefficients of h that are kept is $N+1$, with N even. In some applications, leakage into higher frequencies of the actual amplitude response, as shown in Figure 33.12, may not be serious. If it is required to reduce this effect, then this can be done using a windowing technique similar to that used when computing the DFT. Consider the graph of the truncated filter coefficients in

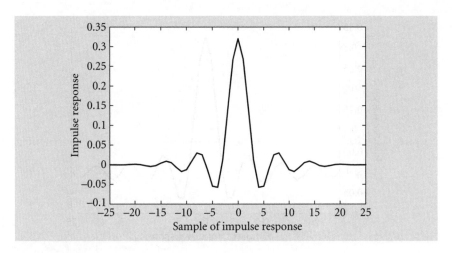

Figure 33.13 Windowed impulse response.

Figure 33.11. It can be seen that the gradient of $\{h[p]\}$ at $p=\pm25$ is finite so abruptly making $h[p]=0$ outside those limits introduces a discontinuity of slope. These discontinuities can be reduced by multiplying the set $\{h[p]\}$ by window function $\{w[p]\}$ to give a modified set of coefficients

$$h'[p] = w[p]h[p].$$

One example of a window function is

$$w[p] = \frac{1}{2}\left[1 + \cos\left(\frac{2\pi p}{N}\right)\right] \quad p = -\frac{N}{2}, \ -\frac{N}{2}+1, \ \dots, \ \frac{N}{2}-1, \frac{N}{2} \tag{33.33}$$

which is the Von Hann window we previously met in (18.22).

It can be seen that $w[p]$ goes to zero at both the beginning of the data ($p=-N/2$) and at the end of the data ($p=N/2$). The graph of the modified coefficients $\{h'[p]\}$ is shown in Figure 33.13.

The amplitude response of the windowed FIR filter is shown as a dashed line in Figure 33.12. It can be seen that the amplitude response for frequencies greater than 80 Hz is reduced compared to the case when no windowing was used; this makes the amplitude response sharper. There is another effect that occurs, which is clearer if we plot the amplitude response in non-logarithmic units as in Figure 33.14.

With no windowing there are oscillations within the pass-band of the filter ($0 \le f \le 80$ Hz), which is a direct result of truncation of the impulse responses to 51 samples. Windowing solves this problem and the amplitude response is flatter within the pass-band.

Let us now apply FIR filtering to the fetal ECG example in Figure 33.1. As a first step it is convenient to use the FFT method in §33.2 to eliminate the mains interference. Subsequently we will use the FIR filter method to reduce the effects of noise. When the 50 Hz is removed, then the signal is as seen in Figure 33.15. Note that whilst the effect of mains has been eliminated from the signal, there is still white noise present. If we now apply the FIR low-pass filter, (33.31), using 51 samples of the impulse response with and without using a von Hann window, then the filtered signals are shown in Figure 33.16.

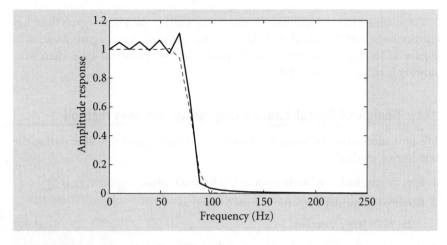

Figure 33.14 Amplitude response for designed digital filter. No windowing (full curve); windowing (dashed curve).

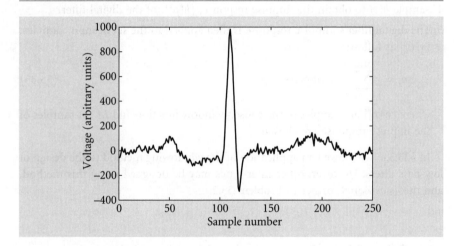

Figure 33.15 ECG signal with the mains frequency removed.

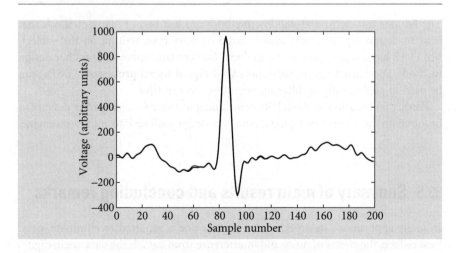

Figure 33.16 Filtered ECG signal. No windowing (full curve); windowing (dashed curve).

Comparing Figures 33.15 and 33.16, we can see that the FIR low-pass filter has further improved the signal to noise ratio for the underlying signal. As seen in Figure 33.16 there is no significant difference for this particular case when windowing is either used or not used.

33.3.5 Design of digital filters using the windowing method

We now summarize the general procedure to design digital FIR filters using the windowing method:

1 Specify required amplitude response, $|H(f)|$ and phase response, $\phi\ (f)$.

2 Required frequency response is then formed as

$$H(f) = |H(f)| \exp(i\phi(f)). \tag{33.34}$$

3 Determine the inverse Fourier transform of $H(f)$ to obtain the impulse response, $h(t)$.

4 Sample $h(t)$ to obtain the impulse response, $\{h[p]\}$, of the digital filter.

5 The digital filter's impulse response is then applied to the set of input samples, $\{x[n]\}$ as follows:

$$y[n] = \sum_{p=-M/2}^{M/2} w[p]h[p]x[n-p] \tag{33.35}$$

where $\{w[p]\}$ are samples of the chosen window function and $M+1$ samples of the impulse response are chosen.

In §33.3.4, we showed an application of the windowing method to the design of low-pass filters. However, other filter types may be designed using this method, and this is explored further in Problem 33.2.

33.4 Other filter design methods

The Fourier transform and window methods are just two of many methods that exist to design digital filters; whole textbooks have been written on the subject. MATLAB has a 'signal-processing toolbox' that contains many digital filter design methods. Specialized microprocessors called **digital signal processors** (DSPs) can be used to perform digital filtering operations in real time.

Further information on digital filters including software for digital filter design can be found on the internet by typing *digital filter design methods* into any search engine.

33.5 Summary of main results and concluding remarks

In many applications in experimental physics one is required to eliminate, or at least reduce, the effects of noise and interference from data. If the data are in digital form, then one can design the appropriate digital filter to enhance the underlying signal. In this chapter, we have described two approaches to applying a digital filter to sampled data.

In one approach we compute the Fourier transform of the data, for example using the DFT. Noise and interference effects are identified by prior knowledge of their characteristic frequencies and the DFT is put to zero at these frequencies. An inverse DFT is then carried out on the modified DFT to obtain the filtered signal. This method has the advantage that it can be adapted to the situation in hand; for example when filtering several sets of data where the bandwidths of the underlying signals are different for each dataset, in which case it is appropriate to get rid of a different set of frequencies for each set of data. This method does have the disadvantage when rapid online applications are required where it may not be feasible to carry out a DFT on windows of the data.

In another approach, one defines the frequency response, $H(f)$, by

$$Y(f) = H(f)X(f)$$

where $X(f)$ and $Y(f)$ are the Fourier transforms of the input and output signals respectively (33.16). One chooses an $H(f)$ that eliminates specified frequencies in the input signal. The impulse response, $h(t)$, is then found by application of the inverse Fourier transform to $H(f)$. Next, $h(t)$ is sampled and the filter implemented using (33.32). This second type of filter is defined for general sets of input signals with an assumed common bandwidth.

It is important not to cause unacceptable distortion to the underlying signal, which can occur if the amplitude response is not sharp enough or the phase is non-linear in frequency. There are countless digital filter design filter methods that are available through computer packages and on the web. It is common practice to experiment with applying different filters on simulated signals to ensure that the required performance is achieved in filtering off noise and interference while, at the same time, not distorting the underlying signal.

The following problems explore further the application of the Fourier transform and FIR window techniques.

Problems

33.1 In §33.2, we looked at he case of eliminating 50 Hz interference from an ECG signal when using the Fourier transform method to filter the data. The sampling frequency chosen was 500 Hz and 250 samples were processed.

(i) Would the results have been any better or worse if, say, 256 samples of the data were analysed? Give reasons for your answer.

(ii) In general, if one wishes to eliminate interference of frequency f_i Hz, and the sampling frequency is f_s Hz, then what is the relation between f_i, f_s, and the number of samples, N, such that the interference is completely eliminated?

33.2 It is required to design a digital FIR band-pass filter with lower cut-off frequency 100 Hz and an upper cut-off frequency 200 Hz. The frequency response of the filter is shown below.

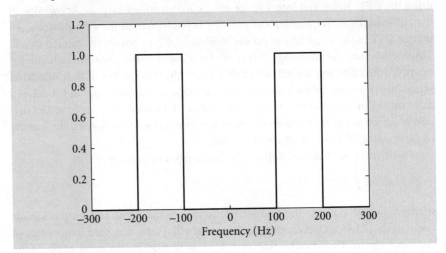

(i) Using (33.27), derive an expression for the impulse response, $h(t)$.

(ii) If the sampling frequency is set at 500 Hz, derive an expression for the coefficients $\{h[p]\}$ for the impulse response for this filter. Evaluate specifically $h[0]$.

Introduction to estimation theory

In many areas of experimental physics, measurement noise contaminates the signal of interest. It is important to reduce the effects of noise in order to extract more information about the signal. Conventional techniques used to address this problem are described in Chapter 33. However, sometimes the model of the underlying signal we wish to estimate is not so well defined as in the examples that have so far been discussed. In this chapter, we will develop methods to estimate underlying signals using least-squares methods, where the estimates are determined recursively rather than by using batch methods. This can be regarded as a 'real time' implementation of the least-squares technique. We will develop methods to detect and take into account sudden changes in the underlying model in the estimation of the underlying signal of interest.

34.1 Introduction

Consider the signal shown in Figure 34.1. We can see that noise is affecting the data. The underlying model is that the signal is constant over segments of the data, but then undergoes jumps. This sort of problem could occur, for example, in geophysics when we are looking for changes in seismic activity, or in astronomy when monitoring solar flares. In these applications, the sudden changes in signal occur at unpredictable times. The aim here is to estimate the underlying signal in noise and, at the same time, take into account the unpredictable jumps in the data.

The general problem can then be expressed as follows: given some experimental data as a function of time, where the underlying model can change unpredictably, estimate the underlying signal. The least-squares process for solving over-determined sets of equations, described in Chapter 30, is an example of a **batch method** where all the data are processed at once. This is fine if the underlying model does not change over the dataset but will not work in applications such as in Figure 34.1, where the underlying model changes abruptly. What we require is to be able to estimate the underlying signal on a sample-by-sample basis. This is an example of what is known as a **recursive** estimator. If the underlying model

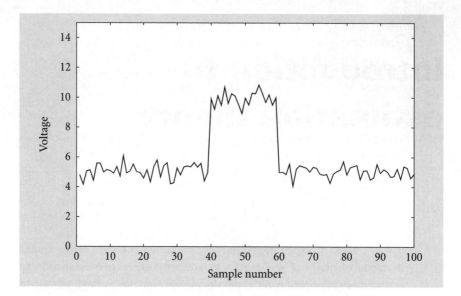

Figure 34.1 Example of a noisy constant signal with jumps.

changes suddenly, then the algorithm should be able to detect this and take it into account in the estimation process.

For simplicity, we will concentrate on the specific case where the underlying signal looks like Figure 34.1, i.e. a signal that is piecewise constant. The basic principles hold for more complicated patterns of underlying signal.

34.2 **Estimation of a constant**

In the simplest type of least-squares problem we are required to estimate the value of an underlying signal that is constant with respect to time, but has been contaminated by noise. A typical example is if we are taking several measurements of, for example, temperature with associated noise and we wish to improve our estimate of the temperature. Now, it is shown in (12.18) that if there are $n-1$ noisy measurements, $\{z[i]\}$ of a physical constant, x, then the least-squares estimator, i.e. the mean value that gives the least variance of the measurements, is just the straight average:

$$\tilde{x}[n-1] = \frac{1}{n-1}[z[n-1] + z[n-2] + \cdots + z[2] + z[1]] \tag{34.1}$$

where the symbol $\tilde{x}$ signifies that we are taking an estimate of x from the measurements $\{z[1], z[2], \ldots, z[n-1]\}$. If we obtain a new measurement at the nth sample point, $z[n]$, then our new estimate of x is given by the average over n measurements:

$$\tilde{x}[n] = \frac{1}{n}[z[n] + z[n-1] + z[n-2] + \cdots + z[2] + z[1]]. \tag{34.2}$$

As we take more measurements, we can add in each new measurement to the average. However, in some applications, where we are carrying out real-time monitoring of a physical constant, it is not always convenient to store an increasing number of measurements. We can rewrite the average to make it more amenable to real-time calculation. First, we rewrite $\tilde{x}[n]$ in (34.2) as follows:

$$\tilde{x}[n] = \frac{1}{n}z[n] + \frac{1}{n}[z[n-1] + z[n-2] + \cdots + z[2] + z[1]]. \tag{34.3}$$

Now, we can rewrite the second term on the right-hand side of (34.3) as

$$\left(\frac{n-1}{n}\right)\left(\frac{1}{n-1}\right)[z[n-1] + z[n-2] + \cdots + z[2] + z[1]]. \tag{34.4}$$

Using (34.1), this can be further rewritten as

$$\left(\frac{n-1}{n}\right)\cdot\left\{\left(\frac{1}{n-1}\right)\cdot[z[n-1] + z[n-2] + \cdots + z[2] + z[1]]\right\}$$
$$= \left(\frac{n-1}{n}\right)\tilde{x}[n-1]. \tag{34.5}$$

Substituting from (34.5) into (34.3), we can relate the average at time point n to the average at time point $n-1$ as

$$\tilde{x}[n] = \left(\frac{n-1}{n}\right)\tilde{x}[n-1] + \frac{1}{n}z[n]. \tag{34.6}$$

Exercise 34.1

Suppose that we have the following four measurements: $z[1] = 3.4$, $z[2] = 3.3$, $z[3] = 3.4$, and $z[4] = 3.1$.

(i) Determine the average value of these measurements, $\tilde{x}[4]$.

(ii) Supposing that a fifth measurement is received with value $z[5] = 3.6$. Determine the average over five measurements, $\tilde{x}[5]$, using both (34.6) and (34.2) and verify that these two values agree with each other.

We can now see that we do not need to keep all the measurements up to time point n in order to work out the average. All we need to do is to take the average, $\tilde{x}[n-1]$, at the previous data point, weight it by $(n-1)/n$, and add the new measurement, $z[n]$, weighted by $1/n$. We refer to (34.6) as a **recursive estimator** of the constant x at time point n; by 'recursive' we mean that the estimate of x at time point n depends only on its previous value and the present measurement. After adding in the measurement $z[n]$, we discard it; we do not need to keep it for further updates of the average. On the other hand, (34.2) is referred to as a **batch estimator** of x in that *all* measurements $\{z[1], z[2], \ldots, z[n]\}$ are kept to form the average. The recursive form is much more convenient for real-time implementation.

Now, (34.6) can be further rewritten as follows:

$$\tilde{x}[n] = \tilde{x}[n-1] - \frac{1}{n}\tilde{x}[n-1] + \frac{1}{n}z[n]. \tag{34.7}$$

Combining the last two terms,

$$\tilde{x}[n] = \tilde{x}[n-1] + \frac{1}{n}(z[n] - \tilde{x}[n-1]). \tag{34.8}$$

This form of the estimator is known as a **predictor–corrector form**. To explain this phrase, suppose that we have estimated our constant x using the first $n-1$ measurements and we have obtained our estimate $\tilde{x}[n-1]$. Now consider the situation just before we receive the nth measurement. We are assuming that our physical constant x does not change between measurements – it would not be a constant otherwise! Hence, at time point n just before we process the measurement $z[n]$, we can consider $\tilde{x}[n-1]$ as the **prediction** of x at time point n given the measurements *up to* and including time point $n-1$. Let us write this as

$$\hat{x}[n] = \tilde{x}[n-1] \tag{34.9}$$

where $\hat{x}$ denotes the predicted value of x at the nth time point before processing the nth measurement. Now what happens when we receive the measurement, $z[n]$? Using (34.9), we can rewrite the estimator (34.8) as

$$\tilde{x}[n] = \hat{x}[n] + \frac{1}{n}(z[n] - \hat{x}[n]). \tag{34.10}$$

The first term on the right-hand side of (34.10) is the **prediction** and the second term is the **correction** to this prediction taking into account the additional information at time point n through the measurement $z[n]$. Equation (34.10) can be further rewritten as

$$\tilde{x}[n] = \hat{x}[n] + \frac{1}{n}r[n] \tag{34.11}$$

where

$$r[n] = z[n] - \hat{x}[n] \tag{34.12}$$

is called the **residual**, which is the difference between the prediction and the new measurement.

> **Exercise 34.2**
> Derive an expression for the residual at $n=4$ in terms of the measurements $z[1]$, $z[2]$, $z[3]$, and $z[4]$.

How does this estimator work in practice? Well, suppose that we are processing measurements $\{z[1], z[2], \ldots, z[n]\}$. When the first measurement, $z[1]$, comes in, we put $n=1$ in (34.10) and our estimate of the average using the first measurement is then given by

$$\tilde{x}[1] = \hat{x}[1] + z[1] - \hat{x}[1] = z[1]. \tag{34.13}$$

That is, we are averaging over one measurement, which is all that we can do at this stage! Next, using (34.9), the prediction of the average at $n=2$, before processing the measurement at $n=2$, is given by

$$\hat{x}[2] = \tilde{x}[1]. \tag{34.14}$$

Now, suppose that we receive the second measurement, $z[2]$. Hence, the estimate of the average value of x, having received this measurement is given from (34.10) by

$$\tilde{x}[2] = \hat{x}[2] + \frac{1}{2}(z[2] - \hat{x}[2]). \tag{34.15}$$

This process is iterated further, predicting to $n=3$ using (34.9) and then, after receiving the measurement at $n=3$, using (34.10) to form the estimate at $n=3$ and so on. One point to note from (34.10) is that as we process more measurements, that is as n increases, then the measurements are weighted less and less through the factor $1/n$. In fact, as $n \to \infty$ in (34.10),

$$\tilde{x}[n] \approx \hat{x}[n].\qquad(34.16)$$

In other words, the estimator is almost ignoring the measurements and just basing its estimate on the prediction. Another way of looking at (34.16) is to say that the estimator has processed so many measurements, it is now so confident that it can ignore the measurements and just rely on the predictions. We shall see later that this could lead to problems in some practical situations.

Now, we need to have a measure as to how accurate is the estimate at the nth sample point. We will never know this exactly, because of the random nature of the noise. Let us first assume that the noise is white, i.e. uncorrelated from sample to sample, and has a Gaussian distribution. The variance, $\tilde{P}[n]$ for the average of n measurements has been derived in (26.5) as

$$\tilde{P}[n] = \frac{\sigma^2}{n}.\qquad(34.17)$$

where σ is the standard deviation of each measurement. Hence, we can characterize the error in the average of x at the nth sample by its standard deviation:

$$\sigma_x[n] = \sqrt{\tilde{P}[n]} = \frac{\sigma}{\sqrt{n}}.\qquad(34.18)$$

As $n \to \infty$, $\sigma_x[n] \to 0$, implying a perfect estimate. As mentioned above, in this limit the estimator is so confident of the predictions from the model that it is effectively ignoring the measurements. Our final algorithm procedure for estimating a constant, recursively, from noisy measurements is given as Algorithm 34.1. We can now apply this to the estimation of a constant in noise. In this simulation, the noise has a Gaussian distribution with standard deviation of 0.5 and the actual value of the constant is 5. In Figure 34.2, the noisy measurements are shown as dashed lines and the estimate as a full line. It can be seen in Figure 34.2 that the estimate is quite noisy initially and has a similar trend to the noise. However, for larger sample values the estimate becomes smoother and smoother and approaches the actual value of the signal, which is 5.

Algorithm 34.1 Estimation of a constant

Initialization
  Input measurement $z[1]$
    $\tilde{x}[1] = z[1]$
For the nth sample, with $n > 1$
  Prediction : $\hat{x}[n] = \tilde{x}[n-1]$
Input measurement : $z[n]$
    Estimation : $\tilde{x}[n] = \hat{x}[n] + \frac{1}{n}(z[n] - \hat{x}[n])$
    Variance of estimate : $\tilde{P}[n] = \frac{\sigma^2}{n}.$
  End loop

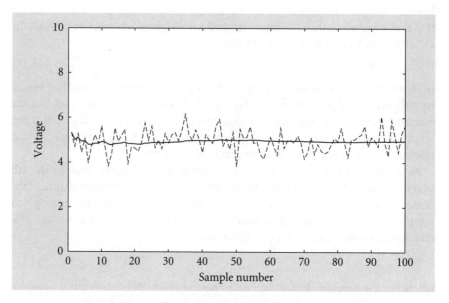

Figure 34.2 Noisy measurements (dashed line) and estimate (full line).

34.3 Taking into account changes in the underlying model

So far in this chapter, we have applied recursive least-squares estimators to noisy data where the underlying signal is a constant. These signals could have been estimated just as well by using the batch least-squares technique described in Chapter 30. However, the recursive techniques have advantages when the underlying signal model changes suddenly, for example when there is a change in seismic activity or if a target tracked by radar suddenly accelerates or manoeuvres.

We cannot use the techniques described above, or batch techniques, in situations where the model of the variation with time of the underlying signal changes at unpredictable times. To give an example, suppose that the signal is represented by the full curve in Figure 34.3 below and we apply the estimator in Algorithm 34.1. The result is given by the dashed line in Figure 34.3. The estimated signal follows the actual one very well up to the sudden change but subsequently does a poor job tracking the signal. The estimated signal does increase, but very slowly. The reason for this is that, in (34.10), the gain $1/n$ becomes smaller for increasing n and this means that the weighting put on the new measurements $z[n]$ after the transition becomes very small. In this case, the approximation in (34.16) holds, namely that the estimated values of the signal are given by the predicted values. However, these predictions are mainly determined by the data before the transition, and, because the new data after the transition has a relatively low weighting, these predictions are inaccurate.

To improve matters there are two things we must do:

- automatically detect the sudden change in the underlying signal
- respond to this sudden change to obtain improved estimates after the change.

We turn our attention first to detecting sudden changes.

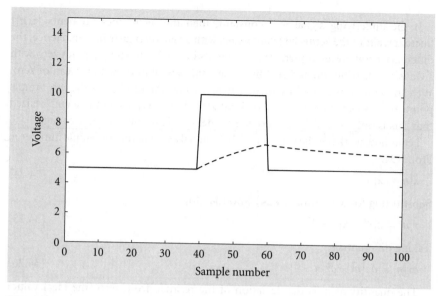

Figure 34.3 Actual signal (full line) and estimate (dashed line).

34.3.1 **Detection of sudden changes**

Let us use (34.11) for the estimator

$$\tilde{x}[n] = \hat{x}[n] + \frac{1}{n} r[n] \tag{34.19}$$

where

$$r[n] = z[n] - \hat{x}[n] \tag{34.20}$$

is the residual. The residual is a measure of the difference between the predicted value of the underlying signal and the measurement. In the ideal case where we have modelled the underlying signal correctly and there is no noise, $r[n]$ is zero. With noise, $r[n]$ will fluctuate about zero.

At this stage we make an assumption about the statistics of the noise. In practice, to a good approximation, if we select a large enough number of noise samples, then they will approximately obey a Gaussian distribution with mean 0. This observation is explained by the central limit theorem (§14.3.1). Using this assumption, we can show that, when the time variation of the signal is a constant when using the first-order estimator, the residual, $r[n]$, has a mean of zero and follows a Gaussian distribution. To see this for $n = 2$, note that the residual at the second time point is given from (34.12) by

$$r[2] = z[2] - \hat{x}[2]. \tag{34.21}$$

From (34.14),

$$\hat{x}[2] = \tilde{x}[1] = z[1]. \tag{34.22}$$

Substituting for $\hat{x}[2]$ from (34.22) into (34.21),

$$r[2] = z[2] - z[1]. \tag{34.23}$$

If the underlying signal is a constant, then the set of measurements $\{z[n]\}$ fluctuates about the actual constant value, with a Gaussian distribution. Hence, the difference between subsequent measurements in (34.23) fluctuates about zero with a Gaussian distribution and $r[2]$ has a Gaussian distribution with a mean of zero. We can iterate this procedure further to show that the residual at the nth sample point, $r[n]$, is Gaussian distributed about zero. However, it should be emphasized that this is only true if the underlying model of a constant does not change.

Let us now work out the variance of the residual as a function of sample number n. From (34.9),

$$\hat{x}[n] = \tilde{x}[n-1] \tag{34.24}$$

Substituting for $\hat{x}[n]$ from (34.24) into (34.20),

$$r[n] = z[n] - \tilde{x}[n-1]. \tag{34.25}$$

Taking differentials,

$$\delta r[n] = \delta z[n] - \delta \tilde{x}[n-1]. \tag{34.26}$$

The quantity $\delta r[n]$ is the deviation of the residual from zero (the ideal value) while $\delta z[n]$ is the deviation of the measurement from the actual value of the underlying signal and $\delta \tilde{x}[n-1]$ is the deviation of the estimate at sample $n-1$ from the actual value of the signal at sample point $n-1$.

From (14.19), the variance of the sum of uncorrelated Gaussian variables is equal to the sum of the variances of the individual variables. In (34.26), $\delta \tilde{x}[n-1]$ is the error in the estimate of the signal at sample $n-1$ before measurement $z[n]$ is processed. Hence, $\delta \tilde{x}[n-1]$ and $\delta z[n]$ are uncorrelated in (34.26).

From the solution to Exercise 34.3, the residual $r[n]$ follows a Gaussian distribution about 0 with a standard deviation given by

Exercise 34.3

From (34.17) and (34.26), show that the variance of the residual is given by $\sigma_r^2 = n/(n-1)\sigma^2$.

$$\sigma_r = \sqrt{\frac{n}{n-1}}\sigma. \tag{34.27}$$

The variations of $r[n]$ about zero are purely due to noise. Suppose now that there is a sudden change in the residual. As an example, we will look at applying the first-order estimator in Algorithm 34.1 to the data in Figure 34.1. In Figure 34.4 we plot the residual as a function of time up to the point when the first jump occurs. The residual fluctuates about 0 until the underlying signal jumps. When this occurs, the measurement, $z[n]$, in (34.12) increases in value significantly, while the prediction, $\hat{x}[n]$, is equal to the estimate at the previous time point, $n-1$ before the jump occurs. This prediction contains no information about the sudden increase in value of the signal, hence the residual significantly increases in value, as observed in Figure 34.4. How can we quantitatively detect this jump in residual as significant? Also, how do we know that other increases in the magnitude of the residual, for example at sample number 14 in Figure 34.4, are not significant? To answer this, recall that we noted above that the residual at time point n is drawn

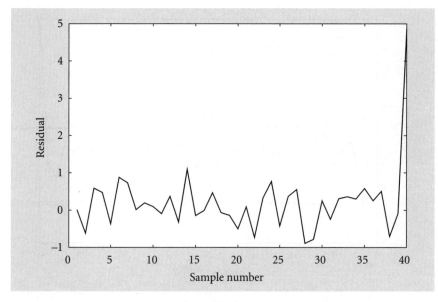

Figure 34.4 Residual up to when the jump occurs.

from a Gaussian distribution with standard deviation σ_r, given by (34.27). One approach is to set thresholds $\pm T$ where

$$T = k\sigma_r \tag{34.28}$$

where k is a constant to be determined. We are going to set k in (34.28) so that if the residual is larger than T or less than $-T$ then a significant jump in the data is deemed to have occurred; remember that the residual could be either positive or negative and at a jump in the signal, the residual could also go large and negative. If we set k too small in (34.28), then T will be too small and random fluctuations due to noise in the residual will cause false jumps to be detected, for example at samples 14 and 28 in Figure 34.4. If k is chosen to be too large, then the jump at sample 40 in Figure 34.4 will not be detected, causing the estimator to be in error. As a compromise, k is usually set to be 2 or 3, for reasons to be discussed later. Let us choose $k = 3$ so that the condition for a jump to have occurred is

$$|r[n]| > 3\sigma_r \tag{34.29}$$

with σ_r given in (34.27). Note that we are taking the magnitude of the residual in (34.29), as it could be either positive or negative. In (34.27), we would need to have an initial estimate of the measurement noise standard deviation, σ, for example estimating it during a segment where there are no changes in the underlying signal.

If we now re-plot the residual up to sample point 40 with the thresholds at $\pm 3\sigma_r$, we obtain Figure 34.5. We can now see that the thresholds are not exceeded until the jump at sample point 40. Note that we define the residual as zero at $n = 1$. Before the sudden change, all the residual values are within the $\pm 3\sigma_r$ limits. However, at the point of the transition, the residual suddenly jumps in magnitude and passes through the upper threshold.

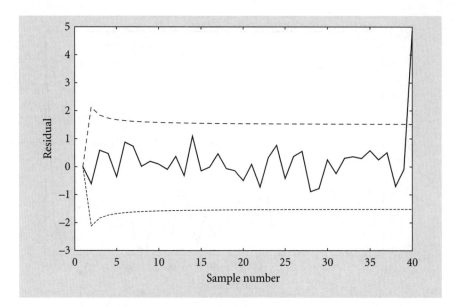

Figure 34.5 Residual with upper and lower threshold.

How can we be so sure that thresholds will not be exceeded by the residual due to the random fluctuations? The answer is that we cannot be 100% sure. However, note that the residual at each time point has a Gaussian distribution centred about zero. For Gaussian distributions, the probability that the value exceeds 3 times the standard deviation is 0.3% (Table 14.1). Or, put another way, the probability that the residual has this value due to noise alone is so small (less than 0.3%) that we say that this fluctuation is due to a change in the underlying signal. Of course, we may be wrong here and the deviation may be due to a rogue noise value, but the probability of this is so small that we deem this possibility to be negligible.

34.3.2 Response to detection of sudden changes

Now that we have detected that a 'significant' change in the underlying signal has occurred, we now have to decide how to respond to this detection. Clearly, we cannot continue to implement the estimator as we were doing up to the detection. There are many possible ways that one could respond to this detection. A simple way to respond to the detection is to re-initialize the estimator. This is done as follows. Suppose that a jump is detected at sample point M. The estimator is then initialized as follows:

$$\tilde{x}[M] = z[M]. \tag{34.30}$$

For $n \geq M + 1$:

$$\hat{x}[n] = \tilde{x}[n - 1], \tag{34.31}$$

$$\tilde{x}[n] = \hat{x}[n] + \frac{1}{[n - (M - 1)]}(z[n] - \hat{x}[n]). \tag{34.32}$$

What we are saying here is that the underlying model has changed, so we start from the sample point where we detected the change and estimate the new value of

Algorithm 34.2 Estimation of a constant in the presence of jumps

Input threshold constant k in (34.28): k
Initialization
   Input measurement z[1]

   $\tilde{x}[1] = z[1]$

   Manoeuvre index j = 1
For the nth sample, with n > 1
   Increment manoeuvre index: j = j + 1
Prediction : $\hat{x}[n] = \tilde{x}[n - 1]$
   Input measurement : z[n]
   Residual : $r[n] = z[n] - \hat{x}[n]$

   Standard deviation of residual: $\sigma_r = \sqrt{\dfrac{j}{j-1}}\sigma$
Reset manoeuvre index if manoeuvre detected
   If $|r[n]| > k\sigma_r$ then j = 1

Estimation : $\tilde{x}[n] = \hat{x}[n] + \dfrac{1}{j}(z[n] - \hat{x}[n])$

Variance of estimate : $\tilde{P}[n] = \dfrac{\sigma^2}{j}$.
End loop

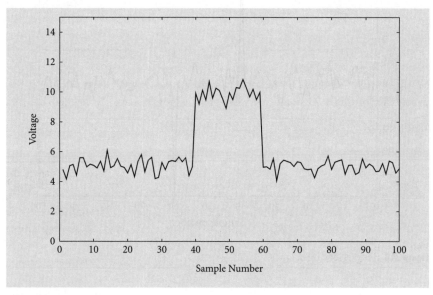

Figure 34.6 Noisy data.

the constant. The algorithm for the estimation of a constant in the estimation of a jump is a combination of Algorithm 34.1 and (34.30)–(34.32), and is summarized in Algorithm 34.2.

Now we can implement Algorithm 34.2 for the data in Figure 34.1. The data are shown again for reference in Figure 34.6. The standard deviation of the measurement noise error is equal to 0.5. If we apply the estimator and jump detector combined, then significant jumps in the data are detected at sample points 40 and

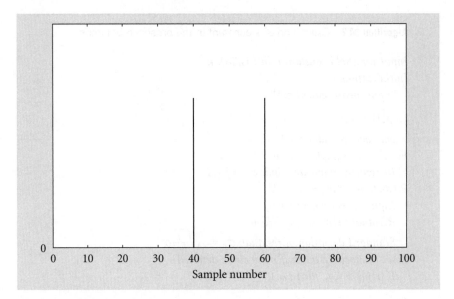

Figure 34.7 Detection of jumps in data.

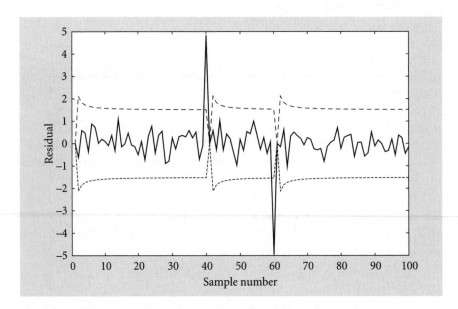

Figure 34.8 Residual and $\pm 3\sigma_r$ limits.

60 as shown in Figure 34.7. The residual as a function of time, before, during and after the jumps is shown in Figure 34.8. Note that the jump at sample point 60 is detected by the residual becoming large and negative and crossing the lower threshold.

The estimated signal is shown as a dashed curve in Figure 34.9, and the actual signal is shown as a full line. Comparing Figure 34.9 with Figure 34.3, we can see that the incorporation of a jump detector (Algorithm 34.2) has significantly improved the performance of the estimator in Algorithm 34.1.

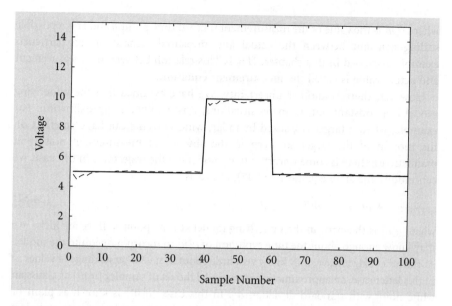

Figure 34.9 Actual signal (full line) and estimated signal (dashed line).

34.4 **Further methods**

In this chapter, we have described the application of recursive least-squares methods to fit data to the simplest model of an underlying signal – that is, the signal is a constant with possible jumps in values at unpredictable times. The difference between this and the model discussed in Chapter 30 is that now the model of the underlying signal may change at unpredictable times, and it is important to be able to detect such changes and compensate for them in some way. We have discussed one of the simplest methods, which is to re-initialize the estimator.

If we take the example of the estimation of a constant, then the estimation of the underlying signal, (34.6), can be rewritten as

$$\tilde{x}[n] = (1 - K[n])\hat{x}[n] + K(n)z[n] \tag{34.33}$$

where

$$K(n) = \frac{1}{n} \tag{34.34}$$

and we have used (34.9) to replace $\tilde{x}[n-1]$ by $\hat{x}[n]$.

Now $K(n)$ determines how much you weight the measurements at each time point. For small n, $K(n)$ is close to 1, reflecting that we have not processed enough measurements to be sure what the underlying constant signal is. The underlying model of the signal can be written as

$$x_a[n] = x_a[n-1] \tag{34.35}$$

where $x_a[n]$ is the **actual** value of the signal at time point n. The time evolution of the underlying model is known as the **state equation**.

The relation between the measurement and actual signal can be written as

$$z[n] = Hx_a[n] + v[n] \tag{34.36}$$

where $v[n]$ is the value of the measurement noise at time point n and H is a possible scaling constant between the actual and measured signals; in the particular example discussed in this chapter, $H = 1$. This relation between the measurement and actual value is called the **measurement equation**.

However, there is another uncertainty. We have assumed that the underlying model is a constant but, in many situations, this is only an approximation. For example, when a target is tracked by radar, wind currents can have an effect on the motion of the target so, even in the absence of measurement noise and manoeuvres, there is some uncertainty in modelling the trajectory. In this case, we can modify the state equation, (34.35), above as

$$x_a[n] = x_a[n-1] + q[n] \qquad (34.37)$$

where $q[n]$ is the error in the underlying model at time point n. If we are lucky, we may know enough about the time evolution of $q[n]$ to incorporate it into the model. At the other extreme, we may know very little about it, save what its range of values is; in this latter case, an approximation is to model the set of samples $\{q[n]\}$ as Gaussian white noise with standard deviation σ_q^2. In this case, $\{q[n]\}$ is known as **plant** or **process** noise. Of course, this is an approximation, but it is often better to include it than just leaving this factor out altogether. In this case, the gain factor $K[n]$ is no longer given by (34.34) but instead becomes a function of both σ_q^2 and the measurement noise variance σ^2. The effect of incorporating plant noise as in (34.37) is to increase $K[n]$ over the case when plant noise is ignored, i.e. more emphasis is being put on the measurements than the predictions because there is more uncertainty in the predictions due to the modelling errors reflected in the plant noise. Remember that the predictions will be in more error, as they exclude the effects of plant noise. It can also be shown that $K[n] \to m$ as $n \to \infty$, where $m > 0$. Thus, one is always putting some weight on the measurements because one cannot completely trust the predictions due to uncertainties in the time evolution of the underlying signal.

The resultant estimator, which takes the form of (34.33) with subsidiary equations for $K[n]$, is called a **Kalman filter**. When applied to estimating a constant, the resultant Kalman filter is called a **first-order estimator** because only one variable, the constant, is being estimated. In tracking an aircraft travelling at constant velocity using radar, the resultant Kalman filter is said to be **second-order** because now we are estimating both position and velocity. In this case, the estimation equation (34.33) becomes a matrix equation. The above procedure of incorporating uncertainty in the signal model can be extended to second- and higher-order estimators. Further discussion about Kalman filters is beyond the scope of this book; an excellent introduction to the subject is given by Candy (2005).

34.5 Concluding remarks

In many applications encountered in experimental physics, we are required to monitor a signal where the underlying model changes at unpredictable times, for example in the study of solar flares or in seismological applications. Implicit in the estimation process is the filtering of any underlying noise to provide an estimate that is hopefully closer to the actual signal than the measurement.

Before implementing an estimator, the first step is to derive a model for the time variation of the underlying signal. We need to ask ourselves questions about the time evolution of the underlying signal. For example, is it essentially a constant in time, with possible jumps in the data, like solar flare activity? Or is the underlying signal varying linearly with time, for example if one is tracking an airplane or a car travelling at constant velocity? The variation with time may be much more complicated: for example, the underlying signal may evolve with time sinusoidally.

In the real world the evolution of the signal with time will not follow our model exactly. There will be variations, for example a car suddenly accelerating. We can detect these variations by using jump or manoeuvre detection techniques and correct for them accordingly. However, if in a certain situation we are detecting such jumps for most of the time, then this implies that the underlying model is inaccurate. In critical situations, for example when tracking aircraft, it may not be possible to change models during monitoring. This emphasizes the importance of modelling the underlying signal first, before applying the estimator. Thus knowledge of the physics of the underlying signal is important in order for the estimator to perform to an acceptable standard. Candy (1986, p. 123) gives an example of monitoring the evaporation of plutonium nitrate in a storage tank. It is important to model the evaporation process accurately before the Kalman filter is applied, in order to estimate the solution mass and density as a function of time. Approximations are necessary to make the problem tractable, and it is important to determine that such approximations do not significantly affect the performance of the final designed estimator.

Problems

All these examples are based on using the program estimateconstant. The underlying signal is assumed to look like the full curve in Figure 34.3 where the signal starts off as a constant, then jumps in value, then jumps down again to its initial value. The basic method is based on Algorithm 34.2.

The input parameters to this program are as follows (the variables in brackets do not correspond to the names of the variables in the program):

- Number of samples in the signal, N
- Actual standard deviation of the noise added on to the underlying signal, $SDACT$
- Initial value of the signal, $INIT$
- Jump in value, $JUMP$
- Sample number when the jump occurs, $JUMPSTART$
- Sample number when the jump ends, $JUMPEND$
- Assumed standard deviation of the noise, σ, (Equation (34.27)), $SDASS$

- Threshold for jump detector (Equation (34.28)), k

For the MATLAB program, two figures are produced.

- Figure 1: Actual signal (red), estimated signal (blue), and measurements (green).
- Figure 2: Sample numbers where manoeuvres are detected.

For the Fortran and C programs two files are produced:

- ESTOUT.DAT which contains three columns of data: (1) the actual signal, (2) the estimated signal, and (3) the measurements
- JUMPOUT.DAT which indicates where the jumps in the data are detected.

The information in these files can be plotted out using a graphics package, for example GNUPLOT.

34.1 In this problem, we are going to look at the performance of the first-order estimator when detecting jumps. Put in the following parameters: $N=100$, $SDACT=1$, $INIT=5$, $JUMP=5$, $JUMPSTART=45$, $JUMPEND=55$, $SDASS = 1$, $k=3$.

Carry out the simulation a few times. For each simulation, a different set of random numbers is produced, obeying the same statistics (Gaussian noise). Comment on the results. Example results are shown in the solutions.

34.2 For the simulation in Problem 34.1, investigate the effects of keeping $SDACT$ at 1 and increasing $SDASS$ to above 1 and then decreasing it to below 1. Comment on your results.

34.3 For the simulation in Problem 34.2, investigate the effect of increasing k to values above 3 and decreasing it to values below 3. Comment on the results. Example results are shown in the solutions.

34.4 For the simulation in Problem 34.1, put $JUMP$ to 0, so that the signal remains at a constant value of 5 for all time. Change k to 2. Now run the simulation a few times and comment on the results. Example results are shown in the solutions.

Linear programming and optimization

Linear programming is part of a general area of applicable mathematics known as operational research. It has its origins at the time of World War II, where it was developed to apply scientific methods to the planning of military operations, and was developed further in the postwar period when it was applied to industrial and commercial decision-making. However, it also has scientific applications and so is of interest to engineers and research scientists. You will learn how to find the values of variables, linked by linear relationships and with linear constraints, which give an optimum value, either maximum or minimum, of some target function. You will also be introduced to 'hill-climbing' methods of finding the maxima or minima of non-linear functions of variables. Finally, we describe the Lagrange multiplier method which enables maxima and minima to be found for non-linear functions with general constraints.

35.1 Basic ideas of linear programming

A very simple problem will illustrate the principle of linear programming without introducing any mathematical complication.

EXAMPLE 35.1

A paint manufacturer has a stock of 10 000 litres of red paint and 12 000 litres of green paint. He can sell red paint for £2.00 per litre and green paint for £2.50 per litre. By mixing green and red paints in the ratio 2:1 he can make brown paint that sells at £2.40 per litre. In what form should he sell the paint to maximize his income?

We assume that he uses x litres of red paint, and hence $2x$ litres of green, to make brown. His income, in pounds, is then

$$I = 2.00(10\ 000 - x) + 2.50(12\ 000 - 2x) + 2.40 \times 3x$$
$$= 50\ 000 + 0.2x. \tag{35.1}$$

This is plotted in Figure 35.1. The straight line representing the income steadily rises as the amount of red converted to brown (x) increases but there is a

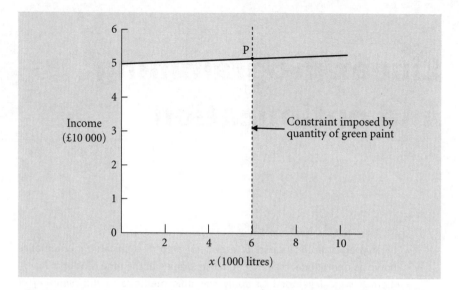

Figure 35.1 Income as a function of the amount of red paint converted to brown.

constraint imposed by the amount of paint available. This constraint is imposed by the green paint since $2x$ cannot be greater than $12\,000$ litres. Thus the maximum income is obtained by converting all the green paint into brown – the point P in the figure – with $x = 6000$ litres. The paint is sold in the form of $4\,000$ litres of red and $18\,000$ litres of brown to give an income of £51 200.

This example illustrates the general principle of linear programming. It involves transforming an optimization process into one or more linear equations that are solved subject to a number of constraints. A second, and slightly more complicated problem, will highlight other features of the method.

EXAMPLE 35.2

Three basic colours of paint, together with the quantities available in litres, are:

red (5000)   yellow (4000)   blue (3000).

The basic paints sell for £2.00 per litre but can be combined to give two other colours, with the proportions of red, yellow, and blue and the price as shown in Table 35.1.

We shall now find the maximum income in this situation. Since there are two products of the basic ingredients we take the amounts of these made, in litres, as x for the orange paint and y for the green paint. From this the amount of red used is $0.4x + 0.2y$ but since the amount of red available is 5000 litres the inequality relationship

$$0.4x + 0.2y \leq 5000 \tag{35.2a}$$

must be satisfied. Similarly, from the quantities of yellow and blue available,

$$0.4x + 0.4y \leq 4000 \tag{35.2b}$$

Table 35.1 The price of paint

Colour	Red	Yellow	Blue	Price per litre
Dark orange	0.4	0.4	0.2	£2.40
Dark green	0.2	0.4	0.4	£2.50

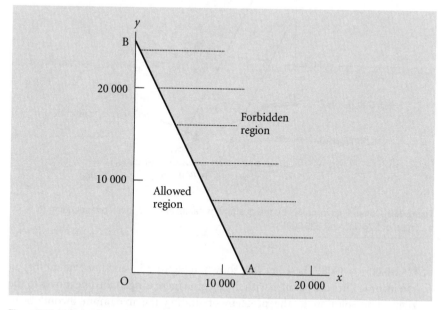

Figure 35.2 Values of x and y allowed by the quantity of red paint.

and
$$0.2x + 0.4y \le 3000. \tag{35.2c}$$
Inequality (35.2a) is represented graphically in Figure 35.2, which shows the line
$$0.4x + 0.2y = 5000.$$

Values of x and y allowed by the quantity of red paint are restricted to the region within the triangle OAB. Figure 35.3 shows the effect of adding the constraints imposed by (35.2b) and (35.2c). It is clear that the allowed regions of x and y are now constrained to be within the quadrilateral OPQR, a region entirely dictated by the quantities of yellow and blue paint. The quantity of red paint imposes no extra constraint in this case.

We now consider the income when quantities x of orange paint and y of green paint are produced. This is

$$
\begin{aligned}
I &= 2.00(5000 - 0.4x - 0.2y) + 2.00(4000 - 0.4x - 0.4y) \\
&\quad + 2.00(3000 - 0.2x - 0.4y) + 2.40x + 2.50y \\
&= 24\,000 + 0.40x + 0.50y. \tag{35.3}
\end{aligned}
$$

For an income of £26 000 the values of x and y are on the line $0.4x + 0.5y = 2000$, while for an income of £28 000 they must be on the line

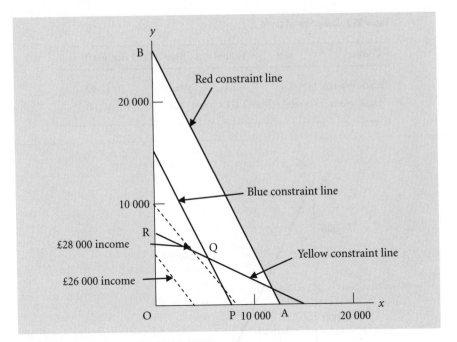

Figure 35.3 Constraints imposed by the quantity of paints and the lines corresponding to incomes of £26 000 and £28 000.

$0.4x + 0.5y = 4000$. These lines are shown in Figure 35.3 and it is evident that all the income lines are parallel, with the income increasing as the line moves to the right. The solution to the problem of finding the maximum income is to determine how far to the right the income line can move and still contain an allowed point (x, y). By inspection, this is seen to be when the income line passes through the point Q. The coordinates of this point are given by the intersections of the limiting yellow and blue lines

$$0.4x + 0.4y = 4000$$

and

$$0.2x + 0.4y = 3000$$

which give $(x, y) = (5000, 5000)$. Substituting these values in (35.3) gives the maximum income as £28 500.

If the prices of the orange and green paint were different from the ones used in Example 35.2, this would affect the slope of the income lines. In Figure 35.4 we show some possible relationships between the slope of income lines and the quadrilateral OPQR.

Depending on the slope of the income line, the optimum point for maximum income can be any of the three non-origin points of the quadrilateral. Another situation that can arise is if the income line is parallel to one of the sides of the quadrilateral. The optimum position of the income line is then when it is coincident with the side of the quadrilateral and any (x, y) on this side gives the maximum income. In such a case the solution is said to be **degenerate**, a term met in quantum mechanics (§21.5.2) when there were several eigenstates with the same energy.

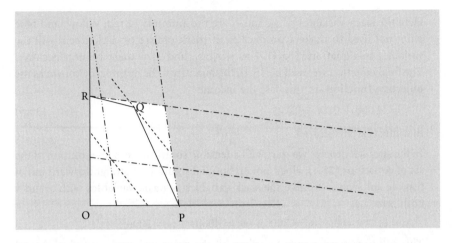

Figure 35.4 Income lines displaced to go through an apex of the allowed-region quadrilateral.

So far the examples we have considered have been simple enough to allow a two-dimensional graphical representation of the linear equations and the constraints. It was by visual inspection of Figure 35.3 that we decided that the relevant maximum-income point was Q. However, we can envisage problems with, say, 10 components and 20 products where graphical representation would be impossible and for which a totally analytical approach would be necessary.

Exercise 35.1
A florist has 300 red carnations that can be sold for 20 pence each and 500 white carnations that sell for 16 pence each. Bunches of 5 red plus 5 white carnations sell for £1.90 per bunch. How should the florist sell the carnations to make the maximum income, and what is that income?

35.2 **Simplex method**

The **simplex method** is a very powerful technique for solving linear programming problems of any complexity and is usually applied as a standard computer program. Here it will be illustrated by the problem in §35.1 involving the three basic paints, the inequality relationships for which are given in (35.2).

EXAMPLE 35.3

The first step is to convert the inequality relationships to equalities by introducing what are called **slack variables**. Thus the three relationships (35.2) are written as

$$0.4x + 0.2y + s_1 = 5000, \tag{35.4a}$$

$$0.4x + 0.4y + s_2 = 4000, \tag{35.4b}$$

$$0.2x + 0.4y + s_3 = 3000. \tag{35.4c}$$

Here the slack variables s_1, s_2, and s_3 are the amounts of red, yellow, and blue paint *not* used to make a product paint (dark orange or dark green). All the variables in Equations (35.4) – x, y, s_1, s_2, and s_3 – must be non-negative. The final equation we need is (35.3), the quantity to be optimized, known as the **objective function**, in this case the income

$$I = 24\,000 + 0.4x + 0.5y \qquad (35.5)$$

that must be maximized.

In the simplex process we start with a **feasible solution**, i.e. some solution of the set of equations (35.4), albeit not an optimal solution. A straightforward initial feasible solution is to take the slack variables as **basic variables** with x and y both zero, i.e.

$$x = 0, \quad y = 0, \quad s_1 = 5000, \quad s_2 = 4000, \quad s_3 = 3000.$$

This solution corresponds to selling all the paint unmixed, and gives a total income of £24 000. The equations are now rewritten in a form that expresses the basic variables in terms of the others – thus

$$s_1 = 5000 - 0.4x - 0.2y, \qquad (35.6a)$$
$$s_2 = 4000 - 0.4x - 0.4y, \qquad (35.6b)$$
$$s_3 = 3000 - 0.2x - 0.4y. \qquad (35.6c)$$

It is clear from (35.5) that increasing either, or both, of x and y will increase the income, and of the two it is better to increase y because it has a larger coefficient. The problem now is to find out how much y can be increased without contravening the condition that all the variables must be non-negative. To determine this we examine the equations (35.6). From (35.6a) it is found that $y \le 25\,000$ otherwise s_1 would become negative. Similarly we find from (35.6b) $y \le 10\,000$ and from (35.6c) $y \le 7500$. This means that the maximum that y can take while still satisfying all the inequality relationships is 7500, and from (35.6c) that gives this restriction

$$y = 7500 - 0.5x - 2.5s_3. \qquad (35.7a)$$

This value of y is now substituted in (35.6a), (35.6b), and (35.5). With the revised set of basic variables on the left-hand side we have, in addition to (35.7a)

$$s_1 = 3500 - 0.3x + 0.5s_3 \qquad (35.7b)$$

and

$$s_2 = 1000 - 0.2x + s_3 \qquad (35.7c)$$

with objective function

$$I = 27\,750 + 0.15x - 1.25s_3. \qquad (35.7d)$$

We have now moved to a new feasible solution with $y = 7500$, $s_1 = 3500$, $s_2 = 1000$, and both x and s_3 zero. On Figure 35.3 we have moved to the point R. With this situation the value of y and x indicate that 7500 litres of green paint and no orange paint have been produced. The amount of red paint unused (given by s_1) is 3500 litres, unused yellow paint (given by s_2) is 1000 litres, and all the blue paint has been used, since $s_3 = 0$.

Examination of (35.7) shows that the income has not been maximized since it can be further increased by increasing the value of x. The amount by which x

can be increased is restricted by Equations (35.7). These indicate a maximum value for x of 15 000, 3500/0.3, and 5000 respectively so the operative constraint is that from (35.7c) from which

$$x = 5000 - 5s_2 + 5s_3. \tag{35.8a}$$

Substituting this value in (35.6a), (35.6b), and (35.7)

$$y = 5000 + 2.5s_2 - 5.0s_3, \tag{35.8b}$$

$$s_1 = 2000 + 1.5s_2 - s_3, \tag{35.8c}$$

$$I = 28\,500 - 0.75s_2 - 0.5s_3. \tag{35.9}$$

The form of the objective function indicates that the maximum income has been achieved since the coefficients of the variables in it are negative. The maximum income is £28 500 and the values of the variables are:

$x = 5000$, giving the amount of orange paint

$y = 5000$, giving the amount of green paint

$s_1 = 2000$, giving the amount of red paint

$s_2 = s_3 = 0$, indicating that no yellow or blue paint is left.

The final step moved the feasible solution from R to Q in Figure 35.3. Both these points, and the starting point, are apices of the quadrilateral ORQP, and it is a characteristic of the simplex method that it moves from apex to apex in achieving the optimum solution. This is easily visualized in two dimensions. If three products were produced from the original paints then, with three variables x, y, and z the equations would be representing planes in three dimensions and we can envisage apices in this case, like the corners of a cube. In more than three dimensions we cannot envisage what is equivalent to an apex, but such points exist in multidimensional space and can be described mathematically.

The simplex method is systematic and readily programmed. However, such programs are fairly complex and must allow for problems giving inequalities where sums of variables are greater than rather than less than some limit.

Exercise 35.2
Write down a set of equations, analogous to (35.4a–c) and the objective function for the problem in Exercise 35.1, and proceed to a solution by the simplex method.

35.3 Non-linear optimization; gradient methods

Situations occur in scientific and other contexts where it is required to find the maximum (or minimum) of a function in some multidimensional space. If there is only one maximum then a straightforward and methodical process exists to find where it is. The principle of this is best explained by a two-dimensional example. Figure 35.5 shows contours of constant value of a function $f(x, y)$ where the objective is to find the maximum at point P.

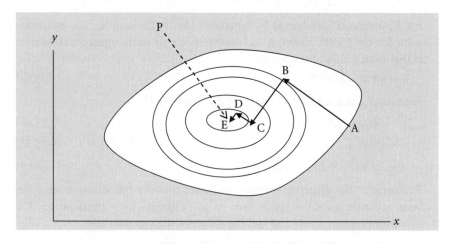

Figure 35.5 Steepest-ascent determination of the maximum of a two-dimensional function.

To begin the process a point in the plane is selected, close to the maximum if possible but otherwise just a random point. This is shown as A in Figure 35.5. The direction in which the gradient of the function is a maximum is then found, shown as AB in the figure. This direction will be perpendicular to the contour line through A. The point B is that where the function is a maximum along that line and is where it is parallel to the contour at the point. Now the direction of greatest gradient is found at B, which is perpendicular to the contour at B and hence perpendicular to AB. The process is repeated. Point C is where the function is a maximum along BC, CD is the direction of greatest gradient at C and D the point where the function has the highest value along CD. The next step goes to point E, close to the maximum of the function, and steps may be repeated until the maximum point is reached within the required precision.

This method is quite general and may be applied to finding a minimum value by simply going downhill rather than uphill at each stage. The extension of this process of maximization to three or more dimensions is straightforward, although not as easily illustrated as the two-dimensional case. We shall now see how to carry out this process in two dimensions if the analytical function $f(x, y)$ is available and deduce from this the general approach for a multidimensional problem.

35.4 Gradient method for two variables

The direction of the shift of variables at each stage is that of the greatest rate of change of the function $f(x, y)$, which is the direction of ∇f as defined in §16.2. If the total shift is $(\delta x, \delta y)$ then the individual shifts are given by

$$\delta x = k\frac{\partial f}{\partial x} \quad \text{and} \quad \delta y = k\frac{\partial f}{\partial x} \tag{35.10}$$

where k is of magnitude that gives the total shift from A to B or from B to C in Figure 35.5. Finding the value of k is the difficult part of the process.

We now illustrate a possible process that can be applied in the two-dimensional case if the starting point of the maximization process is not too distant from the maximum point. We start at the point A(x, y) with the extremum at B$(x+\delta x, y+\delta y)$ not too distant. The condition for the extremum is

$$\left(\frac{\partial f}{\partial x}\right)_{x+\delta x, y+\delta y} = 0 \tag{35.11a}$$

and

$$\left(\frac{\partial f}{\partial y}\right)_{x+\delta x, y+\delta y} = 0. \tag{35.11b}$$

Adapting (11.34) to the first partial derivative we write

$$\left(\frac{\partial f}{\partial x}\right)_{x+\delta x, y+\delta y} = \frac{\partial f}{\partial x} + \frac{\partial^2 f}{\partial x^2}\delta x + \frac{\partial^2 f}{\partial x \partial y}\delta y = 0 \tag{35.12a}$$

where the expansion on the right-hand side has only been taken to first-order terms in small quantities and all the partial derivatives on the right-hand side are at the point (x, y). Similarly

$$\left(\frac{\partial f}{\partial y}\right)_{x+\delta x, y+\delta y} = \frac{\partial f}{\partial y} + \frac{\partial^2 f}{\partial y \partial x}\delta x + \frac{\partial^2 f}{\partial y^2}\delta y = 0. \tag{35.12b}$$

These linear equations in δx and δy can be solved to give

$$\delta x = \frac{\frac{\partial f}{\partial y}\frac{\partial^2 f}{\partial x \partial y} - \frac{\partial f}{\partial x}\frac{\partial^2 f}{\partial y^2}}{\frac{\partial^2 f}{\partial x^2}\frac{\partial^2 f}{\partial y^2} - \left(\frac{\partial^2 f}{\partial x \partial y}\right)^2} \tag{35.13a}$$

and

$$\delta y = \frac{\frac{\partial f}{\partial x}\frac{\partial^2 f}{\partial x \partial y} - \frac{\partial f}{\partial y}\frac{\partial^2 f}{\partial x^2}}{\frac{\partial^2 f}{\partial x^2}\frac{\partial^2 f}{\partial y^2} - \left(\frac{\partial^2 f}{\partial x \partial y}\right)^2}. \tag{35.13b}$$

Then, starting from the new value of (x, y) corresponding to point B, the process could be repeated to find point C and so on. This looks like, and is, a very complicated process and extending it into more than two dimensions is not really feasible. A more practical approach is needed and will now be described.

35.5 A practical gradient method for any number of variables

We now consider the problem of finding the maximum for a function of n variables, $f(x_1, x_2, \ldots, x_n)$ starting from point $(_0x_1, _0x_2, \ldots, _0x_n)$. For the maximum benefit the direction of shift from the starting point must be in the direction of ∇f so that the shift in each variable is proportional to the corresponding component of ∇f. This gives

$$\delta x_i = k\frac{\partial f}{\partial x_i},$$

(35.14)

for the variable x_i. The 'length' of the shift is defined as

$$d = \sqrt{\sum_i (\delta x_i)^2} = k\sqrt{\sum_i \left(\frac{\partial f}{\partial x_i}\right)^2}.$$

(35.15)

If the length of the shift is chosen then this gives the value of k as

$$k = \frac{d}{\sqrt{\sum_i \left(\frac{\partial f}{\partial x_i}\right)^2}}.$$

(35.16)

The shifted values of the variables are now

$$_1x_i = {}_0x_i + k\frac{\partial f}{\partial x_i}.$$

(35.17)

We must now test that the shift actually increases the value of f. If in the step from A in Figure 35.4 one goes well beyond the point B then the value of f could be below the value at A. If this happens then the value of d is halved and the test repeated until the new value of f is greater than the original value.

This whole process is repeated until the value of f converges to some degree of precision that is required, which can be determined by the magnitude of the difference of successive values of f. Normally this method is performed on a computer, but here we illustrate the first two steps of a simple two-dimensional example.

EXAMPLE 35.4

We wish to find the maximum of the function

$$f(x, y) = \sin(2x + 3y) - \sin(x - y).$$

(35.18)

The first thing to be done is find the form of the partial derivatives, which are

$$\frac{\partial f}{\partial x} = 2\cos(2x + 3y) - \cos(x - y)$$

(35.19a)

and

$$\frac{\partial f}{\partial y} = 3\cos(2x + 3y) + \cos(x - y).$$

(35.19b)

Taking the starting point as $(x_0, y_0) = (0, 0)$,

$$\left(\frac{\partial f}{\partial x}\right)_0 = 1 \quad \text{and} \quad \left(\frac{\partial f}{\partial y}\right)_0 = 4.$$

With $d = 0.5$, from (35.16), $k = 0.5/\sqrt{17} = 0.121268$ and hence, from (35.14),

$$x_1 = x_0 + k\frac{\partial f}{\partial x} = 0.121\,268 \quad \text{and} \quad y_1 = y_0 + k\frac{\partial f}{\partial y} = 0.485\,071.$$

The value of f for these values of x and y is 1.347 783, higher than the original value of zero.

From these latest values of x and y and still with $d = 0.5$, we find

$$\left(\frac{\partial f}{\partial x}\right)_1 = -1.187\,774 \quad \text{and} \quad \left(\frac{\partial f}{\partial y}\right)_1 = 0.554\,715$$

giving

$$k = 0.381\,420,$$
$$x_2 = 0.121\,268 + 0.381\,420 \times (-1.187\,774) = -0.331\,773,$$
$$y_2 = 0.485\,071 + 0.381\,420 \times 0.554\,715 = 0.696\,650.$$

The new value of f is now 1.846 080, again an increase over the previous value. If the value of f had been smaller then that would have been a sign that the shift d was too great and a smaller value would need to be tried.

This problem has been programmed and is outlined as GRADMAX in Appendix 25. By changing the functions in the program it could be run for any two-variable problem. The user inputs the initial values of x and y and an initial shift d. Another input quantity is a tolerance, the accuracy with which the maximum value of the function is required. The output of a version of this program is given in Table 35.2. The results for the first two steps differ slightly from those given by the above calculations, since the computer is working to more significant figures.

The maximum value is 2.0, the values of x and y being found making the first sine argument in f equal to $\pi/2$ and the second sine argument equal to $-\pi/2$. Clearly other solutions are possible, for example that found by solving the simultaneous equations

$$2x + 3y = 5\pi/2 \quad \text{and} \quad x - y = \pi/2.$$

In this case, because of the periodic nature of the sine functions there are an infinite number of solutions for finding the maximum of f and the one found is usually that closest to the starting point if a modest shift is used. There are many situations where the function has many local maxima although there may be only one **global maximum**, i.e. the **absolute maximum value** of the function. If the starting point for the refinement is near one of the local maxima, that is the point to which the

Table 35.2 Steps in running the program GRADMAX

Step	x	y	d	f
1	0.121268	0.485071	0.500000	1.347784
2	−0.333563	0.696645	0.500000	1.846072
3	−0.390332	0.939688	0.250000	1.863804
4	−0.569149	0.764975	0.250000	1.887575
5	−0.477359	0.997444	0.250000	1.888223
6	−0.632658	0.801672	0.250000	1.899212
7	−0.576443	0.913319	0.125000	1.996586
8	−0.636580	0.930343	0.062500	1.998592
9	−0.628280	0.943582	0.015625	1.999994
10	−0.628723	0.942711	0.000977	2.000000

refinement will converge. A useful strategy in such circumstances is to select several starting points and refine from each of them in turn. This gives a good chance of finding the global maximum, but does not guarantee that it will be found.

35.6 Optimization with constraints – the Lagrange multiplier method

Problems occasionally arise where one is interested in the maximum or minimum (optimal values) of a function but where some constraint is imposed. For example, consider the problem of finding the optimal values of the function

$$f(x, y) = 3x^3 + 5xy + y^3 - 2x \tag{35.20a}$$

under the condition

$$x - y = 0. \tag{35.20b}$$

In this case from (35.20b) one finds that $y = x$ and substituting this in (35.20a) gives

$$f(x, y) = g(x) = 4x^3 + 5x^2 - 2x. \tag{35.21}$$

The extrema of $g(x)$ are now found in the normal way:

$$\frac{dg}{dx} = 12x^2 + 10x - 2 = 0,$$

giving $x = 1/6$ and $x = -1$. To find which extremum is which, we find

$$\frac{d^2g}{dx^2} = 24x + 10.$$

For $x = 1/6$ the second derivative is positive and hence this point is a minimum corresponding to $(x, y) = (1/6, 1/6)$. For $x = -1$ the second derivative is negative and hence the point is a maximum corresponding to $(x, y) = (-1, -1)$.

This example is simple because it was possible to find y in terms of x and so convert $f(x, y)$ into a function of a single variable for which the extrema were to be found. However the constraining function may be very complex so that it is not feasible to express either y as a function of x or x as a function of y. The method of **Lagrange multipliers** gives a way of finding optimal values in such cases.

We now take the general case of finding optimal values of $f(x, y)$ under the constraint that $g(x, y) = 0$. Figure 35.6 shows contours of constant values of $f(x, y)$ and also the line $g(x, y) = 0$.

At the point representing the extremum the line $g(x, y) = 0$ must be tangential to the contour of $f(x, y)$ corresponding to the optimal value. If this were not so then the line would cross the contour and then points on $g(x, y) = 0$ on either side of the intersection point would be greater and smaller than the value at the intersection point – which means that it could not be an extremum. The implication of this is that at the extremum point the gradients of the $f(x, y)$ and $g(x, y)$ must be parallel or

$$\left(\frac{\partial f}{\partial x} \mathbf{i} + \frac{\partial f}{\partial y} \mathbf{j} \right) = \lambda \left(\frac{\partial g}{\partial x} \mathbf{i} + \frac{\partial g}{\partial y} \mathbf{j} \right).$$

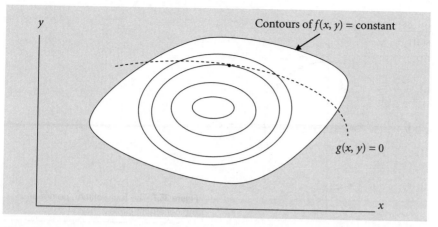

Figure 35.6 Contours of $f(x, y) = $ constant (full lines) and the line $g(x, y) = 0$ (dashed line).

From this we find

$$\frac{\partial f}{\partial x} = \lambda \frac{\partial g}{\partial x} \quad \text{and} \quad \frac{\partial f}{\partial y} = \lambda \frac{\partial g}{\partial y}. \tag{35.22}$$

The quantity λ is known as the **Lagrange multiplier**. The equations (35.22) do not completely define the problem because, for example, exactly the same equations would be found for an extremum on the line $g(x, y) = c$, which would give a different solution. Hence the complete set of equations that must be solved to find the extremum values is

$$\frac{\partial f}{\partial x} = \lambda \frac{\partial g}{\partial x}, \quad \frac{\partial f}{\partial y} = \lambda \frac{\partial g}{\partial y}, \quad \text{and} \quad g(x, y) = 0. \tag{35.23}$$

We now illustrate the method with an example.

EXAMPLE 35.5

We are required to find the extrema of the function $f(x, y) = 4x^2 - 4xy + y^2$, subject to the condition $g(x, y) = x^2 + y^2 - 36 = 0$

First we find the partial derivatives

$$\frac{\partial f}{\partial x} = 8x - 4y, \quad \frac{\partial f}{\partial y} = -4x + 2y, \quad \frac{\partial g}{\partial x} = 2x, \quad \text{and} \quad \frac{\partial g}{\partial y} = 2y.$$

Hence the set of equations to be solved to find the positions of the extrema is:

$$8x - 4y - 2\lambda x = 0 \tag{35.24a}$$

$$-4x + 2y - 2\lambda y = 0 \tag{35.24b}$$

$$x^2 + y^2 - 36 = 0. \tag{35.24c}$$

From (35.24a) and (35.24b) we find

$$\lambda = \frac{8x - 4y}{2x} = \frac{-4x + 2y}{2y}$$

which gives

$$2x^2 + 3xy - 2y^2 = (2x - y)(x + 2y) = 0$$

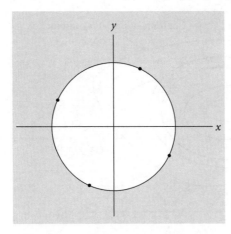

Figure 35.7 The four solution points on the constraint circle $x^2 + y^2 = 36$.

or

$$x = y/2 \quad \text{and} \quad x = -2y. \tag{35.25}$$

Putting the first of these relationships between x and y into (35.24c) gives

$$\frac{5}{4}y^2 = 36 \quad \text{or} \quad y = \pm\frac{12}{\sqrt{5}}.$$

The second relationship gives

$$5y^2 = 36 \quad \text{or} \quad y = \pm\frac{6}{\sqrt{5}}.$$

From this we find the following solutions:

$$(x, y) = \left(\frac{6}{\sqrt{5}}, \frac{12}{\sqrt{5}}\right), \left(-\frac{6}{\sqrt{5}}, -\frac{12}{\sqrt{5}}\right), \left(-\frac{12}{\sqrt{5}}, \frac{6}{\sqrt{5}}\right), \left(\frac{12}{\sqrt{5}}, -\frac{6}{\sqrt{5}}\right).$$

The constraint equation, $g(x, y) = 0$, is a circle of radius 6 units centred on the origin and the positions of the extrema on this circle are shown in Figure 35.7. The values of $f(x, y)$ for the four points are

$$f\left(\frac{6}{\sqrt{5}}, \frac{12}{\sqrt{5}}\right) = 0, \quad f\left(-\frac{6}{\sqrt{5}}, -\frac{12}{\sqrt{5}}\right) = 0, \quad f\left(-\frac{12}{\sqrt{5}}, \frac{6}{\sqrt{5}}\right) = 180,$$

$$\text{and} \quad f\left(\frac{12}{\sqrt{5}}, -\frac{6}{\sqrt{5}}\right) = 180$$

and it is clear in this case that the first two points are minima and the last two maxima.

35.6.1 Identifying maxima and minima

In the example just dealt with the positions of the extrema points and the pattern of $f(x, y)$ values at these points indicated without doubt the nature of the points. Finding which optimal points are maxima and which minima can often be decided in a rather *ad hoc* way for each particular problem. In cases of difficulty a general recipe that can be followed is now given for the optimal point (x, y).

1 Select a value $x + \Delta x$ where Δx is a very small quantity.

2 Set up an equation $h(y_d) = g(x + \Delta x, y_d) = 0$

3 Solve $h(y_d) = 0$ for y_d by one of the methods described in Chapter 31. The solution should be little different from y and the point $(x + \Delta x, y_d)$ is on the constraint curve close to the optimal point (x, y).

4 Calculate $f(x + \Delta x, y_d)$. If this is greater than $f(x, y)$ then (x, y) is a minimum point; if it is less then (x, y) is a maximum point. One could guard against the possibility that (x, y) was a point of inflection by repeating the calculation with $-\Delta x$ replacing Δx. If it gave an opposite indication to that given by Δx then we are dealing with a point of inflection.

Problems

35.1 A manufacturer has 2000 kg of red yarn, 2000 kg of yellow yarn, and 1000 kg of blue yarn. With these yarns he can make three types of cloth – A, B, and C – with fractional components of the colours as follows:

	red	yellow	blue
A	0.25	0.5	0.25
B	0.5	0.25	0.25
C	0.25	0.25	0.5

He sells his cloth by weight for the following prices: A, £10.00/kg, B £12.00/kg, and C £15.00/kg. Unused yarn can be sold for £6.00/kg. What is the maximum income he can make, and how much of each cloth does he make?

35.2 Modify the program GRADMAX outlined in Appendix 25 to find the maximum, both in magnitude and position, of the following function:

$$Z = 3\cos\{2\pi(x + y + 0.3)\} + 2\sin\{2\pi(x + 3y - 0.1)\}.$$

This function is periodic in x and y with a repeat distance of 1 in each direction. Take as a starting point $(x, y) = (0.5, 0.5)$ and an initial refinement shift 0.1. A tolerance of 0.0001 is suitable for this problem.

35.3 We wish to make an enclosed cylindrical tin can that has a capacity of 1 litre with the minimum amount of sheet metal. What are the dimensions of the can? Solve the problem using the Lagrange multiplier approach.

36 Laplace transforms

In Chapter 17 we were introduced to the Fourier transform and found that it has a wide range of applicability in physics and other branches of science. Now we consider another kind of transform, related to the Fourier transform, which also has wide applicability. It is a powerful way of solving linear constant-coefficient differential equations, particularly when used with extensive standard tables of Laplace transforms that are widely available. They constitute a very useful tool for the solution of some kinds of problems in science and engineering – for example, in the design of control systems – but also more generally.

36.1 Defining the Laplace transform

The Laplace transform of a function $f(t)$ is given by

$$F(s) = \mathcal{L}\{f(t)\} = \int_0^\infty f(t)e^{-st}dt, \tag{36.1}$$

where the symbol $\mathcal{L}$ indicates the Laplace transform of the function that follows.

As an example we consider the Laplace transform of $f(t) = c$, where c is a constant. Then

$$F(s) = \mathcal{L}(c) = c \int_0^\infty e^{-st}dt = \frac{c}{s}. \tag{36.2}$$

Using the rules for integration described in Chapter 5 it is possible to find Laplace transforms for many functions $f(t)$ quite straightforwardly. However, because of their usefulness as a mathematical tool, tables of Laplace transforms are widely available.

Table 36.1 gives the Laplace transforms for a selection of common functions. By its use it is possible to find Laplace transforms for a number of composite functions not appearing directly in the table. Consider for example the function

$$f(t) = e^{-\alpha t}t \sin bt. \tag{36.3}$$

From the table we find

$$\mathcal{L}\{t \sin bt\} = \frac{2bs}{(s^2 + b^2)^2}. \tag{36.4}$$

Table 36.1 Laplace transforms for some standard functions

$f(t)$	$F(s)$	$f(t)$	$F(s)$
1	$1/s$	$\sinh bt$	$\dfrac{b}{s^2 - b^2}$
t^n	$\dfrac{n!}{s^{n+1}}$	$\cosh bt$	$\dfrac{s}{s^2 - b^2}$
$e^{\alpha t}$	$\dfrac{1}{s - \alpha}$	$t \sin bt$	$\dfrac{2bs}{(s^2 + b^2)^2}$
$\sin bt$	$\dfrac{b}{s^2 + b^2}$	$t \cos bt$	$\dfrac{s^2 - b^2}{(s^2 + b^2)^2}$
$\cos bt$	$\dfrac{s}{s^2 + b^2}$	Step function $u(t)$ $f(t) = 1,\ t \geq 0$ $f(t) = 0,\ t < 0$	$\dfrac{1}{s}$
$t^n e^{-\alpha t}$	$\dfrac{n!}{(s + \alpha)^{n+1}}$	$\delta(t)$	1
$* f(t)\, e^{-\alpha t}$	$F(s + \alpha)$	$\delta(t - d)$	e^{-sd}

We now note the entry in Table 36.1 at the bottom of the first column, marked with a $*$. This says that the Laplace transform of $e^{-\alpha t}f(t)$ is obtained from the Laplace transform of $f(t)$ by replacing s by $s + \alpha$. Hence

$$\mathcal{L}\left\{e^{-\alpha t} t \sin bt\right\} = \frac{2b(s + \alpha)}{\left\{(s + \alpha)^2 + b^2\right\}^2}. \tag{36.5}$$

Finally, we note that Laplace transforms satisfy the rules

$$\mathcal{L}\{f_1(t) + f_2(t)\} = \mathcal{L}\{f_1(t)\} + \mathcal{L}\{f_2(t)\} \tag{36.6a}$$

and

$$\mathcal{L}\{kf(t)\} = k\mathcal{L}\{f(t)\}. \tag{36.6b}$$

indicating that Laplace transforms are **linear transforms**, which is obvious from the definition (36.1).

It should be noted that Table 36.1 gives only a small selection of functions, sufficient for the preszent purpose, and that much more extensive tables are available.

Exercise 36.1
Find the Laplace transforms of (i) $\cos 4t$; (ii) $\cosh 4t$; (iii) t^3, (iv) $t^3 e^{-2t}$; (v) $e^{-t}\cosh t$ (vi) $\cos 2t + \sin 2t$.

36.2 Inverse Laplace transforms

Another aspect of Laplace transforms that has important applications is the ability to determine **inverse Laplace transforms**, which means to deduce what the original function was, given its Laplace transform. In mathematical form the inverse transform appears as

$$f(t) = \mathcal{L}^{-1}\{F(s)\} \tag{36.7}$$

and the problem is to find $f(t)$ given $F(s)$.

It is clear that Table 36.1 can provide the answers directly in simple cases. Given that $F(s) = 1/s$ we can easily deduce that $f(t) = 1$. However, to exploit fully the facility provided by a table of Laplace transforms it is necessary to use the following rules that express the linearity of inverse Laplace transforms:

$$\mathcal{L}^{-1}\{F_1(s) + F_2(s)\} = \mathcal{L}^{-1}\{F_1(s)\} + \mathcal{L}^{-1}\{F_2(s)\} \tag{36.8a}$$

and

$$\mathcal{L}^{-1}\{kF(s)\} = k\mathcal{L}^{-1}\{F(s)\}. \tag{36.8b}$$

 Exercise 36.2

Find the inverse Laplace transforms of (i) $\dfrac{4}{s}$; (ii) $\dfrac{1}{s-2}$; (iii) $\dfrac{s}{s+4}$;

(iv) $\dfrac{6}{s^2+1}$; (v) $\dfrac{1}{(s+2)^3}$; (vi) $\dfrac{8s}{(s^2+1)^2} + \dfrac{1}{(s+1)^3}$.

We now give a number of examples that will illustrate the various processes that can be used to find inverse Laplace transforms.

EXAMPLE 36.1

Find the inverse Laplace transform of

$$\frac{1}{(s+1)(s+2)}. \tag{36.9}$$

Using partial fractions (§5.4),

$$\frac{1}{(s+1)(s+2)} = \frac{1}{s+1} - \frac{1}{s+2}.$$

Hence

$$\mathcal{L}^{-1}\left(\frac{1}{(s+1)(s+2)}\right) = \mathcal{L}^{-1}\left(\frac{1}{s+1}\right) - \mathcal{L}^{-1}\left(\frac{1}{s+2}\right) = e^{-t} - e^{-2t}. \tag{36.10}$$

EXAMPLE 36.2

Find the inverse Laplace transform of

$$\frac{1}{s^2+4s+5}. \tag{36.11}$$

Here we can use the process of 'completing the square' (§5.3.1) to give

$$\mathcal{L}^{-1}\left(\frac{1}{s^2 + 4s + 5}\right) = \mathcal{L}^{-1}\left(\frac{1}{(s + 2)^2 + 1}\right).$$

We now consider the solution in two stages. First we note that

$$\mathcal{L}^{-1}\left(\frac{1}{s^2 + 1}\right) = \sin t$$

and then, from the starred entry in Table 36.1, we note that replacing s by $s + 2$ multiplies the original function by e^{-2t}. Hence

$$\mathcal{L}^{-1}\left(\frac{1}{s^2 + 4s + 5}\right) = e^{-2t} \sin t. \tag{36.12}$$

EXAMPLE 36.3

Find the inverse Laplace transform of

$$\frac{6 - s}{s^2 + 4s + 5}. \tag{36.13}$$

We write

$$\frac{6 - s}{s^2 + 4s + 5} = \frac{6 - s}{(s + 2)^2 + 1} = -\frac{s + 2}{(s + 2)^2 + 1} + \frac{8}{(s + 2)^2 + 1}.$$

Hence

$$\mathcal{L}^{-1}\left(\frac{6 - s}{s^2 + 4s + 5}\right) = \mathcal{L}^{-1}\left\{\frac{s + 2}{(s + 2)^2 + 1}\right\} + 8\mathcal{L}^{-1}\left\{\frac{1}{(s + 2)^2 + 1}\right\}$$

$$= e^{-2t} \cos t + 8e^{-2t} \sin t. \tag{36.14}$$

These examples illustrate the general principles of finding inverse Laplace transforms. The function to be transformed is manipulated to give components that correspond to entries in tables of Laplace transforms – and it must be stressed again that tables far more extensive than Table 36.1 are available. The processes by which the functions are manipulated are often similar to those used on functions to find integrals, as are described in Chapter 5.

Exercise 36.3

Find the inverse Laplace transform of $\dfrac{1}{(s + 2)(s + 3)}$.

36.3 Solving differential equations with Laplace transforms

Laplace transforms offer a powerful technique of solving linear constant-coefficient differential equations. In order to explain this process it is first necessary to describe how to find the Laplace transforms of derivatives of a function.

36.3.1 The Laplace transform of derivatives

We first consider how to find the Laplace transform of $f'(t)$, given by

$$\mathcal{L}\{f'(t)\} = \int_0^\infty f'(t)e^{-st}dt. \tag{36.15}$$

We use integration by parts (§5.5) to give

$$\mathcal{L}\{f'(t)\} = |f(t)e^{-st}|_0^\infty + s\int_0^\infty f(t)e^{-st}dt = sF(s) - f(0), \tag{36.16}$$

where $s > 0$ is assumed.

Now we find the Laplace transform of $f''(t)$ given by

$$\mathcal{L}\{f''(t)\} = \int_0^\infty f''(t)e^{-st}dt. \tag{36.17}$$

Again using integration by parts and (36.17),

$$\mathcal{L}\{f''(t)\} = |f'(t)e^{-st}|_0^\infty + s\mathcal{L}\{f'(t)\} = s^2F(s) - sf(0) - f'(0). \tag{36.18}$$

The pattern established by (36.16) and (36.18) leads to a general formula for the Laplace transform of the nth derivative $f^n(t)$, which is

$$\mathcal{L}\{f^n(t)\} = s^nF(s) - s^{n-1}f(0) - s^{n-2}f'(0) - \cdots - sf^{n-2}(0) - f^{n-1}(0). \tag{36.19}$$

Exercise 36.4

Find the Laplace transform of $f''(t) - 2f'(t) + f(t)$.

36.3.2 Solving a differential equation

The best way of illustrating this technique of solving differential equations is with actual examples. In what follows the function to be found is $x(t)$, the Laplace transform of which is $X(s)$.

EXAMPLE 36.4

Solve the differential equation

$$\frac{dx}{dt} = 4x - e^t \text{ with boundary condition } x(0) = 1.$$

Taking the Laplace transform of each side, using (36.16) and Table 36.1,

$$sX(s) - x(0) = 4X(s) - \frac{1}{s-1}$$

which comes to

$$X(s) = \frac{1}{s-4} - \frac{1}{(s-1)(s-4)} = \frac{1}{3(s-1)} + \frac{2}{3(s-4)}. \tag{36.20}$$

Now to find $x(t)$ we find the inverse Laplace transform of each side, giving

$$x(t) = \frac{1}{3}e^t + \frac{2}{3}e^{4t}. \qquad (36.21)$$

It will be noticed that the boundary condition is automatically included in the solution.

EXAMPLE 36.5

Solve the differential equation

$$\frac{d^2x}{dt^2} - 2\frac{dx}{dt} + 5x = \cos 2t \text{ with boundary conditions } x(0) = x'(0) = 0$$

Taking the Laplace transform of each side,

$$s^2 X(s) - sx(0) - x'(0) - 2sX(s) + 2x(0) + 5X(s) = \frac{s}{s^2 + 4}$$

or, putting in the boundary conditions

$$X(s) = \frac{s}{(s^2 + 4)(s^2 - 2s + 5)}. \qquad (36.22)$$

By the use of partial fractions we find

$$X(s) = \frac{1}{17}\left(\frac{s - 8}{s^2 + 4} - \frac{s - 10}{s^2 - 2s + 5}\right)$$

$$= \frac{1}{17}\left\{\frac{s}{s^2 + 4} - \frac{8}{s^2 + 4} - \frac{s - 1}{(s - 1)^2 + 4} + \frac{9}{(s - 1)^2 + 4}\right\}.$$

Taking the inverse Laplace transform of each side,

$$x(t) = \frac{1}{17}\left(\cos 2t - 4\sin 2t - e^t \cos 2t + \frac{9}{2}e^t \sin 2t\right), \qquad (36.23)$$

which is the required solution.

Laplace transforms provide a powerful and very elegant method of solution of linear constant-coefficient differential equations.

Exercise 36.5

Solve the differential equation

$$\frac{dx}{dt} = e^t\left(\cos 2t + \frac{1}{2}\sin 2t\right)$$

subject to the boundary condition $x(0) = 0$.

36.4 **Laplace transforms and transfer functions**

We have seen that Fourier transforms have important practical applications (e.g. Chapters 17 and 18) and the same is true for Laplace transforms. This should not be surprising, since there is a strong relationship between the two kinds of transform. Thus (17.30) is

$$X(\omega) = \int_{-\infty}^{\infty} x(t) \exp(-i\omega t) \, dt.$$

If it is accepted that the function $x(t)$ is zero for $t \leq 0$, which is true for many engineering applications, and if s replaces $i\omega$ then (17.29) becomes the Laplace transform equation (36.1).

The applications of Laplace transforms to signal processing, control theory, and communication theory are very extensive. Here we shall just consider some simple applications involving a concept referred to as the **transfer function** to illustrate the practical nature of Laplace transform theory. We shall be using the concept of a signal processor without defining exactly what it is in a physical sense. In general, if the input and output signals are electrical, then it would be a collection of circuit elements.

36.4.1 **A simple signal processor**

In Figure 36.1 we show in symbolic form the action of a simple signal processor.

The ingoing signal, which is zero for $t \leq 0$, is $u(t)$ and the outgoing signal is $y(t)$. It is a linear processing system, which means that if the ingoing signal is increased by a factor k but keeps the same form then the same will be true for the outgoing signal. We now represent the outgoing signal as the solution of the differential equation

$$ay'(t) + y(t) = Ku(t). \tag{36.24}$$

It was shown in §36.3 that the solution of linear differential equations with constant coefficients can be found through the use of the inverse Laplace transform and the inverse Laplace transforms are indicated as $U(s)$ and $Y(s)$ in Figure 36.1. Taking the Laplace transform of both sides of (36.24),

$$asY(s) - y(0) + Y(s) = KU(s). \tag{36.25}$$

Since $u(0) = 0$ then $y(0) = 0$ (no signal in, no signal out) so rearranging (36.25)

$$Y(s) = \frac{K}{1 + as} U(s). \tag{36.26}$$

Figure 36.1 A simple signal-processing procedure.

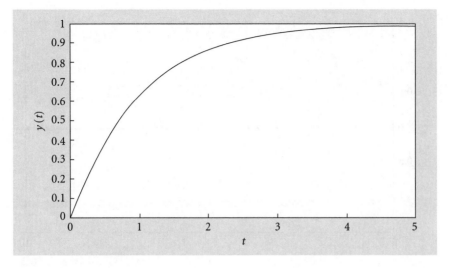

Figure 36.2 The output signal for $u(t)=1$.

The transfer function is defined as

$$G(s) = \frac{Y(s)}{U(s)} = \frac{K}{1+as}. \tag{36.27}$$

It has no particular role to play in the present analysis with a single signal-processor, but its usefulness will be demonstrated later.

We now consider a particular form of input, a step function defined by

$$u(t) = 0 \qquad t \le 0$$
$$u(t) = 1 \qquad t > 0. \tag{36.28}$$

With this form of input $U(s) = 1/s$ so that, from (36.26),

$$Y(s) = \frac{K}{s(1+as)} = \frac{K}{s} - \frac{K}{1/a+s}. \tag{36.29a}$$

From Table 36.1 this gives

$$y(t) = K - Ke^{-t/a} = K(1 - e^{-t/a}). \tag{36.29b}$$

The response as a function of time is shown in Figure 36.2 for $K = a = 1$. There is a transient response where the output steadily climbs to $y(t) = 1$ after which the output is steady, so that the output is just a scaled version of the input. The steady-state output is proportional to K and if a is made larger the transient time is greater and vice versa.

Another interesting case is where the input is a pure sine wave, for example,

$$u(t) = \sin 2t$$

giving

$$U(s) = \frac{2}{s^2 + 4}$$

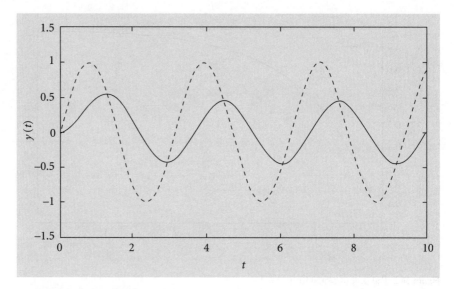

Figure 36.3 The solution (full line) given by (36.33) with $a = K = 1$ together with the input signal $\sin 2t$ (dashed line).

and hence

$$Y(s) = \frac{2K}{(s^2 + 4)(1 + as)} = \frac{2K}{4a^2 + 1}\left(\frac{1}{s^2 + 4} - \frac{as}{s^2 + 4} + \frac{a^2}{1 + as}\right). \qquad (36.30)$$

From Table 36.1 we find

$$y(t) = \frac{K}{4a^2 + 1}\left(\sin 2t - 2a\cos 2t + 2ae^{-t/a}\right). \qquad (36.31)$$

The last term in the bracket in the right-hand side of (36.31) is a transient term that will disappear after some time. The other terms can be combined as follows:

$$\sin 2t - 2a\cos 2t = \sqrt{4a^2 + 1}\left(\frac{1}{\sqrt{4a^2 + 1}}\sin 2t - \frac{2a}{\sqrt{4a^2 + 1}}\cos 2t\right). \qquad (36.32)$$

Writing

$$\frac{1}{\sqrt{4a^2 + 1}} = \cos\delta \text{ and } \frac{2a}{\sqrt{4a^2 + 1}} = \sin\delta$$

(note that $\sin^2\delta + \cos^2\delta = 1$) we now have

$$y(t) = \frac{K}{\sqrt{4a^2 + 1}}\sin(2t - \delta) + \frac{2aK}{4a^2 + 1}e^{-t/a}. \qquad (36.33)$$

The form of the input and output is shown in Figure 36.3 for $a = K = 1$. The phase shift (-1.107 radians) is clearly seen as is the effect of the early transient. The input is of the form $\sin\omega t$ with $\omega = 2$ and if another value of ω had been chosen then both the amplitude and the phase shift would be different. If the input had been a mixture of components with different frequencies then in the output the relative amplitudes and the phase shift would all be different from what they were in the input and the waveforms of input and output would be different. The way that passing a signal through a piece of equipment changes its form is, naturally, of critical interest to those engaged in signal processing.

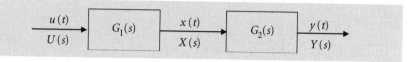

Figure 36.4 Effect or two signal processors in series.

36.4.2 Signal processors in series

We now consider what the effect is of using the output of one signal processor as the input of another, illustrated in Figure 36.4.

The first signal processor modifies the incoming signal, $u(t)$, to give an outgoing signal, $x(t)$ and the form of $x(t)$ may be found by solving a differential equation of the form (36.24). Then $x(t)$ is the ingoing signal for the second processor and plays the role of $u(t)$ in (36.24) to find the final output, $y(t)$. The two signal processors may be different in the way they act, requiring different differential equations of the form (36.24) and if $x(t)$ is fairly complicated then finding the effect of the two signal processors in series might be difficult. This is where the idea of the transfer function comes into its own. With the Laplace transforms as indicated in Figure 36.4 we have

$$X(s) = G_1(s)U(s) \tag{36.34a}$$

and

$$Y(s) = G_2(s)X(s) = G_2(s)G_1(s)U(s) \tag{36.34b}$$

so that the overall transfer function for the pair of processors in series, G_s, is given by the product of the individual transfer functions, i.e.

$$G_s(s) = G_1(s)G_2(s). \tag{36.35}$$

Let us see how this works out in practice with the action of the first signal processor described by

$$x'(t) + x(t) = u(t) \tag{36.36a}$$

and the second by

$$2y'(t) + y(t) = x(t). \tag{36.36b}$$

With the usual assumption that $u(0) = x(0) = y(0) = 0$, then from (36.27)

$$G_1(s) = \frac{1}{1+s} \quad \text{and} \quad G_2(s) = \frac{1}{1+2s}$$

so that

$$G_s(s) = G_1(s)G_2(s) + \frac{1}{(1+s)(1+2s)} = \frac{2}{1+2s} - \frac{1}{1+s}. \tag{36.37}$$

We now take an input signal, $u(t) = \sin 2t$ so that $U(s) = 2/(s^2 + 4)$. This gives

$$Y(s) = G_s(s)U(s) = \frac{4}{(1+2s)(s^2+4)} - \frac{2}{(1+s)(s^2+4)}. \tag{36.38}$$

By the use of partial fractions we find

$$\frac{4}{(1+2s)(s^2+4)} = \frac{4}{17}\left(\frac{4}{1+2s} + \frac{1-2s}{s^2+4}\right) \tag{36.39a}$$

and

$$\frac{2}{(1+s)(s^2+4)} = \frac{2}{5}\left(\frac{1}{1+s} + \frac{1-s}{s^2+4}\right). \tag{36.39b}$$

Substituting from (36.39) into (36.38), rationalizing and putting in a form suitable for comparing with Table 36.1 we find

$$Y(s) = \frac{8}{17}\frac{1}{\frac{1}{2}+s} - \frac{2}{5}\frac{1}{1+s} - \frac{7}{85}\frac{2}{s^2+4} - \frac{6}{85}\frac{s}{s^2+4}$$

giving

$$y(t) = \frac{8}{17}e^{-t/2} - \frac{2}{5}e^{-t} - \frac{7}{85}\sin 2t - \frac{6}{85}\cos 2t. \tag{36.40}$$

The form of this solution is shown in Figure 36.5.

The output finally settles down to a sinusoidal wave shifted in phase from the input signal. Once again, for a simple sinusoidal input the amplitude and phase of the output would be frequency dependent so that, if the input contained components of more than one frequency, the forms of the input and output signals would be different.

This example shows the usefulness of the transfer function as a means of dealing with processors in series where the overall transfer function is the product of the individual transfer functions for each of the processors. We now consider what happens if the processors are in parallel.

 Exercise 36.6
Two signal processors, whose actions are described by (36.36) are connected in series. Find the output for a step-function input signal $u(t) = 1$.

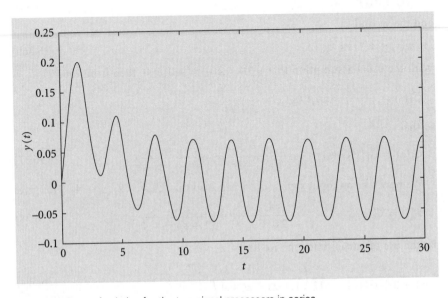

Figure 36.5 Form of solution for the two signal processors in series.

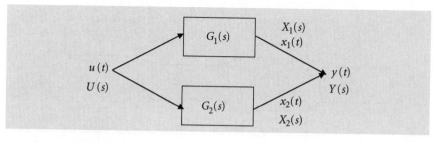

Figure 36.6 Two signal processors in parallel.

36.4.3 **Signal processors in parallel**

We consider an arrangement of processors as in Figure 36.6 where the same signal, $u(t)$ is fed into two processors in parallel and then the two individual outputs, $x_1(t)$ and $x_2(t)$ are combined to give a composite final output $y(t)$.

This is far simpler to deal with than the case of processors in series. By the transfer-function approach

$$X_1(s) = G_1(s)U(s) \tag{36.41a}$$

and

$$X_2(s) = G_2(s)U(s) \tag{36.41b}$$

so that

$$Y(s) = X_1(s) + X_2(s) = \{G_1(s) + G_2(s)\}U(s).$$

The components are similar to those in (36.40) but with different coefficients.

Thus for processors in parallel the overall transfer function, $G_p(s)$, is just the sum of the individual transfer functions or

$$G_p(s) = G_1(s) + G_2(s). \tag{36.42}$$

EXAMPLE 36.6

For the processors considered in §36.4.1 with input signal $u(t) = \sin 2t$ we find

$$Y(s) = \frac{1}{(1+s)(s^2+4)} + \frac{1}{(1+2s)(s^2+4)}.$$

By the use of partial fractions, and rationalization of the result we find

$$Y(s) = \frac{2}{5}\frac{1}{1+s} + \frac{4}{17}\frac{1}{\frac{1}{2}+s} + \frac{22}{85}\frac{2}{s^2+4} - \frac{54}{85}\frac{s}{s^2+4}$$

which gives

$$y(t) = \frac{2}{5}e^{-t} + \frac{4}{17}e^{-t/2} + \frac{22}{85}\sin 2t - \frac{54}{85}\cos 2t. \tag{36.42}$$

For systems consisting of many signal-processing units connected together in various ways the simplest way to determine their combined effect is to find the transfer function for the whole system using (36.35) for processors in series and

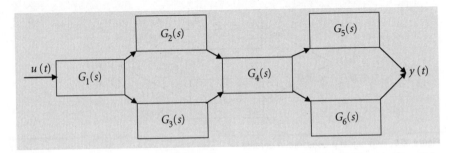

Figure 36.7 System of six signal processors.

(36.42) for processors in parallel. A composite system with six processors is shown in Figure 36.7. By combining results from (36.35) and (36.42) the overall transfer function of the system is found as

$$G(s) = G_1(s)\{G_2(s) + G_3(s)\}G_4(s)\{G_5(s) + G_6(s)\}. \tag{36.43}$$

Exercise 36.7

Two signal processors, whose actions are described by (36.36) are connected in parallel. Find the output for a step-function input signal $u(t) = 1$.

Problems

36.1 Find the Laplace transforms of (i) $\cos(a + t)\,e^{-t}$; (ii) $\cosh t \sinh t$; (iii) $\cos^2 t$.

36.2 Find the inverse Laplace transforms of

(i) $\dfrac{1}{s(s+1)(s+2)}$; (ii) $\dfrac{1}{s^4 - b^4}$; (iii) $\dfrac{s^2}{s^4 - b^4}$.

36.3 Solve the differential equation

$$x''(t) - 2x'(t) + x(t) = \cos t - \sin t, \quad x(0) = 1, \quad x'(0) = 1.$$

36.4 The output of a signal processor is described by the solution of $3y'(t) + y(t) = u(t)$. What is the transfer function for this processor? If the input is of the form $u(t) = e^{-t}\sin t$, then find the form of the output.

36.5 Two signal processors, the outputs of which are described by the solutions of $3y'(t) + y(t) = u(t)$ and $y'(t) + 3y(t) = u(t)$ are arranged in series. Find the output for a step function input $u(t) = 1$.

Networks

Many important systems have the characteristic that something flows on pathways between particular locations – gas through pipelines or messages through telephone wires. Such a system is usually referred to as a network – the rail network, for example – and a branch of mathematics called network theory, or sometimes graph theory, has been developed to handle network problems. Sometimes these problems crop up in a scientific context, but here you will be introduced to network theory through examples that occur in everyday life.

37.1 Graphs and networks

In mathematics a **graph** is a set of points, called **nodes** connected by links called **edges**. Figure 37.1 shows a simple graph in which the nodes are numbered and the edges may be described by the nodes they connect.

There are many physical systems that have the characteristics of a graph. A road network is a set of nodes (cities, towns, and villages) connected by edges (roads); a telephone network is a set of nodes (telephones) connected by edges (wires) – although in this case there is a complicated hierarchical structure involving local, regional, and national telephone exchanges. In these problems of practical interest, where traffic of some kind – vehicles or messages, for example – flow along the edges, the system is normally called a **network**. Networks are of great importance in everyday life, and sometimes impinge on the work of a scientist. Getting the best out of a network in terms of a desired outcome presents interesting challenges and we shall explore some of these here. They are just examples of many applications of network theory.

37.2 Types of network

In the graph shown in Figure 37.1 the nodes are just numbered locations connected by edges of non-specified nature or properties. Some pairs of nodes are connected and some are not, and there is no particular overall description that one could give

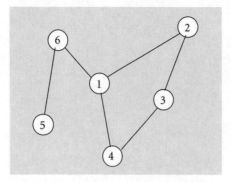

Figure 37.1 Simple graph.

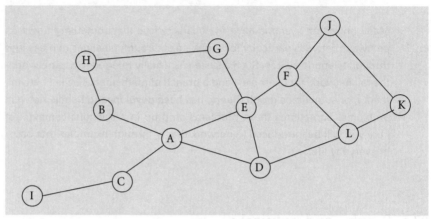

Figure 37.2 Road system connecting 12 population centres.

concerning the nature of the connections. However, there are networks with specific structures and we now describe these.

37.2.1 **A general network and directionality**

Consider a small country with 12 major centres of population connected by a system of major roads. This is illustrated in Figure 37.2 with the population centres lettered from A to L.

All the roads may be traversed in both directions, which makes it an **undirected network**. This is in contrast to a **directed network** in which movement is restricted to only one direction in each edge. An example of this would be an oil pipeline system in which the oil flows from the oilfield to the oil terminal and not the other way round. There are many examples of **mixed networks** in which some edges allow passage in both directions while others only allow passage in one direction – for example a town road system in which some streets are designated 'one-way'.

37.2.2 **Trees**

For the road network illustrated in Figure 37.2 it is required to indicate the 'best' routes, i.e. the quickest, between the capital city A and other population centres. This leads to the network shown in Figure 37.3.

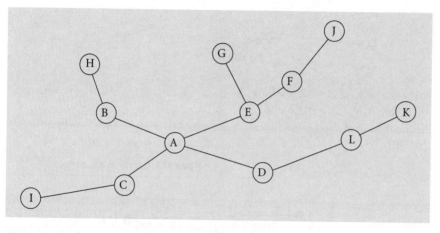

Figure 37.3 'Best' routes from A to all other population centres.

This structure is described as a **tree** and it has the characteristic that there is a unique pathway from any node to any other node. Thus to go from G to K the edges traversed must be (GE) (EA) (AD) (DL) (LK). The description 'tree' comes from the structure of a tree to the trunk of which branches are connected, to the branches of which twigs are connected, and to the twigs of which leaves are connected. The path from any leaf to any other leaf is unique. However, a network of the 'best' routes from some other centre, say L, would be different from Figure 37.3 but still be in the form of a tree.

 Exercise 37.1
Draw a possible 'best-route' system from centre L to all other centres and confirm that it has the form of a tree.

37.2.3 **Complete networks**

A **complete network** is a network in which all pairs of nodes are connected, as in Figure 37.4. An example of a complete network is the airline connections between a number of major cities – for example, New York, Washington, Boston, Los Angeles, San Francisco, and Chicago – all of which have direct links to all the others.

 Exercise 37.2
How many edges are there in a complete network containing 7 nodes?

37.2.4 **Bipartite networks**

We now consider a case where there are two sets of nodes – set A and set B. The nodes within set A are not connected in any way, nor are those in set B. However, there are various connections between individual nodes in set A and those of set B. We show this in Figure 37.5. Such a network is described as **bipartite**; examples would be transatlantic connections between various coastal cities in the Americas

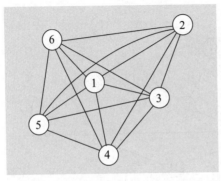

Figure 37.4 Complete network.

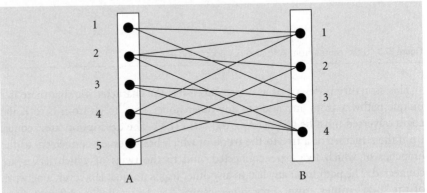

Figure 37.5 Bipartite network.

with ports in Europe or electrical connections between the terminals of one terminal box to the terminals of another.

A useful way of representing the connections of a bipartite network is by means of a matrix, M, where element $m_{i,j} = 1$ if A_i is connected to B_j but is zero otherwise. Thus the system in Figure 37.5 can be represented as

$$M_{AB} = \begin{bmatrix} 1 & 0 & 1 & 0 \\ 1 & 1 & 0 & 1 \\ 0 & 0 & 1 & 1 \\ 1 & 0 & 0 & 1 \\ 0 & 1 & 1 & 0 \end{bmatrix}. \tag{37.1}$$

An extension of the bipartite pattern is if there were three junction boxes, some terminals of box A being joined to some in box B and some of those of box B being joined to some in box C. We show this in Figure 37.6.

A matrix representation of the links between terminal boxes B and C is M_{BC}, but what we are going to examine is the matrix product $M_{AB}M_{BC}$:

$$M_{AB}M_{BC} = \begin{bmatrix} 1 & 0 & 1 & 0 \\ 1 & 1 & 0 & 1 \\ 0 & 0 & 1 & 1 \\ 1 & 0 & 0 & 1 \\ 0 & 1 & 1 & 0 \end{bmatrix} \begin{bmatrix} 0 & 0 & 1 & 1 \\ 0 & 1 & 0 & 1 \\ 1 & 1 & 1 & 0 \\ 1 & 0 & 0 & 1 \end{bmatrix} = \begin{bmatrix} 1 & 1 & 2 & 1 \\ 1 & 1 & 1 & 3 \\ 2 & 1 & 1 & 1 \\ 1 & 0 & 1 & 2 \\ 1 & 2 & 1 & 1 \end{bmatrix}. \tag{37.2}$$

This product matrix has elements other than 0 and 1, and what these elements represent is the number of *different* paths from a terminal of A to a terminal of C

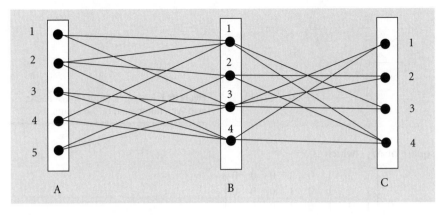

Figure 37.6 Three linked terminal boxes.

via a terminal of B. Representing the product matrix as M_{AC}, then since $m_{1,2} = 1$ this means that there is just one path from A_1 to C_2, the path $A_1B_3C_2$. Similarly, since $m_{4,2} = 0$ there is no path linking A_4 to C_2 and an examination of Figure 37.6 shows this to be true. Now we look for the three links indicated by $m_{2,4} = 3$. Inspection shows that these are $A_2B_1C_4$, $A_2B_2C_4$, and $A_2B_4C_4$.

We see here that matrix algebra can be used to interpret the properties of networks. If, instead of 3 terminal boxes there were 10 and instead of 4 or 5 terminals per box there were 100, then finding the number of links between terminals on the first and last terminal box is not something that could easily be done by inspection but could be done systematically via a matrix approach.

Exercise 37.3
Write down the matrix corresponding to connections from terminal box C to terminal box A in Figure 37.6.

37.2.5 Links in a molecule

Consider the simple molecule of acetic acid, the atomic linkages of which are shown in Figure 37.7.

It is clear that this conventional representation of a molecule has the characteristics of a network and, not taking account of the different atoms involved the structure can be represented by a matrix of linkages:

$$M_{AA} = \begin{bmatrix} 0 & 0 & 0 & 1 & 0 & 0 & 0 & 0 \\ 0 & 0 & 0 & 1 & 0 & 0 & 0 & 0 \\ 0 & 0 & 0 & 1 & 0 & 0 & 0 & 0 \\ 1 & 1 & 1 & 0 & 1 & 0 & 0 & 0 \\ 0 & 0 & 0 & 1 & 0 & 1 & 1 & 0 \\ 0 & 0 & 0 & 0 & 1 & 0 & 0 & 0 \\ 0 & 0 & 0 & 0 & 1 & 0 & 0 & 1 \\ 0 & 0 & 0 & 0 & 0 & 0 & 1 & 0 \end{bmatrix}. \tag{37.3}$$

There is a 1 wherever a bond between atoms exists and 0 otherwise. By the nature of the problem the matrix has to be symmetric – i.e. $m_{i,j} = m_{j,i}$. We now look at the

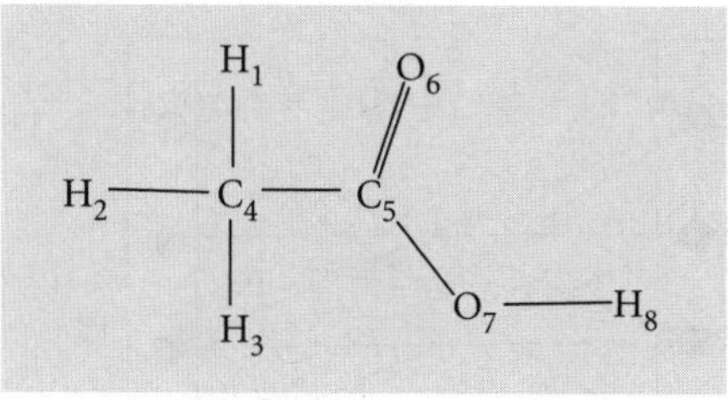

Figure 37.7 Molecular structure of acetic acid with numbering of the atoms.

square of M_{AA} which is

$$M_{AA}^2 = \begin{bmatrix} 1 & 1 & 1 & 0 & 1 & 0 & 0 & 0 \\ 1 & 1 & 1 & 0 & 1 & 0 & 0 & 0 \\ 1 & 1 & 1 & 0 & 1 & 0 & 0 & 0 \\ 0 & 0 & 0 & 4 & 0 & 1 & 1 & 0 \\ 1 & 1 & 1 & 0 & 3 & 0 & 0 & 1 \\ 0 & 0 & 0 & 1 & 0 & 1 & 1 & 0 \\ 0 & 0 & 0 & 1 & 0 & 1 & 2 & 0 \\ 0 & 0 & 0 & 0 & 1 & 0 & 0 & 1 \end{bmatrix}. \tag{37.4}$$

The interpretation of (37.4) is that it represents the number of two-bond paths between pairs of atoms. Representing the elements of M_{AA}^2 as $_2m_{i,j}$ then, since $_2m_{1,2} = 1$ there must be one two-bond path between H_1 and H_2 – which is $H_1C_4H_2$. Similarly there is no *two-bond* path between the neighbouring atoms H_1 and C_4. The diagonal elements represent the number of two-bond paths between an atom and itself, which is another way of describing the number of its neighbours. Thus C_4 is linked to itself by the four two-bond paths $C_4H_1C_4$, $C_4H_2C_4$, $C_4H_3C_4$, and $C_4C_5C_4$.

Although the interpretation of the linkages in acetic acid is amenable to visual inspection, there are other, much more complicated, systems where we wish to know how many one-removed or two-removed neighbours there are, and a matrix representation of the linkage system provides a systematic approach to solving problems of this kind.

> **?** **Exercise 37.4**
> Express the pattern of edges connecting the apices of a tetrahedron in matrix form. Square this matrix and interpret the elements of the squared matrix.

37.3 Finding cheapest paths

In planning a journey, say to deliver goods from one point to another, there is usually some criterion to be satisfied. It may be that the quickest route is required and this may involve paying tolls for both travelling along the fastest roads or crossing bridges. Another criterion may be cost. Travelling on minor roads may not involve tolls but the journey may be longer, costing more both in fuel, salaries, and wear-and-tear on the vehicles. Here we shall consider finding the cheapest path and

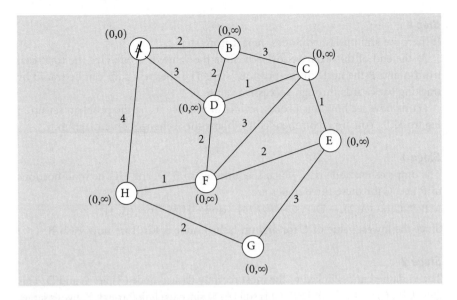

Figure 37.8 Initial diagram for the cheapest-route algorithm.

to each edge of the road network there is allocated a cost that takes all elements of expenditure into account. Such a network of roads is shown in Figure 37.8.

The problem we shall solve is that of finding the cheapest path from node A to all other nodes. It is a task best done with pencil and paper for small systems and a computer program for larger systems. The progress of the algorithm involves several steps at each of which a diagram similar to Figure 37.8 is an outcome, with minor changes from one diagram to the next. What we do here is first to describe the algorithm and then to present a concise version of the changes at each step and invite you to write those changes on a reproduction of Figure 37.8.

Step 1

Draw the network with nodes indicated by letters and the edges that link nodes marked by the cost of traversing that edge. The objective in this case is to find the cheapest routes from A to all other nodes. Each node is marked by a pair of quantities (P, C) where P eventually indicates the previous node on the cheapest path and C the cost up to that point. Initially mark point A as (0, 0) and all other points as (0, ∞). Then put a slash through point A to indicate that its (P, C) is finalized. This is the initial state indicated in Figure 38.8.

Step 2

Find the unslashed nodes (i) directly connected to each slashed node (j). For each of these unslashed nodes find $C'_i = C_j + c_{i,j}$ where $c_{i,j}$ is the cost of the edge linking node i to node j. Where there are two or more slashed nodes linked to an unslashed node then accept the lower or lowest value of C'_i. Change C_i to C'_i and P to the j that gave C'_i.

Step 3

Now slash the node with the lowest value of C_i.

Step 4

If there are still unslashed nodes, repeat from step 2.

At the end of this exercise for each node the value of C will give the total cost from A and P the node in the previous step. The cheapest path can be traced by tracking backwards through successive values of P.

Let us now see how this process works in practice using the problem set up in Figure 37.8. You are recommended to follow these changes on a diagram.

Stage 1

The only slashed node is A, which is connected to B, D, and H. The modifications of P and C for these three nodes are

$$B(P, C) = (A, 2), \quad D(P, C) = (A, 3) \quad \text{and} \quad H(P, C) = (A, 4).$$

Since the lowest value of C for any unslashed node is for B we now slash B.

Stage 2

The unslashed nodes linked to the most recently slashed node (B) are C and D. This gives $C(P, C) = (B, 5)$ but $D(P, C)$ is still $(A, 3)$ since the link through B gives a larger value of C. Since the lowest value of C for any unslashed node is for D, we now slash D.

Stage 3

The unslashed nodes linked to the most recently slashed node (D) are C and F. We now make $F(P, C) = (D, 5)$ and we change $D(P, C)$ to $(D, 4)$ since this value of C is less than in the previous $(B, 5)$. Two of the Cs for unslashed nodes, those of H and C, are both equal to 4 and since these are equal lowest we slash both these nodes.

Stage 4

The unslashed nodes linked to the most recently slashed nodes are: E linked to C, F linked to H and C and G linked to H. We find $E(P, C) = (C, 5)$ and $G(P, C) = (H, 6)$. F is previously labelled $(D, 5)$ but its new linkages are $(H, 5)$ and $(C, 7)$. We leave it unchanged since the paths via C and H are of equal cheapness and we must choose one or the other. Now we slash F and C, for each of which $C = 5$.

Stage 5

The slashed node connected to the most recently slashed node is G, linked to E. We already have $G(P, C) = (H, 6)$ and since the link through E is more expensive we leave it unchanged. The last node, G, is slashed and the process is complete.

Examination of the final diagram gives the following for the cheapest routes from A:

To B: Cost 2 by direct edge

To C: Cost 4 via D

To D: Cost 3 by direct edge

To E: Cost 5 via D and C

To F: Cost 5 via D

To G: Cost 6 via H

To H: Cost 4 by direct edge.

For very complex systems this hands-on approach is not very practicable and a computer program NETWORK is outlined in Appendix 26 that enables solutions for large systems to be found.

37.4 **Critical path analysis**

When planning a project with various stages a certain logical ordering of activities must be taken into account. For example, in constructing a conventional brick building one cannot construct the roof until the walls have been built – although that rule may not apply when the building is based on a steel framework. It is also necessary to avoid delays due to the non-availability of components. If roof tiles are available within one day of ordering then their supply will not be a critical factor. On the other hand if it takes 6 weeks from ordering to supply then they must be ordered in sufficient time for them to be on site when required.

Let us consider what must occur if a decision is made to build a small factory. The following processes, not in any particular order, must be gone through:

A   Planning
B   Ordering basic building materials (and delivery)
C   Pouring foundations
D   Building walls
E   Constructing roof
F   Installing services – electricity and water
G   Plastering walls
H   Installing interior furniture and fittings
I   Painting
J   Ordering machinery (and delivery)
K   Installing machinery
L   External works.

There are some obvious sequences that must occur – walls cannot be plastered before they are built – but others are not so obvious. The machinery may be on a long delivery date and these may need to be ordered even as the planning is taking place.

We now construct a flow diagram, Figure 37.9, in which nodes represent instants of time and directed edges represent the activity that has occurred between those instants. The time required for each activity, in days, is marked in parentheses. After the walls are built the installation of services and the building of the roof can both proceed and the dummy edge is put in to give continuity of flow.

In Figure 37.9 the nodes are numbered in such a way that the higher number is always at the arrow-point end of the edge, which means that as time passes so the numbers increase. For this simple project diagram this can be done by inspection but, for more complex diagrams, algorithms exist to give the numbering system.

We now wish to know how long the project is going to take and to do this we need to determine the **critical path**, the path following the edges going from the

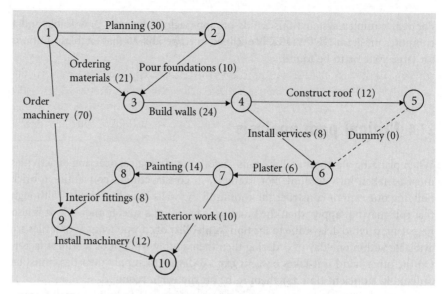

Figure 37.9 Project flow diagram representing the construction of a factory.

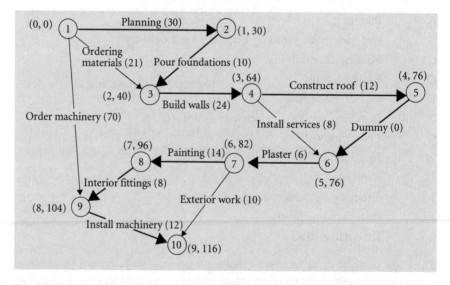

Figure 37.10 The values of (P, C) are shown at each node together with the critical path.

start, point 1, to the end, point 10, that has the longest duration. To do this we use the following algorithm which eventually gives each node a label (P, C) where P is the number of the previous node along the critical path and C is the duration, in days, to reach that point.

Step 1: Label the start point (0, 0).

Step n with n from 2 to 10: For all the edges incident on point n find $C_i + c(i, n)$ where i is the node at the previous end of the edge and c(i, n) the duration of the edge. Make C_n equal to the largest sum and $P_n = i$, the previous node that gave the largest sum.

At the end of this exercise the value of C_{10} will be the required time for the project and the critical path will be found by tracing backwards the values of P.

Figure 37.10 shows the critical path as a thick line. It will be seen that there are some non-critical activities; the delivery of the machinery could be up to 34 days late before it affects the timetable and, similarly, installing services and doing the exterior work are non-critical.

Of course the planned timings of activities can never be guaranteed and can be affected by bad weather, illness of critical workmen, delays in delivery, or just misjudgement by the planners. A technique called PERT (Programme Evaluation and Review Technique) has been designed to try to assess the uncertainties of timing in carrying out the project. For each activity there is designated an expected time, t_e, a best time if things went better than expected, t_b, and a worst time, t_w, allowing for the worst possible circumstances of weather or other delaying problems. Putting every activity at its best time gives a project duration, T_b, and, similarly T_w and T_e are the project durations if all activities are at their worst times or expected times respectively.

The overall expected time for the project is then assessed as

$$T_P = \frac{T_w + T_b + 4T_e}{6}.$$
(37.5)

Problems

37.1 Express the pattern of edges connecting the apices of a cube in matrix form. Square this matrix and interpret the elements of the squared matrix.

37.2 A road system between numbered population centres is shown below, with the costs of traversing each road. By using the program NETWORK find the

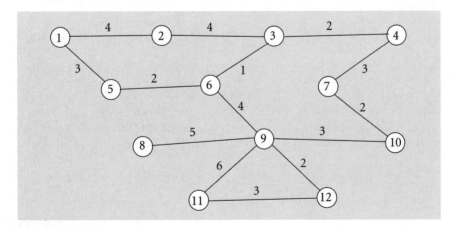

Figure P37.1 A road system with costs.

cheapest routes from (i) centre 1, (ii) centre 7, and (iii) centre 12 to all other centres.

37.3 For the project described in Figure 37.9 the best, expected and worst times, in days, for each activity are as follows:

	T_b	T_e	T_w
Planning	28	30	32
Pouring foundations	8	10	15
Constructing roof	10	12	20
Plastering walls	5	6	8
Painting	12	14	18
Installing machinery	10	12	15
Ordering materials	18	21	25
Building walls	20	24	30
Installing services	7	8	10
Interior fittings	7	8	10
Ordering machinery	60	70	100
Exterior work	8	10	14

Use the PERT technique to find the best estimate of time for the completion of the project.

Simulation with particles

Many physical systems can be simulated by following the behaviour of collections of particles. Sometimes there is a one-to-one correspondence between simulation particles and particles in the actual physical system. More often the number of actual particles is so large that, for computational purposes, they must be represented by a much smaller number of simulation particles that a computer can handle. In this chapter you will learn about many simulations using particles – planetary systems, clusters of stars, the motion of electrons in an electron-microscope lens, molecules in a liquid, and charged particles in a plasma.

38.1 Types of problem

In Chapter 19 we introduced the idea that numerical methods of solving ordinary differential equations can be used to deal with problems involving the motions of individual particles. In the case of the driven oscillator there was only one particle involved. However, by using a set of coupled differential equations, as described in (19.10) and (19.11), that can be extended to any number of bodies, we can solve many-body problems such as, for example, the motions of a system of interacting planets orbiting a star.

These are representative of a class of problems where the number of bodies is small and there is a one-to-one correlation between the bodies in the system and those in the simulation. Examples of few-body problems which fall into this category are:

- individual electron orbits in magnetic and/or electric fields, e.g. in the lens system of an electron microscope
- planetary motion with either a single planet or several planets
- the evolution of small stellar clusters.

In particle-interaction problems it is often necessary to consider interactions between pairs of bodies. Assuming that interactions are symmetrical, so that the force on particle i due to particle j, $\mathbf{F}_{i,j}$ is equal to $-\mathbf{F}_{j,i}$ and that particles do not exert forces on themselves then the total number of pair interactions is $\frac{1}{2}N(N-1)$, where N is the number of bodies. Thus for a cluster of 100 stars there are 4950 pair interactions, which can easily be handled by computers of modest power. This astronomical example is clearly an easy-to-handle few-body system, but astronomy also offers much larger systems, for example, globular stellar clusters with 10^5–10^6 stars. Here the number of pair-interactions is in the range 5×10^9 to 5×10^{11} which is clearly beyond the capacity of most readily available computers on a reasonable timescale. There are systems which are larger still in terms of the number of particles they contain, such as:

- liquids containing approximately 10^{16} molecules in $1\,\mu g$
- galaxies containing 10^{11} stars
- plasmas – intimate mixtures of positive ions, electrons, and neutral atoms – which are the predominant state of matter in the universe
- electrons in solids – of interest in semiconductor device simulation, for example.

Pair-interactions cannot be considered in systems of this size. The study of such systems involves approximations which, while making the problem manageable, still give results of sufficient precision to be useful.

38.2 Binary systems

The simplest many-body problem is that of a star binary system. With finite mass of both bodies the two stars are both in orbit around their stationary centre of mass. The situation is illustrated in Figure 38.1a where the stars, S_1 and S_2, with masses M_1 and M_2 are separated by a distance r. The centre of mass is at O, so that

$$OS_1 = rM_2/(M_1 + M_2) \tag{38.1a}$$

and

$$OS_2 = rM_1/(M_1 + M_2). \tag{38.1b}$$

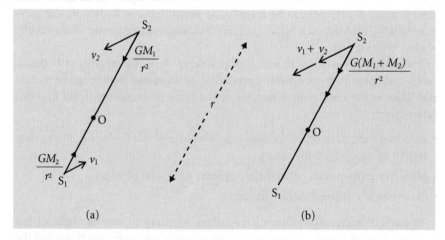

Figure 38.1 (a) Two bodies orbiting around their centre of mass. (b) Applying an acceleration that brings S_1 to rest.

The accelerations, shown in Figure 38.1a, have magnitudes GM_2/r^2 and GM_1/r^2 respectively. However, from the representational point of view it is an advantage to transform the problem so that one of the stars is at rest at the origin, as described in §10.7. This can be done by applying a uniform acceleration to the whole system which brings one of the stars to rest and then taking its position as the origin. This is shown in Figure 38.1b where S_1 is brought to rest. Bringing S_1 to rest requires that the acceleration of S_2 now has magnitude $G(M_1 + M_2)/r^2$. For a two-body system this is equivalent to using a **reduced mass frame** where the mass $M_1 M_2/(M_1 + M_2)$ is subjected to a force with magnitude $GM_1 M_2/r^2$ corresponding to

$$\text{acceleration} = \frac{\text{force}}{\text{reduced mass}} = \frac{GM_1 M_2}{r^2} \div \frac{M_1 M_2}{M_1 + M_2} = \frac{G(M_1 + M_2)}{r^2}.$$

The concept of applying a uniform acceleration to bring one body to rest is also applicable in many-body systems.

One way of testing the accuracy of the computational process is to check on the conservation of energy, momentum, and angular momentum. When the origin is shifted to one of the bodies the conservation of momentum cannot be checked because the assumption has been built in that the centre of mass is given by equations (38.1). If the velocity of S_2 relative to S_1 is $\mathbf{v}_{12}$, then the velocities of S_1 and S_2 relative to the centre of mass are

$$\mathbf{v}_1 = -\frac{M_2}{M_1 + M_2} \mathbf{v}_{12} \tag{38.2a}$$

and

$$\mathbf{v}_2 = \frac{M_1}{M_1 + M_2} \mathbf{v}_{12}. \tag{38.2b}$$

This implies that the momentum relative to the centre of mass, $M_1 \mathbf{v}_1 + M_2 \mathbf{v}_2$, is always zero.

The total energy of the system is given by

$$E = \frac{1}{2} M_1 v_1^2 + \frac{1}{2} M_2 v_2^2 - \frac{GM_1 M_2}{r} \tag{38.3}$$

or, from (38.2), this can be expressed as

$$E = \frac{1}{2} \frac{M_1 M_2}{M_1 + M_2} v_{12}^2 - \frac{GM_1 M_2}{r} \tag{38.4}$$

which quantity should be conserved during the progress of the computation.

The motion of a binary system is two-dimensional and defining it in the x-y plane the angular momentum vector is in the z direction and, from (10.37), is of magnitude

$$H_z = M_1 \left(v_{x1} y_1 - v_{y1} x_1 \right) + M_2 \left(v_{x2} y_2 - v_{y2} x_2 \right), \tag{38.5}$$

where v_{x1} is the x component of v_1 and x_1 the x coordinate of S_1 relative to O as origin with corresponding definitions for the other terms in the equation. Using (38.2) this becomes

$$H_z = \frac{M_1 M_2}{M_1 + M_2} \left(v_{x12} y - v_{y12} x \right) \tag{38.6}$$

where the velocity components and coordinates are now those of S_2 relative to S_1.

The program GRAVBODY (Appendix 27) is a general program for integrating the motion of any small number of bodies under mutual gravitational attractions. The integration is carried out by the four-step Runge–Kutta routine (§19.3) and automatic step-length control is included to keep the error down to a required tolerance. The units used are convenient for problems involving the motions of astronomical bodies with solar-system-scale parameters. The unit of length is 1 AU (1.496×10^{11} m), that of mass is one solar mass ($1\,M_\odot = 1.99 \times 10^{30}$ kg) and that of time is 1 year (3.156×10^7 s). The gravitational constant in these units becomes $4\pi^2$. In establishing initial boundary conditions for the calculation, positions relative to some origin must be given in units of AU and velocities in units of AU year^{-1}. The program suggests 10^{-6} AU, approximately 150 km, as a suitable tolerance for calculations where distances of a few AU are involved, but a suitable tolerance would be larger or smaller for systems of larger or smaller extent.

The program was run for two bodies simulating the motions of the Sun and Jupiter, not a binary-star system but similar in kind. The information input into the program is given in Table 38.1.

The program was run with initial timestep 0.01 years and for a duration of 100 years. Three different tolerances were used – 10^{-6} AU, as recommended by the program, 10^{-3} AU, and 10^{-2} AU. The variations of total energy and angular momentum are shown in Figure 38.2. It is clear that over the duration of the simulation the recommended tolerance keeps both energy and angular momentum sensibly constant. For a tolerance 10^{-3} there is a slight departure from constancy, but for 10^{-2} there is a large and unacceptable variation. For very long simulation times even the recommended tolerance would be found to be unsatisfactory and either a smaller tolerance, or a better integration technique, would be required.

Exercise 38.1

(i) Jupiter has a mass of 0.001 $M_\odot$ and its innermost major satellite, Io, is in a circular orbit of radius 2.8209×10^{-3} AU. What is the speed of Io in its orbit in units AU year^{-1}?

(ii) The next major satellite, Europa, has a circular orbit with period exactly twice that of Io. What is the radius of its orbit in AU and the speed in its orbit in units AU year^{-1}?

Table 38.1 Input into GRAVBODY for a Jupiter orbit

Property	Sun	Jupiter
Mass	1.000	0.001
x	0	0
y	0	5.2
z	0	0
$\dot{x}$	0	-2.75674
$\dot{y}$	0	0
$\dot{z}$	0	0

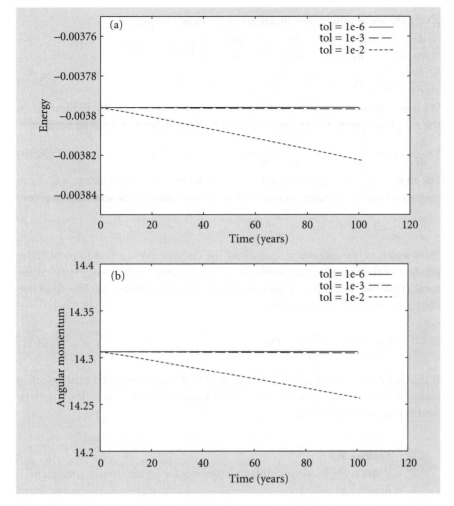

Figure 38.2 Variation of (a) energy and (b) angular momentum for three different tolerances with GRAVBODY run for the Jupiter–Sun problem.

38.3 **An electron in a magnetic field**

For an electron in a magnetic field the basic equation of motion is

$$m\frac{d^2\mathbf{r}}{dt^2} = e\mathbf{v} \times \mathbf{B} \tag{38.7}$$

where e is the charge of the electron, $\mathbf{v}$ its velocity, and $\mathbf{B}$ the magnetic field. This is factorized into the pair of equations

$$\frac{d\mathbf{r}}{dt} = \mathbf{v} \tag{38.8a}$$

and

$$\frac{d\mathbf{v}}{dt} = \frac{e}{m}\mathbf{v} \times \mathbf{B} \tag{38.8b}$$

giving the set of differential equations in terms of components

$$\frac{dx}{dt} = v_x, \quad \frac{dy}{dt} = v_y, \quad \frac{dz}{dt} = v_z, \quad \frac{dv_x}{dt} = \frac{e}{m}\left(v_y B_z - v_z B_y\right),$$

$$\frac{dv_y}{dt} = \frac{e}{m}\left(v_z B_x - v_x B_z\right), \quad \frac{dv_z}{dt} = \frac{e}{m}\left(v_x B_y - v_y B_x\right). \tag{38.9}$$

We now take a system with a field in the z direction $\mathbf{B} = (0, 0, B_z)$ and motion in the x-y plane, giving $\mathbf{v} = (v_x, v_y, 0)$. The set of equations to be solved becomes

$$\frac{dx}{dt} = v_x, \quad \frac{dy}{dt} = v_y, \quad \frac{dv_x}{dt} = \frac{e}{m}v_y B_z, \quad \frac{dv_y}{dt} = -\frac{e}{m}v_x B_z. \tag{38.10}$$

For an electron moving with speed v in a circular orbit perpendicular to a uniform magnetic field $\mathbf{B}$, equating central force to mass $\times$ centripedal acceleration gives

$$\frac{v^2}{r} = \frac{e}{m}vB \quad \text{or} \quad \frac{v}{r} = \omega = \frac{e}{m}B$$

from which ω, the angular frequency of the circular motion is seen to be independent of the radius of the circular motion. This angular frequency is called the **local cyclotron frequency**, usually represented by the symbol Ω. The final two equations in (38.10) now become

$$\frac{dv_x}{dt} = \Omega v_y \quad \text{and} \quad \frac{dv_y}{dt} = -\Omega v_x. \tag{38.11}$$

It should be noted that Ω could be dependent both on position and time. We now consider the workings of a simple electron microscope lens where the magnetic field is position dependent.

Exercise 38.2
What is the local synchrotron frequency for an electron in a magnetic field 0.1 T?

38.3.1 The electron microscope lens

We now consider a form of magnetic field that operates just in two dimensions and illustrates the general principle behind lens design for electron microscopes. In Figure 38.3 we show the paths of electrons leaving both a point at the origin and also a point $(0, y)$. There is a magnetic field along the z direction with magnitude

$$B_z = cy \tag{38.12}$$

where c is a constant. The trajectories of these electrons can be followed with the computer program ELECLENS (Appendix 28) which uses a predictor–corrector Euler method to solve the differential equations given by (38.10) and (38.11). We take 1 keV electrons, for which the speed of motion is $1.785 \times 10^7 \, \text{ms}^{-1}$. It is evident that electrons starting at the origin and moving along the x-axis will not be deflected because they are in zero field. We now take electrons starting at the origin at angles 0.01, 0.02, and 0.03 radians. These trajectories all pass through the same point at various equispaced positions on the x axis. Next we start the electrons from $(0, 0.001)$ with trajectories making small angles with each other. These also

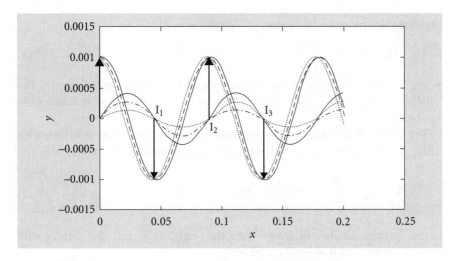

Figure 38.3 Simulation of magnetic focusing with ELECLENS showing the formation of a series of images.

have common points which have the same x coordinates as those found for trajectories from the origin. A one-dimensional object situated on the y axis would form an inverted image at I_1, an upright image at I_2 and alternating inverted and upright images along the x direction. An actual lens would not have the field over the whole x range but restricted to a narrow range of x and ELECLENS has provision for defining the range within which the field operates. This gives behaviour more akin to that of a normal optical lens and enables enlarged images to be formed. This behaviour is explored in Problem 38.2.

38.4 *N*-body problems

In §38.2 a system of two bodies with gravitational interactions was considered. If there are N bodies interacting gravitationally then, as for the two-body system, the centre of mass does not accelerate and the system of equations to be solved is

$$\frac{\mathrm{d}\mathbf{r}_i}{\mathrm{d}t} = \mathbf{v}_i, \quad \frac{\mathrm{d}\mathbf{v}_i}{\mathrm{d}t} = \sum_{\substack{j=1 \\ j \neq i}}^{N} \frac{Gm_j(\mathbf{r}_j - \mathbf{r}_i)}{\left|\mathbf{r}_j - \mathbf{r}_i\right|^3} \tag{38.13}$$

for $i = 1$ to N. For $N = 2$ there is an analytical solution, as there also is for some special three-body problems. For larger values of N numerical methods must be used. However, before we explore N-body systems any further we shall first derive an important theorem, the **virial theorem**, that applies to systems of interacting bodies.

38.4.1 Virial theorem

The virial theorem applies to any system of interacting particles with pair-conservative interactions for which the distribution of particles, in a general macroscopic sense, remains constant. The theorem is

$$2T + \Omega = 0 \tag{38.14}$$

where T is the translational kinetic energy and Ω is the potential energy. The validity of the theorem will be demonstrated for a system of gravitationally interacting particles.

The ith body has mass m_i, coordinates (x_i, y_i, z_i), and velocity components $(\dot{x}_i, \dot{y}_i, \dot{z}_i)$. We define the **geometrical moment of inertia** of the system as

$$I = \sum_{i=1}^{N} m_i r_i^2 = \sum_{i=1}^{N} m_i(x_i^2 + y_i^2 + z_i^2), \tag{38.15}$$

where r_i is the distance of mass i from the origin. Differentiating I twice with respect to time and dividing by 2 gives

$$\frac{1}{2}\ddot{I} = \sum_{i=1}^{N} m_i(\dot{x}_i^2 + \dot{y}_i^2 + \dot{z}_i^2) + \sum_{i=1}^{N} m_i(x_i \ddot{x}_i + y_i \ddot{y}_i + z_i \ddot{z}_i). \tag{38.16}$$

The first term on the right-hand side is $2T$. The second term can be transformed by noting that $m_i \ddot{x}_i$ is the total force on body i due to all other bodies in the x-direction so that

$$m_i x_i \ddot{x}_i = \sum_{\substack{j=1 \\ j \neq i}}^{N} \frac{G m_i m_j}{r_{ij}^2} x_i \frac{x_j - x_i}{r_{ij}} \tag{38.17}$$

where r_{ij} is the distance between particle i and particle j. The last factor on the right-hand side of (38.17) gives the component in the x-direction. Inserting (38.17) into (38.16) and combining contributions of pairs of particles we find

$$\sum_{i=1}^{N} m_i(x_i \ddot{x}_i + y_i \ddot{y}_i + z_i \ddot{z}_i) = -\sum_{\text{pairs}} G m_i m_j \frac{(x_i - x_j)^2 + (y_i - y_j)^2 + (z_i - z_j)^2}{r_{ij}^3}$$

$$= -\sum_{\text{pairs}} \frac{G m_i m_j}{r_{ij}} = \Omega. \tag{38.18}$$

Equation (38.16) now appears as

$$\frac{1}{2}\ddot{I} = 2T + \Omega \tag{38.19}$$

and if the system stays within the same volume and retains the same general distribution of particles then I is constant, $\dot{I} = 0$ and hence $\ddot{I} = 0$ and the virial theorem is verified.

The virial theorem applies to a wide range of systems where the particles can be stars in a cluster of stars or molecules within a single star. If a cluster of N stars, each of the same mass M, has a uniform distribution within a spherical volume of radius R then the kinetic energy of the stars is

$$T = \frac{1}{2}NM\langle V^2 \rangle \tag{38.20}$$

where $\langle V^2 \rangle$ is the mean-square speed of the stars. The potential energy will, to a first approximation, be that of a sphere of radius R, mass NM, and uniform

density, which is

$$\Omega = -\frac{3G(NM)^2}{5R}. \text{ (see Exercise 38.4)} \tag{38.21}$$

Applying the virial theorem gives the root-mean-square speed of the stars as

$$\langle V^2 \rangle^{1/2} = \left(\frac{3GNM}{5R}\right)^{1/2}. \tag{38.22}$$

If a cluster of 300 stars with mean mass $\frac{1}{2}M_\odot$ uniformly occupied a spherical volume of radius 10^{16} m, then the mean square speed of those stars would be

$$\langle V^2 \rangle^{1/2} = \left(\frac{3 \times 6.67 \times 10^{-11} \times 300 \times 10^{30}}{5 \times 10^{16}}\right)^{1/2} \text{ m s}^{-1} = 1.096 \text{ km s}^{-1}.$$

It is important to be clear that what the virial theorem actually means. It applies to the collection of molecules in a star or stars in a cluster since, by and large, stars and clusters evolve slowly so that $\dot{I} = 0$ (and hence $\ddot{I} = 0$) is a good approximation. On the other hand if there happened to be a system that was expanding or contracting in such a way that $\dot{I} = $ constant then again the virial theorem would be valid. What the theorem does *not* say, which is sometimes assumed by those that use it, is that a system with $2T + \Omega = 0$ will necessarily stay in a compact state. The conservation of energy says that

$$E = T + \Omega \tag{38.23}$$

is a constant so we can write the virial theorem as

$$\ddot{I} = E + T. \tag{38.24}$$

If the system is in a state for which $E + T = 0$ then $\dot{I}$ can be finite so that the system changes its configuration. This will change the potential energy and hence the kinetic energy so that the right-hand side of (38.24) changes and $\ddot{I}$ is no longer equal to zero. Thus the system can evolve without limit, expanding or contracting, with the only constraints imposed by the constancy of energy and angular momentum.

Exercise 38.3
Show that the virial theorem is valid for a planet in a circular orbit around the Sun.

Exercise 38.4
Show that a uniform sphere of mass M and radius R has gravitational potential energy $V = -3GM^2/5R$. (Hint: Consider building up the sphere in thin shells, bringing the material to do so from infinity.)

38.4.2 Evolution of a stellar cluster

A program CLUSTER (Appendix 29) is available that follows the evolution of a cluster of stars up to 20 in number – but this can be increased by changing the dimensions of some arrays. The stars are initially placed in random positions within a sphere of a chosen radius and the positions are then adjusted so that the

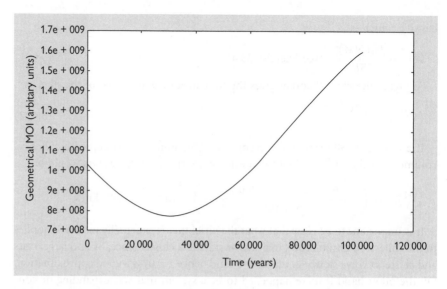

Figure 38.4 The variation of the geometrical moment of inertia for a system originally satisfying the virial theorem.

centre of mass is at the origin. They are given velocities that give a mean velocity zero for the complete cluster but with a mean square value that satisfies the virial theorem. All stars are given the same mass, which may be chosen.

In one run of CLUSTER 20 stars, each of solar mass, were placed within a sphere of radius 10^4 AU. After a simulated time of 100 000 years the estimated energy is $-0.472\,134\,9$ units compared with the initial value of $-0.472\,071\,2$ and the angular momentum is $9\,295.974\,2$ units compared with the initial value of $9\,295.974\,4$. This near-constancy of the conserved quantities indicates that the computation is almost certainly giving sensible results over the period of the simulation. If the constancy was not found then it would be necessary either to change the tolerance, taken as 10^{-3} in the calculation, or to use a better integrational algorithm.

The value of the geometrical moment of inertia was also monitored and is shown in Figure 38.4. At first the value falls, corresponding to a general collapse of the system but this is then reversed and considerable expansion is indicated. Actually what happens is that the centre of the system moves inwards while external stars move outwards. This underlines the point that a system that satisfies the virial theorem will not necessarily stay in a compact form.

38.5 **Molecular dynamics**

The general principle behind molecular dynamics is very simple. It involves studying the properties of a system, which could be a liquid, for example, by following the dynamics of a number of single particles, the ith of which has position $\mathbf{r}_i$ and velocity $\mathbf{v}_i$ and is subject to a total force $\mathbf{F}_i$. The motion of each particle is computed by the numerical integration of the kinematic equation of motion

$$\frac{\mathrm{d}\mathbf{r}_i}{\mathrm{d}t} = \mathbf{v}_i \tag{38.25}$$

and Newton's second law

$$\frac{d\mathbf{v}_i}{dt} = \frac{\mathbf{F}_i}{m_i}, \tag{38.26}$$

where m_i is the mass of the particle and the force $\mathbf{F}_i$ is the sum of the external forces and the mutual interactions.

The molecular dynamics procedure is particularly applicable where the forces are short-range, which allows the summation of the forces on a particle to be substantially curtailed. In addition, to reduce the total number of independent particles that have to be considered, the cell model, as described in §23.7.1, can be used. Such a procedure for finding the equation of state of a liquid is now described.

38.5.1 Equation of state from molecular dynamics

A molecular dynamics program FLUIDYN (Appendix 30), using a cell model, is available. It is based on the Lennard–Jones potential in the form (23.27). The constants ε and σ for the Lennard–Jones potential and the atomic mass given in FLUIDYN are those for argon, but constants for other inert gases and for nitrogen are given in Table 23.3 and may be substituted for the argon values. The cell contains 125 molecules which are initially placed on a regular grid and then slightly displaced in a random fashion. For an actual fluid the molecules would have a distribution of speeds given by the Maxwell–Boltzmann distribution (Chapter 22), which is what occurs when a system of particles undergoes perfectly elastic interactions. However, in FLUIDYN the particles are given an initial speed equal to the root-mean-square speed appropriate to their temperature but in random directions. By the time the particles have interacted for 50 timesteps the velocity distribution approaches the correct form.

The integration of the equations to follow the motions of the particles is carried out by the Runge–Kutta method (§19.3) and at each step after the first 50 the contributions to the virial term in (23.22) are found and accumulated. The appropriate value of the virial term to use to determine the equation of state is the average found for all the steps after the first 50. In addition, after each step the radial distribution function, $\rho(r)$, as described in §23.7.5, is found. Input to the program consists of the temperature and the value of V^* for which the radial distribution is required. The output file EOS.DAT contains values of $\{V^*, PV/(NkT)\}$ for V^* from 0.85 to 1.8 in steps of 0.05 and the output file RADIAL.DAT contains the radial density distribution. These are shown for argon at a temperature 329 K with the radial distrution for $V^* = 1$ in Figure 38.5. These are similar to the results shown in Figures 23.14 and 23.15, although different in detail.

38.6 Modelling plasmas

Plasma is often called the fourth state of matter, the others being solid, liquid, and gas. A plasma consists of mixtures of ions and electrons, together with some neutral atoms if the material is only partially ionized. For many-electron atoms

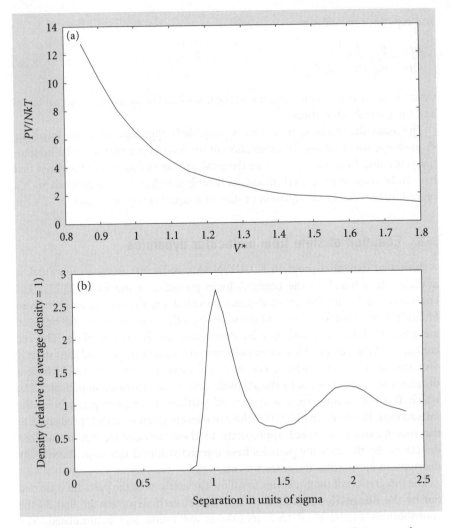

Figure 38.5 (a) The equation of state and (b) the radial distribution function from a run of FLUIDYN.

there can be different degrees of ionization which will, in general, depend on the temperature and the density of the material.

There will be quasi-neutrality in the plasma with

$$n_e = Zn_i \tag{38.27}$$

where n_e and n_i are the electron and ion density respectively and Z the average ion charge number, i.e. the average number of electrons lost by ionization per atom. Approximate neutrality in any region of the plasma is due to the large space-charge fields which are generated by any significant inbalance of charges which act to reduce the inbalance. Most plasmas are at temperatures above 10^4 K and temperatures of interest go up to 10^8 K and beyond, which is the thermal regime for fusion research.

For one class of problems, described by the term **magnetohydrodynamics**, the motions of the electrons and ions can be described in terms of currents and behaviour is dominated by the forces on the current elements due to magnetic fields, both imposed externally and also generated by the currents themselves. As the currents can be regarded as flowing in closed circuits there is little or no charge separation in such a system. The behaviour of such a system is governed by fluid dynamics and by Maxwell's equations of electromagnetism. However, we shall be dealing with a class of problems in which electrostatic fields, partly externally applied and partly space-charge fields, are dominant. For the one-dimensional systems we shall consider, current loops do not exist and there are no current-generated magnetic fields. The equations which describe the motion of the jth individual particle, with mass m_i and charge q_i placed in a combination of an electric field $\mathbf{E}$ and a magnetic field, $\mathbf{B}$, are

$$m_j \frac{d\mathbf{v}_j}{dt} = q_j \left(\mathbf{E} + \mathbf{v}_j \times \mathbf{B} \right) \tag{38.28a}$$

$$\mathbf{E} = -\nabla\phi \tag{38.28b}$$

and the electrostatic form of Poisson's equation (§20.6)

$$\nabla^2\phi = -\frac{\rho}{\varepsilon_0} \tag{38.28c}$$

where $\mathbf{v}_j$ is the velocity of the particle, ϕ the electric potential, and ρ the local electron density. As previously indicated, it is assumed here that any magnetic field is imposed externally and not generated by currents within the plasma.

Since the number of charged particles involved in even a tiny region of a plasma is extremely large, there is no way of modelling such systems using individual particles, one for each ion or electron. In a plasma, motions are coherent over the whole system so that we are interested in the collective behaviour of a plasma, the way that streams of electrons and ions move relative to one another or themselves. Collective effects mean that the individual charges are lost in a continuum description which enables us to simplify the problem by the use of **superparticles** which represent large numbers of electrons or ions and possess the combined mass and the combined charge of the group of particles they represent. For a good simulation, if the continuous nature of the charge distribution is to be well represented, then it is necessary that there should be many superparticles. For a particular model the number of superparticles being used might typically be in the range 10^4–10^7 with each superparticle representing 10^7–10^9 electrons or ions.

Even with the reduction in number by the use of superparticles, dealing with the simulation by particle–particle interactions is still not practicable since the inverse-square forces are long range and cannot be terminated at a certain distance from the particles – as was the case for the Lennard–Jones potential. Nevertheless it is evident that distant particles do not have to be dealt with individually, but that a reasonably small cluster of them will have almost the same effect as the combined charge at the centre of mass of the cluster. Although in the method to be described the field experienced by individual particles is *not* calculated by dividing the combined charge of a cluster by the square of its mean distance, it is instructive to see what errors might come about from such an approximation. In Figure 38.6 we show in two dimensions a square box of unit side with centre at coordinate (m, n). By a random number generator 10 equal charges are positioned in the box and the components of field at the origin due to the 10 individual charges are found. For each value of (m, n) 100 different random configurations are generated and the

Table 38.2 Testing the approximation that 10 charges in a unit-side square box may be replaced by the combined charge at the centre of the box. The coordinates of the box centre are (m, n). Approx. E_x and Approx. E_y are the field components with the combined charge at the box centre. Mean E_x and σ_x are the mean and standard deviation of the x-component of the field for 100 random arrangements of charges; there are similar quantities for the y-component

m	n	Approx. E_x	Mean E_x	σ_x	Approx. E_y	Mean E_y	σ_y
10	10	0.035 4	0.035 4	0.000 6	0.035 4	0.035 4	0.000 6
10	6	0.063 1	0.063 1	0.001 1	0.037 8	0.037 8	0.000 8
10	2	0.094 3	0.094 3	0.001 8	0.018 9	0.018 8	0.001 0
8	8	0.055 2	0.055 4	0.001 1	0.055 2	0.055 3	0.001 1
8	4	0.111 8	0.112 0	0.002 5	0.055 9	0.055 9	0.001 7
8	0	0.156 2	0.156 3	0.003 8	0.000 0	− 0.000 1	0.001 9
6	6	0.098 2	0.098 5	0.002 5	0.098 2	0.098 3	0.002 5
6	2	0.238 2	0.237 8	0.007 4	0.079 1	0.078 9	0.004 3
4	4	0.221 0	0.222 3	0.008 7	0.221 0	0.221 7	0.008 5
4	0	0.625 0	0.628 3	0.030 0	0.000 0	− 0.001 2	0.015 8
3	3	0.392 8	0.396 7	0.020 8	0.392 8	0.395 3	0.020 2
3	1	0.948 7	0.960 8	0.060 2	0.316 2	0.316 9	0.035 1
2	2	0.883 9	0.902 7	0.072 7	0.883 9	0.898 1	0.069 8
2	1	1.788 8	1.843 0	0.175 1	0.894 4	0.909 0	0.108 0
2	0	2.500 0	2.564 1	0.242 6	0.000 0	− 0.012 9	0.141 9
1	1	3.535 5	3.874 6	0.707 5	3.535 5	3.848 8	0.674 0
1	0	10.000 0	10.979 8	2.122 2	0.000 0	− 0.187 9	1.570 4

mean and standard deviation of the components of field at the origin are calculated. These results are shown in Table 38.2 compared with the approximate field components calculated by assuming that all the 10 charges are at the centre of the box. It will be seen that there is a systematic difference between the mean and approximate values which can be summarized by the statement that 'the average of inverse squares is greater than the inverse square of the average'. This effect is small for the larger distances but, as appears from a comparison of the mean and approximate values, can be as much as 10% for very small distances. The standard deviations show that the random fluctuations can be even larger, as much as 20% or even more, at close distances. The errors would be smaller, but similar in size, if the combined charge was taken at the centre of mass of the charges rather than at the centre of the box.

The way in which the distribution of electrons and ions is transformed into electric fields in the method to be described is not by the direct process we have used in this test. However, one part of the method does involve the approximation that all particles act as though they were placed at the nearest point of a grid and the errors we have found here will still appear, albeit in a less transparent way. For the errors to be small the cells defined by the grid should contain around 100 particles. Another condition to be satisfied, if the coherent motions of electrons are to be properly represented, is that the cell size should be small compared

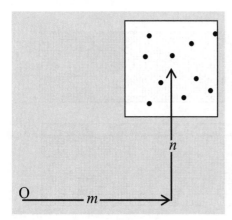

Figure 38.6 10 equal charges randomly placed in a square box of unit side, with the box centre at (m, n) with respect to the origin.

with the characteristic coherence length – the average distance apart of two ions with correlated motions. Later we shall be describing the **cloud-in-cell** method in which each particle is spread out over several cells, a procedure that reduces the errors.

The kinds of forces acting on a charged particle can be divided into two parts: those due to distant charges, which give a slowly varying field in the vicinity of the particle and so contribute to collective motion of the particles in its region, and those due to nearby particles which cause it to move relative to its neighbours in an uncorrelated way. The correlated motions are governed by a timescale which is known as the **plasma period**, t_p, which is the timescale for natural oscillations in the plasma. Consider a slab of plasma as illustrated in Figure 38.7a that is then subjected to a brief electric field which slightly separates the charges and creates two thin layers of opposite polarity, each of thickness x as shown in Figure 38.7b. An infinite sheet of uniform charge density σ per unit area gives rise to a uniform field $\sigma/2\varepsilon_0$. Since the surface density of charge in each layer has magnitude nex and they reinforce one another within the plasma the total field is nex/ε_0 in an upwards direction. Each electron, of mass m, experiences a force $-ne^2x/\varepsilon_0$ and thus undergoes simple harmonic motion with angular frequency, known as the plasma frequency,

$$\omega_p = \left(\frac{ne^2}{m\varepsilon_0}\right)^{1/2} \tag{38.29}$$

corresponding to a plasma period

$$t_p = 2\pi\left(\frac{m\varepsilon_0}{ne^2}\right)^{1/2}. \tag{38.30}$$

The distance within which the uncorrelated particle–particle interactions are considered as important is called the **Debye length**, denoted by λ_D. If we consider the **Debye sphere**, a sphere of radius λ_D surrounding the particle, then the effect of the distribution of charge beyond a distance λ_D is substantially reduced by screening. To understand the role of the screening consider a small test charge, q, at the origin surrounded by the hot plasma. The ions, being heavy, will have low velocities and so their contribution to the charge density will fluctuate very little. The electrons, being of much lower mass, will have high velocities and hence their

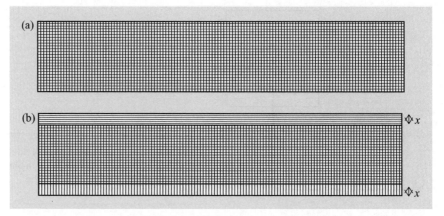

Figure 38.7 (a) A neutral plasma with equal positive ion charge (horizontal lines) and negative electron charge (vertical lines). (b) Two slabs of charge, each of thickness x, due to separation of the charges.

contribution to the charge density will fluctuate much more. This means that the mean local charge density will fluctuate around zero, the mean for a plasma, and hence the local electric potential will also fluctuate. The situation is that the local time-averaged density of electrons will depend on the local potential through the influence of the Boltzmann distribution and, in its turn, the potential will depend on the local charge density through the influence of the Poisson equation. The local time-averaged electron density will be

$$n = n_e \exp\left(\frac{e\phi}{kT}\right) \tag{38.31}$$

which gives a local space charge

$$\rho(r) = (n_e - n)e = n_e e\left\{1 - \exp\left(\frac{e\phi}{kT}\right)\right\} \approx -\frac{ne^2\phi}{kT} \tag{38.32}$$

assuming a hot plasma (the usual state of affairs) so that $kT \gg e\phi$.

The Poisson equation is given by (38.28c) and with the spherically symmetric form of the Laplacian operator, (16.39), we have

$$\frac{1}{r^2}\frac{d}{dr}\left(r^2\frac{d\phi}{dr}\right) = -\frac{1}{\varepsilon_0}\rho(r) = \frac{ne^2}{\varepsilon_0 kT}\phi. \tag{38.33}$$

The solution of this differential equation is subject to the boundary condition that as $r \to 0$ so $\phi \to q/4\pi\varepsilon_0 r$, which gives the solution, which you should check, as

$$\phi = \frac{q}{4\pi\varepsilon_0 r}\exp\left(-\frac{r}{\lambda_D}\right) \tag{38.34a}$$

where

$$\lambda_D = \sqrt{\frac{\varepsilon_0 kT}{e^2 n}} \tag{38.34b}$$

is the Debye length. For $r \ll \lambda_D$ the field is due to the test charge, but for r of order λ_D or greater the field is essentially removed by the screening effect of the plasma charges. Carrying this concept over into what an individual charged particle experiences within the plasma what we infer is that other charged particles within a Debye length contribute an effective field while at larger distances their effects are screened out by the collective behaviour of many particles. Indeed over these lengths only *collective* behaviour, in which the particles behave coherently, usually in the form of a plasma wave, can occur.

Particle behaviour in the short-range (individual) regime is characterized by collisional effects, distinguished from those in gases by the importance of the simultaneous interaction of the test particle with many other bodies. Now the question is to decide how important the induced motions due to local particle–particle interactions are compared to the collective motions due to the more distant particles. To do this we consider an interaction between two particles, each with magnitude of charge e, as shown in Figure 38.8.

The distance of nearest approach of the two bodies is D, B is at rest and the speed of A is V. It is shown in Appendix 31 that the effect of the interaction is to give to A a component of velocity along the direction CB of magnitude

$$\delta V = \frac{e^2}{2\pi\varepsilon_0 mDV} \tag{38.35}$$

where m is the mass of particle A. This corresponds to an angular deflection of its path by

$$\delta\theta = \frac{e^2}{2\pi\varepsilon_0 mDV^2}. \tag{38.36}$$

A collision timescale, t_c, is defined which is the expected time for a particle to be deflected through an angle of $\pi/2$. Since the individual deflections are randomly oriented in space then N deflections, each with a root-mean-square magnitude of deviation $\langle\delta\theta^2\rangle^{1/2}$, will give an expected total deviation of magnitude $N^{1/2}\langle\delta\theta^2\rangle^{1/2}$. Based on this relationship it is shown in Appendix 31 that the ratio t_c/t_p is of the order of the number of superparticles in the Debye sphere. If the number of superparticles in the Debye sphere is very large, which it normally is, then the modelling of a plasma may be carried out over several plasma periods

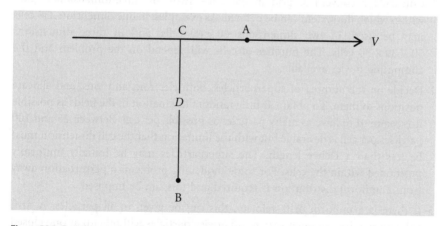

Figure 38.8 Relative motion of two charged particles with B at rest.

without collision effects being important. Under these conditions the system can be regarded as collisionless and the modelling done on the basis of calculating smoothed-out average fields in the vicinity of each particle.

Exercise 38.5
For an electron density of $10^{25} \, \mathrm{m}^{-3}$ and a temperature $10^5 \, \mathrm{K}$ in a plasma find (i) the plasma frequency, (ii) the plasma period, and (iii) the Debye length.

38.7 Collisionless particle-in-cell model

Plasmas can be modelled in one, two, or three dimensions. The physical reality is that any plasma must be three-dimensional, so when we study a two-dimensional plasma, say in the x-y plane, we are assuming that all sections of constant z are similar and also that the system has infinite extent in the z direction. The latter condition can never be true, but may be a reasonable assumption if the z dimension is large compared with the scale of the fluctuations in particle densities. Each superparticle in a two-dimensional simulation represents a rod of charge and mass density per unit distance parallel to z. Whereas point charges give a $1/r^2$ force law for fields, charge rods give a $1/r$ law although this relationship is not explicitly used in the computations. For a one-dimensional plasma each super-particle represents an infinite sheet of charge and mass density per unit area perpendicular to x. Infinite charge sheets give a uniform field independent of distance from the sheet.

In the **collisionless particle-in-cell (PIC) model** the particles are placed in a mesh, which is usually cubical in three dimensions, square in two dimensions, and equispaced in one dimension. What is needed from the distribution of charges is the electric field at the position of each electron or ion so that its acceleration can be found and its motion followed. The steps in the process for purely electric fields are:

1 Construct a convenient grid in the one-, two-, or three-dimensional space within which the system can be defined. As examples, in one dimension 250 cells may be used, in two dimensions 60×60 cells, and in three dimensions $30 \times 30 \times 30$ cells. The number of cells will depend on the problem and the computing power available.

2 Decide on the number of superparticles, both electrons and ions, and allocate positions to them. To obtain as little random fluctuation in the fields as possible it is required to have as many particles as possible per cell. Between 25 and 100 particles per cell is desirable but with the limitation that the cell dimension must be less than a Debye length. The superparticles may be initially uniformly positioned within the cells. For some modelling problems a perturbation away from a uniform distribution is required and this can be imposed.

3 If there is a uniform drift velocity, this can be given to all particles. A drift velocity requires an open system otherwise particles will pile up at one closed

end and a particle vacuum will be created at the other. Where there is a drift velocity then periodic boundary conditions are usually used as described for liquid models in §23.7.1. Another possibility is for the boundary to reflect the particles back into the system. The size of the space being modelled must be larger than the scale of the largest significant feature in the phenomenon being investigated.

4 A 'cold' plasma is one in which there is no random thermal motion of the particles. This is very rarely used, as numerical energy generation heats the particles because the Debye length is very small. For the more usual 'warm' plasma model, velocities must be chosen randomly from a Maxwellian distribution and allocated to the particles in random fashion. Due to statistical fluctuations with small numbers of superparticles, 'hot spots' or 'cold spots' may develop within the plasma and in some computer algorithms correction procedures are built in to avoid this.

5 Assign the charge of each superparticle to the nearest cell, or cells, and calculate the charge densities within the cells, associated with the centre of the cell.

6 Using the densities at cell centres Poisson's equation, (38.28c), is solved to find the potential at each grid point using techniques similar to those described in §20.7.

7 From the potentials, field components are found at the grid points by a finite-difference approximation

$$E_x(i) = -\frac{\phi(i+1) - \phi(i-1)}{2\Delta x} \tag{38.37}$$

(see the one-dimensional example that follows).

8 Move the superparticles by numerical solution of (38.28a). This can be done by any suitable method. For complex problems a combination of computer storage and time will often dictate the method to be used. The timestep for the integration should be about 0.03 or so of the shortest time of interest in the system – which in the electrostatic case we are considering here would be the cell crossing time or, less likely, the plasma period.

9 Calculate and store any characteristics of the plasma which are required. Output numerically or graphically at required intervals.

10 If the total simulation time is not exceeded, then return to step 5.

38.7.1 Design of a one-dimensional plasma program

Now we shall describe a simple one-dimensional program for investigating aspects of plasma behaviour. One simplification that can be imposed is in the way that the ions are handled. Although in general ions and electrons must be treated as distinct kinds of particles with different charge-to-mass ratios it is obvious that, because of their large mass, ions move much less than do electrons. For some kinds of problem it is sufficient to consider the ions as a stationary background just contributing to the net charge density and this is the assumption for this program.

The one-dimensional cell structure for a total of N cells is set up as shown in Figure 38.9 where the centres of cells are at

$$x_{i+1/2} = \left(i + \frac{1}{2}\right)\Delta x \qquad 0 \leq i \leq N - 1 \tag{38.38a}$$

and the boundaries of cells at

$$x_i = i\Delta x \qquad 0 \leq i \leq N. \tag{38.38b}$$

For one dimension, Poisson's equation can be simplified to give fields directly from charge density by

$$\nabla^2 \phi = \frac{d^2\phi}{dx^2} = -\frac{dE}{dx} = -\frac{\rho}{\varepsilon_0}. \tag{38.39}$$

In finite-difference form this gives for the field gradient at a cell centre

$$\frac{E_{i+1} - E_i}{\Delta x} = \frac{\rho_{i+1/2}}{\varepsilon_0}$$

where $\rho_{i+1/2}$ is the charge density at the cell centre. Rearranging (38.39) gives

$$E_{i+1} = E_i + \frac{\rho_{i+1/2}\Delta x}{\varepsilon_0} \tag{38.40a}$$

and the field at a cell centre, if required, is then given as

$$E_{i+1/2} = \frac{1}{2}(E_i + E_{i+1}). \tag{38.40b}$$

The recurrence relationship (38.40a) for the fields at centres can be initiated either by some field boundary condition imposed by the problem, e.g. $E_0 = 0$, or by some known potential drop across the system such as

$$\phi_N - \phi_0 = -\sum_{i=0}^{N-1} E_{i+1/2}\Delta x = 0. \tag{38.41}$$

The charges at cell centres are found from the positions of the individual particles by

$$\rho_{i+1/2} = \sum_j w_{i+1/2}(x_j)q_j \tag{38.42}$$

where the summation is over all the particles, particle j has coordinate x_j and charge q_j and where the weighting function $w_{i+1/2}(x_j)$ gives the proportion of the

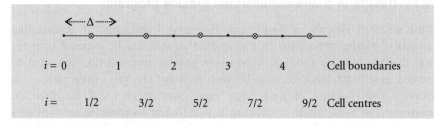

Figure 38.9 A one-dimensional cell structure with cell boundaries at integral values and cell centres at half-integral values.

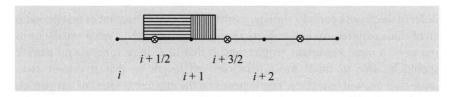

Figure 38.10 The cloud-in-cell method. The horizontally shaded part of the charge is allocated to cell-centre $i + 1/2$ and the vertically shaded part to cell-centre $i + 3/2$.

charge of a particle at position x_j that is allocated to the centre of cell $i + 1/2$. There are various weighting schemes the simplest being to allocate all the charge to the nearest cell centre for which the weighting function is

$$w_{i+1/2}(x) = 1 \quad \text{if} \quad |x - x_{i+1/2}| < \frac{1}{2}\Delta x$$
$$w_{i+1/2}(x) = 0 \quad \text{otherwise.} \tag{38.43}$$

The cloud-in-cell method is another simple scheme where the charge is assumed to be uniformly distributed over a length Δx and each part of the charge is then allocated to the nearest cell centre. This is illustrated in Figure 38.10; the horizontally-shaded portion of the charge is allocated to cell-centre $i + \frac{1}{2}$ and the vertically-shaded portion to cell-centre $i + \frac{3}{2}$. The weighting function is

$$w_{i+1/2}(x) = 1 - |x - x_{i+1/2}|/\Delta x \quad \text{if} |x - x_{i+1/2}| < \Delta x$$
$$w_{i+1/2}(x) = 0 \quad \text{otherwise.} \tag{38.44}$$

Since changes in velocity depend on the positions of the particles through the fields they generate and changes of position depend upon velocities it is sometimes convenient to solve the equations of motion by what is known as a **leapfrog process**, although this is not used in the program described in the following section. For the jth electron (the ions are not moved) this is of the form

$$x_j^n = x_j^{n-1} + v_j^{n-1/2}\Delta t \tag{38.45a}$$

and

$$v_j^{n+1/2} = v_j^{n-1/2} + \frac{q_j}{m_j}E_j^n\Delta t, \tag{38.45b}$$

where the superscript n represents the number of timesteps from the beginning of the calculation. The 'leapfrog' designation of this process comes from the fact that positions and velocities are found at times that interleave each other.

To find the field at any particle position the same weighting functions used for the charge density, described in (38.43) and (38.44), *must* be used so that

$$E_j^n = \sum_{i=0}^{N-1} w_{i+1/2}(x_j^n)E_{i+1/2} \tag{38.46}$$

where $E_{i+1/2}$ is found from E_i and E_{i+1} as indicated by (38.40b).

With computations of this kind the use of an appropriate timestep is very important. Since the natural period of disturbances in the plasma will be of the order of the plasma period a timestep which is some small fraction of that period is an obvious requirement and stability requires that $\Delta t \leq 0.2 t_p$. What usually turns out to be a more important requirement is that in a single timestep no particle should be able to travel more than one cell length so that it cannot cross more than one cell boundary. This depends on the maximum speed of any particle and for a one-dimensional case we can safely set $v_{\max}$ equal to four times the root-mean-square speed of the electrons. The limitation of the timestep by this condition may be found to be given by the relationship

$$\Delta t < 0.25 \frac{\Delta x}{v_{\max}} = 0.25 \frac{\Delta x}{\lambda_{\mathrm{D}}} \frac{1}{\omega_p}. \tag{38.47}$$

Since the scale length of events in the plasma will be of order λ_{D} and the cell length must be less than the Debye length, (38.47) will then be the critical relationship for determining the timestep.

38.7.2 Program PLASMA1

Program PLASMA1 (Appendix 32) models the dynamic boundary of a plasma layer. The initial ion density is set up as shown in Figure 38.11 where the density is constant for the first X_1 Debye lengths then falls linearly over the next X_2 Debye lengths and then is zero to the end of the mesh which has a total extent of X_t Debye lengths. For the program as given $X_t = 10$, $X_1 = 2.5$, and $X_2 = 0.5$. The number of cells in the mesh is 100 so each cell is one-tenth of a Debye length. The temperature of the plasma is set at 10^4 K and the density of ions in the constant-density part of the plasma is 10^{25} m^{-3}. The program generates 5000 electron positions which give the required distribution and also assigns velocities to these electrons following a Maxwell–Boltzmann distribution that, in one dimension, is a normal distribution with zero mean and standard deviation $(kT/m)^{1/2}$. To keep the plasma electrically neutral on average, particles which leave the mesh region are returned. Any leaving at $x = 0$, at the dense plasma boundary, are returned to $x = 0$ with positive velocities chosen from a Maxwell–Boltzmann distribution. Those leaving at the other boundary, $x = X_t$, are reflected back into the system with reversed velocity.

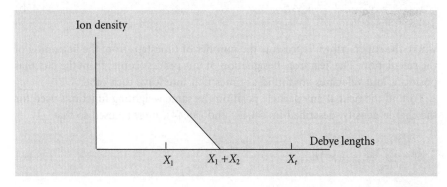

Figure 38.11 An initial configuration of ion density for PLASMA1.

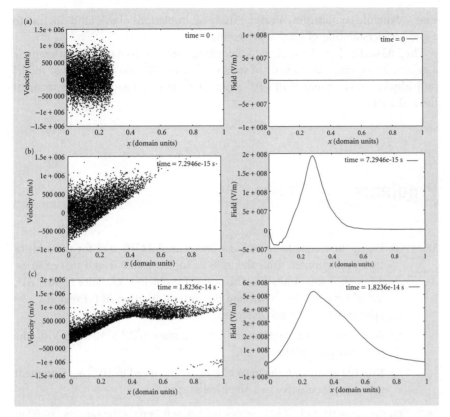

Figure 38.12 The position velocity and position field diagrams at (a) $t = 0$, (b) $t = 7.2946 \times 10^{-15}$ s, and (c) $t = 1.8326 \times 10^{-14}$ s.

The results of this kind of model are best understood from graphical output and we show a number of stages in Figure 38.12. The left-hand diagrams show the position–velocity coordinates of each particle and the right-hand diagrams show the variation of field throughout the mesh. Since each particle represents an infinite sheet of charge, which gives a constant field independent of distance from the sheet, the field at a point is just a measure of the total inbalance of charge up to that point. From Figure 38.12a, which shows the situation at $t = 0$, it is clear that the net charge seen by a test particle at the right hand end of the mesh is zero, so there is no net field. As the test particle moves towards the origin this situation is still true even when it enters that part of the region initially occupied by the plasma since the electron and ion distributions are initially identical.

The left-hand diagram of Figure 38.12b shows the transient effect of the faster thermal electrons leaving the dense plasma and moving into the vacuum, leading to a positive field everywhere outside the ion boundary. This positive field also exists in most of the ion region except near the left-hand reflecting boundary where a local accumulation of electrons reverses the field. Eventually, as seen in Figure 38.12c electrons are reflected back from outside the mesh on the right hand side, seen as the lower negative-velocity band. As time progresses so this band moves upwards as slower electrons move into and across the vacuum region. Eventually this fills and the particle distribution and field settle down into a nearly quiescent state – dynamic equilibrium. A careful study of simulations of this kind can lead to a good understanding of plasma behaviour.

The PIC method can be applied to modelling other kinds of system, for example a galaxy. In a galaxy the number of stars is of order 10^{11}, far too many to model individually, and a system of 10^4–10^5 'superstars' can give a good representation of the real system.

Problems

38.1 (i) Using the results from Exercise 38.1, run GRAVBODY for the orbit of Io around Jupiter. Take the mass of Jupiter as 0.001 and that of Io as 4.5×10^{-8} solar units. For this scale of problem a tolerance of 10^{-7} AU, initial timestep 10^{-6} year, and simulation duration of 0.1 year is appropriate. View the outcome graphically.

(ii) Add Europa to the system in (i) with a mass of $2.5 \times 10^{-8} M_\odot$ View the outcome graphically.

(iii) Run (ii) again with the mass of Europa enhanced to $10^{-6} M_\odot$ Comment on what you observe in a graphical output.

38.2 Run the program ELECLENS with a field $B_z = 0.7y$ over the range $x = 0.02$ to 0.03 m. This means changing the values of $x1$ and $x2$ in the program. Trace rays from the origin and from (0, 0.001) and hence find the magnification produced by this magnetic lens for 1 keV electrons..

38.3 Run CLUSTER with 20 stars, each of solar mass, initially within a spherical distribution of radius 10^4 AU. Use an initial timestep of 1 year, a duration of 10^5 years for the simulation and a tolerance of 10^{-3} AU. From the output file, PROPERTY.DAT, plot the virial term, $2T + \Omega$, against time. This will have some sharp peaks corresponding to near approaches of stars and consequent sharp increases of kinetic energy. Choose a time where the virial term is reasonably constant and from the values in PROPERTY.DAT, using a finite difference approach, deduce a value of $\ddot{I}$ for that time. Confirm the relationship $\frac{1}{2}\ddot{I} = 2T + \Omega$.

38.4 Modify FLUIDYN to find PV/NkT for xenon at 500 K and 1000 K for values of V^* from 0.85 to 1.8 and plot your results. The values of σ and ε/k are given in Table 23.3 and the atomic mass of xenon is 131.3. Also plot the radial density function for these temperatures for $V^* = 1$. Explain the differences at the two temperatures. MATLAB users should use the values of V^* in the provided program, otherwise running times will be excessive.

38.5 Run PLASMA1 with the following parameters: XT = 12, IT = 100, NT = 5000, $X_1 = 3$, $X_2 = 3$, DENS = 10^{24} m^{-3}, TEMP = 2×10^4 K. In addition, the number of timesteps between outputs should be changed to 750. Produce graphical output for the velocity of the electrons and the field within the space for times close to 6×10^{-14} s and 1×10^{-13} s and comment upon your results.

Chaos and physical calculations

The time development of some physical systems is so sensitively dependent on some parameters of the system that the pattern of their evolution is indeterminate and apparently chaotic. However, even for these so-called chaotic systems there are underlying patterns of order.

39.1 **The nature of chaos**

When we plan a trip to the countryside or the beach we usually check the weather forecasts either in newspapers or, more commonly these days, on the web. They give the weather outlook for a few days ahead, but experience indicates that the further ahead is the forecast the more likely it is to be faulty. Why is this?

In theory, if we had a complete knowledge of worldwide atmospheric conditions at any instant then it should be possible to solve sets of partial differential equations to describe the atmospheric conditions, and hence the weather, at any place at any time in the future. The knowledge required to do this is considerable – the density, temperature, humidity, and velocity of the air at densely spaced points over the whole atmosphere plus knowledge of ground terrain over the whole surface of the Earth. Such knowledge is only crudely determined in practice but, nevertheless, in principle long-range weather forecasting seems to be feasible.

At the end of the 1950s computer technology, although crude by today's standards, had improved to the point where a one-day forecast took less than one day of computation, which made forecasting a borderline useful activity. The original programs were not very successful since, although they predicted a plausible weather pattern at future times, it was not the weather that actually occurred. One of the meteorologists working in this field, Edward Lorenz, decided to repeat a calculation he had done earlier. To save time, he started the calculation using the printed parameters from the previous run at some intermediate time. He was surprised to find that the new run gave completely different results from the previous one. He found that the reason was that the printed numbers had fewer significant digits than those in the computer, so his starting point had differed slightly from the corresponding point in the original computer run. This was bad news for meteorologists. It meant that the outcome of their calculations depended critically on very small differences in the starting parameters – differences so small that there was no chance of measuring quantities to that accuracy. So began the new mathematical theory of **chaos**. It led to a description of chaos known as the **butterfly phenomenon**. This puts forward the proposition that a butterfly flapping its wings slightly changes the movement of air in its vicinity. Over the course of time this changes the development of the state of the atmosphere to such an extent that in some distant part of the globe a tornado develops that would not have occurred had the butterfly either not flapped its wings at all or flapped them a few seconds later.

39.2 **An example from population studies**

In §24.1 the biological field of population studies was used as a practical way of introducing the concepts of eigenvectors and matrix eigenvalues. The same area of biology offers a stunning example of the occurrence of chaos. We consider a population that, when birth-rates and natural death-rates are taken into account, gives a factor of increase of population r per generation. This means that the population at the $(i+1)$th generation, x_{i+1}, is related to the population at the ith generation, x_i, by

$$x_{i+1} = rx_i. \tag{39.1}$$

The conclusion from (39.1) that, for $r > 1$, the population would increase without limit is clearly impossible since large populations impose pressure on food resources and overcrowding can lead to the propagation of disease. To model the adverse factor associated with a large population, the propagation model is changed to

$$x_{i+1} = rx_i(1 - \alpha x_i) \tag{39.2}$$

the final term providing a brake on growth. However, the behaviour of (39.2) as r is changed is rather peculiar. For $r=2$, $x_0 = 1\,000$ and $\alpha = 1.0 \times 10^{-5}$ the pattern of population size as a function of generation is shown in Figure 39.1a. The population rises from the original value to a stable value which is maintained generation after generation. Increasing r maintains this pattern until quite suddenly the pattern changes as shown in Figure 39.1b for $r=3$. The population oscillates between a higher and lower value and eventually it settles down to a two-value state. The onset of this bifurcation is quite sudden, with values of r differing by an infinitesimal amount giving behaviours like Figure 39.1a and Figure 39.1b. On increasing r further quite suddenly another bifurcation occurs, as shown in Figure 39.1c for $r=3.5$. Now in the steady state the population takes on four values forming a periodic pattern of four-generation units. As r increases further so further bifurcations occur giving a periodicity of 8 generations, 16 generations, and so on with the increments of r giving the next bifurcation becoming ever smaller. Eventually a value of r is reached beyond which the development of population becomes chaotic and completely unpredictable. This is shown for $r=4$ in Figure 39.1d . The butterfly effect also becomes apparent at this stage. Increasing r to 4.000 001, a difference of growth factor r that could not even be estimated, gives a completely different pattern of

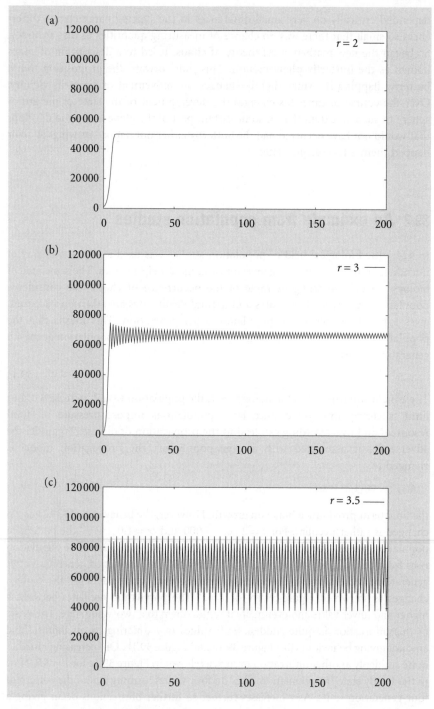

Figure 39.1 (a) Equation (39.2) with $x_0 = 1000$ and $\alpha = 1.0 \times 10^{-5}$ for (a) $r = 2$, (b) $r = 3$, (c) $r = 3.5$, (d) $r = 4$, and (e) $r = 4$ and $r = 4.000\,001$.

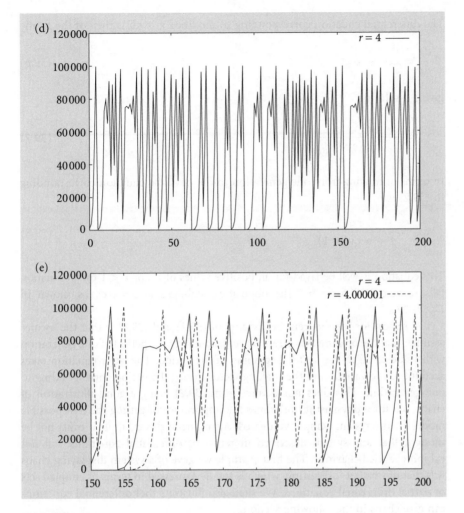

Figure 39.1 (*Contd.*)

development – still chaotic, but with different chaos. This is shown in Figure 39.1e, which shows the populations for the two values of r in some detail between generations 150 and 200.

The illustration of the chaotic behaviour of the very simple equation (39.2) is interesting, although it must be said that it is inappropriate to the problem of modified population growth which it is said to represent. The differential equation representing population growth, corresponding to (39.1), is

$$\frac{\mathrm{d}x}{\mathrm{d}t} = \lambda x, \tag{39.3}$$

giving a solution

$$x = x_0 e^{\lambda t}. \tag{39.4}$$

If t_g is the period of one generation then this gives, equivalent to (39.1),

$$x_{i+1} = rx_i \tag{39.5a}$$

where

$$r = \exp(\lambda t_g). \tag{39.5b}$$

The differential equation corresponding to modified growth is then of the form

$$\frac{dx}{dt} = \lambda x(1 - \alpha x) \tag{39.6}$$

giving a solution

$$x_t = \frac{e^{\lambda t} x_0}{1 + \alpha x_0 (e^{\lambda t} - 1)}. \tag{39.7}$$

In terms of moving from one generation to the next, the equation corresponding to (39.2) is

$$x_{i+1} = \frac{r x_i}{1 + \alpha x_i (r - 1)}. \tag{39.8}$$

This equation is well behaved for all positive values of r. For $r < 1$ the population falls to zero and for $r > 1$ the population stabilizes at $x = 1/\alpha$, as shown in Figure 39.2.

The reason for the different behaviours of (39.2) and (39.8) is that the former assumes that the braking factor remains constant over the period of one generation with value equal to that at the beginning of the period. The latter equation takes account of the variation of the braking factor *during* the period corresponding to one generation. This does not detract from (39.2) being a good illustration of chaos, but for a physicist it underlines the importance of getting the best possible model for the system being investigated. Again, from this example it must not be thought that any system described in terms of differential equations will not exhibit chaotic behaviour. The first example we gave of a system displaying chaos was that of weather forecasting, which is totally expressed in terms of coupled sets of partial differential equations. We shall also illustrate that differential equations can give chaos in the following section.

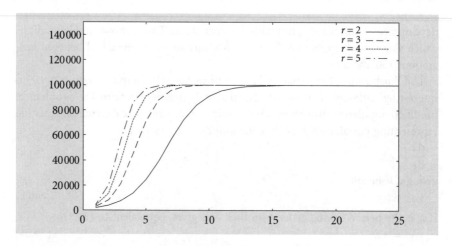

Figure 39.2 Results from (39.8) with $r = 2$, 3, 4, and 5.

39.3 **Other aspects of chaos**

The possible occurrence of chaos in physical calculations is something the physicist should be aware of, although, fortunately, it does not often occur in practice. For example, astronomers wishing to know the future state of the solar system should realize that the solar system is essentially chaotic so that tiny errors in defining the present state of the system lead to complete unpredictability of the state of the system in a period of time that is short by geological standards. It is not even certain that the solar system is stable, in the sense that planets will continue to move close to their present orbits over the future lifetime of the solar system – that is until the Sun expands to the red giant stage and swallows the inner planets.

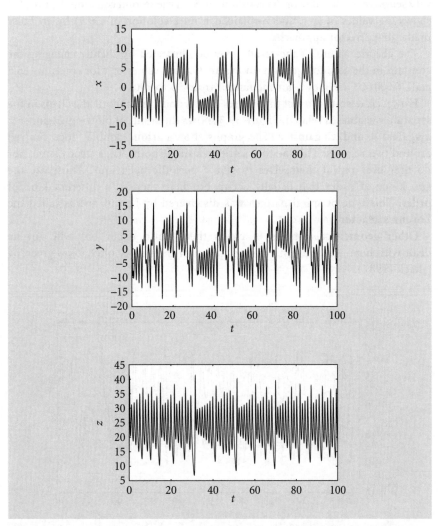

Figure 39.3 The values of x, y, and z as functions of t obtained by the solution of (39.9).

There are other aspects of chaos theory that are fascinating and of great interest to mathematicians but do not directly affect the work of the physicist. Here we shall just describe one of these that shows that in some cases, deeply submerged in chaotic behaviour, there is some semblance of order. The case we take is the set of coupled differential equations

$$\frac{dx}{dt} = 3(y - x)$$

$$\frac{dy}{dt} = -xz + 26.5x - y \qquad\qquad (39.9)$$

$$\frac{dz}{dt} = xy - z.$$

The values of x, y, and z that come from solving these equations are chaotic and very sensitive to the value of the constant, 26.5, in the second equation. Figure 39.3 shows the values of x, y, and z obtained from a solution of (39.9) by an Euler predictor–corrector approach.

The chaotic nature of the results may be illustrated by slightly changing the constant in the second equation. In Figure 39.4 the values of x for constants 26.5 and 26.500 001 are shown for the range t from 50 to 100.

However, despite the fact that values of x, y, and z are chaotic the chaos in the separate variables is related. Figure 39.5 shows the result of plotting y against x, z against y, and x against z. The graphs show curious double loops centred around two regions. They neither settle to a single point, i.e. a steady state, nor do they ever repeat themselves to give a periodic behaviour. Those are the two kinds of order that usually occur, but here there is a different kind of order. This type of spiral pattern was discovered by Lorenz and is called the **Lorenz attractor**.

Other geometrical aspects of chaos theory also occur but will not be dealt with here. A good general description of the theory of chaos is given by Gleick (1987).

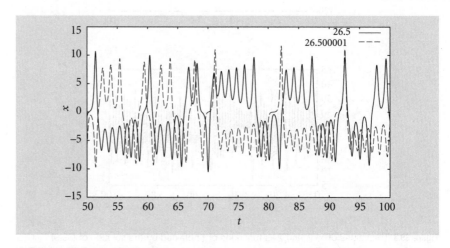

Figure 39.4 A comparison of values of x when the constant in the second equation of (39.9) is changed from 26.5 to 26.500 001.

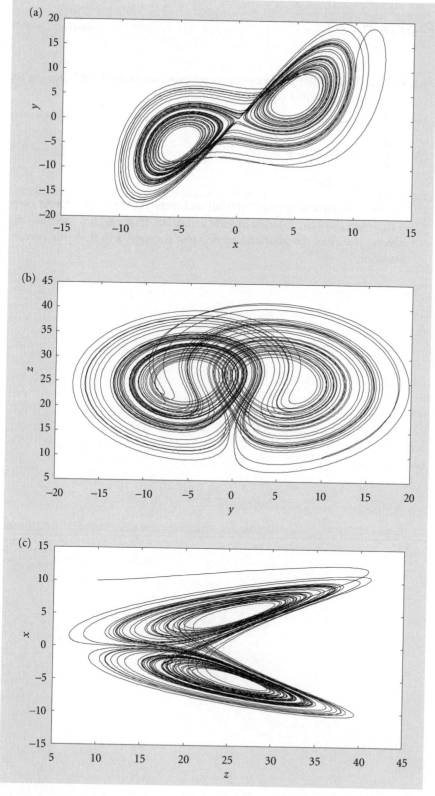

Figure 39.5 Lorenz attractors for (a) y against x, (b) z against y, and (c) x against z.

Problem

39.1 Write a simple Euler predictor–corrector program to solve the following set of coupled differential equations.

$$\frac{dx}{dt} = 10(y - x)$$

$$\frac{dy}{dt} = x(28 - z) - y$$

$$\frac{dz}{dt} = xy - 8z/3.$$

Use an integration step length of 0.001 and make 100 000 steps. Output every 5 steps will show the solution in sufficient detail. Plot x against t to show the chaotic nature of the solution and also plot y against x to show the Lorenz attractor.

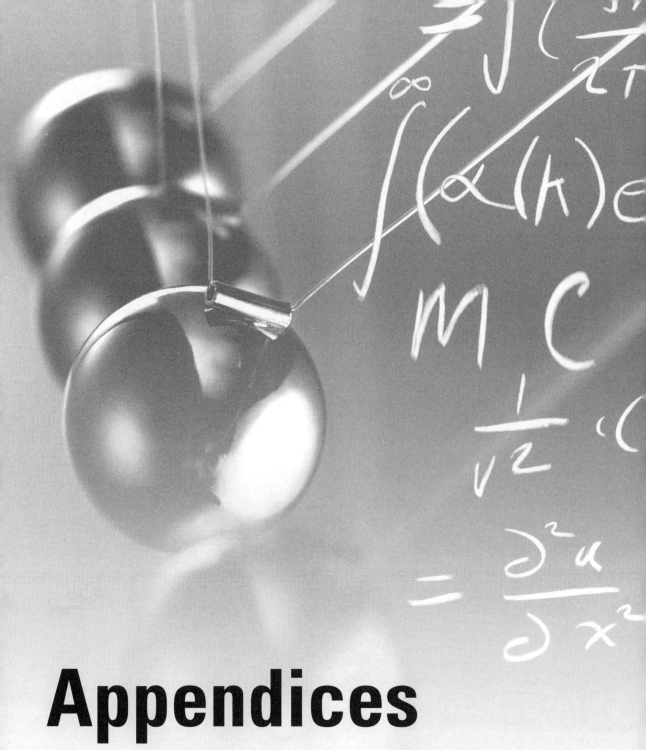

Appendices

Appendix 1 **Table of integrals**

For brevity the constant of integration has been excluded.

$f(x)$	$\int f(x)dx$	$f(x)$	$\int f(x)dx$
$x^n \ (n \neq -1)$	$\dfrac{x^n}{n+1}$	$\ln x$	$x \ln x - x$
$\dfrac{1}{x}$	$\ln x$	$\sinh x$	$\cosh x$
$e^{cx} \ (c \neq 0)$	$\dfrac{1}{c}e^{cx}$	$\cosh x$	$\sinh x$
$c^x \ (c > 0)$	$\dfrac{1}{\ln c}c^x$	$\tanh x$	$\ln(\cosh x)$
$\sin x$	$-\cos x$	$\coth x$	$\ln(\sinh x)$
$\cos x$	$\sin x$	$\operatorname{sech}^2 x$	$\tanh x$
$\tan x$	$\ln(\sec x)$	$\operatorname{cosech}^2 x$	$-\coth x$
$\cot x$	$\ln(\sin x)$	$\dfrac{1}{\sqrt{x^2 + c^2}}$	$\sinh^{-1}\left(\dfrac{x}{c}\right)$
$\sec^2 x$	$\tan x$	$\dfrac{1}{\sqrt{x^2 - c^2}}$	$\cosh^{-1}\left(\dfrac{x}{c}\right)$
$\operatorname{cosec}^2 x$	$-\cot x$	$\dfrac{1}{c^2 - x^2}$	$\dfrac{1}{c}\tanh^{-1}\left(\dfrac{x}{c}\right)$
$\dfrac{1}{\sqrt{c^2 - x^2}}$	$\sin^{-1}\left(\dfrac{x}{c}\right)$	$\dfrac{1}{x^2 - c^2}$	$\dfrac{1}{c}\coth^{-1}\left(\dfrac{x}{c}\right)$
$\dfrac{1}{c^2 + x^2}$	$\dfrac{1}{c}\tan^{-1}\left(\dfrac{x}{c}\right)$	$\dfrac{f'(x)}{f(x)}$	$\ln(f(x))$

Appendix 2 **Inverse Fourier transform**

In this appendix the inverse Fourier transform relation in (17.41b) is derived starting with the Fourier transform relation, (17.30):

$$X(\omega) = \int_{-\infty}^{\infty} x(t) \exp(-i\omega t)dt \tag{A2.1}$$

First multiply both sides by the factor $\exp(i\omega t')$, where t' is a time variable that is different from t giving

$$X(\omega) \exp(i\omega t') = \exp(i\omega t') \int_{-\infty}^{\infty} x(t) \exp(-i\omega t)dt = \int_{-\infty}^{\infty} x(t) \exp(i\omega(t' - t))dt \tag{A2.2}$$

Now integrate both sides over the frequency ω between the limits

$$-\frac{W}{2} \leq \omega \leq \frac{W}{2} \tag{A2.3}$$

$$\int_{-W/2}^{W/2} X(\omega) \exp(i\omega t')d\omega = \int_{-\infty}^{\infty} x(t)\left\{ \int_{-W/2}^{W/2} \exp(i\omega(t' - t))d\omega \right\}dt = \int_{-\infty}^{\infty} x(t)I(W)dt \tag{A2.4}$$

where

$$I(W) = \int_{-W/2}^{W/2} \exp(i\omega(t' - t))d\omega. \tag{A2.5}$$

Evaluate the integral $I(W)$ by integrating over ω:

$$I(W) = \left[\frac{\exp(i\omega(t'-t))}{i(t'-t)}\right]_{-W/2}^{W/2} = \left[\frac{\exp\left(i\frac{W}{2}(t'-t)\right) - \exp\left(-i\frac{W}{2}(t'-t)\right)}{i(t'-t)}\right]$$

$$= \frac{2\sin\left(\dfrac{W(t'-t)}{2}\right)}{(t'-t)} \tag{A2.6}$$

Substituting for $I(W)$ from (A2.6) into (A2.4):

$$\int_{-W/2}^{W/2} X(\omega)\exp(i\omega t')d\omega = 2\int_{-\infty}^{\infty} x(t)\frac{\sin\left(\dfrac{W(t'-t)}{2}\right)}{t'-t}dt. \tag{A2.7}$$

We now take the limit as $W \to \infty$ on both sides of (A2.7):

$$\int_{-\infty}^{\infty} X(\omega)\exp(i\omega t')d\omega = 2\int_{-\infty}^{\infty} x(t)\left[\frac{\sin\left(\dfrac{W(t'-t)}{2}\right)}{t'-t}\right]_{W\to\infty} dt. \tag{A2.8}$$

To evaluate the limit in the term on the right-hand side of (A2.8), we use (17.36a), which is the definition of a δ-function:

$$\delta(x) = \frac{1}{\pi}\left[\frac{\sin\left(\dfrac{xT}{2}\right)}{x}\right]_{T\to\infty}. \tag{A2.9}$$

Substituting $x \to t' - t$, $T \to W$ in this formula, we obtain the result:

$$\delta(t'-t) = \frac{1}{\pi}\left[\frac{\sin\left(\dfrac{(t'-t)W}{2}\right)}{t'-t}\right]_{W\to\infty} \tag{A2.10}$$

or

$$\left[\frac{\sin\left(\dfrac{(t'-t)W}{2}\right)}{t'-t}\right]_{W\to\infty} = \pi\delta(t'-t). \tag{A2.11}$$

Inserting the result (A2.11) on the right-hand side of (A2.8):

$$\int_{-\infty}^{\infty} X(\omega)\exp(i\omega t')d\omega = 2\pi\int_{-\infty}^{\infty} x(t)\delta(t'-t)dt. \tag{A2.12}$$

Using the property in (17.38) for δ-functions, on the right-hand side of (A2.12):

$$\int_{-\infty}^{\infty} x(t)\delta(t'-t)dt = x(t') \tag{A2.13}$$

or

$$\int_{-\infty}^{\infty} X(\omega)\exp(i\omega t')d\omega = 2\pi x(t'). \tag{A2.14}$$

Since the variable t' was arbitrary it can be replaced by any variable we want, so let us replace $t' \to t$ in which case (A2.14) becomes, after rearranging:

$$x(t) = \frac{1}{2\pi}\int_{-\infty}^{\infty} X(\omega)\exp(i\omega t)d\omega, \tag{A2.15}$$

which is the required result.

Appendix 3 **Fourier transform of a sampled signal**

Let the continuous signal be $x(t)$. If the sampling frequency is f_s, then the time between samples is given by

$$T = \frac{1}{f_s} \tag{A3.1}$$

The sampling function can be represented as a periodic train of pulses, $p(t)$ with period T, width τ and height $1/\tau$ as illustrated in Figure A3.1: later we shall take the limit $\tau \to 0$. The sampled signal can then be approximated as

$$x_s(t) = x(t)p(t). \tag{A3.2}$$

The function $p(t)$ can be expanded as a Fourier series:

$$p(t) = \frac{1}{T} + \frac{2}{\pi} \sum_{n=1}^{\infty} \frac{1}{n\tau} \sin\left(\frac{\pi n\tau}{T}\right) \cos\left(\frac{2\pi nt}{T}\right) \tag{A3.3}$$

where T is the time between pulses.

Substituting for $p(t)$ from (A3.3) into (A3.2) we find

$$x_s(t) = \frac{1}{T} x(t) + \frac{2}{\pi} \sum_{n=1}^{\infty} \left[\frac{1}{n\tau} \sin\left(\frac{\pi n\tau}{T}\right) \right] x(t) \cos\left(\frac{2\pi nt}{T}\right). \tag{A3.4}$$

Taking the limit $\tau \to 0$ where the sampling is 'ideal', i.e. the samples have zero width,

$$x_s(t) = \frac{1}{T} x(t) + \frac{2}{\pi} \sum_{n=1}^{\infty} \left[\frac{1}{n\tau} \sin\left(\frac{\pi n\tau}{T}\right) \right]_{\tau \to 0} x(t) \cos\left(\frac{2\pi nt}{T}\right). \tag{A3.5}$$

Now, using L'Hôpital's rule, §4.7.2,

$$\left[\frac{1}{n\tau} \sin\left(\frac{\pi n\tau}{T}\right) \right]_{\tau \to 0} = \frac{\pi}{T} \tag{A3.6}$$

giving

$$x_s(t) = \frac{1}{T} x(t) + \frac{2}{T} \sum_{n=1}^{\infty} x(t) \cos\left(\frac{2\pi nt}{T}\right). \tag{A3.7}$$

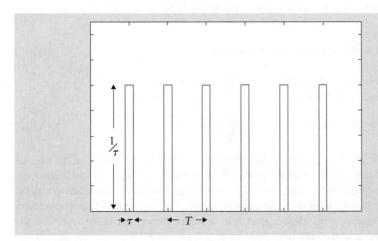

Figure A3.1 Sampling function, $p(t)$.

Substituting

$$T = \frac{1}{f_s}$$

$$x_s(t) = f_s x(t) + 2f_s \sum_{n=1}^{\infty} x(t) \cos(2\pi n f_s t) \tag{A3.8}$$

Taking Fourier transforms of both sides of this equation:

$$X_s(f) = f_s X(f) + 2f_s \sum_{n=1}^{\infty} \int_{-\infty}^{\infty} [x(t) \cos(2\pi n f_s t)] \exp(-i2\pi f t) dt. \tag{A3.9}$$

Now we use (6.12a), with $x = 2\pi n f_s t$, to expand the cosine in the above integral as a sum of complex exponentials:

$$\cos(2\pi n f_s t) = \frac{1}{2} [\exp(i2\pi n f_s t) + \exp(-i2\pi n f_s t)]. \tag{A3.10}$$

Substituting from (A3.10) into (A3.9):

$$X_s(f) = f_s X(f) + f_s \sum_{n=1}^{\infty} \int_{-\infty}^{\infty} x(t) [\exp(i2\pi n f_s t) + \exp(-i2\pi n f_s t)] \exp(-i2\pi f t) dt \tag{A3.11}$$

or

$$X_s(f) = f_s X(f) + f_s \sum_{n=1}^{\infty} \int_{-\infty}^{\infty} x(t) \exp(-i2\pi (f - n f_s) t) dt$$

$$+ f_s \sum_{n=1}^{\infty} \int_{-\infty}^{\infty} x(t) \exp(-i2\pi (f + n f_s) t) dt. \tag{A3.12}$$

The basic definition of the Fourier transform of the analogue signal $x(t)$ is given by (17.28):

$$X(f) = \int_{-\infty}^{\infty} x(t) \exp(-i2\pi f t) dt. \tag{A3.13}$$

Substituting $f \rightarrow f - n f_s$ into this expression,

$$X(f - n f_s) = \int_{-\infty}^{\infty} x(t) \exp(-i2\pi (f - n f_s) t) dt \tag{A3.14}$$

which is the integral in the second term on the right-hand side of (A3.12).
Similarly,

$$X(f + n f_s) = \int_{-\infty}^{\infty} x(t) \exp(-i2\pi (f + n f_s) t) dt \tag{A3.15}$$

which is the integral in the third term on the right-hand side of (A3.12).
Hence, using (A3.14) and (A3.15), (A3.12) can be rewritten as:

$$X_s(f) = f_s X(f) + f_s \sum_{n=1}^{\infty} X(f - n f_s) + f_s \sum_{n=1}^{\infty} X(f + n f_s). \tag{A3.16}$$

The first term on the right-hand side represents the Fourier transform of the original continuous signal. For the example in Figure 18.10, this is the band centred at zero frequency. The nth term in the first summation is the continuous Fourier transform centred at frequency $n f_s$. The nth term in the second summation is the continuous Fourier transform centred at frequency $-n f_s$. Hence, apart from a scaling factor, f_s, (A3.16) represents the Fourier transform of the sampled signal shown in Figure 18.10.

Appendix 4 **Derivation of the discrete and inverse discrete Fourier transforms**

The DFT assumes that the underlying signal is periodic. First consider a periodic, *continuous* signal, $x(t)$; we will sample it later. As $x(t)$ is periodic, it can be expanded as a Fourier series. In particular, the complex Fourier coefficient for the pth harmonic is given, from §17.14, as

$$c_p = \frac{1}{2}(a_p - ib_p) \tag{A4.1}$$

where, from (17.75) and (17.76):

$$a_p = \frac{2}{T} \int_0^T x(t) \cos(p\omega_0 t) \, dt \tag{A4.2}$$

and

$$b_p = \frac{2}{T} \int_0^T x(t) \sin(p\omega_0 t) \, dt. \tag{A4.3}$$

This gives

$$c_p = \frac{1}{2}\left[\frac{2}{T} \int_0^T x(t) \cos(p\omega_0 t) \, dt - i\frac{2}{T} \int_0^T x(t) \sin(p\omega_0 t) \, dt\right] = \frac{1}{T} \int_0^T x(t) \exp(-ip\omega_0 t) \, dt. \tag{A4.4}$$

Making the substitution $\omega_0 = 2\pi f_0$ where f_0 is the fundamental frequency in Hz,

$$c_p = \frac{1}{T} \int_0^T x(t) \exp(-2\pi i p f_0 t) \, dt. \tag{A4.5}$$

In the above form, the expression for the complex Fourier coefficient is only appropriate for *analogue* systems, as one is taking the *continuous* signal $x(t)$ and performing an integral. We now consider how to determine c_p for a *sampled signal*.

The sampled function may be considered as the product of the continuous function $x(t)$ with a set of δ-functions of the form $\delta(t - n\tau)$ for $0 \leq n \leq N - 1$, where there are N samples in one period of duration T. Hence the times of the samples are of the form $n\tau$ and we can replace the integral in (A4.5) by a summation over all the samples to give

$$c_p = \frac{1}{T} \sum_{n=0}^{N-1} x(n\tau) \exp(-2\pi i p f_0 n\tau). \tag{A4.6}$$

Substituting $f_0 = 1/T$ and $\tau = 1/f_s$,

$$c_p = \frac{1}{T} \sum_{n=0}^{N-1} x(n\tau) \exp\left(-\frac{2\pi i n p}{T f_s}\right). \tag{A4.7}$$

Since a period T of the signal corresponds to N samples,

$$N = T f_s. \tag{A4.8}$$

Substituting for $T f_s$ from (A4.8) into (A4.7) and making the notational changes,

$$T c_p \rightarrow X[p] \quad \text{and} \quad x(n\tau) \rightarrow x[n]$$

we obtain the required result, (18.6):

$$X[p] = \sum_{n=0}^{N-1} x[n] \exp\left(-i\frac{2\pi n p}{N}\right). \tag{A4.9}$$

To obtain the inverse discrete Fourier transform relationship, (18.8), we multiply both sides of (A4.9) by $\exp(2\pi i n' p/N)$ where n' is an integer. This gives

$$X[p]\exp\left(\frac{2\pi i n' p}{N}\right) = \sum_{n=0}^{N-1} x[n]\exp\left(-\frac{2\pi i n p}{N}\right)\exp\left(\frac{2\pi i n' p}{N}\right) = \sum_{n=0}^{N-1} x[n]\exp\left(\frac{2\pi i(n'-n)p}{N}\right).$$

(A4.10)

Now sum both sides of this equation over p between $p=0$ and $N-1$:

$$\sum_{p=0}^{N-1} X[p]\exp\left(\frac{2\pi i n' p}{N}\right) = \sum_{p=0}^{N-1}\sum_{n=0}^{N-1} x[n]\exp\left(\frac{2\pi i(n'-n)p}{N}\right)$$

(A4.11)

or

$$\sum_{p=0}^{N-1} X[p]\exp\left(\frac{2\pi i n' p}{N}\right) = \sum_{n=0}^{N-1} x[n]\sum_{p=0}^{N-1}\exp\left(\frac{2\pi i(n'-n)p}{N}\right).$$

(A4.12)

The summation over p on the right-hand side is a geometrical series (§3.2) of N terms, the first of which is 1 and with the constant factor $\exp(2\pi i(n'-n)/N)$. The sum is

$$S = \frac{1-\left[\exp\left(\dfrac{2\pi i(n'-n)}{N}\right)\right]^N}{1-\exp\left(\dfrac{2\pi i(n'-n)}{N}\right)} = \frac{1-\exp(2\pi i(n'-n))}{1-\exp\left(\dfrac{2\pi i(n'-n)}{N}\right)}.$$

(A4.13)

Now, in the numerator of S,

$$\exp(i[2\pi(n'-n)]) = \cos(2\pi(n'-n)) + i\sin(2\pi(n'-n)).$$

(A4.14)

Since n and n' are integers, then so is $n'-n$. Hence $\cos(2\pi(n'-n))=1$ and $\sin(2\pi(n'-n))=0$, in which case, from (A4.14):

$$1-\exp\{2\pi i(n'-n)\} = 0.$$

(A4.15)

In general, for $n \neq n'$, in the denominator of (A4.13), $\exp(2\pi i(n'-n)/N) \neq 1$ and hence the denominator is non-zero. Therefore

$$S = 0 \quad \text{for} \quad n' \neq n.$$

(A4.16)

The only exception to this is when $n'=n$. In this case, $\exp(2\pi i(n'-n)/N)=1$ and the denominator of S in (A4.13) is also zero. Hence, for $n'=n$, $S=0/0$, which is undefined. However, we now use l'Hôpital's rule, (§4.7.2), to find the limit of S when $n'=n$. Differentiating the numerator and denominator of (A4.13) with respect to n',

$$S = \left[\frac{\dfrac{d}{dn'}\{1-\exp\{2\pi i(n'-n)\}\}}{\dfrac{d}{dn'}\left\{1-\exp\left\{\dfrac{2\pi i(n'-n)}{N}\right\}\right\}}\right]_{n'\to n} = \left[\frac{-2\pi i\exp\{2\pi i(n'-n)\}}{-\dfrac{2\pi i}{N}\exp\left\{\dfrac{2\pi i(n'-n)}{N}\right\}}\right]_{n'\to n} = N.$$

(A4.17)

It can be seen that the summation over p on the right-hand side of (A4.12) is only non-zero when $n=n'$, when it equals N so that (A4.12) becomes

$$\sum_{p=0}^{N-1} X[p]\exp\left(\frac{2\pi i n' p}{N}\right) = N x[n']$$

(A4.18)

or

$$x[n'] = \frac{1}{N}\sum_{p=0}^{N-1} X[p]\exp\left(\frac{2\pi i n' p}{N}\right).$$

(A4.19)

We can use any variable other than n' for the time index, so if we change $n' \rightarrow n$, we obtain the result in (18.8) for the inverse discrete Fourier transform.

Appendix 5 **Program OSCILLATE**

Initialization

Set up standard values of $m, f, k, F, \omega, x_0, u_0$

Fix values of array $w(1) = 0, w(2) = 0.5, w(3) = 0.5, w(4) = 1$

Set up files for output as later requested

Make provision for change of standard variables as requested by user

Input a

Set integration timestep, h, **as** 0.01 **of period corresponding to** ω

Point A:

Loop to integrate coupled differential equations in batches of 1000 timesteps

    Loop for ith timestep to calculate values of $\delta_{j,x}$ **and** $\delta_{j,u}$ **for** j **from 1 to 4.**

$$\delta_{j,x} = hf_x\{x_i + w(j)\delta_{j-1,x}, u_i + w(j)\delta_{j-1,u}\} \ (f_x \text{ is rhs of } 19.12b)$$
$$\delta_{j,u} = hf_u\{x_i + w(j)\delta_{j-1,x}, u_i + w(j)\delta_{j-1,u}\} \ (f_u \text{ is rhs of } 19.12a)$$

    End loop

    Store $x_{i+1} = x_i + \frac{1}{6}(\delta_{1,x} + 2\delta_{2,x} + 2\delta_{3,x} + \delta_{4,x})$

    Store $u_{i+1} = u_i + \frac{1}{6}(\delta_{1,u} + 2\delta_{2,u} + 2\delta_{3,u} + \delta_{4,u})$

End loop

Output if requested

    Is output required of some or all of previous batch of output?

    If 'yes' then output to files and move to next section

    If 'no' move to next section

Continue the calculation?

    If 'no' then stop

    If 'yes' then back to point A to calculate next batch.

Appendix 6 **Program EXPLICIT**

This program differs from HEATCRNI only in the routine CYCLE. For graphical output six files are available. If a seventh graphical output is requested it overwrites the first, and so on.

Intitialization

    Set up standard values of $Blen$ **(length of bar),** $\kappa, c,$ **and** ρ

    Fix temperatures at the bar ends as functions of time

Make provision for change of standard variables as requested by user

Input N, **the number of segments in the bar and determine** Δx

Read in initial temperatures at internal points

Read in timestep, h **from which** r **is calculated or** r **directly**

Choose graphical or printed output

    Select number of timesteps between outputs, m

Point A:

Loop for m **timesteps**

To routine CYCLE to advance temperatures in one timestep

For internal points

$$\theta_j = r(\theta_{j-1} + \theta_{j+1}) + (1 - 2r)\theta_j$$

End loop

Printed or file output

Continue the calculation?
   If 'no' then stop
   If 'yes' then back to point A to calculate next batch

Appendix 7 **Program HEATCRNI**

This program differs from EXPLICIT only in the routine CYCLE. For graphical output six files are available. If a seventh graphical output is requested it overwrites the first, and so on.

Initialization
   Set up standard values of $Blen$ **(length of bar),** κ, c **and** ρ
   Fix temperatures at the bar ends as functions of time
Make provision for change of standard variables as requested by user
Input N, **the number of segments in the bar and determine** Δx
Read in initial temperatures at internal points
Read in timestep, Δt, **from which** r **is calculated or** r **directly**
Choose graphical or printed output
   Select number of timesteps between outputs, m

Point A
Loop for m timesteps
To routine CYCLE to advance temperatures in one timestep
   For internal points set up table of coefficients for

$$-r\theta_{i-1, j+1} + 2(1 + r)\theta_{i, j+1} - r\theta_{i+1, j+1} = r\theta_{i-1, j} + (1 - 2r)\theta_{i, j} + r\theta_{i+1, j}$$

   Solve for temperatures at time $j + 1$ **from tridiagonal matrix coefficients**
End loop

Printed or file output

Continue the calculation?
   If 'no' then stop
   If 'yes' then back to point A to calculate next batch.

Appendix 8 **Program SIMPLATE**

This solves for the thermal equilibrium in a plate with insulated top and bottom surfaces. Boundaries may be at fixed temperatures, insulated or exchanging heat with the outside.

The plate is defined on a square mesh contained within a rectangular region with n_r rows and n_c columns. As provided 12 rows and columns are designated, but the user may change this. The plate must be defined within a 10×10 mesh so that ghost points can also be accommodated. All sides of the plate must be along the principal directions of the mesh. Each side of the plate must contain at least three points.

Initialization
> **Define plate on a character grid by:**
> **Numerical value for fixed-temperature boundary point**
> I = **an insulated boundary point**
> E = **boundary point exchanging heat with outside**
> U = **point inside the plate whose temperature is to be determined**
> **Read in values of K and S for boundary E-points (equation 20.30).**
> **Set up the form of Q_s, the heat generation as function of position on plate**

Set maximum number of iterations, n_I, for determining temperatures
Set tolerance for temperature difference between temperature refinement steps

Loop – temperature determination step
> **Loop through plate points on rectangular grid covering plate**
> **If I, E, or U point then examine four neighbouring points giving 'star'**
> **If neighbouring point is:**
> > **Constant temperature or U, I or E point then note temperature**
> > **Ghost point related to I surface find temperature of corresponding interior point**
> >
> > **Ghost point related to E surface find temperature from corresponding interior**
> > > **point and K and S values.**
> >
> > **Find Q_s**
> **Use neighbouring point temperatures to update temperature of point under consideration**
> **End loop**
Test for maximum change of temperature during this loop. If less than tolerance then stop

End loop
Maximum number of iterations exceeded so stop

Appendix 9 **Program STRING1**

A string of mass m per unit length, of length l, and under tension T is fixed at both ends. It is subjected to an initial displacement and then released. The program finds subsequent displacements as a function of time.

The displacements for every IQ timesteps can be given in numerical form, either through the printer or on the screen. If required six data files, STRING0.dat to STRING5.DAT, giving values of (x, y), can be produced starting from timestep M1 and then every L1 timesteps thereafter. These can be used for graphical output later.

Standard values of m, l, T, and r are provided but may be changed by the user. The program runs for a simulated 10 ms or for 6 data-file outputs, whichever is shorter.

> **Input number of divisions, N, of the string. Hence find Δx and Δt from value of r.**
> **Input initial displacements, y, at $N - 1$ internal points of string. This can be done through the keyboard or by analytical formula.**
> **Read in IQ for screen or printer output and M1 and L1 if data files are required.**

> **Loop – timesteps**
> > **Loop interior points**
> > > **For first step** $\quad z_i = r(y_{i+1} + y_{i-1})/2 + (1 - r)y_i$
> > > $$u_i = z_i$$

For subsequent steps
Loop interior points

$$y_i = z_i$$

$$z_i = u_i$$

End loop

$$u_i = r(z_{i+1} + z_{i-1}) + 2(1 - r)z_i - y_i$$

End loop

Is number of timesteps divisible by IQ?
   If 'no' then go to A
Printer output?
   If 'no' then go to B
Output (x, y) on printer
Go to A
Point B   **Output (x, y) on screen**
Point A   **Is number of timesteps of the form $M1 + kL1$ where k is an integer?**
                  If 'no' go to C
                  Write output file STRINGk.DAT
                  If $k = 6$ then stop
Point C   **If total time ≥ 10 ms then stop**
End loop

Appendix 10 **Program DRUM**

A circular drum skin of mass m per unit area, of radius S and with tautness T is fixed at its boundaries. It is given a circularly symmetric initial displacement and then released. The program finds subsequent circularly symmetric displacements as a function of time.

The displacements for every IQ timesteps can be given in numerical form, either through the printer or on the screen. If required six data files, DRUM0.dat to DRUM5.DAT, giving values of (x, y) can be produced starting from timestep M1 and then every L1 timesteps thereafter. These can be used for graphical output later.

Standard values of m, S, T, and r are provided but may be changed by the user. The program runs for a simulated 10 ms or for 6 data-file outputs, whichever is shorter.

Input number of divisions, N, of the radius. Hence find Δs and Δt from value of r.
Input initial displacements, y, at $N - 1$ points along the radius.
Read in IQ for screen or printer output and M1 and L1 if data files are required.

Loop – timesteps
   Loop interior points
   For first step
For centre point $z_1 = ry_2 + (i - r)y_1$
                           $u_1 = z_1$

Loop for other points $z_1 = \dfrac{r}{2}\left\{ y_{i+1}\left(1 + \dfrac{\Delta s}{2s_i}\right) + y_{i-1}\left(1 - \dfrac{\Delta s}{2s_i}\right)\right\} + (1 - r)y_i$

   $u_i = z_i$

End loop
   For subsequent steps
Loop interior points

$$y_i = z_i$$

$$z_i = u_i$$

End loop

For centre point $\qquad u_1 = 2rz_2 + 2(1 - r)z_1 - y_1$

Loop for other points

$$u_i = r\left\{ z_{i+1}\left(1 + \frac{\Delta s}{2s_i}\right) + z_{i-1}\left(1 - \frac{\Delta s}{2s_i}\right)\right\} + 2(1 - r)z_i - y_i$$

    End loop

End loop

Is number of timesteps divisible by IQ?

    If 'no' then go to A

Printer output?

    If 'no' then go to B

Output (x, y) on printer

    Go to A

Point B   **Output (x, y) on screen**

Point A   **Is number of timesteps of the form $M1 + kL1$ where k is an integer?**

        If 'no' go to C

        Write output file DRUMk.DAT

        If $k = 6$ then stop

Point C   **If total time $\geq$ 10ms then stop**

End loop

Appendix 11 **Program SHOOT**

This program integrates for a perturbed particle in a box from $y = 0$ to $y = 1$ with a trial value of E. The TISWE solved is given by (21.109) and the coupled differential equations must be in the form (21.111). The Runge–Kutta method of solution is used.

Read in trial value of E

Set weights $w(1) = 0$, $w(2) = 0.5$, $w(3) = 0.5$, $w(4) = 1$

Set up boundary conditions: $y = 0$

                                 $\psi = 0$

                         and $\chi = 1$ (arbitrary slope)

Set integration step length as $h = 0.01$

Loop from $i = 0$ to 99

$y = hi$

    Loop from $j = 1$ to 4

    $d\psi(j) = h\{\chi + w(j)d\chi(j-1)\}$

    $d\chi(j) = h[0.26224 + \{20.0 \sin(\pi(y + w(j)h) - E\}$

    End loop

$\psi = \psi + \{d\psi(1) + 2d\psi(2) + 2d\psi(3) + d\psi(4)\}/6$

$\chi = \chi + \{d\chi(1) + 2d\chi(2) + 2d\chi(3) + d\chi(4)\}/6$

Write (y, ψ) to file

End loop

    Print final ψ to check boundary value

End program.

Appendix 12 **Program DRUNKARD**

A modified random-walk program in which unit steps are restricted to be along the x and y directions. After the first step the probabilities of following steps are assigned as follows:

P(1) probability that in same direction as previous step
P(2) probability that in opposite direction as previous step
$P(3) = P(4)$ probabilities of turning right or turning left.
P(1) and P(2) are set in the program, as is the number of Monte Carlo trials, nr, set at 10^6, but these may be changed by the user before running the program.

> **Calculate cumulative probabilities**
> $cum(1) = P(1)$
> $cum(2) = P(1) + P(2)$
> $cum(4) = 1$
> $cum(3) = (cum(2) + cum(4))/2.0$
>
> **Read in, n, number of steps for the random walk**
>
> **Loop from $i = 1$ to nr (number of trials)**
> **Set** $sumx = 1$ **(arbitrary first step in x direction)**
> $sumy = 0$
> **Loop from $j = 1$ to $n - 1$ (giving n steps, including the first already taken)**
>   **Set $id = 0$**   **In what follows $id = 0$ gives step in direction $+x$**
>   **$id = 1$ gives step in direction $-x$**
>   **$id = 2$ gives step in direction $+y$**
>   **$id = 3$ gives step in direction $-y$**
>   **Generate $r(uv(0,1))$ by random number generator routine**
> $iadd = 3$
>   **If $cum(3) > r$** $iadd = 1$
>   **If $cum(2) > r$** $iadd = 2$
>   **If $cum(1) > r$** $iadd = 0$
>   **$id = (id + iadd)$ modulo 4**
>   **If $id = 0$** $sumx = sumx + 1$
>   **If $id = 1$** $sumx = sumx - 1$
>   **If $id = 2$** $sumy = sumy + 1$
>   **If $id = 3$** $sumy = sumy - 1$
>   **End loop**
> $d(i) = sqrt(sumx^2 + sumy^2)$
> **End loop**
>   **Output the r.m.s distance of travel** $\left\{ \sum_{i=1}^{nr} d(i)^2 / nr \right\}^{1/2}.$

Appendix 13 **Program POLYMER**

A modified random-walk program in which unit steps are restricted to be along the x and y directions. A step that produces an intersection in the path is repeated up to three more times to attempt to avoid the intersection but if still unsuccessful then the trial is abandoned. The mean length between the ends of the polymer are calculated from the results of the successful trials. After the first step the probabilities of following steps are assigned as follows:

P(1) probability that in same direction as previous step
P(2) probability that in opposite direction as previous step $(= 0)$

$P(3) = P(4)$ probabilities of turning right or turning left.

$P(1)$ and $P(2)$ are set in the program, as is the number of Monte Carlo trials, nr, set at 10^5, but these may be changed by the user before running the program.

Input n, the number of units of the polymer.

Calculate cumulative probabilities

$cum(1) = P(1)$

$cum(2) = P(1) + P(2)$

$cum(4) = 1$

$cum(3) = (cum(2) + cum(4))/2.0$

Clear a table linktab(i, j) for i and $j = -100$ to 100 (or equivalent) in which (0, 0) represents the starting point. An unoccupied point in the x-y plane has linktab(i, j) $= 0$ while an occupied point has linktab(i, j) $= 1$.

Set $nsum = 0$ ($nsum$ gives the number of successful trials)

Set $sumd = 0$ ($sumd$ gives $\sum d^2$ where d is the length of a successful trial)

Loop A from $i = 1$ to nr (number of trials)

Set $ix = 1$ (arbitrary first step in x direction)

** $iy = 0$**

** $id = 0$**

In what follows $id = 0$ gives step in direction $+x$

** $id = 1$ gives step in direction $-x$**

** $id = 2$ gives step in direction $+y$**

** $id = 3$ gives step in direction $-y$**

Loop B from $j = 1$ to $n - 1$ (giving n steps, including the first already taken)

** $itry = 0$**

Point C $itry = itry + 1$

**   If $itry > 4$ goto End loop A**

Generate r ($uv(0, 1)$) by random number generator routine

	$iadd = 3$
If $cum(3) > r$	$iadd = 1$
If $cum(2) > r$	$iadd = 2$
If $cum(1) > r$	$iadd = 0$

** $id = (id + iadd)$ modulo 4**

If $id = 0$

**   If linktab($ix + 1, iy$) $= 1$ goto C**

** $ix = ix + 1$**

**   linktab(ix, iy) $= 1$**

**   goto End loop B**

If $id = 1$

**   If linktab($ix, iy + 1$) $= 1$ goto C**

** $iy = iy + 1$**

**   linktab(ix, iy) $= 1$**

**   goto End loop B**

If $id = 2$

**   If linktab($ix - 1, iy$) $= 1$ goto C**

** $ix = ix - 1$**

**   linktab(ix, iy) $= 1$**

**   goto End loop B**

If $id = 3$

**   If linktab($ix, iy - 1$) $= 1$ goto C**

** $iy = iy - 1$**

linktab$(ix, iy) = 1$
End loop B
$sumd = sumd + sqrt(ix^2 + iy^2)$
$nsum = nsum + 1$
End loop A

Output number of successful trials nsum
Output the r.m.s. distance between ends of polymer $\{sumd/nsum\}^{1/2}.$

Appendix 14 **Program METROPOLIS**

This is a cell-structure Monte Carlo program in which the cell contains 125 atoms (or molecules). The atoms are first placed on a $5 \times 5 \times 5$ regular grid. Thereafter new configurations are generated by moving atoms, one at a time, in random directions by up to $d/20$, where d is a cell edge.

Before information is taken from the system a number, *nstart* (set at 50 000) atom moves are made to allow the system to settle down. Thereafter every 10th configuration is sampled to provide both a contribution to the virial term and contributions to the radial density function. The total number of generated configurations is *ntot* (set at 100 000). The information for the radial density function is in file RADIST.DAT.

Set constants (σ, ε) for the Lennard–Jones potential, Boltzmann
constant, and temperature.
The range of the Lennard–Jones potential is set at 2.4σ where the force falls to
1% of its central value.
Read in the reduced volume V^*
Place molecules within the cell on a uniform grid
Set up potential energy table – pot(i, j) contains the potential energy due to
atoms i and j.
Initialize a table tab to zero that will contain number of inter-atom distances in
bins of size $\sigma/20$.
Initialize *virial* to zero
Loop i from 1 to *ntot*
Choose a random atom, a random moving distance and a random direction.
Move atom.
In array tempot(j) insert potential energy due to moved atom and all other atoms.
Find deltaphi, the change in potential energy due to the move.
If deltaphi is negative then accept atomic move and update potential energy
table
If deltaphi is positive then calculate $z = \exp(-\text{deltaphi}/kT)$ and generate r $\{uv(0, 1)\}$
If $z > r$ then accept atomic move and update potential energy table. Otherwise
retain original atomic position and original potential energy table.
If $i > $ nstart and i is 10m, where m is an integer, then calculate contributions to
(i) the virial term
and (ii) the radial distribution function. All distances between
pairs of atoms are calculated and recorded in bins of size $\sigma/20$ for
distances up to 2.4σ.
End loop
Calculate mean of *virial*, $vir = 10 \times virial/(ntot - nstart)$
Output $1 + \dfrac{vir}{125kT}$
Write table tab to file.
Stop.

Appendix 15 **Program REACTOR**

This program calculates the multiplication factor of a spherical homogeneous nuclear reactor of radius r consisting of a mixture of graphite and uranium with an atomic ratio U/C $= a$. The uranium is enriched by a factor e from the natural uranium composition of $U_{235} : U_{238} = 1 : 138$. The Monte Carlo method is used to follow the path of neutrons through the reactor. Experimental values are used for the cross sections of scattering, absorption, and fission. It is assumed that after 114 scattering events by moderator atoms the neutrons have thermal energy and may thereafter either be purely absorbed or give fission by interaction with U_{235}.

Input the characteristics of the reactor r, a, and e and the trial number of neutrons, nn.

Constant $barn = 10^{-28}$

Calculate concentrations of U_{235}, $conc35$, and U_{238}, $conc38$, based on a concentration of carbon, $concc = 1.0 \times 10^{29}$, corresponding to a graphite density $\sim 2000\,\text{kg m}^{-3}$.

Calculate scattering probability per unit length for carbon, $sigsc$, and for all scattering events, $sigst$.

> $sigsc = 4.8 \times barn \times concc$
>
> $sigst = sigsc + (10.0 \times conc35 + 8.3 \times conc38) \times barn$

Calculate the absorption collision probability for U_{238} from Table 23.5. This requires the total scattering cross section (in barns) per U_{238} atom, sigovn.

> $sigovn = sigst/conc38/barn$

Use interpolation in Table 23.5 to find the absorption cross section for U_{238}, $siga38$. If $sigovn > 2000$ then

> $siga38 = 15.936 - 21072.0/sigovn$

Calculate the following:

> **Total interaction probability for U_{238} + moderator for fast neutrons**
>
> $sigtfast = sigst + (3.2 \times 10^{-3} \times concc + siga38 \times conc38) \times barn$
>
> **Total interaction probability for U_{235} + U_{238} + moderator for slow neutrons**
>
> $sigtslow = sigst + (3.2 \times 10^{-3} \times concc + 694.0 \times conc35 + 2.73 \times conc38) \times barn$
>
> **Probability per unit length for fission for thermal neutrons**
>
> $sigf = 582.0 \times conc35 \times barn$
>
> **Ratio of probability of carbon scattering to all scattering**
>
> $scatcovt = sigsc/sigst$
>
> **Ratio of probability of scattering/all interactions for fast neutrons**
>
> $scaototf = sigst/sigtfast$
>
> **Ratio of probability of scattering/all interactions for slow neutrons**
>
> $scaotots = sigst/sigtslow$
>
> **Ratio of probability of fission/slow absorption**
>
> $fisoslow = sigf/(sigtslow - sigst)$

Start of Monte Carlo trials

> **Number of fission neutrons $fisn = 0$**

Loop $i = 1$ to nn

> **Number of moderator scattering events $noscat = 0$**

Find initial distance of neutron from centre of vessel as $ran \times r$ where ran is a random selection from a distribution $P(x) = x\,sin(\pi x)$. This is found by the rejection method with a rectangular comparator function of height 0.6.

> $x = ran \times r$
>
> $y = 0$
>
> $z = 0$

Point C

Generate θ and ϕ for random direction of motion

Generate expected path length to next interaction
 if *noscat* < 114 *avpath* $= 1/sigtfast$
 if *noscat* ≥ 114 *avpath* $= 1/sigtslow$

Generate actual pathlength
 ran $= uv(0, 1)$
 path $= - avpath \times log(ran)$

Position of next interaction
 $x = x + sin(\theta)cos(\phi) \times path$
 $y = y + sin(\theta)sin(\phi) \times path$
 $z = z + cos(\theta) \times path$

Test for absorption by wall
 if $x^2 + y^2 + z^2 \geq r^2$ **goto end of loop**

Test for scattering
 ran $= uv(0, 1)$
 if *noscat* < 114
  if *ran* $<$ *scaototf* **goto A**
  goto B
 if *noscat* ≥ 114
  if *ran* $<$ *scaotots* **goto A**
  goto B

Point A

Scattering has taken place. Test for moderator scattering
 ran $= uv(0, 1)$
 if *ran* $<$ *scatcovt* **then** *noscat* $=$ *noscat* $+ 1$
 goto C

Point B

Absorption has taken place. If with fast neutron begin the next trial
 if *noscat* < 114 **goto end of loop**

Absorption of a slow neutron has taken place. Test for fission
 ran $= uv(0, 1)$
  if *ran* $<$ **fisoslow then** *fisn* $=$ *fisn* $+ 2.47$

End of loop

Calculate multiplication factor
 factor $=$ *fisn/nn*

Output *r, a, e,* **factor.**

Stop.

Appendix 16 **Program LESLIE**

A program for applying a Leslie matrix for up to 10 generation groups for any number of generations. The principal eigenvalue and eigenvector can be found from the output.

Input number of generations, *ng*
Input initial number within each generation, *pop(i)* for *i* from 1 to *ng*
Input fertility factors, *f(i)* for *i* from 1 to *ng*
Input survival factors, *s(i)* for *i* from 1 to *ng* $- 1$
Initialize number of generations, *nogen* $= 0$
Point A   **Input number of generations to run, *ig***
Loop for *i* from 1 to *ig*
 Clear temporary array *pp(ii)* $= 0$, for *ii* from 1 to *ng*

Calculate new number of infants

$pp(1) = \sum_{j=1}^{ng} f(j) \times pop(j)$

Calculate new number for other groups

$pp(ii) = s(ii-1) \times pop(ii-1)$

Find the fractional change in population for each group

$fr(ii) = pp(ii)/pop(ii)$ **for** ii **from 1 to** ng

Update populations

$pop(ii) = pp(ii)$ **for** i **from 1 to** ng

End loop

Output $pp(i)$ **and** $fr(i)$ **for** i **from 1 to** ng

Do you wish to continue (y/n)?

Input ans

If ans = y goto A

Stop.

Appendix 17 **Eigenvalues and eigenvectors of Hermitian matrices**

This appendix will show that the eigenvalues of an Hermitian matrix are real and the eigenvectors are orthogonal. First it should be made clear what is meant by orthogonal eigenvectors, taking into account that the elements of the eigenvectors may be complex. If two eigenvectors, x_1 and x_2, are orthogonal then the *scalar product*

$$\left(x_1^T\right)^* x_2 = 0 \tag{A17.1}$$

where the first term on the left-hand side is the complex conjugate of the transpose of x_1. Thus if

$$x_1 = \begin{bmatrix} x_{11} \\ x_{12} \\ x_{13} \end{bmatrix} \quad \text{and} \quad x_2 = \begin{bmatrix} x_{21} \\ x_{22} \\ x_{23} \end{bmatrix}$$

then

$$\left(x_1^T\right)^* x_2 = \begin{bmatrix} x_{11}^* & x_{12}^* & x_{13}^* \end{bmatrix} \begin{bmatrix} x_{21} \\ x_{22} \\ x_{23} \end{bmatrix} = x_{11}^* x_{21} + x_{12}^* x_{22} + x_{13}^* x_{23}. \tag{A17.2}$$

If $x_2 = x_1$ then the scalar product is

$$\left(x_1^T\right)^* x_2 = |x_{11}|^2 + |x_{12}|^2 + |x_{13}|^2, \tag{A17.3}$$

which will equal unity if the eigenvector is normalized. For normalized eigenvectors we can write

$$\left(x_i^T\right)^* x_j = \delta_{ij}, \tag{A17.4}$$

where δ_{ij} is the Kronecker delta function.

Two relationships will be used in the derivation of the properties of Hermitian matrices.

Relationship 1

If a and b are two vectors of the same order then $a^T b = b^T a$. For two order-3 vectors

$$a^T b = \begin{bmatrix} a_1 & a_2 & a_3 \end{bmatrix} \begin{bmatrix} b_1 \\ b_2 \\ b_3 \end{bmatrix} = a_1 b_1 + a_2 b_2 + a_3 b_3 = \begin{bmatrix} b_1 & b_2 & b_3 \end{bmatrix} \begin{bmatrix} a_1 \\ a_2 \\ a_3 \end{bmatrix} = b^T a.$$

Relationship 2

If A is an $n \times n$ matrix and b is a $n \times 1$ vector then $(Ab)^T = b^T A^T$. Thus

$$Ab = \begin{bmatrix} a_{11} & a_{12} & a_{13} \\ a_{21} & a_{22} & a_{23} \\ a_{31} & a_{23} & a_{33} \end{bmatrix} \begin{bmatrix} b_1 \\ b_2 \\ b_3 \end{bmatrix} = \begin{bmatrix} a_{11}b_1 + a_{12}b_2 + a_{13}b_3 \\ a_{21}b_1 + a_{22}b_2 + a_{23}b_3 \\ a_{31}b_1 + a_{32}b_2 + a_{33}b_3 \end{bmatrix}$$

and

$$b^T A^T = \begin{bmatrix} b_1 & b_2 & b_3 \end{bmatrix} \begin{bmatrix} a_{11} & a_{21} & a_{31} \\ a_{12} & a_{22} & a_{32} \\ a_{13} & a_{23} & a_{33} \end{bmatrix} = (Ab)^T.$$

Eigenvalues are real

We now consider the product

$$\left(x_1^T\right)^* H x_1 = \left(x_1^T\right)^* \lambda_1 x_1 = \lambda_1 \left(x_1^T\right)^* x_1 \tag{A17.5}$$

where H is an Hermitian matrix and x_1 is an eigenvector. Then, remembering that Hx_1 is just a vector and using first Relationship 1 and then Relationship 2 we find also that

$$\left(x_1^T\right)^* H x_1 = (Hx_1)^T x_1^* = x_1^T H^T x_1^*$$

However, from the form of an Hermitian matrix we have (8.19) that gives $H^T = H^*$ so that

$$\left(x_1^T\right)^* H x_1 = \lambda_1 \left(x_1^T\right)^* x_1 = x_1^T H^* x_1^* = \lambda_1^* x_1^T x_1^*. \tag{A17.6}$$

Using Relationship 1 with $a = b = x_1$ we find $\left(x_1^T\right)^* x_1 = x_1^T x_1^*$ so that

$$\lambda_1 \left(x_1^T\right)^* x_1 = \lambda_1^* \left(x_1^T\right)^* x_1 \tag{A17.7}$$

showing that $\lambda_1 = \lambda_1^*$ so that λ_1 is real.

Eigenvectors are orthogonal

We now start with

$$\left(x_1^T\right)^* H x_2 = \lambda_2 \left(x_1^T\right)^* x_2 \tag{A17.8}$$

From Relationships 1 and 2

$$\left(x_1^T\right)^* H x_2 = (Hx_2)^T x_1^* = x_2^T H^T x_1^* = x_2^T H^* x_1^* = \lambda_1 x_2^T x_1^*.$$

From Relationship 1 with $a = x_1^*$ and $b = x_2$ we find

$$x_2^T x_1^* = \left(x_1^T\right)^* x_2. \tag{A17.9}$$

Combining results from (A17.7), (A17.8) and (A17.9) we find

$$\lambda_2 \left(x_1^T\right)^* x_2 = \lambda_1 \left(x_1^T\right)^* x_2. \tag{A17.10}$$

If $\lambda_1 \neq \lambda_2$ then (17.10) shows that $\left(x_1^T\right)^* x_2 = 0$, which means that the eigenvectors are orthogonal. However, if $\lambda_1 = \lambda_2$ then the eigenvectors need not be orthogonal. This is the case when there is degeneracy. As indicated in §21.5.2 when there is n-fold degeneracy there can be found n linearly-independent eigenfunctions from which n linearly-independent mutually orthogonal eigenfunctions can be produced by Gram–Schmidt orthogonalization.

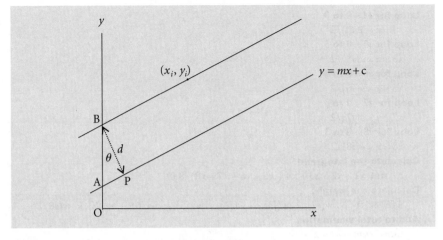

Figure A18.1 The line $y = mx + c$ and the parallel line through (x_i, y_i).

Appendix 18 **The distance of a point from a line**

Given a point (x_i, y_i) it is required to find its distance from the line $y = mx + c$. The line and the point are shown in Figure A18.1.

The intercept of the line on the y axis, $OA = c$ and the intercept of the parallel line through the point is $OB = c'$. The equation for the parallel line is

$$(y - y_i) = m(x - x_i) \qquad (A18.1)$$

since this clearly has slope m and passes through (x_i, y_i). From (A18.1), $c' = y_i - mx_i$ and hence $BA = c' - c = y_i - mx_i - c$.

The angle $ABP = \theta$ in Figure A18.1 is also the slope angle of the line so that $m = \tan \theta$. The distance of the point from the line is $d (= BP)$ and is given by

$$d = BA \cos \theta = \frac{y_i - mx_i - c}{\sec \theta} = \frac{y_i - mx_i - c}{\sqrt{1 + \tan^2 \theta}} = \frac{y_i - mx_i - c}{\sqrt{1 + m^2}},$$

which is the required result.

Appendix 19 **Program MULGAUSS**

A multiple integration program designed to give integral (29.39) with three-point integration in each of the eight dimensions. It can provide a model for more general cases involving n-point Gauss integration in m dimensions and with different integrands.

> **Set up a and w values for 3-point Gauss integration**
> $sum = 0$
> **Loop for $i1 = 0$ to 3**
> $\quad x1 = \pi \times a(i1)/2$
> **Loop for $i2 = 0$ to 3**
> $\quad x2 = \pi \times a(i2)/2$
> **Loop for $i3 = 0$ to 3**
> $\quad x3 = \pi \times a(i3)/2$

Loop for $i4 = 0$ **to 3**
   $x4 = \pi \times a(i4)/2$
Loop for $i5 = 0$ **to 3**
   $x5 = \pi \times a(i5)/2$
Loop for $i6 = 0$ **to 3**
   $x6 = \pi \times a(i6)/2$
Loop for $i7 = 0$ **to 3**
   $x7 = \pi \times a(i7)/2$
Loop for $i8 = 0$ **to 3**
   $x8 = \pi \times a(i8)/2$
Calculate the integrand
   $z = sin\{(x1 + x2 + x3 + x4 + x5 + x6 + x7 + x8)/8.0\}$
Calculate the weight
   $wt = w(i1) \times w(i2) \times w(i3) \times w(i4) \times w(i5) \times w(i6) \times w(i7) \times w(i8)$
Add to total summation
   $sum = sum + wt \times z$
End loop
End loop
End loop
End loop
End loop
End loop
End loop
End loop
Multiply sum by 'volume' in eight-dimensional space
$Integral = sum \times (\pi/2)^8$
Output integral
Stop

Appendix 20 **Program MCINT**

A Monte Carlo integration program designed to give integral (29.39). Random coordinates in each of the eight dimensions are found to produce a large number of integrand terms from which an average integrand can be found. The standard deviation of the final estimate is also found. This program can provide a model for more general cases involving m dimensions and with different integrands.

Input the number of trials, n
$sum = 0$
$sum2 = 0$

Loop for $i = 1$ **to** n
Generate random number r $uv(0, 1)$
   $x1 = r$
Generate random number r $uv(0, 1)$
   $x2 = r$
Generate random number r $uv(0, 1)$
   $x3 = r$
Generate random number r $uv(0, 1)$
   $x4 = r$

Generate random number r $uv(0, 1)$
$x5 = r$
Generate random number r $uv(0, 1)$
$x6 = r$
Generate random number r $uv(0, 1)$
$x7 = r$
Generate random number r $uv(0, 1)$
$x8 = r$
Calculate the integrand
$z = sin\{\pi \times (x1 + x2 + x3 + x4 + x5 + x6 + x7 + x8)/16.0\}$
Add to total summations
$sum = sum + z$
$sum2 = sum2 + z^2$
End loop

Multiply average integrand by *'volume'* **in eight-dimensional space to find integral**
$Integral = sum \times (\pi/2)^8/n$
Output integral
Calculate mean of sample $av = sum/n$
Calculate mean-square of sample values $av2 = sum2/n$
Calculate variance of sample members $sampvar = av2 - av^2$
Calculate variance of sample mean $varmean = sampvar/n$
Calculate standard deviation of estimate $sd = \sqrt{varmean} \times (\pi/2)^8$
Output sd
stop

Appendix 21 **Program GS**

This program is specifically designed to solve the equations (30.12). It may easily be adapted to solve other sets of equations.

Set $x1 = x2 = x3 = x4 = 0$
Carry out a cycle of refinement
Point A
$xx1 = 4 - x2/3 + x3/3 - 2x4/3$
$xx2 = 3.5 - xx1/4 + x3/2 - x4/4$
$xx3 = -1.6 - 0.4xx1 + 0.2xx2 + 0.2x4$
$xx4 = 11.0/3.0 - xx1/3 - xx2/3 - xx3/3$
Output $xx1, xx2, xx3, xx4$
Test for convergence
if $((xx1 - x1)^2 + (xx2 - x2)^2 + (xx3 - x3)^2 + (xx4 - x4)^2 < 10^{-9})$ **goto B**
$x1 = xx1$
$x2 = xx2$
$x3 = xx3$
$x4 = xx4$
goto A
Point B
Output $xx1, xx2, xx3, xx4$.
Stop

Appendix 22 **Second moments for uniform and Gaussian noise**

From (32.18), the second moment for noise with general probability distribution $p(x)$ is given by:

$$\bar{x}_2 = \int_{-\infty}^{\infty} x^2 p(x)\, dx. \tag{A22.1}$$

For uniform noise, the probability distribution is given by (32.12):

$$p(x) = \begin{cases} \dfrac{1}{2A} & \text{for} \quad -A \leq x \leq A \\ 0 & \text{otherwise.} \end{cases} \tag{A22.2}$$

We can see that $p(x)$ is non-zero only for the range of x: $-A \leq x \leq A$. Hence, substituting for $p(x)$ from (A22.2) into (A22.1):

$$\bar{x}_2 = \int_{-A}^{A} x^2 \frac{1}{2A}\, dx = \frac{1}{2A} \int_{-A}^{A} x^2\, dx. \tag{A22.3}$$

Carrying out the integration:

$$\bar{x}_2 = \frac{1}{2A} \left[\frac{x^3}{3} \right]_{-A}^{A} = \frac{1}{2A} \left[\frac{A^3}{3} - \left(-\frac{A^3}{3} \right) \right] = \frac{1}{2A} \cdot \frac{2A^3}{3} = \frac{A^2}{3} \tag{A22.4}$$

For Gaussian noise, the probability distribution is given by (32.11):

$$p(x) = \frac{1}{\sqrt{2\pi}\sigma} \exp\left(-\frac{x^2}{2\sigma^2} \right). \tag{A22.5}$$

Substituting for $p(x)$ from (A22.5) into (A22.1):

$$\bar{x}_2 = \int_{-\infty}^{\infty} x^2 \frac{1}{\sqrt{2\pi}\sigma} \exp\left(-\frac{x^2}{2\sigma^2} \right) dx. \tag{A22.6}$$

Now it is convenient here to change variables as follows:

$$z^2 = \frac{x^2}{2\sigma^2} \tag{A22.7}$$

in which case

$$z = \frac{x}{\sqrt{2}\sigma} \tag{A22.8}$$

Hence

$$dz = \frac{dx}{\sqrt{2}\sigma}. \tag{A22.9}$$

From (A22.9),

$$dx = \sqrt{2}\sigma\, dz. \tag{A22.10}$$

Using (A22.7) to (A22.10), (A22.6) can be rewritten as:

$$\bar{x}_2 = \frac{1}{\sqrt{2\pi}\sigma} \int_{-\infty}^{\infty} 2\sigma^2 z^2 \exp(-z^2) \sqrt{2}\sigma\, dz. \tag{A22.11}$$

Cancelling factors, this expression can be simplified to:

$$\bar{x}_2 = \frac{2\sigma^2}{\sqrt{\pi}} \int_{-\infty}^{\infty} z^2 \exp(-z^2)\, dz. \tag{A22.12}$$

The integral can now be evaluated by parts. Note however, that in (A22.12), we cannot put $u = z^2$ and $dv = \exp(-z^2)dz$. This is because in this case $v = \int \exp(-z^2)dz$ which cannot be integrated analytically. Instead, in (A22.12), we put

$$u = z \tag{A22.13}$$

and

$$dv = z \exp(-z^2)\, dz. \tag{A22.14}$$

In this case, from (A22.13),

$$du = dz, \tag{A22.15}$$

and from (A22.14)

$$v = \int z \exp(-z^2)\, dz \tag{A22.16}$$

which can be evaluated as

$$v = -\frac{1}{2}\exp(-z^2) \tag{A22.17}$$

Using integration by parts:

$$\bar{x}_2 = \frac{2\sigma^2}{\sqrt{\pi}}\left[-\frac{z}{2}\exp(-z^2)\right]_{-\infty}^{\infty} + \frac{2\sigma^2}{\sqrt{\pi}}\frac{1}{2}\int_{-\infty}^{\infty}\exp(-z^2)\, dz. \tag{A22.18}$$

Now the first term in (A22.18) goes to zero as $z \to \pm\infty$. Hence, after cancellation of factors, we are left with the second term in (A22.18):

$$\bar{x}_2 = \frac{\sigma^2}{\sqrt{\pi}}\int_{-\infty}^{\infty}\exp(-z^2)\, dz. \tag{A22.19}$$

Substituting for z and dz from (A22.8) and (A22.9) into (A22.19):

$$\bar{x}_2 = \frac{\sigma^2}{\sqrt{\pi}}\int_{-\infty}^{\infty}\exp\left(-\frac{x^2}{2\sigma^2}\right)\frac{1}{\sqrt{2}\sigma}\, dx = \sigma^2\int_{-\infty}^{\infty}\frac{1}{\sqrt{2\pi}\sigma}\exp\left(-\frac{x^2}{2\sigma^2}\right)\, dx. \tag{A22.20}$$

Now, from (A22.5), the integrand $\frac{1}{\sqrt{2\pi}\sigma}\exp\left(-\frac{x^2}{2\sigma^2}\right) = p(x)$ is the probability distribution function for Gaussian noise. Hence

$$\bar{x}_2 = \sigma^2\int_{-\infty}^{\infty}p(x)\, dx. \tag{A22.21}$$

Now, from (32.14), for any probability distribution function:

$$\int_{-\infty}^{\infty}p(x)\, dx = 1. \tag{A22.22}$$

Substituting this result in (A22.21), we obtain our final expression for the second moment of Gaussian noise:

$$\bar{x}_2 = \sigma^2. \tag{A22.23}$$

Appendix 23 **Convolution theorem**

We start off with the following convolution relation:

$$y(t) = \int_{-\infty}^{\infty}h(\tau)x(t-\tau)\, d\tau. \tag{A23.1}$$

Taking Fourier transforms of both sides of this equation:

$$Y(f) = \int_{-\infty}^{\infty}\left\{\int_{-\infty}^{\infty}h(\tau)x(t-\tau)\, d\tau\right\}\exp(-i2\pi ft)\, dt. \tag{A23.2}$$

Rewrite the complex exponential as follows:

$$\exp(-i2\pi ft) = \exp(-i2\pi f\tau - i2\pi ft + i2\pi f\tau) = \exp(-i2\pi f\tau)\exp(-i2\pi f(t-\tau)). \tag{A23.3}$$

Substituting from (A23.3) into (A23.2) and rearranging terms:

$$Y(f) = \int_{-\infty}^{\infty} \left\{ \int_{-\infty}^{\infty} h(\tau)x(t - \tau) \exp(-i2\pi f\tau) \exp(-i2\pi f(t - \tau)) \, dt \right\} d\tau. \qquad \text{(A23.4)}$$

Changing variables,

$$t' = t - \tau \quad \text{and} \quad dt' = dt. \qquad \text{(A23.5)}$$

Note that $dt' \neq dt - d\tau$ because t and τ are independent variables and, when performing the integral over t in curly brackets, τ is kept constant. Equation (A23.4) now appears as:

$$Y(f) = \int_{-\infty}^{\infty} \int_{-\infty}^{\infty} h(\tau)x(t') \exp(-i2\pi f\tau) \exp(-i2\pi ft') \, dt' \, d\tau. \qquad \text{(A23.6)}$$

This expression may be rearranged as follows:

$$Y(f) = \left\{ \int_{-\infty}^{\infty} h(\tau) \exp(-i2\pi f\tau) \, d\tau \right\} \left\{ \int_{-\infty}^{\infty} x(t') \exp(-i2\pi ft')) \, dt' \right\}. \qquad \text{(A23.7)}$$

The first integral on the right hand side of (A23.7) is $H(f)$ and the second integral is $X(f)$. Hence we obtain the convolution theorem (33.16):

$$Y(f) = H(f)X(f). \qquad \text{(A23.8)}$$

Appendix 24 **Output from a filter when the input is a cosine**

Combining (33.16) and (33.17), the Fourier transform, $Y(f)$, of the output from a filter is given by:

$$Y(f) = \alpha(f) \exp(i\phi(f))X(f) \qquad \text{(A24.1)}$$

where $\alpha(f)$ is the amplitude response of the filter, $\phi(f)$ is the phase response, and $X(f)$ is the Fourier transform of the input signal. Now specialize to the case where the input signal is a cosine:

$$x(t) = A\cos(2\pi f_0 t). \qquad \text{(A24.2)}$$

From (17.39b) the Fourier transform of a cosine is

$$X(f) = \frac{A}{2}\delta(f - f_0) + \frac{A}{2}\delta(f + f_0). \qquad \text{(A24.3)}$$

Substituting for $X(f)$ from (A24.3) into (A24.1):

$$Y(f) = \frac{A}{2}\alpha(f) \exp(i\phi(f))\delta(f - f_0) + \frac{A}{2}\alpha(f) \exp(i\phi(f))\delta(f + f_0). \qquad \text{(A24.4)}$$

We can think of the δ-function $\delta(f - f_0)$ as picking out the product term $\alpha(f) \exp(i\phi(f))$ at $f = f_0$. Similarly in (A24.4) we can think of the δ-function $\delta(f + f_0)$ as picking out the product term $\alpha(f)\exp(i\phi(f))$ at $f = -f_0$.

Hence we replace $\alpha(f) \rightarrow \alpha(f_0)$ and $\phi(f) \rightarrow \phi(f_0)$ in the first term of (A24.4) and $\alpha(f) \rightarrow \alpha(-f_0)$ and $\phi(f) \rightarrow \phi(-f_0)$ in the second term, which gives

$$Y(f) = \frac{A}{2}\alpha(f_0) \exp(i\phi(f_0))\delta(f - f_0) + \frac{A}{2}\alpha(-f_0) \exp(i\phi(-f_0))\delta(f + f_0). \qquad \text{(A24.5)}$$

Using the symmetry properties of amplitude and phase response, (33.18) and (33.19), we may write:

$$\phi(-f_0) = -\phi(f_0) \quad \text{and} \quad \alpha(f_0) = \alpha(-f_0). \qquad \text{(A24.6)}$$

Substituting for $\phi(-f_0)$ and $\alpha(-f_0)$ from (A24.6) into (A24.5):

$$Y(f) = \frac{A}{2}\alpha(f_0)\exp(i\phi(f_0))\delta(f-f_0) + \frac{A}{2}\alpha(f_0)\exp(-i\phi(f_0))\delta(f+f_0). \qquad (A24.7)$$

We now take the inverse Fourier transform of both sides of (A24.7). On the right-hand side, the only functions of frequency, f, are $\delta(f-f_0)$ and $\delta(f+f_0)$. Now, in the solutions to Problem 17.4 it was shown that the Fourier transform of $\exp(2\pi i f_0 t)$ is $\delta(f-f_0)$ and the Fourier transform of $\exp(-2\pi i f_0 t)$ is $\delta(f+f_0)$. We can turn these statements around and say that the inverse Fourier transforms of $\delta(f-f_0)$ and $\delta(f+f_0)$ are given as follows:

$$\delta(f-f_0) \rightarrow \exp(2\pi i f_0 t) \qquad (A24.8a)$$

and

$$\delta(f+f_0) \rightarrow \exp(-2\pi i f_0 t) \qquad (A24.8b)$$

Using these expressions and taking inverse Fourier transforms of both sides of (A24.7),

$$y(t) = \frac{A}{2}\alpha(f_0)\exp(i\phi(f_0))\exp(2\pi i f_0 t) + \frac{A}{2}\alpha(f_0)\exp(-i\phi(f_0))\exp(-2\pi i f_0 t). \qquad (A24.9)$$

Factorizing this expression and combining the two products of exponentials:

$$y(t) = \frac{A}{2}\alpha(f_0)[\exp(i[\phi(f_0) + 2\pi f_0 t]) + \exp(-i[\phi(f_0) + 2\pi f_0 t])]. \qquad (A24.10)$$

Finally, using the identity

$$\cos(x) = \frac{1}{2}[\exp(ix) + \exp(-ix)]$$

we obtain our final expression for the output from the filter, when the input is a cosine of amplitude A and frequency f:

$$y(t) = A\alpha(f_0)\cos(2\pi f_0 t + \phi(f_0)). \qquad (A24.11)$$

Appendix 25 **Program GRADMAX**

This uses a gradient method to maximize the function given by $phi(x, y)$. It requires functions $phi(x, y)$ and those for the derivatives $dphix(x, y)$ and $dphiy(x, y)$.

Initialize
Set counter for number of steps $kount = 0$
Set starting point x **and** y
Set initial step length d
Set tolerance tol
Start calculation
   $t = phi(x, y)$
Point A
   $a = dphix(x, y)$
   $b = dphiy(x, y)$
Point B
Find the constant k
   $k = d/sqrt(a*a + b*b)$
Find x **and** y **at the end of the step**
   $xn = x + ka$
   $yn = y + kb$

Find the value of the function at the end of the step
  $tn = phi(xn, yn)$
If $tn < t$ then $d = d/2$ and return to B. Otherwise
  $kount = kount + 1$
  Output $kount, x, y, d, tn$
Check to see if tolerance is reached
  If ($|tn - t| < tol$ then stop. Otherwise go to A.

Appendix 26 **Program NETWORK**

This program solves the 'cheapest path' problem. The input is a list of edges linking nodes and their costs. Each edge needs only to be put in once but duplication will not affect the program. The user may specify the node from which the cheapest paths to all other nodes is required. The output is the cost to each other node and a description of the cheapest path.

Read in number of nodes, n

Loop i from 1 to n
  read j, a node linked to i or 1 000 if no more nodes linked
  if ($j > 999$) go to end of loop
  read in $c(i, j)$ and $c(j, i) = c(i, j)$
End loop
Point A
  Input node m from which costs are required, or 1 000 to end calculation
  If ($m > 999$) stop
Clear arrays for previous node p, cost to that node, k, and slash indicator, s
Loop i from 1 to n
  $p(i) = 0$ $k(i) = 100\ 000$ $s(i) = 0$
End loop

  $k(m) = 0$ $s(m) = 1$ **(node m is slashed)**
Update all costs from slashed nodes
Point B
  Loop $i = 1$ to n
  if ($s(i) = 0$) go to end of loop
  $t = i$
  Loop $j = 1$ to n
    if ($j = t$ or $s(j) = 1$ or $c(t, j) = 0$) go to end of loop
    $kk = k(t) + c(t, j)$
    if ($kk > k(j)$) go to end of loop
      $k(i) = kk$ $p(j) = t$
  End loop
  End loop
Find the lowest unslashed node
  $less = 100\ 000$ $t = 0$

  Loop for $i = 1$ to n
  if ($s(i) = 1$ or $k(i) > less$) go to end of loop
    $less = k(i)$ $t = i$
  End loop
  if ($t = 0$) goto C
  $s(t) = 1$
  go to B

Point C   **Analysis is complete. Output results**

    Loop $i = 1$ to n
        if $(i = m)$ go to end of loop
        print $k(i)$ the cost from m to i
    Trace path backwards from i to m
        $t = i$
Point D
        $nn = p(t)$
        if $(nn = m)$ print m and go to end of loop
            print nn
            $t = nn$
        go to D
    End loop

Go back to beginning to input new initial node
    go to A.

Appendix 27 **Program GRAVBODY**

Initialization
    Set number of timesteps between output, $iout$
    Input number of bodies, their masses and position and velocity components
    Input initial timestep, h, and duration of the simulation, $totime$
    Input tolerance
    If a body is to be chosen as origin give $norig =$ number of body. Otherwise $norig = 0$
Determine initial energy and angular momentum with respect to centre of mass
    Find position of centre of mass and its velocity
    Find positions and velocities of all bodies relative to the centre of mass
    Calculate energy (38.3) and angular momentum (38.5)
Point A
Begin integration
    Advance positions and velocities of all bodies using Runge–Kutta for two steps with timestep h
From same initial position advance positions and velocities of all bodies using Runge–Kutta with timestep $2h$
If the greatest distance between two estimates of position for any body is greater than tolerance then halve the timestep and return to A
Otherwise accept position after two steps of h and double the timestep for the next integration. Also output energy and angular momentum to check for constancy
Output
    If number of timesteps = integer × $iout$ then write positions of bodies to a datafile.
    If total time of simulation > $totime$ then stop
    Otherwise return to A.

Appendix 28 **Program ELECLENS**

This program maps the path of an electron in the x-y plane starting at point $(0, y)$ and moving at a small angle, *theta*, to the x-axis. There is a magnetic field along z with strength $B = Cy$.

Initialization
Set x_1, x_2, the limits of the magnetic field
   Input initial y
   Input initial speed, v_0, and angle *theta*
   Input final value of x, $xlim$
   Input constant C
   Fix timestep, h, to give shift of 0.001 m of x for each timestep
   Set initial values
   $x = 0$
   $y = 0$
   $u = v_0 \cos(theta)$
   $v = v_0 \sin(theta)$
The integration is by Euler predictor corrector.
Point A
First approximations
   $yt = y + vh$
   $omega = Cye/m$
If $x < x_1$ or $x < x_2$, omega $= 0$
   $ut = u + omega \times v \times h$
   $vt = v - omega \times u \times h$
Final estimate
   $x = x + 0.5(u + ut)h$
   $y = y + 0.5(v + vt)h$
   $omega = 0.5C(y + yt)e/m$
If $x < x_1$ or $x > x_2$, omega $= 0$
   $u = u + 0.5 \times omega(v + vt)h$
   $v = v - 0.5 \times omega(u + ut)h$
Store values of (x, y)
If $x > xlim$ then stop
   Otherwise goto A.

Appendix 29 **Program CLUSTER**

The heart of this program is similar to GRAVBODY. The initial setting up of the system of bodies and the information derived from the computed positions is the main difference.

Initialization
   Set number of timesteps between output, *iout*
   Input initial timestep, h, and duration of the simulation, *totime*
   Input tolerance
   Input number of stars, the mass of each star (all equal) and radius of the cluster
Setting up initial system
   Set up random positions of stars in spherical region
Find potential energy of star arrangement and hence r.m.s. speed from virial theorem
Using a random-number generator find random directions of motion of stars
Find centre of mass and velocity of centre of mass
Subtract com coordinates from star coordinates and so make com the origin
Subtract com velocity from star velocities and so bring com to rest
Multiply all star velocities by constant factor to obtain rms speed indicated by the virial theorem
   Calculate energy (38.3) and angular momentum (38.5)
Point A
Begin integration

Advance positions and velocities of all bodies using Runge–Kutta for two steps with timestep h

From same initial position advance positions and velocities of all bodies using Runge–Kutta with timestep $2h$

If the greatest distance between two estimates of position for any body is greater than tolerance then halve the timestep and return to A

Otherwise accept position after two steps of h and double the timestep for the next integration. Also output energy and angular momentum to check for constancy

Calculate energy (38.3) and angular momentum (38.5)

Output

If number of steps = integer $\times$ $iout$ then store positions of all stars. Also calculate and store the moment-of-inertia factor, energy and angular momentum

If total time $>$ $totime$ then stop

Otherwise goto A.

Appendix 30 **Program FLUIDYN**

In this cell-structure fluid dynamics program 125 molecules are placed on a regular grid and then randomly displaced by up to 0.05 of the original spacing, d, in each direction. The initial speeds of the molecules, again in random directions, are all equal to the rms speed appropriate to the temperature. The motions of the molecules, under the influence of Lennard–Jones forces, are followed using the Runge–Kutta process. After 50 timesteps, to allow a proper distribution of speeds to be established, the molecular configuration is sampled every timestep to find the virial term and also the radial density function.

Initialization

Within the program there are set the Lennard–Jones constants and the atomic mass of argon and the number of timesteps to completion, $ntot$. These can be changed.

Input temperature, T

Input value of V^* for which radial distribution function is required

Loop for k from 1 to 20

Set normalized volume $V^* = 0.8 + 0.5k$

Calculate size of cubical cell of dimension $5d$

Set up molecules on a regular grid with coordinates $(0.5, 1.5, 2.5, 3.5, 4.5)d$ in each principal direction. For each molecule displace each coordinate randomly by up to $0.05d$.

Calculate rms speed of molecules and give each molecule this speed in a random direction.

The equations of motion are solved by Runge–Kutta with variable step, h. This finds the maximum speed of any molecule, v_{max}, and takes $h = 0.05d/v_{max}$.

Loop for 50 timesteps

Find h

Solve for motion of molecules by Runge–Kutta. For each pair of molecules, i and j, find the closest distance from i to j or any of its 'ghosts' in neighbouring cells. If this distance is less than 2.4σ then the forces of i on j and j on i are taken into account.

End loop

Loop for ntot $-$ 50 timesteps

Find h
Solve for motion by Runge–Kutta.
For each pair of molecules find contribution to virial term and to radial distribution function in bins and store
End loop
If V^* has required value then store radial distribution in file RADIAL.DAT
    Find virial term and store PV/NkT in file EOS.DAT
End k loop

Stop.

Appendix 31 **Condition for collisionless PIC**

In Figure A31.1 an interaction between an electron at C and a singly charged ion at O is shown. The electron at C experiences an acceleration towards the ion at O equal to

$$a = \frac{e^2}{4\pi\varepsilon_0 m (D\sec\theta)^2} \tag{A31.1}$$

where D is the closest approach of the interaction. By symmetry the net change of velocity of the electron due to the interaction will be in the direction AO and the component of the velocity change in that direction in time dt will be

$$v_\perp = a\cos\theta\,dt = \frac{e^2\cos\theta}{4\pi\varepsilon_0 m D^2\sec^2\theta}\,dt. \tag{A31.2}$$

We also have $s = D\tan\theta$ so that

$$ds = v\,dt = D\sec^2\theta\,d\theta. \tag{A31.3}$$

Substituting from (A31.3) into (A31.2) gives

$$dv_\perp = \frac{e^2}{4\pi\varepsilon_0 m D v}\cos\theta\,d\theta. \tag{A31.4}$$

Hence the total change in velocity along AO due to the passage of the electron is

$$\Delta v_\perp = \frac{e^2}{4\pi\varepsilon_0 m D v}\int_{-\pi/2}^{\pi/2}\cos\theta\,d\theta = \frac{e^2}{2\pi\varepsilon_0 m D v}. \tag{A31.5}$$

This will cause a deviation in the path of the electron:

$$\Delta\phi = \frac{\Delta v_\perp}{v} = \frac{e^2}{2\pi\varepsilon_0 m D v^2}. \tag{A31.6}$$

The deviations due to close interactions will be in completely random directions and the expected deviation after N interactions at speed v will be

$$\langle\Delta\phi^2\rangle^{1/2} = N^{1/2}\frac{e^2}{2\pi\varepsilon_0 m v^2}\left\langle\frac{1}{D^2}\right\rangle^{1/2}. \tag{A31.7}$$

The analysis is simplified by making the approximation

$$\left\langle\frac{1}{D^2}\right\rangle^{1/2} \approx \left\langle\frac{1}{D}\right\rangle$$

since the two quantities will differ by a factor of order unity and the conclusions we draw will still be valid.

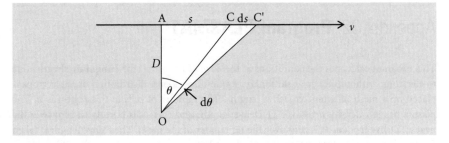

Figure A31.1 An electron at C moving at speed v relative to a singly charged ion at O.

We assume a maximum value of D, D_{max}, for a close interaction. The total target area for a close interaction is thus πD_{max}^2 and for a closest approach between D and $D + dD$ the target area is $2\pi D\, dD$. From this the value of $\langle 1/D \rangle$ is found as

$$\left\langle \frac{1}{D} \right\rangle = \frac{1}{\pi D_{max}^2} \int_0^{D_{max}} \frac{1}{D} \times 2\pi D\, dD = \frac{2}{D_{max}}. \tag{A31.8}$$

A more precise analysis, without the simplification, involves finding $\langle 1/D^2 \rangle$ and requires the introduction of a minimum distance for an interaction, D_{min}. This is taken as the **Landau length**, the distance of an interaction that will give a deviation of $\pi/2$.

From (A31.7) and (A31.8) we can now find the expected number of close interactions required to give a deviation of $\pi/2$ which will happen in time t_c. This is

$$N_c = \frac{\pi^4 \varepsilon_0^2 m^2 D_{max}^2 v^4}{e^4}. \tag{A31.9}$$

As the electron travels through the plasma every ion within a distance D_{max} will give a close interaction. To give N_c close interactions when the number density of ions is n requires a distance of travel L for the electron where

$$N_c = \pi D_{max}^2 Ln \tag{A31.10}$$

or

$$L = \frac{\pi^3 \varepsilon_0^2 m^2 v^4}{e^4 n},$$

requiring a time

$$t_c = \frac{L}{v} = \frac{\pi^3 \varepsilon_0^2 m^2 v^3}{e^4 n}. \tag{A31.11}$$

From (38.30), which gives an expression for the plasma period, writing $mv^2 = 3kT$ and using (38.34b), the expression for the Debye length, we find

$$\frac{t_c}{t_p} = \frac{\pi^2 3^{3/2}}{2} n\lambda_D^3, \tag{A31.12}$$

which is about six times the number of particles in the Debye sphere. This result applies to the number of superparticles in the Debye sphere of the model since the root-N statistics we used will be relevant to the model and not to the plasma being modelled.

Appendix 32 **Program PLASMA1**

This program calculates the conditions at the edge of a plasma (the **Langmuir sheath**) due to electrons, with ions taken as stationary. A one-dimensional distribution of nt electrons is placed on a mesh of extent xt Debye lengths. The number of cells in this region is it. The plasma has ion density $dens$ over $x1$ Debye lengths and then falls linearly to zero over the next $x2$ Debye lengths. It is zero over the remainder of the mesh. The Debye length is taken for the density $dens$. Integration is carried out by a simple Euler approach with timestep governed by thermal speed and cell length and also by plasma frequency. Positions and velocities of the electrons are output for the initial distribution and after every 500 timesteps thereafter for 2 500 timesteps. These are in files V0.dat to V5.DAT. At the same intervals the potential at cell edges is output in files E0.DAT to E5.DAT.

> **Parameters** $dens$, xt, $x1$, $x2$, it, nt **and temperature** $temp$ **are set in the program. These may be changed for other simulations.**
>
> **Calculate plasma frequency, plasma period, t_p, Debye length and number of electrons per superelectron**
>
> **Assign positions of superelectrons to give required distribution**
>
> **Calculate charge density due to ions**
>
> **Assign velocities to superelectrons from normal distribution with rms value** $\sqrt{kT/m}$
>
> **Find maximum speed** v_{max}
>
> **Set integration timestep as minimum of** $0.05t_p$ **and** $0.05 \times celllength/v_{max}$
>
> **Point A**
>
> **Find field at cell boundaries**
>
> **Assign superelectron charges to nearest cell centres. Take E at vacuum end of plasma equal zero and use equivalent of (38.40a) to find field at other cell boundaries.**
>
> **If number of timesteps is integer** $\times 500$ **then output velocities in file Vx.DAT and field in file Ex.DAT where** $x =$ **the integer**
>
> **If number of timesteps** $= 2\,500$ **then stop**
>
> **Loop over all superparticles**
>
> > **Interpolate field at superparticle position from field at cell edges**
> >
> > **Advance velocity of superparticle by simple Euler**
> >
> > **Advance position of superparticle by simple Euler**
>
> **If new position gives** $x < 0$ **then move particle to** $x = 0$ **and give positive velocity taken from normal distribution as done previously**
>
> **If new position gives particle beyond vacuum boundary then reflect both position and velocity in the boundary**
>
> **End loop over superparticles**
>
> > **Go to A.**

References and further reading

References

Abramowitz, M. and Stegun, I. A. (1964) *Handbook of Mathematical Functions, Graphs and Mathematical Tables.* National Bureau of Standards, Washington, DC. (Reprinted by Dover Books, New York and also widely available on the Internet.)

Candy, J. V. (1986) *Signal processing: the model-based approach.* McGraw-Hill, New York

Gleick, James (1988) *Chaos: Making a new science.* Penguin, London. (A book written in a popular style.)

Press, W.H., Flannery, B.P., Teukolsky, S.A. and Vetterling, W.T. *Numerical recipes: the art of scientific computing.* This is the title of a series of books published by Cambridge University Press, each of which deals with a different computer language. Versions are available for C^{++}, C, FORTRAN77, FORTRAN90, and Pascal. There is also a version for BASIC by J. C. Sprott. For full details refer to *www.numerical-recipes.com.*

Shannon, C.E. (1948) A mathematical theory of communication. *Bell System Technical Journal,* **27**, 379–423; 623–656.

Woolfson, M. M. and Pert, G. J. (1999) *An introduction to computer simulation.* Oxford University Press, Oxford.

Further reading

Many of the topics covered by chapters of this book form the material of either complete, or large sections of, specialist texts. For those wishing to explore more, or more advanced, aspects of what is presented here the following textbooks may be found useful.

Brigham, E. Oran (1988) *The fast Fourier transform and its applications.* Prentice-Hall, Englewood Cliffs, NJ.

Candy, J. V. (2005) *Model-based signal processing.* Wiley, New York.

Greenhow, R. C. (1990) *Introductory quantum mechanics.* Institute of Physics Publishing, Bristol.

Hirst, Ann (1995) *Vectors in two or three dimensions.* Butterworth-Heinemann, Oxford.

Keisler, H. Jerome. *Elementary calculus; an approach using infinitesimals.* Online edition, http://www.math.wisc.edu/ ∼ keisler/calc.html

McClellan, J. H., Schafer, R. W. and Yoder, M. A. (2003) *Signal processing first.* Pearson/Prentice Hall, London.

Meddins, Robert (2000) *Introduction to signal processing.* Newnes, London.

Ockendon, John (2003) *Applied partial differential equations.* Oxford University Press, Oxford.

Roy, A. E. (2005) *Orbital motion,* 4th edition. Institute of Physics Publishing, Bristol.

Shannon, Claude E. and Weaver, Warren (1963) *The mathematical theory of communication.* University of Illinois Press, Champaign, IL.

Suli, Endre and Mayers, D. (2003) *An introduction to numerical analysis.* Cambridge University Press, Cambridge.

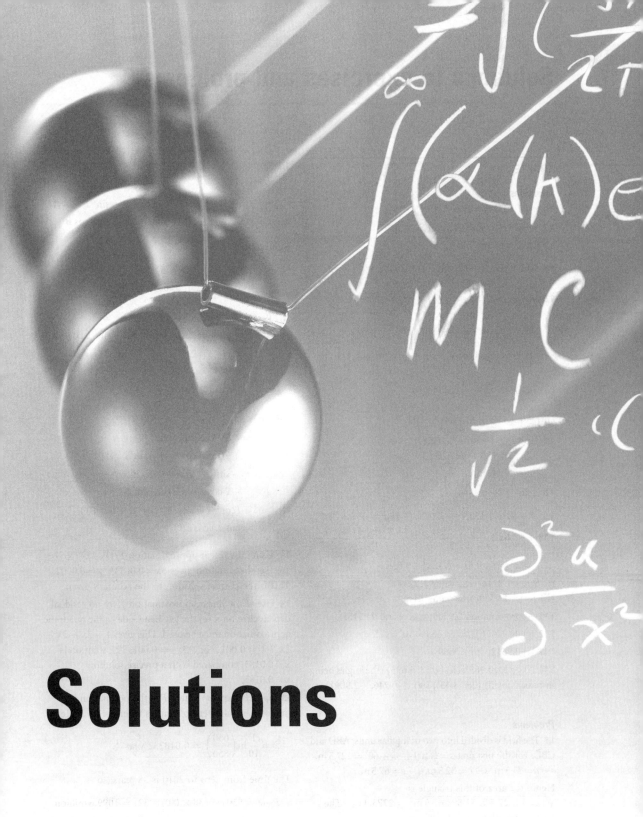

Solutions

Solutions to exercises and problems

CHAPTER 1

Exercises

1.1 $c = 7.832$, $A = 39.67°$, $C = 90.33°$.

1.2 $A = 38.62°$, $B = 48.51°$, $C = 92.87°$.

1.3 13.42 square units.

1.4 (i) 0.9659; (ii) 0.2588; (iii) 0.2588; (iv) 0.9659.

1.5 (i) $\dfrac{1}{2}\cos(A - B) + \dfrac{1}{2}\cos(A + B)$

$$= \frac{1}{2}(\cos A \cos B + \sin A \sin B$$

$$+ \cos A \cos B - \sin A \sin B) = \cos A \cos B;$$

(ii) $\dfrac{1}{2}\cos(A - B) - \dfrac{1}{2}\cos(A + B)$;

(iii) $\dfrac{1}{2}\sin(A + B) + \dfrac{1}{2}\sin(A - B)$.

1.6 (i) 3.732; (ii) 0.2679; (iii) 0.2679; (iv) 3.732.

1.7 $-2 \sin\dfrac{A + B}{2}\sin\dfrac{A - B}{2}$.

1.8 $2\sin\dfrac{A - B}{2}\cos\dfrac{A + B}{2}$.

1.9 $a^6 + 6a^5 b + 15a^4 b^2 + 20a^3 b^3 + 15a^2 b^4 + 6ab^5 + b^6$

1.10 $1 + 0.6x + \dfrac{0.6 \times (-0.4)}{1 \times 2}x^2 + \dfrac{0.6 \times (-0.4) \times (-1.4)}{1 \times 2 \times 3}$

$$\times x^3 = 1 + 0.6x - 0.12x^2 + 0.056x^3$$

(i) Series 1.2244, true value 1.2237
(ii) Series 1.0589, true value 1.0589.

1.11 Approximation 0.9500, true value 0.9515.

1.12 (i) 24 976; (ii) $\lambda = 4.489 \times 10^{-7}\,\text{s}^{-1}$.

1.13 (i) 2.9512; (ii) 2.3569

1.14 (i) $(\sqrt{13}, 0.983)$; (ii) $(\sqrt{13}, 5.695)$ (the angles being in radians); (iii) $(2.5, 4.33)$; (iv) $(-0.746, -2.906)$

Problems

1.1 The field is divided into two triangular units, ABD and CBD. For the first unit $s = \frac{1}{2}(100 + 95 + 60) = 127.5\,\text{m}$, $s - a = 27.5\,\text{m}$, $s - b = 32.5\,\text{m}$, $s - c = 67.5\,\text{m}$.
Hence the area of this triangle is
$\sqrt{127.5 \times 27.5 \times 32.5 \times 67.5}\,\text{m}^2 = 2773.4\,\text{m}^2$. The area of the other triangle is
$\sqrt{120.5 \times 30.5 \times 25.5 \times 64.5}\,\text{m}^2 = 2458.6\,\text{m}^2$, giving a total area of $5232.0\,\text{m}^2 = 0.5232$ hectares.

1.2 (i) $\sin^2(A + B) = \sin^2 A \cos^2 B + \sin^2 B \cos^2 A$
$$+ 2\sin A \cos A \sin B \cos B$$
$$\sin^2(A - B) = \sin^2 A \cos^2 B + \sin^2 B \cos^2 A$$
$$- 2\sin A \cos A \sin B \cos B$$

Subtracting,

$$\sin^2(A + B) - \sin^2(A - B)$$
$$= 4\sin A \cos A \sin B \cos B = \sin 2A \sin 2B$$

(ii) From (i), $\sin^2(A + B) + \sin^2(A - B)$
$$= 2\sin^2 A \cos^2 B + 2\sin^2 B \cos^2 A$$
$$= 2(1 - \cos^2 A)\cos^2 B$$
$$+ 2(1 - \cos^2 B)\cos^2 A$$
$$= 4(1 - \cos^2 A \cos^2 B)$$

(iii) $\cos^2(A - B) - \sin^2(A + B)$
$$= \cos^2 A \cos^2 B + \sin^2 A \sin^2 B$$
$$+ 2\cos A \cos B \sin A \sin B$$
$$- \sin^2 A \cos^2 B - \sin^2 B \cos^2 A$$
$$- 2\cos A \cos B \sin A \sin B$$
$$= (\cos^2 A - \sin^2 A)(\cos^2 B - \sin^2 B)$$
$$= \cos 2A \cos 2B$$

1.3 Using the linear approximation $1/(1 - x)^3 = 1 + 3x$, the equation gives $5x = 0.001$ or $x = 0.0002$. This is the correct solution to the accuracy given.

1.4 Use of the linear approximation gives no residual dependence on x on the left-hand side so the quadratic approximation must be used. This gives $1 - 2x + 3x^2 + 2x - 1 = 0.001$ or $3x^2 = 0.001$. The solution is $x = 0.0183$, compared with a precise solution $x = 0.0185$.

1.5 $P_t = P_0 \exp(\lambda t)$ where P is the population in millions and the zero of time is taken as 1972. Hence

$$\lambda = \frac{1}{10}\ln\left(\frac{660}{550}\right) = 0.018232\ \text{year}^{-1}.$$

The time from zero to 2010 is 38 years, so

$$P_{2010} = 550\exp(38 \times 0.018232) = 1099.6\ \text{million}.$$

1.6 $\ln(1.5) = 0.4055$; $\ln(1.4) = 0.3365$;
$\ln(1.3) = 0.2624$; $\ln(1,2) = 0.1823$; $\ln(1.1) = 0.09531$;

$\ln(1.01) = 0.009950$; $\ln(1.001) = 0.0009995$; $\ln(1.0001) = 0.000099995$.

The conclusion is that as $x \to 0$ so $\ln(1+x) \to x$.

CHAPTER 2

Exercises

2.1 0.230 sr.

2.2 $ml^{-1}t^{-1}$.

2.3 All elements of the equation have dimensions $ml^2 t^{-4} c^{-1}$.

Problems

2.1 Force has the dimensions of mass $\times$ acceleration or $[mlt^{-2}]$. Hence pressure, which is force per unit area, has dimensions $[mlt^{-2}] \div [l^2] = [ml^{-1}t^{-2}]$.

Energy has dimensions force $\times$ distance or $[mlt^{-2}] \times [l] = [ml^2t^{-2}]$. Hence energy density has dimensions $[ml^2t^{-2}] \div [l^3] = [ml^{-1}t^{-2}]$, which are the same as those for pressure.

2.2 We express the relationship in the form $P_B = CB^{\alpha} \times \mu_0^{\beta}$, where B is the magnetic field, and C is a numerical constant. Dimensionally,

$$[ml^{-1}t^{-2}] = [mt^{-2}c^{-1}]^{\alpha} \, [mlt^{-2}c^{-2}]^{\beta}.$$

Equating dimensions,

for **m**: $\alpha + \beta = 1$

for **l**: $\beta = -1$

for **t**: $2\alpha + 2\beta = 2$

for **c**: $-\alpha - 2\beta = 0$.

There are more equations than unknowns, but the equations are consistent and give the solution $\alpha = 2$ and $\beta = -1$. The actual physical expression is

$$P_B = \frac{B^2}{2\mu_0}.$$

2.3 The relationship is written in the form

$$\lambda = C\rho^{\alpha} c^{\beta} G^{\delta}$$

or, dimensionally,

$$[l] = [ml^{-3}]^{\alpha} [lt^{-1}]^{\beta} [m^{-1}l^3t^{-2}]^{\delta}.$$

Equating dimensions

for **m**: $\alpha - \delta = 0$

for **l**: $-3\alpha + \beta + 3\delta = 1$

for **t**: $-\beta - 2\delta = 0$

giving the solution $\alpha = -\frac{1}{2}, \beta = 1, \delta = -\frac{1}{2}$. The physical relationship was first derived by the astrophysicist James Jeans in 1919 and is

$$\lambda = \left(\frac{\pi}{\gamma G \rho}\right)^{1/2} c$$

where γ is the ratio of the specific heats at constant pressure and constant volume for the gas.

2.4 The expression is written in the form

$$a = Ce^{\alpha} \varepsilon_0^{\beta} c^{\gamma} m_e^{\delta}$$

or dimensionally

$$[l] = [ct]^{\alpha} [m^{-1}l^{-3}t^4c^2]^{\beta} [lt^{-1}]^{\gamma} [m]^{\delta}.$$

Equating dimensions,

for **m**: $-\beta + \delta = 0$

for **l**: $-3\beta + \gamma = 1$

for **t**: $\alpha + 4\beta - \gamma = 0$

for **c**: $\alpha + 2\beta = 0$

The solution of these equations is $\alpha = 2$, $\beta = -1$, $\gamma = -2$, and $\delta = -1$. The derived form of this theoretical equation is

$$a = \frac{e^2}{4\pi\varepsilon_0 c^2 m_e}.$$

CHAPTER 3

Exercises

3.1 1600.

3.2 (i) 7.9375; (ii) 7.96875. The difference, $0.03125 = 1/32$.

3.3 Group as $1 - \left(\dfrac{1}{2} + \dfrac{1}{3}\right) + \left(\dfrac{1}{4} + \dfrac{1}{5}\right) - \left(\dfrac{1}{6} + \dfrac{1}{7}\right) + \cdots$.

Following the first term successive pairs of brackets give a negative contribution, so the sum must be less than 1. After the first bracket successive pairs of brackets give a positive contribution, so the sum must be greater than $1 - \left(\dfrac{1}{2} + \dfrac{1}{3}\right) = \dfrac{1}{6}$.

Problems

3.1 The distance travelled by the ball, in metres, is

$$D = 1 + 2\{0.8 + (0.8)^2 + (0.8)^3 + \cdots\}$$

$$= 1 + 2\frac{1}{1 - 0.8} = 11.$$

3.2 The total potential energy per ion is, from (3.17),

$$\Phi = -\frac{e^2}{4\pi\varepsilon_0 r} \times 2 \times \left(1 - \frac{1}{2} + \frac{1}{3} - \frac{1}{4} + \cdots\right)$$

$$= -\frac{e^2}{4\pi\varepsilon_0 r} \times 2 \ln 2.$$

Hence the Madelung constant is $2 \ln 2$.

3.3 (i) Applying the ratio test,

$$\left(\frac{a_{t+1}}{a_t}\right)_{t \to \infty} = \left\{\frac{(t+1)^5}{(t+1)!} \frac{t!}{t^5}\right\}_{t \to \infty}$$

$$= \left\{\frac{1}{t+1} \left(\frac{t+1}{t}\right)^5\right\}_{t \to \infty} = 0.$$

Hence the series is convergent.

(ii) Applying the ratio test,

$$\left(\frac{a_{t+1}}{a_t}\right)_{t\to\infty} = \left(\frac{e^{t+1}}{(t+1)!}\frac{t!}{e^t}\right)_{t\to\infty} = \left(\frac{e}{t+1}\right)_{t\to\infty} = 0.$$

Hence the series is convergent.

(iii) Applying the ratio test again,

$$\left(\frac{a_{t+1}}{a_t}\right)_{t\to\infty} = \left(\frac{t(t+1)}{(t+1)(t+2)}\right)_{t\to\infty} = 1,$$

which does not indicate either convergence or divergence. However, $\frac{1}{t(t+1)} = \frac{1}{t} - \frac{1}{t+1}$, so that the sum of the series to t terms is

$$S_t = \left(1 - \frac{1}{2}\right) + \left(\frac{1}{2} - \frac{1}{3}\right)$$
$$+ \left(\frac{1}{3} - \frac{1}{4}\right) + \cdots + \left(\frac{1}{t} - \frac{1}{t+1}\right).$$

Hence $S_\infty = 1$ and the series is convergent.

(iv) Applying the ratio test,

$$\left(\frac{a_{t+1}}{a_t}\right)_{t\to\infty} = \left(\frac{(t+1)^{t+1}}{(t+1)!}\frac{t!}{t^t}\right)_{t\to\infty}$$

$$= \left\{\left(\frac{t+1}{t}\right)^t\right\}_{t\to\infty}$$

$$= \left\{\left(1 + \frac{1}{t}\right)^t\right\}_{t\to\infty} = e \text{ from } (1.37).$$

Hence the series is divergent.

3.4 Applying the ratio test,

$$\left(\frac{a_{t+1}}{a_t}\right)_{t\to\infty} = \left(\frac{(t+1)^{t+1}x^{t+1}}{t!}\frac{(t-1)!}{t^t x^t}\right)_{t\to\infty}$$

$$= \left\{x\left(1 + \frac{1}{t}\right)^{t+1}\right\}_{t\to\infty} = ex. \text{ from } (1.37)$$

The convergence range is $-\dfrac{1}{e} < x < \dfrac{1}{e}$.

CHAPTER 4

Exercises

4.1

$$\frac{dy}{dx} = \left\{\frac{\tan(x+\delta x) - \tan x}{\delta x}\right\}_{\delta x\to 0}$$

$$= \left\{\frac{1}{\delta x}\left(\frac{\tan x + \tan\delta x}{1 - \tan x \tan\delta x} - \tan x\right)\right\}_{\delta x\to 0}$$

$$= \left\{\frac{1}{\delta x}\frac{\tan\delta x(1 + \tan^2 x)}{1 - \tan x \tan\delta x}\right\}_{\delta x\to 0}$$

$$= \sec^2 x\left\{\frac{\tan\delta x}{\delta x(1 - \tan x \tan\delta x)}\right\}_{\delta x\to 0}.$$

Since $\tan\delta x \to \delta x \to 0$ as $\delta x \to 0$, then $dy/dx = \sec^2 x$.

4.2 Writing $y = \sin u$, $u = \exp v$, and $v = \tan x$,

$$\frac{dy}{dx} = \frac{dy}{du}\frac{du}{dv}\frac{dv}{dx} = \cos u \exp v \sec^2 x$$

$$= \cos\{\exp(\tan x)\}\exp(\tan x)\sec^2 x.$$

4.3

$$\frac{dy}{dx} = (x + e^x \sin x)\cos x + \sin x(1 + e^x \cos x + e^x \sin x).$$

4.4

$$\frac{dy}{dx} = b(c + dx)e^x + d(a + bx)e^x + (a + bx)(c + dx)e^x.$$

4.5

$$\frac{dy}{dx} = \frac{(x + \cos x)(1 + \cos x) - (x + \sin x)(1 - \sin x)}{(x + \cos x)^2}$$

$$= \frac{1 + \cos x - \sin x + x(\cos x + \sin x)}{(x + \cos x)^2}.$$

4.6 $x = -3$, maximum and $x = 2$, minimum.

4.7 Writing

$$(1+x)^n = a_0 + a_1 x + a_2 x^2 + a_3 x^3 + \cdots + x^m + \cdots,$$

then

$$n(1+x)^{n-1} = a_1 + 2a_2 x + 3a_3 x^2 + \cdots + mx^{m-1} + \cdots$$

and

$$n(n-1)x^{n-2} = 2a_2 + 3 \times 2a_3 x + \cdots + m(m-1)x^{m-2} + \cdots.$$

Substituting $x = 0$ in these equations gives successive coefficients leading to $(1+x)^n = \sum_{i=1}^n {}^nC_i x^i$, the normal binomial expansion.

4.8 0.

Problems

4.1

(i) The chain rule with $u = x^3$ and $y = \sin u$ gives $dy/dx = 3x^2 \cos(x^3)$.

(ii) The chain rule with $u = \sin x$ and $y = \exp(u)$ gives $dy/dx = \cos x \exp(\sin x)$.

(iii) The chain rule with $u = 1 + x^4$ and $y = u^4$ gives $dy/dx = 16x^3(1 + x^4)^3$.

(iv) The product rule with $f_1 = x$ and $f_2 = \sin x$ gives $dy/dx = x \cos x + \sin x$.

(v) The product rule is used with $f_1 = x$ and $f_2 = \sin(x^2)$. To find df_2/dx requires the chain rule with $u = x^2$ and $f_2 = \sin u$. This gives

$$\frac{dy}{dx} = 2x^2 \cos(x^2) + \sin(x^2).$$

(vi) The product rule is used with $f_1 = \sin(1 + x^2)$ and $f_2 = \cos x$. To find df_1/dx requires the chain rule with $u = 1 + x^2$ and $f_1 = \sin u$. This gives

$$\frac{dy}{dx} = -\sin(1 + x^2)\sin x + 2x \cos x \cos(1 + x^2).$$

(vii) The product rule, in the form (2.19) is used with $f_1 = x^2$, $f_2 = \sin(x^2)$, $f_3 = \cos(2x)$. The chain rule is required to find df_2/dx and df_3/dx. The solution is

$$\frac{dy}{dx} = 2x\sin(x^2)\cos(2x) + 2x^3\cos(x^2)\cos(2x) - 2x^2\sin(x^2)\sin(2x).$$

(viii) The product rule is used with $f_1 = x^3$ and $f_2 = \sin(1 - \cos x)$. To find df_2/dx requires the use of the chain rule with $u = 1 - \cos x$ and $f_2 = \sin u$. The solution is

$$\frac{dy}{dx} = x^3\sin x\cos(1 - \cos x) + 3x^2\sin(1 - \cos x).$$

(ix) The quotient rule is used with $u = 1 + x^2$ and $v = 1 + x^3$. This gives

$$\frac{dy}{dx} = \frac{2x(1 + x^3) - 3x^2(1 + x^2)}{(1 + x^3)^2} = \frac{x(2 - 3x - x^3)}{(1 + x^3)^2}.$$

(x) The quotient rule is used with $u = x^2$ and $v = (1 + x)\cos x$. The product rule is used for dv/dx with $f_1 = 1 + x$ and $f_2 = \cos x$. This gives

$$\frac{dy}{dx} = \frac{2x(1 + x)\cos x - x^2\{\cos x - (1 + x)\sin x\}}{(1 + x)^2\cos^2 x}.$$

4.2 (i) Using the quotient rule with $y = \tan x = \sin x/\cos x$ we find

$$\frac{dy}{dx} = \frac{\cos x \times \cos x - \sin x \times (-\sin x)}{\cos^2 x} = \frac{1}{\cos^2 x} = \sec^2 x$$

since $\cos^2 x + \sin^2 x = 1$.

(ii) $y = \cot x = \dfrac{\cos x}{\sin x}$ so that

$$\frac{dy}{dx} = \frac{\sin x \times (-\sin x) - \cos x \times \cos x}{\sin^2 x} = -\frac{1}{\sin^2 x}$$
$$= -\mathrm{cosec}^2 x.$$

(iii) $y = (\cos x)^{-1}$ so that

$$\frac{dy}{dx} = -(\cos x)^{-2} \times (-\sin x) = \tan x\sec x.$$

(iv) $y = (\sin x)^{-1}$ so that

$$\frac{dy}{dx} = -(\sin x)^{-2} \times \cos x = -\cot x\,\mathrm{cosec}\,x.$$

4.3 The distribution is of the form $P(v) = Cv^2\exp(-\alpha v^2)$. Applying the product rule with $f_1 = v^2$ and $f_2 = \exp(-\alpha v^2)$ and finding df_2/dv from the chain rule with $u = -\alpha v^2$ and $f_2 = \exp(u)$ gives, with $C = 1$,

$$\frac{dP}{dv} = 2v\exp(-\alpha v^2) - 2\alpha v^3\exp(-\alpha v^2).$$

Equating this to zero for an extremum gives $v = 1/\sqrt{\alpha} = \sqrt{2kT/m}$. Other solutions are $v = 0$ and $v = \infty$. Differentiating again,

$$\frac{d^2 P}{dv^2} = 2\exp(-\alpha v^2)\left(1 - 5\alpha v^2 + 2\alpha^2 v^4\right).$$

Since $\alpha v^2 = 1$ for the solution of interest this gives negative $d^2 P/dv^2$ and hence a maximum.

4.4 We have

$$R = ABC\exp\left(-\frac{E}{kT} - \frac{E_G^{1/2}}{E^{1/2}}\right).$$

This will be a maximum when $y = \dfrac{E}{kT} + \dfrac{E_G^{1/2}}{E^{1/2}}$ is a minimum. Taking $dy/dE = 0$ gives

$$E = \left(\frac{1}{2}E_G^{1/2}kT\right)^{2/3}. \text{ We find}$$

$$E = \left(\frac{\sqrt{4.93 \times 10^5 \times 1.602 \times 10^{-19}} \times 1.38 \times 10^{-23} \times 2 \times 10^7}{2}\right)^{2/3} \text{ J}$$

$$= 1.1457 \times 10^{-15}\text{J} = 7.15 \text{ keV}.$$

$(1 \text{ eV} = 1.602 \times 10^{-19}\text{ J})$.

4.5 $f(x) = \frac{1}{1+x}$, $f'(x) = -\frac{1}{(1+x)^2}$, $f''(x) = \frac{2}{(1+x)^3}$, $f'''(x) = -\frac{3!}{(1+x)^4}$ or $f(0) = 1$, $f'(0) = -1$, $f''(0) = 2!$, $f'''(0) = -3!$ and in general $f^n(0) = (-1)^n n!$. Substituting into (2.39) gives

$$\frac{1}{1+x} = 1 - x + x^2 - x^3 + x^4 - \cdots + (-1)^n x^n + \cdots.$$

From the binomial theorem,

$$(1+x)^{-1} = 1 - x + \frac{-1\times-2}{1\times 2}x^2 + \frac{-1\times-2\times-3}{1\times 2\times 3}x^3 + \cdots$$

which gives the same result as the Maclaurin expansion.

4.6 The Maclaurin expansion for $\ln(1 + x)$, with remainder, is

$$\ln(1+x) = x - \frac{x^2}{2} + \frac{x^3}{3} - \cdots + (-1)^{n-1}\frac{x^n}{n} + (-1)^n\frac{\eta^{n+1}}{n+1}.$$

The maximum magnitude of η^{n+1} is 1, so that to get the required precision requires $1/(n + 1) = 0.001$ or $n = 999$.

4.7 The Taylor series is

$$\sin x = \sin\left(\frac{\pi}{4}\right) + \cos\left(\frac{\pi}{4}\right)\left(x - \frac{\pi}{4}\right) - \frac{\sin(\pi/4)(x-\pi/4)^2}{2!}$$
$$- \cdots + E_n(x)$$

where the remainder term is

$$E_n(x) = \frac{1}{(n+1)!}\left[\frac{d^{n+1}(\sin x)}{dx}\right]_{x=\eta}\left(x - \frac{\pi}{4}\right)^{n+1}.$$

Depending on the value of n the derivative is of the form $\pm\sin x$ or $\pm\cos x$ so that with x in the range 0 to $\pi/2$ its maximum magnitude is 1. Hence for the required degree of accuracy over the range we require $\frac{1}{(n+1)!}\left(\frac{\pi}{2} - \frac{\pi}{4}\right)^{n+1} \leq 0.001$. By trial and error this requires $n = 6$, or seven terms in the series.

4.8 (i) We repeatedly differentiate both numerator and divisor until one or other or both of them are non-zero at the limit (note results in Problem 4.2):

$$\left[\frac{x^3}{\tan^2 x}\right]_{x\to 0} = \left[\frac{3x^2}{2\tan x\sec^2 x}\right]_{x\to 0}$$
$$= \left[\frac{6x}{2\sec^4 x + 4\tan^2 x\sec x}\right]_{x\to 0} = 0$$

since $\sec(0) = 1$.

(ii) $\left[\dfrac{1-x^n}{(1-x)^n}\right]_{x\to1} = \left[\dfrac{-nx^{n-1}}{-n(1-x)^{n-1}}\right]_{x\to1} = \infty.$

CHAPTER 5

Exercises

5.1 (i) $\tan x + C$; (ii) $\frac{1}{5}(1+x)^5 + C$.

5.2 (i) $\frac{1}{2}\tan^2 x + C$; (ii) $\ln(\sin x) + C$; (iii) $-\dfrac{1}{\sin x}+C$.

5.3 With $t = \tan\frac{1}{2}\theta$ we have

$$I = \int \frac{2dt/(1+t^2)}{1+2t/(1+t^2)} = 2\int \frac{dt}{(1+t)^2}$$

$$= -\frac{2}{1+t} + C = -\frac{2}{1+\tan\frac{1}{2}\theta} + C.$$

5.4 $\ln(1+x) + \tan^{-1}x + C$.

5.5 (i) $x\tan x + \ln(\cos x) + C$;
(ii) $-x^2\cos x + 2x\sin x + 2\cos x + C$.

5.6 (i) $\dfrac{1}{3}\sin^3 x - \dfrac{1}{5}\sin^5 x + C$;

(ii) $-\dfrac{1}{3}\cos^3 x + \dfrac{1}{5}\cos^5 x + C$.

5.7 (i) $\dfrac{2}{3}$; (ii) 4.

Problems

5.1 (i) Completing the square gives $\displaystyle\int \frac{1}{1+(x+2)^2}dx.$

Substituting $x+2 = u$, the integral is

$$\int \frac{1}{1+u^2}du = \tan^{-1}u + C = \tan^{-1}(x+2) + C.$$

(ii) Substituting $\sin x = u$ gives $\cos x\,dx = du$. Hence

$$\int\cot x\,dx = \int\frac{\cos x}{\sin x}dx = \int\frac{1}{u}du = \ln u + C = \ln(\sin x) + C.$$

(iii) Substituting $\cos x = u$ gives $-\sin x\,dx = du$.
Hence

$$\int\tan x\,dx = \int\frac{\sin x}{\cos x}dx = -\int\frac{1}{u}du$$

$$= -\ln u + C = -\ln(\cos x) + C.$$

(iv) Substituting $x/2 = u$ gives

$$\int\sqrt{\frac{1}{4-x^2}}dx = \int\frac{1}{\sqrt{1-u^2}}du = \sin^{-1}u + C$$

$$= \sin^{-1}\left(\frac{x}{2}\right) + C.$$

(v) $\displaystyle\int \frac{1}{\sqrt{3+2x-x^2}}dx = \int\frac{1}{\sqrt{4-(x-1)^2}}dx.$

Substituting $x-1 = u$ gives

$$\int\frac{1}{\sqrt{4-u^2}}du = \sin^{-1}\left(\frac{u}{2}\right) + C = \sin^{-1}\left(\frac{x-1}{2}\right) + C.$$

5.2 (i) $\displaystyle\int\frac{1}{2x^2+3x+1}dx = \int\frac{1}{(2x+1)(x+1)}dx$

$$= \int\left\{\frac{2}{2x+1} - \frac{1}{x+1}\right\}dx$$

$$= \ln(2x+1) - \ln(x+1) + C$$

$$= \ln\left(\frac{2x+1}{x+1}\right) + C.$$

(ii) We note that differentiating the divisor gives
$2x+6$ and we write

$$\int\frac{x+2}{x^2+6x+8}dx = \frac{1}{2}\int\frac{2x+6}{x^2+6x+8}dx - \int\frac{1}{x^2+6x+8}dx.$$

The first integral is solved by the substitution
$x^2 + 6x + 8 = u$ and equals $\frac{1}{2}\ln(x^2+6x+8) + C$.
For the second integral.

$$\int\frac{1}{x^2+6x+8}dx = \int\frac{1}{(x+3)^2-1}dx.$$

Substituting $x+3 = u$ gives

$$\int\frac{1}{u^2-1}du = \frac{1}{2}\int\left(\frac{1}{u-1} - \frac{1}{u+1}\right)du$$

$$= \frac{1}{2}\ln\left(\frac{u-1}{u+1}\right) + C = \frac{1}{2}\ln\left(\frac{x+2}{x+4}\right) + C.$$

Putting together all components of the solution gives

$$\int\frac{x+2}{x^2+6x+8}dx = \ln\sqrt{\frac{(x^2+6x+8)(x+4)}{(x+2)}} + C.$$

5.3 Substituting $\tan\frac{1}{2}\theta = t$ gives

$$I = \int\frac{1-2(1-t^2)/(1+t^2)}{1+2(1-t^2)/(1+t^2)}\frac{2}{1+t^2}dt$$

$$= 2\int\frac{3t^2-1}{3-t^2}\frac{1}{1+t^2}dt.$$

Using partial fractions,

$$\frac{3t^2-1}{3-t^2}\frac{1}{1+t^2} = \frac{2}{3-t^2} - \frac{1}{1+t^2}$$

$$= \frac{1}{\sqrt3}\left(\frac{1}{\sqrt3-t} + \frac{1}{\sqrt3+t}\right) - \frac{1}{1+t^2}.$$

Hence

$$I = \frac{2}{\sqrt3}\int\left(\frac{1}{\sqrt3-t} + \frac{1}{\sqrt3+t}\right)dt - 2\int\frac{1}{1+t^2}dt$$

$$= \frac{2}{\sqrt3}\ln\left(\frac{\sqrt3+t}{\sqrt3-t}\right) - 2\tan^{-1}t + C$$

$$= \frac{2}{\sqrt3}\ln\left(\frac{\sqrt3+\tan\frac{1}{2}\theta}{\sqrt3-\tan\frac{1}{2}\theta}\right) - \theta + C.$$

5.4 (i) Integrate by parts with

$$U = x^3\left(\frac{dU}{dx} = 3x^2\right) \text{ and } \frac{dV}{dx} = \cos x\ (V = \sin x).$$

This gives $x^3 \sin x - 3 \int x^2 \sin x\, dx$. For this new integral we take

$$U = x^2 \left(\frac{dU}{dx} = 2x\right) \text{ and } \frac{dV}{dx} = \sin x \ (V = -\cos x).$$

This gives $-x^2 \cos x + 2 \int x \cos x\, dx$. One more stage of integration by parts with $U = x$ and $dV/dx = \cos x$ gives

$$\int x \cos x\, dx = x \sin x - \int \sin x\, dx = x \sin x + \cos x + C.$$

Assembling all the components of the solution gives

$$\int x^3 \cos x\, dx = x^3 \sin x + 3x^2 \cos x - 6x \sin x - 6\cos x + C.$$

(ii) From (5.52),

$$I_4 = \frac{1}{4}\cos^3 x \sin x + \frac{3}{4}I_2 = \frac{1}{4}\cos^3 x \sin x + \frac{3}{4}\left(\frac{1}{2}\cos x \sin x + \frac{1}{2}I_0\right).$$

$I_0 = \int dx = x + C$. This gives

$$\int \cos^4 x\, dx = \frac{1}{4}\cos^3 x \sin x + \frac{3}{8}\cos x \sin x + \frac{3}{8}x + C.$$

(iii) From (5.57),

$$I_{5,3} = -\frac{1}{8}\cos^6 x \sin^2 x + \frac{1}{4}I_{5,1}$$

$$I_{5,1} = \int \cos^5 x \sin x\, dx = -\frac{1}{6}\cos^6 x + C.$$

Hence

$$\int \cos^5 x \sin^3 x\, dx = -\frac{1}{8}\cos^6 x \sin^2 x - \frac{1}{24}\cos^6 x + C.$$

5.5 The area under the curve between $x=0$ and $x=2$ is

$$A_2 = \int_0^2 (x^2 - 5x + 6)\, dx = \left|\frac{1}{3}x^3 - \frac{5}{2}x^2 + 6x\right|_0^2$$

$$= \frac{8}{3} - 10 + 12 = \frac{14}{3}.$$

Similarly the area between $x=0$ and $x=3$ is

$A_3 = 9 - \frac{45}{2} + 18 = \frac{9}{2}$. A_3 is less than A_2 since the curve is below the x-axis between $x=2$ and $x=3$. The curve is shown in Figure S5.1.

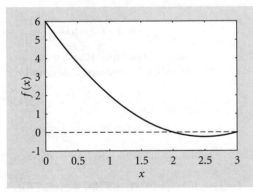

Figure S5.1 The curve $y = x^2 - 5x + 6$ between $x=0$ and $x=3$.

5.6 (i) The final velocity when the motor stops is

$$V = \int_0^{450} a\, dt = 10 \int_0^{450} \frac{1}{1 - 0.002t}\, dt.$$

Making the substitution $1 - 0.002t = u$ gives

$$V = -5000 \int_1^{0.1} \frac{1}{u}\, du = -5000|\ln u|_1^{0.1}$$

$$= -5000 \ln 0.1 = 11\,513\,\text{m s}^{-1}.$$

Notice the limits of u corresponding to the t limits and also that the final speed is somewhat greater than the escape speed from the Earth's surface, $11.2\,\text{km s}^{-1}$.
(ii) Use integration by parts with

$$U = \ln x \left(\frac{dU}{dx} = \frac{1}{x}\right) \text{ and } \frac{dV}{dx} = 1(V = x).$$

This gives $\int \ln x\, dx = x \ln x - \int dx = x \ln x - x + C$.
(iii) The general expression for v as a function of t is $V = -5000 \ln(1 - 0.002t)$, which is found by replacing 450 by t in the first integral. Hence the height when the motor stops is

$$H = \int_0^{450} V\, dt = -5000 \int_0^{450} \ln(1 - 0.002t)\, dt.$$

Substituting $1 - 0.002x = u$

$$H = 5000 \times 500 \int_1^{0.1} \ln u\, du$$

$$= 2.5 \times 10^6 |u \ln u - u|_1^{0.1}$$

$$= 2.5 \times 10^6 (0.1 \ln 0.1 - 0.1 + 1)$$

$$= 1.674 \times 10^6\,\text{m} = 1674\,\text{km}.$$

CHAPTER 6

Examples

6.1 (i) $4 - i$; (ii) $4 - 6i$; (iii) $5 + i$; (iv) $13 - 13i$
6.2 (i) $\sqrt{2}\exp\left(i\frac{\pi}{4}\right)$; (ii) $\sqrt{2}\exp\left(i\frac{3\pi}{4}\right)$
(iii) $\sqrt{13}\exp(-0.983i)$; (iv) $\sqrt{13}\exp(0.588i)$.
6.3 (i) $-0.832 + 1.819i$; (ii) $-1.248 - 2.728i$;
(iii) $-0.654 + 0.757i$.
6.4 (i) $24e^{-3i}$; (ii) $\frac{3}{2}e^{5i}$.
6.5 $\cos^3 \theta = \frac{3}{4}\cos\theta + \frac{1}{4}\cos 3\theta$; $\sin^3 \theta = \frac{3}{4}\sin\theta - \frac{1}{4}\sin 3\theta$.
6.6 $\cos 3\theta = \cos\theta(\cos^2 \theta - 3\sin^2 \theta)$;
$\sin 3\theta = \sin\theta(3\cos^2 \theta - \sin^2 \theta)$.
6.7 Cube roots are:
$2\exp(\pi i/3)$; $2\exp(\pi i)(= -2)$; $2\exp(5\pi i/3)$.
6.8 Both sides equal $7 - i$.
6.9 $\frac{1}{5} - \frac{3}{5}i$
6.10 $\frac{1}{2}e^x(\sin x - \cos x)$
6.11 $\frac{1}{Z} = \frac{1}{i\omega L} - \frac{\omega C}{i}$, giving $Z = i\frac{\omega L}{1 - \omega^2 LC}$. With $\omega = 100\,\pi\,\text{rad s}^{-1}$ the impedance

$$|Z| = \left|\frac{\omega L}{1 - \omega^2 LC}\right| = 348.6\ \Omega.$$

Problems

6.1 (i) Given that the solutions are complex, they are of the form $-b \pm i\sqrt{4ac - b^2}/2a$. The product of the two solutions is

$$\left(\frac{-b}{2a}\right)^2 + \frac{4ac - b^2}{4a^2} = \frac{c}{a}.$$

(ii) The solutions are

$$\frac{-3 \pm i\sqrt{40 - 9}}{4} = -0.75 \pm \frac{\sqrt{31}}{4}i.$$

The modulus of the solutions is

$$r = \sqrt{0.75^2 + \frac{31}{16}} = 1.5811.$$

The argument of one solution is $\tan^{-1}(\sqrt{31}/4 - 0.75) = 2.0650$ radians (remember sine is $+$ve and cosine $-$ve). Hence the solutions are $1.5811\exp(2.0650i)$ and $1.5811\exp(-2.0650i)$.

6.2 With $z = \exp i\theta$ we have

$$\cos^6\theta = \frac{1}{2^6}\left(z + \frac{1}{z}\right)^6 = \frac{1}{2^6}\left\{\left(z^6 + \frac{1}{z^6}\right)\right.$$
$$\left. + 6\left(z^4 + \frac{1}{z^4}\right) + 15\left(z^2 + \frac{1}{z^2}\right) + 20\right\}.$$

From (6.19a) this gives

$$\cos^6\theta = \frac{1}{64}(2\cos 6\theta + 12\cos 4\theta + 30\cos 2\theta + 20)$$
$$= \frac{1}{32}(\cos 6\theta + 6\cos 4\theta + 15\cos 2\theta + 10).$$

6.3 $\exp i4\theta = (\cos\theta + i\sin\theta)^4$
$$= \cos^4\theta + 4i\sin\theta\cos^3\theta + 6i^2\sin^2\theta\cos^2\theta$$
$$+ 4i^3\sin^3\theta\cos\theta + i^4\sin^4\theta$$
$$= (\cos^4\theta - 6\sin^2\theta\cos^2\theta + \sin^4\theta)$$
$$+ i(4\sin\theta\cos^3\theta - 4\sin^3\theta\cos\theta).$$

Equating real and imaginary parts,
$$\cos 4\theta = \cos^4\theta - 6\cos^2\theta\sin^2\theta + \sin^4\theta \quad \text{and}$$
$$\sin 4\theta = 4\cos\theta\sin\theta(\cos^2\theta - \sin^2\theta).$$

6.4 $2 + 3i = \sqrt{13}\exp(0.98279 + 2\pi n)$.

Thus $(2 + 3i)^{1/5}$.
for $n = 0$: $13^{1/10}\exp(0.196\,558i) = 1.267\,51 + 0.252\,40i$
for $n = 1$: $13^{1/10}\exp(1.453\,195i) = 0.151\,64 + 1.283\,47i$
for $n = 2$: $13^{1/10}\exp(2.709\,832i) = -1.173\,79 + 0.540\,83i$
for $n = 3$: $13^{1/10}\exp(3.966\,469i) = -0.877\,08 - 0.949\,22i$
for $n = 4$: $13^{1/10}\exp(5.223\,106i) = 0.631\,731 - 1.127\,48i$.

6.5 (i) $\dfrac{2 - i}{3 + 2i} = \dfrac{(2 - i)(3 - 2i)}{(3 + 2i)(3 - 2i)} = \dfrac{4 - 7i}{13}.$

(ii)
$$\frac{3 - i}{(1 + i)(2 - i)} = \frac{3 - i}{3 + i} = \frac{(3 - i)^2}{10} = \frac{8 - 6i}{10} = \frac{4 - 3i}{5}.$$

6.6 The complex conjugate is

$$\frac{z_1^* z_2^* + z_3^* z_4^*}{(z_1^* + z_2^*)(z_3^* + z_4^*)}.$$

The original expression is

$$\frac{(1 + i)^2 + (2 + i)^2}{4(1 + i)(2 + i)} = \frac{3 + 6i}{4(1 + 3i)}$$
$$= \frac{3(1 + 2i)(1 - 3i)}{40} = \frac{3(7 - i)}{40}.$$

The complex conjugate expression is

$$\frac{(1 - i)^2 + (2 - i)^2}{4(1 - i)(2 - i)} = \frac{3 - 6i}{4(1 - 3i)}$$
$$= \frac{3(1 - 2i)(1 + 3i)}{40} = \frac{3(7 + i)}{40}.$$

6.7 Not including the integration constant until the final step,

$$\int\cos x\sin x\,e^x\,dx = \frac{1}{2}\int\sin 2x\,e^x\,dx$$
$$= \frac{1}{2}i\left[\int\exp\{(1 + 2i)x\}dx\right]$$
$$= \frac{1}{2}i\left[\frac{1}{1 + 2i}\exp\{(1 + 2i)x\}\right]$$
$$= \frac{1}{2}i\left[\frac{e^x(\cos 2x + i\sin 2x)}{1 + 2i}\right]$$
$$= \frac{1}{2}i\left[\frac{e^x(\cos 2x + i\sin 2x)(1 - 2i)}{5}\right]$$
$$= \frac{1}{10}e^x(\sin 2x - 2\cos 2x) + C.$$

6.8 The angular frequency of the applied voltage is $\omega = 2\pi\nu = 100\pi$. Hence the resistor and capacitor in series have impedance

$$z_{RC} = 5 - \frac{i}{100\pi \times 0.01} = 5 - 0.3183i\ \Omega.$$

The inductor has impedance $z_L = 100\pi \times 0.01i = 3.1416i\ \Omega$. The combined impedance, Z, is given by

$$\frac{1}{Z} = \frac{1}{5 - 0.3183i} + \frac{1}{3.1416i} = \frac{5 + 0.3183i}{25.101} - 0.3183i$$
$$= 0.1992 - 0.3056i = 0.3648\exp(-0.9931i).$$

The current through the main circuit is, in amperes,

$$I = \frac{V}{Z} = 100\exp(i100\pi t) \times 0.3648\exp(-0.9931i)$$
$$= 36.48\exp\{i(100\pi t - 0.9931)\}.$$

CHAPTER 7

Exercises

7.1 1300 m.

7.2 $y = \sqrt{(x+1)^2 + C}$.

7.3 $\ln \sqrt{x^2 + y^2} - \tan^{-1}\frac{y}{x} = C$.

7.4 $y = C\exp(-\frac{1}{2}x^2)+1$.

7.5 (i) 0.6164 Hz; (ii) $3.873\,\mathrm{s}^{-1}$.

7.6 $0.968\,\mathrm{s}^{-1}$.

7.7 3.331 m.

7.8 $\omega_r = 10^3\,\mathrm{s}^{-1}$, $I_0 = 10\,\mathrm{A}$.

Problems

7.1 When $t = \infty$, $d^3 x/dt^3 = 0$ since the factor e^{-t} is dominant. The basic differential equation is $d^3 x/dt^3 = te^{-t}$. Integrating gives

$$\frac{d^2 x}{dt^2} = -te^{-t} - e^{-t} + C.$$

Since $a = 0$ when $t = 0$, $C = 1$. Integrating again,

$$\frac{dx}{dt} = te^{-t} + 2e^{-t} + t + C'.$$

Since $dx/dt = 0$ when $t = 0$ this gives $C' = -2$. Finally, integrating again,

$$x = -te^{-t} - 3e^{-t} + \frac{1}{2}t^2 - 2t + C''.$$

Since $x = 0$ when $t = 0$ we find $C'' = 3$. Hence

$$x = -te^{-t} - 3e^{-t} + \frac{1}{2}t^2 - 2t + 3.$$

7.2 (i) Separating variables,

$$\int \frac{dy}{1+y^2} = \int \frac{dx}{1+x} \text{ or } \tan^{-1} y = \ln(1+x) + C.$$

This gives $y = \tan\{\ln(1+x) + C\}$.

(ii) Separating variables,

$$\int \frac{dy}{1-y} = \int \frac{dx}{x} \text{ or } -\ln(1-y) = \ln x + \ln C.$$

The constant of integration is put in this form for convenience since it leads to $y = 1 - (1/Cx)$.

7.3 (i) $\dfrac{dy}{dx} = \dfrac{xy}{x^2 - y^2} = \dfrac{y/x}{1 - (y/x)^2}$.

Substituting $u = y/x$,

$$u + x\frac{du}{dx} = \frac{u}{1-u^2} \text{ or } x\frac{du}{dx} = \frac{u^3}{1-u^2}.$$

Hence

$$\int \frac{1-u^2}{u^3}\,du = \int \frac{du}{u^3} - \int \frac{du}{u} = \int \frac{dx}{x},$$

which gives $-(1/2u^2) - \ln u = \ln x + C$. Substituting back to y,

$$\ln y + \frac{1}{2}\frac{x^2}{y^2} + C = 0.$$

(ii) Substituting $x = X+1$ and $y = Y+1$ we have

$$\frac{dY}{dX} = \frac{X+Y}{X+4Y}.$$

Now substituting $u = Y/X$ gives

$$u + X\frac{du}{dX} = \frac{1+u}{1+4u}$$

or

$$X\frac{du}{dX} = \frac{1-4u^2}{1+4u}.$$

On rearranging, this gives

$$\int \frac{dX}{X} = \int \frac{1+4u}{1-4u^2}\,du = \int \frac{1+4u}{(1-2u)(1+2u)}\,du$$

$$= \frac{3}{2}\int \frac{1}{1-2u}\,du - \frac{1}{2}\int \frac{1}{1+2u}\,du.$$

Integrating and expressing the integration constant in a convenient form,

$$\ln X = -\frac{3}{4}\ln(1-2u) - \frac{1}{4}(1+2u) + \ln C.$$

This simplifies to

$$X(1-2u)^{3/4}(1+2u)^{1/4} = C$$

or

$X^4 (1-2u)^3 (1+2u) = K$, where K is a constant. Substituting $X = x-1$, $u = Y/X$ and $Y = y-1$ we find the solution $(x-2y+1)^3 (x+2y-3) = K$.

7.4

(i) $\dfrac{dy}{dx} + yx^2 = 2x^2$. Integrating factor $\exp(\int x^2\,dx) = \exp(x^3/3)$. Hence

$$y\exp\left(\frac{x^3}{3}\right) = 2\int x^2 \exp\left(\frac{x^3}{3}\right)dx = 2\exp\left(\frac{x^3}{3}\right) + C$$

or $y = 2 + C\exp(-x^3/3)$

(ii) $\dfrac{dy}{dx} + y\sin x = \sin x$. Integrating factor $\exp(\int \sin x\,dx) = \exp(-\cos x)$. Hence

$$y\exp(-\cos x) = \int \sin x \exp(-\cos x)dx$$

$$= \exp(-\cos x) + C.$$

This gives $y = 1 + C\exp(\cos x)$.

7.5

(i) $v\dfrac{dv}{dx} = \dfrac{dx}{dt}\dfrac{dv}{dx} = \dfrac{dv}{dt} = \dfrac{d^2 x}{dt^2}$

(ii) $v\dfrac{dv}{dx} = x - 2v^2$.

(iii) Substitute $K = \frac{1}{2}v^2$, then $dK/dx + 4K = x$.

(iv) The integrating factor is $\exp(4x)$, giving

$$K\exp(4x) = \int x\exp(4x)\,dx$$

$$= \frac{1}{4}x\exp(4x) - \frac{1}{16}\exp(4x) + C.$$

Substituting back to v,

$$v^2 = \frac{1}{2}x - \frac{1}{8} + C\exp(-4x).$$

With boundary condition $v=0$ when $x=0$,

$$v^2 = \frac{1}{2}x + \frac{1}{8}\{\exp(-4x) - 1\}$$

For $x=1$,

$$v^2 = \frac{1}{2} + \frac{1}{8}\{\exp(-4) - 1\} = 0.377\,\text{m}^2\,\text{s}^{-2} \quad \text{or}$$

$$v = 0.614\,\text{m s}^{-1}$$

7.6 The expressions (7.50) and (7.51) are of the form
$C\exp(-pt)$ and $A\exp\{-(p-q)t\} + B\exp\{-(p+q)t\}$
respectively, where p and q are positive quantities. The
ratio of the magnitudes is $\dfrac{|C|}{|A\exp(qt)+B\exp(-qt)|}$. As t
increases so the first term in the divisor increases
without limit while the second term tends to zero.
Hence, for some value of t the magnitude of (7.51)
exceeds the magnitude of (7.50).

7.7 Without air resistance,

$$v = \frac{\omega}{2\pi} = \frac{1}{2\pi}\sqrt{\frac{k}{m}} = \frac{1}{2\pi}\sqrt{\frac{1}{1}} = 0.159\,155\,\text{cycles s}^{-1}.$$

With air resistance we note that $\dfrac{f}{4m^2} < \dfrac{k}{m}$ so we have
subcritical damping and the frequency is

$$\frac{1}{2\pi}\left(\frac{k}{m} - \frac{f^2}{4m^2}\right)^{1/2} = \frac{1}{2\pi}\sqrt{\frac{k}{m}}\left(1 - \frac{f^2}{4km}\right)^{1/2}.$$

Using the binomial theorem approximation (1.35) this
becomes $\dfrac{1}{2\pi}\sqrt{\dfrac{k}{m}}\left(1 - \dfrac{f^2}{8km}\right)$. Hence the frequency is
less than that of the undamped vibration by an amount

$$-\frac{1}{2\pi}\sqrt{\frac{k}{m}}\frac{f^2}{8km} = -\frac{1}{2\pi}\sqrt{\frac{1}{1}}\frac{10^{-4}}{8\times1\times1}$$

$$= -1.99\times10^{-6}\,\text{cycles s}^{-1}.$$

Notice in this case that if the binomial theorem
approximation had not been used eight-figure
accuracy would have been required to give the
difference of frequency to the precision given here.

7.8 The basic circuit equation is

$$R\frac{dI}{dt} + \frac{I}{C} = -V_0\omega\sin\omega t \quad\text{or}\quad \frac{dI}{dt} + \frac{I}{RC} = -\frac{V_0\omega}{R}\sin\omega t.$$

The integrating factor is

$$\exp\left(\int\frac{1}{RC}\,dt\right) = \exp\left(\frac{t}{RC}\right).$$

Hence

$$\exp\left(\frac{t}{RC}\right)I = -\frac{V_0\omega}{R}\int\sin\omega t\exp\left(\frac{t}{RC}\right)dt.$$

Solving the integral on the right-hand side by parts
gives

$$\exp\left(\frac{t}{RC}\right)I = -\frac{V_0\omega}{R(1+R^2C^2\omega^2)}\left\{RC\exp\left(\frac{t}{RC}\right)\sin\omega t\right.$$

$$\left. -\omega R^2C^2\exp\left(\frac{t}{RC}\right)\cos\omega t + A\right\}$$

where A is the constant of integration, or

$$I = -\frac{V_0\omega C}{1+R^2C^2\omega^2}\left(\sin\omega t - RC\omega\cos\omega t + A\exp\left(-\frac{t}{RC}\right)\right).$$

The final term involving the integration constant is a
transient, since it disappears for large t. The remainder
can be put in the form

$$I = -\frac{V_0\omega C}{\sqrt{1+R^2C^2\omega^2}}\left(\sin\omega t\frac{1}{\sqrt{1+R^2C^2\omega^2}}\right.$$

$$\left. -\cos\omega t\frac{RC\omega}{\sqrt{1+R^2C^2\omega^2}}\right)$$

$$= -\frac{V_0\omega C}{\sqrt{1+R^2C^2\omega^2}}\sin(\omega t - \delta)$$

where $\tan\delta = \dfrac{-RC\omega}{1}$ and the signs of $\sin\delta$ and $\cos\delta$ are
given by the numerator and divisor respectively.

CHAPTER 8

Exercises

8.1 $\begin{bmatrix} 13 & 10 & 15 \\ 20 & 18 & 16 \\ 9 & 8 & 8 \end{bmatrix}$.

8.2 $\begin{bmatrix} 3+i & 5-2i \\ 3+2i & 7-i \end{bmatrix}$.

8.3 $F = 2.0\,\text{m}$. The answer is independent of the
thickness of the lens.

8.4 $L(v)L(-v) = \beta^2\begin{bmatrix} 1 & -v \\ -\frac{v}{c^2} & 1 \end{bmatrix}\begin{bmatrix} 1 & v \\ \frac{v}{c^2} & 1 \end{bmatrix}$

$$= \beta^2\begin{bmatrix} 1-\frac{v^2}{c^2} & 0 \\ 0 & 1-\frac{v^2}{c^2} \end{bmatrix} = I.$$

8.5

$$A_{13} = \begin{vmatrix} 2 & 3 & -1 \\ 1 & 0 & 4 \\ 2 & 1 & 0 \end{vmatrix};\ A_{23} = -\begin{vmatrix} 1 & 2 & 1 \\ 1 & 0 & 4 \\ 2 & 1 & 0 \end{vmatrix}.$$

8.6 40.

8.7 $|A_1| = 7$ and $|A_2| = 5$ so that $|A_1||A_2| = 35$.

$$A_1 A_2 = \begin{bmatrix} 7 & -4 \\ 7 & 1 \end{bmatrix} \text{ and } |A_1 A_2| = 35.$$

8.8 $\dfrac{1}{5} \begin{bmatrix} 2 & -1 \\ -1 & 3 \end{bmatrix}$.

8.9

$$\begin{bmatrix} x \\ y \end{bmatrix} = \frac{1}{5} \begin{bmatrix} 2 & -1 \\ -1 & 3 \end{bmatrix} \begin{bmatrix} a \\ b \end{bmatrix} = \frac{1}{5} \begin{bmatrix} 2a - b \\ -a + 3b \end{bmatrix}.$$

(i) $\begin{bmatrix} x \\ y \end{bmatrix} = \begin{bmatrix} 1 \\ 1 \end{bmatrix}$; (ii) $\begin{bmatrix} x \\ y \end{bmatrix} = \begin{bmatrix} 6/5 \\ 2/5 \end{bmatrix}$; (iii) $\begin{bmatrix} x \\ y \end{bmatrix} = \begin{bmatrix} 2/5 \\ 4/5 \end{bmatrix}$.

Problems

8.1 The product of the matrices gives

$$\begin{bmatrix} 41 & 19 & 41 \\ 15 & 9 & 15 \end{bmatrix}.$$

8.2 The matrix product describing the passage of a ray, initially parallel to the optical axis, through the four surfaces and one gap is

$$\begin{bmatrix} d_{\text{final}} \\ \alpha_{\text{final}} \end{bmatrix}$$

$$= \begin{bmatrix} 1 & 0 \\ \frac{1}{R_4}(1 - \mu_2) & \mu_2 \end{bmatrix} \begin{bmatrix} 1 & 0 \\ \frac{1}{R_3}\left(1 - \frac{1}{\mu_2}\right) & \frac{1}{\mu_2} \end{bmatrix} \begin{bmatrix} 1 & -t \\ 0 & 1 \end{bmatrix}$$

$$\times \begin{bmatrix} 1 & 0 \\ \frac{1}{R_2}(1 - \mu_1) & \mu_1 \end{bmatrix} \begin{bmatrix} 1 & 0 \\ \frac{1}{R_1}\left(1 - \frac{1}{\mu_1}\right) & \frac{1}{\mu_1} \end{bmatrix} \begin{bmatrix} d \\ 0 \end{bmatrix}$$

Substituting $R_1 = -R_2 = R_3 = -R_4 = 1.0$ m, $t = 0.1$ m, $\mu_1 = 1.5$ m, and $\mu_2 = 1.4$ m,

$$\begin{bmatrix} d_{\text{final}} \\ \alpha_{\text{final}} \end{bmatrix}$$

$$= \begin{bmatrix} 1 & 0 \\ 0.4 & 1.4 \end{bmatrix} \begin{bmatrix} 1 & 0 \\ 0.2857 & 0.7143 \end{bmatrix} \begin{bmatrix} 1 & -0.1 \\ 0 & 1 \end{bmatrix}$$

$$\times \begin{bmatrix} 1 & 0 \\ 0.5 & 1.5 \end{bmatrix} \begin{bmatrix} 1 & 0 \\ 0.3333 & 0.6667 \end{bmatrix} \begin{bmatrix} d \\ 0 \end{bmatrix} = \begin{bmatrix} 0.9d \\ 1.72d \end{bmatrix}.$$

The focal length is given by $d/\alpha_{\text{final}} = 1/1.72\,\text{m} = 0.581$ m.

8.3 The product of Lorentz matrices describing observations in the frame O''' in terms of observations in frame O is

$$\begin{bmatrix} x''' \\ t''' \end{bmatrix} = \beta\beta'\beta'' \begin{bmatrix} 1 & -v'' \\ -\frac{v''}{c^2} & 1 \end{bmatrix} \begin{bmatrix} 1 & -v' \\ -\frac{v'}{c^2} & 1 \end{bmatrix} \begin{bmatrix} 1 & -v \\ -\frac{v}{c^2} & 1 \end{bmatrix} \begin{bmatrix} x \\ t \end{bmatrix}.$$

Evaluating the product of matrices gives

$$\begin{bmatrix} x''' \\ t''' \end{bmatrix} = \begin{bmatrix} 1 + \frac{vv' + v'v'' + v''v}{c^2} & -(v + v' + v'') - \frac{vv'v''}{c^2} \\ -\frac{v + v' + v''}{c^2} - \frac{vv'v''}{c^4} & 1 + \frac{vv' + v'v'' + v''v}{c^2} \end{bmatrix} \begin{bmatrix} x \\ t \end{bmatrix}.$$

Taking out a factor equal to the diagonal elements gives the form of a Lorentz matrix corresponding to a

speed of O''' relative to O:

$$v_{O'''O} = \frac{c^2(v + v' + v'') + vv'v''}{c^2 + vv' + v'v'' + v''v}.$$

For $v = v' = v'' = 0.5c$ we find $v_{O'''O} = 0.9286c$.

8.4 The original determinant is

$$\begin{vmatrix} 2 & 1 & 3 & 4 & 1 \\ 1 & 2 & 3 & 2 & 1 \\ 1 & 1 & 5 & 2 & 2 \\ 3 & 0 & 0 & 2 & 1 \\ 1 & -1 & 2 & 0 & -1 \end{vmatrix}.$$

Now subtract factors of the last column from other columns to make their first row elements equal zero. This gives

$$\begin{vmatrix} 0 & 0 & 0 & 0 & 1 \\ -1 & 1 & 0 & -2 & 1 \\ -3 & -1 & -1 & -6 & 2 \\ 1 & -1 & -3 & -2 & 1 \\ 3 & 0 & 5 & 4 & -1 \end{vmatrix} = \begin{vmatrix} -1 & 1 & 0 & -2 \\ -3 & -1 & -1 & -6 \\ 1 & -1 & -3 & -2 \\ 3 & 0 & 5 & 4 \end{vmatrix}.$$

Now subtract factors of the first column from the other columns to make their first row elements equal to zero:

$$\begin{vmatrix} -1 & 0 & 0 & 0 \\ -3 & -4 & -1 & 0 \\ 1 & 0 & -3 & -4 \\ 3 & 3 & 5 & -2 \end{vmatrix} = - \begin{vmatrix} -4 & -1 & 0 \\ 0 & -3 & -4 \\ 3 & 5 & -2 \end{vmatrix}.$$

Finally, use the second column to make the top row element of the first column equal to zero:

$$- \begin{vmatrix} 0 & -1 & 0 \\ 12 & -3 & -4 \\ -17 & 5 & -2 \end{vmatrix} = - \begin{vmatrix} 12 & -4 \\ -17 & -2 \end{vmatrix} = 92.$$

8.5 The cofactors of the matrix are:

$A_{11} = 3,\ A_{12} = 5,\ A_{13} = -1$;
$A_{21} = -4,\ A_{22} = 1,\ A_{23} = 9$;
$A_{31} = 9,\ A_{32} = -8,\ A_{33} = -3$.

The determinant of the matrix is

$$\begin{vmatrix} 3 & 3 & 1 \\ 1 & 0 & 3 \\ 2 & -1 & 1 \end{vmatrix} = \begin{vmatrix} 9 & 0 & 4 \\ 1 & 0 & 3 \\ 2 & -1 & 1 \end{vmatrix} = \begin{vmatrix} 9 & 4 \\ 1 & 3 \end{vmatrix} = 23.$$

The inverse matrix is the adjoint matrix divided by the determinant, which is

$$\frac{1}{23} \begin{bmatrix} 3 & -4 & 9 \\ 5 & 1 & -8 \\ -1 & 9 & -3 \end{bmatrix}.$$

To check,

$$\frac{1}{23} \begin{bmatrix} 3 & -4 & 9 \\ 5 & 1 & -8 \\ -1 & 9 & -3 \end{bmatrix} \begin{bmatrix} 3 & 3 & 1 \\ 1 & 0 & 3 \\ 2 & -1 & 1 \end{bmatrix} = \begin{bmatrix} 1 & 0 & 0 \\ 0 & 1 & 0 \\ 0 & 0 & 1 \end{bmatrix}.$$

8.6 The solution vector is given by

$$\begin{bmatrix} x \\ y \\ z \end{bmatrix} = \begin{bmatrix} 1 & -1 & -1 \\ 2 & 1 & -2 \\ 1 & -2 & 1 \end{bmatrix}^{-1} \begin{bmatrix} 4 \\ 3 \\ 5 \end{bmatrix}.$$

The cofactors of the matrix are

$A_{11} = -3, A_{12} = -4, A_{13} = -5;$
$A_{21} = 3, A_{22} = 2, A_{23} = 1;$
$A_{31} = 3, A_{32} = 0, A_{33} = 3$

and the determinant is

$$\begin{vmatrix} 1 & -1 & -1 \\ 2 & 1 & -2 \\ 1 & 2 & 1 \end{vmatrix} = \begin{vmatrix} 0 & 0 & -1 \\ 0 & 3 & -2 \\ 2 & 1 & 1 \end{vmatrix} = -1 \begin{vmatrix} 0 & 3 \\ 2 & 1 \end{vmatrix} = 6.$$

Hence

$$\begin{bmatrix} x \\ y \\ z \end{bmatrix} = \frac{1}{6} \begin{bmatrix} -3 & 3 & 3 \\ -4 & 2 & 0 \\ -5 & 1 & 3 \end{bmatrix} \begin{bmatrix} 4 \\ 3 \\ 5 \end{bmatrix} = \begin{bmatrix} 2 \\ -5/3 \\ -1/3 \end{bmatrix}.$$

CHAPTER 9

Examples

9.1 $\frac{1}{\sqrt{29}}(3\mathbf{i} + 2\mathbf{j} - 4\mathbf{k}).$

9.2 $l = 3/\sqrt{29}, m = 2/\sqrt{29}, n = -4/\sqrt{29}.$

9.3 (i) 1; (ii) $3\mathbf{i} + \mathbf{j} - 4\mathbf{k}.$

9.4 $78.90°.$

9.5 11 cubic units.

9.6 (i) $7\sqrt{2}$ in units of problem; (ii) $-\mathbf{i}/\sqrt{2} + \mathbf{j}/\sqrt{2}.$

9.7 $\sqrt{56}$ units in a direction with unit vector
$\frac{1}{\sqrt{14}}(-2\mathbf{i} + 3\mathbf{j} + \mathbf{k}).$

9.8

$$\begin{vmatrix} 1 & 1 & 1 \\ 2 & -2 & 3 \\ 3 & 7 & 2 \end{vmatrix} = 0.5. (\mathbf{i}+\mathbf{j}+\mathbf{k}) - (2\mathbf{i}-2\mathbf{j}+3\mathbf{k}) = 3\mathbf{i}+7\mathbf{j}+2\mathbf{k}.$$

9.9 The lines are: $\mathbf{u} = (4t - 1)\mathbf{i} + 2t\mathbf{j} + (6t - 2)\mathbf{k}$ and $\mathbf{v} = (4s - 1)\mathbf{i} + (4s - 1)\mathbf{j} + \mathbf{k}$ (other forms are possible). Common point where $s = t = -\frac{1}{2}$ giving position of intersection as $\mathbf{i} + \mathbf{j} + \mathbf{k}$ (your solution should give this).

9.10 Vector form: $\mathbf{r} = (s + 1)\mathbf{i} + (2s + 2t - 1)\mathbf{j} + (2 - t)\mathbf{k}$ (other forms possible).

Cartesian form: $2x - y - 2z = -1$ (your solution should give this).

9.11 $-4/\sqrt{3}$. Since d is positive the point and origin are on the opposite sides.

9.12 $61.9°.$

9.13 Position of intersection $\frac{5}{3}\mathbf{i} + \frac{5}{3}\mathbf{j} + \frac{5}{3}\mathbf{k}.$

9.14 (1, 1, 2).

9.15 5 and 3 in the units of the problem.

9.16

$$\frac{d\hat{\mathbf{r}}}{d\theta} = \hat{\theta}, \quad \frac{d^2\hat{\mathbf{r}}}{d\theta^2} = \frac{d}{d\theta}\left(\frac{d\hat{\mathbf{r}}}{d\theta}\right) = -\hat{\mathbf{r}},$$

$$\frac{d^3\hat{\mathbf{r}}}{d\theta^3} = \frac{d}{d\theta}\left(\frac{d^2\hat{\mathbf{r}}}{d\theta^2}\right) = -\frac{d\hat{\mathbf{r}}}{d\theta}.$$

Hence $\frac{d^3\hat{\mathbf{r}}}{d\theta^3} + \frac{d\hat{\mathbf{r}}}{d\theta} = 0.$

9.17 (i) Component along radius vector $= \sqrt{2} \, \text{m s}^{-1}.$

(ii) $\omega = \frac{\text{tangential speed}}{\text{distance}} = \frac{\sqrt{2}}{10} \, \text{rads s}^{-1}.$ This is due to the central force and is $\mathbf{F}_r = -0.1\hat{\mathbf{r}}$, indicating direction.

$$-0.1 = m\left(\frac{d^2r}{dt^2} - r\omega^2\right) \quad \text{or}$$

$$\frac{d^2r}{dt^2} = -\frac{0.1}{m} + r\omega^2 = 0.15 \, \text{m s}^{-2}.$$

Problems

9.1 The magnitudes of the vectors are $\sqrt{3}, \sqrt{6}$ and $\sqrt{6}.$

$\cos\theta_{ab} = \frac{0}{\sqrt{3} \times \sqrt{6}}$ hence $\theta_{ab} = 90°;$ $\cos\theta_{bc} = \frac{1}{\sqrt{6} \times \sqrt{6}}$ hence $\theta_{bc} = 80.4°;$ $\cos\theta_{ca} = \frac{-2}{\sqrt{6} \times \sqrt{3}}$ hence $\theta_{ca} = 118.1°.$

9.2

$$\mathbf{a} \times \mathbf{b} = \begin{vmatrix} \mathbf{i} & \mathbf{j} & \mathbf{k} \\ 1 & 1 & 1 \\ 1 & 1 & -2 \end{vmatrix} = -3\mathbf{i} + 3\mathbf{j}.$$

$$\theta_{ab} = \sin^{-1}\left(\frac{\sqrt{3^2 + 3^2}}{\sqrt{3} \times \sqrt{6}}\right) = 90°.$$

$$\mathbf{b} \times \mathbf{c} = \begin{vmatrix} \mathbf{i} & \mathbf{j} & \mathbf{k} \\ 1 & 1 & -2 \\ -2 & 1 & -1 \end{vmatrix} = \mathbf{i} + 5\mathbf{j} + 3\mathbf{k}.$$

$$\theta_{bc} = \sin^{-1}\left(\frac{\sqrt{35}}{6}\right) = 80.4°$$

$$\mathbf{c} \times \mathbf{a} = \begin{vmatrix} \mathbf{i} & \mathbf{j} & \mathbf{k} \\ -2 & 1 & -1 \\ 1 & 1 & 1 \end{vmatrix} = 2\mathbf{i} + \mathbf{j} - 3\mathbf{k}.$$

$$\theta_{ca} = \sin^{-1}\left(\frac{\sqrt{14}}{\sqrt{18}}\right) = 61.9° \, \text{or} \, 118.1°.$$

The answers found agree with those found in problem 9.1. For θ_{ca} both the cosine and sine were required to give an unambiguous solution.

9.3 Both vectors have unit magnitude with an angle $\phi - \theta$ between them. Hence

$\mathbf{a} \cdot \mathbf{b} = \cos(\phi - \theta) = \cos\theta\cos\phi + \sin\theta\sin\phi$

$$\mathbf{a} \times \mathbf{b} = \sin(\phi - \theta)\mathbf{k} = \begin{vmatrix} \mathbf{i} & \mathbf{j} & \mathbf{k} \\ \cos\theta & \sin\theta & 0 \\ \cos\phi & \sin\phi & 0 \end{vmatrix}$$

$= (\cos\theta\sin\phi - \sin\theta\cos\phi)\mathbf{k}$

which is the required relationship with sign reversal.

9.4

$$\mathbf{a} \cdot \mathbf{b} \times \mathbf{c} = \begin{vmatrix} 1 & 1 & 1 \\ 1 & 1 & -2 \\ -2 & 1 & -1 \end{vmatrix} = 9$$

$$\mathbf{a} \times (\mathbf{b} \times \mathbf{c}) = \begin{vmatrix} \mathbf{i} & \mathbf{j} & \mathbf{k} \\ a_x & a_y & a_z \\ b_y c_z - b_z c_y & b_z c_x - b_x c_z & b_x c_y - b_y c_x \end{vmatrix}$$

$$= \begin{vmatrix} \mathbf{i} & \mathbf{j} & \mathbf{k} \\ 1 & 1 & 1 \\ 1 & 5 & 3 \end{vmatrix} = -2\mathbf{i} - 2\mathbf{j} + 4\mathbf{k}$$

9.5 The intrinsic angular momentum

$$\mathbf{L}_I = \mathbf{r} \times \mathbf{v} = \begin{vmatrix} \mathbf{i} & \mathbf{j} & \mathbf{k} \\ 1 & -1 & 2 \\ 2 & 1 & 1 \end{vmatrix} = -3\mathbf{i} + 3\mathbf{j} + 3\mathbf{k} \text{ with}$$

magnitude $\sqrt{27}$. Angle with axis is $\cos^{-1}\left(\frac{3}{\sqrt{27}}\right)$

$= \cos^{-1}\left(\frac{1}{\sqrt{3}}\right) = 54.7°$.

9.6 For $\mathbf{b}, \mathbf{c}, \mathbf{d}$, $\begin{vmatrix} -1 & -5 & 1 \\ 1 & 3 & -2 \\ 2 & 1 & 1 \end{vmatrix} = -15$, hence linearly

independent.

For $\mathbf{a}, \mathbf{c}, \mathbf{d}$, $\begin{vmatrix} 1 & -1 & 1 \\ 1 & 3 & -2 \\ 2 & 1 & 1 \end{vmatrix} = 5$, hence linearly

independent.

For $\mathbf{a}, \mathbf{b}, \mathbf{d}$, $\begin{vmatrix} 1 & -1 & 1 \\ -1 & -5 & 1 \\ 2 & 1 & 1 \end{vmatrix} = 0$, hence linearly

dependent.

For $\mathbf{a}, \mathbf{b}, \mathbf{c}$, $\begin{vmatrix} 1 & -1 & 1 \\ -1 & -5 & 1 \\ 1 & 3 & -2 \end{vmatrix} = 10$, hence linearly

independent.

9.7

$\mathbf{r}_l = (1-m)(\mathbf{i}+\mathbf{j}+\mathbf{k})+m(\mathbf{i}+2\mathbf{j}+\mathbf{k})=\mathbf{i}+(1+m)\mathbf{j}+\mathbf{k}$

$\mathbf{r}_p = (1-s-t)(2\mathbf{i}-3\mathbf{j}+\mathbf{k})+s(\mathbf{i}-\mathbf{j}-\mathbf{k})+t(2\mathbf{i}-3\mathbf{k})$

$= (2-s)\mathbf{i}+(-3+2s+3t)\mathbf{j}+(1-2s-4t)\mathbf{k}$.

For the intersection $2 - s = 1$, $-3 + 2s + 3t = 1 + m$, and $1 - 2s - 4t = 1$, the solution of which is $m = -7/2$, $s = 1$, and $t = -1/2$. Substituting the value for m in $\mathbf{r}_l$ gives the intersection point $\mathbf{i} - \frac{5}{2}\mathbf{j} + \mathbf{k}$.

9.8 $\mathbf{r}_p = x\mathbf{i} + y\mathbf{j} + z\mathbf{k}$. Hence $x = 2 - s$, $y = -3 + 2s + 3t$ and $z = 1 - 2s - 4t$. Eliminating s and t gives the equation $2x + 4y + 3z = -5$. The distance to the plane from the point $(1, 1, 1)$ is $p = \frac{-5-2-4-3}{\sqrt{29}} = -2.60$.

Since the sign of p is the same as that of d the point $(1, 1, 1)$ is on the same side of the plane as the origin.

9.9 The unit normal to the plane in Problem 9.7 is $\hat{\mathbf{p}}_1 = \frac{2\mathbf{i}+4\mathbf{j}+3\mathbf{k}}{\sqrt{29}}$. The second plane is $\mathbf{r} = (1 - s - t) \times (\mathbf{i} + \mathbf{j} + \mathbf{k}) + s(2\mathbf{i} + \mathbf{j} + \mathbf{k}) + t(\mathbf{i} + 2\mathbf{j} + \mathbf{k})$ $= (1 + s)\mathbf{i} + (1 + t)\mathbf{j} + \mathbf{k}$.

Treating these equations in the usual way we would write $x = 1 + s$, $y = 1 + t$ and $z = 1$. This plane is $z = 1$. To see this, think of the last equation as $z = 1 + 0s + 0t$. The unit normal to this plane is $\hat{\mathbf{p}}_2 = \mathbf{k}$. The angle between the two planes is

$\theta = \cos^{-1}(\hat{\mathbf{p}}_2 \cdot \mathbf{p}_2) = \cos^{-1}\left(\frac{3}{\sqrt{29}}\right) = 56.1°$.

9.10 $\mathbf{r} = (3 + t - t^2)\mathbf{i} + 2t\mathbf{j} + 3t^2\mathbf{k}$

$\mathbf{v} = \frac{d\mathbf{r}}{dt} = (1 - 2t)\mathbf{i} + 2\mathbf{j} + 6t\mathbf{k}$,

$|\mathbf{v}| = \sqrt{40t^2 - 4t + 5}$.

$\frac{d\mathbf{v}}{dt} = -2\mathbf{i} + 6\mathbf{k}$, $\left|\frac{d\mathbf{v}}{dt}\right| = \sqrt{40}$

CHAPTER 10

Exercises

10.1 $\pm 83.6°$.

10.2 $(3.33, \pm 2.98)$.

10.3 $101.54°$.

10.4 $(-6.00, 9.80)$

10.5 (i) $\frac{x^2}{1} - \frac{y^2}{4} = 1$; (ii) $\frac{4}{\sqrt{5}\cos\theta \pm 1}$, depending on which focus as origin.

10.6 (i) 3.5×10^{11} m; (ii) 1.65×10^{12} m; (iii) 1.32×10^{34} kg m^2 s^{-1}; (iv) 5.44×10^8 s $= 17.24$ years.

10.7 (i) -8.393×10^8 J kg^{-1}; (ii) 7.95×10^{10} m.

10.8 (i) 1.5×10^{15} m^2 s^{-1}; (ii) 0.888.

10.9 (i) 1.384×10^7 m s^{-1}; (ii) 1.384×10^{-6} m^2 s^{-1}; (iii) 5.456 m^3 s^{-2}; (iv) 3.651; (v) $74.1°$; (vi) $31.8°$.

Problems

10.1 From the points given,

$$\frac{1.886^2}{a^2} + \frac{1}{b^2} = 1 \text{ and } \frac{1.5^2}{a^2} + \frac{1.984^2}{b^2} = 1.$$

Solving for $1/a^2$ and $1/b^2$ gives $a = 2.00$ and $b = 3.00$. Actually the major axis is along y in this case. We also have $e = (1 - (4/9))^{1/2} = 0.745$.

10.2 From Figure S10.1,

$$QP = 2q - r\cos\theta = 2q - \frac{2q\cos\theta}{1 + \cos\theta} = \frac{2q}{1 + \cos\theta} = r.$$

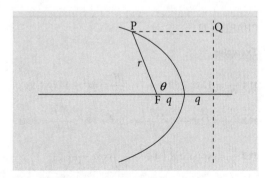

Figure S10.1 A parabola and the line $x = 2q$.

10.3 From Figure 10.6,

$$(F_1Q)^2 = (x - ae)^2 + y^2 = (x - ae)^2$$
$$+ (e^2 - 1)(x^2 - a^2) = (a - ex)^2.$$

Since $ae \leq x$ then $a \leq x/e < ex$, so for a positive value $F_1Q = ex - a$. Similarly $F_2Q = ex + a$. Hence $F_2Q - F_1Q = 2a$.

10.4 From the given perihelion $a(1 - e) = 2.3$ and from the $60°$ position

$$3.0 = \frac{a(1 - e^2)}{1 + e\cos 60°} = \frac{2.3(1 + e)}{1 + 0.5e}$$

from which $e = 0.875$. Hence

$$a = \frac{2.3}{(1 - 0.875)} = 18.4 \text{ AU}.$$

10.5 Working in metres and taking $1\,\text{AU} = 1.5 \times 10^{11}\,\text{m}$ we have

$$x = 1.8 \times 10^{11}\,\text{m}, \quad y = 1.2 \times 10^{11}\,\text{m}$$
$$\dot{x} = 2 \times 10^4 \cos 55° = 1.147 \times 10^4 \text{ m s}^{-1} \text{ and}$$
$$\dot{y} = 2 \times 10^4 \sin 55° = 1.638 \times 10^4 \text{ m s}^{-1}.$$

Hence

$$H = \begin{vmatrix} \mathbf{i} & \mathbf{j} & \mathbf{k} \\ 1.8 & 1.2 & 0 \\ 1.147 & 1.638 & 0 \end{vmatrix} \times 10^{15} = 1.572 \times 10^{15}\mathbf{k}$$

$$E = -\frac{GM}{\sqrt{x^2 + y^2}} + \frac{1}{2}(\dot{x}^2 + \dot{y}^2)$$
$$= -4.167 \times 10^8 \text{ J kg}^{-1}.$$

$$a = -\frac{GM}{2E} = 1.601 \times 10^{11} \text{ m} = 1.07 \text{ AU}.$$

$$e = \left(1 + \frac{2EH^2}{G^2M^2}\right)^{1/2} = 0.940.$$

10.6 For $\beta = \pi/2 \cos \beta = 0$, so that $8\pi\varepsilon_0 E_\alpha b = zZq_e^2$ or $b = zZq_e^2/8\pi\varepsilon_0 E_\alpha$. For a proton $z = 1$, for silver $Z = 47$ and $E_\alpha = 3.204 \times 10^{-13}$ J. This gives $b = 1.69 \times 10^{-14}$ m.

CHAPTER 11

Examples

11.1 $\dfrac{\partial f}{\partial x} = 12x^2 + 12xy + 7y^2; \dfrac{\partial f}{\partial y} = 6x^2 + 14xy + 9y^2.$

11.2 $\dfrac{\partial^2 f}{\partial x^2} = 72x^2 + 10y^2, \dfrac{\partial^2 f}{\partial y^2} = 10x^2 + 24y^2, \dfrac{\partial^2 f}{\partial x\partial y} = 20xy.$

11.3 $\dfrac{df}{dt} = \exp(\sin t)\left\{\left(1 + \dfrac{2}{t} + \dfrac{3}{t^2}\right)\cos t - \dfrac{2}{t^2}\left(1 + \dfrac{3}{t}\right)\right\}.$

11.4 $\dfrac{dy}{dx} = -\dfrac{b^4 x^3}{a^4 y^3}.$

11.5
$$\left(\frac{\partial P}{\partial V}\right)_T = -\frac{RT}{(V - b)^2} + \frac{2a}{TV^3},$$
$$\left(\frac{\partial P}{\partial T}\right)_V = \frac{R}{V - b} + \frac{a}{T^2 V^2}.$$

Hence

$$\left(\frac{\partial V}{\partial T}\right)_P = \frac{\dfrac{R}{V - b} + \dfrac{a}{T^2 V^2}}{\dfrac{RT}{(V - b)^2} - \dfrac{2a}{TV^3}}$$

or something equivalent.

11.6 $f(1.1, 0.9) = 0.880$. True value $0.882\,09$.

11.7 Minimum at $x = y = -\dfrac{1}{2}$.

Problems

11.1 (i) $\dfrac{\partial f}{\partial x} = 3x^2 + 12xy + 2y^2 + 2;$

$$\frac{\partial f}{\partial y} = 6x^2 + 4xy + 3y^2 + 3$$

$$\frac{\partial^2 f}{\partial x^2} = 6x + 12y; \quad \frac{\partial^2 f}{\partial y^2} = 4x + 6y;$$

$$\frac{\partial^2 f}{\partial x\partial y} = 12x + 4y.$$

(ii) $\dfrac{\partial f}{\partial x} = \dfrac{2}{y}\sin\left(\dfrac{x}{y}\right)\cos\left(\dfrac{x}{y}\right) = \dfrac{1}{y}\sin\left(\dfrac{2x}{y}\right);$

$$\frac{\partial f}{\partial y} = -\frac{x}{y^2}\sin\left(\frac{2x}{y}\right)$$

$$\frac{\partial^2 f}{\partial x^2} = \frac{2}{y^2}\cos\left(\frac{2x}{y}\right); \quad \frac{\partial^2 f}{\partial y^2} = \frac{2x}{y^3}\sin\left(\frac{2x}{y}\right) + \frac{2x^2}{y^4}\cos\left(\frac{2x}{y}\right)$$

$$\frac{\partial^2 f}{\partial x\partial y} = -\frac{1}{y^2}\sin\left(\frac{2x}{y}\right) - \frac{2x}{y^3}\cos\left(\frac{2x}{y}\right).$$

Here we use $\dfrac{d}{dx}(\sin^{-1}x) = \dfrac{1}{\sqrt{1 - x^2}}.$

$$\frac{\partial f}{\partial x} = \frac{1}{y}\frac{1}{\sqrt{1 - x^2/y^2}} = \frac{1}{\sqrt{y^2 - x^2}}; \quad \frac{\partial f}{\partial y} = -\frac{x}{y}\frac{1}{\sqrt{y^2 - x^2}}$$

$$\frac{\partial^2 f}{\partial x^2} = \frac{x}{(y^2 - x^2)^{3/2}}; \quad \frac{\partial^2 f}{\partial y^2} = \frac{x}{y^2}\frac{1}{\sqrt{y^2 - x^2}} + \frac{x}{(y^2 - x^2)^{3/2}}$$

$$\frac{\partial^2 f}{\partial x\partial y} = -\frac{y}{(y^2 - x^2)^{3/2}}.$$

11.2 From Problem 11.1,

$$\frac{\partial f}{\partial x} = \frac{1}{\sqrt{y^2 - x^2}} = \frac{1}{\sqrt{e^{2t} - \sin^2 t}} \text{ and}$$

$$\frac{\partial f}{\partial y} = -\frac{x}{y}\frac{1}{\sqrt{y^2 - x^2}} = -\frac{e^{-t}\sin t}{\sqrt{e^{2t} - \sin^2 t}}.$$

We also have $dx/dt = \cos t$ and $dy/dt = e^t$. Hence

$$\frac{df}{dt} = \frac{\cos t}{\sqrt{e^{2t} - \sin^2 t}} - \frac{e^{-t} \sin t}{\sqrt{e^{2t} - \sin^2 t}} \times e^t$$

$$= \frac{\cos t - \sin t}{\sqrt{e^{2t} - \sin^2 t}}.$$

11.3 $\dfrac{\partial f}{\partial x} = 27x^2 + 16xy - 3y$, $\dfrac{\partial f}{\partial y} = 8x^2 - 3x - 4y^3$,

$$\frac{dy}{dx} = -\frac{\partial f/\partial x}{\partial f/\partial y} = -\frac{27x^2 + 16xy - 3y}{8x^2 - 3x - 4y^3}.$$

For the point $(1, 2)$ $\dfrac{dy}{dx} = \dfrac{53}{27}$.

11.4 $P = \dfrac{RT}{V}$ so that $\left(\dfrac{\partial P}{\partial V}\right)_T = -\dfrac{RT}{V^2}$,

$V = \dfrac{RT}{P}$ so that $\left(\dfrac{\partial V}{\partial T}\right)_P = \dfrac{R}{P}$,

$T = \dfrac{PV}{R}$ so that $\left(\dfrac{\partial T}{\partial P}\right)_V = \dfrac{V}{R}$. Hence

$$\left(\frac{\partial P}{\partial V}\right)_T \left(\frac{\partial V}{\partial T}\right)_P \left(\frac{\partial T}{\partial P}\right)_V = -\frac{RT}{V^2}\frac{R}{P}\frac{V}{R} = -\frac{RT}{PV} = -1.$$

11.5 $f = \exp(x^2 + y^2)$, $\dfrac{\partial f}{\partial x} = 2x \exp(x^2 + y^2)$,

$$\frac{\partial f}{\partial y} = 2y \exp(x^2 + y^2),$$

$$\frac{\partial^2 f}{\partial x^2} = (2 + 4x^2) \exp(x^2 + y^2),$$

$$\frac{\partial^2 f}{\partial y^2} = (2 + 4y^2) \exp(x^2 + y^2),$$

$$\frac{\partial^2 f}{\partial x \partial y} = 4xy \exp(x^2 + y^2).$$

At $x = 1, y = 1$, $f = \exp 2$, $\dfrac{\partial f}{\partial x} = \dfrac{\partial f}{\partial y} = 2\exp 2$,

$\dfrac{\partial^2 f}{\partial x^2} = \dfrac{\partial^2 f}{\partial y^2} = 6\exp 2$ and $\dfrac{\partial^2 f}{\partial x \partial y} = 4\exp 2$

$f(1.1, 0.9) = \exp 2[1 + \{2 \times 0.1 + 2 \times (-0.1)\} + 0.5$
$\times \{6 \times (0.1)^2 + 2 \times 4 \times 0.1 \times (-0.1) + 6 \times (0.1)^2\}]$
$= 7.5368.$

The correct value is 7.5383.

11.6 The lines are $\mathbf{r}_1 = (1 - s)\mathbf{u} + s\mathbf{v} = (2 - s)\mathbf{i}$
$+ (2 - 3s)\mathbf{j} + (2s - 1)\mathbf{k}$ and $\mathbf{r}_2 = (1 - t)\mathbf{p} + t\mathbf{q}$
$= (3 - t)\mathbf{i} - \mathbf{j} + (1 - 2t)\mathbf{k}$.

The square of the distance between points defined by s
and t is $D^2 = (1 - t + s)^2 + (3 - 3s)^2 +$
$(2s + 2t - 2)^2$.
The condition $\partial(D^2)/\partial s = 0$ leads to $14s + 3t = 12$
and $\partial(D^2)/\partial t = 0$ leads to $3s + 5t = 5$. The solution
of the linear simultaneous equations gives $s = 45/61$
and $t = 34/61$. Substituting these values in the
expression for D^2 gives $D = \sqrt{8784}/61 = 1.536$.

CHAPTER 12

Examples

12.1 (i) $1/52 + 1/52 = 1/26$; (ii) $1/52 \times 1/2 = 1/104$.

12.2 $1/(10 \times 9 \times 8) = 1/720$.

12.3 $3! \, 7!/10! = 1/120$.

12.4 Probability that all birth-months different for n
individuals is $12!/(12 - n)!12^n$. This is less than 0.5
for $n = 5$.

12.5 (i) 0.25; (ii) 0.75; (iii) 1;

(iv) $P(D|NC) = \dfrac{P(NC|D)P(D)}{P(NC)} = \dfrac{1}{3}$.

12.6 (i) 17.81; (ii) 19; (iii) 42.06; (iv) 6.49.

12.7 $C = \left(\frac{1}{2.7^2} + \frac{1}{4.2^2}\right)^{-1} = 5.158$.

Hence standard deviation of best estimate = 2.27.

$$w_1 = \frac{C}{V_1} = \frac{5.158}{2.7^2} = 0.7076. \text{ Similarly,}$$

$$w_2 = 0.2924 \, (w_1 + w_2 = 1)$$

Best estimate $= 0.7076 \times 15.2 + 0.2924 \times 16.1 = 15.46$.

Problems

12.1 There are 36 (6×6) different ways in which the dice
can fall, each way having a probability of 1/36. Six of these
give rise to a total of $7 - 1 + 6, 2 + 5, 3 + 4, 4 + 3, 5 + 2$,
and $6 + 1$ – and the combined probability of these
mutually exclusive events is $6 \times 1/36 = 1/6$.

The probability of a head with a coin is 0.5 and since
with two coins the outcomes are independent then the
probability of two heads is $1/2 \times 1/2 = 1/4$. The
outcomes from the coins and dice are independent so
the combined probability of having a sum of 7 and two
heads is $1/6 \times 1/4 = 1/24$.

12.2 The number of ways of selecting 4 from 15
without regard to order is

$$\frac{15!}{4! \, 11!} = \frac{15 \times 14 \times 13 \times 12}{4 \times 3 \times 2 \times 1} = 1365.$$

12.3 The number of ways that n Sumodians could have
birthdays, without restriction is 100^n. Of these the
number of ways that the birthdays could be all different is

$$100 \times 99 \times 98 \times \cdots \times (101 - n) = \frac{100!}{(100 - n)!}.$$

Hence the condition that there is a 50% probability
that at least two Sumodians have the same birthday is
given by $\dfrac{100!}{(100 - n)!100^n} < 0.5$ or, taking natural
logarithms,

$$\ln(100!) - \ln\{(100 - n)!\} - n\ln(100) - \ln(0.5) < 0.$$

Using Stirling's approximation and rearranging this gives

$$(100 - n)\ln\left(\frac{100}{100 - n}\right) - n - \ln(0.5) < 0.$$

The minimum value of n to satisfy this relationship is
found by trial and error, which is illustrated in the
following table. Incrementing n first by steps of 5
indicated a solution between 10 and 15 and then unit
increments gave the solution as 12.

n	Expression
5	0.566
10	0.176
15	−0.493
11	0.065
12	−0.058

12.4 The prior probability of faulty screws $P(F) = 0.0001$. The posterior probability that a screw is faulty and tested as faulty is $P(T|F) = 0.99$. The probability that a screw is tested as faulty is $P(T) = 0.00011$. Hence the probability that a screw tested as faulty really is faulty is

$$P(F|T) = \frac{P(T|F)P(F)}{P(T)} = \frac{0.99 \times 0.0001}{0.00011} = 0.90.$$

12.5 The mean-square rainfall for each district is found from $\overline{x^2} = \sigma^2 + \bar{x}^2$. The mean-square daily rainfall in the six districts, in units mm^2, is:

5.1746, 6.8429, 2.2850, 4.1513, 6.0176, 10.9888.

The mean square daily rainfall for the whole country is the average of these, $5.9100\,\text{mm}^2$. The mean daily rainfall is the average of the means for the districts, 2.2283 mm. Hence the whole-country standard deviation of daily rainfall is

$$\sigma = \sqrt{5.9100 - 2.2283^2}\,\text{mm} = 0.972\,\text{mm}.$$

Rounded off to the accuracy of the data provided, the whole-country average and standard deviation are 2.23 mm and 0.97 mm respectively.

12.6 The relative weights, using inverse-variance weighting, are

$$w_1 : w_2 : w_3 = \frac{1}{(0.020)^2} : \frac{1}{(0.0071)^2} : \frac{1}{(0.0112)^2}$$

$$= 2500 : 19837 : 7972.$$

The best estimate of g is

$$g_{\text{best}} = \frac{9.841w_1 + 9.8152w_2 + 9.8231w_3}{w_1 + w_2 + w_3}$$

$$= 9.8193\,\text{m s}^{-2}.$$

The standard deviation of the best estimate is

$$\sigma_{\text{best}} = (w_1 + w_2 + w_3)^{-1/2} = 0.0057\text{m s}^{-2}.$$

CHAPTER 13

Exercises

13.1 (i) $\frac{1}{4}$; (ii) $\frac{11}{3}$.

13.2 $\frac{5}{3}$.

13.3 $\frac{1}{4}$.

13.4

$$J\left(\frac{x,y}{x',y'}\right) = \begin{vmatrix} \dfrac{\partial x}{\partial x'} & \dfrac{\partial x}{\partial y'} \\ \dfrac{\partial y}{\partial x'} & \dfrac{\partial y}{\partial y'} \end{vmatrix} = \begin{vmatrix} \cos\alpha & -\sin\alpha \\ \sin\alpha & \cos\alpha \end{vmatrix}$$

$$= 1. \text{ Hence elemental area is } dx'dy'.$$

13.5 $(3.742, 105.50°, 33.69°)$.

13.6 $(1.414, 2.449, −2.828)$.

13.7 $(\sqrt{8}, 60°, −2.828)$.

13.8 61.

13.9 $\pi^2/16$.

13.10 0.733 kg.

13.11 $\frac{1}{3}Ml^2$ (note in comparing with 13.38 that the length of the rod here is l).

13.12 $a^5 : b^5$

13.13 $\frac{3}{2}MR^2$.

13.14 $\frac{5}{4}MR^2$.

Problems

13.1 (i) $I = \int_0^1 \left|\frac{1}{3}xy^3 + \frac{1}{2}x^2y^2\right|_0^3 dx = \int_0^1 \left(9x + \frac{9}{2}x^2\right)dx$

$$= \left|\frac{9}{2}x^2 + \frac{3}{2}x^3\right|_0^1 = 6$$

(ii) $I = \int_0^1 \left|-x\exp(-xy)\right|_0^2 dx = \int_0^1 \{x - x\exp(-2x)\}dx$

$$= \left|\frac{1}{2}x^2 + \frac{1}{2}x\exp(-2x) + \frac{1}{4}\exp(-2x)\right|_0^1$$

$$= \frac{1}{4} + \frac{3}{4}\exp(-2) = 0.352$$

(iii) $I = \int_0^1 \left|-\frac{1}{1 + r\sin\theta}\right|_0^{\pi/2} dr = \int_0^1 \left(1 - \frac{1}{1 + r}\right)dr$

$$= |r - \ln(1 + r)|_0^1 = 1 - \ln 2 = 0.307.$$

13.2

$$M = 5000 \int_0^{0.1} \int_0^{0.2} (x + 0.1)(y + 0.1)\,dy\,dx$$

$$= 5000 \times \left|\frac{1}{2}x^2 + 0.1x\right|_0^{0.1} \times \left|\frac{1}{2}y^2 + 0.1y\right|_0^{0.2}$$

$$= 5000 \times 0.015 \times 0.04 = 3.0\,\text{kg}.$$

13.3 $\begin{vmatrix} \cos\theta & -r\sin\theta & 0 \\ \sin\theta & r\cos\theta & 0 \\ 0 & 0 & 1 \end{vmatrix} = r.$ Hence the elemental volume is $r\,dr\,d\theta\,dz$.

13.4 Rewrite density relationship as $\rho = \rho_0(1 + r^2\sin^2\theta\cos^2\phi)$. Then mass is

$$M = \rho_0 \int_0^{2\pi} \int_0^\pi \int_0^1 (1 + r^2\sin^2\theta\cos^2\phi)r^2$$
$$\times \sin\theta\,dr\,d\theta\,d\phi.$$

The 1 in the bracket corresponds to a density ρ_0 over the sphere and hence contributes $4\pi\sigma_0/3$. The remaining contribution can be factorized, giving

$$\rho_0 \int_0^1 r^4 \mathrm{d}r \int_0^\pi \sin^3 \theta \, \mathrm{d}\theta \int_0^{2\pi} \cos^2 \phi \, \mathrm{d}\phi$$

$$\int_0^1 r^4 \mathrm{d}r = \frac{1}{5}, \quad \int_0^\pi \sin^3 \theta \, \mathrm{d}\theta$$

$$= \int_0^\pi (1 - \cos^2 \theta) \sin \theta \, \mathrm{d}\theta$$

$$= \left| -\cos\theta + \frac{1}{3}\cos^3\theta \right|_0^\pi = \frac{4}{3}; \quad \int_0^{2\pi} \cos^2 \phi \, \mathrm{d}\phi$$

$$= \int_0^{2\pi} \frac{1}{2}(1 + \cos 2\phi)\mathrm{d}\phi = \left| \frac{1}{2}\phi + \frac{1}{2}\sin 2\phi \right|_0^{2\pi} = \pi.$$

Hence the mass is

$$M = \rho_0 \left(\frac{4}{3}\pi + \frac{1}{5} \times \frac{4}{3} \times \pi \right) = \frac{8}{5}\pi\rho_0.$$

13.5 The areal density of the plate (mass per unit area) is $\rho t_0 \left\{ 1 - \left(\frac{r}{R}\right)^2 \right\}$. For the purpose of integration it can be considered in annular rings of radius r and width $\mathrm{d}r$. The mass is

$$M = 2\pi\rho t_0 \int_0^R \left\{ 1 - \left(\frac{r}{R}\right)^2 \right\} r \, \mathrm{d}r$$

$$= 2\pi\rho t_0 \left| \frac{1}{2}r^2 - \frac{1}{4}\frac{r^4}{R^2} \right|_0^R = \frac{1}{2}\pi\rho t_0 R^2.$$

The moment of inertia is

$$I_m = 2\pi\rho t_0 \int_0^R \left\{ 1 - \left(\frac{r}{R}\right)^2 \right\} r^3 \, \mathrm{d}r$$

$$= 2\pi\rho t_0 \left| \frac{1}{4}r^4 - \frac{1}{6}\frac{r^6}{R^2} \right|_0^R = \frac{1}{6}\pi\rho t_0 R^4.$$

This gives $I_m = \frac{1}{3}MR^2$.

13.6 The simplest approach is to consider the Moon as the sum of two components – a sphere of radius 1740 km and density 3300 kg m^{-3} plus a sphere of radius 400 km with a density of 4700 kg m^{-3}. Hence the total mass is

$$M = \frac{4}{3}\pi \left\{ 3300(1.74 \times 10^6)^3 + 4700(4.00 \times 10^5)^3 \right\}$$
$$= 7.4080 \times 10^{22} \text{ kg}.$$

The moment of inertia is

$$I_m = 0.4 \times \frac{4}{3}\pi \left\{ 3300(1.74 \times 10^6)^5 \right.$$
$$\left. + 4700(4.00 \times 10^5)^5 \right\}$$
$$= 8.8268 \times 10^{34} \text{ kg m}^2.$$

Hence the moment of inertia factor

$$\alpha = \frac{8.8268 \times 10^{34}}{7.4080 \times 10^{22} \times (1.74 \times 10^6)^2} = 0.394.$$

13.7 The arrangement of spheres is shown in Figure S13.1. The moment of inertia of each sphere about an axis perpendicular to the rod through O, the centre of the rod is, using the parallel-axis theorem,

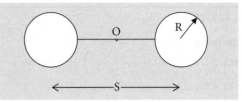

Figure S13.1 Uniform spheres connected by a light rod.

$$I_S = 0.4MR^2 + M\left(\frac{1}{2}S\right)^2.$$

For both spheres it is

$$I_{2S} = 0.8MR^2 + 0.5MS^2.$$

Hence

$$0.8MR^2 + 0.5MS^2 = 5MR^2 \text{ or } S = \sqrt{8.4}R.$$

13.8 (i) First we find the moment of inertia about an axis in the plane through its midpoint and parallel to one of its sides – as shown in Figure S13.2. The contribution of the strip, distant x from the axis and of thickness $\mathrm{d}x$ is $\mathrm{d}I_y = \sigma a x^2 \mathrm{d}x$, where σ is the mass per unit area of the plate. Hence the total moment of inertia about the axis is

$$I_y = \sigma a \int_{-a/2}^{a/2} x^2 \mathrm{d}x = \frac{1}{12}\sigma a^4 = \frac{1}{12}Ma^2$$

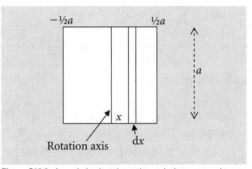

Figure S13.2 An axis in the plane through the centre along y.

where M is the total mass of the plate. The moment of inertia about an axis through the centre in the x-direction has the same value so, using the perpendicular-axis theorem for the axis through the centre perpendicular to the plate

$$I_z = I_x + I_y = \frac{1}{6}Ma^2.$$

(ii) A parallel axis through a corner is distant $a/\sqrt{2}$ from the axis through the centre. Hence the moment of inertia about an axis perpendicular to the plate through a corner is

$$I_C = I_z + \frac{1}{2}Ma^2 = \frac{2}{3}Ma^2.$$

CHAPTER 14

Exercises

14.1 $P = \dfrac{1}{\sqrt{2\pi \times 20^2}} \exp\left(-\dfrac{11^2}{2 \times 20^2}\right) \times 0.1 = 1.71 \times 10^{-3}$.

14.2 0.02275.

14.3 Mean $= 0$, standard deviation $= 10$.

14.4 0.0879.

14.5 Average $= 1$, standard deviation $= 0.913$.

14.6 Average $= 250\,000$, standard deviation $= 433.0$, probability $= 0.0105$.

14.7 1.93×10^{-3}.

14.8 (i) 0.189; (ii) 0.315; (iii) 0.496.

Problems

14.1 (i) (a) $\bar{r} = np = 2.5$, $P(r) = {}^5C_r(0.5)^5$.

$P(0) = 0.031\,25$, $P(1) = 0.156\,25$, $P(2) = 0.312\,50$, $P(3) = 0.312\,50$, $P(4) = 0.156\,25$, $P(5) = 0.031\,25$.

$m(3) = \sum_{r=0}^{5} P(i)(r - 2.5)^3 = 0$. The odd moment of any distribution symmetrical about the mean is zero.

(b) $\bar{r} = np = 4, P(r) = {}^5C_r(0.8)^r(0.2)^{5-r}$.

$P(0) = 0.000\,32$, $P(1) = 0.006\,40$, $P(2) = 0.051\,20$, $P(3) = 0.204\,80$, $P(4) = 0.409\,60$, $P(5) = 0.327\,68$.

$m(3) = \sum_{r=0}^{5} P(i)(r - 4)^3 = -0.48$.

(ii) $m(n + 2) = \dfrac{1}{\sqrt{2\pi\sigma^2}} \int_{-\infty}^{\infty} x^{n+2} \exp\left(-\dfrac{x^2}{2\sigma^2}\right) dx$.

Integrating by parts gives

$m(n + 2) = \dfrac{1}{\sqrt{2\pi\sigma^2}} \left\{ \left| -\sigma^2 x^{n+1} \exp\left(-\dfrac{x^2}{2\sigma^2}\right) \right|_{-\infty}^{\infty} \right.$

$\left. + \sigma^2(n + 1) \int_{-\infty}^{\infty} x^n \exp\left(-\dfrac{x^2}{2\sigma^2}\right) dx \right\}$.

By symmetry the first term in the curly brackets is zero and the remaining term gives
$m(n + 2) = \sigma^2(n + 1)m(n)$. Hence
$m(10) = 9 \times 7 \times 5 \times 3 \times 1 \times \sigma^{10} m(0)$. Since $m(0) = 1$ (normalized distribution), $m(10) = 945\sigma^{10}$.

14.2 The expected mean number of successes in 1000 trials $= 1000 \times 0.2 = 200$, with standard deviation $\sqrt{1000 \times 0.2 \times 0.8} = 12.65$.

The binomial distribution of the number of successes is approximated by a normal distribution and, since 240 represents the range 239.5–240.5 of the normal distribution, the required probability is the area under the curve beyond $239.5 = (39.5/12.65)\sigma = 3.12\sigma$ from the mean. By linear interpolation from Table 14.1 this is 9.12×10^{-4}.

14.3 From (14.40), the required probability is

$$P = \frac{10!}{3!2!2!1!1!1!} \left(\frac{1}{6}\right)^{10} = 0.0025.$$

14.4 The probabilities of the numbers of photons arriving at the detector will be governed by the Poisson distribution with mean 1.2. The probability that the experiment will *fail* is given by

$P(0) + P(1) + P(2)$

$= e^{-1.2}\left(\dfrac{(1.2)^0}{0!} + \dfrac{(1.2)^1}{1!} + \dfrac{(1.2)^2}{2!}\right) = 0.8794$.

Hence the probability of success is
$1 - 0.8794 = 0.1206$.

CHAPTER 15

Exercises

15.1 $\cosh^3 x = \dfrac{1}{8}(e^x + e^{-x})^3 = \dfrac{1}{8}(e^{3x} + 3e^x + 3e^{-x} + e^{-3x})$

$= \dfrac{1}{4}\cosh 3x + \dfrac{3}{4}\cosh x$.

15.2 (i) and (ii) $2\cosh x \sinh x = \sinh 2x$
(iii) $2\tanh x \operatorname{sech}^2 x$.

15.3 $\cosh x - \sinh x = 1 - x + \dfrac{x^2}{2!} - \dfrac{x^3}{3!} + \dfrac{x^4}{4!} - \cdots = e^{-x}$
from (4.34) with $x = -1$.

15.4 (i) $\dfrac{1}{2}\sinh^{-1}(2x) + C$; (ii) $\dfrac{1}{2}\cosh^{-1}(2x) + C$.

Problems

15.1 Taking $A = \dfrac{X + Y}{2}$ and $B = \dfrac{X - Y}{2}$ gives
$X = \dfrac{A + B}{2}$ and $Y = \dfrac{A - B}{2}$.

(i) $\sinh\left(\dfrac{X + Y}{2}\right) + \sinh\left(\dfrac{X - Y}{2}\right)$

$= 2\sinh\dfrac{X}{2}\cosh\dfrac{Y}{2}$

$= 2\sinh\left(\dfrac{A + B}{2}\right)\cosh\left(\dfrac{A - B}{2}\right)$.

(ii) $\cosh\left(\dfrac{X + Y}{2}\right) + \cosh\left(\dfrac{X - Y}{2}\right)$

$= 2\cosh\dfrac{X}{2}\cosh\dfrac{Y}{2}$

$= 2\cosh\left(\dfrac{A + B}{2}\right)\cosh\left(\dfrac{A - B}{2}\right)$.

15.2 (i) $\dfrac{d}{dx}\left(\dfrac{\sinh x}{1 + x^2}\right) = \dfrac{\cosh x}{1 + x^2} - \dfrac{2x\sinh x}{(1 + x^2)^2}$.

(ii) $\dfrac{d}{dx}(\sinh x \cos x) = \cosh x \cos x - \sinh x \sin x$.

(iii) Substituting $u = \sinh x$ then writing $y = \tan^{-1} u$ we have $dy/du = 1/(1 + u^2)$. Hence

$\dfrac{dy}{dx} = \dfrac{dy}{du}\dfrac{du}{dx} = \dfrac{1}{1 + \sinh^2 x}\cosh x = \dfrac{\cosh x}{\cosh^2 x} = \operatorname{sech} x$.

15.3 (i) We write $\tanh x = a_0 + a_1 x + a_2 x^2 + a_3 x^3 + a_4 x^4 + a_5 x^5 + \cdots$. Putting $x = 0$ gives $a_0 = 0$. Differentiate both sides:

$$\text{sech}^2 x = a_1 + 2a_2 x + 3a_3 x^2 + 4a_4 x^3 + 5a_5 x^4 + \cdots$$

Putting $x = 0$ gives $a_1 = 1$. Differentiate both sides:

$$-2\text{sech}^2 x \tanh x = 2a_2 + 6a_3 x + 12a_4 x^2 + 20a_5 x^3 + \cdots.$$

Putting $x = 0$ gives $a_2 = 0$. Differentiate both sides:

$$4\text{sech}^2 x \tanh^2 x - 2\text{sech}^4 x$$
$$= 6a_3 + 24a_4 x + 60a_5 x^2 + \cdots.$$

Putting $x = 0$ gives $a_3 = -\frac{1}{3}$. Differentiate both sides:

$$-8\text{sech}^2 x \tanh^3 x + 8\text{sech}^4 x \tanh x$$
$$+ 8\text{sech}^4 x \tanh x = 24a_4 + 120a_5 x + \cdots.$$

Putting $x = 0$ gives $a_4 = 0$. Differentiate both sides. Various terms involving $\tanh x + 16\text{sech}^6 x = 120a_5 + \cdots$. Putting $x = 0$ gives $a_5 = 2/15$.

(ii) Since $\int \dfrac{1}{1-x^2}\,dx = \tanh^{-1} x + C$ we have

$$\frac{d}{dx}(\tanh^{-1} x) = \frac{1}{1-x^2}.$$

We write

$$\tanh^{-1} x = a_0 + a_1 x + a_2 x^2 + a_3 x^3 + a_4 x^4 + a_5 x^5 + \cdots.$$

Putting $x = 0$ gives $a_0 = 0$. Differentiate both sides.

$$\frac{1}{1-x^2} = a_1 + 2a_2 x + 3a_3 x^2 + 4a_4 x^3 + 5a_5 x^4 + \cdots.$$

Putting $x = 0$ gives $a_1 = 1$. Differentiate both sides:

$$\frac{2x}{(1-x^2)^2} = 2a_2 + 6a_3 x + 12a_4 x^2 + 20a_5 x^3 + \cdots$$

Putting $x = 0$ gives $a_2 = 0$. Differentiate both sides:

$$\frac{2}{(1-x^2)^2} + \frac{8x^2}{(1-x^2)^3} = 6a_3 + 24a_4 x + 60a_5 x^2 + \cdots.$$

Putting $x = 0$ gives $a_3 = \frac{1}{3}$. Differentiate both sides:

$$\frac{8x}{(1-x^2)^3} + \frac{16x}{(1-x^2)^3} + \frac{48x^3}{(1-x^2)^4} = 24a_4 + 120a_5 x + \cdots.$$

Putting $x = 0$ gives $a_4 = 0$. Differentiate both sides:

$$\frac{24}{(1-x^2)^3} + \frac{48x^2}{(1-x^2)^4} + \frac{144x^2}{(1-x^2)^4} + \frac{384x^4}{(1-x^2)^5} = 120a_5 + \cdots.$$

Putting $x = 0$ gives $a_5 = \frac{1}{5}$.

15.4 (i) $\int \dfrac{1}{\sqrt{x^2 + 4x + 3}}\,dx = \int \dfrac{1}{\sqrt{(x+2)^2 - 1}}\,dx.$

Substitute $u = x + 2$, giving $du/dx = 1$:

$$\int \frac{1}{\sqrt{u^2 - 1}}\,du = \cosh^{-1} u + C = \cosh^{-1}(x+2) + C.$$

(ii) $\int \dfrac{1}{\sqrt{x^2 + 4x + 5}}\,dx = \int \dfrac{1}{\sqrt{(x+2)^2 + 1}}\,dx.$

Substitute $u = x + 2$, giving $du/dx = 1$:

$$\int \frac{1}{\sqrt{u^2 + 1}}\,du = \sinh^{-1} u + C = \sinh^{-1}(x+2) + C.$$

(iii) $\displaystyle \int e^x \cosh^2 x\,dx = \frac{1}{2}\int e^x (1 + \cosh 2x)\,dx$

$$= \frac{1}{2}\int e^x \left\{ 1 + \frac{1}{2}\left(e^{2x} + e^{-2x}\right) \right\}dx$$

$$= \frac{1}{2}\left(e^x + \frac{1}{6}e^{3x} - \frac{1}{2}e^{-x}\right) + C.$$

(iv) Substitute $x = \sinh u$, then $\sinh^{-1} x = u$, $\sqrt{1 + x^2} = \cosh u$ and

$$\frac{dx}{du} = \cosh u.$$

$$I = \int_0^{\sinh^{-1} 1} u \cosh^2 u \sinh u\,du.$$ Integrate by parts with $dv/du = \cosh^2 u \sinh u$:

$$I = \left| \frac{1}{3}u \cosh^3 u \right|_0^{\sinh^{-1} 1} - \frac{1}{3}\int_0^{\sinh^{-1} 1} \cosh^3 u\,du.$$

If $\sinh u = 1$ then $\cosh u = \sqrt{2}$

$$I = \frac{2\sqrt{2}}{3}\sinh^{-1} 1 - \frac{1}{3}\int_0^{\sinh^{-1} 1}\left(\cosh u + \sinh^2 u \cosh u\right)du$$

$$= \frac{2\sqrt{2}}{3}\sinh^{-1} 1 - \frac{1}{3}\left|\sinh u + \frac{1}{3}\sinh^3 u\right|_0^{\sinh^{-1} 1}$$

$$= \frac{2\sqrt{2}}{3}\sinh^{-1} 1 - \frac{4}{9}.$$

15.5 Weight function $\tanh(100\alpha) = 0.5$, then $100\alpha = 0.5493$ and $\alpha = 5.493 \times 10^{-3}$.

CHAPTER 16

Exercises

16.1 $\nabla\phi = (8x - 2y)\mathbf{i} + (2y - 2x + z)\mathbf{j} + (2z + y)\mathbf{k}$.

16.2 (i) $\mathbf{F}$ along radius vector so work done only along BC and DA, which cancel.

(ii) Transverse force so work only along AB and CD where force is parallel to the motion. Net work done by the field $= -\dfrac{K \times AB}{OA} + \dfrac{K \times CD}{OD} = 0$ since $\dfrac{AB}{OA} = \dfrac{CD}{OD}$.

16.3 $2x + 2y + 2z$.

16.4 (i) $\dfrac{a^3 \sigma}{3\varepsilon_0 r^2}$; (ii) $\dfrac{r\sigma}{3\varepsilon_0}$

16.5 (i) $-(k_x^2 + k_y^2 + k_z^2)\sin k_x x \sin k_y y \sin k_z z$
$= -(k_x^2 + k_y^2 + k_z^2)\psi$.

(ii) $(6\sin\theta + \cot\theta\cos\theta - \sin\theta - \text{cosec }\theta)\cos\phi$.

16.6 0.

Problems

16.1 For $\mathbf{A} = x\mathbf{i} + y\mathbf{j} + 2z\mathbf{k}$ we have $\nabla \cdot \mathbf{A} = 1 + 1 + 2 = 4$. Hence

$$\oiint \nabla \cdot \mathbf{A}\,dv = 4 \times \text{volume of cube} = 4 \times 8 = 32.$$

The component $V_x = x$ and has the same value, 1, everywhere on the face centred $(1, 0, 0)$. Hence the flux through this face is $1 \times \text{area of face} = 4$.

The flux through the face $(-1, 0, 0)$ *out of the cube* is $-V_x(-1, 0, 0) \times 4 = 4$. Hence the net flux out of the cube through the x faces $= 8$. By similar reasoning the net flux through the y faces is 8 and that through the z faces is 16. Hence $\oint_S \mathbf{V} \cdot d\mathbf{A} = 32$ and the divergence theorem holds.

16.2

$$W_{AB} = \int_D^{D+a} \frac{K}{x^3} dx = \left| -\frac{K}{2x^2} \right|_D^{D+a} = \frac{K}{2}\left(\frac{1}{D^2} - \frac{1}{(D+a)^2} \right),$$

$$W_{AC} = \int_0^b \frac{Kx}{(D^2+x^2)^2} dx = \frac{K}{2}\left| -\frac{1}{D^2+x^2} \right|_0^b$$
$$= \frac{K}{2}\left(\frac{1}{D^2} - \frac{1}{D^2+b^2} \right),$$

$$W_{CD} = \int_0^a \frac{K(D+x)}{\left\{(D+x)^2+b^2\right\}^2} dx = \frac{K}{2}\left| -\frac{1}{(D+x)^2+b^2} \right|_0^a$$
$$= \frac{K}{2}\left\{ \frac{1}{D^2+b^2} - \frac{1}{(D+a)^2+b^2} \right\},$$

$$W_{DB} = -\frac{K}{2}\left\{ \frac{1}{(D+a)^2} - \frac{1}{(D+a)^2+b^2} \right\},$$

$W_{AC} + W_{CD} + W_{DB} = W_{AB}$, as is required to show a conservative field.

16.3 From (16.47),

$$\nabla \times \mathbf{V} = \left(\frac{\partial V_z}{\partial y} - \frac{\partial V_y}{\partial z} \right)\mathbf{i} + \left(\frac{\partial V_x}{\partial z} - \frac{\partial V_z}{\partial x} \right)\mathbf{j}$$
$$+ \left(\frac{\partial V_y}{\partial x} - \frac{\partial V_x}{\partial y} \right)\mathbf{k}.$$

Hence

$$\nabla \times (\nabla \times \mathbf{V}) = \begin{vmatrix} \mathbf{i} & \mathbf{j} & \mathbf{k} \\ \frac{\partial}{\partial x} & \frac{\partial}{\partial y} & \frac{\partial}{\partial z} \\ \frac{\partial V_z}{\partial y} - \frac{\partial V_y}{\partial z} & \frac{\partial V_x}{\partial z} - \frac{\partial V_z}{\partial x} & \frac{\partial V_y}{\partial x} - \frac{\partial V_x}{\partial y} \end{vmatrix}.$$

The x component is

$$\frac{\partial^2 V_y}{\partial y \partial x} - \frac{\partial^2 V_x}{\partial y^2} - \frac{\partial^2 V_x}{\partial z^2} + \frac{\partial^2 V_z}{\partial z \partial x}.$$

We now calculate the x component of the right-hand side:

$$\nabla(\nabla \cdot \mathbf{V}) = \left(\mathbf{i}\frac{\partial}{\partial x} + \mathbf{j}\frac{\partial}{\partial y} + \mathbf{k}\frac{\partial}{\partial z} \right)\left(\frac{\partial V_x}{\partial x} + \frac{\partial V_y}{\partial y} + \frac{\partial V_z}{\partial z} \right),$$

and the x component is

$$\frac{\partial^2 V_x}{\partial x^2} + \frac{\partial^2 V_y}{\partial x \partial y} + \frac{\partial^2 V_z}{\partial x \partial z}.$$

$$\nabla^2 \mathbf{V} = \left(\frac{\partial^2}{\partial x^2} + \frac{\partial^2}{\partial y^2} + \frac{\partial^2}{\partial z^2} \right)(V_x\mathbf{i} + V_y\mathbf{j} + V_z\mathbf{k}),$$

and the x component is

$$\frac{\partial^2 V_x}{\partial x^2} + \frac{\partial^2 V_x}{\partial y^2} + \frac{\partial^2 V_x}{\partial z^2}.$$

The x component of $\nabla(\nabla \cdot \mathbf{V}) - \nabla^2 \mathbf{V}$

$$= \frac{\partial^2 V_y}{\partial x \partial y} + \frac{\partial^2 V_z}{\partial x \partial z} - \frac{\partial^2 V_x}{\partial y^2} - \frac{\partial^2 V_x}{\partial z^2}.$$

Since

$$\frac{\partial^2 V_y}{\partial x \partial y} = \frac{\partial^2 V_y}{\partial y \partial x} \quad \text{and} \quad \frac{\partial^2 V_z}{\partial x \partial z} = \frac{\partial^2 V_z}{\partial z \partial x}$$

the x components of the two sides are identical. By symmetry, the same will be true for the other components.

16.4

$$\nabla \cdot \nabla \times \mathbf{V} = \left(\frac{\partial}{\partial x}\mathbf{i} + \frac{\partial}{\partial y}\mathbf{j} + \frac{\partial}{\partial z}\mathbf{k} \right) \cdot \left\{ \left(\frac{\partial V_z}{\partial y} - \frac{\partial V_y}{\partial z} \right)\mathbf{i} \right.$$
$$\left. + \left(\frac{\partial V_x}{\partial z} - \frac{\partial V_z}{\partial x} \right)\mathbf{j} + \left(\frac{\partial V_y}{\partial x} - \frac{\partial V_x}{\partial y} \right)\mathbf{k} \right\}$$

$$= \frac{\partial^2 V_z}{\partial x \partial y} - \frac{\partial^2 V_y}{\partial x \partial z} + \frac{\partial^2 V_x}{\partial y \partial z} - \frac{\partial^2 V_z}{\partial y \partial x}$$
$$+ \frac{\partial^2 V_y}{\partial z \partial x} - \frac{\partial^2 V_x}{\partial z \partial y} = 0$$

$$\nabla\phi = \frac{\partial\phi}{\partial x}\mathbf{i} + \frac{\partial\phi}{\partial y}\mathbf{j} + \frac{\partial\phi}{\partial z}\mathbf{k}.$$

Hence

$$\nabla \times \nabla\phi = \begin{vmatrix} \mathbf{i} & \mathbf{j} & \mathbf{k} \\ \frac{\partial}{\partial x} & \frac{\partial}{\partial y} & \frac{\partial}{\partial z} \\ \frac{\partial\phi}{\partial x} & \frac{\partial\phi}{\partial y} & \frac{\partial\phi}{\partial z} \end{vmatrix} = \left(\frac{\partial^2\phi}{\partial y \partial z} - \frac{\partial^2\phi}{\partial z \partial y} \right)\mathbf{i}$$
$$+ \left(\frac{\partial^2\phi}{\partial z \partial x} - \frac{\partial^2\phi}{\partial x \partial z} \right)\mathbf{j} + \left(\frac{\partial^2\phi}{\partial x \partial y} - \frac{\partial^2\phi}{\partial y \partial x} \right)\mathbf{k} = 0.$$

16.5

$$\nabla \times \mathbf{V} = \begin{vmatrix} \mathbf{i} & \mathbf{j} & \mathbf{k} \\ \frac{\partial}{\partial x} & \frac{\partial}{\partial y} & \frac{\partial}{\partial z} \\ -y^3 & x^3 & 0 \end{vmatrix} = (3x^2 + 3y^2)\mathbf{k},$$

$$d\mathbf{A} = dx\,dy\,\mathbf{k},$$

$$\oiint (\nabla \times \mathbf{V}) \cdot d\mathbf{A} = \int_{-a}^{a}\int_{-a}^{a}(3x^2 + 3y^2)\,dx\,dy = 8a^4.$$

Along AB, $\mathbf{V} = a^3\mathbf{i} + x^3\mathbf{j}$ and $d\mathbf{r} = dx\mathbf{i}$. Hence

$$\int_{-a}^{a} \mathbf{V} \cdot d\mathbf{r} = \int_{-a}^{a} a^3 dx = 2a^4.$$

Along BC, $\mathbf{V} = -y^3\mathbf{i} + a^3\mathbf{j}$ and $d\mathbf{r} = dy\mathbf{j}$. Hence

$$\int_{-a}^{a} \mathbf{V} \cdot d\mathbf{r} = \int_{-a}^{a} a^3 dy = 2a^4.$$

Along CD, $\mathbf{V} = -a^3\mathbf{i} + x^3\mathbf{j}$ and $d\mathbf{r} = -dx\mathbf{i}$. Hence

$$\int_{-a}^{a} \mathbf{V} \cdot d\mathbf{r} = \int_{-a}^{a} a^3 dx = 2a^4.$$

Along DA, $\mathbf{V} = -y^3\mathbf{i} - a^3\mathbf{j}$ and $d\mathbf{r} = -dy\mathbf{j}$. Hence

$$\int_{-a}^{a} \mathbf{V} \cdot d\mathbf{r} = \int_{-a}^{a} a^3 dy = 2a^4.$$

Summing along the four sides, $\oint \mathbf{V} \cdot d\mathbf{r} = 8a^4.$

Exercises

17.1 The amplitude of the cosine is 3, and its frequency is 1.5 Hz, so the amplitude spectrum looks like Figure S17.1. The constant phase of the cosine is $\pi/2$ radians, so the phase spectrum looks like Figure S17.2.

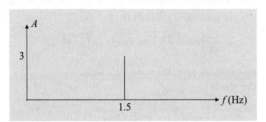

Figure S17.1

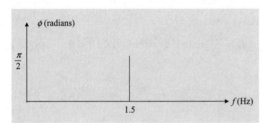

Figure S17.2

17.2 From the amplitude spectrum, the frequency of the cosine is 20 Hz and the amplitude is 1. From the phase spectrum, the phase is -2.6 radians. Combining all this information, $s(t) = \cos(2\pi(20)t - 2.6)$.

17.3 From (17.43), multiplying the amplitude spectrum by $2/T$ means that the peak heights correspond to the amplitudes of the cosines. From the amplitude spectrum, there is a cosine of frequency 3 Hz of amplitude 2 and a cosine of frequency 5 Hz of amplitude 1. From the phase spectrum, the 3 Hz cosine has a phase of -2 radians whilst the 5 Hz cosine has a phase of 1 radian. Hence the signal is
$s(t) = 2\cos(2\pi(3)t - 2) + \cos(2\pi(5)t + 1)$.

17.4 (i) The period $= 0.01$ s, hence the fundamental frequency is given by $f_0 = 1/0.01 = 100$ Hz.
(ii) 3. (iii) $5\cos(2\pi(100)t) + 3\sin(2\pi(100)t)$
(iv) 0 (there are no terms with frequency 200 Hz).
(v) $-4\sin(2\pi(300)t)$.

17.5 The solutions to Exercises 17.5 and 17.6 rely on the result that if a function that has odd symmetry is integrated between equal but opposite limits, then the integral is zero. For example, suppose that a general function $f(t)$ looks like Figure S17.3. The definite

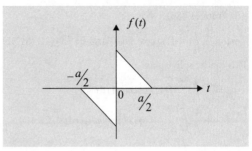

Figure S17.3

integral between $-a/2$ and $a/2$ is the total area under $f(t)$ between these two limits. It can be seen that the area between $-a/2$ and 0 is equal and opposite to the area between 0 and $a/2$. These two areas cancel each other out and so the integral between $-a/2$ and $a/2$ is zero for $f(t)$ odd. From (17.77),
$$a_0 = \frac{1}{T} \int_{-T/2}^{T/2} f(t)\,\mathrm{d}t.$$
From our above discussion, as $f(t)$ is odd, hence $a_0 = 0$.
From (17.78),
$$a_m = \frac{2}{T} \int_{-T/2}^{T/2} f(t) \cdot \cos(m\omega_0 t)\,\mathrm{d}t.$$
Now $\cos(x) = \cos(-x)$, hence the cosine is an even function of time.
If $f(t)$ is odd, then $f(t) \cdot \cos(m\omega_0 t)$ is odd (odd $\times$ even $=$ odd) and so $a_m = 0$ for $f(t)$ odd.

17.6 From (17.79),
$$b_m = \frac{2}{T} \int_{-T/2}^{T/2} f(t) \sin(m\omega_0 t)\,\mathrm{d}t.$$
If $f(t)$ is even, then $f(t) \cdot \sin(m\omega_0 t)$ is odd (even $\times$ odd $=$ odd), hence the integral is zero i.e. for $f(t)$ even, $b_m = 0$.

17.7 (i) $\dfrac{a_1^2}{2} + \dfrac{a_3^2}{2} = \dfrac{1}{2}\left(\dfrac{4}{\pi}\right)^2 + \dfrac{1}{2}\left(\dfrac{4}{3\pi}\right)^2 = 0.90$ (90% of the total power).

(ii) $\dfrac{a_1^2}{2} + \dfrac{a_3^2}{2} + \dfrac{a_5^2}{2} = \dfrac{1}{2}\left(\dfrac{4}{\pi}\right)^2 + \dfrac{1}{2}\left(\dfrac{4}{3\pi}\right)^2 + \dfrac{1}{2}\left(\dfrac{4}{5\pi}\right)^2 = 0.93$
(93% of the total power).

(iii) $\dfrac{a_1^2}{2} + \dfrac{a_3^2}{2} + \dfrac{a_5^2}{2} + \dfrac{a_7^2}{2} = \dfrac{1}{2}\left(\dfrac{4}{\pi}\right)^2 + \dfrac{1}{2}\left(\dfrac{4}{3\pi}\right)^2 + \dfrac{1}{2}\left(\dfrac{4}{5\pi}\right)^2 +$
$\dfrac{1}{2}\left(\dfrac{4}{7\pi}\right)^2 = 0.95$ (95% of the total power).

Problems

17.1 (i) The mean value of the signal over a period is zero, hence from (17.77), $a_0 = 0$. The signal has even symmetry, $f(t) = f(-t)$, hence from (17.82), $b_n = 0$ for all n.

(ii) From (17.86)

$$a_n = \frac{4}{T} \int_0^{T/2} f(t) \cos(n\omega_0 t) \, dt. \qquad (S17.1)$$

This can be split up as

$$a_n = \frac{4}{T} \int_0^{T_1/2} f(t) \cos(n\omega_0 t) \, dt$$
$$+ \frac{4}{T} \int_{T_1/2}^{T/2-T_1/2} f(t) \cos(n\omega_0 t) \, dt$$
$$+ \frac{4}{T} \int_{T/2-T_1/2}^{T/2} f(t) \cos(n\omega_0 t) \, dt.$$

Substituting $f(t) = A$ in the first integral, $f(t) = 0$ in the second, and $f(t) = -A$ in the third:

$$a_n = \frac{4A}{T} \int_0^{T_1/2} \cos(n\omega_0 t) \, dt$$
$$- \frac{4A}{T} \int_{T/2-T_1/2}^{T/2} \cos(n\omega_0 t) \, dt.$$

Substituting $\omega_0 = 2\pi/T$ and integrating:

$$a_n = \frac{4A}{T} \cdot \frac{T}{2\pi n} \left| \sin\left(\frac{2\pi n t}{T}\right) \right|_0^{T_1/2}$$
$$- \frac{4A}{T} \cdot \frac{T}{2\pi n} \left| \sin\left(\frac{2\pi n t}{T}\right) \right|_{T/2-T_1/2}^{T/2}.$$

Putting in the limits and cancelling terms:

$$a_n = \frac{2A}{\pi n} \left[\sin\left(\frac{\pi n T_1}{T}\right) - 0 \right]$$
$$- \frac{2A}{\pi n} \left[\sin(n\pi) - \sin\left(\frac{2\pi n}{T}\left\{ \frac{T}{2} - \frac{T_1}{2}\right\}\right) \right].$$

Now $\sin(n\pi) = 0$ for all integers n, hence the above equation can be rewritten as

$$a_n = \frac{2A}{\pi n} \left[\sin\left(\frac{\pi n T_1}{T}\right) + \sin\left(n\pi - \frac{\pi n T_1}{T}\right) \right]. \qquad (S17.2)$$

Using (1.13), the second term may be expanded as

$$\sin\left(n\pi - \frac{\pi n T_1}{T}\right) = \sin(n\pi)\cos\left(\frac{\pi n T_1}{T}\right)$$
$$- \cos(n\pi)\sin\left(\frac{\pi n T_1}{T}\right)$$

and since $\sin(n\pi) = 0$,

$$a_n = \frac{2A}{n\pi} \left[\sin\left(\frac{\pi n T_1}{T}\right) - \cos(n\pi)\sin\left(\frac{\pi n T_1}{T}\right) \right]. \qquad (S17.3)$$

(i) For even harmonics, put $n = 2p$, where p is an integer. In this case

$$a_{2p} = \frac{A}{p\pi} \left[\sin\left(\frac{\pi 2p T_1}{T}\right) - \cos(2p\pi)\sin\left(\frac{\pi 2p T_1}{T}\right) \right].$$

Now $\cos(2p\pi) = 1$, hence $a_{2p} = 0$, that is all even harmonics are absent. This can also be seen graphically. For the example when $A = 1$, $T_1 = 1$ s, and $T = 2$ s, the second harmonic amplitude is given from (S17.1) by:

$$a_2 = \frac{2}{1} \int_0^1 f(t) \cos\left(2\pi \times 2 \times \frac{1}{2} t \right) dt$$
$$= 2 \int_0^1 f(t) \cos(2\pi t) \, dt. \qquad (S17.4)$$

The signal $f(t)$ is shown as a full curve and the term $\cos(2\pi t)$ in (S17.4) as a dashed curve in Figure S17.4. If we multiply these two functions together, we obtain the curve in Figure S17.5 as the integrand in (S17.4). Now a_2 is the net area under the curve in the above diagram which, by inspection, is zero. Hence a_2 in (S17.4) is zero. The cancellation of positive and negative areas takes place for all even harmonics.

(ii) Substituting $T_1 = T/3$ in (S17.3),

$$a_n = \frac{2A}{n\pi} \left[\sin\left(\frac{\pi n}{3}\right) - \cos(n\pi) \cdot \sin\left(\frac{\pi n}{3}\right) \right].$$

From (i), all even harmonics are zero.
Put $n = 1$:

$$a_1 = \frac{2A}{\pi} \left[\sin\left(\frac{\pi}{3}\right) - \cos(\pi) \sin\left(\frac{\pi}{3}\right) \right]$$
$$= \frac{2A}{\pi} \times 2 \times \sin\left(\frac{\pi}{3}\right) = \frac{2A\sqrt{3}}{\pi}. \qquad (S17.5)$$

Put $n = 3$:

$$a_3 = \frac{2A}{3\pi} [\sin(\pi) - \cos(3\pi) \cdot \sin(\pi)] = 0. \qquad (S17.6)$$

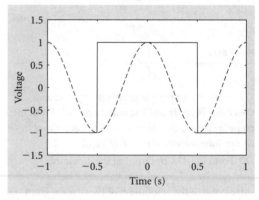

Figure S17.4

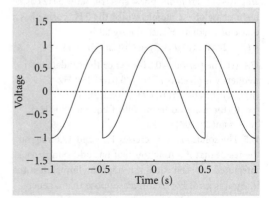

Figure S17.5

Put $n=5$:

$$a_5 = \frac{2A}{5\pi}\left[\sin\left(\frac{5\pi}{3}\right) - \cos(5\pi)\sin\left(\frac{5\pi}{3}\right)\right]$$

$$= \frac{2A}{5\pi} \times 2 \times \sin\left(\frac{5\pi}{3}\right) = -\frac{2A\sqrt{3}}{5\pi}. \qquad \text{(S17.7)}$$

Put $n=7$:

$$a_7 = \frac{2A}{7\pi}\left[\sin\left(\frac{7\pi}{3}\right) - \cos(7\pi)\sin\left(\frac{7\pi}{3}\right)\right]$$

$$= \frac{2A}{7\pi} \times 2 \times \sin\left(\frac{7\pi}{3}\right) = \frac{2A\sqrt{3}}{7\pi}. \qquad \text{(S17.8)}$$

Put $n=9$:

$$a_9 = \frac{2A}{9\pi}[\sin(3\pi) - \cos(9\pi)\cdot\sin(3\pi)] = 0. \quad \text{(S17.9)}$$

When $T_1 = T/3$, not only are the even harmonics zero, but also those harmonics that are a multiple of 3. Hence, the first non-zero harmonic after the fundamental is the fifth.

(iii) From (17.103), the *total* power for an even signal is given by $P_{\text{TOT}} = \frac{2}{T}\int_0^{T/2}[f(t)]^2\,dt$. Substituting for $f(t)$ and splitting up the integral,

$$P_{\text{TOT}} = \frac{2}{T}\int_0^{T_1/2} A^2\,dt + \frac{2}{T}\int_{T/2-T_1/2}^{T/2} A^2\,dt$$

where we have used the fact that
$f(t) = 0 \quad \text{for} \quad \frac{T_1}{2} \leq t \leq \frac{T}{2} - \frac{T_1}{2}$
Performing the integrations

$$P_{\text{TOT}} = \frac{2}{T}|A^2 t|_0^{T_1/2} + \frac{2}{T}|A^2 t|_{T/2-T_1/2}^{T/2}$$

$$= \frac{2}{T}A^2\frac{T_1}{2} + \frac{2}{T}A^2\frac{T}{2} - \frac{2}{T}A^2\left(\frac{T}{2} - \frac{T_1}{2}\right) \qquad \text{(S17.10)}$$

$$= \frac{2A^2}{3}.$$

From Parseval's theorem, (17.102), the power contained in the fundamental and fifth harmonics combined is given by

$$P_{1+5} = \frac{a_1^2}{2} + \frac{a_5^2}{2}.$$

Substituting a_1 from (S17.5) and a_5 from (S17.7),

$$P_{1+5} = \frac{1}{2}\left[\frac{2A\sqrt{3}}{\pi}\right]^2 + \frac{1}{2}\left[\frac{2A\sqrt{3}}{5\pi}\right]^2 = \frac{156}{25\pi^2}A^2. \qquad \text{(S17.11)}$$

The percentage power in the fundamental and fifth harmonics is given by $R = P_{1+5}/P_{\text{TOT}} \times 100\%$. Substituting for P_{1+5} from (S17.11) and P_{TOT} from (S17.10),

$$R = \frac{156A^2}{25\pi^2} \cdot \frac{3}{2A^2} \times 100\% = 94.8\%.$$

17.2 (i) $F_D(\omega) = \int_{-\infty}^{\infty}\frac{df(t)}{dt}\exp(-i\omega t)dt$. Integrate by parts with $\frac{dv}{dt} = \frac{df(t)}{dt}$ and $u = \exp(-i\omega t)$.

This gives $v = f(t)$ and $du/dt = -i\omega\exp(-i\omega t)$. Hence

$$F_D(\omega) = [\exp(-i\omega t)f(t)]_{-\infty}^{\infty}$$

$$- \int_{-\infty}^{\infty} f(t)(-i\omega)\exp(-i\omega t)dt.$$

Putting in the limits, $f(t) \to 0$ as $(t) \to \pm\infty$.

$$F_D(\omega) = \int_{-\infty}^{\infty}\frac{df(t)}{dt}\exp(-i\omega t)dt$$

$$= i\omega\int_{-\infty}^{\infty} f(t)\exp(-i\omega t)dt = i\omega F(\omega).$$

(ii) $V_L(t) = L\dfrac{dI_L(t)}{dt}$.

Take Fourier transforms of both sides of this equation:

$$_TV_L(\omega) = L\int_{-\infty}^{\infty}\frac{dI_L(t)}{dt}\exp(-i\omega t)dt. \qquad \text{(S17.12)}$$

Using the result of part (i) with $f(t) = I_L(t)$,

$$\int_{-\infty}^{\infty}\frac{dI_L(t)}{dt}\exp(-i\omega t)dt = i\omega \,_TI_L(\omega). \qquad \text{(S17.13)}$$

Hence, substituting from (S17.13) into (S17.12),
$_TV_L(\omega) = i\omega L_T\, I_L(\omega)$.

(iii) $I_C(t) = C\dfrac{dV_C(t)}{dt}$.

Take Fourier transforms of both sides of this equation:

$$_TI_C(\omega) = C\int_{-\infty}^{\infty}\frac{dV_C(t)}{dt}\exp(-i\omega t)\,dt. \qquad \text{(S17.14)}$$

Using the result of part (i) with $f(t) = V_C(t)$,

$$\int_{-\infty}^{\infty}\frac{dV_C(t)}{dt}\exp(-i\omega t)dt = i\omega \,_TV_C(\omega). \qquad \text{(S17.15)}$$

Hence, substituting from (S17.15) into (S17.14),
$_TI_C(\omega) = i\omega C_T\, V_C(\omega)$ and hence, solving for $_TV_C(\omega)$,

$$_TV_C(\omega) = \frac{1}{i\omega C}\,_TI_C(\omega).$$

(iv) The results of parts (ii) and (iii) are that one can write the following general relation for the Fourier transform of the voltage across and current through a capacitor and inductor as $V(\omega) = I(\omega) \cdot Z(\omega)$, where $Z(\omega) = i\omega L$ for an inductor and $Z(\omega) = 1/i\omega C$ for a capacitor.

In Chapter 6 it was shown how these impedances could be used to solve circuit problems when the input is a cosine (AC) signal. The result of this question is that one can also use these impedances to solve circuit problems when the input is non-sinusoidal.

17.3 (i) From §17.8 the total energy is given by

$$E_{\text{TOT}} = \int_{-\infty}^{\infty}[f(t)]^2 dt. \qquad \text{(S17.16)}$$

Now $f(t) = 4u(t)\exp(-t)$, where

$$u(t) = \begin{cases} 1 & t \geq 0 \\ 0 & t < 0. \end{cases}$$

Hence

$$f(t) = \begin{cases} 4\exp(-t) & \text{for } t \geq 0 \\ 0 & \text{for } t < 0. \end{cases}$$

Substituting this expression into (S17.16),

$$E_{TOT} = \int_0^\infty [4\exp(-t)]^2 dt$$
$$= 16 \int_0^\infty \exp(-2t)dt. \qquad \text{(S17.17)}$$

Now

$$\int_0^\infty \exp(-2t)dt = -\frac{1}{2}|\exp(-2t)|_0^\infty = -\frac{1}{2}[0-1] = \frac{1}{2}.$$

Hence, in (S17.17), $E_{TOT} = 16 \times \frac{1}{2} = 8$.

(ii) From (17.30) substituting f for x,

$$F(\omega) = \int_{-\infty}^\infty f(t)\exp(-i\omega t)\, dt.$$

Substituting for $f(t)$,

$$F(\omega) = \int_0^\infty 4\exp(-t)\exp(-i\omega t)\, dt.$$

Combining the exponentials,

$$F(\omega) = 4 \int_0^\infty \exp\{-(1+i\omega)t\}dt$$
$$= -\frac{4}{1+i\omega}|\exp\{-(1+i\omega)t\}|_0^\infty$$
$$= -\frac{4}{1+i\omega}|\exp(-i\omega t)\exp(-t)|_0^\infty$$

Now $\exp(-i\omega t)$ is not defined as $t \to \infty$; all that can be said that it has a magnitude of 1 for all t, i.e. it is always finite. However, we do know that $\{\exp(-t)\}_{t\to\infty} = 0$, Hence $\{\exp(-i\omega t)\exp(-t)\}_{t\to\infty} = 0$. Putting in the lower limit,

$$F(\omega) = -\frac{4}{1+i\omega}[0-1] = \frac{4}{1+i\omega}.$$

(iii) From §17.8 the total energy is given in terms of the Fourier transform of the signal as

$$E_{TOT} = \frac{1}{\pi}\int_0^\infty |F(\omega)|^2 d\omega, \text{ where}$$

$|F(\omega)|^2 = F(\omega)F^*(\omega)$. Now, from the answer to part (ii), $F(\omega) = \frac{4}{1+i\omega}$. The complex conjugate, from (6.36), is $F^*(\omega) = \frac{4}{1-i\omega}$.

Hence

$$|F(\omega)|^2 = \frac{4}{(1+i\omega)} \times \frac{4}{(1-i\omega)} = \frac{16}{1+i\omega-i\omega-i^2\omega^2}.$$

Using the relation $i^2 = -1$, we can thus write $|F(\omega)|^2 = \frac{16}{1+\omega^2}$. Substituting this into the above equation for E_{TOT},

$$E_{TOT} = \frac{16}{\pi}\int_0^\infty \frac{1}{1+\omega^2}d\omega = \frac{16}{\pi}|\tan^{-1}\omega|_0^\infty = \frac{16}{\pi}\left[\frac{\pi}{2}-0\right] = 8,$$

in agreement with the answer to part (i).

(iv) The energy associated with the frequency band is given by (17.49) as $E(\omega_1, \omega_2) = \frac{1}{\pi}\int_{\omega_1}^{\omega_2} |F(\omega)|^2 d\omega$. Substituting the Fourier transform for the filtered signal, putting $\omega_1 = 0$ and $\omega_2 = \omega_c$, $E(0, \omega_c) = \frac{1}{\pi}\int_0^{\omega_c} |F(\omega)|^2 d\omega$ where $\omega_c = 2\pi f_c$. Substituting

$$|F(\omega)|^2 = \frac{16}{1+\omega^2} \text{ as found in part (iii),}$$

$$E(0, \omega_c) = \frac{16}{\pi}\int_0^{\omega_c} \frac{d\omega}{1+\omega^2} = \frac{16}{\pi}\tan^{-1}\omega_c.$$

Now, we have been asked to choose ω_c such that $E(0, \omega_c) = 0.95 \times E_{TOT} = 0.95 \times 8 = 7.6$. Hence

$$\frac{16}{\pi}\tan^{-1}(\omega_c) = 7.6 \text{ or } \tan^{-1}(\omega_c) = \frac{7.6\pi}{16} = 1.4923,$$

which gives $\omega_c = \tan(1.4923) = 12.71$ rad s^{-1}. In terms of cyclical frequency, $f_c = \omega_c/2\pi = 2.02$ Hz.

17.4 (i)

$$S_r(f) = \frac{1}{2}[\delta(f+f_d) + \delta(f-f_d)] \qquad \text{(S17.18)}$$

$\delta(f+f_d)$ is a peak at $f = -f_d$ and $\delta(f-f_d)$ is a peak at $f = f_d$. Hence the Fourier transform of $S_r(f)$ looks like

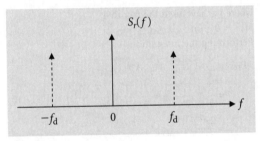

$$S_r(f)$$

$$-f_d \qquad 0 \qquad f_d \qquad \rightarrow f$$

Figure S17.6

the curve shown in Figure S17.6, with the δ-functions shown as dotted vertical lines.

(ii) Replacing $f_d \to -f_d$ in (S17.18), the Fourier transform of $\cos(-2\pi f_d t)$ becomes

$$S_r(f) = \frac{1}{2}[\delta(f-f_d) + \delta(f+f_d)]$$
$$= \frac{1}{2}[\delta(f+f_d) + \delta(f-f_d)],$$

i.e. the same as for $\cos(2\pi f_d t)$. This is not a surprising result, as from a property of cosines $\cos(-2\pi f_d t) = \cos(2\pi f_d t)$.

(iii) The Fourier transform of $A\cos(2\pi f_d t)$ is given by

$$S_R(f) = A \int_{-\infty}^\infty \cos(2\pi f_d t)\exp(-2\pi i f t)dt$$
$$= \frac{A}{2}[\delta(f+f_d) + \delta(f-f_d)], \qquad \text{(S17.19)}$$

the Fourier transform of $A\sin(2\pi f_d t)$ is given by

$$S_I(f) = A\int_{-\infty}^{\infty} \sin(2\pi f_d t)\exp(-2\pi ift)dt$$

$$= \frac{A}{2}i[\delta(f+f_d) - \delta(f-f_d)], \qquad (S17.20)$$

and the Fourier transform of $s(t) = A\exp(2\pi if_d t)$ is given by

$$S(f) = A\int_{-\infty}^{\infty} \exp(2\pi if_d t)\exp(-2\pi ift)dt.$$

Expanding $A\exp(2\pi if_d t)$ as in (P17.4),

$$S(f) = A\int_{-\infty}^{\infty} [\cos(2\pi f_d t)\exp(-2\pi ift)dt$$

$$+ i\sin(2\pi f_d t)\exp(-2\pi ift)]dt.$$

Splitting this expression up into two integrals,

$$S(f) = A\int_{-\infty}^{\infty} \cos(2\pi f_d t)\exp(-2\pi ift)\,dt$$

$$+ iA\int_{-\infty}^{\infty} \sin(2\pi f_d t)\exp(-2\pi ift)\,dt.$$

From (S17.19) and (S17.20), this expression can be rewritten as

$$S(f) = S_R(f) + iS_I(f).$$

Substituting for $S_R(f)$ from (S17.19) and $S_I(f)$ from (S17.20),

$$S(f) = \frac{A}{2}[\delta(f+f_d) + \delta(f-f_d)] + i\frac{A}{2}i[\delta(f+f_d) - \delta(f-f_d)].$$

Using the identity $i^2 = -1$,

$$S(f) = \frac{A}{2}[\delta(f+f_d) - \delta(f+f_d)] + \frac{A}{2}[\delta(f-f_d) + \delta(f-f_d)].$$

Hence

$$S(f) = A\delta(f - f_d) \qquad (S17.21)$$

which is a single peak at $f = f_d$. The negative peak has disappeared (Figure S17.7).

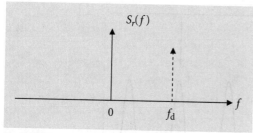

Figure S17.7

To find the Fourier transform of $A\exp(-2\pi if_d t)$, replace $f_d \to -f_d$ in (S17.21):

$$S(f) = A\delta(f + f_d)$$

which is a single peak at $f = -f_d$.

Hence, we have solved the problem of distinguishing between positive and negative radial velocities of the target. This sort of processing is also used in medical applications of Doppler ultrasound.

The above result may seem odd, as it is often stated that the amplitude spectrum has even symmetry about DC and that if there is a peak at f_d Hz, then there is another peak at $-f_d$ Hz. However, this is only true if the input signal to the Fourier transform is *real*, which would be the case if the signal $s_r(t)$ in (P17.1) is fed in. However, if the input signal is made *complex*, as in (P17.4), then the amplitude spectrum no longer has this symmetry.

CHAPTER 18

Exercises

18.1 $X[p+lN] = \sum_{n=0}^{N-1} x[n]\exp\left(-2\pi i\frac{n(p+lN)}{N}\right)$

$$= \sum_{n=0}^{N-1} x[n]\exp\left(-2\pi i\frac{np}{N} - 2\pi inl\right)$$

$$= \sum_{n=0}^{N-1} x[n]\exp\left(-2\pi i\frac{np}{N}\right)\exp(-2\pi inl) = X(p)$$

since n and l are integers and hence $\exp(-2\pi inl) = 1$.

$$X[-p] = \sum_{n=0}^{N-1} x[n]\exp\left\{2\pi i\frac{np}{N}\right\} = X[p]^*.$$

18.2 $p = 40$.

18.3 (i) 256; (ii) Peak at $p = 0$; (iii) 0 Hz (DC), since sampling the peaks is similar to that in Figure 18.18.

18.4 (i) 16; (ii) 13.333.

18.5 Case (ii), because peak occurs at non-integer value of p.

18.6 Let the percentage of multiplications required by the FFT compared to that required by the DFT be R where $R = \dfrac{N\log_2(N)}{N^2}\times 100\%$. Hence R for various N is as shown in Table S18.1.

The larger the number of samples, N, the greater the saving in computation time when using the FFT compared to using the DFT.

Table S18.1 R for various N

N	$\log_2(N)$	$N\log_2 N$	N^2	R
16	4	64	256	25%
256	8	2 048	65 536	3%
8192	13	106 496	6.7×10^7	0.16%
131 072	17	2.2×10^6	1.72×10^{10}	0.01%

Problems

18.1 (i) From (18.4) and the discussion in §18.1, the Fourier transform of a sampled signal is equal to the Fourier transform of the corresponding continuous signal along with additional bands centred at integer multiples of the sampling frequency. Hence, for a

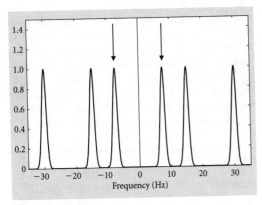

Figure S18.1

sampling frequency of 22.5 Hz, the Fourier transform of the sampled signal is as shown in Figure S18.1.

(ii) The bands corresponding to the Fourier transform of the original continuous signal are indicated by the arrows; these bands do not overlap with the additional bands due to sampling hence it should be possible, in theory, to recover the original continuous signal from its samples for this value of the sampling frequency.

(iii) The Fourier transform of the sampled signal for this value of the sampling frequency is as shown in Figure S18.2.

(iv) As for the previous value of the sampling frequency, the bands corresponding to the Fourier transform of the original continuous signal are indicated by the arrows; these bands do not overlap with the additional bands due to sampling hence it should be possible, in theory, to recover the original continuous signal from its samples for this value of the sampling frequency.

This finding appears to contradict the sampling theorem, which states that the sampling frequency should be at least twice the highest frequency component of the continuous signal – in this case 20 Hz – in order that aliasing will not occur. However, there are exceptions and this signal is one of them. The amplitude spectrum in Figure P18.1 is zero between 0 and 5 Hz and its bandwidth is also 5 Hz. Hence, when a sampling frequency of 10 Hz is chosen, the sidebands due to aliasing exactly fit into the gap between 0 Hz and 5 Hz and corresponding gaps at higher frequencies. However, this is a strict value that should be used. If f_s is increased to 10.1 Hz. say, then aliasing will occur again.

18.2 From (18.17), the relation between the frequency, f_p, number of samples processed, N, frequency point p and sampling frequency f_s is given by

$$p = \frac{Nf_p}{f_s}.$$

In this question, $f_p = 1000$ Hz and $N = 32$.

(i) It should be remembered that, for a cosine with frequency f_p, there are two peaks in the amplitude spectrum, one at $+f_p$ and one at $-f_p$. From the above equation, the peaks at $f_p = \pm 1000$ Hz, occur at values of p given by $p = \pm \frac{32 \times 1000}{4000} = \pm 8$. If we implement (18.6) on a computer and plot $|X[p]|$ as a function of p we obtain the amplitude spectrum shown in Figure S18.3, in agreement with our calculation.

(ii) The sampling frequency has now been increased to 5000 Hz. In this case, from (18.17), the positive frequency peak occurs at

$$p = \frac{32 \times 1000}{5000} = 6.4.$$

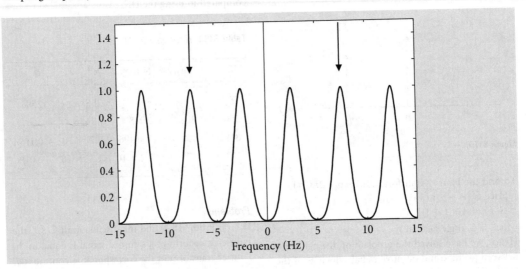

Figure S18.2

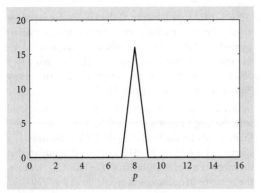

Figure S18.3

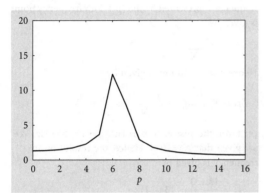

Figure S18.4

Since p is non-integer, from the discussion in §18.2.4, leakage will occur and the DFT will be non-zero for all values of p with a peak between $p = 6$ and $p = 7$; this is verified from the computed amplitude spectrum shown in Figure S18.4.

(iii) The sampling frequency has now been reduced to 600 Hz. From (18.17), the frequency peaks occur at values of p equal to

$$p = \pm \frac{32 \times 1000}{600} = \pm 53.33.$$

Both these values are outside the required range $0 \le p \le 16$.

Using the periodicity property, (18.14), the peak at $p = 53.33$ also occurs at $p = 53.33 - 32 = 21.33$. The peak at $p = -53.33$ also occurs at $-53.33 + 32 = -21.33$ and also at $-21.33 + 32 = 10.67$, which is the only value between 0 and 16. This value is non-integer, hence leakage is occurring. Hence the DFT will be non-zero for all values of p with a peak between $p = 10$ and $p = 11$; this is verified from the computed amplitude spectrum (Figure S18.5).

But a problem has now occurred here. From (18.13), $p = 10.67$ corresponds to a frequency given by

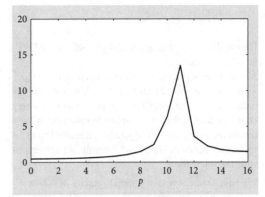

Figure S18.5

$$f = \frac{10.67 \times 600}{32} = 200 \text{ Hz}.$$

However, we know that the underlying frequency is 1000 Hz. In practice, we would not know the signal's frequency and we would be underestimating this frequency by a factor of 5.

The answer to this anomaly is that the underlying signal has a frequency of 1000 Hz and from the sampling theorem, the minimum sampling frequency should be 2000 Hz. We have chosen a sampling frequency less than this value, 600 Hz, and hence we are undersampling the signal, as in Example 18.2 in §18.2.4. In this case, the information obtained from the DFT will be erroneous, as we have seen. Hence, it is important to sample above the Nyquist frequency in order to obtain accurate enough information on the underlying frequency or frequencies of the signal.

18.3 (i) The maximum radial velocity is $+1000 \text{ m s}^{-1}$ and the minimum radial velocity is -1000 m s^{-1}. Hence, the maximum Doppler shift frequency is given by

$$f_d^{max} = \frac{2 \times 1000 \times 10^9}{3 \times 10^8} = 6.667 \text{ kHz}$$

and the minimum Doppler shift frequency is given by

$$f_d^{min} = \frac{2 \times (-1000) \times 10^9}{3 \times 10^8} = -6.667 \text{ kHz}.$$

(ii) The maximum deviation of the Doppler shift from the radar's carrier frequency is 6.667 kHz. Hence to measure this shift digitally, the minimum sampling frequency allowed by the sampling theorem is given by

$$f_s^{min} = 2 \times 6.667 \times 10^3 = 13.333 \text{ kHz}.$$

(iii) The Doppler frequency corresponding to a radial velocity of 600 m s^{-1} towards the radar is given by

$$f_d = \frac{2 \times 600 \times 10^9}{3 \times 10^8} = 4 \text{ kHz}.$$

From (18.17), the frequency point p in the DFT corresponding to a frequency of f_d is given by

$$p = \frac{4000}{13333} \times 1024 = 307.2.$$

Hence, there is a peak at around $p = 307$ in the DFT amplitude spectrum.

As we are processing the complex signal $\exp(2\pi i f_d t)$, with $f_d > 0$ because the target is travelling towards the radar, there are no negative frequency peaks to worry about. In fact, if the DFT amplitude spectrum is computed using the FFT algorithm, the amplitude spectrum shown in Figure S18.6 results, in agreement with the calculation.

We have seen above that a radial velocity of $600\ \text{m s}^{-1}$ towards the radar corresponds to a Doppler frequency shift of $+4\ \text{kHz}$, which corresponds to a value of p in the amplitude spectrum of approximately 307. Hence, if the target was travelling *away* from the radar with a radial velocity of $600\ \text{m s}^{-1}$, $V = -600\ \text{m s}^{-1}$, this would correspond to a Doppler shift of $-4\ \text{kHz}$, which would correspond to a value of p in the DFT amplitude spectrum of -307.

Now we have seen that in applications where the signal is *real* (rather than complex) the negative frequencies carry the same information as the positive frequencies, so the DFT amplitude spectrum is usually displayed for values of p between 0 and $N/2$. However, in the Doppler application where the input signal is made complex, both positive and negative frequencies are important; positive frequencies signify that the target is travelling towards the radar, negative that it is travelling away from the radar. Hence, the DFT in this case is displayed for value of p between 0 and $N-1$, as values of p between $N/2 + 1$ and $N - 1$ correspond to negative frequencies, see §18.2.3.

Now, from the periodicity property, (18.14), a p value of -307 corresponds to a p value of

$-307 + 1024 = 717$. Hence, if the DFT is displayed for values of p between 0 and 1023, then a peak will occur around $p = 717$ for the case where the target is travelling away from the radar, which is indeed observed if we compute the amplitude spectrum using Equation (18.6) – see Figure S18.7.

In this case, peaks between 513 and 1023 are associated with the target travelling away from the radar. Equivalently we could display the DFT amplitude spectrum for $-512 \leq p \leq 512$, with $p < 0$ signifying that the target is travelling away from the radar. In this case, the negative frequency peak would be at $p = -307$.

(iv) From (18.13), the frequency represented by the pth frequency point in the DFT is given by $f_p = \frac{p f_s}{N}$ and the frequency represented by the $(p+1)$th frequency point is given by

$$f_{p+1} = \frac{(p+1)f_s}{N}.$$

Hence, the resolution is given by

$$\delta f_d = f_{p+1} - f_p = \frac{(p+1)f_s}{N} - \frac{p f_s}{N} = \frac{f_s}{N}.$$

(v) From the answer to part (ii), $f_s = 13.333\ \text{kHz}$. We are given that $N = 1024$. Hence, the resolution in frequency is given by

$$\delta f_d = \frac{13\,333}{1024} = 13.02\ \text{Hz}.$$

Let δv be the corresponding resolution in radial velocity. Now, we are given that the Doppler frequency, f_d, is given by $f_d = 2Vf_0/c$. Taking increments on both sides,

$$\delta f_d = \frac{2(\delta V)f_0}{c}.$$

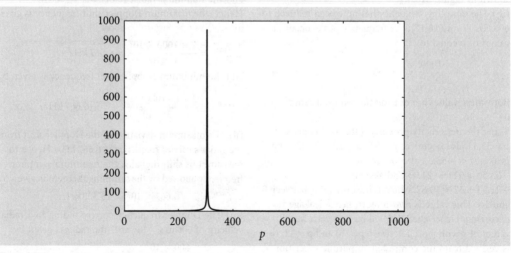

Figure S18.6

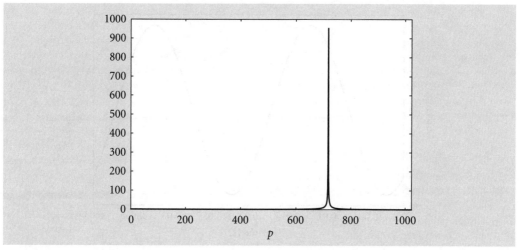

Figure S18.7

Solving this equation for δV and putting in numbers,

$$\delta V = \frac{c(\delta f_d)}{2f_0} = \frac{3 \times 10^8 \times 13.02}{2 \times 10^9} = 1.95 \text{ m s}^{-1}.$$

CHAPTER 19

Exercises

19.1 The true solution is $y = -\cos x$ which, for $x = 0.2$, gives $y = -0.980$. Two steps with $h = 0.1$ give $y = -0.990$. One step with $h = 0.2$ gives $y = -1.000$, i.e. error $\propto h$.

19.2 EPC gives $Y = -1.000$ and $y = -0.980$.

19.3 The true solution is $y = e^x + x + 1$ which, for $x = 2$, gives $y = 10.3891$. Application of the Numerov method gives $Y_2 = 10.1548$ and then $y_2 = 10.3814$.

Problems

19.1 Taking $V = V_0 \exp(i\omega t)$ gives $I = \frac{V_0 \exp(i\omega t)}{R - i(1/\omega C)}$ with $\omega = 2\pi\nu = 100\pi$ radians s^{-1}. Hence the magnitude of I is given by

$$|I| = \frac{|V_0 \exp(i\omega t)|}{|R - i(1/\omega C)|} = \frac{|V_0|}{\sqrt{R^2 + (1/\omega^2 C^2)}}$$

$$= \frac{100}{\sqrt{20^2 + [1/(100\pi \times 10^{-4})^2]}} \text{ A} = 2.660 \text{ A}.$$

There follows the outline of a program to solve the first-order ODE

$$\frac{dI}{dt} = \frac{\omega V_0}{R} \cos(\omega t) - \frac{1}{RC} I$$

with boundary condition $I = 0$ when $t = 0$. The driving potential has been taken as $V = V_0 \sin(\omega t)$.

Program CIRCUIT

Initialization
Set up values of V_0, C, R and ω
Set integration timestep, h, as 0.01 of period
Set boundary conditions $t = 0$ and $I = 0$
Loop to integrate over 1 000 200 timesteps. The first 1 000 000 will remove transients.
Loop for ith timestep
Initial gradient of current,
$$\text{grad}_{\text{init}} = \frac{1}{R}\left(\omega V_0 \cos \omega t - \frac{I_i}{C}\right)$$
Euler estimate of current $I_{\text{Eu}} = I_i + \text{grad} \times h$
Estimate of final gradient
$$\text{grad}_{\text{fin}} = \frac{1}{R}\left\{\omega V_0 \cos \omega(t + h) - \frac{I_{\text{Eu}}}{C}\right\}$$
EPC estimate of final current
$$I_{i+1} = I_i + \tfrac{1}{2}h(\text{grad}_{\text{init}} + \text{grad}_{\text{fin}})$$
Is $i > 1\,000\,000$?
If 'no' then continue to loop
If 'yes' then write result to file and continue to loop
End loop
Stop

This program uses timesteps of 0.01 times the driving frequency and initially runs one million timesteps to eliminate the transient. It then outputs (t, I) to a data file for 200 timesteps. This output file in graphical form is shown in Figure S19.1.

Examination of the data file shows an amplitude of 2.664 A, the difference between this result and the analytical result being due to errors in the numerical process.

19.2 The standard parameters in OSCILLATE that are retained are: $m = 0.001$ kg, $k = 100$ N m^{-1}, $F = 0.01$ N and $\omega = 300$ radians s^{-1}. The outputs for cycles

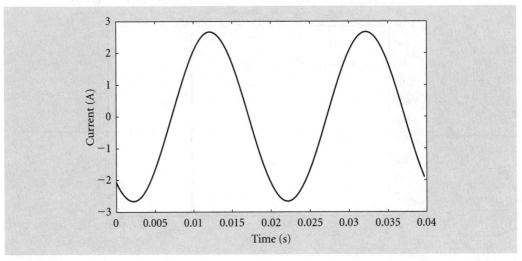

Figure S19.1

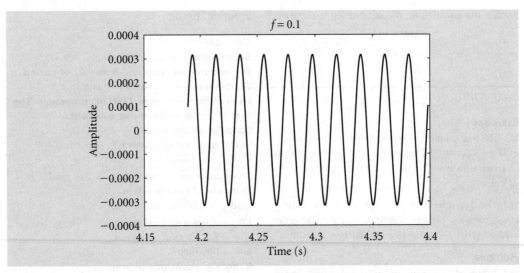

Figure S19.2

201–210 for $f = 0.1\,\mathrm{N\,s\,m^{-1}}$ and $f = 1.0\,\mathrm{N\,s\,m^{-1}}$ are shown in Figures S19.2 and S19.3.

The analytical results from (7.59) give amplitudes 3.162×10^{-4} m and 3.331×10^{-5} m, which are consistent with results from the plots.

CHAPTER 20

Exercises

20.1 (i) $A = 1$, $B = 0$, $C = 1$; $B^2 < 4AC$, hence elliptic.
(ii) $A = 1$, $B = 1$, $C = 0$; $B^2 > 4AC$, hence hyperbolic.
(iii) $A = 1$, $B = 2$, $C = 1$; $B^2 = 4AC$, hence parabolic.

20.2 (i) 2.0208, $\sec^2 45° = 2.0000$; (ii) 4.0575, $2\sec^2 45° \tan 45° = 4.0000$.

20.3 $3.75\,\mathrm{Ks^{-1}}$.

20.4 (i) $r = 0.37$ (stable); (ii) $r = 0.74$ (unstable); (iii) and (iv) $r = 0.074$ (stable). (iv) will give the most precise results.

20.5 $x_1 = 1$, $x_2 = 2$, $x_3 = 3$, $x_4 = 4$.

20.6 $\dfrac{\partial^2 \theta}{\partial x^2} = -\dfrac{\partial^2 \theta}{\partial y^2}$.

20.7 $300 + 400 + 500 + 600 - 4\theta = 0$, hence $\theta = 450\,\mathrm{K}$.

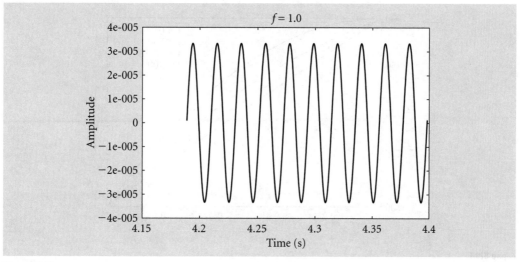

Figure S19.3

20.8 For the wave equation to hold, then for some value of c

$$\frac{\partial^2 \eta}{\partial t^2} = c_1^2 f_1''(x - c_1 t) + c_2^2 f_2''(x - c_2 t) = c^2 f_1''(x - c_1 t)$$
$$+ c^2 f_2''(x - c_2 t) \text{ or } (c_1^2 - c^2) f_1''(x - c_1 t)$$
$$+ (c_2^2 - c^2) f_2''(x - c_2 t) = 0.$$

Since f_1 and f_2 are independently variable functions, the only way that this equation can be true is if $c_1 = c_2 = c$.

Problems

20.1 The first derivative of $\sin(x)$ is $\cos(x)$ and for $x = \pi/4$ this is 0.707107. The values found by applying (20.3), (20.5) and (20.6), with errors in parentheses are:

(i)(a) 0.607 022 ($-0.100\,085$), (ii)(a) 0.791 090 (0.083 983), (iii)(a) 0.699 056 (0.008 051), (i)(b) 0.658 871 ($-0.048\,236$), (ii)(b) 0.751 308 (0.044 201), (iii)(b) 0.705 090 (0.002 017).

The ratios of errors a/b are 2.07, 1.90, and 3.99 – close to the theoretical values 2, 2, and 4 that come from ignoring higher powers of h.

The second derivative of $\sin(x)$ is $-\sin(x)$ and for $x = \pi/4$ this is -0.707107. Applying (20.7) gives $a - 0.703\,090$ (0.004 017) $b - 0.706\,168$ (0.000 939) with error ratio a/b of 4.28.

20.2 and **20.3** It is necessary to make $r \leq \frac{1}{2}$. For $r = \frac{1}{2}$ the time interval

$$\Delta t = r \frac{\rho c (\Delta x)^2}{\kappa} = \frac{1}{2} \frac{3400 \times 800 \times (0.05)^2}{400} s = 8.5 \text{ s}.$$

A sensible time interval to use in the calculation is 5 s. Applying the program EXPLICIT with graphical output gives the result shown below. Applying

HEATCRNI with $\Delta t = 50$ s, corresponding to $r = 2.94$, gave a very reasonable result hardly distinguishable from that with EXPLICIT (Figure S20.1).

20.4 To give the required heating, the function HEAT must be of the form

$$Q = 1.0 \times 10^7 xy(1 - x)(1 - y).$$

The input is of the form

400	300	300	300	300	300	300	300	300
500	U	U	U	U	U	U	U	E
500	U	U	U	U	U	U	U	E
500	U	U	U	U	U	U	U	E
500	U	U	U	U	U	U	U	E
500	U	U	U	U	U	U	U	E
500	U	U	U	U	U	U	U	E
500	U	U	U	U	U	U	U	E
500	U	U	U	U	U	U	U	E
500	I	I	I	I	I	I	I	I

with $K = 15$ and $S = 3 \times 10^3$. The output solution is

400	300	300	300	300	300	300	300	300
500	424	404	401	401	399	395	389	384
500	483	482	487	490	488	480	469	462
500	517	536	551	559	559	550	537	527
500	537	569	594	607	609	601	588	577
500	547	587	617	634	639	633	621	611
500	549	592	624	644	651	648	639	631
500	547	589	621	641	651	651	645	639
500	545	584	616	636	647	648	645	642

20.5 The initial profile was entered at 19 calculated initial points. The plotted output is shown in

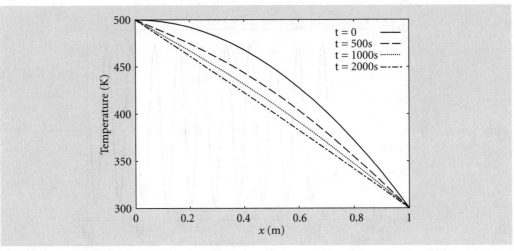

Figure S20.1

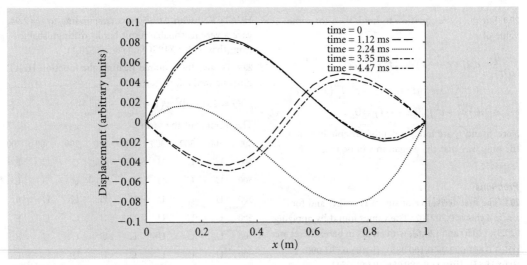

Figure S20.2

Figure S20.2 and shows approximately one period of the motion.

20.6 The initial displacements, were entered from calculated values. The output of DRUM was as shown in Figure S20.3.

CHAPTER 21

Exercises

21.1 (i) 3.98×10^{-19} J; (ii) 1.325×10^{-27} kg m s^{-1}.

21.2 127.9 nm.

21.3 37.6 eV. Electrons contained in atoms of typical dimensions ~ 0.1 nm have energies of a few eV.

21.4 51.2 MeV. Alpha-particles, contained in nuclei of typical dimensions $\sim 10^{-15}$ m have energies of order MeV.

21.5 Probability is

$$\frac{2}{L} \int_0^{L/4} \sin^2 \frac{\pi x}{L} \mathrm{d}x = 0.25 - \frac{1}{2\pi} = 0.091.$$

21.6 Nine-fold degenerate with integer sets (7, 7, 1), (7, 1, 7), (1, 7, 7). (5, 5, 7), (5, 7, 5), (7, 5, 5), (9, 3, 3), (3, 9, 3), (3, 3, 9).

21.7 Zero. The general wave function is antisymmetric about $x = L/2$ whereas the ground-state function is symmetrical around that point.

21.8 $\frac{1}{2}\hbar\omega = 5.27 \times 10^{-20}$ J $= 0.329$ eV.

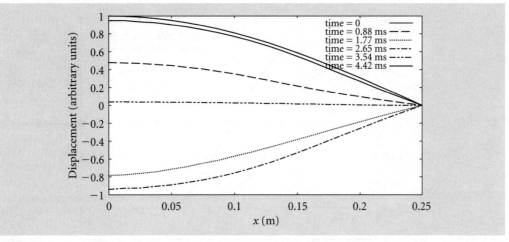

Figure S20.3

21.9 Six-fold degenerate with integer sets $(2, 0, 0)$, $(0, 2, 0)$, $(0, 0, 2)$, $(0, 1, 1)$, $(1, 0, 1)$, and $(1, 1, 0)$.

21.10 $\Delta v = \dfrac{\hbar}{2m\Delta x} = 5.8 \times 10^6 \, \text{m s}^{-1}$.

21.11 $T = \dfrac{4}{4 + k_1^2 a^2}$.

Problems

21.1 The TISWE in this case is $\dfrac{d^2\psi}{dx^2} = -\dfrac{2m}{\hbar^2}(E - V)$ where V is the constant potential within the box. The solution will be of the form

$$E - V = \frac{n^2\pi^2\hbar^2}{2mL^2} \quad \text{or} \quad E = \frac{n^2\pi^2\hbar^2}{2mL^2} + V.$$

For the ground state $n = 1$ and $E_1 = 1.828 \times 10^{-18}\,\text{J} = 11.40\,\text{eV}$.

For the first excited state $n = 2$ and $E_2 = 6.350 \times 10^{-18}\,\text{J} = 39.60\,\text{eV}$.

21.2 The normalized eigenfunctions for a cubical box of side L are of the form

$$u(n_x, n_y, n_z) = \left(\frac{2}{L}\right)^{3/2} \sin\left(\frac{\pi n_x x}{L}\right) \sin\left(\frac{\pi n_y y}{L}\right) \sin\left(\frac{\pi n_z z}{L}\right)$$

with energy eigenvalues

$$E(n_x, n_y, n_z) = \frac{\pi^2\hbar^2}{2mL^2}(n_x^2 + n_y^2 + n_z^2).$$

The energy given has six-fold degeneracy with (n_x, n_y, n_z) values $(1, 2, 3)$, $(1, 3, 2)$, $(2, 3, 1)$, $(2, 1, 3)$, $(3, 1, 2)$, $(3, 2, 1)$.

21.3 The normalization constant, C, for the mixed state is given by

$$C^2 \int_{-\infty}^{\infty} \exp\left(-\frac{2x^2}{a^2}\right) dx = 1.$$

From the normalized normal distribution (14.13) we find

$$C^2 = \sqrt{\frac{2}{\pi a^2}} \quad \text{or} \quad C = \left(\frac{2}{\pi a^2}\right)^{1/4}.$$

The overlap integral with the ground state eigenfunction is

$$A = \left(\frac{2}{\pi a^2}\right)^{1/4} \left(\frac{m\omega}{\pi\hbar}\right)^{1/4} \int_{-\infty}^{\infty} \exp\left\{-\left(\frac{m\omega}{\hbar} + \frac{1}{a^2}\right)x^2\right\} dx.$$

Again from (14.13) we find

$$\int_{-\infty}^{\infty} \exp(-Kx^2) dx = \sqrt{\pi/K}.$$

Hence

$$A = \left(\frac{2m\omega}{\pi^2\hbar a^2}\right)^{1/4} \sqrt{\frac{\pi}{\left(\frac{m\omega}{\hbar} + \frac{1}{a^2}\right)}} = 0.820.$$

The probability of measuring the ground state is $A^2 = 0.672$.

20.4 (i) For $E > V$,

$$T = \frac{4k_1^2 k_2^2}{(k_1^2 - k_2^2)^2 \sin^2(k_2 a) + 4k_1^2 k_2^2}.$$

$$k_1 = \sqrt{\frac{2mE}{\hbar^2}} = 1.698 \times 10^{10}\,\text{m}^{-1} \text{ and}$$

$$k_2 = \sqrt{\frac{2m(E - V)}{\hbar^2}} = 5.121 \times 10^9\,\text{m}^{-1}$$

giving $T = 0.647$.

(ii) For $E < V$,

$$T = \frac{4k_1^2 k_3^2}{(k_1^2 + k_3^2)^2 \sinh^2(k_3 a) + 4k_1^2 k_3^2}$$

$$k_3 = \sqrt{\frac{2m(V - E)}{\hbar^2}} = 5.121 \times 10^9\,\text{m}^{-1}$$

giving $T = 0.517$.

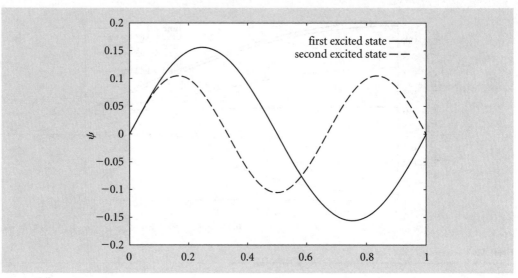

Figure S21.1

(iii) For $E = V$, $k_2 = k_3 = 0$. Dividing top and bottom of the equation in (i) by k_2 gives

$$T = \frac{4k_1^2}{(k_1^2 - k_2^2)^2 \frac{\sin^2(k_2 a)}{k_2^2} + 4k_1^2}.$$

In the limit $k_2 \to 0$ we have

$$\frac{\sin^2(k_2 a)}{k_2^2} = a^2 \quad \text{so that} \quad T = \frac{4k_1^2}{k_1^4 a^2 + 4k_1^2} = 0.581.$$

The same result is obtained from the formula for T in (ii).

21.5 The energies for the first and second excited states for zero potential energy in the box are 150.5 eV and 338.7 eV respectively. These provide convenient starting points for an investigation using SHOOT. The first excited state of the problem of interest has energy 164 eV. A plot of ψ is shown in Figure S21.1. The figure also shows ψ for the second excited state that has energy 352 eV.

CHAPTER 22

Exercises

22.1 $u_{max} = 1$; $\frac{P(2)}{P(1)} = 4e^{-3} = 0.199$.

22.2 $v_{esc} = 2654 \text{ m s}^{-1}$, $v_{rms} = 298 \text{ m s}^{-1}$.

22.3 $2.10 \times 10^{-14} \text{ J}$.

Problems

22.1 Using (22.18) and the given mean energy

$$\bar{E} = \int_0^\infty E P(E) dE = 2\pi \left(\frac{1}{\pi kT} \right)^{3/2}$$

$$\times \int_0^\infty E^{3/2} \exp\left(-\frac{E}{kT} \right) dE = \frac{3}{2} kT$$

we have

$$\overline{E^2} = \int_0^\infty E^2 P(E) dE$$

$$= 2\pi \left(\frac{1}{\pi kT} \right)^{3/2} \int_0^\infty E^{5/2} \exp\left(-\frac{E}{kT} \right) dE.$$

Integrating by parts,

$$\overline{E^2} = \left| -2\pi kT \left(\frac{1}{\pi kT} \right)^{3/2} E^{5/2} \exp\left(-\frac{E}{kT} \right) \right|_0^\infty$$

$$+ 5\pi kT \left(\frac{1}{\pi kT} \right)^{3/2} \int_0^\infty E^{3/2} \exp\left(-\frac{E}{kT} \right) dE.$$

The first term on the right-hand side is zero while the second term is $\frac{5}{2} kT \times \bar{E} = \frac{15}{4} k^2 T^2$. The standard deviation of the kinetic energy is

$$\sigma_E = \sqrt{\overline{E^2} - \bar{E}^2} = \sqrt{\frac{15}{4} - \frac{9}{4}} kT = \sqrt{\frac{3}{2}} kT.$$

22.2 The escape speed from the exosphere is

$$v_{esc} = \sqrt{\frac{2GM_{Ven}}{R_{exo}}} = 10.00 \text{ km s}^{-1}.$$

For H,

$$v_p(H) = \sqrt{\frac{2kT}{\mu_H}} = \sqrt{\frac{2 \times 1.38 \times 10^{-23} \times 400}{1.67 \times 10^{-27}}}$$

$$= 2.57 \text{ km s}^{-1}.$$

Similarly, for D, $v_p(D) = 1.818 \text{ km s}^{-1}$. For H, $u_{esc}(H) = v_{esc}/v_p(H) = 3.89$ and from Figure 22.2 the proportion of escaping H atoms, P_H, is $10^{-6.0}$.

Similarly $u_{esc}(D) = 5.50$ and $P_D = 10^{-12.4}$. Hence the required ratio is $10^{-6.0}/10^{-12.4} = 2.5 \times 10^6$.

22.3 The reaction efficiency function is

$$R(E) = E^{3/2} P(E) = CE^2 \exp\left(-\frac{E}{kT}\right)$$

where C is a constant dependent on T. For a maximum,

$$\frac{dR(E)}{dE} = 2CE \exp\left(-\frac{E}{kT}\right) - \frac{C}{kT} E^2 \exp\left(-\frac{E}{kT}\right) = 0$$

giving $E = 2kT$ or

$$\frac{E}{\bar{E}} = 2kT \div \frac{3}{2}kT = \frac{4}{3}.$$

CHAPTER 23

Exercises

23.1 Averages found, with theoretical values in parentheses:

$\bar{x} = 0.4844\,(0.5000)$, $\quad \overline{x^2} = 0.3300\,(0.3333)$,

$\overline{x^3} = 0.2514\,(0.2500)$.

23.2 The 25 trials gave the following square distances, assuming unit steps. An X represents an unsuccessful trial.

13 5 17 17 13 9 13 17 13 5   5   9   13

X   5 8 17 5   5 5 X   9   13 13 13

The root-mean-square for the 23 successful trials is 3.24. $5^{0.681} = 2.99$.

23.3 The ratio of volumes is $\pi/6 = 0.524$. Of the 33 sums of squares 16 are less than unity, or a proportion 0.485.

23.4 $\int_0^X e^x dx = e^X - 1$. If r is from uv(0,1) then transform to $X = \ln(1 + r)$.

23.5 Comparing (23.27) and (23.29), $a = 4\varepsilon\sigma^{12}$ and $b = 4\varepsilon\sigma^6$. Substituting in (23.28) gives the required result.

23.6 From Table 23.3, for argon $\sigma = 0.345$ nm and $\varepsilon = 120k = 1.656 \times 10^{-21}$. Substituting in the expression gives $F(r) = 1.111 \times 10^{-9}$ N. The contribution to the virial term is $\frac{F(r)r}{3} = 1.111 \times 10^{-19}$ J.

Problems

23.1 Changing $P(1)$ to 0.333333 and $P(2)$ to zero in DRUNKARD gave the following results:

No. of steps (n)	1	4	9	16	25	36	49
r.m.s. path (d)	1	2.552	4.065	5.516	6.958	8.408	9.823

A plot of log(d) against log(n), shown in Figure S23.1, indicated a best slope of 0.604 giving $d = n^{0.604}$.

23.2 Changing $P(1)$ to 0.5 in POLYMER gave the following results:

No. of steps (n)	1	4	9	16	25	36	49
r.m.s. path (d)	1	2.90	5.06	7.42	9.99	12.85	15.82

A plot of log(d) against log(n), shown in Figure S23.2, indicated a best slope of 0.718, giving $d = n^{0.718}$.

23.3 For this distribution

$$f(X) = \frac{2}{3} \int_0^X (1 + x)\, dx = \frac{2}{3}\left(X + \frac{1}{2}X^2\right).$$

Equating this to r from a uv(0,1) gives
$$X = -1 + \sqrt{1 + 3r}.$$

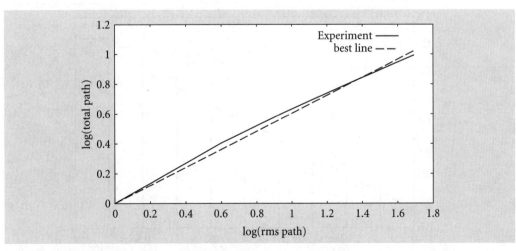

Figure S23.1 A log–log plot.

The following outline program generates the scaled histogram:

Clear histogram array
  Loop for i = 1 to 10
    $h(i) = 0$
  End loop
Generate 100 000 variables
  Loop for i = 1 to 100 000
    Generate r from uv(0, 1)
    $x = 1 + \sqrt{1 + 3r}$
    $k = \text{int}(10x) + 1$
    $h(k) = h(k) + 1$
  End loop
Normalize histogram
  Loop for i = 1 to 100 000
    $h(i) = 0.0001h(i)$
  End loop
Output on printer or file

A program based on this algorithm gave the histogram plotted in Figure S23.3, which is shown together with $p(x)$.

23.4 The maximum of the distribution is 16/9 so the uniform-variate comparator function has this height. An outline program to use the rejection method is

Clear histogram array
  Loop i from 1 to 10
    $h(i) = 0$
  End loop
Generate variables, enter in histogram and
  find total accepted
    sum = 0
  Loop i from 1 to 1 000 000
    Generate r_1 from uv(0, 1)
    Generate r_2 from uv(0, 1)
    if $16r_2/9 > 12r_1^2(1 - r_1)$ go to end of loop

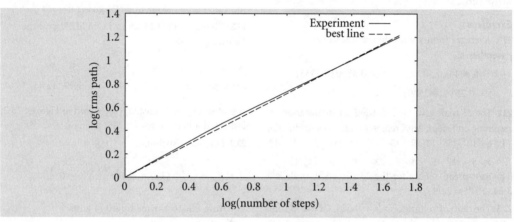

Figure S23.2 A log–log plot for a biased POLYMER random walk.

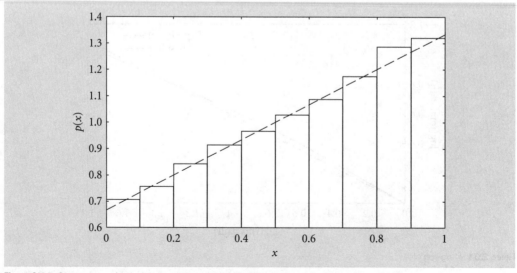

Figure S23.3 Comparison of calculated and theoretical distributions for $p(x) = 2(1 + x)/3$.

```
        k = int(10r₁) + 1
        h(k) = h(k) + 1
        sum = sum + 1
     End loop
Proportion rejected
     rej = sum/10 ⁶
Normalize histogram
     Loop i from 1 to 10
        h(i) = 10 h(i)/sum
     End loop
Output histogram on printer or in file.
```

The result of running a program based on this algorithm is illustrated in Figure S23.4 together with the distribution $p(x)$, shown as a dashed line. The area under the comparator function is 16/9 so that, since the distribution is normalized, the proportion of

rejected trials is $1 - 9/16 = 7/16 = 0.4375$. The value from a program run was 0.4384.

23.5 For nitrogen parameters the equation of state is as shown in Figure S23.5.

23.6

(i) For $U/C = 0.002$ and $U/C = 0.005$ the results gave Figure S23.6. Varying the U/C ratio, keeping other parameters constant gave Figure S23.7.

(ii) If there is too little uranium then, clearly, there are fewer U_{235} atoms available to give fission and, in the limit, with no uranium the multiplication factor would be zero. However, increasing the uranium without limit does not continuously increase the multiplication factor. Absorption by U_{238} is competing with the slowing of neutrons by the moderator and if there is too much uranium the

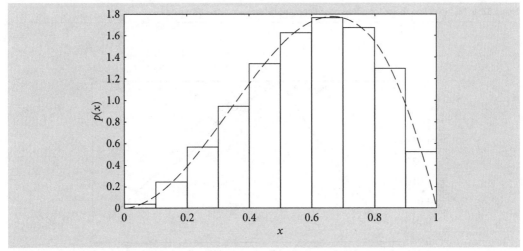

Figure S23.4 Comparison of calculated and theoretical distributions for $p(x) = 12x^2(1-x)$.

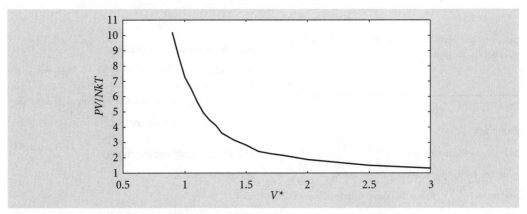

Figure S23.5 Calculated equation of state for nitrogen from the METROPOLIS program.

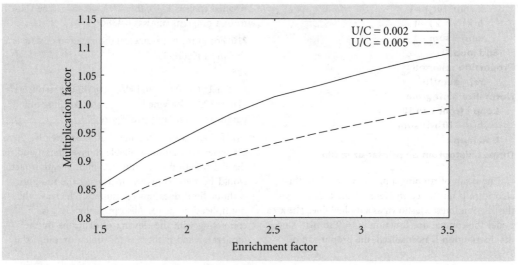

Figure S23.6 Multiplication factor for different enrichment factors and two U/C ratios

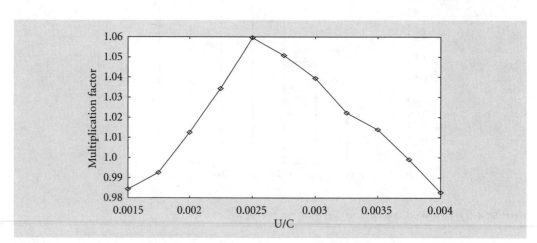

Figure S23.7 Multiplication factors for different U/C and constant enrichment factor.

neutrons are absorbed before they are slowed down sufficiently to give fission.

CHAPTER 24

Exercises

24.1 The Leslie matrix is $L = \begin{bmatrix} 0.6 & 0.4 \\ 0.8 & 0 \end{bmatrix}$. After 4 and 5 generations the population vectors are $n_4 = \begin{bmatrix} 8176 \\ 6976 \end{bmatrix}$ and $n_5 = \begin{bmatrix} 7696 \\ 6541 \end{bmatrix}$. From this the estimate of eigenvalue is 0.939 and the ratio $Y : O = 1.177 : 1$.

24.2 The characteristic equation is $\begin{vmatrix} 0.6 - \lambda & 0.4 \\ 0.8 & -\lambda \end{vmatrix} = 0$, which gives $\lambda_1 = 0.9403$ and $\lambda_2 = -0.3403$. For λ_1 the first equation gives $-0.3403Y + 0.4O = 0$ or $Y : O = 1.1754$, corresponding to what was found in Exercise 24.1. For λ_2 the eigenvector is $\begin{bmatrix} -0.4254 \\ 1 \end{bmatrix}$.

24.3 The diagonalizing matrix is $Q = \begin{bmatrix} 1.1754 & -0.4254 \\ 1 & 1 \end{bmatrix}$ and the inverse is

$$Q^{-1} = \frac{1}{1.6008} \begin{vmatrix} 1 & 0.4254 \\ -1 & 1.1754 \end{vmatrix}.$$

24.4

$$Q^{-1}LQ = \frac{1}{1.6008}\begin{bmatrix} 1 & 0.4254 \\ -1 & 1.1754 \end{bmatrix}\begin{bmatrix} 0.6 & 0.4 \\ 0.8 & 0 \end{bmatrix}$$

$$\times \begin{bmatrix} 1.1754 & -0.4254 \\ 1 & 1 \end{bmatrix}$$

$$= \frac{1}{1.6008}\begin{bmatrix} 1 & 0.4254 \\ -1 & 1.1754 \end{bmatrix}$$

$$\times \begin{bmatrix} 1.10524 & 0.14476 \\ 0.94032 & -0.34032 \end{bmatrix}$$

$$= \begin{bmatrix} 0.9403 & 0 \\ 0 & -0.3403 \end{bmatrix}.$$

24.5 (i) $\begin{bmatrix} \ddot{x}_1 \\ \ddot{x}_2 \end{bmatrix} = -\frac{T}{ma}\begin{bmatrix} 2 & -1 \\ -1 & 2 \end{bmatrix}\begin{bmatrix} x_1 \\ x_2 \end{bmatrix}.$

(ii) $\lambda_1 = 1$ with eigenvector $\begin{bmatrix} 1 \\ 1 \end{bmatrix}$ and $\lambda_2 = 3$ with eigenvector $\begin{bmatrix} 1 \\ -1 \end{bmatrix}$.

(iii) See Figure S24.1 for the diagrams showing the normal modes for λ_1 and λ_2.

For λ_1

For λ_2

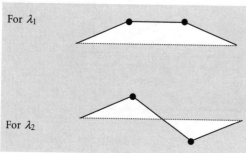

Figure S24.1

Problems

24.1 The characteristic equation is $\begin{vmatrix} 1-\lambda & 1 & 1 \\ 0 & 2-\lambda & 1 \\ 0 & 0 & 3-\lambda \end{vmatrix} = 0,$

which gives $(1-\lambda)(2-\lambda)(3-\lambda) = 0$ or $\lambda_1 = 1$, $\lambda_2 = 2$, and $\lambda_3 = 3$.

For λ_1 the three equations from the characteristic equation give $x_2 + x_3 = 0$, $x_2 + x_3 = 0$, and $2x_3 = 0$.

These give $x_2 = x_3 = 0$ and an eigenvector $\begin{bmatrix} 1 \\ 0 \\ 0 \end{bmatrix}$.

For λ_2 the three equations from the characteristic equation give $-x_1 + x_2 + x_3 = 0$, $x_3 = 0$, and $x_3 = 0$. These give $x_1 = x_2$ and $x_3 = 0$ and an eigenvector $\begin{bmatrix} 1 \\ 1 \\ 0 \end{bmatrix}$.

For λ_3 the three equations from the characteristic equation give $-2x_1 + x_2 + x_3 = 0$, $-x_2 + x_3 = 0$, and $0 = 0$. These give $x_1 = x_2 = x_3$ and an eigenvector $\begin{bmatrix} 1 \\ 1 \\ 1 \end{bmatrix}$.

24.2 Running LESLIE with the original survival and fertility factors gave the following relative numbers, normalized to unity for infants, with the generation growth factor 1.00664: I 1.0000, J 0.8941, Y 0.7549, A 0.6000, E 0.4172. With the modified fertility factors due to culling the generation growth factor is 0.99157 and the population profile is: I 1.0000, J 0.9077, Y 0.7781, A 0.5885, E 0.3858.

24.3 The accelerations are given by

$$\ddot{x}_1 = -\frac{T}{2ma}(2x_1 - x_2)$$

$$\ddot{x}_2 = -\frac{T}{ma}(2x_2 - x_1 - x_3) = -\frac{T}{2ma}(4x_2 - 2x_1 - 2x_3)$$

$$\ddot{x}_3 = -\frac{T}{2ma}(2x_3 - x_2)$$

or

$$\begin{bmatrix} \ddot{x}_1 \\ \ddot{x}_2 \\ \ddot{x}_3 \end{bmatrix} = -\frac{T}{2ma}\begin{bmatrix} 2 & -1 & 0 \\ -2 & 4 & -2 \\ 0 & -1 & 2 \end{bmatrix}\begin{bmatrix} x_1 \\ x_2 \\ x_3 \end{bmatrix}.$$

We now find the eigenvalues for the matrix

$$\begin{vmatrix} 2-\lambda & -1 & 0 \\ -2 & 4-\lambda & -2 \\ 0 & -1 & 2-\lambda \end{vmatrix} = 0.$$

This gives $\lambda^3 - 8\lambda^2 + 16\lambda - 8 = (\lambda - 2)(\lambda^2 - 6\lambda + 4) = 0$, leading to the eigenvalues $\lambda_1 = 2$, $\lambda_2 = 3 + \sqrt{5}$ and $\lambda_3 = 3 - \sqrt{5}$. Substituting for λ_1 the equations giving the eigenvector are

$$-x_2 = 0$$

$$-2x_1 + 2x_2 - 2x_3 = 0$$

$$-x_2 = 0$$

giving the eigenvector $\{1, 0, -1\}$. Similarly, λ_2 and λ_3 give eigenvectors $\{1, -(\sqrt{5}+1), 1\}$ and $\{1, \sqrt{5}-1, 1\}$ respectively. The general motions of the masses are described by

$$x_1 = c_1 \exp\left(i\sqrt{\frac{T\lambda_1}{2ma}}t\right) + c_2 \exp\left(i\sqrt{\frac{T\lambda_2}{2ma}}t\right)$$
$$+ c_3 \exp\left(i\sqrt{\frac{T\lambda_3}{2ma}}t\right),$$

$$x_2 = -\left(\sqrt{5}+1\right)c_2 \exp\left(i\sqrt{\frac{T\lambda_2}{2ma}}t\right)$$
$$+ \left(\sqrt{5}-1\right)c_3 \exp\left(i\sqrt{\frac{T\lambda_3}{2ma}}t\right),$$

$$x_3 = -c_1 \exp\left(i\sqrt{\frac{T\lambda_1}{2ma}}t\right) + c_2 \exp\left(i\sqrt{\frac{T\lambda_2}{2ma}}t\right)$$
$$+ c_3 \exp\left(i\sqrt{\frac{T\lambda_3}{2ma}}t\right).$$

The normal modes are similar to those shown in Figures 24.2, 24.3, and 24.4 except that the ratio of the amplitude of the centre mass is $\sqrt{5}+1$ times as great as

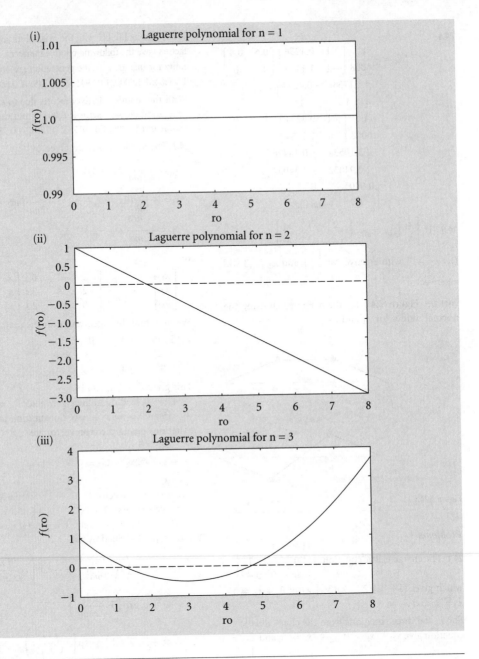

Figure S25.1

that of the others in the Figure 24.3 configuration and $\sqrt{5} - 1$ times as great for the Figure 24.4 configuration.

CHAPTER 25

Exercises

25.1 (i) 0, 0; (ii) $6\hbar^2$, $-2\hbar$; (iii) $6\hbar^2$, $2\hbar$.

25.2 $\lambda = c/\nu$. From (25.28) with $n = 2$ and $m = 3$, $\lambda_{max} = 656.5$ nm; with $n = 2$ and $m = \infty$, $\lambda_{min} = 364.7$ nm.

25.3 See Figure S25.1.

25.4

$$[\hat{s}_y, \hat{s}_z] = \frac{1}{4}\hbar^2 \left(\begin{bmatrix} 0 & -i \\ i & 0 \end{bmatrix} \begin{bmatrix} 1 & 0 \\ 0 & -1 \end{bmatrix} \right.$$

$$\left. - \begin{bmatrix} 1 & 0 \\ 0 & -1 \end{bmatrix} \begin{bmatrix} 0 & -i \\ i & 0 \end{bmatrix} \right)$$

$$= \frac{1}{4}\hbar^2 \left(\begin{bmatrix} 0 & i \\ i & 0 \end{bmatrix} - \begin{bmatrix} 0 & -i \\ -i & 0 \end{bmatrix} \right)$$

$$= \frac{1}{2}\hbar^2 \begin{bmatrix} 0 & i \\ i & 0 \end{bmatrix} = i\hbar\hat{s}_x$$

$$[\hat{s}_z, \hat{s}_x] = \frac{1}{4}\hbar^2 \left(\begin{bmatrix} 1 & 0 \\ 0 & -1 \end{bmatrix} \begin{bmatrix} 0 & 1 \\ 1 & 0 \end{bmatrix} \right.$$
$$\left. - \begin{bmatrix} 0 & 1 \\ 1 & 0 \end{bmatrix} \begin{bmatrix} 1 & 0 \\ 0 & -1 \end{bmatrix} \right)$$
$$= \frac{1}{4}\hbar^2 \left(\begin{bmatrix} 0 & 1 \\ -1 & 0 \end{bmatrix} - \begin{bmatrix} 0 & -1 \\ 1 & 0 \end{bmatrix} \right)$$
$$= \frac{1}{2}\hbar^2 \begin{bmatrix} 0 & 1 \\ -1 & 0 \end{bmatrix} = i\hbar\hat{s}_y$$

25.5

$$\frac{\sqrt{3}}{2}\begin{bmatrix} 1 \\ 0 \end{bmatrix} + \frac{1}{2}\begin{bmatrix} 0 \\ 1 \end{bmatrix} = \begin{bmatrix} \sqrt{3}/2 \\ 1/2 \end{bmatrix}. \text{ Probability} = \frac{1}{4}.$$

Problems

25.1 The spherical harmonic $Y_2^1 = C \sin\theta\cos\theta e^{i\phi}$.

$$\hat{L}^2 Y_2^1 = -C\hbar^2 \left\{ \frac{1}{\sin\theta}\frac{\partial}{\partial\theta}\sin\theta(\cos^2\theta - \sin^2\theta) \right.$$
$$\left. - \frac{\sin\theta\cos\theta}{\sin^2\theta} \right\} e^{i\phi}$$
$$= -C\hbar^2 \left\{ \frac{1}{\sin\theta}(\cos^3\theta - 2\sin^2\theta\cos\theta \right.$$
$$\left. - 3\sin^2\theta\cos\theta) - \frac{\cos\theta}{\sin\theta} \right\} e^{i\phi}$$
$$= -C\hbar^2 \left\{ \frac{1}{\sin\theta}(\cos\theta - 6\sin^2\theta\cos\theta) \right.$$
$$\left. - \frac{\cos\theta}{\sin\theta} \right\} e^{i\phi}$$
$$= 6C\hbar^2\sin\theta\cos\theta e^{i\phi} = 2(2+1)\hbar^2 Y_2^1.$$

The spherical harmonic $Y_1^1 = C\sin\theta e^{i\phi}$.

$$\hat{L}^2 Y_1^1 = -C\hbar^2 \left\{ \frac{1}{\sin\theta}\frac{\partial}{\partial\theta}(\sin\theta\cos\theta) - \frac{\sin\theta}{\sin^2\theta} \right\} e^{i\phi}$$
$$= -C\hbar^2 \left\{ \frac{1}{\sin\theta}(\cos^2\theta - \sin^2\theta) - \frac{1}{\sin\theta} \right\} e^{i\phi}$$
$$= -C\hbar^2 \left\{ \frac{1}{\sin\theta}(1 - 2\sin^2\theta) - \frac{1}{\sin\theta} \right\} e^{i\phi}$$
$$= 2C\hbar^2\sin\theta e^{i\phi} = 1(1+1)\hbar^2 Y_1^1.$$

25.2 The overlap integral is of the form

$$C \int_0^{2\pi}\int_0^\pi (\sin\theta\cos\theta e^{-i\phi} \times \sin\theta e^{i\phi})\sin\theta \, d\theta \, d\phi$$
$$= C \int_0^\pi \sin^2\theta\cos\theta \, d\theta \int_0^{2\pi} d\phi.$$

The second integral equals 2π but the first equals $\left|\frac{1}{3}\sin^3\theta\right|_0^\pi = 0$. Hence these spherical harmonics are orthogonal.

25.3 For $u_{100} = Ce^{-r/a_B}$ the radial density function is $S(r) = 4\pi C^2 r^2 e^{-2r/a_B}$. To find the maximum,

$$\frac{dS}{dr} = 4\pi C^2\left(2re^{-2r/a_B} - 2\frac{r^2}{a_B}e^{-2r/a_B}\right) = 0.$$

The solutions $r = 0$ and $r = \infty$ correspond to minima and $r = a_B$ is a maximum (see Figure 25.2a).

25.4 Passing through the first field blocks out a fraction 0.5 of the atoms. Thereafter for each of the following fields a fraction $\cos^2(30°) = 3/4$ of the atoms pass through in the spin-up condition. Hence the intensity is reduced by a factor $0.5 \times (0.75)^5 = 0.119$.

CHAPTER 26

Exercises

26.1 For A, $\bar{h} = 178.6$ cm, $V_h = 77.44$ cm^2. For B, $\bar{h} = 187.4$ cm, $V_h = 25.98$ cm^2.

26.2 For A, $\langle\mu_A\rangle = 178.6$ cm, $\langle V_h\rangle = \frac{10}{9}\times 77.44 = 86.04$ cm^2.

For B, $\langle\mu_B\rangle = 187.4$ cm, $\langle V_h\rangle = \frac{8}{7}\times 25.98 = 29.69$ cm^2.

26.3 $\langle\sigma_P\rangle = \sqrt{\frac{0.485\times 0.515}{10^4}} = 0.005$. The claim is $3\langle\sigma_P\rangle$ from the result of the sample. From Table 14.1 the probability of this, or a greater difference, is 0.00135. Hence the claim cannot be supported.

26.4 $\sigma_d = \sqrt{\frac{77.44 + 25.98}{16}} = 2.54$ and $t = \frac{187.4 - 178.6}{2.54} = 3.46$. Since this is a two-tail problem we look at the table in the column headed 0.05. For 16 degrees of freedom the interpolated table value is 1.75, so the difference is significant.

26.5 $\sigma_\mu = \sqrt{\frac{77.44}{9}} = 2.93$ cm and $t = \frac{186 - 178.6}{2.93} = 2.53$. This is a one-tail problem and from the table, with 9 degrees of freedom and a significance level 0.1 the value of t is 1.38. Hence the claim cannot be supported.

26.6

	Observed		Expected	
	A	B	A	B
Pass	81	59	70	70
Fail	19	41	30	30

$$\chi^2 = \frac{121}{70} + \frac{121}{70} + \frac{121}{30} + \frac{121}{30} = 11.52.$$

This is a highly significant result.

Problems

26.1 For the sample data provided, $\mu_s = 2904.8$ kg m^{-3} and $\sigma_s = 16.64$ kg m^{-3}. The estimated standard deviation of sample means is

$$\sigma_{\mu_s} = \frac{\sigma_s}{\sqrt{n-1}} = \frac{16.64}{3} = 5.55 \text{ kg m}^{-3}.$$

The null hypothesis is that the density of the mineral is 2914 kg m^{-3} and since we are comparing the sample average with some claimed value a one-tail t-test is appropriate. The value of t is given by

$$t = \frac{2914 - 2904.8}{5.55} = 1.66.$$

For 9 degrees of freedom and a significance level 0.1 the interpolated value from Table 26.1 is 1.385. Since the calculated value of t is greater than this tabulated value we may reject the claimed density at the 0.1 level of significance.

26.2 For the two samples

$$n_1 = 10, \mu_1 = 9.8189 \, \text{m s}^{-2}, \sigma_{s1} = 0.006\,59 \, \text{m s}^{-2}$$
$$n_2 = 10, \mu_2 = 9.8232 \, \text{m s}^{-2}, \sigma_{s1} = 0.008\,66 \, \text{m s}^{-2}.$$

The null hypothesis is that the two locations have the same acceleration due to gravity and we are testing at the significance level 0.1 whether the difference between the two sample means is significant. This requires a two-tail test that means that the column of interest in the Student t-test table corresponds to a significance level of 0.05 with $n_1 + n_2 - 2 = 18$ degrees of freedom. The estimated variance of the difference of sample means is

$$\sigma_d^2 = \frac{\sigma_{s1}^2 + \sigma_{s2}^2}{18} = 6.579 \times 10^{-6}$$

giving $\sigma_d = 0.00256$. The value of t from the samples is thus

$$t = \frac{9.8232 - 9.8189}{0.00256} = 1.68.$$

The tabulated value for the appropriate significance level and degrees of freedom is 1.74. Since the value from the samples is less than this, the difference is not significant.

26.3 The expected number of each digit is 100, so

$$\chi^2 = \frac{1}{100}(225 + 64 + 169 + 121 + 16 + 81 + 81 + 144 + 256 + 169) = 11.26.$$

For 9 degrees of freedom and a significance level of 0.1 the value of chi-squared is, from Table 26.2, 14.7. Since chi-squared for the data is less than this, we cannot reject the null hypothesis that the random digit generator is working properly.

CHAPTER 27

Exercises

27.1 For the pairs of points $m_1 = 1.5, c_1 = -0.5$; $m_2 = 1.2, c_2 = -0.2$; $m_3 = 1.0, c_3 = 1.0$ with means $\bar{m} = 1.233$ and $\bar{c} = 0.1$. This line and points are shown in Figure S27.1.

27.2 $y = 1.184x + 0.053$.

27.3 0.993.

27.4 (i) $k = 1$; (ii) $A = 5$; (iii) 1.1935; (iv) 0.022.

27.5 Convert to $\log w = \log a + 3\log l$. The function to minimize is $S = \sum (\log w - \log a - 3\log l)^2$. Making $\partial S / \partial a = 0$ gives $\log a = \langle \log w \rangle - 3\langle \log l \rangle$ from which $\log a = -3.685$ and $a = 2.065 \times 10^{-4}$. The predicted values of w are, in kg wt, 0.21, 18.8, and 697.

Problems

27.1 The basic equation giving the distance fallen, s, against time, t, is $s = \frac{1}{2}gt^2$ or $t = \sqrt{2s/g}$. However, if the zero of time measurement is time t_0 after the release of the object then this becomes

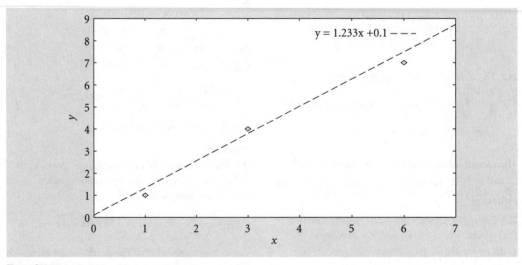

Figure S27.1

$$t + t_0 = \sqrt{\frac{2s}{g}} \quad \text{or} \quad t = \sqrt{\frac{2}{g}}\sqrt{s} - t_0.$$

To find g requires the slope of the line, which is

$$\sqrt{\frac{2}{g}} = \frac{\langle\sqrt{s}t\rangle - \langle\sqrt{s}\rangle\langle t\rangle}{\langle s\rangle - \langle\sqrt{s}\rangle^2}.$$

Table S27.1

s	$\sqrt{s}$	t	$\sqrt{s}t$
0.3	0.548	0.000	0.000
0.4	0.632	0.037	0.023
0.5	0.707	0.073	0.052
0.6	0.775	0.109	0.084
0.7	0.837	0.128	0.107
0.8	0.894	0.161	0.144
0.55	0.732	0.0847	2.0683

Table S27.1 gives the required averages. This gives

$$\sqrt{\frac{2}{g}} = \frac{0.0683 - 0.732 \times 0.0847}{0.55 - (0.732)^2} = 0.444 \quad \text{or}$$

$$g = 10.1 \, \text{m s}^{-2}.$$

The calculation can also be carried out by writing a simple computer program and the result found is then $9.4 \, \text{m s}^{-2}$ – very different from the previous result. The reason for this is that not enough significant figures were used in deriving the averages in the table. Quantities such as $\langle x^2\rangle - \langle x\rangle^2$ and $\langle xy\rangle - \langle x\rangle\langle y\rangle$ are the small differences of two much larger numbers. For example, in this case $\langle s\rangle = 0.55$ and $\langle\sqrt{s}\rangle^2 =$ 0.536 060 4. The difference is 0.0139 396 rather than 0.014 18 derived from the table values, that is about 2% less. Similarly, using seven significant figures, $\langle\sqrt{s}t\rangle - \langle\sqrt{s}\rangle\langle t\rangle$ equals 0.006 434 65 rather than 0.006 300 obtained from the table, about 2% more. The quotient of the two numbers from the table thus differs by almost 4% from the result given by a more precise calculation and, since this quotient is squared to give g, the final table value is about 7% greater than the precise value.

The lesson to be drawn here is that where these 'difference of large numbers' calculations are involved it is sensible to calculate with as much precision as possible and truncate the final result to reflect the quality of the data used. If simple computer programs are written to perform the calculations – and this usually takes less time than using a hand calculator – then, normally, all the precision required is automatically obtained.

27.2 For this problem the straight-line relationship is between n, the number of the ring, and r_n^2, the square of the ring radius. Since the straight line passes through the origin it is of the form $y = mx$ and the least-squares condition is to minimize

$$S = \sum_{i=1}^{n} (y_i - mx_i)^2$$

Since $\dfrac{\mathrm{d}S}{\mathrm{d}m} = -2\sum_{i=1}^{n} x_i(y_i - mx_i)$ this leads to

$$m = \frac{\langle xy\rangle}{\langle x^2\rangle} \quad \text{or,} \quad \text{for this case,} \quad 2R\lambda = \frac{\langle nr^2\rangle}{\langle n^2\rangle},$$

where R is the radius of the convex lens. There are no 'differences of two large numbers' involved in this calculation. The averages are $\langle nr^2\rangle = 1.895 \times 10^{-4}$ and $\langle n^2\rangle = 379.2$, giving $\lambda = 5.00 \times 10^{-7}$ m.

27.3 The correlation coefficient for the first two rows is found from

$$r(t, t+5) = \frac{\langle s(t)s(t+5)\rangle - \langle s(t)\rangle\langle s(t+5)\rangle}{\sigma_t \sigma_{t+5}}$$

where $s(t_i)$ is the signal at time t_i and σ_t is the standard deviation of the set of signals given in the top row. This gives a correlation coefficient of -0.262 which shows a fairly weak negative correlation and no possible period of 5 s or near. On the other hand $r(t, t+8) = 0.624$, a fairly strong positive correlation indicating a good likelihood of an 8 s periodicity in the stream of signals.

27.4 The basic equation for a simple pendulum is

$$P = 2\pi\sqrt{\frac{l}{g}} \quad \text{or} \quad P^2 = 4\pi^2 \frac{l}{g}.$$

If l_0 is the length of the cord and x the distance of the mass from its end, then

$$P^2 = -\frac{4\pi^2}{g}x + \frac{4\pi^2}{g}l_0.$$

Thus the slope of the LSSL line linking P^2 and x gives g and then the intercept gives l_0. The value of k is given by

$$k = \frac{\varepsilon_{P^2}}{\varepsilon_x} = \frac{0.20}{0.05} = 4$$

and this is used in (27.29) to give the slope m, which is found to be $-4.063\,50$. From (27.22) the intercept is 79.7459. Hence

$$\frac{4\pi^2}{g} = 4.06350 \quad \text{or} \quad g = 9.72 \, \text{m s}^{-2} \quad \text{and}$$

$$\frac{4\pi^2 l_0}{g} = 79.7459, \quad \text{giving } l_0 = 19.62 \, \text{m}.$$

27.5 The activity of the source is related to time by $r = Ce^{-\lambda t}$ or, taking natural logarithms, $\ln(r) = -\lambda t + \ln(C)$. A plot of $\ln(r)$ against t gives a straight line of slope $-\lambda$. We shall take the unit of time as the day, so that λ will be in units d^{-1}. We have

$$-\lambda = \frac{\langle \ln(r)t\rangle - \langle \ln(r)\rangle\langle t\rangle}{\langle t^2\rangle - \langle t\rangle^2} = -0.973 \, \text{d}^{-1}.$$

The half-life is thus $\tau = \ln(2)/\lambda = 0.71$ d. Inspection of the data shows that it takes less than one day to halve the activity.

CHAPTER 28

Exercises

28.1 10.56 cm.

28.2 (i) f is given by $(0.1)^2 - 0 - (d/df)(0.1f)^2 = 0$ which gives $f = 0.5$. The error is then $0.5 \times 0.1^2 + 0.5 \times 0.0^2 - (0.05)^2 = 0.0025$.

(ii) f is given by $(1.0)^2 - (0.9)^2 - (d/df)(0.9 + 0.1f)^2 = 0$ which gives $f = 0.5$. The error is then $0.5 \times 1.0^2 + 0.5 \times 0.9^2 - (0.95)^2 = 0.0025$.

28.3 $a = \dfrac{14.3 + 4.9 - 2 \times 9.9}{2 \times 2^2} = -0.075$,

$b = \dfrac{14.3 - 4.9}{2 \times 2} = 2.35$, $c = 9.9$.

Hence $y = -0.075 \times (0.3)^2 + 2.35 \times 0.3 + 9.9 = 10.60$ cm. Compare Exercise 28.1.

Problems

28.1 (i) The linear interpolation in this notation is

$$y = \frac{x_{i+1} - x}{x_{i+1} - x_i} y_i + \frac{x - x_i}{x_{i+1} - x_i} y_{i+1}.$$

For $x = 2/3$, $x_i = x_0 = 0$; $y_i = y_0 = 0$; $x_{i+1} = x_1 = 1.0$; $y_{i+1} = y_1 = 1.5$. This gives $y = 1.00$.
For $x = 1.7$, $x_i = x_1 = 1.0$; $y_i = y_1 = 1.5$; $x_{i+1} = x_2 = 2.0$; $y_{i+1} = y_2 = 2.0$. This gives $y = 1.85$.

(ii) For $x = 2/3$ the nearest tabulated x value is $i = 1$. From (7.8) the coefficients of the parabola are:

$$a = \frac{0.0 + 2.0 - 2 \times 1.5}{2 \times 1.0^2} = -0.5;$$

$$b = \frac{2.0 - 0.0}{2 \times 1.0} = 1.0; \quad c = 1.5.$$

This gives $y = -0.5(2/3 - 1.0)^2 + 1.0(2/3 - 1.0) + 1.5 = 1.111$. For $x = 1.7$ the nearest tabulated x value is $i = 1$ and parabola coefficients are:

$$a = \frac{1.5 + 3.5 - 2 \times 2.0}{2 \times 1.0^2} = 0.5;$$

$$b = \frac{3.5 - 1.5}{2 \times 1.0} = 1.0; \quad c = 2.0$$

This gives $y = 0.5(1.7 - 2.0)^2 + 1.0(1.7 - 2.0) + 2.0 = 1.745$.

(iii) From (28.17a) and (28.18)

$$2y_0' + y_1' = 4.5 \tag{S28.1}$$

$$y_2' + 2y_3' = 4.5 \tag{S28.2}$$
$$y_0' + 4y_1' + y_2' = 6 \tag{S28.3}$$
$$y_1' + 4y_2' + y_3' = 6. \tag{S28.4}$$

From (S28.1), $y_1' = 4.5 - 2y_0'$. $\tag{S28.5}$

From (S28.3), $y_2' = 6 - y_0' - 4y_1'$
$$= 7y_0' - 12. \tag{S28.6}$$

From (S28.4), $y_3' = 6 - y_1' - 4y_2'$
$$= 49.5 - 26y_0'. \tag{S28.7}$$

From (S28.2), $7y_0' - 12 + 99 - 52y_0' = 4.5$
or $45y_0' = 82.5$. $\tag{S28.8}$

From (S28.8), $y_0' = 1.8333$.

From (S28.5), $y_1' = 0.8333$.

From (S28.6), $y_2' = 0.8333$.

From (S28.7), $y_3' = 1.8333$.

The value $x = 2/3$ is in the first segment including a natural boundary condition for which equations (28.16) are appropriate. These give $a_0 = 0$, $b_0 = 1.8333$, $c_0 = 0$, and $d_0 = -0.3333$. Hence

$$y = 0 + 1.8333 \times (2/3) + 0 - 0.3333 \times (2/3)^3$$
$$= 1.123.$$

The value $x = 1.7$ is in the segment $i = 1$ for which the equations (28.13) are appropriate. These give $a_1 = 1.5$; $b_1 = 0.8333$; $c_1 = -1.0$, and $d_1 = 0.6667$. Hence $y = 1.5 + 0.8333 \times 1.7 - 1.0 \times 1.7^2 + 0.6667 \times 1.7^3 = 1.822$.

28.2 Linear interpolation in the table gives

$$f = 0.27 \times 0.003\ 578 + 0.63 \times 0.008\ 050 +$$
$$0.03 \times 0.006\ 573 + 0.07 \times 0.014\ 789$$
$$= 0.007\ 270.$$

The function tabulated is $f(x, y) = x^{1.5}y^2$ and the true value $f(2.1, 2.7) = 0.007\ 015$.

CHAPTER 29

Exercises

29.1 (i) $I_4 = 0.3437$, $\varepsilon_4 = 0.0104$; (ii) $I_8 = 0.3359$, $\varepsilon_8 = 0.0026$. $\varepsilon_4/\varepsilon_8 = 4$ as expected.

29.2 (i) $I_4 = 0.3333$, $\varepsilon_4 = 0$; (ii) $I_8 = 0.3333$, $\varepsilon_8 = 0$. Simpson's rule gives a precise answer for any function of cubic power or less.

29.3 The Romberg table is

$T_1 = 0.7500$
$\qquad S_2 = 0.7833$
$T_2 = 0.7750 \qquad\qquad R_4 = 0.7855$
$\qquad S_4 = 0.7854$
$T_4 = 0.7828$

29.4 The analytical solution is $\pi/4 = 0.7854$.
Since the range of the integral is 0 to 1 the integral is

$$I = w_1 \frac{1}{1 + \alpha_1^2} + w_2 \frac{1}{1 + \alpha_2^2} + w_3 \frac{1}{1 + \alpha_3^2} = 0.7853,$$

where the w and α come from Table 29.4.

29.5 The table of values of xy^2 is

y	x 0	$\frac{1}{2}$	1
0	0.000	0.000	0.0
$\frac{1}{2}$	0.000	0.125	0.250
1	0.000	0.500	1.000

For the trapezium rule the weights of rows and columns have the pattern 1 2 1 with the overall multiplication factor $\frac{1}{4}h^2 = 1/16$. The estimate of the integral is 0.1875.

For Simpson's rule the weights have the pattern 1 4 1 and the overall multiplication factor is $h^2/9 = 1/36$. The estimate of the integral is 1/6, the correct value since the function involves powers of x and y below 4.

Problems

29.1 The trapezium rule results are as follows:

No of intervals	Estimate	Error
1	0.75	0.035 398
2	0.775	0.010 398
4	0.782 794	0.002 604
8	0.784 747	0.000 651
16	0.785 235	0.000 163

The error for two intervals is markedly greater than 1/4 of that for one interval. For larger values of h the assumption made in (29.13) that only the lowest power of h needs to be considered breaks down.

29.2 The Simpson rule results are obtained from the trapezium rule results by $S_{2n} = (4T_{2n} - T_n)/3$. These results are:

Number of intervals	Trapezium	Simpson
1	0.75	
2	0.775	0.783 333
4	0.782 794	0.785 392
8	0.784 747	0.785 398
16	0.785 235	0.785 398

The Simpson results for 8 and 16 intervals are correct to 6 decimal places.

29.3 The modifications of the programs were just to change π to $\pi/2$ wherever it appeared. The result from MULGAUSS was 0.055229. Those from MCINT with various numbers of points were:

10 000 points   0.055 291 $\sigma = 0.000\,109$
100 000 points   0.055 228 $\sigma = 0.000\,034$
1 000 000 points 0.055 225 $\sigma = 0.000\,011$.

29.4 An adaptation of the program MULGAUSS gave the estimate 0.535 757.

CHAPTER 30

Exercises

30.1 (i) Linearly dependent; (ii) independent; (iii) incompatible.

30.2 From the second equation, (a) $x = 1 + y + z$. Substituting in other equations, (b) $6y + 6z = 0$ and (c) $7y + 6z = 1$. From (b), $y = -z$ which, substituted in (c), gives $z = -1$. Then from (b) $y = 1$ and then from (a) $x = 1$.

30.3 (i) $x = -1$, $y = 1$; (ii) $x = -0.83$, $y = 0.85$. The equations are poorly conditioned.

30.4 The equations are set up as $x = 0.1392 - 0.1y$ and $y = 0.298 - 2x/9$. The stages in the solution are:

y	x
0	0
0.1392	0.2671
0.1125	0.2730
0.1119	0.2731
0.1119	0.2731.

30.5 In matrix form the equation is

$$\begin{bmatrix} 4 \\ 1 \\ 3 \end{bmatrix} [x] = \begin{bmatrix} 10 \\ 2 \\ 7 \end{bmatrix}.$$

The solution is $(A^T A)^{-1} A^T b$:

$$A^T A = \begin{bmatrix} 4 & 1 & 3 \end{bmatrix} \begin{bmatrix} 4 \\ 1 \\ 3 \end{bmatrix} = 26 \text{ and } (A^T A)^{-1}$$

$$= 1/26.$$

$$A^T b = \begin{bmatrix} 4 & 1 & 3 \end{bmatrix} \begin{bmatrix} 10 \\ 2 \\ 7 \end{bmatrix} = 63.$$

Hence the least-squares solution is $63/26 = 2.42$.

Problems

30.1 From the first equation

$$x = 4 - 2y + 3z/2 \tag{S30.1}$$

Substituting this in the other equations,

$$-y + z/2 = -1 \tag{S30.2}$$

and $6y + 6z = -9$ (S30.3)

From (S30.2),

$$y = 1 + z/2 \tag{S30.4}$$

and substituting this in (S30.3) gives

$$3z = -3. \tag{S30.5}$$

From (S30.5), $z = -1.0$, then from (S30.4) $y = 0.5$ and from (S30.1) $x = 1.5$.

30.2 The equations are expressed as

$$x_1 = \frac{19}{8} + \frac{x_2}{8} - \frac{x_3}{8} + \frac{x_4}{8} - \frac{x_5}{4}$$

$$x_2 = -2 + \frac{x_1}{4} + \frac{x_3}{4} + \frac{x_4}{4} + \frac{x_5}{4}$$

$$x_3 = 2 - \frac{x_1}{7} - \frac{2x_2}{7} + \frac{2x_4}{7} + \frac{x_5}{7}$$

$$x_4 = -\frac{11}{9} - \frac{2x_1}{9} + \frac{x_2}{9} + \frac{x_3}{9} - \frac{2x_5}{9}$$

$$x_5 = 3 + \frac{x_1}{5} + \frac{x_2}{5} + \frac{x_3}{5} + \frac{x_4}{5}.$$

A modified form of GS gives the output:

2.3750	− 1.4063	2.0625	− 1.6771	3.2708
0.9141	− 0.8574	2.1025	− 2.0139	3.0291
0.9960	− 0.9716	1.9926	− 2.0032	3.0028
1.0034	− 1.0011	1.9993	− 2.0016	3.0000

$$0.9998 \quad -1.0006 \quad 1.9998 \quad -2.0000 \quad 2.9998$$
$$1.0000 \quad -1.0001 \quad 2.0000 \quad -2.0000 \quad 3.0000$$
$$1.0000 \quad -1.0000 \quad 2.0000 \quad -2.0000 \quad 3.0000$$
$$1.0000 \quad -1.0000 \quad 2.0000 \quad -2.0000 \quad 3.0000$$

30.3 The coefficient matrix and right-hand-side vector are:

$$A = \begin{bmatrix} 1 & 2 \\ 2 & -1 \\ 5 & 1 \end{bmatrix} \text{ and } b = \begin{bmatrix} 5 \\ 0 \\ 6 \end{bmatrix}.$$

We find

$$A^T A = \begin{bmatrix} 30 & 5 \\ 5 & 6 \end{bmatrix},$$

$$(A^T A)^{-1} = \frac{1}{155} \begin{bmatrix} 6 & -5 \\ -5 & 30 \end{bmatrix} \text{ and } A^T b = \begin{bmatrix} 35 \\ 16 \end{bmatrix}$$

giving

$$\begin{bmatrix} x \\ y \end{bmatrix} = (A^T A)^{-1} A^T b = \frac{1}{155} \begin{bmatrix} 130 \\ 305 \end{bmatrix} = \begin{bmatrix} 26/31 \\ 61/31 \end{bmatrix}.$$

30.4 Inserting the approximate values on the left-hand side we find

$$F_1' = -0.609, \ F_2' = -0.092, \ F_3' = -1.206.$$

The equations to be solved are

$$-\sin x_1 \, \Delta x_1 - \sin x_2 \, \Delta x_2 = -0.012$$
$$-2 \sin 2x_1 \, \Delta x_1 - 2 \sin 2x_2 \, \Delta x_2 = -0.020$$
$$-3 \sin 3x_1 \, \Delta x_1 - 3 \sin 3x_2 \, \Delta x_2 = -0.031.$$

Substituting for the initial estimates of x_1 and x_2,

$$0.9490\Delta x_1 + 0.3817\Delta x_2 = 0.012$$
$$1.1969\Delta x_1 - 1.4111\Delta x_2 = 0.020$$
$$-1.7147\Delta x_1 + 2.7678\Delta x_2 = 0.031.$$

$$A^T A = \begin{bmatrix} 5.2734 & -6.0727 \\ -6.0727 & 9.7976 \end{bmatrix},$$

$$(A^T A)^{-1} = \frac{1}{14.789} \begin{bmatrix} 9.7976 & 6.0727 \\ 6.0727 & 5.2734 \end{bmatrix}$$

and $A^T b = \begin{bmatrix} -0.0178 \\ 0.0622 \end{bmatrix}.$

Hence

$$\begin{bmatrix} \Delta x_1 \\ \Delta x_2 \end{bmatrix} = (A^T A)^{-1} A^T b = \begin{bmatrix} 0.014 \\ 0.015 \end{bmatrix}.$$

The new estimates of x_1 and x_2 are 1.264 and 2.765, leading to

$$F_1' = -0.628, \ F_2' = -0.088, \ F_3' = -1.223.$$

The new F_2' actually gives a slightly greater discrepancy between the left and right-hand sides of the second equation but overall there is better agreement. The value of $\sum_{i=1}^{3} (F_i - F_i')^2$ is reduced from 1.505×10^{-3} to 8.21×10^{-4}.

CHAPTER 31

Exercises

31.1 A plot of the function (Figure S31.1) shows a solution near $x = 2$.

31.2 1.1142 using $x = 1/\sin x$ for the fixed-point iteration.

31.3 With $g(x) = 1/\sin(x)$ we have

$$g'(x_s) = -\frac{\cos(x_s)}{\sin^2(x_s)} = -0.763 \text{ with } x_s = 1$$

which satisfies the condition for convergence.

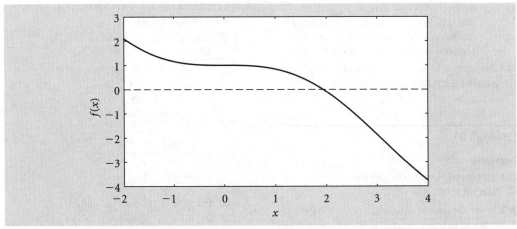

Figure S31.1 A plot of $1 + \sin x - x$.

31.4 Successive approximations are: 1.11473, 1.11416, 1.11416.

31.5 $f'(x_s) = \sin(1.11416) + 1.11416\cos(1.11416)$
$\qquad = 1.3889,$

$f''(x_s) = 2\cos(1.11416) - 1.11416\sin(1.11416)$
$\qquad = -0.1181.$

The starting error, 0.11416, is much less than $2|1.3889/0.1181|$.

Problems

31.1 We write the iteration equation as $x = -\sqrt{\frac{1}{3}x^3 e^x + 1}$. Note the minus sign in taking the square root since we know that the solution is close to -1. Testing for convergence,

$$g'(x) = -\frac{x^2 e^x + \frac{1}{3}x^3 e^x}{2\sqrt{\frac{1}{3}x^3 e^x + 1}}$$

giving $g'(-1) = -0.1309$. Hence the iteration equation is convergent. A program outline to find the solution follows.

Program SOLVE

Read in initial approximation x

Point A $xn = -\sqrt{\frac{1}{3}x^3 e^x + 1}$

  Print xn
  If $((xn - x)^2 < 10^{-9})$ go to B
  $x = xn$
Go to A
Point B **Print xn**
Stop

A version of this program gave the following successive approximations:
$\qquad -0.936\,68, \quad -0.944\,79, \quad -0.943\,78,$
$\qquad -0.943\,90, \quad -0.943\,89, \quad -0.943\,89.$

31.2 For the Newton–Raphson process we have

$$x_{n+1} = x_n - \frac{x_n^3 e^{x_n} + 3(1 - x_n^2)}{3x_n^2 e^{x_n} + x_n^3 e^{x_n} - 6x_n}.$$

The program **SOLVE** in Problem 31.1 can be modified by substituting this expression at point A. Successive approximations found are:
$-0.94538, \quad -0.94389, \quad -0.94389.$

CHAPTER 32

Exercises

32.3 $r[2] = r[-2] = 4$; $r[3] = r[-3] = 3$; $r[4] = r[-4] = 2$; $r[5] = r[-5] = 1$.

Problems

32.1 We are given that

$$r_{xy}[p] = \sum_{m=0}^{M-1} x[m]y[m+p] \tag{S32.1}$$

and

$$y[m] = s_1[m] + s_2[m] + s_3[m] + \cdots + s_N[m]. \tag{S32.2}$$

Substituting for $y[m]$ from (S32.2) into (S32.1):

$$r_{xy}[p] = \sum_{m=0}^{M-1} x[m]\{s_1[m+p] + s_2[m+p] + s_3[m+p] + \cdots + s_N[m+p]\}.$$

This can be rewritten as:

$$r_{xy}[p] = \sum_{m=0}^{M-1} x[m]s_1[m+p] + \sum_{m=0}^{M-1} x[m]s_2[m+p]$$
$$+ \sum_{m=0}^{M-1} x[m]s_3[m+p] + \cdots$$
$$+ \sum_{m=0}^{M-1} x[m]s_N[m+p].$$

Hence

$$r_{xy}[p] = r_{xs_1}[p] + r_{xs_2}[p] + r_{xs_3}[p] + \ldots + r_{xs_N}[p],$$

as required.

32.2 (i) We know that the incoming signal is given by

$$y[m] = s_1[m] + s_2[m] + s_3[m] + \cdots + s_N[m] + n[m].$$

This is cross-correlated with the template $\{x_1[m]\} = \{s_1[m]\}$. Now $r_{x_1 y}[p] = r_{s_1 y}[p]$ as $\{x_1[m]\} = \{s_1[m]\}$. Using the result of Problem 32.1, with $x = s_1$:

$$r_{s_1 y}[p] = r_{s_1 s_1}[p] + r_{s_1 s_2}[p] + r_{s_1 s_3}[p] + \cdots$$
$$+ r_{s_1 s_N}[p] + r_{s_1 n}[p].$$

Now assume that the noise does not correlate with the template $\{s_1[m]\}$ so that the last term $r_{s_1 n}[p] = 0$. In this case:

$$r_{s_1 y}[p] = r_{s_1 s_1}[p] + r_{s_1 s_2}[p] + r_{s_1 s_3}[p] + \cdots$$
$$+ r_{s_1 s_N}[p]. \tag{S32.3}$$

(ii) Now suppose that the cross-correlation between the signal transmitted by satellite i does not correlate with the signal transmitted by satellite j. In this case

$$r_{s_i s_j}[p] = 0 \ (i \neq j) \tag{S32.4}$$

Hence, in (S32.3), all the terms on the right-hand side vanish except for the first and we have:

$$r_{s_1 y}[p] = r_{s_1 s_1}[p]. \tag{S32.5}$$

This result is of great practical importance in GPS. In the GPS receiver, suppose that the signal $\{s_i[m]\}$ is transmitted by each satellite i. (S32.5) can be generalized as follows:

$$r_{s_iy}[p] = r_{s_is_i}[p]. \tag{S32.6}$$

Hence, by correlating each template against the incoming signal, we obtain as the output the autocorrelation function for that template. The time T_i when the autocorrelation function for template i reaches a maximum gives us the distance between the receiver and the ith satellite and by processing the delays from all N satellites in view, an estimation of the user's position can be made. The set of codes $\{s_i\}$ is specially designed so that for each code the autocorrelation function is as low in side lobes as possible, and the cross-correlation function between pairs of codes should be as close to zero as possible.

It should be noted that (S32.6) is derived assuming zero correlations between the templates from different satellites. In practice, for finite segments of code, this ideal cross-correlation (S32.4) can only be approximately obtained. Another assumption is that there is negligible cross-correlation between the templates from each satellite and noise. Again, for finite segments of code this is only approximately the case, but in practice this cross-correlation is small enough that it can be ignored.

32.3

$$y[m] = s[m] + n[m] \tag{S32.7}$$

(i) Now the autocorrelation function of y is given by the general expression (32.29) with x replaced by y:

$$r_{yy}[p] = \sum_{m=-\infty}^{\infty} y[m]y[m+p].$$

Substituting for y from (S32.7):

$$r_{yy}[p] = \sum_{m=-\infty}^{\infty} (s[m] + n[m])(s[m+p] + n[m+p]).$$

Expanding out the brackets:

$$r_{yy}[p] = \sum_{m=-\infty}^{\infty} s[m]s[m+p] + \sum_{m=-\infty}^{\infty} n[m]s[m+p]$$
$$+ \sum_{m=-\infty}^{\infty} s[m]n[m+p] + \sum_{m=-\infty}^{\infty} n[m]n[m+p]. \tag{S32.8}$$

Now, the first term is the autocorrelation function of s at lag p: $r_{ss}[p]$. The second and third terms are the cross-correlation functions between signal and noise which can be assumed to be zero. The last term is the autocorrelation function of noise, $r_{nn}[p]$. Referring to (32.55), this is zero for $p \neq 0$ and is non-zero for $p=0$. Hence

$$r_{yy}[p] = r_{ss}[p] \text{ for } p \neq 0$$

$$r_{yy}[0] = r_{ss}[0] + r_{nn}[0] \text{ for zero lag } p = 0.$$

(ii) $s[m] = s[m+N]$

Now, the autocorrelation function of $s[m]$ is defined according to (32.29) with x replaced by s:

$$r_{ss}[p] = \sum_{m=-\infty}^{\infty} s[m]s[m+p].$$

At lag $p+N$:

$$r_{ss}[p+N] = \sum_{m=-\infty}^{\infty} s[m]s[m+p+N] \tag{S32.9}$$

From the assumed periodicity of s:

$$s[m+p+N] = s[m+p] \tag{S32.10}$$

Substituting for $s[m+p+N]$ from (S32.10) into (S32.9):

$$r_{ss}[p+N] = \sum_{m=-\infty}^{\infty} s[m]s[m+p] \tag{S32.11}$$
$$= r_{ss}[p].$$

(iii) The solution to part (i) shows that, for non-zero lag, the autocorrelation function of the noisy signal y is equal to the autocorrelation function of the underlying signal s. The solution to part (ii) shows that if a signal has a period of N samples, then the autocorrelation function has the same periodicity – see (S32.11).

Hence, the autocorrelation function of the noisy signal y is, for non-zero lag, identical to the autocorrelation function of the underlying signal s and has the same period as s. Therefore, in theory, by calculating the autocorrelation function of y and looking at its periodicity, we can determine the period of the underlying signal s; this is despite the fact that noise could be so large in power that we do not know what the underlying signal looks like. In practice, because the data are of finite duration, this result will only approximately hold.

CHAPTER 33

Exercises

33.1 (i) (a) $0 \leq p \leq 4$ and $196 \leq p \leq 199$; (b) $4 \leq p \leq 40$ and $160 \leq p \leq 196$.

(ii) $p = 10, 20, 30, 170, 180,$ and 190.

33.2 (i) From (33.9), the signal at the nth sample is given by

$$x[n] = A\cos\left(2\pi f \frac{n}{f_s}\right).$$

Substituting $f = 50$ and $f_s = 120$ we obtain for the nth sample of x:

$$x[n] = A\cos\left(2\pi(50)\frac{n}{120}\right) = A\cos\left(\frac{5\pi n}{6}\right).$$

Now replace $n \rightarrow n - 10$:

$$x[n - 10] = A\cos\left(\frac{5\pi(n - 10)}{6}\right)$$

$$= A\cos\left(\frac{5\pi n}{6} - \frac{25\pi}{3}\right).$$

Comparing the above two expressions we can see that $x[n] \neq x[n - 10]$. Hence, for this choice of sampling frequency, $y[n] = x[n] - x[n - 10] \neq 0$ and the mains frequency is not eliminated.

(ii) Putting $f = 60$ Hz and $f_s = 120$ Hz in (33.9),

$$x[n] = A\cos\left(2\pi(60)\frac{n}{120}\right) = A\cos(\pi n).$$

Hence

$$x[n - 10] = A\cos(\pi(n - 10)) = A\cos(\pi n - 10\pi)$$
$$= A\cos(\pi n).$$

For this choice of sampling frequency $y[n] = x[n] - x[n - 10] = 0$ and in the United States the mains frequency would be eliminated by this filter.

Problems

33.1 (i) In §33.2, we had the following parameters when using the DFT method of filtering: $N = 250$, $f = 50$ Hz, $f_s = 500$ Hz.

If we carry out a DFT on the data, then from (33.2) a peak will occur in the spectrum at the following value of p:

$$p = \frac{Nf}{f_s} = \frac{250 \times 50}{500} = 25.$$

Thus, the mains interference occurs at $p = 25$ and is *not present for any other values of p*. Hence, by putting the DFT to zero for this value of p, we are eliminating *all the mains energy*.

Now, let us look what happens when we increase the number of samples to 256. In this case, from (33.2),

$$p = \frac{Nf}{f_s} = \frac{256 \times 50}{500} = 25.6.$$

We can now see that the value of p is *real* rather than *integer*. From our discussion in §18.2.4, the non-integer value of p implies that *leakage* is taking place. Put another way the data interval does not contain an integer number of cycles of the mains interference. Thus, mains energy now occurs at several values of p as demonstrated in Figure S33.1.

Comparing Figure S33.1 with Figure 33.2 it can be seen that when N is increased to 256, the mains peak is now spread across several values of p. This causes a problem when filtering off the mains. For example, suppose that we filter from 45 to 55 Hz, which, from Figure S33.1, appears to contain most of the mains peak. From (33.2), the 45 Hz peak corresponds to a value $p = (256 \times 45)/500 = 23.04$, which can be rounded down to 23. The peak at 55 Hz corresponds to a value of p given by $p = (256 \times 55)/500 = 28.16$, which can be rounded down to 28. If we carry out the DFT filtering procedure on the data, putting to zero the complex DFT for $23 \leq p \leq 28$ and for $228 \leq p$ 233, then Figure S33.2 gives the resulting filtered signal (full line) compared with the original signal (dashed line).

We can see that, although mains frequency has been reduced, the signal looks very distorted, particularly at the ends. The signal itself has significant frequency components between 45 and 55 Hz and these have been eliminated along with the mains. Also, due to leakage, the mains frequency has not been completely eliminated, which was the case when N was 250 samples when the mains occurred at a single integer value of p in the spectrum.

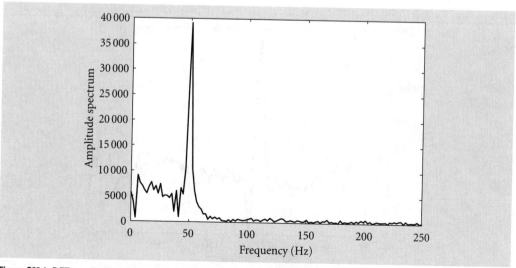

Figure S33.1 DFT amplitude spectrum of signal in Figure 33.1, with $N = 256$.

(ii) If we wish to eliminate an interfering frequency f_i then, in order for leakage not to occur, and for the peak to occur at one value of p only:

$$\frac{Nf_i}{f_s} = \text{integer}.$$

33.2 (i) The impulse response, $h(t)$, is found from the frequency response, $H(f)$ from the following inverse Fourier transform relation:

$$h(t) = \int_{-\infty}^{\infty} H(f) \exp(i2\pi ft) df. \qquad (S33.1)$$

From the figure in problem 33.2, the integrand is 1 for the following frequency ranges: $-200\ \text{Hz} \le f \le -100\ \text{Hz}$, $100\ \text{Hz} \le f \le 200\ \text{Hz}$ and the integrand is zero for frequencies outside these ranges. Hence, from (S33.1):

$$h(t) = \int_{-200}^{-100} \exp(i2\pi ft) df + \int_{100}^{200} \exp(i2\pi ft) df. \qquad (S33.2)$$

Integrating these expressions,

$$h(t) = \left[\frac{\exp(i2\pi ft)}{i2\pi t}\right]_{-200}^{-100} + \left[\frac{\exp(i2\pi ft)}{i2\pi t}\right]_{100}^{200}. \qquad (S33.3)$$

Putting in the limits,

$$h(t) = \frac{\begin{array}{c}\exp(-i2\pi(100)t) - \exp(-i2\pi(200)t) \\ + \exp(i2\pi(200)t) - \exp(i2\pi(100)t)\end{array}}{i2\pi t}. \qquad (S33.4)$$

Regrouping the terms in the numerator:

$$h(t) = \frac{\begin{array}{c}[\exp(i2\pi(200)t) - \exp(-i2\pi(200)t)] \\ -[\exp(i2\pi(100)t) - \exp(-i2\pi(100)t)]\end{array}}{i2\pi t}. \qquad (S33.5)$$

Using the identity

$$\sin(x) = \frac{1}{2i}[\exp(ix) - \exp(-ix)] \qquad (S33.6)$$

we can rewrite (S33.5) as

$$h(t) = \frac{2i\sin(2\pi(200)t) - 2i\sin(2\pi(100)t)}{i2\pi t} \qquad (S33.7)$$

$$= \frac{\sin(2\pi(200)t) - \sin(2\pi(100)t)}{\pi t}.$$

This expression can also be rewritten as

$$h(t) = h_1(t) - h_2(t) \qquad (S33.8)$$

where

$$h_1(t) = \frac{\sin(2\pi(200)t)}{\pi t} \qquad (S33.9)$$

and

$$h_2(t) = \frac{\sin(2\pi(100)t)}{\pi t}. \qquad (S33.10)$$

Comparing (S33.9) and (S33.10) with (33.29), we can see that $h_1(t)$ is the impulse response of a low-pass filter with cut-off of 200 Hz and $h_2(t)$ is the impulse response of a low-pass filter with cut-off of 100 Hz. In other words, the output from a bandpass filter with lower and upper cut-offs of 100 Hz and 200 Hz can be obtained by finding the output from a low-pass filter with cut-off of 200 Hz and subtracting the output from a low-pass filter with cut-off of 100 Hz. To illustrate this, the frequency response of an (ideal) low-pass filter with zero phase response and cut-off 200 Hz is shown in Figure S33.3 The frequency response of the filter with a cut-off of 100 Hz is as shown in Figure S33.4.

If we subtract the frequency response in Figure S33.4 from the frequency response in Figure S33.3, we obtain the bandpass filter's frequency response as shown in the figure in problem 33.2.

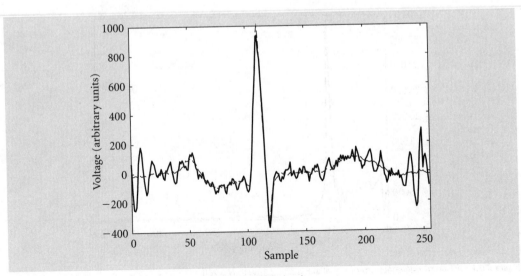

Figure S33.2 Filtered (full curve) and original (dashed curve) ECG signals.

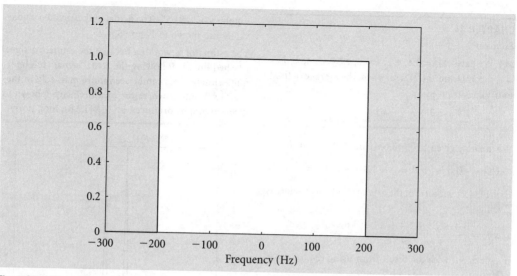

Figure S33.3 Frequency response for low-pass filter with cut-off of 200 Hz.

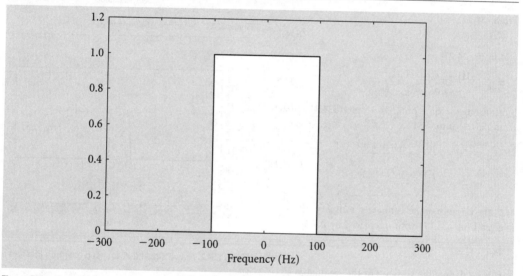

Figure S33.4 Frequency response for low-pass filter with cut-off of 100 Hz.

The pth sample corresponds to a time p/f_s where $f_s = 500$ Hz is the sampling frequency. Hence, making the replacement $t \rightarrow p/500$ in (S33.7), we obtain for the pth sample of the amplitude response:

$$h[p] = \frac{\sin(2\pi(200)(p/500)) - \sin(2\pi(100)(p/500))}{(\pi p/500)}$$

or

$$h[p] = 500\frac{[\sin((4\pi p/5)) - \sin((2\pi p/5))]}{\pi p}. \quad \text{(S33.11)}$$

When evaluating $h[0]$, if we put $p = 0$ in (S33.11), we obtain 0/0 which is indeterminate. Using l'Hôpital's rule,

$$h[0] = 500 \lim_{p \to 0} \left\{ \frac{[\sin(4\pi p/5) - \sin(2\pi p/5)]}{\pi p} \right\}$$

$$= 500 \lim_{p \to 0} \left\{ \frac{(d/dp)[\sin(4\pi p/5) - \sin(2\pi p/5)]}{(d/dp)(\pi p)} \right\}.$$

Carrying out the differentials,

$$h[0] = 500 \lim_{p \to 0} \left\{ \frac{(4\pi/5)\cos(4\pi p/5) - (2\pi/5)\cos(2\pi p/5)}{\pi} \right\}.$$

$$= 500 \left[\frac{(4\pi/5) - (2\pi/5)}{\pi} \right] = 200.$$

CHAPTER 34

Exercises

34.1 We have $z[1] = 3.4$, $z[2] = 3.3$, $z[3] = 3.4$, and $z[4] = 3.1$. Using (34.2) with $n = 4$, the average of these four values is given by

$$\tilde{x}[4] = \frac{3.4 + 3.3 + 3.4 + 3.1}{4} = 3.3.$$

We are now given $z[5] = 3.6$. Using (34.6) with $n = 5$:

$$\tilde{x}[5] = \frac{4}{5}\tilde{x}[4] + \frac{1}{5}z[5] = \frac{4}{5} \times 3.3 + \frac{1}{5} \times 3.6 = 3.36.$$

Using (34.2) with $n = 5$, the average of the five values is given by

$$\tilde{x}[5] = \frac{3.4 + 3.3 + 3.4 + 3.1 + 3.6}{5} = 3.36.$$

in agreement with the result from using (34.6).

34.2 We have from (34.20) with $n = 4$:

$$r[4] = z[4] - \hat{x}[4]. \tag{S34.1}$$

From (34.24),

$$\hat{x}[4] = \tilde{x}[3] \tag{S34.2}$$

and from (34.2)

$$\tilde{x}[3] = \frac{z[1] + z[2] + z[3]}{3}. \tag{S34.3}$$

Substituting for $\tilde{x}[3]$ from (S34.3) into (S34.2) and $\hat{x}[4]$ from (S34.2) into (S34.1):

$$r[4] = z[4] - \frac{(z[1] + z[2] + z[3])}{3}$$

$$= \frac{3z[4] - z[1] - z[2] - z[3]}{3}. \tag{S34.4}$$

34.3 The variance of the estimate at sample point $n - 1$ is given from (34.17) with n replaced by $n - 1$:

$$\tilde{P}[n - 1] = \frac{\sigma^2}{n - 1}$$

and, if we denote the variance of the residual as σ_r^2, then summing the variances of $\delta\tilde{x}[n - 1]$ and $\delta z[n]$, as they are uncorrelated:

$$\sigma_r^2 = \sigma^2 + \tilde{P}[n - 1] = \sigma^2 + \frac{\sigma^2}{n - 1} = \frac{n}{n - 1}\sigma^2.$$

Problems

Note that in the following solutions the graphs shown are not the graphs produced by MATLAB. The graphs shown here are in black and white and, for clarity, show the actual (full line) and the estimated (dashed line) signals.

34.1 When the jump detector and estimator are working together, the sort of results we obtain are shown in Figure S34.1. This graph shows that the estimator plus jump detector, Algorithm 34.2, does indeed detect the start and end of the jumps at sample numbers 45 and 55. The actual and estimated signal values are shown in Figure S34.2.

The estimator is working because the estimated signal (dashed line) is following the actual signal (full line) quite closely. This can be seen more markedly in the MATLAB output where the measurements (shown in green in graphs produced by MATLAB), look noisy

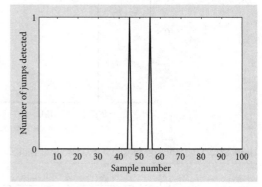

Figure S34.1 Detection of jumps in data.

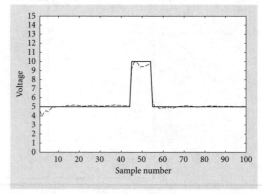

Figure S34.2 Actual signal (full line) and estimate (dashed line).

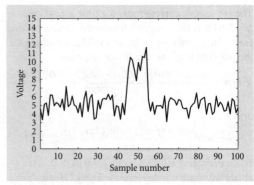

Figure S34.3 Noisy data.

compared to the estimator; the measurements are shown in Figure S34.3.

It should be emphasized that the above results are for one set of random numbers produced by MATLAB. Subsequent runs will produce different results depending on the set of random numbers used. The above should be regarded as typical results. Occasionally, you may notice that false jumps are detected at sample points other than at 45 and 55. Why do you think that happens? (Hint: From the section on the Gaussian distribution, what is the probability that a noise voltage is greater than three times the standard deviation?)

34.2 SDASS is the *assumed* standard deviation of the measurement noise. In practice, we do not know exactly what the noise standard deviation is, but we can estimate it, for example analysing parts of the measurements that consist of noise only and no clean signal. If SDASS, the noise standard deviation, is overestimated by the user, then, for a given threshold parameter k, (34.28), the threshold to detect a jump is increased. The danger here is that genuine jumps in the underlying signal are treated as noise by the algorithm, because the magnitude of the residual does not exceed the threshold in (34.28). For example, suppose that SDASS is increased to 2, with the other parameters kept the same as in Problem 34.1. Figures S34.4 and S34.5 show typical results for the number of jumps detected and the estimated and actual voltages.

It can be seen that the threshold in (34.28) is so large, because of the increase in SDASS, that the jumps at 45 and 55 have not been detected and the estimated signal, shown dashed in Figure S34.5, no longer follows the measurement. If, on the other hand, SDASS is decreased to below its actual value, then the threshold on the right-hand side of (34.28) decreases and spurious noise is detected as jumps and we obtain a lot of false detections. Typical results for the number of jumps detected and estimated and actual voltages

follow in Figures S34.6 and S34.7 for SDASS $= 0.5$. It can now be seen that several spurious jumps are detected. Whenever a jump is detected the estimator re-initializes and this compromises the ability of the estimator to smooth out the effects of noise.

34.3 If all parameters are kept the same as in Problem 34.1, except that now the threshold parameter k in (34.28) is increased, then the effect is the same as

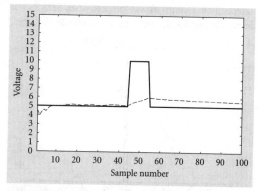

Figure S34.5 Actual signal (full line) and estimate (dashed line).

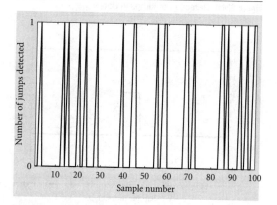

Figure S34.6 Detection of jumps in data.

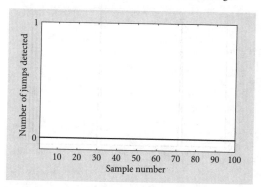

Figure S34.4 Detection of jumps in data

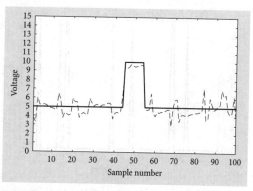

Figure S34.7 Actual signal (full line) and estimate (dashed line).

increasing SDASS, namely that the threshold is increased and the probability is increased that jumps will not be detected. An example is shown in Figures S34.8 and S34.9 for $k = 5$. The jump at sample point 45 is detected too late, at sample point 50, missing the jump at 55, and the estimated signal after the jump does not follow the actual signal.

Decreasing k to say 2 increases the probability of detecting false jumps as was the case when SDASS was decreased and for the same reasons, namely that the

threshold on the right hand side of (34.28) decreases. The results in this case are shown in Figures S34.10 and S34.11. Each time a jump is detected, the estimator is re-initialized, resulting in less smoothing of the estimator and more of a tendency for the estimate to follow the noise.

34.4 An example of the performance of the estimator plus jump detection algorithm (Algorithm 34.2) is shown in Figures S34.12 and S34.13 where the underlying signal is constant with no jumps. In this example, where the signal is constant with no jumps, ideally no jumps should be detected. However, the threshold is low enough so that spurious noise values are causing the residual to cross either the upper or lower thresholds. Hence the estimator is re-initializing itself unnecessarily producing an inferior estimate of the underlying signal.

You should be able to verify that increasing k in (34.28) to 3 reduces the chances of the residual crossing either threshold resulting in an improved estimate of the underlying signal.

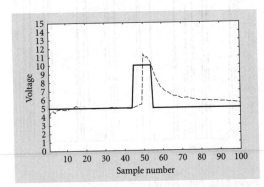

Figure S34.8 Detection of jumps in data.

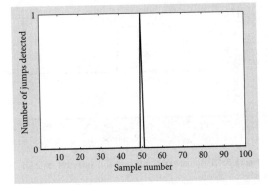

Figure S34.9 Actual signal (full line) and estimate (dashed line).

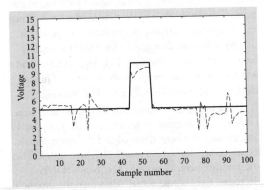

Figure S34.11 Actual signal (full line) and estimate (dashed line).

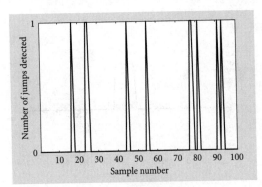

Figure S34.10 Detection of jumps in data.

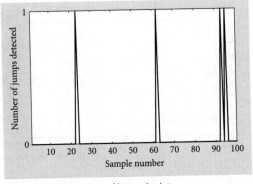

Figure S34.12 Detection of jumps in data.

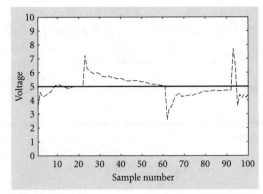

Figure S34.13 Actual signal (full line) and estimate (dashed line).

CHAPTER 35

Exercises

35.1 Let the number of bunches sold be x. Then the constraints are:

for red carnations: $5x \leq 300$
for white carnations: $5x \leq 500$

giving $x \leq 60$. The objective function that must be maximized is

$$I = 0.20(300 - 5x) + 0.16(500 - 5x) + 5x \times 1.90$$
$$= 140 + 0.1x.$$

Income is maximized by having x as large as possible, i.e. $x = 60$, giving an income of £146.

35.2 Introduce slack variables s_1 and s_2. Then

$$s_1 = 300 - 5x, \text{ giving } x \leq 60 \qquad (\text{S35.1})$$
$$s_2 = 500 - 5x, \text{ giving } x \leq 100.$$

The objective function is $I = 140 + 0.1x$
Write (S35.1) as $x = 60 - \frac{1}{5}s_1$, which gives

$s_2 = 200 + s_1$ and

$$I = 140 + 0.1\left(60 - \frac{1}{5}s_1\right) = 146 - \frac{1}{50}s_1.$$

Since the coefficient of s_1 is negative, the maximum income is £146.

Problems

35.1 With slack variables s_1, s_2, and s_3 for red, yellow, and blue and quantities x, y, and z for cloths A, B, and C respectively we have, with basic variables s_1, s_2, and s_3

$$s_1 = 2000 - 0.25x - 0.5y - 0.25z \qquad (\text{S35.2a})$$
$$s_2 = 2000 - 0.5x - 0.25y - 0.25z \qquad (\text{S35.2b})$$
$$s_3 = 1000 - 0.25x - 0.25y - 0.5z \qquad (\text{S35.2c})$$

with the income function to be maximized

$$I = 10x + 12y + 15z + 6(5000 - x - y - z)$$
$$= 30000 + 4x + 6y + 9z.$$

From the coefficients of the income function we increase z by the maximum possible. The tightest constraint on z is given by (S35.2c), which requires $z \leq 2000$. Moving z into the set of basic variables, (S35.2c) becomes

$$z = 2000 - 0.5x - 0.5y - 2s_3 \qquad (\text{S35.3})$$

and substituting this value of z in (S35.2a), (S35.2b), and the income function we find

$$s_1 = 1500 - 0.125x - 0.375y + 0.5s_3 \qquad (\text{S35.3a})$$
$$s_2 = 1500 - 0.375x - 0.125y + 0.5s_3 \qquad (\text{S35.3b})$$
$$I = 48000 - 4.5x + 1.5y - 18s_3.$$

It is clear from the income function that there is only advantage in increasing y and the maximum increase is limited equally by (S35.3a) and (S35.3c), both of which require $y \leq 4000$. From (S35.3a) we have

$$y = 4000 - x/3 + 4s_3/3 - 8s_1/3 \qquad (\text{S35.4a})$$
$$s_2 = 1000 - 5x/12 - s_3/6 + s_1/3 \qquad (\text{S35.4b})$$
$$z = -x/3 - 8s_3/3 + 4s_1/3 \qquad (\text{S35.4c})$$
$$I = 54000 - 5x - 16s_3 - 4s_1.$$

From the coefficients of the income function it is evident that the maximum income is £54 000 and from (S35.4a), (S35.4b) and (S35.4c) we find $y = 4\,000$, $x = z = 0$, and $s_2 = 1\,000$. Thus only cloth B is made and $1\,000\,\mathrm{kg}$ of yellow yarn remains.

35.2 The functions in GRADMAX are modified to give

$$phi(x, y) = 3\cos\{2\pi(x + y + 0.3)\}$$
$$+ 2\sin\{2\pi(x + 3y - 0.1)\}$$

$$dphix(x, y) = -6\pi\sin\{2\pi(x + y + 0.3)\}$$
$$+ 4\pi\cos\{2\pi(x + 3y - 0.1)\}$$

$$dphiy(x, y) = -6\pi\sin\{2\pi(x + y + 0.3)\}$$
$$+ 12\pi\cos\{2\pi(x + 3y - 0.1)\}.$$

The solution is found in 53 steps. The output every 5 steps and for the final step is shown below.

5	0.309 772	0.628 346	1.345 948
10	0.156 593	0.722 742	3.263 695
15	0.041 711	0.723 211	4.041 654
20	0.001 536	0.793 760	4.435 740
25	−0.056 186	0.788 223	4.871 764
30	−0.075 192	0.809 080	4.932 084
35	−0.095 739	0.805 162	4.958 720
40	−0.101 160	0.821 052	4.970 919
45	−0.113 244	0.812 888	4.976 189
50	−0.112 839	0.826 361	4.978 773
53	−0.119 995	0.815 897	4.979 398

35.3 Let the radius be r and the height h. Then the function to be minimized is

$$f(r, h) = 2\pi r^2 + 2\pi rh$$

under the constraint that $g(r, h) = \pi r^2 h - 1000 = 0$, where dimensions are in cm. We have

$$\frac{\partial f}{\partial r} = 4\pi r + 2\pi h, \quad \frac{\partial f}{\partial h} = 2\pi r, \quad \frac{\partial g}{\partial r} = 2\pi rh \text{ and}$$

$$\frac{\partial g}{\partial h} = \pi r^2.$$

The equations to be solved, with a factor π removed where appropriate, are:

$$4r + 2h = 2\lambda rh \qquad \text{(S35.5a)}$$
$$2r = \lambda r^2 \qquad \text{(S35.5b)}$$
$$\pi r^2 h - 1000 = 0. \qquad \text{(S35.5c)}$$

From (S35.5a) and (S35.5b) we find $r = h/2$ and then from (S35.5c) that

$$r = \left(\frac{500}{\pi}\right)^{1/3} \text{cm} = 5.42\,\text{cm} \quad \text{and} \quad h = 2r$$

$$= 10.84\,\text{cm}.$$

CHAPTER 36

Exercises

36.1 (i) $\frac{s}{s^2+16}$; (ii) $\frac{s}{s^2-16}$; (iii) $\frac{6}{s^4}$; (iv) $\frac{6}{(s+2)^4}$;

(v) $\frac{s+1}{(s+1)^2-1}$; (vi) $\frac{s+2}{s^2+4}$.

36.2 (i) 4; (ii) e^{2t}; (iii) $\delta(t) - 4e^{-4t}$; (iv) $6\sin t$;

(v) $\frac{1}{2}t^2 e^{-2t}$; (vi) $4t\sin t + \frac{1}{2}t^2 e^{-t}$.

36.3 $e^{-2t} - e^{-3t}$.

36.4 $(s-1)^2 F(s) + (2-s)f(0) - f'(0)$.

36.5 $\frac{1}{2}e^t \sin 2t$.

36.6 $1 + e^{-t} - 2e^{-t/2}$.

36.7 $2 - e^{-t} - e^{-t/2}$.

Problems

36.1 (i) $\cos(a+t)e^{-t} = \cos a \cos te^{-t} - \sin a \sin te^{-t}$

$$\mathcal{L}\{\cos(a+t)e^{-t}\} = \cos a \frac{s+1}{(s+1)^2+1} - \sin a \frac{1}{(s+1)^2+1}$$

$$= \frac{(s+1)\cos\alpha - \sin\alpha}{(s+1)^2+1}.$$

(ii) $\cosh t \sinh t = \frac{1}{2}\sinh 2t$. Hence $\mathcal{L}(\cosh t \sinh t) = 1/(s^2 - 4)$.

(iii) $\mathcal{L}(\cos^2 t) = \mathcal{L}\left(\frac{1}{2} + \frac{1}{2}\cos 2t\right) = \frac{1}{2s} + \frac{s}{2(s^2+4)}$.

36.2

(i) $\mathcal{L}\left(\frac{1}{s(s+1)(s+2)}\right) = \mathcal{L}\left(\frac{1}{2s} - \frac{1}{s+1} + \frac{1}{2(s+2)}\right)$

$$= \frac{1}{2} - e^{-t} + \frac{1}{2}e^{-2t}.$$

(ii) $\mathcal{L}\left(\frac{1}{s^4 - b^4}\right) = \mathcal{L}\left\{\frac{1}{2b^2}\left(\frac{1}{s^2 - b^2} - \frac{1}{s^2 + b^2}\right)\right\}$

$$= \frac{1}{2b^3}(\sinh bt - \sin bt).$$

(iii) $\mathcal{L}\left(\frac{s^2}{s^4 - b^4}\right) = \mathcal{L}\left\{\frac{1}{2}\left(\frac{1}{s^2 - b^2} + \frac{1}{s^2 + b^2}\right)\right\}$

$$= \frac{1}{2b}(\sinh bt + \sin bt).$$

36.3 Taking the Laplace transform of both sides,

$$(s-1)^2 X(s) - (s-1) = \frac{s-1}{s^2+1}$$

giving

$$X(s) = \frac{1}{(s-1)(s^2+1)} + \frac{1}{s-1}$$

$$= -\frac{1}{2}\frac{s}{s^2+1} - \frac{1}{2}\frac{1}{s^2+1} + \frac{3}{2}\frac{1}{s-1}.$$

Hence $x(t) = -\frac{1}{2}\cos t - \frac{1}{2}\sin t + \frac{3}{2}e^t$.

36.4 Taking Laplace transforms of both sides,

$$3sY(s) - 3y(0) + Y(s) = U(s).$$

With $y(0) = 0$, this gives $(3s+1)Y(s) = U(s)$ or $G(s) = 1/(3s+1)$. For

$$u(t) = e^{-t}\sin t \text{ we have } U(s) = \frac{1}{(s+1)^2 + 1}$$

so that

$$Y(s) = \frac{1}{\{(s+1)^2 + 1\}(3s+1)}$$

$$= \frac{1}{13}\left(\frac{9}{3s+1} - \frac{3s+5}{(s+1)^2+1}\right)$$

$$= \frac{3}{13}\frac{1}{s+\frac{1}{3}} - \frac{3}{13}\frac{s+1}{(s+1)^2+1} - \frac{2}{13}\frac{1}{(s+1)^2+1}.$$

Hence

$$y(t) = \frac{3}{13}e^{-t/3} - \frac{3}{13}e^{-t}\cos t - \frac{2}{13}e^{-t}\sin t.$$

36.5 The individual transfer functions are $1/(3s+1)$ and $1/(s+3)$. The transfer function for the series combination is

$$G(s) = \frac{1}{(3s+1)(s+3)} = \frac{3}{8(3s+1)} - \frac{1}{8(s+3)}.$$

For the input $U(s) = 1/s$ and hence

$$Y(s) = \frac{3}{8}\frac{1}{s(3s+1)} - \frac{1}{8}\frac{1}{s(s+3)}$$

$$= \frac{3}{8}\left(\frac{1}{s} - \frac{3}{3s+1}\right) - \frac{1}{24}\left(\frac{1}{s} - \frac{1}{s+3}\right)$$

$$= \frac{1}{3s} - \frac{3}{8}\frac{1}{s+\frac{1}{3}} + \frac{1}{24}\frac{1}{s+3}.$$

Hence

$$y(t) = \frac{1}{3} - \frac{3}{8}e^{-t/3} + \frac{1}{24}e^{-3t}.$$

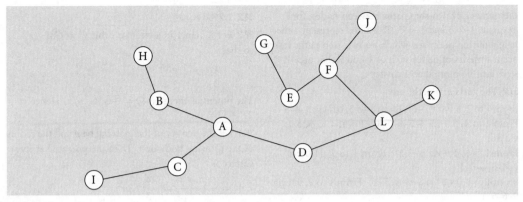

Figure S37.1

CHAPTER 37

Exercises

37.1 A possible best-route system from L is shown in Figure S37.1, but there are other possibilities.

37.2 21.

37.3

$$M_{CB}M_{BA} = \begin{bmatrix} 0 & 0 & 1 & 1 \\ 0 & 1 & 1 & 0 \\ 1 & 0 & 1 & 0 \\ 1 & 1 & 0 & 1 \end{bmatrix} \begin{bmatrix} 1 & 1 & 0 & 1 & 0 \\ 0 & 1 & 0 & 0 & 1 \\ 1 & 0 & 1 & 0 & 1 \\ 0 & 1 & 1 & 1 & 0 \end{bmatrix}$$

$$= \begin{bmatrix} 1 & 1 & 2 & 1 & 1 \\ 1 & 1 & 1 & 0 & 2 \\ 2 & 1 & 1 & 1 & 1 \\ 1 & 3 & 1 & 2 & 1 \end{bmatrix}.$$

37.4 The tetrahedron, with numbered indices, is shown in Figure S37.2.

$$M_{tetrahedron} = \begin{bmatrix} 0 & 1 & 1 & 1 \\ 1 & 0 & 1 & 1 \\ 1 & 1 & 0 & 1 \\ 1 & 1 & 1 & 0 \end{bmatrix} \quad \text{and}$$

$$M_{tetrahedron}^2 = \begin{bmatrix} 3 & 2 & 2 & 2 \\ 2 & 3 & 2 & 2 \\ 2 & 2 & 3 & 2 \\ 2 & 2 & 2 & 3 \end{bmatrix}.$$

The 3s represent the number of connected nodes and the 2s the number of two-step pathways between nodes.

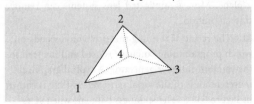

Figure S37.2

Problems

37.1 The numbered cube is shown in Figure S37.3. We have

$$M_{cube} = \begin{bmatrix} 0 & 1 & 0 & 1 & 1 & 0 & 0 & 0 \\ 1 & 0 & 1 & 0 & 0 & 1 & 0 & 0 \\ 0 & 1 & 0 & 1 & 0 & 0 & 1 & 0 \\ 1 & 0 & 1 & 0 & 0 & 0 & 0 & 1 \\ 1 & 0 & 0 & 0 & 0 & 1 & 0 & 1 \\ 0 & 1 & 0 & 0 & 1 & 0 & 1 & 0 \\ 0 & 0 & 1 & 0 & 0 & 1 & 0 & 1 \\ 0 & 0 & 0 & 1 & 1 & 0 & 1 & 0 \end{bmatrix} \quad \text{and}$$

$$M_{cube}^2 = \begin{bmatrix} 3 & 0 & 2 & 0 & 0 & 2 & 0 & 2 \\ 0 & 3 & 0 & 2 & 2 & 0 & 2 & 0 \\ 2 & 0 & 3 & 0 & 0 & 2 & 0 & 2 \\ 0 & 2 & 0 & 3 & 2 & 0 & 2 & 0 \\ 0 & 2 & 0 & 2 & 3 & 0 & 2 & 0 \\ 2 & 0 & 2 & 0 & 0 & 3 & 0 & 2 \\ 0 & 2 & 0 & 2 & 2 & 0 & 3 & 0 \\ 2 & 0 & 2 & 0 & 0 & 2 & 0 & 3 \end{bmatrix}.$$

The 3s on the diagonal indicate the number of other nodes to which each node is linked. The 2s represent

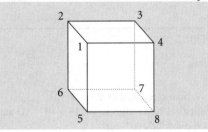

Figure S37.3

the number of two-step paths between nodes, for example, 1–2–3 and 1–4–3. The zeros represent either neighbouring nodes for which no two-step paths can occur or pairs of nodes such as 1 and 7 that are separated by more than two steps.

37.2 The costs and paths are:

From 1 to 2, 4.0 via 2–1   From 1 to 3, 6.0 via 3–6–5–1
From 1 to 4, 8.0 via 4–3–6–5–1   From 1 to 5, 3.0 via 5–1
From 1 to 6, 5.0 via 6–5–1   From 1 to 7, 11.0 via 7–4–3–6–5–1
From 1 to 8, 14.0 via 8–9–6–5–1   From 1 to 9, 9.0 via 9–6–5–1
From 1 to 10, 12.0 via 10–9–6–5–1   From 1 to 11, 14.0 via 11–12–9–6–5–1
From 1 to 12, 11.0 via 12–9–6–5–1
From 7 to 1, 11.0 via 1–5–6–3–4–7   From 7 to 2, 9.0 via 2–3–4–7
From 7 to 3, 5.0 via 3–4–7   From 7 to 4, 3.0 via 4–7
From 7 to 5, 8.0 via 5–6–3–4–7   From 7 to 6, 6.0 via 6–3–4–7
From 7 to 8, 10.0 via 8–9–10–7   From 7 to 9, 5.0 via 9–10–7
From 7 to 10, 2.0 via 10–7   From 7 to 11, 10.0 via 11–12–9–10–7
From 7 to 12, 7.0 via 12–9–10–7
From 12 to 1, 11.0 via 1–5–6–9–12   From 12 to 2, 11.0 via 2–3–6–9–12
From 12 to 2, 7.0 via 3–6–9–12   From 12 to 4, 9.0 via 4–3–6–9–12
From 12 to 5, 8.0 via 5–6–9–12   From 12 to 6, 6.0 via 6–9–12
From 12 to 7, 7.0 via 7–10–9–12   From 12 to 8, 7.0 via 8–9–12
From 12 to 9, 2.0 via 9–12   From 12 to 10, 5.0 via 10–9–12
From 12 to 11, 3.0 via 11–12

37.3 The critical path is the same for the expected, worst, and best timings. The times are
$t_e = 116$ days, $t_b = 100$ days, and $t_w = 148$ days. Rounded up, this gives

$$t_p = \frac{148 + 100 + 4 \times 116}{6} = 119 \text{ days.}$$

SOLUTIONS 38

Exercises

38.1

(i) Io's orbital speed 3.7410 AU year^{-1}.
(ii) Europa's orbital radius is 4.4779×10^{-3} AU and orbital speed 2.9692 AU year^{-1}.

38.2 1.7588×10^{10} s^{-1}.

38.3 For a planet in a circular orbit $v^2 = GM_\odot/r$ so that

$$T = \frac{1}{2}mv^2 = \frac{GM_\odot m}{2r}.$$

The potential energy is $\Omega = -GM_\odot m/r$. Hence $2T + \Omega = 0$.

38.4 Let the density of the material be ρ and the radius of the growing body be r. Then the potential at its surface is

$$V = -\frac{\frac{4}{3}\pi\rho r^3 G}{r} = -\frac{4}{3}\pi G\rho r^2.$$

The mass of a shell of radius r and thickness dr is $4\pi\rho r^2 \mathrm{d}r$ so that the work done *against the field* in bringing this material from infinity and creating the shell is

$$\mathrm{d}\Omega = -\frac{16}{3}G\pi^2\rho^2 r^4 \mathrm{d}r.$$

Hence the total gravitational potential energy of the sphere is

$$\Omega = -\frac{16}{3}G\pi^2\rho^2 \int_0^R r^4 \mathrm{d}r = -\frac{16}{15}G\pi^2\rho^2 R^5 = -\frac{3}{5}\frac{GM^2}{R}$$

since $M = \frac{4}{3}\pi\rho R^3$.

38.5 (i) 1.7840×10^{14} s^{-1}; (ii) 3.5219×10^{-14} s; (iii) 6.901×10^{-9} m.

Problems

38.1 The graphs for the three simulations are shown in Figure S38.1. With the enhanced Europa mass Io's orbit is perturbed away from being circular. In the actual system this happens over a much longer timescale and to a much lesser extent, the eccentricity of Io's orbit being of order 0.0001. The perturbation is enhanced by the 2 : 1 relationship in the period that gives a closest approach of the satellites repeatedly at the same point of Io's orbit. This resonance also affects Europa's orbit. The slight eccentricity in Io's orbit gives a cyclic variation in the tidal stretching due to Jupiter and leads to hysteresis heating in the interior of Io and hence to volcanism.

38.2 The electron paths are shown in Figure S38.2. There is a magnification of about 1.6 but it is clear that not all points come to a focus in the same x-plane. More complicated lens systems can remove this aberration.

38.3 The result of running this problem depends on which programming language is used and the seed for the random-number generator. A typical graphical representation of the virial term versus time is shown in Figure S38.3. The spikiness happens when stars

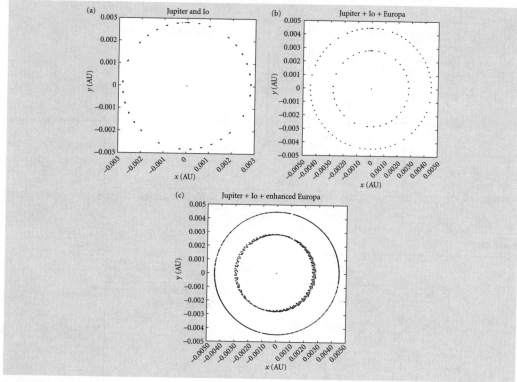

Figure S38.1

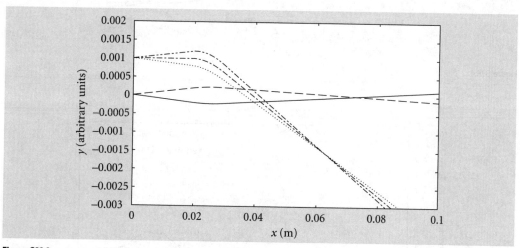

Figure S38.2

approach closely and their speeds increase, so increasing kinetic energy and hence the virial term. This is not seen clearly in an I vs t curve. The term has value about 0.2 for time around 50 000 years.

An excerpt from the table of I is

Time (years)	I (kg m^2)
45 194	8.399×10^8
49 290	8.762×10^8
51 210	8.955×10^8

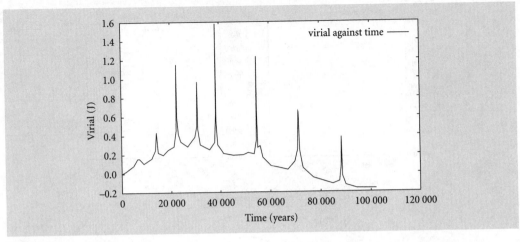

Figure S38.3

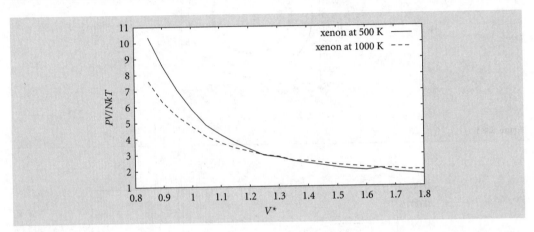

Figure S38.4

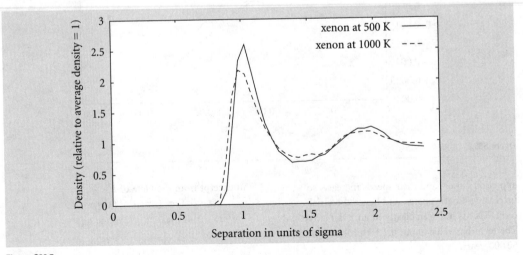

Figure S38.5

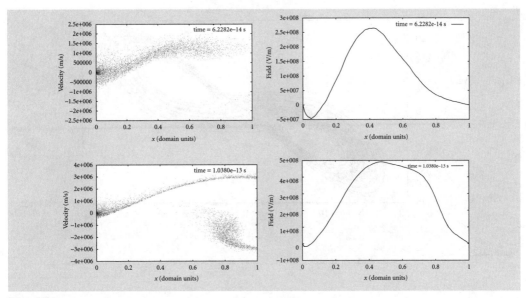

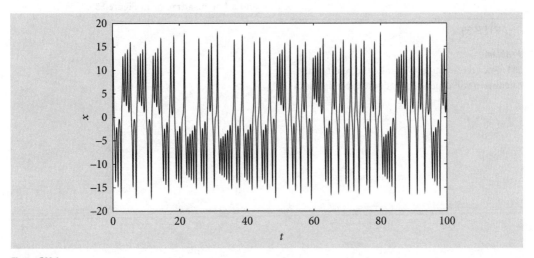

Figure S38.6

Figure S39.1

From the first two rows, using a central difference formula, it may be deduced that for time = 47 242 years $\dot{I} = 8\,862\ \text{kg}\,\text{m}^2\,\text{year}^{-1}$ and that at time = 50 250 years $\dot{I} = 10\,052\ \text{kg}\,\text{m}^2\,\text{year}^{-1}$. Then, from these, again using a central-difference formula, it is deduced that at time 48 796 years $\ddot{I} = 0.39$ or $\frac{1}{2}\ddot{I} = 0.195$. Given the crudeness of the calculation, this is consistent with (38.19).

38.4 The equations of state and radial density distributions are shown in Figures S38.4 and S38.5.

The virial term in the equation of state (23.24) will depend only on the density, i.e. the value of V^*, so that PV/NkT will decrease with temperature, as seen in the above equation-of-state curves. Also at higher temperature the atoms are less rigidly bound and hence the radial density distribution is fuzzier at higher temperatures.

38.5 The velocity and field diagrams are shown in Figure S38.6. At the earlier time electrons are being reflected back towards the original plasma region.

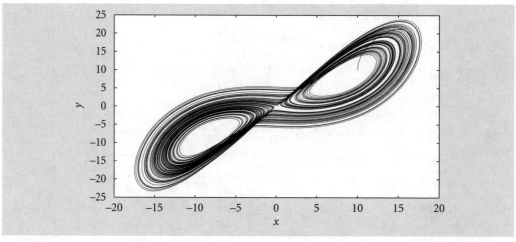

Figure S39.2

However, the positive field slows down the returning electrons, a process that is seen at the later time.

CHAPTER 39

Problem

39.1 Since the solution is chaotic the result of the calculation will depend on both the exact details of the program and the nature of the computer. However, a typical output of x against t is shown in Figure S39.1 and a Lorenz attractor in Figure S39.2.

Index